Ancient Lakes:
Biodiversity, Ecology and Evolution

Advances in
ECOLOGICAL RESEARCH

VOLUME 31

Ancient Lakes: Biodiversity, Ecology and Evolution

Advances in

ECOLOGICAL RESEARCH

Edited by

A. ROSSITER and H. KAWANABE

International Lake Studies, Lake Biwa Museum, Kusatsu, Japan

VOLUME 31

ACADEMIC PRESS

A Harcourt Science and Technology Company

San Diego   San Francisco   New York
Boston   London   Sydney   Tokyo

Academic Press
A Harcourt Science and Technology Company
Harcourt Place, 32 Jamestown Road, London, NW1 7BY, UK
http://www.academicpress.com

Academic Press
A Harcourt Science and Technology Company
525 B Street, Suite 1900, San Diego, California 92101-4495, USA
http://www.academicpress.com

ISBN 0-12-013931-6

A catalogue record for this book is available from the British Library

Typeset by Gray Publishing, Tunbridge Wells, UK
Printed in Great Britain by MPG Books Limited, Bodmin, Cornwall

00 01 02 03 04 05 MP 9 8 7 6 5 4 3 2 1

Contributors to Volume 31

M. AMANO, *Otsuchi Marine Research Center, Ocean Research Institute, University of Tokyo, 2-106-1 Akahama, Otsuchi, Iwate 028-1102, Japan*

P. BÍRÓ, *Balaton Limnological Research Institute of the Hungarian Academy of Sciences, H-8237 Tihany, Hungary*

G.A. BOXSHALL, *The Natural History Museum, Cromwell Road, London SW7 5BD, UK*

P.-F. CHEN, *Institute of Vertebrate Paleontology and Paleoanthropology, Chinese Academy of Sciences, Beijing 100044, China*

C. COCQUYT, *Laboratory of Botany, Department of Biology, University of Ghent, K.L. Ledeganckstraat 35, B-9000 Ghent, Belgium*

A.S. COHEN, *Department of Geosciences, University of Arizona, Tucson, AZ 85721, USA*

K.F. DOWNING, *DePaul University, 243 South Wabash Avenue, Chicago, IL 60604-2302, USA*

H.J. DUMONT, *Laboratory of Animal Ecology, University of Ghent, Ledeganckstraat, 35, B-9000 Ghent, Belgium*

E.A. ERBAEVA, *Scientific Research Institute of Biology, Irkutsk State University, 664003, Irkutsk-3, Lenin St 3, P.B. 24, Siberia, Russia*

E.J. FEE, *Department of Fisheries and Oceans, Winnipeg, Manitoba R3T 2N6, Canada*

G. FRYER, *Institute of Environmental and Natural Sciences, University of Lancaster, Lancaster LA1 4YQ, UK*

D.H. GEARY, *Department of Geology and Geophysics, 1215 W. Dayton Street, University of Wisconsin, Madison, WI 53706, USA*

R.E. HECKY, *National Water Research Institute, P.O. Box 5050, 867 Lakeshore Road, Burlington, Ontario L7R 4A6, Canada*

K. IRVINE, *Department of Zoology, Trinity College, Dublin University, Dublin 2, Ireland*

D. JAUME, *Instituto Mediterráneo de Estudios Avanzados (CSIC-UIB), Ctra. Valldemossa, km 7'5, 07071 Palma de Mallorca, Spain*

R.M. KAMALTYNOV, *Limnological Institute, Siberian Division of the Russian Academy of Sciences, Ulan-Batorskaya 3, 664003 Irkutsk, Russia*

O.M. KOZHOVA, *Scientific Research Institute of Biology, Irkutsk State University, 664003, Irkutsk-3, Lenin St. 3, P.B. 24, Siberia, Russia*

L. KUUSIPALO, *Department of Biology, University of Joensuu, P.O. Box 111, FIN-80101 Joensuu, Finland*

I. MAGYAR, *Department of Geology and Geophysics, 1215 W. Dayton Street, University of Wisconsin, Madison, WI 53706, USA*

K. MASHIKO, *Laboratory of Basic Life Sciences, Teikyo University, Ohtsuka 359, Hachioji, Tokyo 192-0395, Japan*

E. MICHEL, *Institute for Systematics and Population Biology, University of Amsterdam, P.O. Box 94766, 1090-GT Amsterdam, The Netherlands*

N. MIYAZAKI, *Otsuchi Marine Research Center, Ocean Research Institute, University of Tokyo, 2-106-1 Akahama, Otsuchi, Iwate 028-1102, Japan*

H. MORINO, *Department of Environmental Sciences, Ibaraki University, Mito, Ibaraki 310-8512, Japan*

P. MOURGUIART, *IRD (formerly ORSTOM)/UFR Sciences et Technologies Côte Basque, Université de Pau et des Pays de l'Adour, Département d'écologie, Campus universitaire, Parc Montaury, BP 155, 64600 Anglet, France*

P. MÜLLER, *Geological Institute of Hungary, H-1143 Budapest, Stefánia út 14, Hungary*

K. NAKAI, *Lake Biwa Museum, Kusatsu, 1091 Oroshimo-cho, Shiga 525-0001, Japan*

M. NISHINO, *Lake Biwa Research Institute, Uchide-hama 1-10, Otsu, Shiga 520, Japan*

T.G. NORTHCOTE, *Department of Zoology, The University of British Columbia, Vancouver, BC, V6T 2A9, Canada*

L.E. PARK, *Department of Geology, University of Akron, Akron, OH 44325-4101, USA*

G. PATTERSON, *Natural Resources Institute, University of Greenwich, Central Avenue, Chatham Maritime, Chatham, Kent ME4 4TB, UK*

E.A. PETROV, *Limnological Institute, Russian Academy of Sciences, Ulan-Batorskaya 664033, Irkutsk, Russia*

L. RÜBER, *Zoological Museum of the University of Zürich, Winterthurer-strasse 190, 8057 Zürich, Switzerland*

A. ROSSITER, *International Lake Studies, Lake Biwa Museum, 1091 Oroshimo, Kusatsu, Shiga 525-0001, Japan*

G.P. SAFRONOV, *Scientific Research Institute of Biology, Irkutsk State University, 664003, Irkutsk-3, Lenin St. 3, P.B. 24, Siberia, Russia*

O. SEEHAUSEN, *Institute of Evolutionary and Ecological Sciences, University of Leiden, P.O. Box 9516, NL-2300 RA Leiden, The Netherlands*

V.G. SIDELEVA, *Zoological Institute, Russian Academy of Sciences, 199034 St Petersburg, Russia*

J. SNOEKS, *SADC/GEF Lake Malawi/Nyasa Biodiversity Conservation Project, P.O. Box 311, Salima, Malawi*

V.V. TAKHTEEV, *Department of Invertebrate Zoology, Irkutsk State University, Sukhe-Bator Street 5, Irkutsk 664003, Russia*

G.F. TURNER, *School of Biological Sciences, University of Southampton, Bassett Crescent East, Southampton SO16 7PX, UK*

E. VERHEYEN, *Royal Belgian Institute of Natural Sciences, Section of Taxonomy and Biochemical Systematics, Vautierstraat 29, B-1000 Brussels, Belgium*

J.H. WANINK, *Institute of Evolutionary and Ecological Sciences, Rijksuniversiteit Leiden, P.O. Box 9516, 2300 RA Leiden, The Netherlands*

N.C. WATANABE, *Faculty of Education, Kagawa University, Saiwai-cho 1-1, Takamatsu, Kagawa 760, Japan*

K.A. WEST, *Lake Tanganyika Biodiversity Project, B.P. Kigoma Station, P.O. Box 90, Kigoma, Tanzania*

F. WITTE, *Institute of Evolutionary and Ecological Sciences, Rijksuniversiteit Leiden, P.O. Box 9516, 2300 RA Leiden, The Netherlands*

.

Foreword

The diverse contributions in this volume demonstrate very clearly the importance of studies on ancient lakes in advancing biological ideas in many spheres.

From their associated river systems these lakes received and became reservoirs of ancient phylogenetic lines of organisms, especially of invertebrates and fishes, some of which then exploited the new lacustrine habitats and over time evolved into "flocks" of numerous new endemic species, making these lakes hotspots of biodiversity. Their rich faunas make such lakes natural laboratories in which to study the mechanisms of both speciation and community structure, on how the species interact and are able to coexist in the very complex lacustrine communities.

Furthermore, since the United Nations' Second Earth Summit held in Rio de Janeiro in 1992, much emphasis has been placed on the role of biodiversity in maintaining Earth's life-support ecosystems. To understand the mechanisms involved it is necessary to discover how species arise and are lost, which are key species and the interrelationships of the species living together in complex communities: in competition, co-operation or a situation-specific mixture of both. Ancient lakes, which are hotspots of biodiversity hosting large numbers of very similar endemic species, provide exceptional opportunities for finding answers to such questions. They have been contributing such information for many years and, as this volume shows, continue to do so, especially from those lakes with the most diverse biota: Lake Baikal in Russia, and the East African Great Lakes Malawi and Tanganyika, and from others which have lost many of their endemic species, such as Lake Victoria in East Africa and Lake Titicaca in South America.

The oldest lakes (c. 20 My old), Baikal, Tanganyika and its slightly younger sister lake Malawi, are all deep graben lakes, whose faunas have withstood geological and climatic changes through long ages, with fluctuations in lake levels helping to promote the intralacustrine formation of new species. Initial studies of these ancient lake faunas were based on the morphology of specimens from early expeditions brought back to museums; these aroused interest in these lakes as potential laboratories in which to study the processes of evolution (as summarized by Brooks, 1950). The ability to study the ecology and behaviour of fishes and other organisms *in situ* began with the advent of SCUBA (self-contained underwater breathing apparatus). First used *c.* 1978 in the clear warm waters of lakes Malawi and Tanganyika, this has provided unimagined insights into the often subtle facets of behaviour and ecology, and has revolutionized ideas on how biotic interrelationships affect speciation. Studies of evolution also proceed apace. As illustrated by several papers in this volume, among new techniques being brought to bear on the phylogenetic relationships of these endemic species, examination of cichlid scale patterns is producing interesting results, as are DNA studies, not only on fishes but also on ancient lake mollusc and ostracod faunas.

An indication of the increased recent interest in ancient lakes and their faunas was the holding of the first workshop to consider speciation in ancient lakes, organized in Belgium in 1993, which produced very useful papers edited by Martens, Goddeeris and Coulter (1994). Fundamental questions asked then were, what are the origins of these faunas, how old are they, and when and how did the species radiations occur? A need was identified to decipher the roles of abiotic and biotic selection pressures on mechanisms of speciation. Not all of the phylogenetic lines of fish and invertebrates which gained access to these lakes speciated; why did some do so, such as cichlids, certain mollusc and crustacean groups, and others merely remain as relict species? Fryer (1991) had stressed that there is much to be gained by comparative research between different lakes and between taxonomic groups, both fossil and extant, but the 1993 workshop concluded that, while these comparisons may eventually produce unifying hypotheses, the immediate goal should be to understand the multitude of patterns and processes that exist within the various biota, communities and lake basins. The present volume is an important contribution to the ongoing studies along these lines and also includes some papers on paleoancient lakes, research which can throw light on long-term trends in evolutionary and ecological processes in extant ancient lakes.

Recent research on the fish faunas of the East African Great Lakes has produced many surprises, for example, the discovery of a whole new species flock of haplochromine cichlids living in the protection of the rocky islands at the south end of Lake Victoria, analogous to the mbuna rock-dwelling cichlids of Lake Malawi and rock dwellers in Tanganyika. Behavioural experiments on Malawi cichlids have strengthened the hypothesis that sympatric speciation is aided by sexual selection, the females selecting males for their breeding colours. Furthermore, in Lake Victoria it has now been shown that pollution, which alters light penetration into the water, can affect this process, leading to hybridization of some of these cichlids. In Lake Malawi another recent discovery has been of as yet undescribed cichlids living at circumscribed depths out in the open lake, in contrast with Lake Tanganyika where the pelagic waters are patrolled by large piscivorous endemic *Lates*, a genus absent from Lake Malawi. Another recent surprise has been the demonstration, from bottom cores below the open waters of Lake Victoria, that this relatively shallow (< 90 m) lake appears to have dried up a mere *c.* 12 500 years ago. If substantiated, this means the evolution of several hundred cichlid species in the short time since the lake refilled is the fastest rate of evolution on record.

Projects underway, described in detail in many of the papers, are helping us to understand the roles of biogeographical, ecological and behavioural factors in speciation and the coexistence of species in these lacustrine communities. However, conditions are continually changing, as demonstrated in Cohen's paper on the implications of spatial and temporal change in the diversity structure of these lakes, for both theoretical ecology and conservation biology.

We are learning to understand the roles of climatic changes on these faunas. However, most of these ancient lakes are now subject to accelerating anthropogenic influences affecting almost all lakes: sedimentation from clearance of forests and agricultural developments in the lake basins, organic and industrial pollution from the rapidly increasing riparian human populations, overfishing and, perhaps most damaging, the introduction of alien species, plants such as the water hyacinth, fishes which outcompete the endemics and predators which decimate them. This has been demonstrated by the changes to Lake Victoria. It has been estimated that this lake once supported over 300 endemic cichlid species, two-thirds of which vanished in the three decades following the introduction of the large piscivorous Nile perch (*Lates*), and in which eutrophication – influenced by pollution and slight climatic warming – has led to deoxygenation of the bottom waters and many unforeseen changes to the biota, and is now plagued by water hyacinth. These changes in Lake Victoria are utilized here as a "natural experiment" by Wanink and Witte.

Conservation projects are now of vital importance to protect these very special ancient lake biomes, with their unique and rich faunas, which can contribute so much to biological understanding. However, the complex and diverse nature of their ecosystems makes unravelling them difficult, and perhaps the optimal route to quantitative understanding might be through experimental manipulation. The uniqueness of each ancient lake makes this ethically undesirable, and so insights must be drawn from observation and monitoring, and through comparisons with other, younger lakes. Biro's paper summarizing 100 years of continuous studies on Lake Balaton provides such a comparison, and exemplifies the benefits of long-term and comprehensive research. The contents of this volume make it clear that such a research perspective must be applied to the ancient lakes of the world and that such projects must in each case involve the whole lake basin.

R.H. Lowe-McConnell

Preface

Most lakes are geologically emphemeral, with a post-Pleistocene origin and a maximum age of *c*. 10 000–15 000 years, having been formed after the last glaciation: geological activities, climatic effects and sedimentary processes usually conspire to ensure that their lifespan is often even shorter than this. All the more noteworthy, then, are the ancient lakes of the world, the small number of lakes that have a continuous history of at least 100 000 years; most have existed for considerably longer and some are several million years old. The 20 or so members of this select group display a great diversity in gross physical features, geographical location and ambient climate, water chemistry, bathymetry and limnological characteristics. Some are permanently stratified and have an extensive anoxic hypolimnion, whereas others show annual turnover and are oxygenated throughout. Several are very large – four of the five largest (area) lakes in the world are ancient lakes, as are several of the most voluminous and some of the deepest lakes – but the group also includes much smaller water bodies. Some offer many challenges to the biologist and physical limnologist because of their vast areas, great depths, remoteness and inaccessibility. Most ancient lakes have a complicated geological history which, together with their long continuous existence, has had important implications for the evolutionary histories of their floras and faunas, most notably in providing conditions amenable to the production of the high diversity and endemism which make ancient lake biotas of such interest to biologists.

Perhaps the first scientific research into any ancient lake was the mainly limnological studies made by Russian scientists in the Siberian Lake Baikal during the 1870s and around 1900. Studies of the flora and fauna followed. In the African Great Lakes Malawi, Tanganyika and Victoria, the first specimens to reach biologists were freshwater gastropod shells collected from the lake shores by early European explorers around 1860. These few shells were interesting enough to stimulate biology-orientated collecting expeditions, notably those from the Royal Society of London in 1895 and 1899. In the South American Lake Titicaca, the first faunal inventory was made in 1903 and the first major scientific expedition in 1937. From these and other studies the unusual nature of ancient lake biotas soon became apparent to the scientific community, and it was quickly recognized that most are characterized by highly diverse and unique biological communities, whose component taxa often display high degrees of endemism and sometimes include species flocks. It is these biological features of ancient lakes that the chapters in this book investigate.

Although biological research was already underway in several ancient lakes from around the early 1900s, these studies were performed as independent and unrelated efforts. It was only in 1950, with the publication of Brooks' classic *Speciation in Ancient Lakes*, that a synthetic and comparative approach towards ancient lakes studies was first advocated, and the uniqueness and scientific importance of their faunas were widely promulgated. This landmark paper is as stimulating and thought provoking today as when it was published. Thereafter, research proceeded apace in several ancient lakes and some of these studies produced landmark volumes in their own right, for example, Kozhov's 1963 monograph on Lake Baikal and Fryer and Iles' 1972 volume of the cichlid fishes of the African Great Lakes. Strangely, however, with the possible exception of some African Great Lake studies, for the most part research in each lake continued to be conducted in splendid isolation; interlake comparisons were exceedingly few and little serious effort was made to consolidate the academic foundation provided by Brooks.

It was not until 1994, with the edited proceedings *Speciation in Ancient Lakes*, that a serious attempt was made to gather together such disparate research efforts under the mantle of ancient lakes studies. The 40-year hiatus since Brooks' paper necessitated the inclusion in that volume of potted overviews of the geology and biology of several ancient (and some non-ancient) lakes, along with essays on evolutionary theories and principles, etc. This approach served both to restate Brooks' tenet of the scientific importance of ancient lakes and their faunas, and to re-emphasize the relevance of such studies, but in a modern biological context. With this useful volume there now existed for the first time a sound comparative basis from which to pursue studies of ancient lake biotas.

It has already been 6 years since that most recent review volume on ancient lakes appeared, during which time several major developments and advances in knowledge have taken place; progress towards which many of the authors in the present volume have made important contributions.

The idea for this book was conceived after a forum (ICAL*) which addressed the cultural and anthropological perspectives of interrelationships among people, nature and lakes. In the months following that conference several of the biologists who had attended contacted us to lament the missed opportunity to present an up-to-date volume containing the most recent research on ancient lake biotas. Enquiries made to a number of aquatic biologists on several continents produced almost unanimous recommendations

---

*International Conference on Ancient Lakes, held at the Lake Biwa Museum during summer 1997.

that the time was ripe for a volume such as this, addressing recent and current research on ancient lake organisms and their biology.

So, in summer 1998 we decided to press ahead with the present book. After much consultation and deliberation it was concluded that a synthesis-type presentation, which would, of necessity, have repeated much of the contents of the 1994 volume, was to be avoided, and instead an approach was chosen which illustrates the biodiversity of ancient lakes through emphasis on recent and ongoing studies of the organisms themselves. To achieve this, a taxonomically and topically diverse set of chapters has been included prepared by a truly international group of contributors, each a leader in his or her respective theme. It is hoped that this volume, with its hands-on studies, will both complement and advance the contents of the 1994 volume.

Even so, while preparing this book we remained acutely aware of the old publishing adage concerning potential readership, apposite to editors and authors, that "some will read it to find out what they missed, and some will read it to find out what you missed". As a result of the choice of contents, we hope that persons in the former category, and are sure that those in the latter, will derive satisfaction from reading this book. The former because the international mix of expert contributors, the diversity of taxa and topics that these chapters address, and the contemporaneous nature of each chapter's contents, should ensure that every reader will find something novel among the many exciting recent findings and new perspectives described herein; and the latter because it was never the intention to produce an all-encompassing treatise: the book is, of necessity and by design, incomplete.

This situation springs from at least three causes. First, the information available for ancient lakes is inconsistent in quality and amount; whereas accumulated knowledge for certain lakes is sufficient to have merited monographic treatments (e.g. Lake Tanganyika, Coulter, 1991; Lake Titicaca, Dejoux and Iltis, 1992; the Caspian Sea, Kosarev and Yablonskaya, 1994; and Lake Baikal, Kozhova and Izmest'eva, 1998), for several others almost nothing is known. We have therefore focused on those ancient lakes for which recent information is most available and have not included those for which information is sparse (e.g. the Balkan Lake Prespa, the Burmese Lake Inle, the Chinese Lakes Dianchi, Erhai and Fujian-hu, on the Georgian Lake Sevan), or for which new findings are few (e.g. the Balkan Lake Ohrid on the Philippine Lake Lanao).

Secondly, gathered beneath the umbrella of "Ancient Lakes: Biodiversity, Ecology and Evolution" one finds many fields of research, and the topics are simply too diverse and the literature is too voluminous to attempt anything approaching a comprehensive coverage within a single volume. This has necessitated our being selective in what was included. Ancient lakes are often hotspots of biodiversity and homes to unique and unusual flora and fauna, and many contain endemic species flocks. The aim was to emphasize this by focusing on studies of the diversity and lifestyles of organisms, rather than on purely hypothetical and mathematical treatments or essay-type presentations. Similarly, unless pivotal to the theme of the paper or not previously published, we have assiduously avoided lengthy inventories of taxa, as these can be obtained from specialist volumes on each lake or from journal sources, where these exist. We have also not included physical limnological studies, theoretical papers or general reviews of evolutionary and speciation mechanisms, or those concerned with the geological histories of these lakes, despite these topics being included under the rubric of the book's subtitles. There exists already a rich and informative literature on these topics, the scope and clarity of most of which makes the repetition of their contents here in modified form a futile exercise. Instead, we have taken the liberty of assuming at least a rudimentary knowledge of evolution and ecology on the part of the reader; for those who might wish to delve deeper, some useful titles are given at the end of the Introduction.

Finally, for several ancient lakes and taxa, recent or ongoing research efforts are limited and biologists working on them number only two or three. In some such cases, our first, second and third choices of persons to write about particular subjects were unavailable because of other commitments. In a few other instances, at the last minute promised contributions did not materialize. Such problems forced the omission of some topics that we had hoped to include.

Nevertheless, despite these omissions, and also partly because of them, the book serves admirably the two intended purposes: firstly, to highlight the diversity and unique nature of ancient lake biotas, through cutting-edge papers by expert authors with first-hand knowledge of their taxa and topics; and secondly, to demonstrate the importance of ancient lakes as natural laboratories of evolution, ecology and speciation *in situ*. We are confident that the end result is a book which provides an up-to-date insight of what we know, and need to know, about ancient lake biotas. This being so, we are equally confident that the astute reader will find scattered among these pages many potentially fruitful themes for future research on ancient lake biotas.

The diversity of topics and taxa contained in this book made it impossible to render informed editorial judgement without the assistance of numerous referees. We are indebted to all these persons for the meticulous and conscientious manner in which they processed manuscripts, and for their (often copious) thoughtful comments and advice, which usually resulted in significant improvements in the presentation and content of chapters. Our gratitude goes also to the nine anonymous reviewers, whose insightful comments, advice and critiques helped us to sharpen our ideas and objectives, and the focus of the book. It is a pleasure to thank the contributors for their always enthusiastic, often compliant and usually prompt responses to our sometimes draconian editorial requests. We thank the committees and secretariat of the ICAL, and the Shiga Prefectural Government and other foundations for their support of this forum, which serendipitously triggered the genesis of the present volume. A. Rossiter thanks Y. Rossiter for her tolerance, patience and support during the preparation of this book, Dr N.F. Broccoli for his motivation, and T. Picknett and C. Nehammer at Academic Press for their encouragement, advice and professionalism shown throughout the publication process.

It is our greatest hope that this book will raise the profile of ancient lake biology, and stimulate the interest of biologists and encourage them to join the growing and very active band of scientists* utilizing the unparalled opportunities that ancient lakes offer for studies of biodiversity, ecology and evolution. As the contents of this book illustrate, there is a myriad of exciting ecological, evolutionary and taxonomic questions waiting to be tackled; the answers to which have important implications for biology in general. We also hope that this volume will instil in the reader a sense of awe at the truly amazing diversity of life in these lakes, tempered with an awareness of the urgent need for its conservation. It is perhaps appropriate to close this preface by echoing the sense of wonderment at biodiversity so eloquently expressed in the final paragraph of the *Origin*:

> ... *these elaborately constructed forms, so different from each other, and dependent on each other in so complex a manner, have all been produced by laws acting all around us. ... endless forms most beautiful and most wonderful have been, and are being evolved.*

Andrew Rossiter and Hiroya Kawanabe
Lake Biwa Museum

*SIAL (Species in Ancient Lakes) is an international research group dedicated to promoting research on all aspects of ecology, evolution and diversity of organisms in ancient lakes. The group is open to all interested persons. Details may be obtained from the secretary, Dr K. Martens, at martens@kbinirsnb.be.

Contents

PART 1: SPECIES IN ANCIENT LAKES

Unanticipated Diversity: The Discovery and Biological Exploration of Africa's Ancient Lakes

G. FRYER

How Well Known is the Ichthyodiversity of the Large East African Lakes?

J. SNOEKS

The Nature of Species in Ancient Lakes: Perspectives from the Fishes of Lake Malawi

G.F. TURNER

Making Waves: The Repeated Colonization of Fresh Water by Copepod Crustaceans

G.A. BOXSHALL and D. JAUME

The Ichthyofauna of Lake Baikal, with Special Reference to its Zoogeographical Relations

V.G. SIDELEVA

The Benthic Invertebrates of Lake Khubsugul, Mongolia

O.M. KOZHOVA, E.A. ERBAEVA and G.P. SAFRONOV

Biogeography and Species Diversity of Diatoms in the Northern Basin of Lake Tanganyika

C. COCQUYT

PART 2: EVOLUTION AND ENDEMISM

Evolution and Endemism in Lake Biwa, with Special Reference to its Gastropod Mollusc Fauna

M. NISHINO and N.C. WATANABE

Endemism in the Ponto-Caspian Fauna, with Special Emphasis on the Onychopoda (Crustacea)

H.J. DUMONT

Trends in the Evolution of Baikal Amphipods and Evolutionary Parallels with some Marine Malacostracan Faunas

V.V. TAKHTEEV

Insights into the Mechanism of Speciation in Gammarid Crustaceans of Lake Baikal using a Population-genetic Approach

K. MASHIKO

Explosive Speciation Rates and Unusual Species Richness in Haplochromine Cichlid Fishes: Effects of Sexual Selection

O. SEEHAUSEN

Phylogeny of a Gastropod Species Flock: Exploring Speciation in Lake Tanganyika in a Molecular Framework

E. MICHEL

Implications of Phylogeny Reconstruction for Ostracod Speciation Modes in Lake Tanganyika

L.E. PARK and K.F. DOWNING

The Dynamics of Endemic Diversification: Molecular Phylogeny Suggests an Explosive Origin of the Thiarid Gastropods of Lake Tanganyika

K. WEST and E. MICHEL

PART 3: ECOLOGY AND EVOLUTION

Phenetic Analysis, Trophic Specialization and Habitat Partitioning in the Baikal Amphipod Genus *Eulimnogammarus* (Crustacea)

H. MORINO, R.M. KAMALTYNOV, K. NAKAI and K. MASHIKO

Evolutionary Inferences from the Scale Morphology of Malawian Cichlid Fishes

L. KUUSIPALO

Ecological Interactions Among an Orestiid (Pisces: Cyprinodontidae) Species Flock in the Littoral Zone of Lake Titicaca

T.G. NORTHCOTE

Effect of Hydrological Cycles on Planktonic Primary Production in Lake Malawi/Niassa

G. PATTERSON, R.E. HECKY and E.J. FEE

Macrodistribution, Swarming Behaviour and Production Estimates of the Lakefly *Chaoborus edulis* (Diptera:Chaoboridae) in Lake Malawi

K. IRVINE

Age Determination and Growth of Baikal Seals (*Phoca sibirica*)

M. AMANO, N. MIYAZAKI and E.A. PETROV

PART 4: PALAEOBIOLOGY

Ancient Lake Pannon and its Endemic Molluscan Fauna (Central Europe; Mio-Pliocene)

D.H. GEARY, I. MAGYAR and P. MÜLLER

Using Fish Taphonomy to Reconstruct the Environment of Ancient Shanwang Lake

P.-F. CHEN

Historical Changes in the Environment of Lake Titicaca: Evidence from Ostracod Ecology and Evolution

P. MOURGUIART

PART 5: PERSPECTIVES ON CONSERVATION OF
ANCIENT LAKES

Linking Spatial and Temporal Change in the Diversity Structure of Ancient Lakes: Examples from the Ecology and Palaeoecology of the Tanganyikan Ostracods

A.S. COHEN

Conservation of the Endemic Cichlid Fishes of Lake Tanganyika: Implications from Population-level Studies Based on Mitochondrial DNA

E. VERHEYEN and L. RÜBER

The Use of Perturbation as a Natural Experiment: Effects of Predator Introduction on the Community Structure of Zooplanktivorous Fish in Lake Victoria

J.H. WANINK and F. WITTE

Lake Biwa as a Topical Ancient Lake

A. ROSSITER

Long-term Changes in Lake Balaton and its Fish Populations

P. BÍRÓ

Introduction

Life is full of surprises, particularly for those of us who live in our own, self-imposed, academic microcosms. Perhaps most of us are guilty of ascribing a disproportional import to their own topics of study and fields of research, but surprised I was ...

Interest in the unique biological nature of ancient lakes has increased markedly during the past few decades. The scientific importance of their biotas was now, I believed, widely appreciated by biologists, irrespective of their area of research. Since the early explorations of the African Great Lakes brought back specimens of a rich and diverse mollusc and fish fauna, their importance and unique nature had been widely recognized. The species richness of other ancient lakes would be a feature known to all. The fascinating diversity of Lake Baikal, with over 220 taxa of amphipods alone, and the *Orestias* fish flock of Lake Titicaca are surely examples etched into the minds of all biologists, along with the cichlid fish faunas of the African Great Lakes, the sculpins, ostracods and platyhelminths of Lake Baikal, the gammarids of Lake Titicaca, the cyprinids of Lake Lanao, and so on.

However, when I broached the subject of a volume dedicated to the biotas of ancient lakes with several of my biological colleagues, the uninitiated among them expressed only a moderate excitement, coloured with considerable scepticism. Having discounted my initial response that they were incompetent, it soon became apparent that their reservations were founded on two perspectives. First, in their eyes, the concept of ancient lakes considered in a biological context appeared an entirely synthetic construct, a categorization of geological relevance only. Secondly, the more pragmatic and cynical among them questioned whether this was not just yet another example of the increasingly compartmentalized and myopic approach of modern science: a dividing of the cake of knowledge into increasingly small portions, each piece being claimed by specialist researchers in that field. At the table of science, all too often the crumbs of each piece are seldom accessible, or of much interest, to researchers in other fields. In short, they felt that ancient lake biology seemed extremely specialized and of little relevance to biology *per se*.

Therefore, in case the reader harbours similar reservations, at the outset I would like to try to clarify just why ancient lakes and their biotas merit special attention, and why studies on them are of disproportionate importance to an understanding of ecology, evolution and speciation in general.

I. WHY ANCIENT LAKES?

Extant ancient lakes, those with an uninterrupted history which dates back more than 100 000 years (Gorthner, 1994), number but a handful and are scattered over several continents (Martens *et al.*, 1994). From a biological perspective, the grouping together of ancient lakes might seem an unnatural one, since faunistically and limnologically, these lakes are very different. Furthermore, they show almost no physical similarities and are united only by their great age.

However, they share an important biological feature: in ancient lakes freshwater faunas achieve some of their highest levels of biodiversity. The rich faunas that inhabit these lakes are often predominantly of endemic species and sometimes comprise species flocks. This has occurred because the ancient lakes are, in effect, aquatic islands in which a complex of ecology, genetics and evolutionary constraints has shaped in isolation their biotas over hundreds of thousands to millions of years. Accordingly, ancient lakes offer unique opportunities as natural laboratories for meaningful studies of various facets of biodiversity, ecology and evolution.

Therefore, although ancient lakes are few, they are of disproportionate importance to freshwater biology. Their high biodiversity and the high proportion of endemic forms that they contain simultaneously challenge and stimulate the student of ecology, evolution, speciation and diversity, and studies there can have implications for an understanding of such fundamental problems as, what is a species, how does speciation occur, and how do so many species coexist?

II. ANCIENT LAKE STUDIES ARE NOT EXCLUSIVIST

Furthermore, in contrast to the assertions of these colleagues, contemporary biological studies in ancient lakes reverse the trend towards excessive specialization and reductionism in research. As emphasized repeatedly throughout this volume, biodiversity, ecology and evolution are intimately related, such that any meaningful investigation of one cannot afford to ignore inputs from the others. As such, studies on ancient lake biotas are increasingly utilizing findings and techniques from other disciplines, or even better, are collaborative projects made by scientists from different fields: as evinced in the chapters in this volume, this interdependency often extends beyond these three subjects to include the likes of taxonomy, molecular biology, geology, palaeoecology, etc. By integrating insights from different disciplines when analysing biodiversity, speciation and evolution a deeper understanding often results; indeed, without such integration, findings considered in isolation might prove erroneous or somewhat meaningless.

III. WHO SHOULD READ THIS BOOK?

This book is intended to appeal primarily to students, teachers and researchers of aquatic biology, and to all biologists interested in biodiversity, evolution and ecology. The first group should be brought up to pace with goings on in studies in ancient lakes; the second group will, it is hoped, find new information or research topics of interest and to which its talents might be addressed. The contents are also of relevance to conservationists and to those interested in natural history and biodiversity.

Coverage is biased towards where most of the recent work has been performed and contributors were asked to make their chapters topical and stimulating. Standardized interpretations and mere restatement of old facts were discouraged, and instead a diversity of ideas, interpretations and philosophies was encouraged. Finally, it was emphasized that we would not shy away from controversy, as long as it fell within normal ethical and scientific bounds, a gauntlet which several of the contributors picked up with a fervour and gusto. The result is a book that considers a wide range of taxa and offers a rich diversity of perspectives, including contributions which advocate alternative theories, for example, on speciation mechanisms and species definitions. This diversity of perspectives, championed by some of the most active participants in their respective fields of research, gives vigour to studies and discussion on ancient lakes and stimulates new research.

Despite these differences, it is hoped that two conclusions will be unanimous, that the natural laboratories presented by ancient lakes offer unparalleled opportunities to address questions of fundamental import to biodiversity, ecology and evolution, and that these are indeed exciting times for ancient lake studies.

IV. ABOUT THE BOOK

The book is divided into five parts, but we do not claim that this division is anything but arbitrary: part borders are fuzzy and some overlap exists among the parts. This is a reflection of the content in many individual chapters pertaining to more than one of the three subthemes of this book. We do not find this situation in any way problematic and believe that irrespective of which scheme had been chosen, readers would soon have discovered how interrelated the three subtitles of this volume – biodiversity, ecology and evolution, and hence the entire set of chapters – really are. Therefore, these three themes should not be considered separately, but as subsets of a common theme.

The first part, Species in Ancient Lakes, contains seven chapters, which together provide an introduction to the study of species and biodiversity in several ancient lakes.

Fryer provides a historical perspective on the discovery of the biological treasures of the African Great Lakes Malawi, Tanganyika and Victoria.* These lakes, huge though they are, remained unknown to the western world until the middle of the nineteenth century. Ironically, it was a now disproved hypothesis, that the faunas were relicts of a Jurassic marine incursion, that provided the impetus for the first scientific expedition, to Lake Tanganyika, in 1895. The relative recency of discoveries from these lakes is especially noteworthy. As of 1920 only 40 species of cichlid fish were believed to inhabit Lake Malawi, by the 1930s this had risen to about 170 and today it is believed that over 500 species are present (Snoeks; Turner, this volume). More recently, advances in techniques and technology have been equally as rapid as these increasing species estimates, and these are now being utilized in studies of the phylogenetics, ecology and evolution of a variety of ancient lake faunas. Fryer notes wryly that the questions addressed, and often the solutions arrived at, by modern researchers are essentially the same as those of earlier workers. Biodiversity has been persistently underestimated in these lakes, even in supposedly well-known taxa, and much of their species richness remains to be documented, a situation which is common to all ancient lakes.

This underestimation of diversity is a theme continued and elaborated upon for the ichthyofauna of these lakes, particularly the cichlid fishes, by Snoeks. The species richness and incidence of endemism in these fishes is remarkable: Snoeks estimates some 500 species in Victoria, 250 species in Tanganyika and 700 in Malawi, of which 98%, 99% and 99%, respectively, are endemic. For comparison, the entire ichthyofauna of northern Europe, both marine and freshwater fishes, totals *c.* 350 species (Wheeler, 1978), the freshwater ichthyofauna in the whole of Europe west of the former Soviet Union some 358 species (Kottelat, 1997), and perhaps the most revealing comparison of all: the huge Laurentian Great Lakes and their watersheds contain just 142 species of freshwater fishes (Scott and Crossman, 1973), a depauperate fauna which reflects the relatively recent retreat of the Pleistocene ice fields.

---

*Ancient lakes are those over 100 000 years old, and so the inclusion of Lake Victoria in this volume will surely raise a few hackles. The most recent study concluded that the lake was dry a mere 12 400 years ago. From the viewpoint of lake biologists, this is considered as a resetting of the biological clock to zero, and thus the lake is 12 400 years old. However, in light of the magnitude of the disparity between this age and the age of 75 000 years accepted by many, even very recently (e.g. Greenwood, 1984; Mayr 1984), and the implications for this young age for evolution of the lake's cichlids (e.g. Fryer, 1997; but see also Seehausen, Chapter 12, this volume), we have opted for prudence which leads us to view it as *sub judice*. Lake Victoria is therefore included here as an ancient lake pending future studies. By comparison, although Lake Turkana has a history of > 5 My, this is punctuated with desiccation events, several of which have been conclusively demonstrated to have occurred less than 100 000 years ago. Biologically, then, Lake Turkana does not qualify as an ancient lake.

Studies of the African Great Lake cichlids have a long and intensive history, but how complete is our knowledge, and what can be said of the cichlid taxa in other lakes in this region? Snoeks gives a highly personal and far-ranging overview of what is known and what needs to be done in taxonomic studies of this fauna. He notes that our level of knowledge of the inventories, the phylogenies, the taxonomies and the evolution of these fishes is still incomplete, and predicts large future increases in the number of species described from each of these lakes. A concise and cautionary discussion of the problems and pitfalls of taxonomic research on the cichlid fishes is given, and valuable pointers to alleviate potential problems in future studies are provided. He concludes with a set of thoughtful suggestions, based on personal experience, to improve studies of the systematics of these fishes, advice which should be required reading for all those delving into this taxonomic minefield.

Evaluation of species diversity is founded on the accurate identification and enumeration of species. The validity and usefulness of "species" has been argued at length from philosophical and academic viewpoints (see e.g. Lynch, 1989; King, 1993; Ridley, 1993), but the practical difficulties in its usage experienced by hands-on taxonomists are seldom addressed. Turner provides just such a perspective in his chapter, through an insightful discussion of the development of species concepts, both in practice and in principle, based on his taxonomic studies of Lake Malawi cichlids. He provides a fascinating history of the usage of species concepts throughout the history of studies on this lake. Turner's enthusiastic advocation of the specific make recognition system (SMRS) and denouement of several previous species concepts merits special attention, as does his conclusion that male colour differences, which are based on female choice and can be tested experimentally in the laboratory, are the most practical species-specific characters for taxonomists to use.

Turner concludes his chapter with some speculation, some of which might subsequently be proven incorrect, but all of which is presented so as to encourage critical thought. No drab conservatism here, but instead a refreshingly frank and direct approach which will surely stimulate studies to prove or disprove the several themes addressed, and thereby ultimately advance our knowledge.

Turner's chapter includes the memorable caveat that species concepts and speciation mechanisms appropriate to African cichlid fishes are likely to be of little or no relevance for other ancient lake faunas, a conclusion which harks back to one of the reservations voiced in the introductory paragraphs: is the grouping together of ancient lakes biologically meaningful, and is there any relevance in comparative studies of their faunas? Perhaps it is here appropriate to grasp the nettle and acknowledge that because the different lakes are discrete entities, are located in different locations and climes, and are of different ages, different geological histories, and are populated with suites of organisms of phylogenetically disparate origins, seeking commonalities between them does not seem wise. Instead, it would seem more sensible to investigate each lake as a finite entity, and try to understand the biodiversity and ecological intricacies of each as an individual. In short, comparative studies of different faunal groups in different lakes might not be expected to be very productive, studies of taxonomically similar faunas in different ancient lakes should be more meaningful, and studies of faunas within a given lake, where they have evolved and speciated over long periods, will be most informative.

The process and timing of colonization are topics too often ignored or glossed over for ancient lake taxa, but Boxshall and Jaume's chapter addresses precisely this theme. Copepod crustaceans are present in all ancient lakes, where they sometimes comprise species flocks. The diversity and degree of endemism reflect, in part, the lake's age, area effects and colonization history. Based on a biogeographical analysis and a detailed knowledge of copepod biology and systematics, Boxshall and Jaume present a stimulating and testable novel hypothesis on the timing of colonization of fresh waters by this highly diverse group. From within this grand overview of continental-level resolution, they make a preliminary exploration of the hypothesis that species richness within each copepod lineage is directly correlated with the time elapsed since that lineage colonized fresh waters. The relevance and importance of this approach in providing a framework within which to investigate the colonization history and species diversification of copepods in ancient lakes are clear.

Colonization and biogeographical considerations are also at the fore in Sideleva's chapter, which considers the fishes of Lake Baikal. The ichthyofauna of this lake is numerically depauperate, totalling only 58 taxa or *c.* 3% of its total biodiversity, but is of great biological interest and some commercial importance. Based on their biogeographical origin, Sideleva divides these taxa into four categories and discusses the diversity and general ecology of species within each group. The greatest diversity is found within the Cottoidei – 31 species, all endemic, live in the lake, and even endemic families are found in this group – and Sideleva pays special attention to this suborder. An overview of the history and ecological impact of introduced species is also given. Much of this information is presented here for the first time and provides an invaluable introduction to the fascinating ichthyofauna of this lake.

Lake Baikal has a long and continuous history of biological studies, but its ancient Mongolian neighbour Lake Khubsugul is less well known. Preliminary surveys took place in the early 1900s and again in the 1960s, but only very recently have meaningful studies resumed there. In the following, Kozhova *et al.* present the most complete and latest inventory of the invertebrate fauna of this lake, comprising 287 species, which include 19 species newly described from their recent expedition. Many taxa, especially microinvertebrates, remain uninvestigated, and this number will surely increase. The large differences in the diversity, composition and degree of endemicity of the respective invertebrate faunas of Lakes Baikal and Khubsugul reflect the different ages and evolutionary histories of these lakes, despite a common geological past.

Again, this information, much of which is new, appears here in English for the first time.

The 1990s have seen great advances in our knowledge of the diatom communities of the African Great Lakes, and next Cocquyt reports the latest findings from her ongoing studies of this flora in the northern basin of Lake Tanganyika. Biogeographical analyses of the diverse diatom flora (358 taxa) found there show that *c.* 21% of these taxa are restricted to the African continent and that *c.* 8% are endemic to the lake. Analyses of species richness using ecological diversity indices showed the lowest diversity at a highly perturbed locality at the extreme north of the lake, and a distinct north–south trend of increasing diatom diversity and changing species composition. Diatom community composition may thus have a practical application as a valuable bioindicator of environmental perturbation from pollution and sedimentation.

Part two contains eight chapters broadly categorized under Evolution and Endemism.

The first chapter in this part considers the endemic taxa of the Japanese Lake Biwa. From a thorough and thoughtful examination of the fauna and flora of this lake, Nishino and Watanabe identify 50 species, five subspecies and two varieties as endemic to the lake, and advocate species flock status for the 15 endemic members of the gastropod subgenus *Biwamelania*. By clarifying and correcting several of the existing misconceptions and erroneous hypotheses concerning the evolution and origin of these endemic taxa, this chapter comprises a major advance in studies of this lake. Yet, as the authors note, despite the many biological studies conducted there, knowledge of the origin and evolutionary processes of many of the lake's endemic taxa remains incomplete.

The Caspian Sea is actually a lake: indeed, it is the largest lake in the world, with an area of 374 000 km^2 and a volume of 78 200 km^3. It is also an ancient lake (> 5 My old) and contains a rich fauna of some 950 species, about 400 of which are endemic. Dumont gives a fascinating and far-ranging discussion of speciation and endemism in this lake, with special reference to the onychopod crustaceans. He describes the unusual limnological conditions pertaining to this lake, and describes ecological partitioning mechanisms and differences in reproductive strategies (including asexual reproduction) in the onychopods which might have evolved as adaptations to these conditions. Abandonment of sex seems to have evolved here three times independently, and in view of the extreme rarity of such a reproductive strategy, further research on this fauna is clearly warranted.

As intimated in the opening paragraphs, as editors we have not shied away from controversial theories or opinions, being of the opinion that a diversity of perspectives should stimulate ideas, discussion and eventual testing of alternatives. Thus, one of our stated aims for this book was to emphasize diversity, both of species and of ideas and interpretations. Takhteev tackles both of these aims by considering the ecology and biology of Baikal's diverse amphipod fauna, and by interpreting some aspects of their evolution in the light of the uniquely Russian evolutionary theory of nomogenesis. The amphipod crustaceans of Lake Baikal are a renowned example of adaptive radiation and speciation, the diversity of which rivals that of the African cichlid fishes. The 257 species and 74 subspecies, among which only one species is not endemic to the lake, show great ecological diversity and a variety of evolutionary directions, but a cloak of Cyrillic has meant that for many biologists, information on these animals has remained inacessible. Based on his extensive experience of this group, Takhteev briefly discusses the history of studies of this fauna, and describes some of the trenchant diversity in their ecology and morphology. Takhteev's arguments and the examples of similarities in cuticular armaments that he provides in support of Berg's nomogenesis theory will undoubtedly stimulate some heated discussion. Most readers will wish to understand this theory more fully before arriving at an informed opinion as to its validity, and Berg (1969) is recommended.

Mashiko also looks at the Baikal amphipods, but focuses on a single species, demonstrating genetic differences between neighbouring populations separated by the mouth of a major effluent river. He expands on this finding to offer a simple multilocus gene model based on an allopatric speciation scenario which is sufficient to explain the genetic difference observed in this species, and perhaps speciation in certain other taxa. He notes that increased diversity can be produced by microgeographic changes which split an original population and allow allopatric speciation to occur, but that such changes must not be of such severity or scope that the original population is destroyed.

The controversy about the exact age of the lake notwithstanding, the explosive speciation rates in the Lake Victoria cichlids are higher than in any other known vertebrate radiation. Seehausen gives an elegant and comprehensive overview of his chapter on the studies on the evolution of this fauna. He presents a persuasive hypothesis to explain high diversity and explosive speciation in these fishes, whereby disruptive sexual selection on polymorphic male coloration is the driving force via female choice of male coloration maintaining reproductive isolation. These theoretical, experimental and comparative studies are complemented by an elegant physiological investigation of the visual sensitivity of these fishes. Most excitingly, Seehausen's hypothesis, clearly expounded in this chapter, provides the most plausible explanation to date for the possibility of sympatric speciation in lacustrine fish faunas. This chapter exemplifies the advantages accrued by integrating insights from different disciplines to analyse biodiversity, ecology and evolution.

The gastropods of Lake Tanganyika exhibit high endemism, high taxonomic diversity and great morphological disparity. Michel describes her ongoing investigations of two sister-clade genera of thiarid gastropods from this lake. Using molecular data, she constructs a phylogenetic framework as a model system within which to investigate species relationships and to test speciation hypotheses. She then broadens her investigation by using these phylogenies to explore patterns of character evolution in shell morphology, habitat specificity, reproductive strategy, and radular morphology and trophic ecology.

Using phylogenetic analyses to explore evolutionary patterns is also a theme addressed by Park and Downing. Eighty ostracod species, distributed among 25 genera, are known from Lake Tanganyika. Park and Downing examine the application of speciation models to the nine known species of one of these genera, *Gomphocythere*, in this lake, and seven species from elsewhere in Africa. Using characters of hard- and soft-part anatomy to construct phylogenies, they examine two alternative evolutionary scenarios. They find that *Gomphocythere* ostracods have diversified many times, that this has been a mosaic pattern of multiple speciation events, and that within Lake Tanganyika, major fluctuations in lake level have played a major role in shaping the adaptive radiation and speciation of this group. Colonization scenarios are also considered, which incidentally illustrate clearly one fruitful avenue in which the colonization hypothesis for copepods of Boxshall and Jaume might be tested. In summary, this chapter and that of Michel exemplify the evolutionary insights that can be drawn from careful phylogenetic analysis applied to clearly defined questions.

West and Michel use molecular methods to elucidate the phylogeny of 12 of the 18 recognized gastropod genera of Lake Tanganyika. Their study reveals the endemic thiarid gastropods of this lake to be paraphyletic, but that a larger clade including non-endemic forms is monophyletic. Despite using a variety of laboratory techniques and methods of data exploration, it did not prove possible to resolve relationships among five clades revealed within the endemic thiarids. They interpret this as evidence for a rapid burst-like radiation at the time of origin of this fauna. To describe this group of closely related endemics that have radiated *in situ*, the term "superflock" is resurrected. This study illustrates clearly the difficulties attendant in unravelling the phylogenetic details of explosive speciation events and the utility of molecular approaches in moving towards a solution.

The third part of the book includes chapters based mainly around ecology and evolution.

The first chapter of this part echoes the topic of Baikal amphipods addressed in two earlier chapters, but this time from the viewpoint of phylogenetics, ecology and functional morphology. Despite the many studies of their taxonomy, the familial allocation of Baikal taxa is still in a state of flux, a situation ascribed to the taxonomic usage of the extreme and unique specializations of these fauna, which defy direct comparisons with other non-Baikal taxa. There also exists a surprising paucity of quantitative studies of the community ecology of this fauna. Morino *et al.* make a positive contribution towards both of these areas in their studies of the amphipod genus *Eulimnogammarus*. By using morphological characters and overall similarity to construct a phylogenetic tree and not restricting their analyses to the unique morphological characters of the Baikal amphipod carapace, they were able to address the phylogenetics of this genus and its affinities with several other closely related taxa. Studies of the microdistribution pattern revealed a clear habitat segregation between certain taxa, with sister taxa occupying different habitat zones. From detailed comparisons of mouthpart anatomy they then demonstrate large interspecific differences in mouthpart morphologies. However, in most cases these differences were not reflected in trophic ecology, and high overlaps in food were evident. (With commendable frankness, they caution, however, that this might be due to their having used insufficiently fine resolution in their analyses of gut contents.) As such, an explanation for the coexistence and high diversity of amphipods within Baikal's littoral zone remains elusive. Even so, to reveal the patterns of species diversity and the mechanisms for its maintenance the multidiciplinary approach used here seems most promising.

The relationship between phylogeny and trophic niche is also explored by Kuusipalo, but for Malawi cichlid fishes, where she uses the novel approach, pioneered by Lippitsch (1990), of comparing fine-scale morphology. Using this selectively neutral character to investigate phylogenetic relationships among 20 endemic genera, she then examines niche similarity among component taxa and presents evidence for dramatic niche switches on several occasions during the evolution of these fishes. She concludes with a novel hypothesis to explain how vacant niches might have become available to facilitate such switches. Her hypothesis is supported by evidence from paleontological studies, which indicate two mass mortality events: these might have extirpated many populations of fishes and made newly vacated niches available to survivors. Coincidentally, further support for this scenario has come in October 1999, when mass mortalities of cichlids were recorded in Lake Malawi, caused by either tectonic activity or extreme seiching activity (see Patterson *et al.*) bringing deeper anoxic waters up to the epilimnion.

The ecological interactions among closely related species are of great interest to ecologists and evolutionists, and the faunas of ancient lakes offer unique opportunities for such investigations. One such example is the *Orestias* fishes of Lake Titicaca, but despite almost 100 years of broad-based studies of this lake, to date only rudimentary investigations of their ecology have been made. Northcote describes results of his pioneering and illuminating study of interactions among the seven littoral zone members of this species flock. Through meticulous and comprehensive research, he documents quantitatively for the first time significant differences in relative abundances of species within the littoral zone, and also interspecific differences in the temporal, spatial and ontogenetic use of this habitat. Functional morphology is also considered, and species-specific differences in gill raker number, spacing and length, pharyngeal dentition and alimentary canal length are noted. The differences in trophogastric morphology of these *Orestias* species are related to trophic ecology (unlike the *Eulimnogammarus* amphipods of Morino *et al*'s chapter). Northcote finds evidence that all four mechanisms previously proposed to account for coexistence within fish species flocks in ancient lakes – community/ habitat instability, environmental patchiness, temporal segregation, trophic/ spatial partitioning – apply to the studied *Orestias* community. That this study has advanced significantly our knowledge of this species flock is beyond contention. What makes it even more noteworthy is that Northcote's study was not the primary focus of his work on Lake Titicaca, and that these results were collected as incidental data. This lends much weight to his call for future intensive co-operative approaches of study on the Titicaca *Orestias* species flock.

Patterson *et al.*'s chapter has a limnological basis, but is included here because it has extended beyond physical processes to consider effects of planktonic primary production. Limnological studies in tropical lakes have a long history (e.g. Talling and Lemoalle, 1998), and it is thus surprising that this study by Patterson *et al.* is the latest of but an extremely small number to have examined primary production in Lake Malawi. Using data from their 2-year study of chlorophyll and light distribution in Lake Malawi, the authors model primary production in that lake. Their estimated values correlate well with maximal vertical mixing of the water layers, and primary production in the lake's pelagic zone thus seems to be driven by physical factors such as ambient temperatures and wind strength. These factors have a strong seasonal component and, accordingly, seasonal peaks in primary productivity occur in this lake, fuelled by nutrients brought to the epilimnion by wind-induced mixing.

Such pulses in primary productivity in the pelagic zone can impact strongly on secondary consumers (zooplankters, planktivorous fishes) and ripple along the foodweb. More research is called for to elucidate the strength and directions of these cascade effects, but some intriguing insights into the early effects on the foodweb are given in the chapter by Irvine which describes his studies of the life history and ecology of the lakefly *Chaoborus edulis* in Lake Malawi. As might be predicted from the findings of Patterson *et al.*, the maximum biomass of *Chaoborus* larvae occurs following increased seiching in the water column, a process which brings nutrients from the hypolimnion to the epilimnion and which promotes increased phytoplankton and zooplankton production. *Chaoborus* is a predator of zooplankton, and this abundance of zooplankters is manifest by a maximum of standing biomass of *Chaoborus* larvae some 2 months after the annual maximum of zooplankton. *Chaoborus edulis* is also found in Lake Victoria, and Irvine briefly alludes to studies made there in the 1950s (MacDonald, 1956). However, the dramatic changes which have taken place in that lake's ecosystem (see Wanink and Witte's chapter) since that time limit the value of direct comparisons, and Irvine is wisely judicious in drawing conclusions for such comparisons. This chapter ends with a cautionary note on the inadvisability of any attempt to increase fish production in Lake Malawi through modification of the foodweb. It had been suggested by the Food and Agriculture Organization (FAO) that *C. edulis* should be removed by introducing zooplanktivorous clupeids, a suggestion that Irvine demonstrates to have been based on incorrect assumptions and misguided logic on several accounts.

The last chapter in this part is autobiological, dealing with ageing studies of the endemic Baikal seal. However, the findings of this study have several wider implications, which extend into such diverse areas as ecosystem conservation and comparative studies of life histories and reproductive strategies. Amano *et al.* use growth layers in the teeth to age seals. In the course of their study they found no 1–5-year-old individuals, indicating a mass mortality in 1987–1988, 4–5 years before their sampling. This event is now believed to have been caused by distemper or a similar disease in pups, which resulted from reduced immunoresistance caused by accumulated toxins in the mother's lipid-rich milk. Comparisons of age–length data with those of other species reveal that Baikal seals have a relatively slow growth rate, which may be related to their long nursing period and the fact that they breed on fast ice, which is a stable habitat, unlike most other seal species which breed on ice floes and have short nursing periods. Interspecific comparisons of gross morphology indicate a large sexual dimorphism in body length and weight in the Baikal seal, a feature inferred to indicate competition between males in this species. Similar detailed studies on another ancient lake endemic seal, the Caspian seal, might be especially informative.

The fourth part contains three chapter on palaeobiology. Two of these address palaeoancient lakes,* ancient lakes which existed for over 100 000 years but which no longer exist, and the third describes palaeoecological studies on the extant Lake Titicaca.

Geary *et al.* give a lucid and comprehensive account of the geological history of the European Lake Pannon and the evolution of its spectacular mollusc fauna. This included over 900 described species and many endemic genera, a high diversity which is due to its ancestry in both marine and surrounding freshwater systems and the long lifetime of the lake (*c.* 8 My). The geological history of Lake Pannon is well documented and the detail in the level of knowledge is sufficient to provide a context in which to pursue palaeobiological studies. Many insights can be drawn. Environmental changes coincided with major pulses of faunal turnover, and available data suggest that considerable evolution, both gradual within lineage, and rapid cladogenic, occurred between these turnovers. Although many of the hundreds of endemic species appear to have evolved geologically rapidly ($< 10^5$ years), in several

---

*These have been called "ancient long-lived lakes", or "fossil long-lived lakes", which are cumbersome terms. Likewise, ancient lakes have been called "long-lived lakes", and on technical grounds perhaps they should be. However, the term "ancient lake" is fixed in the biological vocabulary, it is defined clearly in this book and its meaning is understood, as is that of "palaeoancient lake", and therefore these terms are retained throughout this volume.

lineages evolution appears to have been anagenetic and geologically gradual, with morphological intermediates spanning 1–2 My or more. This ongoing study, in which intraspecific variability and diversity can be tracked over long time intervals and placed in a palaeoenvironmental context, provides a natural complement to genetically based studies of living organisms, and reveals much about the pace and process of evolution in this fauna.

Chen utilizes evidence from fish fossils and applies deductive skills worthy of Sherlock Holmes to reconstruct the environment of the Chinese Lake Shanwang. This palaeolake contains abundant and diverse Tertiary fish fossil deposits, and Chen interprets various features in their condition of preservation to infer limnological conditions during the lake's history. In its earlier age (lower sedimentary section) fish fossils are intact; a feature indicating an environment of still waters, anoxic or anaerobic conditions, and a temperature of < 15°C in the hypolimnion. In its later age (upper sedimentary section) fish fossils are disarticulated, indicating that they decayed in a hypolimnion warmer than c. 5°C, at depths of < 8–12 m and under aerobic conditions. Chen applies finer resolution to remains in the older section to discern seasonal mortality in the fish fauna caused, it is hypothesized, by overturn of an anoxic hypolimnion. This chapter comprises a truly fascinating example of the insights that can be drawn when techniques and information from different fields are bought to bear on a single problem.

Mourguiart utilizes information from modern ostracod communities ingeniously to reconstruct past environments of Lake Titicaca. His results support a complicated scenario for this lake's history, where major lake-level fluctuations split the basin into three separate paleolakes, a situation which probably influenced the evolution and speciation of many taxa of aquatic organisms. A similar scenario occurred in Lake Tanganyika, where it also had significant evolutionary implications (see the chapters by Park and Downing; Verheyen and Rüber). Ostracod samples collected beyond the Titicaca basin suggest that the Titicaca lineage is part of extensive radiations which span the entire Andes. By applying transfer functions to quantitative information on the response of extant ostracod communities to habitat heterogeneity and environmental variability, Mourguiart is able to investigate interlake differences in ostracod palaeofaunas and attribute them to climatic differences. Attention is focused on the ostracod subfamily Limnocytheridae and it is noted that while several morphs can be distinguished, these are not discrete and overlap is significant. These morphs are examples of high morphological variability and under the control of environmental factors. There is a clear and urgent need for further studies of the remarkable morphological variation, and Mourguiart wisely concludes that while taxonomic investigations remain based on external morphology only, it is impossible to provide any definitive hypotheses for phylogenetic relations within the Titicaca limnocytherid flock.

This sound conclusion does not augur well for those working with exclusively fossil invertebrate material, where hard-part morphology is often all that is preserved, and raises the point that in such taxa some morphological changes may have been ecophenotypic switches. Especially in fossil forms, it is almost impossible to assess whether the separate forms at different times are real species or ecophenotypic variants. In many taxa, regulatory changes in growth gradients may effect evolutionary changes in morphology, but might not have been accompanied by speciation. Conversely, as discussed in this book, many workers on African cichlids commonly use colour differences or mate preferences, or even DNA differences, to distinguish between "species", characters which are not evident in the fossil record. Perhaps a divergence in taxonomy will soon be upon us, where paleontologists will eventually opt for operational taxonomic units or suchlike, and students of extant forms for "species" defined by other means, e.g. the morphological, anatomical, genetic or behavioural approaches represented in this book.

The final part is entitled Perspectives on Conservation of Ancient Lakes. As research progresses and new understanding is achieved, the scientific importance of these lakes is becoming increasingly recognized. However, it is also becoming apparent that these lakes and their faunas are under threat everywhere. This part addresses this problem from a variety of perspectives.

Cohen continues the palaeobiology theme of the last section, here interfacing between the palaeaoecology and modern ecology of ostracod communities to address questions of ultimate relevance to conservation biology in ancient lakes: namely, what is a community, and how do communities change through space and time? Using data from modern communities he shows that ostracod populations are spatially highly variable and that their species composition at a given locality is highly unpredictable. Palaeoecological analyses of sediment cores show species composition to have varied over periods of hundreds of years. Studies of modern Tanganyika fish communities are also briefly considered, but the conclusions drawn from these differ markedly from those on ostracod communities. Cohen concludes that several models of community assembly may be relevant to ancient lake faunas, and that caution is necessary in applying any particular model when making decisions related to conservation or biodiversity management.

In their chapter, Verheyen and Rüber offer a different perspective on conservation, that of utilizing mitochondrial DNA information in the identification of evolutionary significant units (ESU) and the concomitant assessment of conservation priorities for cichlid taxa. They address Tanganyikan species, but their arguments and recommendations have a much wider relevance. The accurate identification and enumeration of taxa are fundamental prerequisites for conservation decisions, and so their discovery of some genetically divergent yet morphologically cryptic species is especially noteworthy. Another finding with important conservation implications comes from phylogeographical studies which indicate that closely related endemic taxa can have markedly different distribution ranges. Verheyen and Rüber recommend using genetic information to complement zoogeographical data when formulating guidelines for protected areas. The themes raised in this chapter relate to the identification of species and to their enumeration. However, while it is easy to dismiss the need for inventories, without a reasonable degree of understanding of the taxonomic distribution of species diversity and the degree of specialization among species, studies of community ecology and the processes and results of evolution, the priorities for conservation, seem inherently flawed.

In spite of their biological importance, our knowledge of ecosystem-level processes in ancient lakes is especially poor. There have been no long-term monitoring studies, and the unique and irreplaceable nature of ancient lakes and their biotas dictates that experimental manipulatory studies there are ethically unfeasible. For most ancient lakes, present and potential dangers are now being recognized and addressed, but for some it is already too late, and restoration to preperturbation conditions is impossible. One such case is Lake Victoria, where eutrophication and the effects of the introduced Nile perch on the native fish fauna provide a salient example of just how delicate and irreplaceable these ecosystems can be. In the years following the population boom of Nile perch over 200 endemic cichlid species disappeared and many others showed dramatic increases in abundance.

Clearly, the introduction of this predator has severely impacted upon the ecology of Lake Victoria, but the opportunity has been utilized to great effect by Wanink and Witte, who use this predator-induced perturbation as an experimental manipulation. The careful studies by these authors and their HEST colleagues before, during and after the Nile perch boom allowed for the detailed investigation of its effects on the zooplanktivorous fish community. Changes were detected in population densities, habitat choice and life histories of the zooplanktivorous cichlid fishes, and these suggested a shift from a competitively structured community to one structured by predation.

Because the deliberate introduction of alien species into ancient lakes as an experimental manipulation is so unethical, any opportunity which does come along to study this process must be utilized to the full. This makes the lack of any quantitative investigation of the impact of introduced bluegill sunfish and largemouth bass on the Lake Biwa fish community all the more puzzling and regrettable. This and other features are considered in the chapter where Rossiter uses Lake Biwa as an example of an ancient lake, briefly documenting what is known of this lake and the ecological insults that it has suffered, and then relating this information to several other ancient lakes. It is evident that much remains to be learned of the biota and ecology of Lake Biwa, and that the ecosystem is suffering from multiple sources of perturbation.

The conclusions drawn are clear and simple: for all ancient lakes more needs to be done, from both a scientific and conservation perspective, to ensure that we understand and safeguard these biological treasures. As events in Lake Victoria have shown, it would be tragic if some of these unique ancient lake faunas were to be lost, especially before being scientifically understood.

The final chapter is unashamedly not about ancient lakes, but the modern Lake Balaton. It is included here as an example of the benefits and insights that can be derived from long-term and comprehensive studies. Drawing on a database which extends back 100 years, Bíró documents the effects of alien introductions, eutrophication and habitat modification on the lake ecosystem, mainly from the perspective of fisheries. He also notes how the effects of eutrophication were, in some cases, ameliorated. For no ancient lake do we have the long time frame or comprehensiveness of this database, and this elegant example provides a model to which all ancient lake studies should aspire.

V. CONCLUSIONS

The chapters in this volume illustrate clearly how much we know and how much more we need to know, about ancient lake biotas. Between them they summarize previous research, describe contemporary findings, explore novel approaches and set out directions for future research on the biotas of ancient lakes. In total, they epitomize the high quality of the science being conducted, and the broad biological implications of these recent findings. As such, the contents address the past, the present and the future, and convey some of the genuine scientific excitement and advances in modern studies on ancient lakes.

The strongest undercurrent to emerge from these chapters is that ecological research on ancient lakes today is dominated by the study of biodiversity and of interactions of populations, against a background of contribution to the study of the mechanisms of evolution. However, to reach fruition, further studies are needed. In order to evaluate species evolutionary histories and to evaluate hypotheses for speciation, accurate geological age estimation and palaeoecological and palaeoclimatological studies are essential. Similarly, there remains an immense amount of work to be done on cataloguing what species occur in these lakes, how the conditions in each lake have affected the evolution of these species, how the mosaic of interactions allows coexistence, and the uniqueness of these faunas.

However, the recent influx of new ideas and perspectives stimulated by interdisciplinary collaborations has already strengthened our understanding of the relationship between ecological and evolutionary theory, and emphasized the importance of studies of ecological communities over geological time scales. Progress is being made, and I can but repeat an earlier statement: these are indeed exciting times for ancient lake studies.

SOURCES OF INFORMATION

This is not an exhaustive list, but contains works that I have found particularly useful and clear.

Geological histories and physical features of several ancient lakes are summarized in Martens *et al.* (1994). More detailed information on particular lakes can be found in dedicated works listed below. For physical features of African ancient lakes, including some palaeoclimatology, see Johnson and Odada (1996); the pelagic zone of Lake Malawi, including limnology and fisheries, Menz (1995); and a useful Tanganyika literature survey, Patterson and Makin (1998). Suggested background sources for evolutionary writings include Ridley (1993), evolution; King (1993), species concepts; Lynch (1989), speciation; Brooks and McLennan (1991), speciation and phylogenies; Brooks (1950), Coulter (1994) and Ribbink (1994), speciation in ancient lakes. Limnology is addressed most thoroughly in the dedicated volumes, but some general limnology of tropical lakes is given by Talling and Lemoalle (1998). Species lists: Lake Tanganyika, Coulter (1991); Lake Baikal, Kozhova and Izmest'eva (1998); Lake Titicaca, Dejoux and Iltis (1992); and Lake Ohrid, Stankovics (1960). Brief summaries and faunal lists to the ordinal level for several lakes are given in Martens *et al.* (1994). Volumes dedicated to particular lakes that I have found useful exist for Lake Tanganyika, Coulter (1991); Lake Biwa, Horie (1994); Lake Ohrid, Stanovics (1960); Lake Titicaca, Dejoux and Iltis (1992); Lake Baikal, Kozhova and Izmest'eva (1998); and Caspian Sea, Kosarev and Yablonskaya (1994) and Dumont (1998).

REFERENCES

Berg, L.S. (1969). *Nomogenesis, or Evolution Determined by Laws.* London.

Brooks, J.L. (1950). Speciation in ancient lakes. *Q. Rev. Biol.* **25**, 30–60, 131–176.

Brooks, D.L. and McLennan, D.A. (1991). *Phylogeny, Ecology, and Behavior.* University of Chicago Press, Chicago, IL.

Coulter, G.W. (ed.) (1991). *Lake Tanganyika and its Life.* Oxford University Press, Oxford.

Coulter, G.W. (1994). Speciation and fluctuating environments, with reference to ancient East African lakes. *Arch. Hydrobiol. Beiheft. Ergebnisse Limnol.* **44**, 127–137.

Dejoux, C. and Iltis, A. (eds) (1992). *Lake Titicaca. A Synthesis of Limnological Knowledge.* Kluwer, Dordrecht.

Dumont, H.J. (1998). The Caspian Lake: history, biota, structure and function. *Limnol Oceanogr* **43**, 44–52.

Fryer, G. (1991). Biological implications of a suggested Lake Pleistocene desiccation of Lake Victoria. *Hydrobiologia* **354**, 177–182.

Fryer, G. and Iles, D.I. (1972). *The Cichlid Fishes of the African Great Lakes: Their Ecology and Evolution.* Oliver and Boyd, Edinburgh.

Gorthner, A. (1994). What is an ancient lake? *Arch. Hydrobiol. Beiheft. Ergebnisse Limnol.* **44**, 97–100.

Greenwood, P.H. (1984). African cichlids and evolutionary theories. In: *Evolution of Fish Species Flocks* (Ed. by A.A. Echelle and I. Kornfield), pp. 141–154. University of Maine at Orono Press, Orono, ME.

Horie, S. (ed.) (1984). Lake Biwa. In: *Monographiae Biologicae,* Vol. 54. Dr W. Junk, The Hague.

Johnson, T.C. and Odada, E.O. (eds) (1996). *The Limnology, Climatology and Paleoclimatology of the East African Lakes*. Gordon and Breach, Amsterdam.

Johnson, T.C., Scholtz, C.A., Talbot, M.R., Kelts, K., Ricketts, R.D., Ngobi, G., Beuning, K., Ssemmanda, I. and McGill, J.W. (1996). Late Pleistocene desiccation of Lake Victoria and rapid evolution of cichlid fishes. *Science* **273**, 1091–1093.

King, M. (1993). *Species Evolution: The Role of Chromosome Change*. Cambridge University Press, Cambridge.

Kosarev, A.N. and Yablonskaya, E.A. (1994). *The Caspian Sea*. SPB, The Hague.

Kottelat, M. (1997). European freshwater fishes. *Biologia* **52**, Suppl. 5.

Kozhov, M. (1963). Lake Baikal and its life. *Monographiae Biologicae*, Vol. 11. Dr W. Junk, The Hague.

Kozhova, O.M. and Izmest'eva, L.R. (1998). *Lake Baikal. Evolution and Biodiversity*. Backhuys, Leiden.

Lippitsch, E. (1990). Scale morphology and squamation patterns in cichlids (Teleostei, Perciformes): a comparative study. *J. Fish Biol.* **37**, 265–291.

Lynch, J.D. (1989). The gauge of speciation: on the frequencies of modes of speciation. In: *Speciation and its Consequences* (Ed. by D. Otte and J.A. Endler), pp. 527–556. Sinauer Associates, Sunderland, MA.

Martens, K., Goddeeris, B.R., and Coulter, G.W. (eds) (1994). Speciation in Ancient Lakes. *Arch. Hydrobiol. Beiheft. Ergebnisse Limnol.* **44.**

Mayr, E. (1984). Evolution of fish species flocks: a commentary. In: *Evolution of Fish Species Flocks* (Ed. by A.A. Echelle and I. Kornfield), pp. 3–11. University of Maine at Orono Press, Orono, ME.

Menz, A. (ed.) (1995). *The Fishery Potential and Productivity of the Pelagic Zone of Lake Malawi/Niassa*. National Resources Institute, Chatham, UK.

Patterson G. and Makin, J. (1998). *The State of Biodiversity in Lake Tanganyika – A Literature Review*. Natural Resources Institute, Chatham, UK.

Ribbink, A.J. (1994). Biodiversity and speciation of freshwater fishes, with particular reference to African cichlids. In: *Aquatic Ecology: Scale, Patterns and Process* (Ed. by P.S. Giller, A.G. Hildrew and D.G. Raffaelli), pp. 261–288. Blackwell Science, Oxford.

Ridley, M. (1993). *Evolution*. Blackwell Science, Oxford.

Scott, W.B. and Crossman, E.J. (1973). *Freshwater Fishes of Canada. Fisheries Research Board of Canada, Bulletin*, Vol. 184. Canadian Government Publishing Center, Ottawa.

Stankovics, S. (1960). The Balkan Lake Ohrid and its living world. *Monographiae Biologicae*, Vol. 9. Dr W. Junk, The Hague.

Talling, J.F. and Lemoalle, J. (1998). *Ecological Dynamics of Tropical Inland Waters*. Cambridge University Press, Cambridge.

Vrba, E.S. (1985). Environment and evolution: alternative causes of the temporal distribution of evolutionary events. *S. Afr. J. Sci.* **81**, 229–236.

Vrba, E.S. (1993). Turnover-pulses, the Red Queen, and related topics. *Am. J. Sci.* **293-A**, 418–452.

Wheeler, A. (1978). *Key to the Fishes of Northern Europe*. Frederick Warne, London.

Unanticipated Diversity: The Discovery and Biological Exploration of Africa's Ancient Lakes

G. FRYER

I. THE DISCOVERY OF AFRICA'S ANCIENT LAKES

So much is now known about the ancient lakes of Africa – Malawi, Tanganyika and Victoria – that it comes almost as a shock to realize that even the existence, and certainly the true whereabouts, of these enormous lakes was unknown to the outside world until after the middle of the nineteenth century. Indigenous peoples had lived around them for millennia but, having little contact with the outside world and no literature, the knowledge that they possessed remained unknown to outsiders. The story of their discovery is inextricably bound up with one of the greatest of geographical mysteries, the search for the source of the Nile: a quest that proved elusive to the Ancient Egyptians, to the later Greek and Roman rulers of Egypt, and to all who subsequently saw that great river as it flowed through the desert and wondered whence it came. The guardian of that secret was the great Sudd region of the Sudan.

The first clues came about by accident. About 2000 years ago a Greek, Diogenes by name, was exploring or trading in the Indian Ocean when his vessel was blown south to somewhere in the vicinity of what is now Dar-es-Salaam. From here he travelled inland until he reached, or heard about from Arabs who lived on the coast, two large lakes and what he described as a range of snow-covered mountains where he believed the Nile arose. This was fortunately recorded by a Syrian geographer, Marinus of Tyre, from whose records Ptolemy produced his famous map in the middle of the second century.

ADVANCES IN ECOLOGICAL RESEARCH VOL. 31
ISBN 0-12-013931-6

This map was to remain the main source of knowledge (or guesswork) about the Great Lakes of Africa for some seventeen centuries.

In 1848 Johann Rebman, a pioneer missionary, reported the existence of the snow-capped Kilimanjaro (and was ridiculed for his pains), but in 1849 Johann Joseph Krapf saw Mount Kenya, also snow-capped, and soon another missionary, J.J. Erdhardt, produced a map showing a large lake that he called Uniamezi. What is more, Arabs, who had already penetrated into the interior, where they traded in slaves and ivory, knew of two large lakes that they called Ujiji and Nyanza. They had indeed circumnavigated Tanganyika (Burton, 1860). Nyanza simply means "lake" or a large expanse of water. Another large lake further south – Maravi or Malawi – had been seen by the Portuguese explorer Caspar da Boccaro as early as 1616 and this appeared on a map by d'Anville in 1727, but of this no certain information existed until the Scottish missionary and explorer David Livingstone and his party trod its shores in 1859.

It was by penetration from the east coast that, like Diogenes many centuries earlier, Richard Burton and John Hanning Speke arrived at, and briefly explored, first Lake Tanganyika early in 1858, then (Speke alone) Lake Victoria which he reached at the end of July 1858, just over a year before Livingstone arrived at Lake Malawi, which he called by the Yao name Nyasa, which again simply means an expanse of water. Burton was disappointed to learn that the Ruzizi, at the north end of Lake Tanganyika, was an affluent river rather than an exit to the Nile, while Speke was convinced from the outset (on tenuous grounds) that Lake Victoria was the source of the Nile. This he was unable to prove until his next expedition when, in 1862, he saw its outlet at Jinja. Both of these findings were of great zoogeographical interest.

II. EARLY FAUNISTIC DISCOVERIES AND A PROBLEM THAT THEY RAISED

These same early explorers provided the first, very fragmentary, information on the faunas of these ancient lakes. The earliest zoological specimens to achieve scientific recognition were shells picked up at the lake shore by Speke during his visit to Lake Tanganyika in 1858 and described by Woodward (1859). The two gastropods concerned (there were also two bivalves) are now known as *Spekia zonata* and *Lavigeria* (= *Edgaria*) *nassa*. On his second expedition Speke collected molluscs from Lake Victoria that were described by Dohrn (1864). Almost at the same time, John Kirk, physician and botanist on Livingstone's expedition of 1858–1864, collected mollusc shells from Lake Malawi, described by Dohrn (1865), and the first fishes from any African Great Lake to receive scientific study. These, reported on by A. Günther (1864), included six cichlids, a number that gave no inkling of the fact that these were the first fruits of the world's richest lacustrine cichlid fauna.

The gastropods collected by Speke were only empty shells, yet they were the genesis of what proved to be a vital stimulus to the early biological exploration of African lakes. Woodward (1859) was struck by their morphological similarities to marine gastropods and remarked that *Spekia zonata* could be taken for a sea shell, while *Lavigeria* reminded him forcibly of a small marine whelk.

During the next decade or so, the explorations of Livingstone and Henry Morton Stanley made the geography of Lake Tanganyika better known, and by 1878 a group of missionaries established themselves at Ujiji. Among them was E.C. Hore, who collected molluscs and other animals, and whose name is commemorated in the beautiful gastropod *Tiphobia horei* (Figure 1). In 1879 Lake Tanganyika was visited by Joseph Thompson, who also visited Lake Malawi, collected molluscs in both lakes and, in 1881, produced a book, *To the Central African Lakes and Back*. Thompson's molluscs, and others collected by F.A. Simons, were studied by E.A. Smith (1877) who, like Woodward before him, drew attention to the marine-like appearance of some of the Tanganyika (but not Malawian) gastropods. This led Thompson to suggest in his book that, in the not very distant past, Lake Tanganyika had been more saline, and later (Thompson, 1882) to put forward the idea that the lake had been derived from a sea that formerly occupied the Congo Basin. This was done in a lecture of which the short published version is only a summary, but it was the first formulation of the suggestion that, when Tanganyika became isolated as a lake, it retained marine elements among its fauna, some of which persisted when, as a result of inflows, it gradually freshened. This suggestion is usually attributed, incorrectly, to later investigators.

III.   THE EXPEDITIONS AND THEORIES OF J.E.S. MOORE

It was the marine-like gastropod shells, later designated as "thalassoid"* by Bourguignat (1885), which proved to be endemic to Lake Tanganyika and to have no counterpart in any other African lake, that provided the initial, and then apparently ever increasing, evidence in support of a marine origin for at least part of the fauna of Lake Tanganyika. The idea received what seemed dramatic support when, in 1883, Böhm discovered a medusa in the lake. This was described as *Limnocnida tanganyicae* (Figure 2) by R.T. Günther (1893), who later (Günther, 1894) gave an account of its anatomy and elaborated the idea that Lake Tanganyika was formerly an arm of the Atlantic Ocean that

---

*According to Moore, who coined the term "halolimnic", often used as a synonym of "thalassoid", the terms mean different things. "Thalassoid" refers to animals like, but not related to, marine forms, whereas "halolimnic" refers to animals of marine origin (Moore, 1906). As will become apparent, most of the animals concerned are indeed thalassoid.

4   G. FRYER

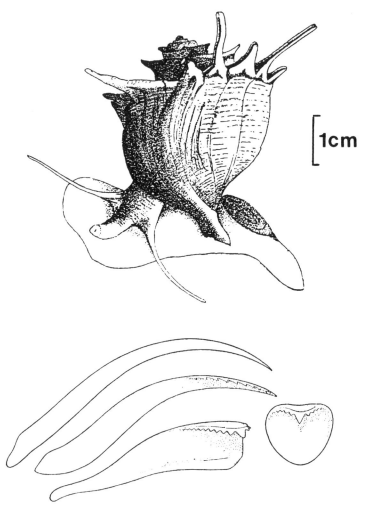

Fig. 1. The thalassoid Tanganyikan gastropod *Tiphobia horei* and some of the teeth of
its radula with which it collects its food. This it does from rich organic sediments over
sandy, silty and muddy bottoms, where it lives at depths of from about 5 m to more than
100 m. From Moore (1903).

extended over the Congo basin and that its waters were later diluted by
inflowing rivers.

The interest raised by these matters proved sufficient to induce the Royal
Society of London to send an expedition to Lake Tanganyika, conducted by
J.E.S. Moore, which left England in 1895, briefly visited Lake Malawi and
devoted several months to the study of Lake Tanganyika. Moore collected
many previously unknown animals and later studied the anatomy of some of

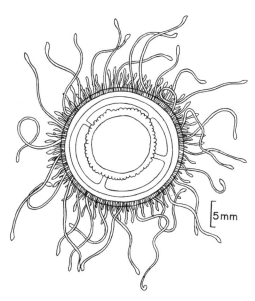

5mm

Fig. 2. The Tanganyikan jellyfish *Limnocnida tanganyicae*, the discovery of which in Lake Tanganyika in 1883 was, like the presence of thalassoid snails, taken as evidence of a marine origin for part of the fauna of the lake. Later, the genus was found in fresh water elsewhere in Africa, and in India, and is now acknowledged as belonging to a group of coelenterates long established in fresh water.

the gastropod molluscs. Of these, some of the prosobranchs* seemed to him to have marine affinities but, as he had difficulties in relating them to modern forms, he concluded that they displayed generalized and archaic attributes, and were deemed to be relics of an older marine fauna. Indeed, some of them seemed so much to resemble fossils of Jurassic age that he concluded (Moore, 1898a–c) that Lake Tanganyika was connected to the sea in Jurassic times. Interest remained so high that a second expedition was despatched in 1899, again conducted by Moore, this time accompanied by Malcolm Fergusson as geologist and surveyor. This concentrated on Tanganyika but visited all three Great Lakes. The resulting faunal comparisons further emphasized the unique nature of Lake Tanganyika. Moore summarized the knowledge accumulated up to that time, and the interpretation he placed upon it, in a book, *The Tanganyika Problem* (1903), in which he pressed the claim that part of the fauna of Tanganyika was derived directly from the sea.

---

*In their study of the phylogeny of gastropods Ponder and Lindberg (1997) reject the use of Prosobranchia, and other familiar higher taxa, on the grounds that they are paraphyletic. Whether this is accepted by malacologists remains to be seen: for present purposes it is convenient and informative to use the familiar name prosobranch.

To Moore, the evidence appeared to be strong. The gastropods in particular not only had a distinctly marine appearance, but the shells of some of them were remarkably like those of marine Jurassic forms, as his illustrations show. Thus, the shell of *Paramelania damoni* is strikingly similar to that of the Jurassic *Purpurina bellona* (Figure 3), that of *Lavigeria nassa* closely resembles that of *Purpurina inflata, Spekia zonata* resembles *Neridomus* (Figure 4), the very different *Chytra kirki* is similar to *Onustus*, and *Bathanalia howesi*, different again, resembles species of *Amberlya*. When one bears in mind that similar gastropods are found in no other African lake, Moore's conclusion seemed plausible. At the time, his belief that the medusa *Limnocnida* was a marine type (although it had by then been found also in the River Niger) also seemed justified and he believed, mistakenly, that the crabs, prawns (better represented in Lake Tanganyika than in the other African ancient lakes) and sponges had marine affinities. What is more, the gymnolaematous polyzoan *Arachnoidea ray-lancesteri*, that he discovered himself, displays marked similarities to the marine *Arachnidium*.

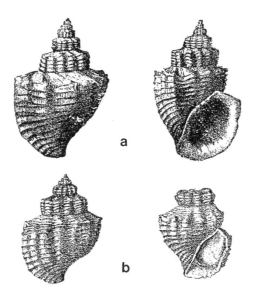

Fig. 3. An example of the convergent similarity of the shells of a living Tanganyikan gastropod mollusc and a marine Jurassic fossil. Moore, who took this as indicating affinity, regarded it as evidence in favour of a theory that the lake was connected to the sea in Jurassic times. (a) The extant *Paramelania damoni*, which achieves a shell length of about 4 cm; (b) the strikingly similar Jurassic *Purpurina bellona*. From Moore (1903).

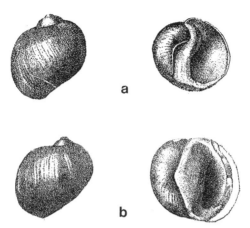

Fig. 4. Shells of (a) *Spekia zonata*, a species of historical interest as it was one of the four molluscs whose shells were brought from Lake Tanganyika by Speke in 1858 and were the first animals from any ancient African lake to be scientifically described; and (b) a convergently similar Jurassic fossil of the genus *Neridomus*. *Spekia zonata* achieves a shell length of about 15 mm. From Moore (1903).

IV.   THE REACTION TO MOORE'S HYPOTHESIS

Moore's ideas did not long remain unchallenged. Indeed, events leading to this challenge had begun before the ideas were promulgated. In 1881 White claimed that the genus *Paramelania*, erected by Smith (1881) for one of the newly discovered gastropods of Tanganyika, was identical with *Pyrgulifera*, a fossil form recorded from a brackish-water formation of Cretaceous age in North America, and pointedly referred to Smith's species as *Pyrgulifera damoni*. Smith (1882) promptly repudiated White's suggestion and claimed, with some justification, that it was unwise on both conchological and geographical grounds to synonymize the two genera. Shortly thereafter, Tausch (1884) reported on some fossil gastropods from Cretaceous deposits in Hungary. These included forms that he attributed to *Pyrgulifera* and some which he regarded as conspecific with the North American *P. humerosa*. Like White, he allocated Smith's Tanganyikan *Paramelania damoni* to *Pyrgulifera*.

After his first expedition, Moore put forward in some detail his hypothesis that Lake Tanganyika represents the remains of a Jurassic sea (Moore, 1898c), and repudiated the suggestions of White and Tausch (and of Oppenheim, who had also studied this genus) that certain Tanganyikan gastropods should be assigned to *Pyrgulifera*. Details need not be given here. Suffice it to say that Moore pointed out that as their evidence involved only one type of shell, the similarities could be explained as mere convergence. His own hypothesis involved several examples and included shells of several different forms, which seemed a telling point in his favour.

Following the publication of Moore's book, Smith (1904) chose the topic of the molluscs of Tanganyika for his presidential address to the Malacological Society and proceeded to attack Moore's conclusions. His criticisms were based largely on detailed comparisons that need not be presented here and that are not always convincing. However, he made the point that the thick-shelled thalassoid nature of various Tanganyika species does not in itself indicate relationship with marine forms, living or extinct. He noted that freshwater gastropods elsewhere may display this attribute, citing *Melania brevis* of Cuban rivers and several other examples. Hudleston (1904) took the same line and rejected Moore's conclusions on zoological, palaeontological and geological grounds. However, all of these criticisms and comparisons were based solely on conchological similarities, whereas Moore, with justification, gave due weight to the anatomy of the animal, i.e. the "soft parts". This led him to claim that the Tanganyika species are ancient.

The dénouement came 2 years later at a meeting of the British Association at which Moore (1906) and Paul Pelseneer (1906) spoke on the gastropods of Lake Tanganyika. Here, Pelseneer settled the matter by a magisterial display of knowledge of the Mollusca. He pointed out that external resemblance of shells is often taxonomically misleading and emphasized that internal organization is the only thing that can throw light on the problem. In comparing the extant Tanganyika species with their alleged Jurassic counterparts, Moore was, of course, only able to use conchological features, but the resemblance between not just one but several pairs of Jurassic and Tanganyikan forms, some of them very different, particularly impressed him. Furthermore, although he was indeed the pioneer investigator of the internal organization of Tanganyikan gastropods, his claim that his anatomical studies revealed affinities between these animals and diverse marine forms was repudiated by Pelseneer, who also noted that certain non-thalassoid freshwater genera, such as *Ampullaria* and *Paludina*, exhibit much more archaic characters than do thalassoid Tanganyikan forms. Pelseneer's conclusion was that all of the thalassoid forms belong to the Melaniidae, an essentially freshwater group, "or to very closely related types". Current practice is to include them in two subfamilies of the Thiaridae (Brown, 1980). This group displays much diversity in shell form but great uniformity in anatomical organization. Thus, although Moore both used conchological comparisons and attempted to deduce the archaic nature of the thalassoid Tanganyikan gastropods from the standpoint of soft-part anatomy, on neither point did he succeed in proving his argument.

It is interesting to see how early emphasis on shell characters contrasts strongly with more recent approaches. Thus, in their analysis of gastropod phylogeny, Ponder and Lindberg (1997) ignore all shell characters except for basic configuration, type of protoconch and structure.

Other support for Moore's theory of a marine element in Lake Tanganyika was also gradually eroded. A linchpin of the argument had already been knocked away when de Guerne (1893, 1894) reported the medusa *Limnocnida tanganyicae* from the River Niger, and in 1903 a medusa, now known to be of a different species, was found in Lake Victoria. The genus is now known from elsewhere in Africa and India. Furthermore, the third Tanganyika expedition, conducted by W.A. Cunnington, which left England in 1904, produced no evidence to support the idea. One objective of this expedition was to see whether there was any supporting botanical evidence, but none was forthcoming.

As for an incursion of the Atlantic Ocean into the Congo basin in Jurassic times, subsequent discoveries have proved Moore, and Thompson and R.T. Günther before him, to be in error, but it should be remembered how little was known of the geological history of the area when the theory was proposed. Even at that time, however, several zoogeographers had suggested that at least land bridges existed between Africa and South America – A. Günther (1880), citing freshwater fishes as an example, was one of them – and Neumayr's book (1889; new edition 1890) included a map showing Africa and South America forming the bulk of a single "Brasilianisch–Æthiopischer Kontinent" and with no trace of any South Atlantic Ocean. The revelations of plate tectonics have now not only disproved Moore's belief but also shown that in Jurassic times, far from invading the Congo basin, the South Atlantic Ocean did not even exist. At that time, Africa and South America were still united on a broad front and it was only towards the end of this period that the physical separation of these continents began.

V.   THE UNRESOLVED ENIGMA OF THE THALASSOID GASTROPODS AND OTHER FAUNAL ELEMENTS OF LAKE TANGANYIKA

Although the alleged marine origin of some of the gastropods of Tanganyika is disproved, these animals still present problems. That they represent the end points of speciation and adaptive radiation of a group of truly freshwater snails is certain. But why are such thalassoid snails present, and in such diversity, in Tanganyika but not in any other ancient (but admittedly younger) African lake? Of the suggestions made, none is entirely convincing. The quasi-oceanic conditions that prevail in Lake Tanganyika and the role of water chemistry can both be rejected as explanations. The bathymetric and shoreline characteristics of Lake Tanganyika and Malawi (and the Siberian Lake Baikal) are remarkably similar, but there is no suggestion of even the incipient development of thalassoid gastropods in Malawi (or in the more ancient Baikal). Beauchamp (1946) suggested that the unusual chemical composition of the water in Tanganyika, including chloride:sulphate and magnesium:calcium ratios that are similar to those in sea water, may have induced thalassoid characteristics, perhaps at times when salinities were higher than at present. However, the present chemical composition of the water is not remarkable among freshwaters as a whole and is well within the range tolerated by freshwater animals in general. Nor is it particularly saline. It is, for example, less saline than that of Lake Edward and has lower concentrations of all major ions (see Talling and Talling, 1965). Although at times it has probably been more saline, its salinity could not have exceeded that tolerated by the diverse freshwater organisms that have long frequented the lake. Nor is there any evidence that high salinities or particular ionic ratios have induced thalassoid shell morphology. The thick shells of some Tanganyikan gastropods are indeed remarkable, as they have achieved their thickness in water that is by no means rich in calcium.

More recently, West *et al.* (1991) and West and Cohen (1994) have suggested that the thick shells of those Tanganyikan snails that coexist with predatory crabs – of which *Platypelphusa armata* in particular has large crushing chelipeds – reflect predator:prey coevolution. Thick shells undoubtedly afford protection against such predation, but could have been acquired for entirely different reasons and the crabs could simply have become adapted for dealing with thick-shelled prey, as have several fishes. Incidentally, the Tanganyikan thalassoid *Tiphobia horei* has a thin shell. Gastropods are also subject to predation by cichlid fishes, some of which have pharyngeal bones specially adapted to crushing mollusc shells and can crush thick, tightly coiled shells, such as those of *Melanoides tuberculata*. Yet, there has been no tendency to develop thick shells as a defensive tactic against the abundant mollusc-crushing cichlids in Lakes Malawi and Victoria, even though there is about twice as much calcium in the water in Malawi as there is in Tanganyika. And of course, a thick shell is no protection against those cichlids that have learned the trick of levering snails out of their shells.

Notwithstanding the demonstration that several key groups of the Tanganyikan fauna are of freshwater and not marine origin, there are in that lake several enigmatic animals that have tantalizing marine affinities. Of these, the gymnolaematous polyzoan *Arachnoidea ray-lancesteri* belongs to an essentially marine group and is one of the very few encrusting gymnolaematous polyzoans known in fresh water. Another gymnolaematous species, of *Victorella,* which is a brackish and freshwater genus, also occurs in Tanganyika. Several harpacticoid copepods, including a flock of at least six endemic species of the diosaccid genus *Schizopera,* also display marine, or at least brackish-water, affinities. The cletodid *Nannopus perplexus* does likewise. Moreover, since 1965 no fewer than three species of parasitic isopods of the genus *Lironeca* have been found in Lake Tanganyika (Fryer, 1965, 1968; Lincoln, 1976). Several other species of *Lironeca* are marine. Thus, mysteries remain.

VI. UNANTICIPATED DIVERSITY: PERSISTENT UNDERESTIMATES OF SPECIES RICHNESS

To provide inventories of species of these large lakes takes much time. However, species richness, especially of cichlid fishes, has been consistently underestimated. In spite of his personal experience, Moore (1903, p. 124) was overconfident when he remarked that "there can be no doubt that we have pretty well exhausted the freshwater fauna of Nyasa, at any rate so far as a knowledge of the different types composing it is concerned". He then listed its cichlid fauna as consisting of 26 species! Following three expeditions, Cunnington (1920) gave its total of cichlids as 40 species. Work by Regan (1921) on collections made by Wood brought this to 84, and by Trewavas (1931, 1935) on the collections of Christy, to about 170. More were discovered in the 1950s, but not until trawling of the deeper water, sampling in open water and a lake-wide exploration of rocky shores were undertaken did its real wealth begin to be appreciated. The rocky shore survey alone (Ribbink *et al.*, 1983) revealed 196 species of the Mbuna group compared with 24 in the review of Trewavas (1935), and this total is incomplete. Work on offshore species at the south end of the lake has revealed yet further species richness (Turner, 1994, 1996). Many species remain to be described. Many sibling species, often undescribed, are also known, and it is probably not too rash to predict that when collections are fully studied and the sibling species are adequately analysed, the total may be nearer to 1000 than to 500 species (see Snoeks, Chapter 2, this volume).

What really shows how easy it is to underestimate species richness is that, although Greenwood resided beside Lake Victoria for several years, did sterling work on the revision of the Victorian cichlid fishes and described 50 new species (see collected papers, 1981), the existence of whole guilds remained unsuspected until the Dutch team led by Frans Witte sampled over a wide area and in a range of habitats. This revealed zooplankton feeders, benthic species, rock-frequenters, and others. Particularly dramatic is the recent discovery of large numbers of rock-frequenting forms – of which there may be as many as 200 species – whose existence was previously unsuspected (Seehausen, 1996). Happily, these live in habitats that provide refuges from the introduced Nile perch *Lates "niloticus"* that has wreaked much destruction among other cichlids of the lake. In Lake Victoria too, sibling species boost the totals. Many remain undescribed and it is tragic that, following the introduction of the Nile perch, some of these are already extinct. Extinction by this predator may also have been the fate of species that remained to be discovered. One can only guess at the total, but 600 or 700 species is probably not an exaggeration.

VII.   MODERN INVESTIGATIONS AND THEIR DEBT TO THE PAST

From an early stage in the biological exploration of these ancient lakes it became apparent that a large proportion of their animals is endemic to a single lake and, after some debate, it became abundantly clear that these endemic faunas are, in the main, the result of extensive intralacustrine speciation, accompanied in many cases by remarkable adaptive radiation in morphology, ecology, behaviour and physiology. This inevitably made them of particular interest to students of evolution, to whom they present many problems in the diverse disciplines involved in modern evolutionary studies.

As the first sentence of this article emphasizes, our knowledge of these faunas is now great. At first it was acquired slowly and with much labour; later, more rapidly and often less arduously. The way in which problems were approached also evolved as local knowledge accumulated. The aim of the early explorers and naturalists was to ascertain what animals inhabited these previously unknown lakes, to enquire where they originated and to discover to what other animals they were related. These objectives are still pursued today, although this tends to be masked by such terminology as molecular genetics, phylogenetic reconstruction, and the like. As more information became available it became possible to put forward hypotheses. Although some of these, such as the suggestions that the fauna of Lake Tanganyika included an element derived from a former incursion of the sea, and that some of its gastropods have been derived from marine Jurassic ancestors, had later to be abandoned, they served as valuable stimuli to their challengers. Today, with greater understanding of Earth history, from the consequences of plate tectonics to detailed, but still incomplete, information on the geological histories and morphometry of individual lakes obtained by seismic surveys, echo-sounding, the analysis of information derived from sediment cores and other techniques, biologists often have a more secure framework within which to consider their results. This does not, however, always facilitate the elucidation of geological events. Thus, a suggested Late Pleistocene desiccation of Lake Victoria, deduced from geophysical data of various kinds (Johnson *et al.*, 1996), appears to be incompatible with the evidence provided by the fauna of the lake (Fryer, 1997).

As faunistic studies, supported by taxonomists in European museums, gradually revealed and named the organisms that inhabited these lakes, it became possible to conduct ecological studies. These showed how the faunas of each lake, the fishes especially, belonged to often well-defined communities and also revealed remarkable cases of convergent evolution among the cichlid fishes of both rocky and sandy shores of Lakes Malawi and Tanganyika. More recently, remarkably similar rocky shore communities have been discovered in Lake Victoria (Seehausen, 1996). To the field worker with some idea of the history of the lakes involved, the convergent nature of these similarities, remarkable as it is, sometimes uncannily so, was self-evident, and it is gratifying that any scepticism from those surveying the picture from afar has now been refuted by those using molecular approaches (e.g. Meyer *et al.*, 1990; Kocher *et al.*, 1993).

Molecular techniques are but some of the new methodologies now being applied to the study of these faunas, and have been very informative.

Sometimes, however, they "discover" what was already known. Thus, in confirming the monophyly of the Lake Victoria haplochromine cichlid flock, Meyer (1993), an expert on the molecular approach, indicated that Fryer and Iles (1972) had concluded otherwise, when in fact they emphatically opted for monophyly.

As examples of advances made by molecular approaches, three suffice. Nishida (1991) showed that there are at least seven old ancestral lineages of cichlid fishes in Lake Tanganyika and suggested that these existed before the lake formed. Sturmbauer and Meyer (1992) showed that the *Tropheus* lineage of Tanganyikan cichlids contains almost twice as much variation as the entire haplochromine flock of Lake Malawi and more than six times as much as the Lake Victoria flock, that it may be about 1.25 My old, and that individual species are in some cases between 200 000 and 750 000 y old. This raises fascinating questions, as the various species of *Tropheus* are very similar in morphology while their Malawian equivalents in the genus *Pseudotropheus*, which are younger, have sometimes undergone speciation in remarkably short periods. Molecular methods also allowed Verheyen *et al.* (1996) to correlate large-scale fluctuations in the level of Lake Tanganyika in Pleistocene times, which at times of low stand gave rise to three separate lakes within the basin, with the evolution of the eretmodine cichlids. This interesting and imaginative combination of disciplines, however, throws up other problems, as the mitochondrially defined clades do not agree with current taxonomy (see Verheyen and Rüber, Chapter 26, this volume).

Valuable as they are, molecular approaches in no way usurp the importance of studies on living animals in nature. Many such studies, particularly on fishes and their behaviour, have now been, and continue to be, conducted in these ancient lakes. Especially in the clear waters of Malawi and Tanganyika, underwater observations of great precision can be made by use of SCUBA (self-contained underwater breathing apparatus). An example of the high resolution achieved by some of these studies is the detailed way in which the complex social structure of the Tanganyikan cichlid *Aulonocranus dewindti* has been elucidated by Rossiter (1994). Likewise, a recent volume on the fish communities in that lake (Kawanabe *et al.*, 1997) gives marvellous insights into the complicated lives led by many of its fishes. Early explorers would have been amazed to know what was going on so close at hand.

Modern approaches are far removed from those of Kirk, Thompson, Simons, Hore, Böhm, Moore, Giraud, Guillemeé, Ancey, Foa, Fülleborn, Cunnington and other early explorers and naturalists, and even further removed from the seemingly trivial act of collecting empty shells by Speke in 1858, yet such early endeavours played a vital role in the investigation of these ancient lakes, as did the hydrographic surveys of Rhoades and Phillips (Malawi) and Stappers (Tanganyika). These early investigators were indeed pioneers and deserve to be remembered as such.

REFERENCES

Beauchamp, R.S.A. (1946). Lake Tanganyika. *Nature* **157**, 183–184.

Bourguignat, J.R. (1885). *Notice prodromique sur les mollusques terrestres et fluviatiles recueillis par M. Victor Giraud dans la règion mèridionale du Lac Tanganika*. V. Tremblay, Paris. pp. 1–100.

Brown, D.S. (1980). *Freshwater Snails of Africa and their Medical Importance*. Taylor and Francis, London.

Burton, R.F. (1860). *The Lake Regions of Central Africa*. Longman, Green, Longman and Roberts, London.

Cunnington, W.A. (1920). The fauna of the African lakes: a study in comparative limnology with special reference to Tanganyika. *Proc. Zool. Soc. Lond.* **1920**, 501–622.

Dohrn, H. (1864). List of shells collected by Capt. Speke during his second journey through Central Africa. *Proc. Zool. Soc. Lond.* **1864**, 116–118.

Dohrn, H. (1865). List of land and freshwater shells of the Zambezi and Lake Nyasa, eastern Tropical Africa, collected by John Kirk. *Proc. Zool. Soc. Lond.* **1865**, 231–234.

Fryer, G. (1965). A new isopod of the genus *Lironeca*, parasitic on fishes of Lake Tanganyika. *Rev. Zool. Bot. Afr.* **71**, 376–384.

Fryer, G. (1968). A new parasitic isopod of the family Cymothoidae from clupeid fishes of Lake Tanganyika: a further Lake Tanganyika enigma. *J. Zool.* **156**, 35–43.

Fryer, G. (1997). Biological implications of a suggested Late Pleistocene desiccation of Lake Victoria. *Hydrobiologia* **354**, 177–182.

Fryer, G. and Iles, T.D. (1972). *The Cichlid Fishes of the Great Lakes of Africa; Their Biology and Evolution*. Oliver and Boyd, Edinburgh.

Greenwood, P.H. (1981). *The Haplochromine Fishes of the East African Lakes. Collected Papers*. Kraus International, Munich.

Guerne, J. de (1893). À propos d'une méduse observée par le Dr. Tautain dans le Niger á Bamakou (Soudan Francais). *Bull. Soc. Zool. Francaise* **1**, 225.

Guerne, J. de (1894). On a medusa observed by Dr. Tautain in the River Niger at Bamakou (French Soudan). *Ann. Mag. Nat. Hist.* **14**, 29–34.

Günther, A. (1864). Report on a collection of reptiles and fishes made by Dr. Kirk in the Zambezi and Nyassa regions. *Proc. Zool. Soc. Lond.* **1864**, 303–314.

Günther, A. (1880). *An Introduction to the Study of Fishes*. A. and C. Black, Edinburgh.

Günther, R.T. (1893). Preliminary account of the freshwater medusa of Lake Tanganyika. *Ann. Mag. Nat. Hist.* **11**, 269–275.

Günther, R.T. (1894). A further contribution to the anatomy of *Limnocnida tanganyicae*. *Q. J. Microsc. Sci.* **36**, 271–293.

Hudleston, W.H. (1904). On the origin of the marine (halolimnic) fauna of Lake Tanganyika. *Geol. Mag. n.s.* **1**, 337–382.

Johnson, T.C., Scholtz, C.A., Talbot, M.R., Kelts, K., Ricketts, R.D., Nagobi, G., Beuning, K., Ssemmanda, I. and McGill, J.W. (1996). Late Pleistocene desiccation of Lake Victoria and rapid evolution of cichlid fishes. *Science* **273**, 1091–1093.

Kawanabe, H., Hori, M. and Nagoshi, M. (eds) (1997). *Fish Communities in Lake Tanganyika*. Kyoto University Press, Kyoto.

Kocher, T.D., Conroy, J.A., McKaye, K.R. and Stauffer, J.R. (1993). Similar morphologies of cichlid fishes in Lake Tanganyika and Malawi are due to convergence. *Mol. Phylogenet. Evol.* **2**, 158–165.

Lincoln, R.J. (1976). A new species of *Lironeca* (Isopoda; Cymothoidae) parasitic on cichlid fishes in Lake Tanganyika. *Bull. Br. Mus. (Nat. Hist.) Zool.* **21**, 329–338.

Meyer, A. (1993). Phylogenetic relationships and evolutionary processes in East African cichlid fishes. *Trends Ecol. Evol.* **8**, 279–284.

Meyer, A., Kocher, T.D., Basasibwaki, P. and Wilson, A.C. (1990). Monophyletic origin of Lake Victoria cichlids suggested by mitochondrial DNA sequences. *Nature* **347**, 550–553.

Moore, J.E.S. (1898a). The marine fauna in Lake Tanganyika and the advisability of further exploration in the great African Lakes. *Nature* **58**, 404–408.

Moore, J.E.S. (1898b). On the zoological evidence for the connection of Lake Tanganyika with the sea. *Proc. R. Soc. Lond.* **62**, 451–458.

Moore, J.E.S. (1898c). On the hypothesis that Lake Tanganyika represents an old Jurassic sea. *Q. J. Microsc. Sci.* **41**, 303–321.

Moore, J.E.S. (1903). *The Tanganyika Problem. An Account of the Researches Undertaken Concerning the Existence of Marine Animals in Central Africa.* Hurst and Blackett, London.

Moore, J.E.S. (1906). Halolimnic faunas and the Tanganyika problem. *Rep. Br. Assoc. Adv. Sci.* **1906**, 601–602.

Neumayr, M. (1890). *Erdgeschichte* (New edn). Bibliographischen Institüt, Leipzig.

Nishida, M. (1991). Lake Tanganyika as an evolutionary reservoir of old lineages of East African cichlid fishes: inferences from allozyme data. *Experientia* **47**, 974–979.

Pelseneer, P. (1906). Halolimnic faunas and the Tanganyika problem. *Rep. Br. Assoc. Adv. Sci.* **1906**, 602.

Ponder, W.S. and Lindberg, D.R. (1997). Towards a phylogeny of gastropod molluscs: an analysis using morphological characters. *Zool. J. Linn. Soc. Lond.* **119**, 83–265.

Regan, T.C. (1921). The cichlid fishes of Lake Nyassa. *Proc. Zool. Soc. Lond.* **1921**, 675–727.

Ribbink, A.J., Marsh, B.A., Marsh, A.C., Ribbink, A.C. and Sharp, B.J. (1983). A preliminary survey of the cichlid fishes of rocky habitats in Lake Malawi. *S. Afr. J. Zool.* **18**, 149–310.

Rossiter, A. (1994). Territory, mating success, and the individual male in a lekking cichlid fish. In: *Animal Societies. Individuals, Interactions and Organisation* (Ed. by P.J. Jarman and A. Rossiter), pp. 43–55. Kyoto University Press, Kyoto.

Seehausen, O. (1996). *Lake Victoria Rock Cichlids–Taxonomy, Ecology and Distribution.* Verduijn Cichlids, Zevenhiuzen, The Netherlands.

Smith, E.A. (1877). On the shells of Lake Nyassa and on a few marine species from Mozambique. *Proc. Zool. Soc. Lond.* **1877**, 712–722.

Smith, E.A. (1881). On a collection of shells from Lakes Tanganyika and Nyassa. *Proc. Zool. Soc. Lond.* **1881**, 276–300.

Smith, E.A. (1882). Tanganyika shells. *Nature* **25**, 218.

Smith, E.G. (1904). Some remarks on the Mollusca of Lake Tanganyika. *Proc. Malacol. Soc.* **6**, 77–104.

Sturmbauer, C. and Meyer, A. (1992). Genetic divergence, speciation and morphological stasis in a lineage of African cichlid fishes. *Nature* **358**, 578–581.

Talling, J.F. and Talling, I.B. (1965). The chemical composition of African lake waters. *Int. Rev. Ges. Hydrobiol. Hydrogr.* **50**, 421–463.

Tausch, L. (1884). Über einige Conchylien aus dem Tanganyika See und desen fossile Verwandte. *Sitzungsberichte Akad. Wissenschaften, Wien, Mathematische-naturwissenschaftliche Classe* **90**, 56–70.

Thompson, J. (1882). On the geographical evolution of the Tanganyika Basin. *Rep. Br. Assoc. Adv. Sci.* **1882**, 622–623.

Trewavas, E. (1931). Revision of the cichlid fishes of the genus *Lethrinops* Regan. *Ann. Mag. Nat. Hist.* **7**, 133–152.

Trewavas, E. (1935). A synopsis of the cichlid fishes of Lake Nyasa. *Ann. Mag. Nat. Hist.* **16**, 65–118.

Turner, G.F. (1994). Fishing and conservation of the endemic fishes of Lake Malawi. *Archiv Hydrobiol. Beiheft. Ergebnisse Limnol.* **44**, 481–494.

Turner, G.F. (1996). *Offshore Fishes of Lake Malawi.* Cichlid Press, Lauenau.

Verheyen, E., Rüber, L., Snoeks, J. and Meyer, A. (1996). Mitochondrial phylogeography of rock-dwelling fishes reveals evolutionary influence of historical lake level fluctuations of Lake Tanganyika, Africa. *Phil. Trans. R. Soc. Lond. B* **351**, 799–805.

West, K. and Cohen, A.S. (1994). Predator-prey co-evolution as a model for the unusual morphologies of the crabs and gastropods of Lake Tanganyika. *Arch. Hydrobiol. Beiheft. Ergebnisse Limnol.* **44**, 267–283.

West, K., Cohen, A.S. and Baron, M. (1991). Morphology and behaviour of crabs and gastropods from Lake Tanganyika, Africa: implications for lacustrine predator-prey co-evolution. *Evolution* **45**, 589–607.

White, C.A. (1881). Tanganyikan shells. *Nature* **25**, 101–102.

Woodward, S.P. (1859). On some new freshwater shells from Central Africa. *Proc. Zool. Soc. Lond.* **1859**, 348–350.

How Well Known is the Ichthyodiversity of the Large East African Lakes?

J. SNOEKS

I. SUMMARY

A review, based on personal experience of problems in systematic research on the fishes of the East African lakes, is given. A characteristic feature of most of these lakes is the high number of endemic cichlid species that they contain. It is estimated that for the three biggest lakes, Victoria, Tanganyika and Malawi/Nyasa, about 1000 species or more are still awaiting scientific description. General and lake-specific taxonomic and phylogenetic problems are discussed. The need for better collaboration on various levels between morphologists and molecular biologists is stressed. Some extrapolations to and suggestions for the future are made.

II. INTRODUCTION

Research on the biodiversity of the large African lakes has been steadily increasing over the last few decades, mainly as a result of a growing scientific interest in the unique biology of the endemic cichlids and an increased awareness of the importance of conservation and sustainable fisheries

ADVANCES IN ECOLOGICAL RESEARCH VOL. 31
ISBN 0-12-013931-6

management in these lakes. Yet, despite some improvement, the basic systematic knowledge of this fish fauna remains rather poor and, consequently, many research programmes experience considerable difficulties because of taxonomic problems.

In this chapter the author will consider some basic taxonomic issues from a rather personal, but not subjective, point of view, and in a perhaps slightly provocative way. This review is eclectic rather than comprehensive, and most issues discussed are highlighted only by reference to examples with which the author is personally familiar.

III.   EAST AFRICAN LAKES AND SPECIES RICHNESS

The numbers of native fish species recorded from the East African lakes are listed in Table 1. Many other comparative tables and figures have been published (e.g. Greenwood, 1981a, 1984, 1991; Coulter et al., 1986; Lowe-McConnell, 1987; Snoeks, 1994; Snoeks et al., 1994; Worthington and Lowe-McConnell, 1994; Pitcher and Hart, 1995), some of which contain species numbers and percentages differing considerably from those given in Table 1. While some data may represent simple typing errors and others may be

Table 1
Indigenous species richness in East African lakes

	Cichlid species		Non-cichlid species		
	n	%	n	%	Total species
Turkana	8	50	36	17	44
Albert	11	36	37	5	48
Edward	60	92	21	5	81
Kivu	16	94	7	0	23
Victoria	500	99	45	16	545
Tanganyika	250	98	75	59	325
Malawi/Nyasa	800	99	45	29	845

Cichlid species numbers of Lakes Edward, Victoria, Tanganyika and Malawi/Nyasa are estimates. Only species occurring naturally in each lake, but not the whole basin, are taken into account; introduced species are excluded. Percentages are of endemism in cichlids and non-cichlids, respectively. Lake Edward figures include Lake George species. Based on Daget et al. (1984, 1986), Coulter et al. (1986), Greenwood (1991), De Vos and Snoeks (1994), Snoeks (1994, 1997), van Oijen (1995), De Vos (1996 and pers. comm), Tweddle (1996 and pers. comm), Seehausen (1996), Seegers (1996) and Snoeks et al. (1997).

considered out of date, the conflict in numbers may also partly reflect the lack of clear annotations as to what "species" have been included, i.e. described versus undescribed, and whether the numbers and percentages refer only to the lakes proper or to the whole drainage basins. Another factor affecting differences may be the inclusion of introduced species.

The fish faunas of the East African lakes are characterized by high numbers of species of endemic cichlids. The total number of described endemic cichlids from these lakes is about one-third of the total of described African freshwater and brackish water fishes of all other families combined, a ratio that may well increase to about one-half, if estimates (Table 1) prove to be accurate.

Among these lakes, Turkana and Albert (Figure 1) are exceptional in having few cichlid species and a low degree of endemicity. Explanations of these features have invoked both abiotic and biotic factors. For example, Greenwood (1991, p. 88) ascribed the low species numbers to "... these water bodies having recently had, or still having, direct connections with a major river system (the Nile)". Correlated with the low cichlid species number, Greenwood also found low levels of adaptive radiation and morphological diversity relative to that seen in the other East African lakes.

Another biotic feature that characterizes the fish fauna of Lakes Albert and Turkana has received less attention, i.e. the ancestor that gave rise to the current species may have been different from the ancestor of the Victoria–Edward–Kivu superflock. The latter lakes only recently came into contact with the Nile system, whereas the former two have had long-term and major contact with the Nile basin (Greenwood, 1979, 1991).

Lakes Victoria, Edward and Kivu belong to the East Coast geographical province (Snoeks *et al.*, 1997) and the ancestor to this flock was probably a riverine *Haplochromis* (*Astatotilapia*)-like species. Lakes Turkana and Albert have at present a Nilotic–Zairean haplochromine fauna (Greenwood, 1979), the common ancestor of which probably was a riverine *Haplochromis* (*Thoracochromis*)-like species. The occurrence of two or three *Haplochromis* (*Thoracochromis*) species in the Victoria–Edward–Kivu system is regarded as a result of recent dispersal (Greenwood, 1979). Possibly, then, the difference in speciation products is simply because the *Haplochromis* (*Astatotilapia*)-like ancestor was more predisposed to life in the ancient lakes than was the *Haplochromis* (*Thoracochromis*)-like species. However, given that the only derived characters found in *Haplochromis* (*Thoracochromis*) are scale pattern features, one might question whether these specializations, if fully present in the ancestor, would have been sufficient to prevent it from producing high species numbers in Lakes Albert and Turkana. It seems likely that factors other than the intrinsic morphological features (bauplan) of the ancestral species have played at least a partial role in the evolution of the low diversity of cichlid species in Lakes Albert and Turkana. Perhaps the lineage was genotypically unable fully to respond evolutionarily to the ecological opportunities present (Greenwood, 1994a), or perhaps the opportunities present in these two lakes

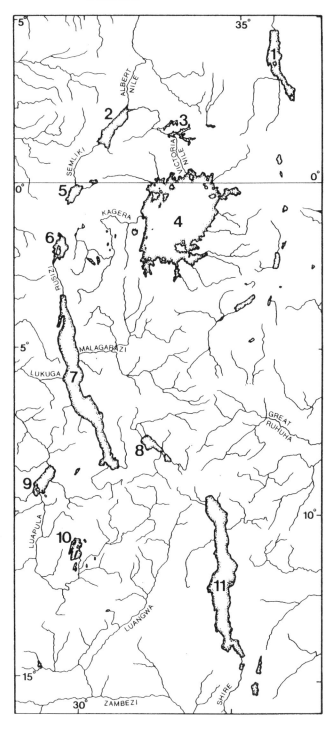

(presence of necessary triggering events) were different from those in other African lakes; or perhaps the answer lies in a mixture of both explanations. Detailed studies are needed to resolve this question. The role of predators in promoting or retarding speciation is still a point of debate (e.g. Fryer, 1965; Lowe-McConnell, 1969, 1987).

In comparison to the cichlids, the degree of endemism in non-cichlids is much lower. Note that the age of the three large lakes is clearly reflected in the percentages of endemism of their non-cichlids. The degree of non-cichlid endemism for Tanganyika diminishes considerably, from 59 to 43%, if the whole basin (Lukuga and all affluent rivers, but only lower Rusizi) is taken into account, while it increases, from 29 to 41% for the basin of Lake Malawi (Lake Malombe and Upper Shire and all affluent rivers included), and from 16 to 31% for the basin of Lake Victoria (Kyoga, Victoria Nile and affluent rivers included). Although these lakes are generally known as "endemic cichlid lakes", an endemic non-cichlid fauna has developed clearly within two of them. Lake Tanganyika has the highest degree of non-cichlid endemism, with eight endemic genera within five different families (De Vos and Snoeks, 1994). Lake Malawi/Nyasa harbours a flock of clariid catfishes, currently comprising 14 *Bathyclarias* species.

IV. GENERAL AND LAKE-SPECIFIC PROBLEMS

A unifying characteristic of the endemic cichlid faunas of the ancient lakes of Africa is that only a fraction of the species in each lake has been described. At present, about 300 endemic species from Lake Malawi have been described and are regarded as valid taxa, which is less than two-fifths of the number of species estimated to be in that lake (Table 1, author's own estimate). In Lake Tanganyika the number of described, valid endemic cichlid species is about 190, which is about three-quarters of a very conservative estimate (Table 1, author's own estimate) of the actual species numbers of this supposedly well-known lake. In Lake Victoria, the recent discovery of many new species in rocky habitats (Seehausen, 1996) reduces the percentage of described species (119 according to Witte and van Oijen, 1995) to about one-quarter of the estimated number.

In the following paragraphs, some more specific features of the research on the systematics of the cichlids of each of the three African Great Lakes, are outlined, comment is made upon their ancestral origin and identification

---

Fig. 1. Map of the East African Lakes region. 1, Lake Turkana; 2, Lake Albert; 3, Lake Kyoga; 4, Lake Victoria; 5, Lake Edward; 6, Lake Kivu; 7, Lake Tanganyika; 8, Lake Rukwa; 9, Lake Mweru; 10, Lake Bangweulu; 11, Lake Malawi/Nyasa. From Snoeks (1994).

problems, and a brief history of their study is provided. A discussion on modes of speciation is beyond the scope of this chapter, and is therefore not included here.

A. Lake Victoria

The haplochromine cichlid fishes of Lake Victoria belong to a superflock with boundaries that encompass other lakes, such as Lakes Kivu, Edward, George, Nabugabo and Kyoga (Greenwood, 1980; Snoeks *et al.*, 1997) and some smaller lakes within the area, such as the Rwandese lakes in the Akagera drainage (pers. obs.). However, there is an indication of a possible monophyletic origin of (most of) the Kivu haplochromines (Lippitsch, 1997; Snoeks *et al.*, 1997). Based on molecular results, the Victoria superflock is presently thought to be monophyletic (Meyer *et al.*, 1990, 1994; but see Greenwood, 1994b). While this might be true, this conclusion is premature in view of the fact that monophyly can only be tested when representatives of the supposed superflock from the other lakes are included in the analysis. To date, this has not been done, although in their analysis Meyer *et al.* (1990) mention *Haplochromis* (*Astatotilapia*) *elegans* Trewavas, 1933, a species endemic to Lakes Edward and George and the Kazinga Channel. However, it is unclear from their data whether they actually examined this species and gave the wrong collection locality (Mwanza Gulf, Tanzania, southern Lake Victoria), or whether they analysed a species that was indeed caught in Lake Victoria, but which was misidentified as *H. elegans*.

It is very difficult to distinguish between most species of the haplochromine superflock of Lake Victoria, and there is also considerable difficulty in deciding whether certain allopatric populations, differing slightly in morphology and colour pattern, are different species or not. Compared with the situation in Lakes Tanganyika and Malawi/Nyasa, however, it is often also very difficult to distinguish between sympatric species, except for highly territorial males. These problems can be partly attributed to the great intraspecific and intra-individual variability in colour pattern, with non-territorial males, females and juveniles of most species having mainly a dull greyish–brownish pattern. A good example of the low level of morphological differentiation between two species is provided by *Haplochromis olivaceus* Snoeks *et al.*, 1990 and *H. crebridens* Snoeks *et al.*, 1990. Territorial males of these species can be distinguished by colour pattern: males of *H. crebridens* are bluish and those of *H. olivaceus* dark olive brown to black. However, all measurements and meristics overlap highly, the greatest difference being found in the lower jaw length, which varies as a percentage of head length from 31.8 to 37.2% (34.2 ± 1.3%, mean ± SD) in *H. olivaceus*, and from 33.7 to 39.4% (36.6 ± 1.4%, mean ± SD) in *H. crebridens*. Other, smaller, differences are present in measurements relating to lower pharyngeal jaw morphology and general body shape (Snoeks, 1991).

Is there anything that makes life easy when studying this group? Yes, there is. Compared with taxonomic studies of cichlids in the other African lakes, the species (re)descriptions are of high quality and provide much detail (cf. Greenwood, 1981b; papers by the HEST team, e.g. Barel *et al.*, 1977; Snoeks, 1994). Surprisingly, however, in contrast to the cichlid fishes of Lakes Tanganyika and Malawi, the taxonomy of the Victoria cichlids has received little attention from aquarists or geneticists.

A special, although sad, feature of Lake Victoria cichlid taxonomy is that about 200 of its species have apparently disappeared within the past few decades (Witte and van Oijen, 1995; Wanink and Witte, Chapter 27, this volume) and hence can only be studied, if at all, by examining preserved specimens in jars.

B. Lake Malawi/Nyasa

Some 300 species of cichlid have been described from Lake Malawi, but more than an equal number remains undescribed. The evolutionary origin of the Malawi cichlids has been extensively studied, but recently, some confusion has crept into the literature concerning the monophyly of "Malawi/Nyasa cichlids". Certainly, stating that the endemic cichlids of Lake Malawi/Nyasa derived from a single founding population (e.g. Klein *et al.*, 1993) is inaccurate; they are paraphyletic. It is only the endemic haplochromine cichlid flock of this lake that is considered to be monophyletic. Several tilapiines (*Oreochromis* spp.) are also endemic, but are not part of the haplochromine clade.

The present level of knowledge indicates that, with the exception of the small and basal *Rhamphochromis* clade, two major clades can be found among the haplochromine endemics (Moran *et al.*, 1994). However, much to the surprise of those who studied these fishes and who almost intuitively distinguish between mbuna and non-mbuna (no monophyly implied for the latter), the practical and readily observable distinction appears to have been refuted, since the mbuna, as formerly understood, are actually paraphyletic (Moran *et al.*, 1994). If one takes the species involved in the latter study as representatives of the genera, then *Alticorpus, Aulonocara* and *Lethrinops* (and hence also the genera *Taeniolethrinops* and *Tramitichromis*) should be called "mbuna" in a phylogenetic sense. The paraphyly found in *Copadichromis* as presently defined is less surprising, as some members of this genus show more affinities with species currently placed in *Nyassachromis* (Konings, 1990a, 1995; but see Stauffer *et al.*, 1993).

The translation of the results of some other molecular studies to species level taxonomy is fraught with difficulties, both practical and theoretical. From their studies of restriction fragment length polymorphisms in mbuna, Moran and Kornfield (1993) postulated the retention of ancestral polymorphism in these fishes. This scheme resulted in an incomplete lineage sorting among species and they concluded that their mitochondrial DNA (mtDNA) gene tree was not congruent with the putative species tree. Bowers (1993), however, using direct mtDNA sequencing, found the population-level mtDNA variation in the Malawian genus *Melanochromis* to be extremely limited, and explained the disjunct distribution of haplotypes as being due to a low signal:noise ratio. In an attempt to resolve the competing hypotheses of whether mbuna mtDNA control region sequence data reflect gene genealogy or species phylogeny, Parker and Kornfield (1997) made a comprehensive analysis of all available data and established that no informative phylogenetic signal is present in the mtDNA control sequence data.

In the author's experience, distinguishing between cichlid species in Lake Malawi/Nyasa is easier than between those of the Victoria superflock, but identifying them is equally difficult. There are three main reasons for this situation: (i) type material is often polyspecific; (ii) data in the literature contain errors which have accumulated over the years; and (iii) keys are either unavailable or only partly workable.

For many species, the most valuable taxonomic observations derive from the 1930s (Trewavas, 1931, 1935) and then even only a few descriptive lines are available for some species. After this pioneering work, mainly by Trewavas but also by Regan, improvement in taxonomic knowledge of Malawi/Nyasa cichlids has come mostly through individual species descriptions by both scientists and aquarists, and have been subject to variable taxonomic standards. Few, although admirable, attempts have been made to review the type material and revise problematic groups, the most useful among which is the classification of some endemic haplochromine taxa (non-mbuna, in the old sense) proposed by Eccles and Trewavas (1989). Lake Malawi/Nyasa shares one characteristic with Lake Tanganyika in that an important source of information comes from popular accounts and scientific publications not principally targeting taxonomic problems (e.g. Ribbink *et al.*, 1983; Konings, 1990b, 1995; Spreinat, 1995; Turner, 1996).

A typical example of the poor knowledge of Malawi/Nyasa cichlids is the genus *Lethrinops*. Although the fishes of this genus have been the subject of some taxonomic revision (Trewavas, 1931; Eccles and Lewis, 1977, 1978, 1979), huge problems remain (see Turner, Chapter 3, this volume). Currently, 24 valid species are recognized (Eccles and Trewavas, 1989), but Konings (1995) mentions eight undescribed shallow-water species, and Turner (1996) lists another 18 additional undescribed deep-water species. This totals about 50 species, the actual taxonomic status of many of which is still questionable. In addition, the ongoing collections of the SADC/GEF project contain several additional new species (pers. obs.). Of all these species, only 15 have been described or redescribed from more than 10 specimens (Table 2). Hence, the morphological variability of most of the species is virtually unknown.

Table 2
Summary of the number of specimens used for the
description or redescription of *Lethrinops* species

Specimens used for (re)description	Species
$> (n)$	(n)
>20	8
>10 <20	7
>5 ≤10	1
>2 ≤5	2
2	2
1	4

C. Lake Tanganyika

Lake Tanganyika is the only African lake with a large number of endemic substratum brooding cichlids. Its great age (9–12 My: Cohen *et al.*, 1993) is often proposed as the reason why the cichlid fishes of this lake show the highest morphological diversity among the African lakes. However, this view ignores the simple fact that it is the polyphyletic origin of its present cichlid species which contributed greatly to its original diversity and which gave rise to its extant species flocks.

In comparison with most other cichlid faunas, the species found in Lake Tanganyika are relatively easy to distinguish among, and compared with the other African lakes there is a relatively good literature record. Aquarists have contributed considerably to the discovery and description of new species, with varying success and, as for Lake Malawi/Nyasa, much useful information is to be found in popular accounts (e.g. Konings, 1988; Brichard, 1989; Konings and Dieckhoff, 1992). However, recent findings suggest that the complexity of the systematics of these fishes may have been grossly underestimated (Snoeks *et al.*, 1994; Verheyen *et al.*, 1996).

One feature recurs in many studies of the systematics, evolution and speciation of Lake Tanganyika cichlids, that of the split of the lake into two or three sub-basins in the past (Figure 2). Indeed, this has probably been the most important event acting on the intralacustrine evolution of these cichlids (Coulter, 1994; Rossiter, 1995). One could consider this event as the foundation upon which other phenomena have been constructed, the results of which are reflected in very complex distribution patterns.

In a recent study of *Lamprologus* sensu lato [the genera *Lamprologus*, *Neolamprologus, Altolamprologus, Variabilichromis* (resurrected by Stiassny,

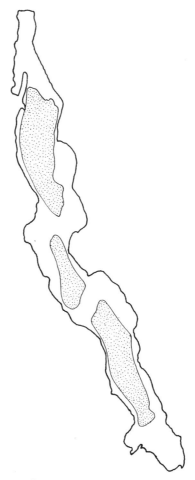

Fig. 2. Map of Lake Tanganyika showing the three palaeo-sub-basins as interpreted from its current 600 m contour. Redrawn from Tiercelin and Mondeguer (1991).

1997) and *Lepidiolamprologus*] in which existing collections and species distributions were reviewed, three major patterns, each with several categories, could be identified (Van Wijngaarden, 1995) as follows.

(1) The effect of the split into sub-basins is difficult to see. This pattern of distribution was evident in two categories:

   (i)   in species with a (nearly) circumlacustrine distribution, *Lamprologus callipterus* Boulenger, 1906, most of the *Lepidiolamprologus* species, etc.;

(ii) in species such as *N. christyi* (Trewavas and Poll, 1952), *N. schreyeni* (Poll, 1974) and *N. wauthioni* (Poll, 1949), that have very limited (single or multiple point) distributions.
(2) The effect of the geographical split can be observed quite clearly. This was evident in three categories:
   (i) in species limited to one or two adjacent sub-basins (Figure 2), e.g. *L. kungweensis* Poll, 1956 and *N. toae* (Poll, 1949) in the north, and *N. leloupi* (Poll, 1948), *V. moorii* (Boulenger, 1898), *N. pulcher* (Trewavas and Poll, 1952) and *N. sexfasciatus* (Trewavas and Poll, 1952) in the south;
   (ii) in species limited to areas that correspond strongly to the former palaeo-shorelines, such as *N. gracilis* (Brichard, 1989), *N. marunguensis* Büscher, 1989 and *N. nigriventris* Büscher, 1992;
   (iii) in species limited to recently submerged areas, especially in the south, e.g. *N. mustax* (Poll, 1978) and *N. prochilus* (Bailey and Stewart, 1977), and only *N. pleuromaculatus* (Trewavas and Poll, 1952) in the north.
(3) Species that are difficult to classify because of their more complex distribution. This category includes species such as *N. mondabu* (Boulenger, 1906), *N. obscurus* (Poll, 1978), *N. brichardi* (Poll, 1974) and *N. savoryi* (Poll, 1949). However, recent studies have identified a relationship between geographical variation in morphology and the historical split in sub-basins in several widespread taxa, such as *N. savoryi* and *N. brichardi* (Louage, 1996).

V. MOLECULAR AND MORPHOLOGICAL STUDIES

Before the advent of molecular techniques, most of the classifications and proposed relationships of East African cichlids (except for Lake Victoria) were based entirely on overall morphological similarities (phenetics). Therefore, it came as no surprise that phylogenetic trees derived from molecular studies differed from some of these earlier assumptions (although some parts have withstood the test remarkably well). Yet, a certain irritation was evident within the taxonomic community when discussing the earlier molecular-based findings. Was there more to this than just human nature, in that no one likes his or her field of research to be criticized? Was it envy that large amounts of money, in sums that traditional taxonomy had rarely attracted, were now being poured into molecular studies? Or was it jealousy that these new techniques swiftly produced results that often found their way into top-ranking journals? In the author's opinion, scientific considerations were involved as well.

First of all, when discussing the new phylogenies that the molecular data produced, the fact that, except for Greenwood's osteology based work on Lake Victoria, morphologists seldom undertook highly time-consuming, detailed phylogenetic studies of the cichlids of the East African Lakes (cf. the review by

Stiassny 1991) was largely ignored. In my opinion, it was this lack of a morphological baseline study, rather than the suspected occurence of homoplasy in the characters traditionally used, that was the basis for the differences in presumed relationships.

Another source of irritation might have been the lack of discussion which characterized earlier molecular-based studies. Because of the eminent pitfall that constitutes homoplasy, in most traditional taxonomic studies results are discussed in a nuanced way, alternative hypotheses are reviewed and conclusions are often presented as tentative. In contrast, several early molecular studies were more cavalier in their approach: less attention was paid to distinguishing between results that were in disagreement with previous classifications but "made sense," or to other results that were in strong conflict with other scientists' "gut feelings". The most likely reason for this was the rigid framework within which molecular data analysis takes place: this left little room for discussion. There was also a lack of awareness on the part of the molecular biologist about the shortcomings and limitations of the techniques used.

An example of this kind of controversy is Sturmbauer and Meyer's molecular study of the problematic genus *Tropheus* in Lake Tanganyika. In their study of 21 *Tropheus* populations, Sturmbauer and Meyer (1992) failed to see the similarities between their classification and those proposed by Konings (1988) (which was completely ignored) and Brichard (1989), and equally did not discuss the dissimilarities. Possible shortcomings of the methods used were not discussed, nor was there any consideration given to alternative explanations for their controversial results, such as the closer relationship of *Tropheus polli* Axelrod, 1977 to *Tropheus* sp. from Bulu (called *T. moorii* Boulenger, 1898, also known as *T.* sp. "kirschfleck") than to *T. annectens* Boulenger, 1900 (see discussion in Konings, 1993). Correctly, Sturmbauer and Meyer (1992) suggested a revision of the taxonomy of *Tropheus* "... which should include molecular data rather than rely solely on taxonomic characters like coloration or the number of spines in the anal fin that are currently used".

Meanwhile, further evidence has been found that both colour patterns and numbers of fin spines are useful characters in *Tropheus* taxonomy (Snoeks *et al.*, 1994, and pers. obs.), and new molecular data have been discussed (Sturmbauer *et al.*, 1997). It is now suspected that *T.* sp. "kirschfleck" and *T. polli* might have hybridized at the Mahale Mountains area. In addition, a shadow of doubt remains as to whether the specimens of *T. annectens* sequenced by Sturmbauer and Meyer indeed belong to this species. To date, *T. annectens* has only been reported from Moba northwards to Mpala, and not from Zongwe (Brichard, 1989), which Sturmbauer and Meyer cite as the source locality of their specimens.

These new molecular data illustrate well the effect of the split of Lake Tanganyika into its sub-basins on the present distributions of certain species. One of the new taxa sequenced (from Cape Kibwesa, Tanzania) was found to belong to the same mtDNA lineage as specimens from Moba and Zongwe

(Democratic Republic of the Congo). Although not all data agree, the presence of this Kibwesa population and its affinities with populations from the opposite shoreline had already been postulated by Konings (1994), based on colour patterns.

Clearly, there is an urgent need for more collaboration between traditional taxonomists and molecular biologists, a fact that few would deny (see also Verheyen, 1994). However, a cautionary note seems in order. The above examples underline the need for those undertaking molecular-based studies to deposit voucher specimens in museum collections, so that identifications can be verified. Indeed, the identification of cichlid specimens used for molecular studies should, ideally, be checked by an experienced cichlid taxonomist **before** the publication of that study, as biologists of other disciplines are not always aware of the difficulties involved in the accurate identification of cichlids.

Although the integration of morphological and molecular research is not easy, the benefits of such an approach are great, as clearly demonstrated by the results of a recent study of the eretmodine cichlid fishes of Lake Tanganyika (Verheyen et al., 1996; see also Verheyen and Rüber, Chapter 26, this volume). Traditional taxonomists had long considered the genus- and species-level taxonomy of this group as having been resolved: the eretmodines consisted of three genera and four species, defined mainly by dentition and head morphology (Poll, 1986). However, the true situation turned out to be more complicated. The monospecific status of the genus *Eretmodus* was first questioned by Konings (1988), who suggested that two species were present. This was confirmed through molecular research by Verheyen et al., (1996). Subsequently, a re-examination of the morphology of the specimens used in the latter study supported Konings' observation that the previously used concept of *Eretmodus cyanostictus* Boulenger, 1898 was incorrect. Examination of head and, especially, mouth morphology revealed the wide-mouthed eretmodine species limited to the northern part of the lake to be clearly different from the true *E. cyanostictus*, which has a smaller mouth and occurs in the southern part of the lake (Konings, 1988; Verheyen et al., 1996). The status of other ectodine taxa also seems in need of revision: differences in colour pattern were found between representatives of the A and B clades identified as *Tanganicodus irsacae* Poll, 1950, and of the two major clades of *Spathodus erythrodon* Boulenger, 1900. Ongoing studies of these fishes include analyses of samples from other areas and an examination of the morphology of those specimens which had an unexpected position within the haplotype tree.

An even more integrated approach has been used to great effect in a study of *Ophthalmotilapia* of Lake Tanganyika with concurrently running molecular and morphometric studies, including a review of information on colour patterns (Hanssens et al., 1999).

VI. THE FUTURE

We appear to have entered the "this is hard to believe" era in the systematics of the cichlids of the East African Lakes. It is hard to believe that perhaps the major part of the endemic cichlid fauna of the Lake Victoria region has evolved since the Late Pleistocene desiccation of the lake (Johnson *et al.*, 1996); so hard, in fact, that some have disputed it (Fryer, 1997). It is equally hard to believe that those rock-dwelling mbuna species that currently have very limited distributions within the southern part of Lake Malawi/Nyasa have evolved *in situ* within the last 1000 years (Owen *et al.*, 1990). It is also hard to believe that some taxa formerly assumed to be non-mbuna are now part of the monophyletic group including the former mbuna.

Of course, "hard to believe" arguments and "gut feelings", however valid they might seem, cannot become the backbone of cichlid systematics. However, in view of the present poor level of knowledge, it must be realized that we have yet to accumulate even the basic information necessary to address adequately many of the existing problems.

A first practical step toward achieving this goal would be to adopt a standardized and scientifically rigorous taxonomy regarding the names used for these fishes. There exists within the East African cichlids a situation, as far as the author is aware, unique amongst vertebrates, in that several hundreds of taxa are reported in the literature by nicknames or cheironyms, owing to the lack of a proper scientific description. Some examples are given in Table 3. The use of these names is not considered good taxonomic practice, but since the system is so well developed (particularly in Lakes Malawi/Nyasa and Victoria cichlids, but less so in Tanganyika species) and has certain benefits, it is currently tolerated. It should be noted that to provide proper scientific descriptions of all of these taxa would entail decades of detailed work by experienced taxonomists.

Within the ongoing SADC/GEF Lake Malawi/Nyasa Biodiversity Conservation Project, a rigid system for names (valid scientific names, synonyms, cheironyms, code names and commercial names) has been adopted for use in the databases. In order to standardize the code names for database use, the present author used the following scheme, building on the conventions used by Konings (1995) for Malawi/Nyasa fishes: genus name (in italics and starting with a capital) + sp. + specific name (between quotes, not in italics and without capital). This results in names such as *Pseudotropheus* sp. "zebra ianth" and *Chalinochromis* sp. "ndobnoi". A serious disadvantage of the use of worknames or codenames is that they are often used inappropriately and without further consideration for valid species as well as for subspecific, and even polyspecific, taxa. It is therefore important that a special effort should be made to ascertain the precise taxonomic status of each taxon. Although it might be argued that the necessary information is often lacking, the advantage

Table 3
Examples of the various styles used for code names of cichlid species of the African lakes

Code name	Lake system
Haplochromis "piceatus-like"	Lake Victoria
Haplochromis "reginus"	Lake Victoria
Haplochromis "velvet black"	Lake Victoria
Otopharynx sp. "auromarginatus jakuta eastern"	Lake Malawi/Nyasa
Stigmatochromis "Spilostichus type"	Lake Malawi/Nyasa
Lethrinops "deep-water albus"	Lake Malawi/Nyasa
Pseudotropheus zebra "ianth"	Lake Malawi/Nyasa
Copadichromis "Verduyni deep blue"	Lake Malawi/Nyasa
Tropheus "kirschfleck"	Lake Tanganyika
Chalinochromis sp. "ndobnoi"	Lake Tanganyika

of using worknames is that they can be allocated before the actual scientific description and without serious nomenclatural implications.

To create some order from the chaos, a more refined and rigorous use of infraspecific and supraspecific categories may have to be encouraged. This is especially true for Lake Malawi/Nyasa, where many scientists, and also hobbyists of various backgrounds, have contributed to the knowledge of the endemic cichlids. Although this might lead to conflicting ideas in a detailed classification of certain taxa, I nevertheless expect that some consensus can be reached for the majority of the groups. This consensus version could then serve as a basis for further refinement, and some such groupings have already been proposed (e.g., Ribbink *et al.*, 1983; Konings, 1995). A further advantage is that while these discussions will be invaluable for specialized taxonomic and phylogenetic studies, they will not necessarily impact upon non-systematic studies.

A. Infraspecific Level

Two infraspecific categories, the subspecies and the race, might be useful in the taxonomy of the cichlids of the African Great Lakes.

A subspecies is "... an aggregate of phenotypically similar populations of a species inhabiting a geographic subdivision of the range of that species and differing taxonomically from other populations of that species" (Mayr and Ashlock, 1991). While this concept appears well suited to the ichthyofauna of the East African lakes, the subspecies category has never been popular in taxonomic studies of these fishes. For example, in the most recent review of the endemic haplochromine taxa of Lake Malawi/Nyasa of Eccles and Trewavas (1989), subspecies were reported in only three species. Similarly, in his taxon review of Lake Tanganyika cichlids, Poll (1986) found isolation between so-called subspecies difficult to prove, and so omitted subspecies altogether and elevated all to the specific level.

Poll (1986) regarded the earlier subspecies as being, in reality, biometrically highly similar species which were assured genetic isolation through their different colour patterns. However, since Poll's (1986) monograph, many more populations of Tanganyika cichlids, often with a slightly different, but population-specific, colour pattern and morphology have been discovered, especially in *Tropheus* and some lamprologine taxa. The degree of subspecific difference recognized in these earlier works, and the degree of differences found now between putative subspecies, are clearly of different magnitudes.

The subspecies is the lowest taxonomic unit recognized by the International Code of Zoological Nomenclature (ICZN) and hence can be utilized within a well-defined and accepted framework. However, the obvious difficulty of objective decision making, stemming partly from an inadequate database and a lack of knowledge, seems to have been, and perhaps will continue to be, the reason for its unpopularity.

Although all infrasubspecific levels are excluded from the provisions of the ICZN, perhaps the use of races (not varieties, which were used by earlier taxonomists in a non-standardized way) could contribute to the taxonomic understanding of these fishes. Names of races are presently used in a geographical context, which is often already implicitly in the mention of the locality of the population, e.g. *Tropheus duboisi* Marlier, 1959 "Bemba", *T. duboisi* "Kigoma" or *Pseudotropheus* sp. "tropheops boadzulu".

B. Supraspecific Level

Two supraspecific categories might be of use in the taxonomy of the African cichlid fishes, the subgenus and the superspecies.

The subgenus would seem an obvious choice for categorization of supraspecific taxa, but except in systematic studies of the Lake Victoria superflock (Greenwood, 1980; van Oijen, 1995) the use of this category has been very limited.

The second alternative category is the superspecies, "...a monophyletic group of closely related and largely or entirely allopatric species" (Mayr and Ashlock, 1991). In taxonomic studies of birds, this category is routinely used, yet it has not found its way into cichlid systematics. Again, a lack of comprehensive knowledge and difficulties in decision making are probably the main reasons for this. The use of superspecies is regulated by the ICZN, Article 6b. An example from Lake Malawi/Nyasa would be: *Pseudotropheus* (superspecies *callainos*) including a.o. *Pseudotropheus* (*callainos*) *callainos* and *Pseudotropheus* (*callainos*) *estherae*.

C. An Eye For Detail

As knowledge of the exceedingly complex taxonomy and phylogeny of the cichlid fishes of the African Great Lakes accumulates, it is becoming increasingly apparent that differences between species are often extremely subtle. In view of this the author recommends the abandonment of ill-defined measurements and meristics, and descriptive statements of the form of imprecise ratios, for example "measurement A, x to y times measurement B". The differences between species may simply be too small (see the example of *H. olivaceus* and *H. crebridens* above) to allow for such inaccurate descriptions. A detailed and accurate description of methods and results is imperative in all morphometrically based taxonomic papers, and will continue to be so in the future.

The need for precision and accuracy in description also applies equally to the data analysis. Principal component analysis is a very popular and highly useful tool in African cichlid taxonomy, and the author considers it the best method for the exploration of large amounts of data. However, while it is easy to use, the interpretation of the results is not always straightforward. For example, increasingly detailed analysis will often reveal differences between populations that may not be specific, but merely a reflection of geographical variation, as was found in a recent study of some lamprologines (Louage, 1996). Hence, there is a clear danger that isolated analyses may lead to incorrect conclusions. In such cases, the inclusion of supplementary data from other sources, such as molecular, ecological and ethological data, is highly informative, as long as one remains aware that these may also show some geographical variation.

D. Phylogeny

The past few decades have witnessed an upsurge of interest in the cichlids of the African Great Lakes and a concomitant increase in the number of species described. More recently, modern DNA methodologies have entered the arena and nowadays many phylogenies of these fishes are based on molecular data. It is clearly an exciting time, yet a note of caution has now been voiced. Since gene trees are suspected of not necessarily corresponding to species trees, in the future, will the phylogeny of these cichlids continue to be based on molecular studies, or will it shift back towards morphology?

In the author's opinion, cichlid studies will continue to be dominated by molecular techniques for some time, simply because the results are obtained much more quickly, and much more time, effort and money is put into this kind of research. No doubt, detailed investigations of the osteological and myological features of these fishes can provide useful information that can contribute significantly to the backbone of any cladogram, as is currently being done for lamprologines (Stiassny, 1997). One can but hope that such studies are undertaken more in the future. However, performing this kind of research below the genus level would be a Herculean task, more so for Lake Malawi Nyasa than for Lake Tanganyika, which might eventually lead to the state of affairs now known for Lake Victoria cichlids. Humphrey Greenwood, one of the best trained ichthyologists of his time, devoted great effort and about 25 years of experience to his studies of the Lake Victoria cichlids, and ended up with groupings, some of them well defined and others doubtfully mono-phyletic, or when supposedly monophyletic only defined by a few apomorphic characters (Greenwood, 1981b). Many had unclear relationships and he was unable to find proof of a monophyletic origin of the flock. This illustrates clearly the complexity and difficulties inherent in morphology-based phyloge-netic studies on the cichlids of the East African lakes.

Therefore, in many cases the input of morphologists into cladograms is currently limited to common sense and not the provision of a set of apomorphic characters.

ACKNOWLEDGEMENTS

The author is much obliged to S. Norris and L. De Vos for providing essential information on some of the non-cichlid taxa within the East African region. E. Verheyen, I. Kornfield, M. Hanssens and A. Rossiter read and commented upon the manuscript. Finally, may those who believe that a discussion of their studies should have been included, but was not, forgive me, as it is hoped those will whose studies are included.

REFERENCES

Barel, C.D.N., Oijen, M.J.P. van, Witte, F. and Witte-Maas, E.L.M. (1977). An introduction to the taxonomy and morphology of the haplochromine Cichlidae from Lake Victoria. *Neth. J. Zool.* **27**, 333–389.
Bowers, N.J. (1993). A revision of the genus *Melanochromis* (Teleostei: Cichlidae) from Lake Malawi, Africa, using morphological and molecular techniques. Unpublished PhD Thesis, Pennsylvania State University, USA.
Brichard, P. (1989). *Pierre Brichard's Book of Cichlids and all Other Fishes of Lake Tanganyika.* T.F.H. Publications, Neptune, NJ.
Cohen, A.S., Soreghan, M.J. and Scholz, C.A. (1993). Estimating the age of formation of lakes, an example from Lake Tanganyika, East-African Rift System. *Geology* **21**, 511–514.
Coulter, G.W. (1994). Speciation and fluctuating environments, with reference to ancient East African lakes. *Arch. Hydrobiol. Beiheft. Ergebnisse Limnol.* **44**, 127–137.

Coulter, G.W., Allanson, B.R., Bruton, M.N., Greenwood, P.H., Hart, R.C., Jackson, P.B.N. and Ribbink, A.J. (1986). Unique qualities and special problems of the African Great Lakes. *Environ. Biol. Fishes* **17**, 161–183.

Daget, J., Gosse, J.P. and Thys van den Audenaerde, D. (eds) (1984). *Check-list of the Freshwater Fishes of Africa*, Vol. I. ORSTOM, Paris, and MRAC, Tervuren.

Daget, J., Gosse, J.P. and Thys van den Audenaerde, D. (eds) (1986). *Check-list of the Freshwater Fishes of Africa*, Vol. II. ISNB, Bruxelles, MRAC, Tervuren, and ORSTOM, Paris.

De Vos, L. (1996). *Rapport final du Centre Regional de Recherche en Hydrobiologie Applique.* IRAZ, Burundi, and KUL, Belgium.

De Vos, L. and Snoeks, J. (1994). The non-cichlid fishes of the Lake Tanganyika basin. *Arch. Hydrobiol. Beiheft. Ergebnisse Limnol.* **44**, 391–405.

Eccles, D.H. and Lewis, D.S.C. (1977). A taxonomic study of the genus *Lethrinops* Regan (Pisces: Cichlidae) from Lake Malawi. Part 1. *Ichthyol. Bull. J.L.B. Smiths. Inst.* **36**, 1–12.

Eccles, D.H. and Lewis, D.S.C. (1978). A taxonomic study of the genus *Lethrinops* Regan (Pisces: Cichlidae) from Lake Malawi. Part 2. *Ichthyol. Bull. J.L.B. Smiths. Inst.* **37**, 1–11.

Eccles, D.H. and Lewis, D.S.C. (1979). A toxonomic study of the genus *Lethrinops* Regan (Pisces: Cichlidae) from Lake Malawi. Part 3. *Ichthyol. Bull. J.L.B. Smiths. Inst.* **38**, 1–25.

Eccles, D.H. and Trewavas, E. (1989). *Malawian Cichlid Fishes. The Classification of Some Haplochromine Genera.* Lake Fish Movies, Herten, Germany.

Fryer, G. (1965). Predation and its effects on migration and speciation in African fishes: a comment. With further comments by P.H. Greenwood, a reply by P.B.N. Jackson and a footnote and postscript by G. Fryer. *Proc. Zool. Soc. Lond.* **144**, 301–322.

Fryer, G. (1997). Biological implications of a suggested Late Pleistocene desiccation of Lake Victoria. *Hydrobiologia* **354**, 177–182.

Greenwood, P.H. (1979). Towards a phyletic classification of the 'genus' *Haplochromis* (Pisces, Cichlidae) and related taxa. Part 1. *Bull. Br. Mus. Nat. Hist. (Zool.)* **35**, 265–322.

Greenwood, P.H. (1980). Towards a phyletic classification of the 'genus' *Haplochromis* (Pisces, Cichlidae) and related taxa. Part 2. *Bull. Br. Mus. Nat. Hist. (Zool.)* **39**, 1–101.

Greenwood, P.H. (1981a). Species flocks and explosive evolution. In: *Chance, Change and Challenge – The Evolving Biosphere.* (Ed. by P.H. Greenwood and P.L. Forey), pp. 61–74. Cambridge University Press and British Museum (Natural History), London.

Greenwood, P.H. (1981b). *The Haplochromine Fishes of the East African Lakes.* British Museum (Natural History), London, and Kraus International Publications, Munich.

Greenwood, P.H. (1984). African cichlids and evolutionary theories. In: *Evolution of Fish Species Flocks (Ed. by A.A. Eschelle and I. Kornfield), pp. 141–154. University of Maine at Orono Press, Orono, ME.*

Greenwood, P.H. (1991). Speciation. In: *Cichlid Fishes. Behaviour, Ecology and Evolution* (Ed. by M. Keenleyside), pp. 86–102. Chapman and Hall, London.

Greenwood, P.H. (1994a). Lake Victoria. *Arch. Hydrobiol. Beiheft. Ergebnisse Limnol.* **44**, 19–26.

Greenwood, P.H. (1994b). The species flocks of cichlid fishes in Lake Victoria and those of other African Great Lakes. *Arch. Hydrobiol. Beiheft. Ergebnisse Limnol.* **44**, 347–354.

Hanssens, M., Snoeks, J. and Verheyen, E. (1999). A morphometric revision of the genus *Ophthalmotilapia* (Teleostei, Cichlidae) from Lake Tanganyika (East Africa). *Zool. J. Linn. Soc. Lond.* **125**, 487–512.

Johnson, T.C., Scholz, C.A., Talbot, M.R., Kelts, K., Ricketts, R.D., Ngobi, G., Beuning, K., Ssemmanda, I. and McGill, J.W. (1996). Late Pleistocene desiccation of Lake Victoria and rapid evolution of cichlid fishes. *Science* **273**, 1091–1093.

Klein, D., Ono, H., O'hUigin, C., Vincek, V., Goldschmidt, T. and Klein, J. (1993). Extensive MHC variability in cichlid fishes of Lake Malawi. *Nature* **364**, 330–334.

Konings, A. (1988). *Tanganyika Cichlids.* Verduyn Cichlids and Lake Fish Movies, Zevenhuizen, The Netherlands.

Konings, A. (1990a). Descriptions of six new Malawi cichlids. *Trop. Fish Hobby.* **38**, 110–129.

Konings, A. (1990b). *Ad Koning's Book of Cichlids and all the Other Fishes of Lake Malawi.* TFH Publications, Neptune City, USA.

Konings, A. (1993). Speciation, DNA, and *Tropheus.* In: *The Cichlids Yearbook*, Vol. 3 (Ed. by A. Konings), pp. 24–27. Cichlid Press, St Leon-Rot, Germany.

Konings, A. (1994). The distribution patterns of *Tropheus* species. In: *The Cichlids Yearbook*, Vol. 4 (Ed. by A. Konings), pp. 6–11. Cichlid Press, St Leon-Rot, Germany.

Konings, A. (1995). *Malawi Cichlids in Their Natural Habitat*, 2nd ed. Cichlid Press, St Leon-Rot, Germany.

Konings, A. and Dieckhoff, H.W. (1992). *Tanganyika Secrets.* Cichlid Press, St Leon-Rot, Germany.

Lippitsch, E. (1997). Phylogenetic investigations on the haplochromine Cichlidae of Lake Kivu (East Africa), based on lepidological characters. *J. Fish. Biol.* **51**, 284–299

Louage, A. (1996). Taxonomische revisie van het *Neolamprologus brichardi* complex (Teleostei:Cichlidae) van het Tanganyikameer (Oost-Afrika). Unpublished MSc Thesis, Katholieke Universiteit Leuven, Belgium.

Lowe-McConnell, R.H. (1969). Speciation in tropical freshwater fishes. *Biol. J. Linn. Soc. Lond.* **1**, 51–75.

Lowe-McConnell, R.H. (1987). *Ecological Studies in Tropical Fish Communities.* Cambridge University Press, Cambridge.

Mayr, E. and Ashlock, P.D. (1991). *Principles of Systematic Zoology.* McGraw-Hill, New York.

Meyer, A., Kocher, T.D., Basasibwaki, P. and Wilson, A.C. (1990). Monophyletic origin of Lake Victoria cichlid fishes suggested by mitochondrial DNA sequences. *Nature* **347**, 550–553.

Meyer, A., Montero, C. and Spreinat, A. (1994). Evolutionary history of the cichlid fish species flocks of the East African great lakes inferred from molecular phylogenetic data. *Arch. Hydrobiol. Beiheft. Ergebnisse Limnol.* **44**, 407–423.

Moran, P. and Kornfield, I. (1993). Retention of an ancestral polymorphism in the mbuna species flock (Teleostei:Cichlidae) of Lake Malawi. *Mol. Biol. Evol.* **10**, 1015–1029.

Moran, P., Kornfield, I. and Reinthal, P. (1994). Molecular systematics and radiation of the haplochromine cichlids (Teleostei:Perciformes) of Lake Malawi. *Copeia* **1994**, 274–288.

Oijen, M.J.P. van (1995). Appendix I. Key to Lake Victoria fishes other than haplochromine cichlids. In: *Fish Stocks and Fisheries of Lake Victoria* (Ed. by F. Witte and Densen van L.T.), pp. 209–300. Samara, Cardigan.

Owen, R.B., Crossley, R., Johnson, T.C., Tweddle, I., Kornfield, I., Davison, S., Eccles, D.H. and Engstrom, D.E. (1990). Major low levels of Lake Malawi and their implications for speciation rates in cichlid fishes. *Proc. R. Soc. Lond. B* **240**, 519–553.

Parker, A. and Kornfield, I. (1997). Evolution of the mitochondrial DNA control region in the mbuna (Cichlidae) species flock of Lake Malawi, East Africa. *J. Mol. Evol.* **45**, 70–83.

Pitcher, T.J. and Hart, P.J.B. (1995). Appendix. Summary of characteristics of major African lakes. In: *The Impact of Species Changes in African Lakes* (Ed. by T.J. Pitcher and P.J.B. Hart), pp. 547–570. Chapman and Hall, London.

Poll, M. (1986). Classification des Cichlidae du lac Tanganika. Tribus, genres et espèces. *Acad. R. Belg. Mém. Classe Sci.* **45**, 1–163.

Ribbink, A.J., Marsh, B.A., Marsh, A.C., Ribbink, A.C. and Sharp, B.J. (1983). A preliminary survey of the cichlid fishes of rocky habitats in Lake Malawi. *S. Afr. J. Zool.* **18**, 149–310.

Rossiter, A. (1995). The cichlid fish assemblages of Lake Tanganyika: ecology, behaviour and evolution of its cichlid flocks. *Adv. Ecol. Res.* **26**, 187–252.

Seegers, L. (1996). The fishes of the Lake Rukwa drainage. *Ann. Mus. R. Cent. Afr. Zool. Sci.* **278**, 1–407.

Seehausen, O. (1996). *Lake Victoria Rock Cichlids: Taxonomy, Ecology and Distribution.* Verduijn Press, Zevenhuizen, The Netherlands.

Snoeks, J. (1991). The use of a standard colour guide and subtle morphological difference in Lake Kivu haplochromine taxonomy. *Ann. Mus. R. Cent. Afr. Zool. Sci.* **262**, 103–108.

Snoeks, J. (1994). The haplochromine fishes (Teleostei, Cichlidae) of Lake Kivu, East Africa: a taxonomic revision with notes on their ecology. *Ann. Mus. R. Afrique Centrale Sci. Zool.* **270**, 1–221.

Snoeks, J. (1997). The non-cichlid fishes of the Lake Malawi/Nyasa system. Annex 1 of the interim review report on systematics and taxonomy for the SADC/GEF Lake Malawi/Nyasa Biodiversity Conservation Project (unpubl.), 5 pp.

Snoeks, J., Rüber, L. and Verheyen, E. (1994). The Tanganyika problem: comments on the taxonomy and distribution patterns of its cichlid fauna. *Arch. Hydrobiol. Beiheft. Ergebnisse Limnol.* **44**, 355–372.

Snoeks, J., De Vos, L. and Thys van den Audenaerde, D. (1997). The ichthyogeography of Lake Kivu. *S. Afr. J. Sci.* **93**, 579–584.

Spreinat, A. (1995). *Lake Malawi Cichlids from Tanzania.* Unterm Hagen, Göttingen.

Stauffer, J.R., Lovullo, T.J. and McKaye, K.R. (1993). Three new sand-dwelling cichlids from lake Malawi, Africa, with a discussion of the status of the genus *Copadichromis* (Teleostei:Cichlidae). *Copeia* **1993**, 1017–1027.

Stiassny, M.L.J. (1991). Phylogenetic intrarelationships of the family Cichlidae: an overview. In: *Cichlid Fishes. Behaviour, Ecology and Evolution* (Ed. by M. Keenleyside), pp. 1–35. Chapman and Hall, London.

Stiassny, M.L.J. (1997). A phylogenetic overview of the lamprologine cichlids of Africa (Teleostei, Cichlidae): a morphological perspective. *S. Afr. J. Sci.* **93**, 513–523.

Sturmbauer, C. and Meyer, A. (1992). Genetic divergence, speciation and morphological stasis in a lineage of African cichlid fishes. *Nature* **358**, 578–581.

Sturmbauer, C., Verheyen, E., Rüber, L. and Meyer, A. (1997). Phylogeographic patterns in populations of cichlid fishes from rocky habitats in Lake Tanganyika. In: *Molecular Systematics of Fishes* (Ed. by T. Kocher and C. Stepien), pp. 93–107. Academic Press, San Diego, CA.

Tiercelin, J.-J. and Mondeguer, A. (1991). The geology of the Tanganyika Trough. In: *Lake Tanganyika and its Life* (Ed. by G.W. Coulter), pp. 7–48. Oxford University Press, London.

Trewavas, E. (1931). A revision of the cichlid fishes of the genus *Lethrinops* Regan. *Ann. Mag. Nat. Hist.* **10**, 133–153.

Trewavas, E. (1935). A synopsis of the cichlid fishes of Lake Nyasa. *Ann. Mag. Nat. Hist.* **10**, 65–118.

Turner, G.F. (1996). *Offshore Cichlids of Lake Malawi.* Cichlid Press, Lauenau, Germany.

Tweddle, D. (1996). Fish survey of Nkhotakota wildlife reserve. *J.L.B. Smith Institute of Ichthyology. Investigative Report* 53, 79 pp.

Van Wijngaarden, C. (1995). Verspreidingspatronen bij *Lamprologus* sensu lato (Teleostei, Cichlidae) in het Tanganyikameer (Oost-Afrika). Unpublished MSc Thesis, Katholieke Universiteit Leuven, Belgium.

Verheyen, E. (1994). Integration of molecular and traditional taxonomic concepts and procedures. *Arch. Hydrobiol. Beiheft. Ergebnisse Limnol.* **44**, 501–503.

Verheyen, E., Rüber, L., Snoeks, J. and Meyer, A. (1996). Mitochondrial phylogeography of rock-dwelling cichlid fishes reveals evolutionary influence of historical lake level fluctuations of Lake Tanganyika, Africa. *Phil. Trans. R. Soc. Lond. B* **351**, 797–805.

Witte, F. and Oijen, M.J.P. van (1995). Appendix III. Biology of haplochromine trophic groups. In: *Fish Stocks and Fisheries of Lake Victoria* (Ed. by F. Witte and W.L.T. van Densen), pp. 321–335. Samara, Cardigan.

Worthington, E.B. and Lowe-McConnell, R. (1994). African Lakes reviewed: creation and destruction of biodiversity. *Envir. Conserv.* **21**, 199–213.

The Nature of Species in Ancient Lakes: Perspectives from the Fishes of Lake Malawi

G.F. TURNER

I. SUMMARY

In this review, the development of species concepts is discussed, both in practice and in principle, applied to the cichlid fishes of Lake Malawi. From 1864 to 1935, museum-based taxonomists with no information on the natural biology of the fishes had no alternative but to apply morphological species definitions. Since then, most descriptions have been carried out by field workers, generally employed on applied fisheries projects, again mostly using morphological criteria. The work of Fryer exemplifies the uncertainty over species concepts during this period, simultaneously demonstrating a profound knowledge of evolutionary theory and biological species concepts, and the difficulties of applying them in practice.

Following the pioneering SCUBA-assisted study of Holzberg in 1978, Ribbink and his co-workers developed a practical species definition for the rocky shore mbuna, based on male breeding colours in addition to behavioural, ecological and morphological features. Such practical species definitions are now used by almost all Malawi cichlid researchers, as well as those working on Lakes Victoria and Tanganyika. Many current workers,

ADVANCES IN ECOLOGICAL RESEARCH VOL. 31
ISBN 0-12-013931-6

including the present author, believe that the most appropriate theoretical basis for this practice lies in the proposition that reproductive isolation between species is often the result of sexual selection by female choice. Rather surprisingly, one implication of this view is that appropriately designed laboratory studies can be used to test the specific status of allopatric populations. In a highly speculative discussion, it is suggested here that speciation mechanisms and species concepts appropriate for African lake cichlids probably have little relevance for other lacustrine endemic taxa.

II. INTRODUCTION

Until recently, most species were described by taxonomists working with long-dead material, and descriptions of species have thus been based largely on anatomical features which remain well preserved. By contrast, most evolutionary biologists working with multicellular organisms believe that interbreeding within a species and reproductive isolation from other species are the key features which permit separate gene pools to accumulate not only selectively neutral differences, but different adaptive features. The purpose of this review is to consider how these practical and theoretical species concepts relate to each other in the context of the species flocks of ancient lakes, in particular the cichlid fish species flocks of the East African Great Lakes, which are arguably the most remarkable of all evolutionary events in present-day lakes. The review will focus principally on the cichlid fishes of Lake Malawi, not only because this is the group with which I am most familiar, but also because there has arguably been a more substantial contribution to theoretical debates by researchers working with these fishes.

III. LAKE MALAWI CICHLIDS: A CASE HISTORY IN TAXONOMIC PRACTICE

A. 137 Years of Progress?

Lake Malawi cichlid research got off to a flying start. Although David Livingstone became the first European known to have set eyes on the lake in 1859, his first visit to the lake was in 1861, accompanied by J. Kirk (Pike, 1968). They sent back fish specimens to the Natural History Museum in London, and within 4 years the first two endemic cichlid species had been described by Günther (1864). These species are now known as *Rhamphochromis longiceps* and *Oreochromis squamipinnis*, representatives of the endemic haplochromine and tilapiine species flocks, respectively. At the time of writing, I probably know these species as well as anyone, yet I can positively identify *O. squamipinnis* only as mature, reproductively active males, although achieving reasonable accuracy with female and immature male specimens over 20 cm total length from the southern part of the lake (Figure 1). Very recently, I came to the conclusion that all of the previous field data on *R. longiceps* refer to a mixture of two species, one of which is probably undescribed. Discounting the possibility of personal spectacular incompetence, what can be responsible for such a striking lack of progress in the last 137 years? To answer this, it is necessary to consider the history of research on Lake Malawi cichlid taxonomy and to review the peculiar nature of the cichlid fish fauna of this lake.

Fig. 1. Immatures of the three "chambo" species, *Oreochromis squamipinnis, O. karongae* and *O. lidole.* Recognition of species of Lake Malawi cichlids is often complicated by the similarity of females and immatures, although mature males are often easily distinguished. In this case, ripe *O. squamipinnis* males have blue or white heads, while those of the other two species are uniformly black.

B. The Museum Years: 1864–1935

For more than 70 years, taxonomic work on Lake Malawi cichlids was carried out by museum-based taxonomists who had never seen the lake, working on collections made by non-biologists. Species were necessarily identified on the basis of anatomical features alone. It was not long before it became widely appreciated that the fish fauna of Lake Malawi was of particular interest. As early as 1896, Boulenger had described the endemic genera *Docimodus* and *Corematodus* to accomodate highly specialized and distinctive fin-biting and scale-eating taxa, respectively. In 1922, Regan increased the number of described species from 38 to 84. He regarded most as belonging to endemic genera or else forming a probably monophyletic group within the large genus *Haplochromis*. By 1935, Trewavas had raised the number of species to 175, but her paper was essentially little more than a key, with the promise of more substantial descriptions to come. Curiously, with the species count virtually doubling with each publication, and African lake cichlids coming to the attention of the major figures in the modern synthesis of Darwinian evolution and Mendelian genetics (e.g. Huxley, 1942; Mayr, 1942), Trewavas' 1935 paper was to mark the end, rather than the beginning, of substantial museum-based alpha-taxonomic research on Malawian cichlids. After describing some tilapiine species in 1941, Trewavas' interests moved elsewhere, and it was not until 54 years later that the full descriptions that she had promised in her synopsis were completed (Eccles and Trewavas, 1989).

C. Fisheries Biologists Go It Alone: 1945–1980

In 1939, Trewavas participated in the first fishery survey of the lake, but this was interrupted by the outbreak of war in Europe. For the first 35 years following the war, virtually all research on Lake Malawi fishes would be funded as applied fisheries projects, with any pure science work largely "spun off". None of these projects had the luxury of operating with identification keys designed for field workers, and most were to find that many of the fish species on which they were working were undescribed and essentially unidentifiable. Many of these researchers carried out their own taxonomic studies, usually focusing on fairly small groups of particular importance to their research: Lowe (1953) for tilapiines, Fryer (1956) for rocky-shore "mbuna", Iles (1960) for zooplanktivores now classed as *Copadichromis*, and Eccles and Lewis (1977–1979) for parts of the genus *Lethrinops*. All produced full formal descriptions. However, none of these projects produced any kind of broad-scale identification guide or key suitable for use in the field, and very little reference material on undescribed species was collected or deposited in properly curated museum collections.

Works dating from this period often refer to the existence of a number of undescribed species, but a substantial amount of ecological and fisheries research was published, virtually all of it apparently referring to formally described species (e.g. Jackson, 1961; Fryer and Iles, 1972; J.L. Turner, 1977a,b). Two factors seem to be at work here.

First, bizarrely, during this time there was a curious tendency to use only formal published scientific names in print. Where an undescribed species is mentioned, it is referred to either as, for example, "undescribed species of the genus *Lethrinops*", or as "*Lethrinops* sp.", neither of which is terribly informative for future researchers. One may surmise that many researchers regarded the use of temporary Anglicized names as unseemly for a scientist who wished to be taken seriously. My own examination of collections at the field museum in Monkey Bay has shown that both Iles (1950s) and Eccles (1960s–1970s) had made considerable progress in identifying species of some groups and had assigned them provisional Latin names, but the formal descriptions were never prepared. There appears to have been a policy that it was better not to publish data on a species than to risk creating a *nomen nudum* by using the provisional name. This has led to a great loss of information, as most of the unpublished data and reference material collected by these and other contemporary workers has now been lost.

Secondly, in many cases, the researchers may not have differentiated between certain similar species. For example, J.L. Turner's (1977a) analysis of changes in size structure of cichlid populations as a result of trawling included a table of 16 species used in the analysis, all given formal scientific names. From my examination of numerous trawl catches from the same area and study of reference material deposited during the surveys of the 1970s, it is likely that nine or 10 of the taxa probably each represents two or more species lumped together. In many cases, these species differ in anatomy, distribution and size at maturity, as well as in male colour. Some of this may simply be the result of bad science or understandable confusion when faced with an overwhelming variety of unknown species, but there may also have been a genuine difficulty with species concepts. This question is probably best investigated through the work of Fryer since, among Malawi cichlid researchers of that era, he made by far the greatest contribution to theoretical matters.

D. Fryer's Ideas on Species and Speciation

Fryer worked for the Joint Fisheries Research Organisation at Nkhata Bay, northern Lake Malawi, from 1953 to 1955. The principal results of this intensely productive period are presented in a paper on the littoral communities of the lake and on the rocky-shore fishes, the mbuna (Fryer, 1959). Fryer's work was the first real field study of Malawian haplochromine cichlids, although SCUBA equipment was not available to him. In his 1959 paper, Fryer did not clearly outline his concept of species, but he seemed to be operating on a morphological species concept. He documented in detail the various colour forms of mbuna at Nkhata Bay. Small differences in coloration of *Pseudotropheus tropheops* and *Petrotilapia* were noted but treated as "random colour variation" which "indicates that in these species specific colouration is unimportant in mating activities" (Fryer, 1959, p. 242). All of these variants are now regarded as separate species (Ribbink *et al.*, 1983; van Oppen *et al.*, 1998). In the case of *P. tropheops*, Fryer synonymized four previously recognized species or subspecies. Their status has still not been resolved. More striking colour variation was regarded as intraspecific colour polymorphism, such as between the orange-blotch and barred female forms of several species, and the blue-barred (BB), orange-blotch (OB), white and blue forms which he considered all to be *Pseudotropheus zebra* (Figure 2). It is now known that the latter two morphs represent another species, *Pseudotropheus callainos*.

Fryer (1959) alluded briefly to the possibility that "homogamy" (assortative mating) might have been important in rapid speciation, as suggested by Kosswig (1947), discussing observations of single-morph aggregations of blue-barred or blue *P. zebra* and predicting (correctly) that this species would be particularly good material for the study of assortative mating. However, he ended up by dismissing the idea on what he admitted was scanty and indecisive evidence. In the course of his discussion of speciation mechanisms (mainly arguing for allopatric speciation by natural selection, with a minor role for drift), he returned twice more to assortative mating. Firstly, he speculated that "peculiarities of courtship behaviour" could prevent interbreeding and that "such a mechanism might be more easily acquired than a cytological sterility barrier" (p. 269), and later discussed how, during secondary contact following speciation, differences in coloration might prevent interbreeding of closely related sympatric species (pp. 273–274). As a possible example of the latter process, he cited the yellow males of *Pseudotropheus livingstonii* (now known to be an undescribed species referred to as *P.* "gold zebra") in contrast to the different colours of the closely related *P. zebra*.

By 1972, Fryer and Iles were still hostile to Kosswig's (1947, 1963) ideas of homogamy (pp. 572–573), perhaps because these ideas were presented as depending on sympatric speciation and monogamous pairing. The former they regarded as implausible, while the latter is obviously inapplicable to the endemic Malawian cichlids, which are all polygynous with no pair bonding. Despite this, in their chapter on taxonomy Fryer and Iles drew attention to two case studies on the difficulty of distinguishing species: the colour forms of *Hemichromis fasciatus* and geographic variation in the bower form of *Tilapia* (now *Oreochromis*) *macrochir* (pp. 54–57). They recognized specific status for the colour forms of the former on the basis of laboratory studies of assortative mating and records of sympatry in the natural habitat, and regarded the status of the latter as unresolved, despite evidence of hybridization in artificial conditions.

(A)

(B)

(C)

(D)

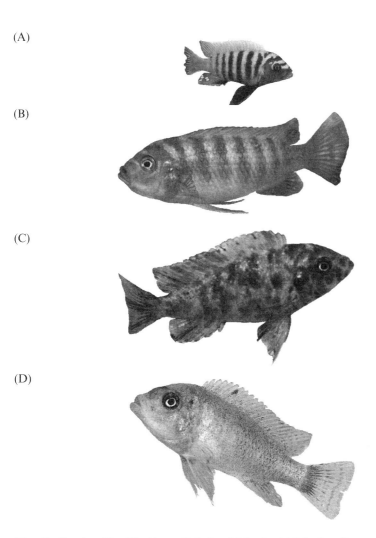

Fig. 2. Species identification of Lake Malawi cichlids is often complicated by intraspecific colour polymorphism. For example, at Nkhata Bay typical blue-barred male *Pseudotropheus zebra* (A) are observed to breed with barred (B), blotched (C) and orange (D) females. Blotched and orange males do occur, but at very low frequency. The matter is further complicated by the presence of very similar barred- and blotched-morph females in the sympatric *Pseudotropheus* "gold zebra", which has yellow and brown barred males.

Overall, it appears that, by the 1970s, Fryer and Iles had accepted Mayr's biological species concept in principle, but no one had pursued the research needed to put it into practice for Malawian cichlids, which were still classified as morphospecies.

E. Ethological Breakthrough: Holzberg and Ribbink

The breakthrough in applying appropriate species concepts to Lake Malawi cichlids came, not from taxonomists, but from ethologists. At Fryer's study site at Nkhata Bay, Holzberg (1978), with the benefit of SCUBA equipment, concluded that *P. zebra* was almost certainly two sympatric species: the BB (blue-barred) males courted with both BB and OB (orange-blotched) females, while the blue (B) males courted with blue or white females. B and BB males defended their territories against males of their own colour, but tolerated overlap with males of the other colour morph. Marsh *et al.* (1981) found similar differences among male colour forms of three sympatric *Petrotilapia* species, even though the females were extremely difficult to distinguish. Non-random mating among these sympatric forms was soon confirmed with allozyme electrophoresis for both complexes (McKaye *et al.*, 1982, 1984). Although slightly complicated by the fact that normally female-limited morphs (such as the OB) were occasionally expressed in males (Holzberg had suggested that this was an "accident" of a hypothetical, and still untested, multi-chromosomal sex- and colour-determining mechanism), the picture emerging by the early 1980s was that male, but not female, colour was diagnostic of reproductively isolated species. On closer inspection, these species were then found to differ in behaviour, ecology and microhabitat preferences, and even to show slight anatomical differences (Holzberg, 1978; Schrœder, 1980; Marsh *et al.*, 1981).

This approach was enthusiastically adopted in a monograph on the mbuna by Ribbink *et al.* (1983). Sympatric male colour forms (except for rare versions of alternative female morphs) were unhesitatingly designated as different species, without any attempt at anatomical study. Ribbink also cut through the tiresome process of formal description by providing Anglicized nicknames for their new species. Subsequent studies have vindicated this approach: the nicknamed species are easy to identify in the field (Konings, 1995; G.F. Turner, pers. obs.), despite the total lack of tables of morphometric ratios and the absence of colourless type specimens in museum collections. Molecular studies have generally confirmed the taxonomic status of sympatric species in these communities (van Oppen *et al.*, 1999).

Paradoxically, although Ribbink's team seem to have classified almost all of the species correctly, they appear to have done so on the basis of a faulty species concept. In the discussion section of his monograph and in later review articles, Ribbink (e.g. 1986) explicitly states that colour (along with behaviour, habitat preferences, etc.) was one of the key traits used in deciding whether taxa were conspecific. Colour was regarded as an integral component of the specific mate recognition system (SMRS), which is under strong stabilizing selection to ensure the efficiency of reproduction.

These latter ideas are derived from the writings of Paterson (e.g. 1980, 1993), who proposed a "recognition concept" in opposition to the more widely accepted biological species concept based on reproductive isolation (e.g. Mayr, 1942). In brief, Paterson's point was initially that the term "isolating mechanisms" implied that the traits causing reproductive isolation were selected for this purpose, which in turn meant that they could not have arisen in allopatry. Since Mayr, Paterson and most other evolutionary biologists believed that most speciation occurred in allopatry, clearly this was misleading terminology. Instead, Paterson proposed that the traits which caused premating isolation were part of a complex of adaptations for efficient location and recognition of conspecific mates (which is surely beyond dispute). However, in later writings he went further and rejected the idea that assortative mating could in any way be the result of sexual selection, and indeed seemed to doubt that mate choice occurred at all (e.g. Paterson, 1993).

To the author's knowledge, what Ribbink did not explicitly mention in his discussions is that in practice he employed only male colour in determination of specific status. The 1983 monograph is full of records of species which are clearly polymorphic for female colour, but as far as I can tell, apart from the rare OB morph males, the only clear example of male colour polymorphism is a cline in the frequencies of white and yellow dorsal fin colour in *Cynotilapia afra* around Likoma Island. Even in this case, Ribbink's writings suggest that he was uncomfortable with the idea of sympatric intraspecific variation in male colour, and he made the plausible suggestion that this variation may be a phenotypic response to environmental conditions.

Ribbink's practice does not seem to be entirely consistent with his own theoretical basis in Paterson's recognition concept, which assumes that speciation is driven by directional natural selection for increased efficiency of mate recognition in allopatric populations. This would surely be expected to lead to monomorphism of female colour within species, but to divergence between species. Furthermore, the SMRS should be under very strong stabilizing selection, except where environmental conditions seriously alter the efficiency of signal transmission. It is difficult to envisage how allopatric populations in different patches of rocky habitat could experience such dramatically different selection pressures. Accordingly, such rapid and repeated diversification of male colour is difficult to explain within the framework of the recognition concept, as Ribbink (1986, 1994) has admitted.

This has been a source of confusion among cichlid researchers, who sometimes seem to think that the recognition concept predicts that divergence of male coloration under sexual selection is an important mode of speciation (e.g. Witte *et al.*, 1997). Apart from Ribbink's own publications, there are few clear statements of Paterson's ideas by cichlid researchers (for exceptions, see Greenwood, 1991; Turner, 1994).

F. Why Should Male Colour be Important in Species Definition?

The use of male colour as a key component of the definition of sympatric species (Ribbink et al., 1983) is entirely consistent with the idea that reproductive isolation among species is largely due to strong directional sexual selection by female choice for conspicuous male coloration. The possible importance of sexual selection in African cichlid speciation has been emphasized by a number of authors (e.g. Lowe-McConnell, 1959; McKaye et al., 1982; Dominey, 1984; McKaye, 1991; Turner, 1994, 1997). Certainly, it would be reasonable to expect to find strong sexual selection by female choice in haplochromine cichlids. There is a hugely male-biased operational sex ratio. A crude minimum estimate for mbuna, assuming a 1:1 sex ratio, might be that half of the mature males are permanently territorial and that each female is receptive all day on only 2 days per year. This would mean that on any one day 90 times as many males as females would be seeking to breed. It has been suggested that multiple mating by females – a single clutch might have as many as six different fathers (Kellogg et al., 1995; Parker and Kornfield, 1996) – might weaken the strength of sexual selection (Ribbink, 1994). However, on the basis of the minimum estimate above, that would still leave 84 territorial males unmated every day for every female that mates.

Furthermore, there is no evidence that competition between males is able to circumvent female choice to any great extent. At Nkhata Bay, it was estimated that 83% of adult male *Pseudotropheus callainos* and 77% of *P. zebra* were territorial, and most of the non-territorial ones were smaller, and probably younger, individuals (M.E. Knight and G.F. Turner, unpublished). It is probable that almost all males are able to obtain a territory. I have never observed males to steal females from each other, and there is no possibility of harem polygyny or male control of scarce resources important to females.

Under conditions of strong sexual selection by female choice, it is to be expected that males are larger and more colourful, have longer fins, and carry out extravagant and energetically expensive courtship rituals. In comparison, females will be smaller and cryptic, have short fins, and expend very little energy in display. All of this is true of virtually all Malawian cichlids studied. Furthermore, in several species at least, females visit a great many males but spawn with relatively few (McKaye, 1991). Male mating success varies with colour (Hert, 1991) and also bower form (McKaye et al., 1990; McKaye, 1991), which itself can be related to the parasite load carried by the male, as predicted by sexual selection theory (Taylor et al., 1998). Studies on Lake Victoria cichlids have shown that females choose males of different colour, but only under conditions where they can see the colour differences, which strongly suggests that other cues are irrelevant, at least in some species (Seehausen and van Alphen, 1998). Mathematical and simulation modelling has demonstrated that it is theoretically plausible for speciation to occur rapidly through the action of sexual selection, even in parapatry (Lande, 1982) or, in some circumstances, in sympatry (Turner and Burrows, 1995).

There are thus good reasons to anticipate that sexual selection acting on male colours (and perhaps other features such as bower form) is likely to play a key role in speciation and the subsequent maintenance of reproductive isolation among sympatric species. It is quite possible that males can accurately recognize conspecific females, and thereby avoid wasting time and energy in courting females which are unresponsive. However, this will be a fairly weak selection pressure, since most conspecific females will be unresponsive anyway most of the time. Initially, males of closely related sympatric species pairs may simply court females of both species, but ultimately mate with whichever females respond to their advances. Where females of two species share the same colour polymorphisms, the task of recognizing conspecific females would seem to be particularly complicated. As such, female colour may often have no influence on assortative mating (Seehausen *et al.* (1999) provide an exception to this generalization). Thus, sexual selection theory provides an explanation of the observed facts that females of different species may be identically coloured, while conspecific females may look totally different.

What light does the recognition concept shed on the matter? I cannot resist including such a superb example of illogical pomposity from Paterson (1993): "Authors frequently attribute to sexual selection phenomena to do with the [SMRS]. Since sexual selection is . . . an intraspecific process, it is clear that the SMRS must first act before sexual selection is possible. Thus, sexual selection is independent of the SMRS. This means that speciation is essentially independent of sexual selection". It would appear that Paterson believes that he has deduced, through pure logic, that interspecific processes must take place prior to intraspecific ones, and that individuals necessarily must therefore first recognize another individual as conspecific, and only then decide whether or not to mate with it.

How, then, might sexual selection lead to speciation? Essentially, this is best understood by ignoring ideas of species recognition. To summarize crudely a long and sophisticated set of arguments (see, for example, Andersson, 1994), sexual selection theory holds that females choose mates on the basis of certain features that indicate their health (which increases the chances of producing healthy offspring), that are attractive to other females (thereby ensuring high mating success for male offspring produced), or that exploit certain sensory/neural pathways evolved for other purposes (which is another way of saying that they look nice or are easy to locate).

Male courtship traits are generally somewhat variable, and this variation has a heritable basis (Turner, 1995, and references therein). Consider a hypothetical species where males produce blue and yellow pigment over the entire body, and in which females prefer males with as much yellow as possible. In many sexually selected species, females have been shown to prefer males with even more spectacular breeding dress than they can actually produce; in other words, preferences are open-ended. However, because predation pressure or simple physical or genetic constraints act to prevent males from expressing the most extreme phenotype possible, it would not be at all surprising to find that males of this hypothetical species were usually slightly greenish-yellow. The more greenish they are, the fewer matings they will achieve.

However, if mutation or recombination were to produce females that preferred less yellow and more blue, these females would tend to mate with the greenish males. This shift could result from something as simple as differential responsiveness of retinal cones sensitive to different wavelengths, or through a neural mechanism concerned with preference. Under these conditions linkage disequilibrium will eventually accumulate: some individuals will contain many genes for yellow and also the genes for preference for yellow, while other individuals will contain many genes for blue and preference for blue. But few, if any, will contain many genes for yellow, together with a preference for blue. If there was enough genetic variability in the males, eventually some of them could end up bright blue. To the new type of females, greenish males, initially preferred to the yellow males, would now be much less attractive than the bright blue males. Similarly, to the original females also, greenish males would again be less attractive than the yellow males. Females which, for example, choose to mate with blue males, but contain many genes for yellow, will tend to produce greenish males, which are unattractive to both types of female. Such genetic combinations will be selected against, and thus green males will be eliminated from the population. With sufficiently choosy females and enough variation in male traits, the species could be split into two reproductively isolated gene pools, one yellow, the other blue. Females would not know that they were selecting males of different species. This would just be a result of divergent mate preferences for different sexually selected traits.

This model is thoroughly sympatric in that both female preferences coexist in an initially undivided population. However, in a species which exists in a number of fairly isolated demes, it might happen that one of the two resulting species will go extinct, either through competition effects or via simple genetic drift. If it is the ancestral form that dies out in a particular deme, but persists elsewhere, the event will appear to have resulted from allopatric speciation. In reality, however, this speciation process is truly sympatric and is almost totally independent of population size.

The above sympatric evolutionary scenario is a simple hypothetical example of the model of Turner and Burrows (1995). As with other models of speciation, this model is highly speculative and may be relevant only in very unusual circumstances. However, the explosive speciation of the Malawian and Victorian haplochromines is an unusual circumstance!

G. Colour Polymorphism

On several occasions above, reference has been made to the occurrence of male colour polymorphism within certain species. If male colour is so important to recognizing species, what is the explanation for the occurrence of markedly different-looking piebald or orange morph males?

Colour polymorphism is widespread in females of maternal mouthbrooding cichlids, where it is probably maintained by some kind of frequency-dependent selection. It is known from haplochromines in lakes Malawi (Ribbink *et al* , 1983), Victoria (Seehausen, 1996a) and Kivu (Snoeks *et al.*, 1989), and in tilapiines from Lake Victoria (Trewavas, 1983). In most of these polymorphic species, the "normal" female morph is dull silvery or brownish, and the "alternative" morphs are either bright orange (O) or pale with dark blotches (OB) (Figure 2). The alternative morph females can be very common, occasionally even being numerically dominant or comprising all of the females in some populations. In contrast, O and OB males are always very rare, but are almost always associated with more numerous females of the same morph.

Several authors have considered the genetic basis for these polymorphisms (e.g. Fryer and Iles, 1972; Holzberg, 1978; Seehausen, 1996b; Seehausen *et al.*, 1999). It is clear that OB/O males rarely appear in broods produced by crossing any female morph with males of the normal morph, whereas broods produced from crosses involving OB or O morph females with males of the same morph produce a large percentage of OB/O males. Why, then, are alternative morph males so scarce?

If females mated at random or preferred males of their own morph, it would be expected that males of the OB/O morph would be much more numerous in the field than they are. Alternative morph males have normal fertility and so sterility can be discounted. It is possible that there is strong natural selection against such males, for example through increased predation, but this seems unlikely, as they seem no more conspicuous than do males of the normal morph. Another explanation is that alternative morph males never mate in nature, since females may prefer to mate with the normal morph males. This has yet to be adequately investigated, but if true it would be compatible with sexual selection theories.

H. The Problems of Allopatric Forms

Although apparent to Fryer (1959) to a lesser extent, one of the most notable things to emerge from the survey by Ribbink *et al.* (1983) was the tremendous degree of intralacustrine allopatric variation in mbuna. Populations on offshore islands or separated by wide sandy bays often differed dramatically in male colour. Some of these populations also differed less strikingly in female colour, body shape, and fine details of behaviour and ecology.

While Ribbink had no difficulty in defining sympatric male colour forms as species, he recognized that there was no similarly clear resolution of allopatric forms. He was tempted to follow Mayr (1969) and designate dubious populations as subspecies, but rejected this for the pragmatic reason that it was often not clear with which of several sympatric species a particular allopatric population should be aligned. In the end, if he could unambiguously align allopatric populations with each other they were regarded as conspecific, even if they differed dramatically in colour. For example, with *Labeotropheus*, there is never more than one slender form (*L. trewavasae*) and one deep-bodied form (*L. fuelleborni*) at any site, and so all populations were assigned to these two species, even though it was recognized that allopatric colour forms might not interbreed in sympatry (Ribbink *et al.*, 1983; p. 154). Where taxa could not be matched up with any one of several species occurring sympatrically, they were tentatively regarded as heterospecific.

Essentially, this policy was adopted by Konings (1995) in his surveys of mbuna and other inshore species of Lake Malawi. With the benefit of repeated surveys, which included the Tanzanian and Mozambican waters inaccessible to Ribbink's team, Konings has recently been able to match up several allopatric forms and has been steadily making minor reductions in the number of "species" described by Ribbink (as Ribbink anticipated, e.g. 1986), while discovering more of his own.

Stauffer and co-workers have adopted a slightly different approach, best exemplified in a recent paper on the *Pseudotropheus zebra* complex (Stauffer *et al.*, 1997). In principle, they designate species on the basis of unique colour patterns, "in conjunction with discontinuity of morphological differences". Allopatric forms which exhibit discontinuities in morphological traits, even if similarly coloured, are treated as distinct species, whereas populations showing clinical variation are accorded conspecific status. For example, several offshore islands harbour populations of *P. zebra* in which males possess orange–red dorsal fins. These populations had previously been regarded as conspecific with each other, and perhaps also with the more widespread blue-dorsal fin *P. zebra* populations (Konings, 1995), or as perhaps representing two species (Ribbink *et al.*, 1983). Stauffer *et al.* formally described three new species of "red-dorsal zebra". One of these differs from the other two in having the black flank bars extending slightly into the dorsal fin, while the other two differ statistically in body shape and some meristic traits. The most notable morphometric difference is in the ratio of snout length to head length, the ranges for the three populations being 29.7–34.2%, 29.1–35.2% and 30.8–37.8%. These populations exist on tiny islands 40–370 km apart, so it would not be too surprising if there were no gene flow between them and small differences in morphometrics and meristics were to accumulate with time.

Designation of allopatric taxa as distinct species has generally been viewed as a matter of opinion, and mine is that these three populations should have been regarded as conspecific until they had been investigated more thoroughly. I believe that it is possible to test specific status of allopatric populations, without undertaking unethical transplantations of fish around the lake. The SMRS is critically context dependent and, viewed under the recognition concept, can only be tested in the field. After all, natural selection for differential efficiency of signal transmission in different locations is essentially regarded as the main process involved in speciation (Paterson, 1993; Ribbink, 1994). However, if sexual selection has been the main engine of speciation, different populations might head off in evolutionary trajectories which will often be essentially random with respect to the environment (with the obvious constraint that a given signal has to be visible). Thus, provided laboratory conditions offer a true choice of males and permit female fishes to differentiate among signals as they would in their natural habitat, should females continue to select mates as they would in the field. Using microsatellite DNA to determine paternity of clutches, such an experimental design was tested recently using three sympatric species of the *P. zebra* complex which have been shown to be reproductively isolated in nature (van Oppen *et al.*, 1999). They also mated 100% assortatively in the laboratory (Knight *et al.*, 1999) and pairs of allopatric taxa are now being tested using the same techniques.

Eventually, laboratory testing should permit researchers to investigate which male traits are important to females in their mating decisions. For example, do females of the *P. zebra* complex choose mates on the basis of the extension of the black flank bars into the dorsal fin? Do they even "care" whether the dorsal fin is red or blue? It will never be possible to test all combinations of populations, but with a few judiciously chosen taxa, it should be possible to achieve a reasonable understanding of which traits define a biological species. Such studies will also provide critical insights into speciation processes.

I. Species Concepts for the Non-mbuna

Since the 1950s, most academic research has focused on the mbuna and to a lesser extent the inshore zooplanktivorous "utaka" *Copadichromis* spp. Much less is known about the pelagic, deep-water benthic and sandy-shore cichlids. In terms of assigning species to formally described taxa, little progress could be made until the publication of Eccles and Trewavas' monograph in 1989, which was largely composed of the full descriptions and illustrations promised in Trewavas' 1935 paper. This timing was especially fortuitous for me, as during 1990–1992 I examined a great number of trawl catches from southern Lake Malawi. From this work I produced a very preliminary field guide to the cichlids of the southern trawling grounds (Turner, 1996), largely following the practice and format of the mbuna guide of Ribbink *et al.* (1983).

Like Ribbink's team, I found that most of the taxa were undescribed and that existing keys based on morphological traits were frequently useless. In general, offshore cichlids, like the mbuna, show few apomorphic anatomical character states, and I was unable to produce a reliable key to their identification. As with the mbuna, many anatomically similar taxa differed markedly in male breeding colours, but unlike mbuna, females of non-mbuna species were not polymorphic. Evidence to date suggests that many non-mbuna cichlids are much less affected by habitat barriers than the mbuna, and it therefore seems probable that there will be far less difficulty in dealing with allopatric populations of dubious status. However, investigation of these species remains complicated, because males in breeding colour occur only sporadically. Male mbuna are usually permanently territorial and occupy most, if not all, of the range of the species. In contrast, it appears that many shallow-water, sandy-shore non-mbuna species are only seasonally territorial, at which time they form breeding arenas in restricted areas of their range. At other times and places, they do not show breeding colours.

Not only is male breeding colour of non-mbuna species more difficult to assay than it is with mbuna, but it may be less important in the maintenance of reproductive isolation. For example, within the genus *Rhamphochromis*, virtually all species have orange pelvic and anal fins, sometimes in both sexes, sometimes only in males. In a few cases, ripe males have orange bellies. However, morphospecies differ little in male coloration.

Although male coloration may be of limited use as a diagnostic character for deep-water species, it is still unclear how much emphasis should be placed on morphology. Most of the diagnoses of described species are heavily influenced by characters associated with feeding: tooth shape, number and arrangement, jaw morphology, pharyngeal bone structure, and the number and shape of gillrakers (Eccles and Trewavas, 1989). There have been very few attempts to investigate intraspecific polymorphisms in such traits in Malawian cichlids, but there are many examples of such polymorphisms in natural populations of neotropical and Victorian cichlids. Futhermore, laboratory studies have shown that diet can have a strong influence on the development of these structures (e.g. Kornfield *et al.*, 1982; Kornfield, 1991; Witte *et al.*, 1997). I strongly suspect that many species diagnoses will become even more difficult once larger samples of individuals from a greater geographic range are studied.

IV. DISCUSSION

The method of Ribbink *et al.* (1983), of defining species on the basis of male colour, has been the most significant step in developing and putting into practice a workable species definition for Malawian cichlids. The importance of male colour in species definitions has also been recognized by researchers working on Lake Victoria cichlids (e.g. Seehausen, 1996a; Witte *et al.*, 1997). It is now clear that there is also a number of allopatric and, in some cases, sympatric, colour forms of Lake Tanganyika cichlids, mainly among maternal mouthbrooding lineages (Snoeks *et al.*, 1994). Some of these colour forms appear to mate assortively (Rossiter, 1995).

Clearly, this critical advance did not come from theoretical constructs, but from a thorough familiarity with the fish, developed through extensive fieldwork. Future progress will likewise depend on empirical studies and much more survey work is necessary. Laboratory studies of assortative mating will be critical, as will studies of sympatric morphotypes using molecular techniques to determine the degree of interbreeding among taxa which cannot be studied either in the laboratory or by using SCUBA surveys of breeding biology.

The development of a sound theoretical basis for species concepts will ultimately depend also on an understanding of the factors responsible for the origin and maintenance of reproductive isolation among species. At present, the main emphasis is likely to be on sexual selection, within which the genetic basis of female preference and male courtship traits will be a key area of investigation. However, the importance of morphological change in influencing assortative mating remains virtually unknown.

Although there is much still to be learned, I will take the opportunity to finish with some wild speculation.

In the majority of the world's lakes, endemic species have probably arisen by allopatric speciation. Initially, this occurred when taxa were isolated within lake basins and subjected to selection regimes different to those present in their ancestral riverine habitats. The initial driving forces have been geographic isolation and natural selection. Genetic bottlenecking and founder effects have probably played little if any role, as most lakes are colonized not by invading migrants, but by organisms that were already present in large numbers in other water bodies within the basin prior to the filling of the lake. In most lakes, this is not the end of the story, as the lacustrine habitat provides both an immense volume and a great many new ecological opportunities not present in rivers and ponds.

The consistent pattern of fish speciation revealed by studies in recently glaciated lakes (e.g. Schluter, 1996) supports the following general scenario. In many cases, a single founding lineage splits sympatrically into several ecologically differentiated morphs, each adapted for a different habitat and/ or trophic niche. Contrary to the assumptions of earlier models of ecological sympatric speciation, in results from empirical investigations, disruptive natural selection against intermediate forms seems unimportant. Reproductive isolation is probably mainly the result of habitat segregation by the different trophic morphs. It is important to note that this is still truly sympatric speciation, because the evolution of the polymorphisms causing reproductive isolation begins in an undivided gene pool. The diversification of most lacustrine fish species flocks was probably influenced by this process.

Allocation of specific status in such taxa will always remain extremely problematic. In most cases, rapidly evolving molecular markers will reveal significant genetic differences from an early stage in the diversification of these new morphs, but the morphs will probably remain interfertile for a long time and would probably often interbreed when forced to coexist in the same habitat. Laboratory studies would seldom reveal assortative mating, as reproductive isolation is almost certainly influenced strongly by the environment. Later, some of these endemic taxa would become specialized to live in patchily distributed habitats. These populations would become isolated, and intralacustrine allopatric speciation by natural selection or drift would then ensue and be accelerated by variations in environmental conditions, such as lake level changes.

Among the cichlids, the substrate spawning and biparental mouthbrooding lineages almost certainly radiated entirely in the various ways just described. Haplochromine cichlids are strikingly aberrant in their accelerated rates of speciation and in the huge numbers of species differing largely in male coloration. Speciation in haplochromines has probably proceeded largely by rapid establishment of reproductive isolation driven by divergent sexual selection through female choice. This process might be possible in sympatry, depending on the genetic basis of female preference rules. Conversely, it might operate only in allopatry, in which case it could not occur until ecological speciation had produced an initial radiation and some of the resulting taxa had developed habitat specializations. Once lineages had thus become prone to intralacustrine allopatry, speciation would have been accelerated. Speciation by sexual selection is probably completed very rapidly, so that ecological competition has often not had time to produce extinction or character displacement. This results in a community in which many taxa are likely to be reproductively isolated, but ecologically equivalent. In the early stages of the radiation, there would have been many empty niches, and most newly formed species would eventually have been able to adapt to one or other of these different niches. The famously plastic cichlid body plan and pharyngeal apparatus would have been important at this stage, as it would have been in the initial radiations of non-haplochromines.

In the later stages of the radiation, there will be little opportunity for new species to find or create new niches. In some niches, the population density of a species will cause significant depletion of critical resources. In these cases, there will be interspecific competition, leading to extinction by competitive exclusion. In other niches, population density will not affect critical resources and densities may be regulated by predation, particularly on juveniles. For example, some mbuna which are confined to rocky shores feed on plankton transported there from all over the lake by currents. It is highly unlikely that their populations could make any impact on the supply of their food resources. In such cases, new species may simply continue to accumulate within niches, although some will be extinguished by stochastic processes.

Haplochromine species are thus properly defined on the basis of reproductive isolation. This will be fairly unambiguous, as speciation will be rapid, and specific status for sympatric taxa or for allopatric taxa will be easy to determine by laboratory studies of assortative mating. While of immense importance to biologists working on these fish faunas, I conclude that species concepts in haplochromine cichlids probably are of little relevance to those appropriate for other taxa.

ACKNOWLEDGEMENTS

The ideas in this manuscript developed from discussions with numerous colleagues, but are all my own fault. The manuscript was improved by comments from two anonymous referees.

REFERENCES

Anderson, M. (1994). *Sexual Selection*. Princeton University Press, Princeton, NJ.

Boulenger, G.A. (1896). Descriptions of new fishes from the Upper Shire River, British Central Africa, collected by Dr Percy Randall, and presented to the British Museum by Sir Harry H. Johnston, K.C.B. *Proc. Zool. Soc. Lond.* 1896, 915–920.

Dominey, W.J. (1984). Effects of sexual selection and life history on species: species flocks in African cichlids and Hawaiian *Drosophila*. In: *Evolution of Fish Species Flocks* (Ed. by A.A. Echelle and I. Kornfield), pp. 231–249. University of Maine at Orono Press, Orono, ME.

Doorn, G.S. van, Noest, A.J. and Hogeweg, P. (1998). Sympatric speciation and extinction driven by environment dependent sexual selection. *Proc. R. Soc. Lond. B* **265**, 1915–1919.

Eccles, D.H. and Lewis, D.S.C. (1977–1979). A taxonomic study of the genus *Lethrinops* Regan (Pisces:Cichlidae) from Lake Malawi. Parts 1–3. *Ichthyol. Bull. J.L.B. Smiths. Inst.* 36–38.

Eccles, D.H. and Trewavas, E. (1989). *Malawian Cichlid Fishes*. Lake Fish Movies, Herten.

Fryer, G. (1956). New species of cichlid fishes from Lake Nyasa. *Rev. Zool. Bot. Afr.* **53**, 81–91.

Fryer, G. (1959). The trophic interrelationships and ecology of some littoral communities of Lake Nyasa with special reference to the fishes, and a discussion of the evolution of a group of rock-frequenting Cichlidae. *Proc. Zool. Soc. Lond.* **132**, 153–281.

Fryer, G. (1991). The evolutionary biology of African cichlid fishes. *Ann. R. Belg. Mus. Cent. Afr. Zool. Sci.* **262**, 13–22.

Fryer, G. and Iles, T.D. (1972). *The Cichlid Fishes of the Great Lakes of Africa: Their Biology and Evolution*. Oliver and Boyd, Edinburgh.

Greenwood, P.H. (1991). Speciation. In: *Cichlid Fishes: Behaviour, Ecology and Evolution* (Ed. by M.H.A. Keenleyside), pp. 86–102. Chapman and Hall, London.

Günther, A. (1864). Report on a collection of reptiles and fishes made by Dr Kirk in the Zambezi and Nyasa regions. *Proc. Zool. Soc. Lond.* 1864, 303–314.

Hert, E. (1991). Female choice based on egg-spots in *Pseudotropheus aurora* Burgess 1976, a rock-dwelling cichlid of Lake Malawi, Africa. *J. Fish Biol.* **38**, 951–953.

Holzberg, S. (1978). A field and laboratory study of the behaviour and ecology of *Pseudotropheus zebra* (Boulenger), and endemic cichlid of Lake Malawi (Pisces: Cichlidae). *Zool. System. Evol. Forschell.* **16**, 171–187.

Huxley, J. (1942). *Evolution: The Modern Synthesis*. Allen and Unwin, London.

Iles, T.D. (1960). A group of zooplankton feeders of the genus *Haplochromis* (Cichlidae) in Lake Nyasa. *Ann. Mag. Nat. Hist.* **13**, 257–280.

Jackson, P.B.N. (1961). *Check List of the Fishes of Nyasaland.* Cambridge University Press, Cambridge.

Kellogg, K.A., Markert, J.A., Stauffer, J.R. and Kocher, T.D. (1995). Microsatellite variation demonstrates multiple paternity in lekking cichlid fishes from Lake Malawi, Africa. *Proc. R. Soc. Lond.* B **260**, 79–84.

Knight, M.E., Turner, G.F., Rico, C., Oppen, M.J.H. van and Hewitt, G.M. (1999). Microsatellite paternity analysis on captive Lake Malawi cichlids supports reproductive isolation by direct mate choice. *Mol. Ecol.* **7**, 1605–1610.

Konings, A. (1995). *Malawi Cichlids in their Natural Habitat.* Cichlid Press, Lauenau, Germany.

Kornfield, I., Smith, D.C., Gagnon, P.S. and Taylor, J.N. (1982). The cichlid fish of Cuatro Cienegas, Mexico: direct evidence of conspecificity among distinct trophic morphs. *Evolution* **36**, 658–664.

Kornfield, I. (1991). Genetics. In: *Cichlid Fishes: Behaviour, Ecology and Evolution* (Ed. by M.H.A. Keenleyside), pp. 103–128. Chapman and Hall, London.

Kosswig, C. (1947). Selective mating as a factor for speciation in cichlid fish of East African Lakes. *Nature* **159**, 604–605.

Kosswig, C. (1963). Ways of speciation in fishes. *Copeia* **1963**, 238–244.

Lande, R.S. (1982). Rapid origin of sexual isolation and character divergence in a cline. *Evolution* **36**, 213–233.

Lowe, R.H. (1953). Notes on the ecology and evolution of the Nyasa fishes of the genus *Tilapia* with a description of *T. saka* Lowe. *Proc. Zool. Soc. Lond.* **122**, 1035–1041.

Lowe-McConnell, R.H. (1959). Breeding behaviour patterns and ecological differences between *Tilapia* species and their significance for evolution within the genus *Tilapia* (Pisces: Cichlidae). *Proc. Zool. Soc. Lond.* **132**, 1–30.

McKaye, K.R. (1991). Sexual selection and the evolution of the cichlid fishes of Lake Malawi, Africa. In: *Cichlid Fishes: Behaviour, Ecology and Evolution* (Ed. by M.H.A., Keenleyside), pp. 241–257. Chapman and Hall, London.

McKaye, K.R., Kocher, T., Reinthal, P. and Kornfield, I. (1982). A sympatric sibling species complex of *Petrotilapia* Trewavas from Lake Malawi analysed by enzyme electrophoresis. *Zool. J. Linn. Soc. Lond.* **76**, 91–96.

McKaye, K.R., Kocher, T., Reinthal, P., Harrison, R. and Kornfield, I. (1984). Genetic evidence for allopatric and sympatric differentiation among color morphs of a Lake Malawi cichlid fish. *Evolution* **38**, 215–219.

McKaye, K.R., Louda, S.M. and Stauffer, J.R. (1990). Bower size and male reproductive success in a cichlid fish lek. *Am. Nat.* **135**, 597–613.

Marsh, A.C., Ribbink, A.J. and Marsh, B.A. (1981). Sibling species complexes in sympatric populations of *Petrotilapia* Trewavas (Cichlidae, Lake Malawi). *Zool. J. Linn. Soc. Lond.* **71**, 253–264.

Mayr, E. (1942). *Systematics and the Origin of Species.* Harvard University Press, Cambridge, MA.

Mayr, E. (1969). *Principles of Systematic Zoology.* McGraw Hill, New York.

Oppen, M.J.H. van, Turner, G.F., Rico, C., Deutsch, J.C., Robinson, R.L., Genner, M.J. and Hewitt, G.M. (1998). Assortative mating among rock-dwelling cichlids supports high estimates of species richness from Lake Malawi. *Mol. Ecol.* **7**, 991–1001.

Payne, R.J.H. and Krakauer, D.C. (1997). Sexual selection, space, and speciation. *Evolution* **51**, 1–9.

Parker, A. and Kornfield, I. (1996). Polygynandry in *Pseudotropheus zebra*, a cichlid fish from Lake Malawi. *Envir. Biol. Fishes* **47**, 345–352.

Paterson, H.E.H. (1980). A comment on "mate recognition systems". *Evolution* **34**, 330–331.

Paterson, H.E.H. (1993). Animal species and sexual selection. In: *Evolutionary Patterns and Processes* (Ed. by D.R. Lees and D. Edmunds), pp. 209–228. Academic Press, London.

Pike, J.G. (1968). *Malawi: A Political and Economic History.* Pall Mall Press, London.

Regan, C.T. (1922). The cichlid fishes of Lake Nyasa. *Proc. Zool. Soc. Lond.* **1921**, 675–727.

Ribbink, A.J. (1986). The species concept, sibling species and speciation. *Ann. R. Belg. Mus. Cent. Afr. Zool. Sci.* **251**, 109–116.

Ribbink, A.J. (1994). Alternative perspectives on some controversial aspects of cichlid fish speciation. *Arch. Hydrobiol. Beiheft. Ergebnisse Limnol.* **44**, 101–125.

Ribbink, A.J., Marsh, B.A., Marsh, A.C., Ribbink, A.C. and Sharp, B.J. (1983). A preliminary survey of the cichlid fishes of rocky habitats in Lake Malawi. *S. Afr. J. Zool.* **18**, 149–310.

Rossiter, A. (1995). The cichlid fishes assemblages of Lake Tanganyika: ecology, behaviour and evolution of its species flocks. *Adv. Ecol. Res.* **26**, 187–252.

Schluter, D. (1996). Ecological speciation in postglacial fishes. *Phil. Trans. R. Soc. Lond.* B **351**, 807–814.

Schrœder, J.H. (1980). Morphological and behavioural differences between the BB/OB and B/W colour morphs of *Pseudotropheus zebra* Boulenger (Pisces: Cichlidae). *Zool. System. Evol. Forschell.* **18**, 69–76.

Seehausen, O. (1996a). *Lake Victoria Rock Cichlids.* Verduijn Cichlids, Zevenhuizen, The Netherlands.

Seehausen, O. (1996b). Polychromatism in rock dwelling Lake Victoria cichlids: types, distributions and observations on their genetics. *The Cichlids Yearbook* **6**, 36–45.

Seehausen, O. and Alphen J.J.M. van (1998). The effect of male coloration on female mate choice in closely-related Lake Victoria cichlids (*Haplochromis nyererei* complex). *Behav. Ecol. Sociobiol.* **42**, 1–8.

Seehausen, O., Alphen, J.J.M. van and Witte, F. (1997). Cichlid fish diversity threatened by eutrophication that curbs sexual selection. *Science* **277**, 1808–1811.

Seehausen, O., Alphen, J.J.M. van and Lande, R. (1999). Colour polymorphism and sex ratio distortion in a cichlid fish as an incipient stage in sympatric speciation by sexual selection. *Ecol. Lett.* **2**, 367–378.

Snoeks J., Rüber, L. and Verheyen. E. (1994). The Tanganyika problem: comments on the taxonomy and distribution patterns of its cichlid fauna. *Arch. Hydrobiol. Beiheft Ergebnisse Limnol.* **44**, 101–125.

Snoeks, J., Coenen, E., de Vos, L. and van den Audenaerde, T. (1989). Genetic polychromatism in Lake Kivu haplochromines. *Ann. R. Belg. Mus. Cent. Afr. Zool. Sci.* **257**, 101–104.

Stauffer, J.R., Bowers, N.J., Kellogg, K.A. and McKaye, K.R. (1997). A revision of the blue–black *Pseudotropheus zebra* (Teleostei: Cichlidae) complex from Lake Malawi, Africa, with a description of a new genus and ten new species. *Proc. Acad. Nat. Sci. Philadel.* **148**, 189–230.

Taylor, M.I., Turner, G.F., Robinson, R.L. and Stauffer, J.R. (1998). Sexual selection, parasites and bower height skew in a bower-building cichlid fish. *Anim. Behav.* **56**, 379–384..

Trewavas, E. (1935). A synopsis of the cichlid fishes of Lake Nyasa. *Ann. Mag. Nat. Hist.* **10**, 294–306.

Trewavas, E. (1941). Nyasa fishes of the genus *Tilapia* and a new species from Portuguese East Africa. *Ann. Mag. Nat. Hist.* **11**, 294–306.

Trewavas, E. (1983). *Tilapiine Fishes of the General Sarotherodon, Oreochromis and Danikilia.* British Museum (Natural History), London.

Turner, G.F. (1994). Speciation in Lake Malawi cichlids: a critical review. *Arch. Hydrobiol. Beiheft. Ergebnisse Limnol.* **44**, 139–160.

Turner, G.F. (1995). The lek paradox resolved? *Trends Ecol. Evol.* **10**, 473–474.

Turner, G.F. (1996). *Offshore Cichlids of Lake Malawi.* Cichlid Press, Lauenau, Germany.

Turner, G.F. (1997). Small fry go big time: why is one fish family so diverse? *New Scientist* **155**, 36–40.

Turner, G.F. (1999). Explosive speciation of African ciclid fishes. In: *Evolution of Biological Diversity.* (Ed. by A.E. Magurran and R.M. May), pp. 113–129. Oxford University Press, Oxford.

Turner, G.F. and Burrows, M.T. (1995). A model of sympatric speciation by sexual selection. *Proc. R. Soc. Lond. B* **260**, 287–292.

Turner, J.L. (1977a). Changes in the size structure of cichlid populations of Lake Malawi resulting from bottom trawling. *J. Fish. Res. Brd Can.* **34**, 232–238.

Turner, J.L. (1977b). Some effect of demersal trawling in Lake Malawi (Lake Nyasa) from 1968 to 1974. *J. Fish Biol.* **10**, 261–271.

Witte, F., Barel, K.D.N. and Oijen, M.J.P. van (1997). Intraspecific variation of haplochromine cichlids from Lake Victoria and its taxonomic implications. *S. Afr. J. Sci.* **93**, 585–594.

Making Waves: The Repeated Colonization of Fresh Water by Copepod Crustaceans

G.A. BOXSHALL and D. JAUME

I. SUMMARY

Twenty-two independent colonizations of fresh and inland continental waters are identified within six of the 10 currently recognized orders of copepods. This number is a minimum estimate and is expected to increase as knowledge of the phylogenetic relationships within copepod families improves. It does not include mere incursions into fresh water, defined as invasions without subsequent diversification (= speciation). The timing of colonization events is estimated, where possible, by inference from biogeographical data. This preliminary analysis supports, in general, a direct relationship between lineage diversity and time elapsed since colonization of fresh water. It is hypothesized that the invasion of South America by the Diaptominae, penetrating from the north, resulted in the displacement of the original calanoid inhabitants of that continent, the *Boeckella* group within the family Centropagidae, except at high altitude and higher latitudes. A similar, but more ancient, invasion by the

Diaptominae is envisaged for Africa, with the original calanoid inhabitants, the Paradiaptominae, now being confined to marginal habitats, such as temporary waters. The analysis indicates a succession of at least four major waves of colonization of inland continental waters, although this may well be an artefact reflecting the indirect method of estimating colonization dates. Taxa currently making incursions into fresh waters may represent the next, or fifth, wave.

II. INTRODUCTION

Ancient lakes are qualitatively different from other lakes, particularly in the relatively high proportion of endemic species that they contain. The key element seems to be time, since the only factor common to all ancient lakes is that they are ancient (i.e. more than 100 000 years old: see discussion in Gorthner, 1994). Copepod crustaceans are present in all ancient lakes, but the levels of diversity and endemism vary and, again, no common factor or general rule is apparent. Copepod diversity and the degree of endemism reflect in part, the age of the lake, area effects and its colonization history. It thus seemed interesting to pursue the concept of time as a factor influencing freshwater copepod diversity in general. A preliminary exploration is presented of the hypothesis that the time elapsed since colonization of fresh water by a lineage of copepods is directly correlated with the species richness of that lineage.

To achieve even a preliminary consideration of this question, it is necessary to make some assumptions. First, it is assumed that knowledge of the systematics of freshwater copepods has improved to the extent that comparing lineage diversity by using absolute numbers of described species is meaningful. This assumption is probably acceptable, although the sampling of some specialist groundwater taxa, such as the Parastenocarididae, is probably uneven geographically and introduces a source of error into the analysis. Secondly, it is assumed that the general patterns of distribution of freshwater taxa contain a high signal: noise ratio, i.e. on balance they reflect historical tectonic events rather than subsequent dispersal events. The present authors also regard this assumption as acceptable, although acknowledging that dispersal of freshwater copepods is an important factor, as testified by the presence of *Paracyclops chiltoni* Thomson on the remote volcanic Easter Island and of diaptomids on the volcanic Galapagos Islands.

In this chapter the number of independent colonizations of fresh water by copepods is identified, then an attempt is made to date some of these colonization events by biogeographical analysis. A colonization of fresh water is defined here as the invasion of inland continental and/or fresh water by the ancestor of a lineage followed by the subsequent diversification of that lineage within continental waters. The subsequent diversification within fresh water requires that the lineage must comprise at least two related taxa. This definition is necessary in order to differentiate between a colonization event and a mere incursion. Several copepod lineages appear to have made incursions into fresh water: such lineages are predominantly coastal marine and estuarine in distribution but their basic euryhalinity allows them to move into fresh water. However, in such cases, no subsequent diversification has occurred within fresh water. The Pseudodiaptomidae is an example of a family that is regarded as having made one or more incursions into fresh water but that has not truly colonized it. Even though any or all of the successful colonization may have been initiated by an original incursion, it is useful to distinguish between them. Each modern incursion into fresh water represents a potential colonization event.

III. COPEPODS OF INLAND CONTINENTAL WATERS

Copepods are abundant and diverse in fresh and inland saline continental waters, occurring as a major component of most planktonic, benthic and groundwater metazoan communities. In addition, they have successfully colonized restricted microhabitats such as phytotelmata, and extreme environments such as hot springs and glacial meltwaters. Copepods of continental waters exhibit an enormous range of lifestyles from small-particle feeders, to predators and parasitic forms, which utilize a wide variety of fishes and a few invertebrates as hosts. In summary, copepods occur in most habitats with sufficient water, including semiterrestrial situations, such as damp moss and leaf litter in humid forests.

Representatives of six of the 10 copepod orders can be found in freshwater habitats (Table 1). The ordinal phylogenetic analysis of Huys and Boxshall (1991) suggested that all orders of copepods originated in marine waters. A minimum of six colonization events is therefore supported; however, in the larger of these orders, it is apparent from available information on phylogenetic relationships that more than one lineage has moved into continental waters independently. The number of independent colonizations is analysed below, order by order.

A. The Order Calanoida

In the order Calanoida, with few exceptions, the fresh and continental water taxa belong to the superfamily Diaptomoidea (= Centropagoidea). This should not be interpreted as evidence that only a single lineage within this superfamily has successfully invaded continental waters. The Centropagidae comprises marine, brackish-water and freshwater species. The type genus *Centropages* Krøyer is marine, comprising 28 species that typically occur in coastal waters, but the family also contains large freshwater genera such as *Boeckella* Guerne and Richard. The phylogenetic relationships between the genera of centropagids are unresolved, but the present authors suggest that the

Table 1
Copepod lineages found in continental waters

Order	Family – lineage	Mar	BrW	FW	Mar/FW
Calanoida	Centropagidae	+ +	+ +	+ +	
	Diaptomidae			+ + +	
	Pseudodiaptomidae	+ +	+ +	+	
	Acartiidae	+ +	+ +	+	
	Temoridae	+ +	+ +	+ +	
	Senecella		+	+	
Harpacticoida	Ectinosomatidae	+ + +	+	+	
	Phyllognathopodidae			+ +	
	Chappuisiidae			+	
	Harpacticidae	+ +	+	+	
	Diosaccidae	+ + +	+ +	+ +	
	Ameiridae	+ +	+ +	+ + +	
	"Canthocamptidae"	+ +	+ +	+ + +	
	Parastenocarididae			+ + +	
	Cletodidae	+ +	+	+	
	Laophontidae	+ + +	+	+	
Cyclopoida	Oithonidae	+ +	+	+	
	Cyclopidae	+ +	+ +	+ + +	
	Ozmanidae			+	
	Lernaeidae			+ + +	
Siphonostomatoida	Caligidae	+ + +	+	+	
	Dichelesthiidae	+			+
	Lernaeopodidae	+ + +		+ + +	+
Poecilostomatoida	Ergasilidae	+ +	+ +	+ + +	+
Gelyelloida	Gelyellidae			+	

+, one to nine species; + +, 10–99 species; + + +, 100 or more species. Mar, Marine; BrW, brackish water; FW, fresh water; Mar/FW, parasitic on anadromous fishes migrating between marine and fresh waters.

family invaded fresh water at least twice. The Centropagidae is the most important family of calanoids in continental waters in Australasia and southern South America. The large group of species contained in *Boeckella* and related genera shows a Gondwana distribution, with representatives even on the smaller fragments of Gondwana such as New Caledonia (Defaye, 1998). Centropagids are absent from African inland waters. Three genera, *Limnocalanus* Sars, *Osphranticum* Forbes and *Sinocalanus* Burckhardt, are present in continental waters in the northern hemisphere. *Osphranticum* is monotypic. *Limnocalanus* comprises two species, both of which occur in marine and fresh waters (Holmquist, 1970). These species are here interpreted as representing incursions into fresh water. *Sinocalanus* currently comprises five species found in estuarine and fresh waters. It may have diversified in fresh water and is here treated as a colonization, rather than as an incursion. It is inferred here that the *Boeckella* group and *Sinocalanus* represent independent colonizations of fresh water. It may be estimated that a minimum of at least two independent colonization events has occurred within the family.

The Diaptomidae is the largest family within the order Calanoida, comprising in excess of 515 valid species from inland waters. Members of the family are primarily small particle feeders, but some are predatory. Most inhabit the water column but some are benthic and a few, small-sized species, inhabit subterranean waters. It is assumed here that the ancestral stock of the Diaptomidae inhabited fresh water and that the family, therefore, represents a single colonization event for the purposes of this analysis. It is possible, however, that the Paradiaptominae and Diaptominae (plus Speodiaptominae and Microdiaptominae) colonized fresh waters independently.

The Pseudodiaptomidae comprises about 50 species of primarily coastal and estuarine species. A small number of species may be found in fresh water and continental waters but this may be considered to be the result of numerous separate incursions into fresh water, rather than representing a true, independent colonization. Members of the Acartiidae, like the pseudodiaptomids, are primarily coastal and estuarine in distribution, but a very small number of species occurs in freshwater habitats. In the absence of information on the phylogenetic relationships of freshwater *Acartia* Dana species, this family is also treated here as having made incursions into fresh water rather than having colonized it.

The Temoridae comprises coastal marine (e.g. *Temora* Baird), estuarine (e.g. some species of *Eurytemora* Giesbrecht) and freshwater (e.g. *Epischura* Forbes and some species of *Eurytemora*) taxa. At least in continental waters, the family is restricted to the northern hemisphere. In Lake Baikal, the temorid *Epischura baikalensis* Sars is usually the dominant copepod in the plankton community. It is assumed that the family represents a single colonization event for the purposes of this analysis, although this may be an underestimate.

The phylogenetic relationships, and hence the classification, of the genus *Senecella* Juday are uncertain. It has been placed in the family Clausocalanidae (see Dussart and Defaye, 1983, as Pseudocalanidae) but was transferred to the Aetideidae by Vyshkvartzeva in 1994. It comprises just two species and, according to its placement in either of these families, appears to represent an independent colonization of fresh water. However, both species can be found in brackish coastal waters and it is possible that this should be treated as two separate incursions into fresh water.

Finally, for completeness, the Fosshageniidae must be mentioned. This monotypic family was established on the basis of material collected from the superficial brackish-water lens overlying the saline water in an anchialine cave in the Caicos Islands (Suárez-Morales and Iliffe, 1996). It was placed in a new superfamily, the Fosshagenioidea, but this taxon is weakly supported and the family probably belongs in the superfamily Diaptomoidea. It is not included as a freshwater taxon in Table 1.

B. The Order Harpacticoida

Within the order Harpacticoida the situation is similar. One large family of more than 210 species, the Parastenocarididae, is exclusively freshwater in distribution, as are two smaller families, the Chappuisiidae (two species) and the Phyllognathopodidae (about 13 species). Lang (1948) indicated that the Chappuisiidae and Phyllognathopodidae may have shared a common ancestor (and this may already have lived in fresh water), but Huys and Iliffe (1998) regarded the former as a specialized lineage within the superfamily Tisboidea. These three families are, therefore, treated as representing three separate colonization events.

The Ectinosomatidae is a primarily marine family, which contains a few estuarine species and even fewer freshwater species, particularly in the genus *Halectinosoma* Lang. This is treated here as representing a single independent colonization event since there may have been some diversification within fresh water.

The Harpacticidae is similar, a primarily marine family with a few freshwater forms, including *Harpacticella inopinata* Sars, one of the most abundant copepods in Lake Baikal. This is also treated here as an independent colonization event.

The Diosaccidae is another predominantly marine family, but it contains numerous freshwater forms, especially in the genus *Schizopera* Sars. There is, for example, a small species flock of at least 10 species described from the ancient Lake Tanganyika. Other species flocks are known, but for the purposes of this chapter, the diosaccids are treated as a single independent colonization event.

The Ameiridae is one of the larger and more diverse families of harpacticoids that inhabit fresh water. The radiation in freshwater habitats is dominated by the group of genera related to *Nitokra* Boeck and particularly to *Nitocrella*

Chappuis. These genera, such as *Nitocrella, Parapseudoleptomesochra* Lang, *Nitocrellopsis* Petkovski and *Stygonitocrella* Petkovski, inhabit karstic systems, inland caves and other subterranean waters (Rouch, 1986). Some ameirids have occupied specialized habitats, such as the hyporheic zone of rivers (e.g. *Psammonitocrella* Rouch), and others have entered into symbiotic relationships with a variety of invertebrate hosts (e.g. species of *Nitokra*). Given the lack of a recent analysis of phylogenetic relationships within the family, the ameirids are treated as a single independent colonization event, but this probably represents an underestimate.

The Canthocamptidae is the largest family of the Harpacticoida, comprising in excess of 600 species, and it is predominantly freshwater in distribution. Determining the evolutionary history, in particular the pattern of colonization of continental waters, of the canthocamptids is grossly obstructed by the lack of a coherent scheme of phylogenetic relationships for the family. The family as currently constituted contains fully marine, even deep-sea forms, such as *Bathycamptus* Huys and Thistle, estuarine forms, such as *Mesochra* Boeck, as well as the mass of true freshwater forms. For this analysis the authors assume only a single colonization event of fresh water, but recognize that this is a minimum estimate for a taxon that is almost certainly polyphyletic.

The Cletodidae and Laophontidae are both predominantly marine families, which contain a small proportion of brackish-water and freshwater taxa. Because the current understanding of phylogenetic relationships is insufficient to allow precise identification of any particular freshwater lineages within these families, in both cases the present discussion is restricted to a minimum assumption of a single colonization event for each family.

C. The Order Cyclopoida

The largest family within the order Cyclopoida is the Cyclopidae. It comprises over 700 species, the great majority of which belongs in the two freshwater subfamilies, Cyclopinae and Eucyclopinae. The small subfamily Euryteinae contains marine and estuarine species only, and the fourth subfamily, the Halicyclopinae, consists predominantly of brackish-water forms belonging to the type genus *Halicyclops* Norman. The Cyclopidae is cosmopolitan in fresh waters and its members exploit a huge variety of habitats from subterranean waters to ancient lakes. Species may be small particle feeders, benthic surface feeders, predators or even specialized symbionts on freshwater sponge hosts. The Cyclopidae is treated as representing a single colonization event.

The Oithonidae contains primarily coastal marine planktonic forms but many species are found in brackish-water habitats and a few occur in fresh water. Some species within the genus *Oithona* Baird are reported exclusively from fresh water. *Limnoithona* Burckhardt species also occur in fresh and brackish waters. Accepting the current taxonomy as valid indicates two independent colonizations of fresh water, one within each of these genera. The existence of a small cluster of *Oithona* species in east-coast drainages of South America (Ferrari and Bowman, 1980) provides evidence of colonization in this lineage, rather than mere incursion into fresh water.

The monotypic Ozmanidae is known from Amazonia. The only species is an endoparasite of a freshwater gastropod mollusc. According to the latest phylogenetic analysis of the Cyclopoida, the Ozmanidae is the sister group of the Lernaeidae (Ho *et al.*, 1998). The Lernaeidae is the second largest family of Cyclopoida found in freshwater habitats. It comprises just over 110 species, all of them parasites of freshwater fishes. For the purposes of this chapter the Ozmanidae and Lernaeidae are treated together as representing a single independent colonization of fresh water following the phylogenetic scheme of Ho *et al.* (1998).

D. The Order Siphonostomatoida

The only members of the Siphonostomatoida found in fresh water are parasites of fishes. One of the two extant monotypic genera in the family Dichelesthiidae parasitizes sturgeons (family Acipenseridae). Since its hosts are anadromous species, migrating between fresh and marine waters, the colonization of fresh water by *Dichelesthium* Abildgaard could probably be classified as passive, with the parasite having been passively carried there on its hosts. The existence of only a single species in fresh water defines this as an incursion into fresh water, rather than a colonization. However, the existence of a Lower Cretaceous fossil copepod, *Kabatarina pattersoni* Cressey and Boxshall, parasitic on the gills of a fossil fish that may have lived in an estuarine habitat (Cressey and Boxshall, 1989), hints at the possibility of colonization by a larger lineage.

One clade only within the large family Lernaeopodidae has radiated in fresh water (Kabata, 1979). The *Salmincola* Wilson lineage comprises 38 species in seven genera. These parasites occur on various freshwater fish hosts, including silurids, cyprinids and coregonids, as well as salmonids and anadromous acipenserids. The original colonization of fresh water by this lineage might also have been passive.

The large family Caligidae contains a single freshwater species, *Caligus lacustris* Steenstrup and Lütken, among several hundreds of marine ones. Another fish parasite, its arrival into fresh water may again be attributed to passive transport via fish hosts and subsequent establishment. It also occurs in brackish waters and in the Black Sea. This is treated as an incursion into fresh water rather than a colonization.

E. The Order Poecilostomatoida

Only one of more than 60 families within the order Poecilostomatoida has successfully colonized inland continental waters. This family, the Ergasilidae, has a unique life cycle: its developmental stages, from nauplius up to adult, occur as free-living members of the plankton community; only after mating do the fertilized adult females seek out and infest hosts for the final, parasitic phase of their life cycle. Most ergasilids are known only from their parasitic females, and these typically utilize fishes as hosts, with a few species on bivalve molluscs. Levels of host specificity appear to be relatively low in the family, partly because of the limitation of the parasitic phase to the postfertilization period of the adult female only. Some ergasilids are fully marine, occurring on coastal fishes, many occur in brackish-water estuaries and lagoons, and many are freshwater. This family is here treated as representing a single colonization event, since the phylogenetic relationships within the group are unresolved and since the complex interactions between host utilization patterns and salinity regimes are not yet understood. It is probably a gross underestimate since several tentatively identified lineages (e.g. *Acusicola* Cressey; see El-Rashidy and Boxshall, 1999) include both marine and freshwater representatives.

F. The Order Gelyelloida

Finally, the order Gelyelloida comprises a single family, the Gelyellidae, one genus and two species. These species occur in European subterranean waters. The family is counted here as an independent colonization of fresh water from a marine origin (Huys and Boxshall, 1991).

In summary, there is evidence of a minimum number of five independent colonizations of continental waters by members of the order Calanoida, 10 by members of the Harpacticoida, four by members of the Cyclopoida, and one each by members of the Siphonostomatoida, Poecilostomatoida and Gelyelloida. The grand total is 22 colonization events. This is a minimal estimate, which will undoubtedly increase as the knowledge of phylogenetic relationships within lineages improves.

IV. BIOGEOGRAPHY AND THE TIMING OF COLONIZATION EVENTS

Using the comparative method the aim was to test whether there is a correlation between the timing of these 22 independent colonization events and the diversity of each taxon in fresh water. The extremely fragmentary nature of the fossil record for the Copepoda (see Cressey and Boxshall, 1989) is a hinderance here, removing one potential set of test data. It is necessary, therefore, to arrive at some estimates of the timing of the original colonization events for each lineage, based on biogeographical data. This method is fraught with difficulties; in particular, the incompleteness of phylogenetic understanding of copepod lineages renders some of the basic data (i.e. the 22 taxonomic lineages enumerated above) doubtful. However, even at this stage it is possible to make a preliminary examination of the correlation between the diversity of a freshwater taxon (estimated as simple species richness) and the time elapsed since initial colonization of fresh water.

A. The Order Calanoida

1. The Family Centropagidae: The Boeckella Group

The species of the *Boeckella* group, within the Centropagidae, occur in fresh and athalassic waters in Australia, New Zealand, southern South America and New Caledonia (Dussart and Defaye, 1983; Defaye, 1998). A single species occurs in Mongolia. Their distribution led Bayly and Morton (1978) and Bayly (1992) to suggest that the lineage first invaded continental waters before the break-up of Gondwana. However, this lineage is not represented in African waters, indicating that this invasion occurred after the separation of Africa from the rest of Gondwanaland (except for South America). The present authors support this suggestion and infer that the *Boeckella* group invaded South America before 120 Mya (= Aptian, Lower Cretaceous) when South America became disconnected from the rest of Gondwanaland (except for Africa).

The *Boeckella* group apparently did not spread into Africa. The progressively deeper incision of the Proto-South Atlantic might have constituted a barrier preventing their spread from South America into Africa until these two continents were completely separated by the South Atlantic, about 95 Mya. Given the lack of a scheme of phylogenetic relationships for the family, the *Boeckella* lineage is defined here as comprising at least those approximately 60 species referred *Boeckella, Pseudoboeckella* Mrázek, *Metaboeckella* Ekman, *Hemiboeckella* Sars and *Parabroteas* Mrázek. This lineage may be even larger, including other continental waters genera such as *Gladioferens* Henry and *Calamoecia* Brady (cf. Maly, 1996). It may be speculated that the centropagids were the only calanoid family represented in fresh water in the southern supercontinent of Gondwana, excluding Africa. Africa was probably never colonized by centropagids, although it is possible that their absence is secondary, the result of regional extinction on that continent.

2. The Family Diaptomidae

The Diaptomidae is the dominant calanoid family in inland waters in Europe, Asia, North America, Africa and northern South America. It does not occur in New Caledonia and New Zealand, and only two species (of *Eodiaptomus*

Kiefer and of *Tropodiaptomus* Kiefer) have been found in Australia. The family comprises about 515 species in four subfamilies, the two largest subfamilies comprising 22 species (the Paradiaptominae) and about 490 species (the Diaptominae). The biology, diversity and evolutionary histories of these two subfamilies differ markedly. Members of the Paradiaptominae are restricted to Africa, with some dispersal outwards to adjacent areas, including southern Europe and the Middle East. They are largely restricted to marginal habitats, such as temporary or inland saline waters. The members of the Diaptominae are extremely widely distributed throughout the range of the family and occupy prime habitats, such as lakes.

It is speculated here that the original colonization of continental waters by the subfamily Diaptominae of the Diaptomidae occurred in the northern supercontinent of Laurasia some time after the break-up of Pangaea around 160 Mya. It is further speculated that the ancestors of the Paradiaptominae were the first calanoids to colonize fresh water on the African plate. The timing of this invasion is difficult to establish. No paradiaptomines have been reported from the neotropical region, suggesting that this invasion occurred only after Africa separated from South America (around 95 Mya). However, the colonization may have been considerably more ancient, if the barriers that prevented the spread of the *Boeckella* group from South America into Africa also acted to prevent the spread of the paradiaptomines in the opposite direction.

A third possibility is that the ancestors of the paradiaptomines may have penetrated into Africa from Laurasia via the Iberian peninsula during the Late Eocene (37 Mya), when Africa was separated from Laurasia by only a narrow passage marking the opening of the Tethys Sea into the Atlantic.

It may be postulated that the presence of diaptomines in northern South America results from a later invasion from North/Central America. After invading from the north, the diaptomines spread rapidly, through the highly interconnected, lowland river systems that make South America unique. They probably replaced the existing calanoid fauna of *Boeckella* group centropagids except at high altitude (i.e. the Andean cordillera) and at high latitude (e.g. Patagonian) areas, where they are still dominant today. The timing of this invasion from the north most probably occurred when North America and South America came back into contact at the closing of the Panama gap in the Pliocene (3 Mya). The remarkable rapidity of the spread through northern South America is attributed here to the exceptional interconnectedness of the major neotropical river systems. Some support for this scenario can be derived from the current state of the systematics of the genus *Notodiaptomus* Kiefer, the dominant and most speciose genus of Diaptominae in South America (Santos Silva *et al.*, 1999). The species of *Notodiaptomus* are notoriously difficult to distinguish on the basis of morphological characters. This is probably not due to especially poor descriptive taxonomy, but is a reflection of the lack of morphological differentiation between recognized species. This, in turn, might be related to the relative recency of the diversification of the genus within South America. The presence of the vagile species *Arctodiaptomus dorsalis* (Marsh) in lowland habitats on the extreme north-western corner of South America is here interpreted as the result of a recent, independent dispersal onto the continent from Central America.

Surprisingly, a somewhat similar scenario can be envisaged for Africa. As mentioned above, the Paradiaptominae can be regarded as the first calanoid group to have colonized African fresh waters. It may be speculated that the diaptomines also invaded the African continental waters, initially from the north, and spread southwards, diversifying and displacing the existing calanoid taxon, the paradiaptomines, except those in marginal habitats, such as saline and temporary waters. The authors regard this invasion as considerably more ancient than the corresponding invasion of South America by representatives of the Diaptominae. First, because the degree of interconnectedness of African river systems is much less (the different African drainage basins are clearly demarcated and faunistically distinct) and the climatic history is different, so the spread of the diaptomines across the continent must have been slower. Secondly, the morphological differentiation between species of the dominant and most speciose genus (*Tropodiaptomus*) is generally greater than between species of *Notodiaptomus*, suggesting an older radiation. A possible date for this colonization is about 17 Mya, when the Arabigo–African and Anatolian plates were connected via the Middle-East landbridge, with great interchange of terrestrial fauna.

The central analysis of this chapter, attempting to identify any correlation between time in fresh water and lineage diversity, treats the Diaptomidae as a single lineage, given that the common ancestor of all subfamilies was almost certainly already an inhabitant of fresh water. However, the postulated displacement of one subfamily lineage within the family by another indicates one of the more obvious possible sources of error in the analysis.

3. The Family Temoridae

According to Dussart and Defaye (1983), the Temoridae of continental waters number about 27 species in three genera, *Eurytemora, Epischura* and *Heterocope* Sars. Most *Eurytemora* species are essentially brackish-water forms, although *E. velox* (Lilljeborg) and *E. lacustris* (Poppe) are freshwater species, and the true diversity of the temorid freshwater lineage is almost certainly considerably less than this figure of 27 species. All are restricted to the higher latitudes of the northern hemisphere and it seems likely that the distribution of this family in continental waters has been profoundly influenced by the Pleistocene glaciations.

4. The "Senecella" Lineage

Senecella calanoides Juday occurs in fresh water in North America, and *S. siberica* Vyshkvartzeva occurs in the brackish waters of the Kara and Laptev

Seas of northern Siberia. The extreme northerly distribution, in areas covered by the ice cap during the Pleistocene glaciation, identifies this as a postglacial colonization.

B. The Order Harpacticoida

1. The Canthocamptidae Lineage and the Family Parastenocarididae

The two large harpacticoid families Canthocamptidae (> 600 species) and Parastenocarididae (> 210 species) both have cosmopolitan distributions, except for the latter's absence from New Zealand (Dussart and Defaye, 1985). Both are presumed here to have inhabited a variety of freshwater habitats on Pangaea. The difference in species richness may be an artefact due to incomplete sampling of the groundwater habitats favoured by parastenocaridids (see Ahnert, 1998), or it may reflect a slower rate of speciation in groundwaters compared with surface waters.

2. The Family Ameiridae

The Ameiridae contains a large number (100 +) of freshwater forms, especially from subterranean waters (Rouch, 1986) and these are distributed widely across Europe, Asia, North America and Africa, especially northern Africa (Dussart and Defaye, 1990), although this may represent sampling bias. Few freshwater ameirids are reported from South America, although typically brackish-water forms, such as some *Nitokra* species, can be found. The Ameiridae have probably colonized continental waters more than once, but the main lineage of subterranean forms probably originated in Laurasia. It seems likely that they colonized Africa early in their radiation.

3. The Family Diosaccidae

The freshwater representatives of the Diosaccidae are found primarily in western Eurasia and in Africa, especially in the Great Lakes Tanganyika and Malawi, where an evolutionary radiation seems to have taken place. These lakes are estimated as having been in existence approximately 9–12 My and 2 Mya, respectively (Cohen *et al.*, 1993; Ribbink, 1994).

4. The Family Harpacticidae

The Harpacticidae contains genera such as *Tigriopus* Norman, which typically live in splash zone pools and are tolerant of extreme variations in temperature and salinity. There is, however, only one colonization of fresh water within the family, and it consists of *Harpacticella* Sars species, which are restricted to the Palaearctic region.

C. The Order Cyclopoida

1. The Family Cyclopidae

The largest family of copepods in fresh water is the Cyclopidae (order Cyclopoida). Members of this family occur on all continents and it is likely that this lineage first invaded and colonized fresh waters prior to the break-up of Pangaea.

2. The Family Oithonidae

The main invasion of the genus *Oithona* within the Oithonidae into fresh water has taken place in Amazonia, where a small cluster of species has been recorded (Ferrari and Bowman, 1980). The extensive brackish-water interface between marine and freshwater habitats in the Amazon probably provided the route by which *Oithona* has colonized neotropical fresh waters, but the possibility that *Oithona* arrived during a marine incursion into the Upper Amazon cannot be excluded (see Lovejoy *et al.*, 1998).

3. The Family Lernaeidae

Unravelling the biogeographical history of parasitic families, such as the cyclopoid family Lernaeidae, is difficult because of the added dimension, the availability of suitable hosts. Ho (1998) demonstrated that the family comprises two monophyletic lineages, the subfamilies Lernaeinae and Lamprogleninae. The 57 species of the Lamprogleninae are found only in Eurasia (mostly from Asia, since this total includes only two species from Europe) and Africa. The 56 species of the Lernaeinae are more widely distributed, with 38 old-world species and 15 new-world species. South America harbours only five species and the South Pacific/Australasia region only two species. Ho (1998) noted that 101 of the 113 species of Lernaeidae are confined to Asia and Africa. This peculiar pattern seems to have resulted from an explosive cladogenesis on the "Indian raft", and this only took place after the ancestral lernaeids colonized the diverse fish family Cyprinidae as hosts. Ho inferred that the ancestral stock of the family invaded Gondwana before the separation of the Indian subcontinental plate from Africa in the Late Cretaceous (about 80 Mya). This invasion presumably occurred after the separation of Africa/India from the rest of Gondwanaland about 95 Mya.

D. The Order Siphonostomatoida

The only colonization of fresh water by siphonostomatoid copepods was by ancestors of the *Salmincola* lineage within the family Lernaeopodidae. Members of this lineage exhibit a boreal distribution, mostly restricted to the higher latitudes of the northern hemisphere. The distribution of this family in continental waters has probably been profoundly influenced both by the Pleistocene glaciations and by the availability of suitable hosts.

E. The Order Poecilostomatoida

The evolutionary history of the Ergasilidae is complex. Ergasilids are found on and around the coastal margins of all continents, with the exception of Antarctica, and they occur in various salinity regimes. It is not possible here to analyse fully the geographical distribution patterns exhibited within the family; however, El-Rashidy (1999) has postulated that most South American ergasilids belong to a single monophyletic lineage. This lineage has invaded and diversified spectacularly in the Amazonian region and has subsequently spread northwards into Central America. The penetration of species of the genus *Acusicola* as far north as Texas indicates that the closure of the Panama gap in the Pliocene may have permitted an intense period of north/south aquatic faunal exchange in both directions. Thatcher (1998) calculated that only a tiny proportion of the possible ergasilid fauna of Amazonia is currently known to science. The biogeographical knowledge of the Ergasilidae is too incomplete to permit its use in the central analysis of this chapter.

F. The Order Gelyelloida

The Gelyelloida are currently reported only from groundwater habitats of north-western Europe, but the authors consider them to be similar to other lineages with a distribution in the northern part of the northern hemisphere.

The estimated colonization dates derived by inference from modern distributions are summarized in Table 2, which also provides approximate estimates of the species richness for each lineage considered.

V. DISCUSSION

This analysis is preliminary. Accurate species richness data are not available for all lineages and there are no fossil data to provide confirmation of colonization dates estimated from modern distribution patterns. Consequently, it is only possible to make a general inference – that the diversity of a lineage appears to be directly proportional to the length of time elapsed since initial colonization of fresh water by that lineage, i.e. the more ancient the colonization, the larger the number of species in the lineage. This inference relies on the assumption, used in the analysis, that tectonic events have shaped modern distributions more than subsequent dispersal events. There is also an obvious link between timing of colonization and the area of habitat available for colonization. The earliest colonists could exploit the whole of Pangaea, but

Table 2

Estimated dates for the colonization of fresh and inland continental waters by copepod lineages and approximate species richness data (number of species in fresh water) for each lineage. The position of the Paradiaptominae is uncertain

Pangaean	Post-Gondwana/ Laurasia split	Post-isolation of Africa/India	Postglacial
Cyclopidae (700+ species)	[Centropagidae] *Boeckella* group (60 species)	Lernaeidae (110 species)	[Centropagidae] *Sinocalanus* group (five species)
Canthocamptidae (600+ species)	Diaptominae (490 species)		[Temoridae] *Eurytemora* group (27 species)
Parastenocarididae (210+ species)	[Ameiridae] *Nitocrella* group (100+ species)	[Diosaccidae] *Schizopera* group (12 species)	*Senecella* (two species)
	Paradiaptominae? (22 species)	Paradiaptominae? (22 species)	Gelyelloida (two species)
			[Lernaeopodidae] *Salmincola* group (38 species)
			[Harpacticidae] *Harpacticella* group (four species)

later colonists had only particular fragments of Pangaea and, therefore, smaller areas available. Unravelling the interplay between time and area effects will be the subject of a future study.

It is surprising that few exceptions have been identified, in the form of ancient but small, relict lineages. One possible candidate must be the family Phyllognathopodidae, which is not included in Table 2. This is a relatively primitive family with a widespread distribution but with few valid species, some of which occur in marginal habitats such as phytotelmata. It apparently contradicts the above general inference; however, its presence in phytotelmata indicates a high dispersal ability. This, in turn, suggests the possibility that the distribution pattern is shaped more by dispersal than by plate tectonics.

The simple estimates of the timing of colonization events generated by interpretation of modern distributions indicate a succession of colonizations. Families such as the Cyclopidae, Parastenocarididae and Canthocamptidae were in the first wave of copepods to colonize and disperse through the varied freshwater habitats of Pangaea. They were followed by a second wave of families, such as the Ameiridae and Diaptomidae (subfamily Diaptominae) invading the northern supercontinent of Laurasia, and the Centropagidae invading the southern supercontinent of Gondwana. Later still came the invasion of Africa/India by the parasitic family Lernaeidae and one or more clades within the Diosaccidae. The Paradiaptominae may also belong in this third wave, or possibly in the second wave. More recently came a fourth wave, but only in the northern part of the northern hemisphere, comprising the *Sinocalanus* group, the Temoridae, *Senecella* species, the Gelyelloida, the *Harpacticella* group and the parasitic *Salmincola* lineage. The restriction of all of these relatively small lineages to the northern part of the Holarctic region suggests that the Pleistocene glaciations may have had a profound effect on their distributions. All may be postglacial invasions.

The taxa identified here as representing modern incursions into fresh water because they have not diversified, for example the Acartiidae, Laophontidae, Cletodidae and the Pseudodiaptomidae, may each be potential colonizers. These taxa may be in the vanguard of the next great wave of copepods to shift from marine into inland continental waters.

The sequence of waves of colonization identified in this analysis is probably an artefact reflecting the indirect method of estimating colonization dates from gross distribution patterns, but it provides a convenient way of grouping the invasions.

The dynamics of the interactions between colonizers and existing faunas is interesting. The hypothesis presented here, that the neotropical diaptomines represent a recent invasion which has swept southwards, displacing the existing *Boeckella* group calanoids in lowland tropical to warm temperate zones, requires testing. To facilitate the testing process the authors predict that the vast majority of South American diaptomines will belong to a single monophyletic lineage (the only likely exceptions being vagile dispersalist species, such as *Arctodiaptomus dorsalis*). Molecular clock dating of this neotropical lineage should indicate a relatively recent radiation (about 3 Mya) to be consistent with this hypothesis. The sister group of this lineage should be found in Central/North America. Similarly, the great majority of African diaptomines should also belong to one or a small number of monophyletic lineages, but the colonization of, and radiation within Africa of the diaptomines may have taken place about 17 Mya. The colonization scenario presented here should provide a framework within which to investigate the colonization history and species diversification of copepods within ancient lakes.

ACKNOWLEDGEMENTS

The authors are grateful to Danielle Defaye (Museum National D'Histoire Naturelle, Paris) and Andrew Rossiter (Lake Biwa Museum) for their informed comments on the ideas presented here. Rony Huys (NHM) provided useful advice on the current state of harpacticoid phylogeny.

REFERENCES

Ahnert, A. (1998). Has the main habitat of *Potamocaris* species been overlooked until now? (Harpacticoida, Parastenocarididae). *J. Mar. Syst.* **15**, 121–125.

Bayly, I.A.E. (1992). The non-marine Centropagidae (Copepoda: Calanoida) of the World. In: *Guides to the Identification of the Microinvertebrates of Continental Waters of the World* 2 (Ed. by H.J. Dumont), pp. 1–30. SPB Academic Publishers, Amsterdam.

Bayly, I.A.E. and Morton, W. (1978). Aspects of the zoogeography of Australian microcrustaceans. *Verh. Int. Verein. Limnol.* **20**, 2537–2540.

Cohen, A.S., Soreghan, M.J. and Scholz, C.A. (1993). Estimating the age of formation of lakes: an example from Lake Tanganyika, East African Rift System. *Geology* **21**, 511–514.

Cressey, R.F. and Boxshall, G.A. (1989). *Kabatarina pattersoni*, a fossil parasitic copepod from a Lower Cretaceous fish, *Cladocyclus gardneri* Agassiz. *Micropalaeontology* **35**, 150–167.

Defaye, D. (1998). Description of the first *Boeckella* (Copepoda, Calanoida, Centropagidae) from New Caledonia. *Crustaceana* **71**, 686–699.

Dussart, B. and Defaye, D. (1983). *Répertoire Mondial Crustacés Copépodes des Eaux Intérieures. I. Calanoïdes*, 224 pp. CNRS, Paris.

Dussart, B. and Defaye, D. (1985). *Répertoire Mondial des Crustacés Copépodes Cyclopoïdes*, 236 pp. CNRS, Paris.

Dussart, B. and Defaye, D. (1990). Répertoire Mondial Crustacés Copépodes des Eaux Intérieures. I. Harpacticoïdes. *Crustaceana* Suppl. **16**, 384 pp.

El-Rashidy, H. (1999). Copepods and grey mullets (Mugilidae). Unpublished PhD thesis, University of London.

El-Rashidy, H. and Boxshall, G.A. (1999). Ergasilid copepods (Poecilostomatoida) from the gills of primitive Mugilidae (grey mullets). *Syst. Parasitol.* **42**, 161–186.

Ferrari, F.D. and Bowman, T.E. (1980). Pelagic copepods of the family Oithonidae (Cyclopoida) from the east coast of Central and South America. *Smiths. Contrib. Zool.* **312**, 1–27.

Gorthner, A. (1994). What is an ancient lake? *Arch. Hydrobiol. Beiheft. Ergebnisse Limnol.* **44**, 97–100.

Ho, J.-s. (1998). Cladistics of the Lernaeidae (Cyclopoida), a major family of freshwater fish parasites. *J. Mar. Syst.* **15**, 177–183.

Ho, J.-s., Conradi, M. and López-González, P. (1998). A new family of cyclopoid copepods (Fratiidae) symbiotic in the ascidian (*Clavelina dellavallei*) from Cádiz, Spain. *J. Zool. Lond.* **246**, 39–48.

Holmquist, C. (1970). The genus *Limnocalanus* (Crustacea, Copepoda). *Z. Zool. Syst. Evol.* **8**, 273–296.

Huys, R. and Boxshall, G.A. (1991). *Copepod Evolution*, 468 pp. The Ray Society, London.

Huys, R. and Iliffe, T.M. (1998). Novocriniidae, a new family of harpacticoid copepods from anchihaline caves in Belize. *Zool. Scripta* **27**, 1–15.

Kabata, Z. (1979). *Parasitic Copepoda of British Fishes*, 468 pp. The Ray Society, London.

Lang, K. (1948). *Monographie der Harpacticiden*, 2 Vols, 1682 pp. Håkan Ohlsson's Bøktryckeri, Lund.

Lovejoy, N.R., Bermingham, E. and Martin, A.P. (1998). Marine incursion into South America. *Nature* **196**, 421–422.

Maly, E.J. (1996). A review of relationships among centropagid copepod genera and some species found in Australasia. *Crustaceana* **69**, 727–733.

Ribbink, A.J. (1994). Lake Malawi. *Arch. Hydrobiol. Beiheft. Ergebnisse Limnol.* **44**, 27–33.

Rouch, R. (1986). Copepoda: les Harpacticoïdes souterrains des eaux douces continentales. In: *Stygofauna Mundi* (Ed. by L. Botosaneanu), pp. 321–355. E.J. Brill, Leiden.

Santos Silva, E.N., Boxshall, G.A. and da Rocha, C.E.F. (1999). The Neotropical genus *Notodiaptomus* Kiefer, 1936 (Calanoida: Diaptomidae): redescription of the type species *Notodiaptomus deitersi* (Poppe, 1891) and designation of a neotype. *Stud. Neotrop. Fauna Envir.* **34**, 114–128.

Suárez-Morales, E. and Iliffe, T.M. (1996). New superfamily of Calanoida (Copepoda) from an anchialine cave in the Bahamas. *J. Crustacean Biol.* **16**, 754–762.

Thatcher, V.E. (1998). Copepods and fishes in the Brazilian Amazon. *J. Mar. Syst.* **15**, 97–112.

Vyshkvartzeva, N.V. (1994). *Senecella siberica* n.sp. and the position of the genus *Senecella* in Calanoida classification. *Hydrobiologia* **292/293**, 113–121.

The Ichthyofauna of Lake Baikal, with Special Reference to its Zoogeographical Relations

V.G. SIDELEVA

I. SUMMARY

Fifty-eight species and subspecies of fishes are found in Lake Baikal, of which 52 taxa are native to the lake and six have been introduced. The fish fauna is of mixed nature and is represented by four groups: Eurasian fishes, Siberian fishes, endemics and introduced fishes. This chapter considers the taxonomic composition of these groups and also comments on their general biology. The fish fauna is dominated, in terms of species diversity, overall abundance and biomass, by endemic taxa of the family Cottoidei. Lake Baikal is the centre of a secondary speculation of cottoid fishes and therefore special attention is given here to this group.

II. INTRODUCTION

Lake Baikal is the deepest lake in the world, reaching 1637 m (Galuzii, 1990). About 70% of the lake area consists of waters deeper than 500 m (Kolokoltzeva, 1961). However, because of renewal of even the abyssal layers, the waters of Lake Baikal are mixed thoroughly and are oxygenated throughout, with concentrations of oxygen of 9 mg l^{-1} at the bottom (Votinzev, 1961). This presence of oxygen even at the deepest layers has allowed for the evolution of a rich abyssal fauna.

The present ichthyofauna of Lake Baikal includes 58 species and subspecies of fishes belonging to 30 genera and 15 families (Table 1), which constitute only

ADVANCES IN ECOLOGICAL RESEARCH VOL. 31
ISBN 0-12-013931-6

c. 3% of the total faunal taxa found in this lake. Of the ichthyofauna, 52 taxa are native to Lake Baikal and six have been introduced. The fish fauna is of mixed nature and is represented by four groups differing in their distribution in the lake, phylogenetic relations with faunas from other fresh waters, the type of distribution and the degree of endemism: Eurasian fishes, Siberian fishes, endemic forms and introduced fishes.

Each of these groups is now considered separately, and details are given of their general biology and ecology.

III. THE FISH FAUNA OF LAKE BAIKAL

A. Eurasian Species

This group contains 12 species and subspecies distributed in five families and is dominated by members of the Cyprinidae (eight species and subspecies), with one species each of Esocidae, Lotidae, Percidae and Cobitidae (Tables 1 and 2). Species of this group are widespread in fresh waters (lakes and rivers) in Eurasia. In Lake Baikal they inhabit the waters of shallow bays (sors) and the mouths of large rivers. From July to September they are found also in the littoral zone of the open lake, where they occupy the shallow waters to depths of 15–20 m.

The highest densities of fishes in this category are seen along the eastern shore of the lake, with its wide shelf. All of the Eurasian taxa spawn in shallow waters of the lake and the major spawning areas of these fishes are also situated in the sors of the eastern shore. Spawning migrations begin in April beneath the ice and spawning takes place over a 10–20 day period in May–June during the earliest period of open water. They feed mainly on benthic invertebrates, such as gammarids, chironomids and molluscs, but some species also feed on fish. Most of these fishes are sought after commercially and the biomass of this group is *c.* 20 000 t, some 7% of the total fish biomass of Lake Baikal (Table 3). The annual catch of Eurasian fishes shows marked fluctuations, for example, in 1994 it was 2115 t (53% of the total commercial fish catch), whereas in 1996 it was 1037 t (26%) (Kalashnikov *et al.*, 1997). This group contains no endemic taxa (Table 2).

B. Siberian Species

The group of Siberian fishes contains nine species and subspecies, distributed within five families: two species of the family Salmonidae, two subspecies of the family Thymallidae, three subspecies of the family Coregonidae, one species of the family Balitoridae and one subspecies of the family Acipenseridae. Outside Lake Baikal members of this group are riverine and are found in an area extending from the Urals to the Far East, but in the lake they inhabit the

Table 1
List of fish taxa occurring in Lake Baikal

Family, species, subspecies	Endemics	Distribution in Lake Baikal			Eurasia	Siberia	Depth (m)
		Shallow water	Deep water	Bays and delta rivers			
Cyprinidae							
Rutilus rutilus lacustris (Pallas)	−	+	−	+	+	+	1–20
Leuciscus leuciscus baicalensis (Dyb.)	−	+	−	+	+	+	1–20
L. idus (L.)	−	−	−	+	+	+	1–20
Phoxinus phoxinus (L.)	−	+	−	+	+	+	1–10
Ph. perenurus (Pallas)	−	−	−	+	+	+	1–10
Gobio gobio cynocephalus Dyb.	−	−	−	+	+	+	1–10
Carassius auratus gibelio (Bloch.)	−	−	−	+	+	+	1–20
Tinca tinca (L.)	−	−	−	+	+	+	1–20
Percidae							
Perca fluviatilis L.	−	+	−	+	+	+	2–20
Cobitidae							
Cobitis melanoleuca Nichols	−	+	−	+	+	+	1–10
Balitoridae							
Barbatula toni (Dyb.)	−	+	−	+	−	+	1–10
Esocidae							
Esox lucius L.	−	+	−	+	+	+	1–20
Lotidae							
Lota lota (L.)	−	+	−	+	+	+	5–60

Table 1 *Continued*

Family, species, subspecies	Endemics	Shallow water	Deep water	Bays and delta rivers	Eurasia	Siberia	Depth (m)
Salmonidae							
Hucho taimen (Pallas)	–	+	–	+	+	+	5–25
Brachymystax lenok (Pallas)	–	+	–	+	–	+	5–25
Thymallidae							
Thymallus arcticus baicalensis (Dyb.)	+	+	–	+	–	+	3–50
Th. arcticus brevipinnis Svetovid.	+	+	–	+	–	+	7–100
Coregonidae							
Coregonus autumnalis migratorius (Georgi)	+	+	+	+	–	+	5–350
C. lavaretus baicalensis (Dyb.)	+	+	–	+	–	+	20–200
C. lavaretus pidschian (Gmelin)	+	+	–	+	–	+	20–150
Acipenseridae							
Acipenser baeri baicalensis Nikolsky	+	–	–	+	–	+	4–200
Cottidae							
Cottus kesslerii Dyb.	+	+	–	+	–	+	1–250
Paracottus knerii (Dyb.)	+	+	–	–	–	+	1–150
Batrachocottus baicalensis (Dyb.)	+	+	–	–	–	–	1–20
B. multiradiatus Berg	+	–	+	–	–	–	50–900
B. talievi Sideleva	+	–	+	–	–	–	100–1000
B. nikolskii Berg	+	–	+	–	–	–	100–1300
Cottocomephorus grewingkii (Dyb.)	+	+	+	–	–	–	15–350

Distribution in Lake Baikal (Shallow water, Deep water, Bays and delta rivers)

Table 1   *Continued*

Family, species, subspecies	Endemics	Distribution in Lake Baikal			Eurasia	Siberia	Depth (m)
		Shallow water	Deep water	Bays and delta rivers			
C. inermis (Jakowlew)	+	+	+	–	–	–	50–500
C. alexandrae Taliev	+	+	+	–	–	–	10–400
Comephoridae							
Comephorus baicalensis (Pallas)	+	–	+	–	–	–	250–1600
C. dybowski Korotneff	+	–	+	–	–	–	150–1600
Abyssocottidae							
Asprocottus herzensteini Erg	+	–	+	–	–	–	25–400
A. intermedius Taliev	+	–	+	–	–	–	200–900
A. platycephalus Taliev	+	–	+	–	–	–	50–800
A. abyssalis Taliev	+	–	+	–	–	–	150–1400
A. pulcher (Taliev)	+	–	+	–	–	–	50–250
A. parmipherus Taliev	+	–	+	–	–	–	50–500
Limnocottus godlewskii (Dyb.)	+	–	+	–	–	–	25–600
L. megalops (Gratzianow)	+	–	+	–	–	–	25–500
L. eurystomus (Taliev)	+	–	+	–	–	–	50–600
L. griseus Taliev	+	–	+	–	–	–	250–1300
L. pallidus Taliev	+	–	+	–	–	–	150–1000
L. bergianus Taliev	+	–	+	–	–	–	100–1000
Abyssocottus korotneffi Berg	+	–	+	–	–	–	400–1600
A. gibbosus Berg	+	–	+	–	–	–	400–1600

Table 1 *Continued*

Family, species, subspecies	Endemics	Distribution in Lake Baikal			Eurasia	Siberia	Depth (m)
		Shallow water	Deep water	Bays and delta rivers			
A. elochini Taliev	+	–	+	–	–	–	200
Cottinella boulengeri (Berg)	+	–	+	–	–	–	400–1600
Procottus jeittelesii (Dyb.)	+	+	–	–	–	–	15–50
P. major Taliev	+	+	+	–	–	–	50–950
P. gurwici Taliev	+	+	–	–	–	–	15–100
Neocottus werestschagini (Taliev)	+	+	–	–	–	–	900–1400
Introduced species							
Cyprinidae							
Abramis brama orientalis Berg	–	–	–	+	+	+	1–15
Cyprinus carpio haematopterus Temm. & Schl.	–	+	–	+	–	+	1–20
Siluridae							
Parasilurus asotus (L.)	–	–	–	+	–	+	1–70
Eleotridae							
Perccottus glenii Dyb.	–	–	–	+	+	+	1–20
Coregonidae							
Coregonus peled (Gmelin)	–	–	–	+	+	+	?
C. albula (L.)	–	–	–	+	+	+	?

Details of distribution in the lake, depth range occupied, endemic status and distribution outside the lake are also given.

Table 2
Taxonomic composition of fishes occurring in Lake Baikal, including introduced forms

Family	Genus number	Species/ subspecies number	% of total species and subspecies number	Endemic species	Endemic subspecies
Cyprinidae	8	10	18	–	–
Percidae	1	1	2	–	–
Cobitidae	1	1	2	–	–
Balitoridae	1	1	2	–	–
Esocidae	1	1	2	–	–
Lotidae	1	1	2	–	–
Thymallidae	1	2	4	–	2
Coregonidae	1	4	7	–	2
Salmonidae	2	2	4	–	–
Acipenseridae	1	1	2	–	–
Cottidae	4	9	12	9	–
Comephoridae	1	2	4	2	–
Abyssocottidae	6	20	35	20	–
Siluridae	1	1	2	–	–
Eleotrididae	1	1	2	–	–
Total	31	58	100	31	4

mouths of cold-water mountain rivers and the littoral zone of the open lake, where they occupy a large depth range of 5–350 m (Figure 1b). However, spawning of these fishes takes place only in rivers and the spawning tactic of most members is to bury their eggs in a gravel/pebble substrate. There is no parental care. This group contains benthic, near-bottom pelagic and pelagic forms. The major part of the diet consists of benthic invertebrates, but adults eat fishes also. The one exception is *Coregonus autumnalis migratorius*, which feeds mainly on mesoplankton and macroplankton (Kontorin and Volerman, 1983).

The families Coregonidae and Thymallidae each contain endemic Baikal subspecies: *C. a. migratorius, C. lavaretus baikalensis* and *C. l. pidschian*, and *Thymallus arcticus baicalensis* and *T. a. brevipinnis*, respectively. The species and subspecies of these two families in Lake Baikal are characterized by a complex intraspecific structure and the presence of a large number of ecological forms. For example, each of the three major populations of *C. a. migratorius* occupies a different habitat: littoral, deep water and pelagic, respectively. Furthermore, each of these populations is further divided into a number of micropopulations, which differ in such features as the habitat and

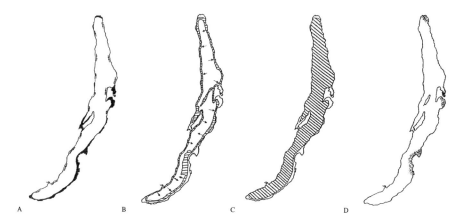

Fig. 1. Distribution of (A) Eurasian, (B) Siberian, (C) Endemic and (D) Introduced
groups of fishes of Lake Baikal. Arrows in B indicate migration trends.

locality occupied, the food taken, mode of feeding, gross morphology and the
rivers used for spawning (Smirnov and Shumilov, 1974), and as such might
represent examples of incipient speciation. The various populations of *C. a.
migratorius* also differ in relative abundance: the pelagic population accounts
for 44% of total species abundance, the littoral population 46% and the deep-
water population 10% (Kalyagin and Maistrenko, 1997).

Several members of this fish group are of commercial importance. The total
biomass of all Siberian fishes is *c.* 34 000 t, which is 13% of the total
ichthyomass in the lake (Table 3). The most valuable commercial species is *C.
a. migratorius*, which has an estimated biomass of 28 000 ± 5000 t (Table 4),
with 2800 t being fished annually (Kalashnikov *et al.*, 1997).

C. Endemic Species

Taxa with endemic subspecific status are not included in this group, but are
categorized according to their species status (Siberian, in all six endemic
subspecies found naturally in the lake). There are 31 endemic species known
from Lake Baikal, all of which are members of the suborder Cottoidei and are
distributed within three families: the Cottidae, the Comephoridae and the
Abyssocottidae (Tables 1 and 2).

World-wide, a total of 73 species of cottoid fishes (19.2%) is known from
fresh waters. With the exception of those found in Lake Baikal, all belong to a
single family, the Cottidae, and are distributed among five genera, *Cottus* (37
species), *Mesocottus* (one species), *Trachidermus* (one species), *Triglopsis* (two
species) and *Leptocottus* (one species) (Goto *et al.*, 1978; Lee *et al.*, 1980;
Reshetnikov, 1998). These freshwater cottoid fishes are exclusively benthic and

Table 3

Taxonomy and quantitative characterization of the fish fauna of Lake Baikal

	Families		Genera		Species		Subspecies		Total	% of endemics	Biomass (10³ tonnes)	% of total biomass
	a	b	a	b	a	b	a	b				
Eurasian species	5	0	11	0	7	0	5	0	12	0	20	7
Siberian species	5	0	6	0	4	0	5	5	9	55	34	13
Endemic species	3	2	11	10	31	31	–	–	31	100	215	80
Introduced species	4	0	5	0	4	0	2	0	6	0	?	<1

a = Total number; b = number of endemics.

Table 4

Total biomass of various categories of fish groups in Lake Baikal

Groups of fish	Biomass (10³ tonnes)	% of total biomass
Eurasian species (12)	20	7
Siberian species (9)	34	13
Coregonus autumnalis migratorius	28	10
Endemic species (31)	215	80
Bottom species (26)	16	6
Pelagic species (2)	185	69

Numbers in parentheses indicate taxa. The coregonid *C. a. migratorius* is the most sought after and valuable target of the commercial fisheries, and is therefore shown here separately.

are found in cold waters of America, Eurasia and Japan. Sixteen freshwater cottoids (excluding the two Baikal species *C. kesslerii* and *Paracottus knerii*) of four genera – *Cottus* (10 species), *Triglopsis* (one species), *Trachidermus* (one species) and *Mesocottus* (one species) – are known from Eurasia.

The family Cottidae is represented in Lake Baikal by two subfamilies, the Cottinae and the Cottocomephorinae. The Cottinae includes only one species, *Cottus* (*Leocottus*) *kesslerii*, whereas the endemic subfamily Cottocomephorinae contains eight species belonging to three genera. The family Comephoridae is endemic to Baikal and includes only two species of one genus *Comephorus*. The family Abyssocottidae is represented by 20 species, distributed within six genera, *Abyssocottus* (three species), *Asprocottus* (six species), *Cottinella* (one species), *Limnocottus* (six species), *Neocottus* (one species) and *Procottus* (three species; cf. Nelson, 1994). As such, 65% of species of Baikal Cottoidei belong to this one endemic family.

The incidence of endemism in the cottoid fauna of Lake Baikal is very high and extends to the family level. Only one cottoid species, *C. kesslerii*, is found in the sors; the other Baikal cottoids are inhabitants of open Baikal, where they occur throughout the entire depth range of the lake, from the shoreline down to depths exceeding 1600 m (Figure 1c).

Only two species of Baikal cottoid fishes occur also outside the lake, *Cottus kesslerii* and *Paracottus knerii*. *Cottus kesslerii* is found in shallow lakes in the Lake Baikal basin and also in water reservoirs of the Angara River. The subspecies *C. kesslerii arachlensis* was described from Lake Arakchlei; another subspecies, *C. k. gussinensis*, inhabits the Selenga River and Gusinoe Lake. *Paracottus knerii* is most abundant in the rivers of the Baikal basin, whereas Lake Agata (Yenisei basin) is inhabited by a subspecies, *P. k. putorania*. Interestingly, the cottoid fishes found in the basin of Lake Baikal and the Angara River flowing from the lake are exclusively Baikal native forms and other freshwater Cottoidei are completely absent. It is evident that a secondary radiation of cottoids has taken place in Lake Baikal and that this has occurred independent of subsequent immigration of other Siberian and Eurasian cottoid taxa.

One characteristic of the distribution of Baikalian Cottoidei is that the maximal number of species (22 species, or 70% of all Baikal cottoid fishes) is found within the upper 400–500 m depths of the lake (Sideleva, 1996) (Figure 2). All species of Cottidae and Abyssocottidae and *c.* 93% of the other endemic species are lithophilous. In these species, the female lays eggs on the underside of a large stone and the male guards the eggs through the entire period of embryogenesis. The exceptions are the two species within the family Comephoridae, *Comephorus baicalensis* (Pallas) and *C. dybowski* Konotreff, which have become ovoviviparous as an adaptation to their pelagic lifestyle.

Baikal cottoid fishes can be grouped according to their habitat utilization of substrate and the water column: a benthic group which contains 26 species (six species of the family Cottidae and all 20 species of the Abyssocottidae); near-bottom pelagic species (three species of the family Cottidae); and pelagic species (two species of the family Comephoridae) (Figure 2).

Benthic species are distributed in Lake Baikal from the shoreline down to maximal depths in excess of 1600 m. This group is taxonomically the most diverse (27 species, 86%) among the Baikal cottoid fishes and can be subdivided into coastal species (seven species, inhabiting depths from 1–250 m), eurybathic species inhabiting a wide range of depths (14 species, which can be further subdivided into two groups which occupy depth ranges of 50–600 m and 100–1300 m, respectively) and abyssal species (six species, restricted to 400–*c*. 1630 m). Benthic species live and reproduce on the bottom, and all (except *C. kesslerii*) have large eggs and benthic larvae. *Cottus kesslerii* is the only Baikal benthic cottid species to have small eggs (diameter 0.8 mm) and pelagic larvae.

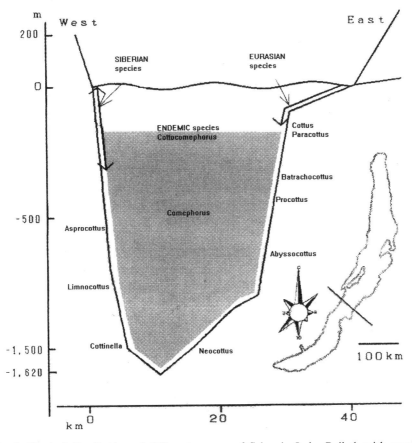

Fig. 2. Vertical distribution of different groups of fishes in Lake Baikal, with special reference to cottoid genera.

Benthopelagic species are represented by the three members of the genus *Cottocomephorus*: these live partly on the bottom and partly in the water column, at depths of 2–500 m. These species reproduce by laying adhesive eggs on the bottom and have pelagic larvae. It is of interest that the benthopelagic species *Cottocomephorus grewinkii* has three discrete spawning populations which differ in the time of spawning. One consequence of this difference is that ambient water temperature at spawning, and therefore the time of embryogenesis, also differs among the three groups. Spawning occurs either in March (beneath the ice at water temperatures of *c.* 0.8–1.0°C), in May in the period of ice melting (water temperature *c.* 1.0–1.5°C) or in August when water temperatures are highest (*c.* 8–12°C).

Pelagic species (two species of Comephoridae) live in the water column at depths of *c.* 150–1630 m. These species show internal fertilization and ovoviviparity, and embryonic development is specialized. Mating and release of larvae, which are pelagic, take place in the water column. The diet of young comephorids consists mostly of mesozooplankton (*Epishura baicalensis*), while adults feed mainly on the pelagic amphipod *Macrohectopus branickii* and comephorid larvae. There is a high degree of interspecific overlap (65%) in the diets of adult *C. baicalensis* and *C. dybowski* (Sideleva, 1996).

The diet of benthic cottoids consists mainly of benthic gammarids and pelagic mesoplankton and microplankton, whereas the diet of near-bottom pelagic species includes benthic and planktonic invertebrates. Twelve types of dietary component have been identified in the gut contents of shallow-water cottoids, while in deep-water benthic species there are only two to four (Sideleva, 1996). Of the 257 species and 74 subspecies of gammarids known from Lake Baikal (Mazepova, 1978; Takhteev, this volume), approximately 70 have been recorded as items in the diet of cottoid fishes (Sideleva and Mekhanikova, 1990) and they are the major dietary items for all types of Lake Baikal cottoids: the percentages of gammarids in the diets of shallow-water, eurybathic and abyssal species range from 44 to 73% (of body mass), 72 to 93% and 92 to 96%, respectively (Sideleva and Mechanikova, 1990).

Although their total biomass constitutes more than 215 000 tonnes, representing 80% of the total fish mass of Lake Baikal, there is no commercial fishery for cottoid fishes. The biomass of cottoid fishes is dominated by the two pelagic comephorid species, with a combined biomass of *c.* 185 000 tonnes (Nagornyi *et al.*, 1984). This amount far exceeds the total biomass all 25 benthic species of the families Cottidae and Abyssocottidae, which reaches 16 000 tonnes. The high abundance and biomass of the comephorids is made possible by the presence in the water column of their major food item, the pelagic amphipod *Macrohectopus branickii*. This amphipod is abundant and widespread throughout the lake, and forms thick layers, mainly at depths of 100–1000 m (Melnik *et al.*, 1993). In Lake Baikal, *M. branickii* occupies an important position in the foodweb, similar to that of oceanic krill (Permitin,

1970), and its high abundance in the pelagic zone has facilitated the evolution of a midwater lifestyle in the two comephorids.

D. Introduced Species

This group comprises six species in four families: the Cyprinidae (two species), Coregonidae (two species), Siluridae (one species) and Eleotridae (one species) (Table 2). The latter two families are the newest members of the ichthyofauna of Lake Baikal. The native habitat of *Cyprinus carpio haematopterus*, *Parasilurus asotus* and *Perccottus glenii* is the Amur River basin. *Abramis brama orientalis* was introduced from Lake Ubinskoye (west Siberia), while *Coregonus pelad* and *C. albula* were introduced from northern water bodies of Siberia. Species of the families Cyprinidae and Siluridae were introduced by humans in the 1930s to 1950s for the purpose of increasing the stock of commercially viable fishes (Ashkaev, 1958). These species inhabit the sors together with Eurasian species (Figure 1d), but have not adapted well to the environment of Lake Baikal. Following their introduction into the lake, their numbers peaked in the 1960s, but thereafter have shown a continuous decline and, currently, both species are present in very low numbers only. The dominant foods of the Cyprinidae in Lake Baikal are benthic invertebrates, gammarids, chironomids and molluscs, and their diet thus overlaps with that of other native Baikal fishes, with which they are trophic competitors. *Parasilurus asotus* is a predator, and its diet consists of cyprinid fishes (Eurasian group), the cottoid *C. kesslerii* (endemic group) and the eleotrid *P. glenii* (acclimatized group). The introduced Cyprinidae and Siluridae are characterized by a slower growth rate, later maturation and a lower annual fecundity (number of spawnings) than is seen in populations within their normal distribution range. For example, *C. carpio haematopterus* in the Amur River spawns three times during the summer, whereas in Lake Baikal it spawns only twice (Skryabin, 1997).

The two coregonid species and the one eleotrid species were introduced into Lake Baikal accidentally. Eggs of *Coregonus peled* and *C. albula* were incubated at fish farms, and newly hatched larvae were mixed with those of Baikal *C. a. migratorius*. Larvae of all three species were then released together into Lake Baikal as part of a supposed *C. a. migratorius* restocking programme.

In 1969 the eleotrid *P. glenii* was accidentally introduced together with eggs of *C. carpio haematopterus* into the Baikal Basin, from where it spread through the Selenga River into Lake Baikal. *Coregonus peled* and *C. albula* are present in Lake Baikal only in very low numbers, but *P. glenii* is very abundant in the delta of the Selenga River, where its density can reach *c.* 100 individuals m^{-2} (Skryabin, 1997). This species reaches maturity at the age of 2 years, when females are *c.* 45 mm in length, and remains reproductively active until at least the age of 8 years. In Baikal, this species spawns twice annually, first in June and then again in late July to early August, with up to 5350 eggs laid in the first clutch. *Perccottus glenii* feeds on the eggs and larvae of Eurasian and Siberian species, and this has a negative effect on the commercial stocks of these fishes. Currently, the population of *P. glenii* is increasing and the species is expanding its range to occupy new areas of the lake. Unusually large numbers are found at depths of 70–75 m (Skryabin, 1997), but why this depth range is favoured by this species is unknown.

In recent years *Salmo trutta* has been recorded in the Irkutsk Reservoir (Shirobokov, 1994), but it has not yet reached Lake Baikal. As a whole, the introduced species have become part of the ichthyocoenoses represented by small Eurasian species, but their abundance is presently low.

IV. CONCLUSIONS

The native fish fauna of Lake Baikal is dominated by endemic forms: fishes in this category comprise 58% of the total number of species and 80% of the total fish biomass. This dominance of endemic forms suggests a high degree of adaptive specialization to the unusual environmental conditions present in this deep and cold-water lake. Each of the three groups of naturally occurring taxa is largely restricted in its use of habitat: Eurasian species are restricted to the sors and to the deltas of large rivers, Siberian species inhabit the outfalls of mountain tributaries and the littoral zone of the open part of Baikal, and endemic species occupy mainly the deep littoral to the abyssal zone.

The taxonomic structure of the ichthyofauna of Baikal reflects peculiar features of its historical development. Eurasian thermophilous species are descendants of the ancient faunas of shallow lakes that were formed and occurred widely in this area in the Neogene (Middle–Upper Miocene). Thus, in the sedimentary deposits of the lake on Olkhon Island (at depths of tens of metres) remains of *Perca, Rutilus* and *Lucioperca* species have been identified (Popova *et al.*, 1989). *Perca* and *Rutilus* are still represented, but *Lucioperca* is now absent from Lake Baikal and its basin. The recent Baikal shallow-water fauna is taxonomically identical with the fauna of Eurasia; however, many genera common in Eurasian waters today have not been recorded from Baikal, either as fossil or as extant forms (Table 5).

It seems most likely that the Siberian species found today in Lake Baikal are representatives of Arctic fauna, assuming that these ancestors probably penetrated Baikal during the glacial period (Berg, 1949). Remains of *Thymallus arcticus* have been found in Pleistocene deposits in caves around Lake Baikal. In Lake Baikal 66% of taxa in this group consist of endemic subspecies. Numerous intraspecific forms, similar to those seen in the glacial Arctic, inhabit Lake Baikal.

Table 5
Dominant modern and fossil Eurasian and Siberian genera of fishes, and their presence in Lake Baikal

Eurasian genera		Siberian genera	
Modern	Fossil, middle–upper Miocene[a]	Modern	Fossil, Pleistocene[b]
Rutilus	*Rutilus*	*Thymallus*	*Thymallus*
Leuciscus	–	*Coregonus*	–
Phoxinus	–	*Acipenser*	–
Gobio	–		
Carassius	–		
Tinca	–		
Perca	*Perca*		
Esox	–		
Lota	–		
Cobitis	–		
–	*Lucioperca*		

[a]Popova *et al.* (1989).
[b]Filippov and Sychevskaya (pers. obs.).

The endemic fish fauna is represented by three families with different degrees of specialization to Baikal conditions, but transitional forms are absent. Only the six species of the endemic cottoid subfamily Cottocomephorinae are morphologically similar to the recent representatives of the genus *Cottus*, which is widespread in fresh waters of Eurasia, North America and Japan, and in which common ancestral forms might be sought. All 22 species of the endemic families Comephoridae and Abyssocottidae are morphologically and ecologically so different from the recent freshwater Cottoidei that it appears impossible to establish ancestral forms. The endemic fauna of fishes is probably the result of intralacustrine evolution from a small number of ancestral forms that invaded Lake Baikal at geologically different times. The presence of 31 species of Cottoidci in Lake Baikal, and only 16 species and subspecies in freshwater bodies elsewhere in Eurasia, indicates Lake Baikal to be the centre of a secondary speciation of cottoid fishes.

REFERENCES

Ashkaev, M.G. (1968). New fish in the Lake Baikal Basin. In: *Fish and Fisheries in the Lake Baikal Basin* (Ed. by M.M. Kozhov and K.I. Misharin), pp. 420–428. Nauka, Irkutsk.
Berg, L.S. (1949). *Essays on the Physical Geography of Baikal, its Nature, and the Genesis of the Organic World.* Academy of Sciences Publications, Moscow (in Russian).

Galuzii, G.I. (1990). Le lac Baikal en sursis. *Recherche* **221**, 628–637.

Goto, A., Nakanishi, H., Utoh, H. and Hamada, K. (1978). A preliminary study of the freshwater fish fauna of rivers in southern Hokkaido. *Bull. Fac. Fish. Hokkaido Univ.* **29**, 118–130 (in Japanese).

Kalashnikov, Y.I., Kalyagin, L.P., Palubis, S.E. and Sokolov, A.V. (1997). Problems of the regulation of management of the Lake Baikal fisheries. In: *Ecological Equivalent Species of Hydrobionths in the Great Lakes of the World*, pp. 13–15. Ulan-Ude, Russia.

Kalyagin, L.P. and Maistrenko, S.G. (1997). The dynamics of distribution of the morphoecological groups of Baikal omul throughout the Baikal expanse. In: *Ecologically Equivalent Species of Hydrobionths in the Great Lakes of the World*, pp. 33–35. Ulan-Ude, Russia.

Kolokoltzeva, E.V. (1961). The morphological characteristics of Lake Baikal. In: *Mesozoic and Cenozoic Lakes of Siberia* (Ed. by N.A. Florensov), pp. 183–192. Nauka, Novosibirsk (in Russian).

Kontorin, V.V. and Volerman, I.V. (1983). *Biological Interactions of Fishes and Seals in Lake Baikal*. Nauka, Novosibirsk (in Russian).

Lee, D.S., Gilbert, C.R., Hocutt, C.H., Jenkins, R.E., McAllister, D.E. and Stauffer, J.R., Jr (eds) (1980). *Atlas of North American Freshwater Fishes*. North Carolina State Museum of Natural History, Raleigh, NC.

Mazepova, G.F. (1978). Baikal fauna, its peculiarities, origins and evolution. In: *Problems of Baikal* (Ed. by G.I. Galuzii and K.K. Votintzev), pp. 181–192. Nauka, Novosibirsk (in Russian).

Melnik, N.G., Timoshkin, O.A., Sideleva, V.G., Pushkin, S.V. and Mamylov, V.S. (1993). Hydroaccoustic measurement of the density of the Baikal macrozooplankter *Macrohectopus branickii*. *Limnol. Oceanogr.* **38**, 425–434.

Nagornyi, V.K., Sideleva, V.G., Volerman, I.B., Kostornov, S.N. and Zavyalova, T.A. (1984). The resource of the Baikal sculpins and their use by omul. In: *Problems in Fisheries Development in the Baikal Basin* (Ed. by D.S. Norenko), pp. 72–79. Nauka, Leningrad.

Nelson, J.S. (1994). *Fishes of the World*, 3rd edn. John Wiley and Sons, New York.

Permitin, Y.E. (1970). The consumption of krill by Antarctic fishes. In: *Antarctic Ecology*, Vol. 1 (Ed. by M.W. Holdgate), pp. 177–182. Academic Press, London.

Popova, S.M., Mats, V.D. and Chernyaeva, G.P. (1989). *Paleolimnological Reconstructions: The Baikal Rift Zone* (Ed. by N.A. Logachev). Nauka, Novosibirsk (in Russian).

Reshetnikov, Yu.S. (ed.) (1998). *Annotated Checklist of Cyclostomata and Fishes of the Continental Waters of Russia*. Nauka, Moscow (in Russian).

Shirobokov, I.I. (1993). On the unpremeditated introduction of the rainbow trout *Onchorhynchus mykiss* into the Irkutsk (L. Baikal). *J. Ichthyol.* **33**, 841–843 (in Russian).

Sideleva, V.G. (1996). Comparative character of the deep-water and inshore cottoid fishes endemic to Lake Baikal. *J. Fish Biol.* **49** Suppl. A, 192–206.

Sideleva, V.G. and Mekhanikova, I.V. (1990). Feeding preference and evolution of the cottoid fishes of Lake Baikal. *Proc. Zool. Inst. USSR Acad. Sci.* **222**, 144–160, (in Russian).

Skryabin, A.G. (1997). Morphological characteristics of *Perccottus glenii* (Eleotridae) in the basin of Lake Baikal. *J. Ichthyol.* **37**, 406–408.

Smirnov, V.V. and Shumilov, I.P. (1974). *Omul of Lake Baikal* (Ed. by B.K. Moskalenko). Nauka, Novosibirsk.

Votinzev, K.K. (1961). Hydrochemistry of Lake Baikal. *Trudy Baikal Limnol. Inst. SO AN SSSR* **20**, 1–312 (in Russian).

The Benthic Invertebrates of Lake Khubsugul, Mongolia

O.M. KOZHOVA, E.A. ERBAEVA and G.P. SAFRONOV

I. SUMMARY

Based on an examination of all relevant literature and including new information from the most recent studies in the lake, the first comprehensive survey is presented of the benthic invertebrate fauna of ancient Lake Khubsugul, a mountainous, cold-water, oligotrophic lake. Lake Khubsugul is located in Mongolia, in the Baikal rift zone, and is connected to Lake Baikal via a river system. The diversity and degree of endemicity of the benthic macroinvertebrate faunas of these two ancient lakes are compared.

The modern Khubsugul fauna comprises at least 287 species and is believed to have resulted from colonization by riverine fauna during the Quaternary period. This is considerably less diverse than the fauna of Lake Baikal. This difference is especially apparent within the Gammaridae, Mollusca, Oligochaeta and Tricladida, which are represented in Baikal by complexes of endemic species. Those taxa found in both lakes are predominantly representatives of General Siberian and Euro-Asian faunas. One species of nematode, two species of oligochaete, three species of chironomid and one

ADVANCES IN ECOLOGICAL RESEARCH VOL. 31
ISBN 0-12-013931-6

species of halacarid are endemic to both lakes. Among the 19 species newly described from Lake Khubsugul, three species of Mermithidae belong to the Baikalian endemic genus *Baicalomermis* and one species of mollusc belongs to the Baikalian endemic subgenus *Achoanomphalus*. These four species, and also perhaps two species of the genus *Gammarus*, are probably endemic to Lake Khubsugul. The remaining 15 newly described species probably also occur in other water bodies.

II. INTRODUCTION

Lake Khubsugul (Kosogol, Khuvsgol) is located in Central Asia in the Sayan mountain region of north Mongolia at an altitude of 1645 m a.s.l. (Figure 1). The lake depression is at the south-western edge of the Baikal rift zone and belongs to the Baikal system of depressions. Geological evidence suggests that the entire Baikal system of depressions developed at the same time; according to Florensov (1960), this was in the Cenozoic, primarily during the Neogene and Quaternary periods. The oldest sedimentary deposits in the Khubsugul depression date back to the Early Quaternary (Pleistocene, 1.6 Mya) (Belova and Kulakov, 1982).

   Three underwater and seven above-water terraces formed by lacustrine–alluvial deposits bear testament to changes in the conditions of the lake during its formation. The Late Cenozoic history of the Khubsugul depression was characterized by a process of intense rock formation and repeated vulcanism. This promoted an increased continental character of the climate, and the appearance of orographical barriers within and beyond the depression.

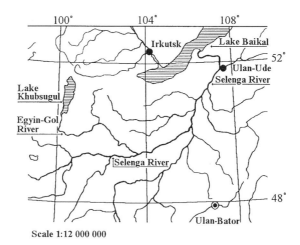

Fig. 1. Map showing the location of Lake Khubsugul and its connection with Lake Baikal via the Egyin-Gol and Selenga rivers.

Belova and Kulakov (1982) distinguish four stages in the change of topography and climate of the Khubsugul depression during the Late Cenozoic:

1. Upper Neogene: "neobaikalian" stage of activation of tectonic movements accompanied by vulcanism. The vegetation contains an admixture of thermophilic elements and climatic conditions are more continental than in the south Baikal trough, with a mean monthly temperature in January of $-10°C$;
2. Early Pleistocene: further activation of tectonic movements, geological uplifting, increased precipitation, effects of erosional processes and a mean monthly temperature in January of $-20°C$;
3. Middle Pleistocene: new activation of tectonic movements with vulcanism. This latest vulcanism has little effect on the main topographical features. With the development of mountains, the climate humidity increases, mountain glaciation develops and outside the glacial zone large lakes are formed. A gradual increase of the climate dryness results in the thawing and degradation of glaciers. Evaporation leads to decreased water flow of rivers and a reduction of the lake's area. Mean January temperatures are only $-25°C$ but, on the whole, the climate is warm;
4. Late Pleistocene: extensive development of mountain glaciation (twofold). The increase in the climate humidity during the Late Quaternary glaciation leads to a rise in lake water levels. During interglacial periods the climate is warm.

Modern Lake Khubsugul is a large, oligotrophic lake with characteristic cold waters, and has a surface area of 2760 km^2 and a water volume of c. 383 km^3. Its maximum depth is 262 m (average 138 m), and its steeply sloping shores mean that almost 85% of its water volume is at depths greater than 50 m. The waters are saturated with oxygen but are poorly mineralized (E_i = 220–230 mg l^{-1}).

The lake is linked to Lake Baikal by the Egyin-Gol River, a tributary of the Selenga River, which feeds into Lake Baikal (Figure 1). It is perhaps unique that two ancient lakes are joined in this manner, and comparisons of the faunas of these lakes may shed light on the roles of colonization processes, lake age and the biology of component species in affecting speciation rates and evolutionary and ecological trends.

III. MATERIALS AND METHODS

The earliest studies of the benthic fauna of Lake Khubsugul were those of Daday (1913), Martynov (1909, 1914), Sovinsky (1915) and Lindholm (1929), and were followed by those of Kozhov (1946, 1962, 1963, 1972) and Kozhov et al. (1965). As such, there exists a reasonably comprehensive literary record of the lake's fauna. More recently, investigations have been carried out by scientists of the Scientific Research Institute of Biology at Irkutsk State University and their Mongolian colleagues, under the auspices of a permanent research project and expedition (Kozhova and Erbaeva, 1984; Kozhova *et al.*, 1994). Information from all of these sources was incorporated into the present chapter.

During the authors' recent surveys, samples of benthic invertebrates were collected from various depths, from the water edge to maximum, and from different parts of the lake, using mainly Petersen grabs, hand nets and scrapers. In the most recent survey over 500 quantitative and qualitative samples were collected. These were preserved and returned to Irkutsk. After preliminary sorting, samples were distributed to systematics specialists from various scientific institutions, for identification of materials.

Comparative data for Lake Baikal were obtained from the detailed compilations of Kozhov (1963) and Kozhova and Izmest'eva (1998), adding new or detailed data where necessary. Sources used for specialist information on certain taxa were: Chironomidae, Linevich (1959, 1964, 1981), Samburova (1982), Wülker *et al.* (1998); Nematoda, Shoshin (1997); Acariformes, Kozhov (1972); and Trichoptera, Ivanov and Menshutkina (1996). The systematic status of certain taxa was based on the following sources: Turbellaria, *Key to Freshwater Invertebrates of Russia and Adjacent Territories* (hereafter KFIRAT) (1994); Nematoda, Gagarin (1993); Mermithidae, Clitellata, Amphipoda and Mollusca, Kozhova *et al.* (1998); Hirudinea, Kozhova *et al.* (1998) and KFIRAT (1994); Acariformes, Ephemeroptera and Plecoptera, KFIRAT (1994); Trichoptera, Spuris (1989); and Chironomidae, Pankratova (1970, 1977, 1983).

IV. RESULTS AND DISCUSSION

A brief summary is presented of the diversity and occurrence of endemism within the main Khubsugul invertebrate fauna, and each group is compared with that found in Lake Baikal.

At present, a total of 287 species of invertebrates has been recorded from Lake Khubsugul. However, several taxa, such as Protozoa, Cnidaria, Rhabdocoela, Rotatoria, Maxillopoda, Ostracoda and Tardigrada, have been subjected to only cursory or no investigation whatsoever, and therefore the number of species recorded from the lake will undoubtedly increase with further study.

A. Parasitic Fauna

In particular, the list does not include known parasitic members of several groups recorded from Lake Khubsugul; such as Protozoa (e.g. a *Mixosporidia* species described from Khubsugul, *Henneguya cerebralis* Pronin, 1972, another protozoan parasite endemic to the Khubsugul subspecies of grayling *Thymallus arcticus nigrescens* Dorogostaisky), Annelida or Arthropoda (e.g. *Salmincola thymalli baicalensis* Messjazeff, 1924). This category of invertebrates was omitted because the present level of knowledge of the Khubsugul parasitic fauna does not permit assessment of its endemic status, nor allow meaningful comparison with the parasitic fauna of Lake Baikal.

B. The Tricladida

Only one species of triclad has been recorded from Lake Khubsugul, the turbellarian *Phagocata* (*Fonticola*) *altaica* (Porfiryeva and Tomilov, 1976). The reasons for this extreme paucity of triclad fauna in Lake Khubsugul are unknown. In stark contrast, Lake Baikal has a rich and diverse triclad fauna, comprising 74 species, all of which (excluding some taxa which occur also in the Angara River) are endemic (Kozhova and Izmest'eva, 1998). This species richness is largely a reflection of the evolutionary radiation and explosive speciation that has occurred within a single family: 37 of these species belong to the family Dendrocoelidae. Livanov and Zabusova (1940) have emphasized that Khubsugul triclads are not closely related to Baikal taxa, and noted that the faunas of these two lakes developed from different sources: the Planariidae in Lake Khubsugul and the Dendrocoelidae in Lake Baikal. The presence of *Phagocata* (*Fonticola*) *altaica* (Livanov and Zabusova, 1940) in Lake Khubsugul and in Lake Teletskoe (Altai) and its tributaries indicates a faunal affinity of these two water bodies. As noted later, this is also true for other aquatic invertebrate groups.

C. The Nematoda

Early investigations by Daday (1913) identified five species of free-living nematodes in Lake Khubsugul, but later studies (Erbaeva and Gagarin, 1980, 1986; Gagarin, 1982; Gagarin and Erbaeva, 1987) have increased this number to 24 species, in 12 families. These include *Eutobrilus peregrinator* Tsalolichin, 1983 and one newly described and endemic species, *Eudorylaimus imitatorius* Gagarin, 1982 (Gagarin and Erbaeva, 1987). Additional species are present in the lake's tributaries. The list of nematodes recorded from Lake Khubsugul will surely increase greatly, as no sampling of small specimens has yet been carried out. In comparison, the free-living nematode fauna of Lake Baikal is more diverse, with 92 species in 20 families, 83 of which are considered as endemic to Lake Baikal (Shoshin, 1997; Kozhova and Izmest'eva, 1998). Only four species of nematode are common to both lakes: *E. peregrinator*, which in Lake Baikal inhabits the open lake and the coastal–sor zone, *Ironus tenuicaudatus* de Man, 1876 and *Dorylaimus stagnalis* Dujardin, 1848, which inhabit the coastal–sor zone, and *Ethmolaimus pratensis* de Man, 1880 from the open lake.

Parasitic nematodes from the family Mermitidae were studied by Rubtsov (1972a, b, 1973), who described 10 new species from Lake Khubsugul, all of which are conditionally considered endemic. He supposed these species to be vicarious to Baikalian endemic species. The Mermitidae of Lake Khubsugul belong to five genera, all of which occur also in Lake Baikal where, according to Bekman (Kozhova and Izmest'eva, 1998), 28 endemic species of Mermithidae in 10 genera are present. One genus, *Baicalomermis* Rubtsov, 1976, contains only taxa found in Lake Khubsugul (three species) and in Lake Baikal (three species). Parasites and their host species often show a coevolutionary interaction, such that parasites are usually host specific. In Lake Khubsugul, the most probable hosts for the Mermithidae are insect larvae. However, as none of the host species of mermitid nematode in Lake Khubsugul is an endemic form, it can be supposed that the mermitids of Khubsugul either are neoendemics (a scenario that is considered unlikely) or belong to a group of poorly known, yet widespread Palaearctic species. As a group, the family Mermithidae has a Holarctic distribution and is poorly known (Rubtsov, 1972a, 1974).

D. The Oligochaeta

The Aphanoneura and Oligochaeta of Lake Khubsugul were described by Semernoy and Tomilov (1972), Semernoy and Akinshina (1980) and Akinshina (1985). The aphanoneuran faunas of these two lakes are very poor: Lake Khubsugul is inhabited by a single species, *Aeolosoma hemprichi* Ehrenberg, 1828, while Lake Baikal is inhabited by just two species, both endemic and also members of the genus *Aeolosoma* (Kozhova and Izmest'eva, 1998).

Twenty-nine species of oligochaete, distributed among 18 genera, are known from Lake Khubsugul, two of which (*Isochaetides tomilovi* Semernoy, 1972 and *Enchytraeus platus* Semernoy, 1972) are endemic to the lake. According to Snimshikova and Akinshina (in Kozhova and Izmest'eva, 1998), the genus *Isochaetides* Hrabê, 1965 in Lake Baikal is fairly diverse and contains, probably, 12 species, 11 of which are endemic. However, this genus is insufficiently studied and seven of these taxa are identified only as "*Isochaetides* sp." The genus *Enchytraeus* Henle, 1837 is distributed in Eurasia and North America, and is not typical of Lake Baikal (only one species, identified as *Enchytraeus* sp., is present). *Enchytraeus platys*, described from Lake Khubsugul, is closely related to species from Lake Issyk-Kul. It is unlikely that either species of oligochaete described from Lake Khubsugul is endemic. In comparison to Lake Khubsgul, the oligochaetes of Lake Baikal are considerably more diverse: 185 species and 20 subspecies (135 species are endemic) are distributed among 45 genera in six families (Kozhova and Izmest'eva, 1998).

Two species, *Nais bekmani* Sokolskaja, 1962, first described from Lake Baikal, and *Teneridrilus* (*Tubifex*) *hubsugulensis* Semernoy et Akinshina, 1980, first described from Lake Khubsugul, are endemic to these two lakes. (*Nais bekmani* also occurs in the Angara River and its reservoirs.)

Nais bekmani is believed to have originated in Lake Khubsugul (Semernoy and Akinshina, 1980). This species is a phytophil and is most abundant at depths of 0–2 m throughout Lake Khubsugul: in Lake Baikal it inhabits open Baikal and coastal–sor zones. The genus *Nais* Müller, 1773 contains a rich assemblage of species, which are widespread and very common in fresh waters. In Lake Baikal this genus is diverse, but represented by many non-endemic species, inhabiting not only open Baikal, but also coastal–sor zones. It can thus be assumed that *N. bekmani* is not a representative of the Baikalian endemic fauna.

Teneridrilus hubsugulensis belongs to a genus represented in Lake Baikal, where its congeners include two endemic species found only in the open lake. In addition to *N. bekmani* and *T. hubsugulensis*, 14 other species of oligochaete, in seven genera, are found in both Lakes Khubsugul and Baikal. This represents more than half of the total number of oligochaete taxa found in Lake Khubsugul. However, whereas *Nais* is the most diverse genus in Lake Khubsugul (nine of 29 species present), in Lake Baikal it is the genus *Baicalodrilus* Holmquist, 1978 (family Tubificidae Vejdovsky, 1884), containing 21 species (all endemic, and most of which occur in open Baikal; only two species are found in the coastal–sor zone), and the genus *Lamprodrilus* Michaelsen, 1901 (family Lumbriculidae Vejdovsky, 1884), comprising 20 species (18 endemic), which show the greatest diversity.

The 53 species of the family Lumbriculidae found in Lake Baikal account for *c.* 30% of the known species throughout the world, and of nine genera of this family, four are endemic to Baikal. In Lake Khubsugul only one species of this family is present, the widespread *Lumbriculus variegatus* (Müller, 1773). The diversity of Baikalian endemic Oligochaeta is a result of intralacustrine evolution within Lake Baikal itself. However, Izosimov (1949) has proposed that the oligochaete fauna of Lake Baikal (at least the Lumbriculidae) is of a relict nature, having evolved in the Tertiary.

In their list of oligochaetes of Lake Khubsugul, Semernoy and Akinshina (1980) noted five palaearctic species found also in Lake Teletskoe (Altai), and several species of the Chinese complex, including a representative of the genus *Monopylephorus* Levinsen, 1883. As noted above, two species, including *E. platys*, are taxonomically close to species from Lake Issyk-Kul (Tuan-Shan) and are found only in Lake Khubsugul, and two species, *N. bekmani* and *T. hubsugulensis*, are found only in Lakes Khubsugul and Baikal. Overall, the Khubsugul oligochaete fauna is unremarkable, and in composition is close to the General Siberian complex.

E. The Hirudinea

Studies of the hirudine fauna of Lake Khubsugul have begun only relatively recently. Eight species in six genera have been described (Lukin and Erbaeva, 1984; Erbaeva et al., 1997). Five of these species occur also in Lake Baikal, in the coastal–sor zone, and one in open Baikal. All six species are widespread Holarctic/Palaearctic forms. Lake Baikal has 18 species of Hirudinea, in 12 genera, of which seven species in two genera are inhabitants of open Baikal and are endemic (Kozhova and Izmest'eva, 1998). In this group there is no evident connection between the Khubsugul fauna and the endemic species complex of Lake Baikal.

F. The Gammaridae

The first information on the gammarid crustaceans of Lake Khubsugul was presented by Dybowsky, who noted the presence of *Gammarus pulex* L., 1755 (= *G. lacustris* Sars, 1863) (cited by Sovinsky, 1915). Subsequently, Bazikalova (1946) established that two species of Gammaridae from the genus *Gammarus* occur in Lake Khubsugul, *G. lacustris* and *G. kozhovi*, the latter being (at that time) a newly described species and endemic to the lake. Later, Safronov (1990) described another new species, *G. hanhi*, from the lake, to bring the total number of species known from there to three. This paucity of gammarids in Lake Khubsugul is in stark contrast to Lake Baikal, where they are the most diverse and group of macroinvertebrates, comprising 254 species and 69 subspecies, distributed among 35 genera; 253 of these species are endemic (Kozhova and Izmest'eva, 1998; cf. Takhteev, this volume).

The only non-endemic gammarid in Lake Baikal is the widely distributed Holarctic *G. lacustris*. This species does not inhabit the open water areas of Lake Baikal, but occurs only in the littoral–sor zone. This fine distinction is used to justify the claim that there is no species of gammarid common to both Lakes Baikal and Khubsugul, despite the hydrographical links between the lakes.

The discrete nature of the gammarid faunas of these two lakes is clearly a consequence of gammarid evolution in Lake Khubsugul having proceeded completely independently of that in Lake Baikal. If *G. kozhovi* Bazikalova, 1946 and *G. hanhi* evolved in Khubsugul (and their distribution in the lake profundal but not in the littoral, as with *G. lacustris* Sars, 1863, confirms this), then the difference in the number of species in Khubsugul and Baikal might reflect the considerably younger age of the Khubsugul fauna. However, the diversity of gammarids in Baikal can be explained not only by its great age, but also by its having a more diverse set of biotopes compared with those in Khubsugul, particularly in terms of zones with depths more than 200 m (cf. Tachteev, this volume).

G. The Acariformes

The Acariformes (Hydrachnidia and Halacaridae) of Lake Khubsugul were first investigated by Daday (1913), but were then ignored until Yankovskaya (1976) established that nine species occurred in Khubsugul itself and an additional 12 species inhabit adjacent water bodies. This list has subsequently been increased by the inclusion of species described by Yankovskaya from recent samples taken in the bays of Lake Khubsugul. The total number of the Hydrachnidia and Halacaridae (27 species) includes species described by Yankovskaya (1976) as found in Lake Khubsugul by Daday (1913): *Hydrachna geographica* O.F. Müller, 1776, *Lebertia tauinsignita* Leb., *Limnesia histrionica* Herm. (= *L. fulgida* Koch.), *Unionicola crassipes* (Müller, 1776) (= *Atax crassipes* Müller) and *Mideopsis orbicularis* O.F. Müller, but excludes species found only in the coastal basins of Lake Khubsugul: *Eylais infundibulifera* Koenike, 1897, *Eylais ussuriensis* Sokolov, *Hydrodroma despiciens* (Müller, 1776), *Hydrachna* (*Dipolhydrachna*) *conjecta* Koenike, 1895, *Piona* (*Piona*) *rotundoides* Thor and *P.* (*P.*) *uncata* Koenike.

Yankovskaya (1976) characterized the hydrachnid and halacarid faunas of Lake Khubsugul as being comprised of genetically different elements. Hydrachnidia of Khubsugul are rheophilic and comprise cold-water stenothermic species, and no true limnophilic species are present. Colonization of Lake Khubsugul by the rheophilic fauna took place in the postglacial period, and is perhaps continuing at present. However, a special place is occupied by the Baikalian endemic *Parasoldanellonyx baicalensis* Sokolov, 1952, which is the sole remnant of the Tertiary fauna. Its presence in Lakes Khubsugul and Baikal is indirect evidence of a common origin of these lakes. Further support for this is provided by the congeneric *P. parviscutatus* (Walter, 1917), which is found in both Lake Khubsugul and open Baikal. Several species of Hydrachnidia found in Lake Khubsugul occur also in Lake Teletskoe (Altai).

The total of 27 species recorded from Lake Khubsugul is a much greater number than is found in Lake Baikal, where there are only eight species and six genera. However, of the eight Baikal species, four are endemic, whereas none of the Khubsugul species is endemic. *Parasoldanellonyx baicalensis*, earlier described as a Baikal endemic, has subsequently been recorded from Lake Khubsugul also. In open Baikal there are five species in the family Halacaridae Murrey, 1976 (Sokolov and Yankovskaya, 1970; Kozhov, 1972; Kozhova and Izmest'eva, 1998). This family includes the endemic species *Pseudosoldanellonyx lohmanneloides* Sokolov et Jankovskaja, 1970 and *Parasoldanellonyx baicalensis* of Lake Khubsugul; a second endemic acariform is *Stygothrombidium* (= *Cerberothrombidium*) *vermiforme* (Sokolov, 1944) in the family Stygothrombiidae Thor, 1935. Two species of Hydrachnidia, *Feltria minuta* Koenike, 1892 and *Pionacercus leukartii* Piersig, inhabit the tributaries of Baikal, and only solitary specimens are found in the coastal zone of the lake (Kozhov, 1972).

H. The Insecta

1. The Ephemeroptera

The Ephemeroptera of Lake Khubsugul comprise 13 species (eight genera, seven families), of which one, *Ameletus eugeni* Sinichenkova et Varykhanova, 1989, is endemic (Baikova and Varychanova, 1978; Varykhanova, 1984, 1989; Erbaeva *et al.*, 1989; Sinichenkova and Varykhanova, 1989). Five species belong to the family Baetidae.

In Lake Baikal, Ephemeroptera are absent from the open waters and are restricted to the coastal areas, where they are represented by 14 species, in six families: Siphlonuridae (one species), Baetidae (three species), Oligoneuriidae (one species), Heptageniidae (three species), Leptophlebiidae (one species) and Ephemerellidae (five species) (Kozhova and Izmest'eva, 1998). The Ephemeroptera fauna of Lake Khubsugul is formed of species which enter the lake from tributaries and colonize the lake littoral, primarily in bays and in areas near river mouths. In Lake Baikal the littoral conditions seem unsuitable for such colonization; this dramatic difference between the faunas of the coastline and the open lake is also seen in the Plecoptera and Trichoptera faunas.

2. The Plecoptera

Plecoptera are represented in Lake Khubsugul by 17 species (10 genera, six families), and in this group diversity in Lake Khubsugul is comparable with that in Lake Baikal. One species is endemic to the lake, *Capnia khubsugulica* Zhiltzova et Varykhanova, 1987 (Varykhanova and Zhiltsova, 1984; Zhiltsova and Varykhanova, 1984, 1987; Erbaeva *et al.*, 1989). None of the species inhabiting Lake Khubsugul is found in open Baikal. Sixteen species of Plecoptera have been recorded from Lake Baikal, but almost all of these occur in the Baikal tributaries, the Angara River and its affluents (Kozhova and Izmest'eva, 1998), and are absent from open Baikal. In the open waters of the lake there are only two endemic species, both of the endemic genus *Baicaloperla* Zapekina-Dulkeit et Zhiltzova, 1973 (family Capniidae) (Kozhova and Izmest'eva, 1998).

3. The Trichoptera

Eighteen species (12 genera, five families) of Trichoptera have been recorded from Lake Khubsugul, all of which are taxa widely distributed throughout the Palaearctic (Martynov 1909, 1914; Levanidova, 1947; Erbaeva *et al.*, 1989; Rozhkova and Varyhanova, 1990). Twelve of these species belong to the family Limnephilidae Kolenati, and it is interesting to note that, like the Baikalian endemic species, the Khubsgul members of this family have lost the ability to fly (Kozhov, 1972). The Khubsugul Trichoptera are taxomonically diverse, and Rozhkova and Varykhanova (1990) have distinguished five zoogeographical groups within this assemblage: Holarctic (two species), Trans-Palaearctic (three species), Trans-Siberian (five species), Eastern-Siberian (six species) and Euro-Siberian (two species). They noted that in Lake Khubsugul the Trichoptera, as well as the Ephemeroptera and Plecoptera, inhabit mainly bays and estuary mouths, and not the open lake. The Trichoptera fauna in these shoreline areas is dominated by species of the genera *Apatania* Kolenati, 1848 and *Limnephilus* Leach, 1815.

In terms of diversity and endemism, the Trichoptera fauna of Khubsgul differs markedly from that of Lake Baikal. In open Baikal 122 species of Trichoptera are present, with five endemic genera and 17 endemic species (Kozhova *et al.*, 1998). A recent revision of this group in Lake Baikal established 14 species and placed them in the family Apataniidae (subfamily Apataniinae) (Ivanov and Menshutkina, 1996). In this scheme, the Baikalian endemics form two tribes, the Baicalinini and the Thamastini. The Baikal shoreline is also home to a diverse Trichoptera fauna: adults of over 100 species have been collected, but for most of these species, the larvae do not inhabit the open waters of the lake. Thus, the Trichoptera fauna of open Baikal differs dramatically both from the fauna of its coasts and from the Khubsugul fauna. However, the composition of the Khubsugul Trichoptera resembles that seen in the tributaries of Baikal, the riverine portions within the Lake Baikal basin.

4. The Chironomidae

The Chironomidae are the most diverse insect group found in Lake Khubsugul, containing 106 species distributed in 52 genera (Linevich, 1964; Erbaeva, 1976; Erbaeva *et al.*, 1989; Makarchenko, 1984). Only one species, *Pseudodiamesa venusta*, which belongs to a common Palaearctic genus, is endemic, and may have evolved from a relict population. It occupies a transitional position between *P. nivosa* Goetghebuer, 1928 and *P. nepalensis* Reiss, 1968 from Nepal. However, it is suspected that larvae and pupae of *P. nivosa*, recorded in Lake Khubsugul, actually belong to *P. venusta* (Makarchenko, 1984).

The diversity of chironomids in Lake Baikal is slightly higher than that in Lake Khubsugul, comprising 139 species in 62 genera (Kozhova and Izmest'eva, 1998). Of these taxa, 13 are endemic, seven of which belong to the endemic subgenus *Baicalosergentia* Linevitsh, 1959. Three species from two genera, *Orthocladius compactus* Linevitch, 1970, *O. gregarius* Linevitch, 1970 and *Sergentia flavodentata* Tshernovskij, 1949, are found only in Lakes Baikal and Khubsugul. All three species have also been recorded in the Baikal effluent Angara River, but here only *O. compactus* is relatively abundant; the other two species are rare (Kozhova *et al.*, 1998).

Unlike in many other insect groups, notably the Ephemeroptera, Plecoptera and Trichoptera, no distinct difference is seen between the chironomid fauna of open Baikal and the coastal–sor zone. Nine Baikalian endemic species inhabit

108 O.M. KOZHOVA *ET AL.*

only the open lake, and four endemic species dwell both in open Baikal and the coastal–sor zone. In total, 67 of the known species live only in open Baikal, 50 species only in the coastal–sor zone, and the remaining 22 species inhabit both regions. Forty-four species are common to Lake Khubsugul and open Baikal, 39 species are common to Khubsugul and the coastal–sor zone, and 65 species are found in both lakes. The highest similarity is observed in the subfamily Diamesinae, the representatives of which are typical of open Baikal (six of the eight species of this subfamily found in Lake Khubsugul live also in open Baikal), and the subfamily Orthocladiinae (23 of 45 species found in Lake Khubsugul inhabit open Baikal). It should be noted that exchange of species between these lakes is possible, as the lakes are only 240 km apart, close enough for wind-assisted migration of flying adults.

I. The Mollusca

A total of 13 molluscan species is known from Lake Khubsugul (Lindholm, 1929; Kozhov, 1946; Tomilov and Dashidorzh, 1965; Sitnikova and Goulden, 1997; present study), among which only *Choanomphalus mongolicus* is endemic. *Kobeltocochlea* (*Pseudobenedictia*) *michnoi*, previously described by Lindholm (1929) (on the basis of a single, empty shell) as endemic to Lake Khubsugul, is not actually found in that lake, but is found in Lake Baikal (Sitnikova, 1988; Sitnikova and Goulden, 1997). Mislabelled samples are the likely source of this error. The identity of another mollusc is more mysterious. Kozhov (1972) noted that among samples taken in the 1959 survey of Khubsugul, Tomilov discovered an empty shell of a gastropod mollusc which was closely reminiscent of species of the genus *Litoglyphus* (= *Fluminicola*). The present authors have found no such molluscs in Khubsugul, nor are there any representatives of this genus in Lake Baikal.

Kozhov (1972) referred *C. mongolicus* and *K. michnoi* to the "Baikalian group" of molluscs in Lake Khubsugul. As seen from the present data, modern classifications retain only one species in this group, *C. mongolicus*, of the subgenus *Achoanomphalus* Lindholm, 1909. This subgenus comprises five species, including *C. angulatus* Dybowsky et Grochmalsky, 1925, a species morphologically very similar to *C. mongolicus* Kozhov, 1946 (Sitnikova and Goulden, 1997).

The other 11 molluscan taxa in Lake Khubsugul belong to species distributed widely in Palaearctic fresh waters. Two of these, *Lymnaea* (*Radix*) *auricularia* (L., 1758) and *L.* (*Peregriana*) *ovata* Draparnaud, 1805, are also present in the coastal-sor zone of Lake Baikal (Kozhova and Izmest'eva, 1998).

The Mollusca are among the most speciose groups of benthic invertebrates found in Lake Baikal, and are considerably more diverse than in Lake Khubsugul, with 169 species, 28 subspecies and 37 genera. The Gastropoda contains 13 endemic genera and 113 endemic species, while the Bivalvia contains 12 endemic species (Kozhova and Izmest'eva, 1998). The mollusc faunas of Lake Khubsugul and of open Baikal differ greatly in their species diversity and composition, and the only similarity is the presence of *C. mongolicus*, which is closely allied to Baikalian species.

V. CONCLUSIONS

Studies of the benthic invertebrate fauna of Lake Khubsugul have an erratic history, with long periods between successive studies. One consequence of this has been an incomplete knowledge of the lake's fauna and that of Lake Baikal by earlier workers, and the occasional incorrect ascribing of endemic status to some taxa at first description. More recently, however, as knowledge of the fauna of both Lakes Khubsugul and Baikal increases, this situation is rapidly improving.

The benthic invertebrate fauna of Lake Khubsugul is rather diverse, and comprises at least 287 species. This total includes 19 newly described species discovered during the authors' recent studies in Lake Khubsugul. To date, they have been recorded only there, and thus can be considered as endemic to the lake.

The most taxonomically diverse group among the Khubsugul endemics is the parasitic Mermithidae, with 10 species, three of which belong to the Baikalian genus *Baicalomermis*. This testifies to a high degree of relatedness of these three species to Baikalian endemics. However, the hosts of the Mermithidae of Khubsugul are not endemic forms. *Teneridrilus hubsugulensis*, first described from Lake Khubsugul, was later also found in Lake Baikal. Further studies will surely result in other species presently referred to as Khubsugul endemics being found in other water bodies, and quite probably, among the Gammaridae only two species, *G. kozhovi* and *G. hanhi*, and among the Mollusca just one species, *C. mongolicus*, are truly endemic to Khubsugul.

The invertebrate fauna of Lake Khubsugul is dominated by stenothermic species and is composed of two groups: limnophilic species widespread in the Palaearctic, and rheophilic species distributed primarily in mountanous regions of Central Asia, Siberia and the Far East. The presence of both typically lacustrine species and species which enter into the lake from tributaries is a feature of Lake Khubsugul's fauna, which differs dramatically from that of Lake Baikal, where an almost complete segregation exists between the fauna inhabiting the open lake and that found in Baikal's tributaries, sor-habitats and surrounding water bodies.

Lake Baikal is characterized by an extremely diverse endemic fauna, particularly of benthic invertebrates, which comprise two genetically and ecologically different complexes. The first complex is typical of open Baikal and is represented in general by Baikalian endemics. The second complex is characteristic of shallow bays (sors), often partly isolated from open Baikal, and of shallow predelta regions, the coastal–sor zone (littoral–sor zone), and is represented by species common to Siberian (Euro-Siberian) water bodies. This complex is often called General Siberian, although it includes representatives of different faunistic groups. Early biologists studying Lake Baikal stressed the difference in the species composition of these complexes and their "immiscibility". The portion of characteristic Baikalian endemics found outside Lake Baikal is insignificant, and these occur mainly in the effluent Angara River and in Siberian lakes of tectonic origin, including Lake Khubsugul.

A special group of seven species (*Eutobrilus peregrinator, Nais bekmani, Teneridrilus hubsugulensis, Orthocladius compactus, O. gregarius, Sergentia flavodentata* and *Parasoldanellonyx baicalensis*) are known only from Lakes Khubsugul and Baikal. However, despite the presence in Lake Khubsugul of these seven species, and species belonging to the Baikalian genus *Baicalomermis* Rubzov, 1976 and subgenus *Achoanomphalus* Lindholm, 1909, the overall species composition in Lake Khubsugul differs markedly from that of open Baikal. Species found in both Lakes Baikal and Khubsugul are mainly inhabitants of the coastal–sor zone of Baikal, not of the open lake. These differences in the diversity, composition and degree of endemicity reflect the different age and evolutionary histories of the two faunas, despite the common geological past of the lakes. During the Quaternary (Pleistocene) period, glaciers had a greater negative influence upon the Khubsugul fauna, owing to the smaller size and lesser depth of the lake, and to its higher position above sea level. In contrast, the greater volume and depth of Lake Baikal allowed for the survival of aquatic animals during the onset of glaciers. The fauna of Lake Khubsugul is thus relatively young compared with that of Baikal. One consequence of this is that in Lake Khubsugul there has been no extensive intralacustrine speciation, whereas in Lake Baikal speciation has been rampant in several groups.

From comparison of the faunal/distributional data with what is known of the geological history of the Khubsugul trough, it can be concluded that the modern fauna of the lake developed in several stages during the Quaternary period, and contains preglacial, glacial and postglacial elements. The fauna is represented mainly by Palaearctic taxa, but elements of Holarctic and Asian faunas (the Far East, China, Nepal), and those with some affinity with Lakes Teletskoye and Issyk-Kul, are also present. A major role in the formation of the present fauna of Lake Khubsugul was played by inhabitants of its tributaries, and some modern rheophilic species penetrate into the lake via this route. This colonization is facilitated by the stone substrate, and the ambient conditions of low temperature and high dissolved oxygen content present in the littoral zone.

It is difficult to account for the presence in Lake Khubsugul of *Parasoldanellonyx baicalensis*, which Yankovskaya (1976) referred to as a Tertiary relict, since geological data testify to the formation of the lake in the Quaternary period. Possibly, such elements may have survived in subterranean basins. There is no clear evidence that river connections (Khubsugul–Egyin-Gol–Selenga–Baikal) played any role in the affinity of several elements in the Khubsugul and Baikal faunas, and the suggested penetration of the oligochaete *Nais bekmani* by such means remains unproven.

ACKNOWLEDGEMENTS

The authors thank Dr C.E. Goulden (Patric Centre for Environmental Research, Benjamin Franklin, USA) for his support in this work and Dr L.R. Izmest'eva for her help in carrying out this project.

REFERENCES

Akinshina, T.V. (1985). The oligochaete fauna of Lake Khubsugul and their distribution. In: *Turn-over of Matter and Energy in Waterbodies*, pp. 100–101. Institute of Limnology Publishing House, Irkutsk (in Russian).

Baikova, O.Ya. and Varykhanova, K.V. (1978). Ephemeroptera of Mongolia. In: *Natural Conditions and Resources of Prikhubsugulye*, pp. 111–121. University Publishing House, Irkutsk and Ulan-Bator (in Russian).

Bazikalova, A.Ya. (1946). Amphipods of Lake Koso Gol (Mongolia). *Doklady Acad. Sci. USSR* **53,** 677–679 (in Russian).

Belova, V.A. and Kulakov, V.S. (1982). Development history of the Khubsugul trough. In: *Late Cenozoic History of Lakes in the USSR*, pp. 80–88. Nauka, Novosibirsk (in Russian).

Daday, E. (1913). Beitrage zur Kennthis der Mikrofauna des Kossogol-Beckens in der Nordwestlichen Mongolei. Mathematische und Naturwissenschaftliche Berichte aus Ungarn. *Hedwigia* **26,** 247–360.

Erbaeva, E.A. (1976). Larvae of Chironomidae in Lake Khubsugul. In: *Natural Conditions and Resources of Prikhubsugulye*, pp. 218–226. University Publishing House, Irkutsk and Ulan-Bator (in Russian).

Erbaeva, E.A. and Gagarin, V.G. (1980). Benthic Nematoda of Lake Khubsugul. In: *Natural Conditions and Resources of Some Regions of the Mongolian People's Republic*, pp. 146–148. Ulan-Bator University Press, Ulan-Bator (in Russian).

Erbaeva, E.A. and Gagarin, V.G. (1986). Fauna of free-living Nematoda of Lake Khubsugul. In: *Natural Conditions and Resources of Some Regions of the Mongolian People's Republic*, pp. 83–84. Ulan-Bator University Press, Ulan-Bator (in Russian).

Erbaeva, E.A., Varychanova, K.V. and Rožkova, N.A. (1989). Wasserinsekten des Chubsugul-Sees in der Nordmongolei. In: *Erforschung Biologische Ressourcen der Mongolischen Volkrepublik*, pp. 69–75. Halle, Saale.

Erbaeva, E.A., Kozhova, O.M. and Safronov, G.P. (1997). *Fauna of Bottom Invertebrates of Lake Khubsugul (Mongolia)*. Report deposited at VINITI 14.01.97, 118-B97. Institute of Biology, Irkutsk State University, Irkutsk (in Russian).

Florensov, N.A. (1960). *Mesozoic and Cenozoic Troughs of Pribaikalye*. Academy of Sciences of the USSR, Moscow (in Russian).

Gagarin, V.G. (1982). New species of free-living nematodes from Mongolia. *Zool. J.* **61,** 1592–1594 (in Russian).

112 O.M. KOZHOVA *ET AL.*

Gagarin, V.G. (1993). *Free-living Nematoda of the Fresh Waters of Russia and Adjacent Territories (Orders Monhysrerida, Araeolaimida, Chromadorida, Enoplida, Mononchida)*. Gidrometisdat, St Petersburg (in Russian).
Gagarin, V.G. and Erbaeva, E.A. (1987). On the fauna of free-living Nematoda in Lake Khubsugul and its tributaries. *Biol. Cont. Wat. Inform. Bull.* **76**, 23–28 (in Russian).
Ivanov, V.D. and Menshutkina, T.V. (1996). Endemic caddisflies of Lake Baikal (Trichoptera; Apataniidae). *Braueria* **23**, 13–28.
Izosimov, V.V. (1949). The Lumbriculidae of Lake Baikal. Unpublished MSc Thesis, University of Kazan (in Russian).
Key to Freshwater Invertebrates of Russia and Adjacent Territories, Vol. 1 (1994). (Ed. by S. Ya Tsallolikhin), Nauka, St Petersburg (in Russian).
Key to Freshwater Invertebrates of Russia and Adjacent Territories, Vol. 3 (1997). (Ed. by E.P. Narchuk, D.V. Tumanov and S. Ya Tsallolikhin), Nauka, St Petersburg (in Russian).
Kozhov, M.M. (1946). Baikalian molluscs in Lake Koso Gol (Mongolia). *Doklady Acad. Sci. USSR* **52**, 369–372. (in Russian).
Kozhov, M.M. (1962). *Biology of Lake Baikal*. Academy of Sciences of the USSR, Moscow (in Russian).
Kozhov, M.M. (1963). *Lake Baikal and its Life*. Junk, The Hague.
Kozhov, M.M. (1972). *Essays on Baikal Science*. East Siberian Publishing House, Irkutsk (in Russian).
Kozhov, M.M., Antipova, N.L. and Nikolaeva, E.P. (1965). On the plankton of Lake Khubsugul. In: *Limnological Investigations of Baikal and some Lakes in Mongolia*, pp. 181–190. Nauka, Moscow (in Russian).
Kozhova, O.M. and Erbaeva, E.A. (1984). Flora and fauna genesis of Lake Khubsugul. In: *Natural Conditions and Resources in Some Regions of the Mongolian People's Republic*, pp. 9–11. University of Bratislava Press, Bratislava (in Russian).
Kozhova, O.M., Izmest'eva, L.R. (eds) (1998). *Lake Baikal. Evolution and Biodiversity*. Backhuys, Leiden.
Kozhova, O.M., Izmest'eva, L.R. and Erbaeva, E.A. (1994). A review of the hydrobiology of Lake Khubsugul (Mongolia). *Hydrobiologia* **291**, 11–19.
Kozhova, O.M., Erbaeva, E.A. and Safronov, G.P. (1998). A comparative analysis of the bottom fauna of Baikal and Khubsugul. *Siber. Ecol. J.* **5**, 391–396 (in Russian).
Levanidova, I.M. (1947). On the fauna of Trichoptera of Lake Koso Gol (Mongolia). *Doklady Acad. Sci. USSR* **55**, 569–591 (in Russian).
Lindholm, W.A. (1929). Die ersten Schnecken (Gastropoda) aus dem See Kosogol in der Nordwest-Mongolei. *Doklady Acad. Sci. USSR* **12**, 315–318.
Linevich, A.A. (1959). New species of mosquitoes of the family Tendipedidae (Diptera) from Lake Baikal. *Entomol. Rev.* **38**, 238–242 (in Russian).
Linevich, A.A. (1964). Tendipedidae (Chironomidae) of Pribaikalye and western Zabaikalye. Unpublished PhD Thesis, University of Leningrad (in Russian).
Linevich, A.A. (1981). *Chironomidae of Baikal and Pribaikalye*. Nauka, Novosibirsk (in Russian).
Livanov, N.A. and Zabusova, Z.I. (1940). Planariae of Lake Teletskoye basin and new data about some other Siberian species. *Proc. Kazan Univ. Soc. Nat. Sci.* **41**, 83–159 (in Russian).
Lukin, E.I. and Erbaeva, E.A. (1984). On the study of leech fauna in Lake Khubsugul (MNR). In: *Ecological Investigations of Lake Baikal and Pribaikalye*, pp. 55–58. University Publishing House, Irkutsk (in Russian).

Makarchenko, E.A. (1984). A new species of Chironomidae of the genus *Pseudodiamesa* Goetghebuer (Diptera, Chironomidae) from Mongolia (Lake Khubsugul). In: *Ecological Investigations of Lake Baikal and Pribaikalye*, pp. 60–65. University Publishing House, Irkutsk (in Russian).

Martynov, A.V. (1909). Trichoptera of Siberia and adjacent regions (1. Phryganeidae and Sericostomatidae). *Ann. Zool. Mus. Imp. Acad. Sci.* **14**, 223–255.

Martynov, A.V. (1914). Die Trichopteren Sibiriens und der angrenzenden Gebiete. III: Subf. Apataniinae (Fam. Limnophilidae). *Ann. Zool. Mus. Imp. Acad. Sci.* **19**, 1–87.

Pankratova, V.Ya. (1970). *Larvae and Pupae of Mosquitoes of the Subfamily Orthocladiinae of the USSR Fauna (Diptera, Chironomidae = Tendipedidae)*. Nauka, Leningrad (in Russian).

Pankratova, V.Ya. (1977). *Larvae and Pupae of Mosquitoes of the Subfamily Podonominae and Tanypodinae of the USSR Fauna (Diptera, Chironomidae = Tendipedidae)*. Nauka, Leningrad (in Russian).

Pankratova, V.Ya. (1983). *Larvae and Pupae of Mosquitoes of the Subfamily Chironominae of the USSR Fauna (Diptera, Chironomidae = Tendipedidae)*. Nauka, Leningrad (in Russian).

Porfiryeva, N.A. and Tomilov, A.A. (1976). Planaria (Tricladida, Paludicola) of Lake Khubsugul. In: *Natural Conditions and Resources of Prikhubsugulye*, pp. 167–178. University Publishing House, Irkutsk and Ulan-Bator (in Russian).

Rozhkova, N.A. and Varykhanova, K.V. (1990). To the knowledge of the caddisflies fauna (Trichoptera) in Lake Khubsugul. In: *Natural Conditions and Resources of Some Regions of the Mongolian People's Republic*, pp. 76–77. Ulan-Bator University Press, Ulan-Bator (in Russian).

Rubtsov, I.A. (1972a). *Water Mermithidae*, Vol. 1. Nauka, Leningrad (in Russian).

Rubtsov, I.A. (1972b). Mermithidae from Lake Khubsugul. In: *Natural Conditions and Resources of Some Regions of Prikhubsugulye*, pp. 106–118. University Publishing House, Irkutsk and Ulan-Bator (in Russian).

Rubtsov, I.A. (1973). Mermithidae of Lake Khubsugul. In: *Natural Conditions and Resources of Prikhubsugulye (MNR)*, pp. 357–382. University Publishing House, Irkutsk and Ulan-Bator (in Russian).

Rubtsov, I.A. (1974). *Water Mermithidae*, Vol. II. Nauka, Leningrad (in Russian).

Safronov, G.P. (1990). Materials on the study of some species of the genus *Gammarus* Fabricius from waterbodies of eastern Siberia and adjacent regions. In: *Natural Conditions and Resources of Some Regions of the Mongolian People's Republic*, pp. 73–75. Ulan-Bator University Press, Ulan-Bator (in Russian).

Samburova, V.A. (1982). Chironomidae. In: *Community State of South Baikal*, pp. 94–104. University Publishing House, Irkutsk (in Russian).

Semernoy, V.P. and Akinshina, T.V. (1980). Oligochaeta of Lake Khubsugul and some other waterbodies in Mongolia. In: *Natural Conditions and Resources of Prikhubsugulye*, pp. 117–134. University Publishing House, Irkutsk (in Russian).

Semernoy, V.P. and Tomilov A.A. (1972). Oligochaeta of Lake Khubsugul (Mongolia). *Biol. Cont. Wat. Inf. Bull.* **16**, 26–29 (in Russian).

Shoshin, A.V. (1997). Fauna and evolution of free-living nematodes in the littoral of South Baikal. Unpublished MSc Thesis, University of St Petersburg (in Russian).

Sinichenkova, N.D. and Varykhanova, K.V. (1989). A new species of Ephemeroptera of the genus *Ameletus* Eaton (Ephemeroptera, Siphlonuridae) from Mongolia. *Entomol. Rev.* **68**, 576–582 (in Russian).

Sitnikova, T.Ya. (1988). New data on the baikalian endemic mollusc *Kobeltocochlea michnoi* Lindholm, 1929 of the family Benedictiidae (Gastropoda, Pectinibranchia). In: *New Studies of the Flora and Fauna of Baikal and its Basin*, pp. 104–108. University Publishing House, Irkutsk (in Russian).

Sitnikova, T.Ya. and Goulden, C.E. (1997). Shell variability of the Hovsgolian *Ch. mongolicus* and comparative study with Baikal *Choanomphalus* (Gastropoda, Planorbidae). In: *Ecologically Equivalent Species of Hydrobionths in the Great Lakes of the World*, pp. 92–93. Buryatian Scientific Centre of the Academy of Sciences of the USSR, Ulan-Ude, Russia.

Sokolov, I.I. (1952). Water mites. Part II. Halacarae. In: *Fauna of the USSR. Arachnida.* Academy of Sciences of the USSR, Moscow (in Russian).

Sokolov, I.I. and Yankovskaya, A.I. (1970). New data on hydrocarynofauna of Baikal. *Izvest. Scient. Res. Inst. Biogeog. Irkutsk State Univ.* **23**, 95–103 (in Russian).

Sovinsky, V.K. (1915). Amphipoda of Lake Baikal (Gammaridae). In: *Zoological Investigations of Lake Baikal*, pp. 1–390. University of Kiev Press, Kiev (in Russian).

Spuris, Z.D. (1989). *A Summary of the Caddisfly Fauna of the USSR.* Zinatne, Riga (in Russian).

Tomilov, A.A. and Dashidorzh, A. (1965). Lake Khubsugul and its capacity for fishery use. In: *Limnological Investigations of Baikal and Some Lakes in Mongolia*, pp. 164–180. Nauka, Moscow (in Russian).

Varykhanova, K.V. (1984). Fauna of Ephemeroptera of Lake Khubsugul and its tributaries. In: *Natural Conditions and Resources of Some Regions of the Mongolian People's Republic.* p. 88. University of Bratislava Press, Bratislava (in Russian).

Varykhanova, K.V. (1989). Zoobenthos of the littoral in Khubsugul and its tributaries. Unpublished MSc Thesis. Irkutsk State University (in Russian).

Varykhanova, K.V. and Zhiltsova, L.A. (1984). On the fauna of Plecoptera in the Lake Khubsugul basin in Mongolia. In: *Natural Conditions and Resources of Some Regions of the Mongolian People's Republic*, pp. 21–22. Ulan-Bator University Press, Ulan-Bator (in Russian).

Wülker, W., Kiknadze, I.I., Kerkis, I.E. and Nevers, P. (1998). Chromosomes, morphology and distribution of *Sergentia baueri*, spec. nov., *S. prima* Proviz, 1997 and *S. coracina* Zett., 1824. *Spixiana* **22**, 69–81.

Yankovskaya, A.I. (1976). Hydracarina (Hydrachnelidae) of Lake Khubsugul. In: *Natural Conditions and Resources of Prikhubsugulye*, pp. 179–217. University Publishing House, Irkutsk and Ulan-Bator (in Russian).

Zhiltsova, L.A. and Varykhanova, K.V. (1984). To the knowledge of Plecoptera in the Lake Khubsugul basin in Mongolia. *Insects Mongolia* **9**, 21–28 (in Russian).

Zhiltsova, L.A. and Varykhanova, K.V. (1987). A new species of the genus *Capnia* Pictet (Plecoptera, Capniidae) from Mongolia. *Entomol. Rev.* **56**, 102–104 (in Russian).

APPENDIX: THE BENTHIC INVERTEBRATES OF LAKE KHUBSUGUL

Taxon	Total no. of species in Khubsugul	No. of species found only in Khubsugul ('endemics' of Khubsugul)	No. of species found only in Khubsugul and in Baikal (and in Angara)	Total no. of species in this genus in Baikal	No. of endemic species in this genus in Baikal
Class TURBELLARIA					
Order TRICLADIDA					
Family Planaridae Stimpson, 1857	1	0	0		
Genus *Phagocata* Leidy, 1847	1	0	0	0	0
Class NEMATODA Rudolphi, 1808	34	11	1		
Order MONCHYSTERIDA Filipjev, 1929					
Family Monchysteridae de Man, 1876	1	0	0		
Genus *Eumonchystera* Andrassy, 1981	1	0	0	4	4
Family Xyalidae Chitwood, 1951	1	0	0		
Genus *Daptonema* Cobb, 1920	1	0	0	0	0
Order ARAEOLAIMIDA de Coninck et Sch. Stekhoven, 1933					
Family Plectidae Örley, 1880	2	0	0	1	
Genus *Plectus* Bastian, 1865	2	0	0		?
Order CHROMADORIDA Chitwood, 1933					
Family Ethmolaimidae Filipjev et Sch. Stekhoven, 1914	2	0	0	5	3
Genus *Etholaimus* de Man, 1880	2	0	0		
Order DIPLOGASTERIDA					
Family Diplogasteridae Micoletzky, 1922	1	0	0	5	5
Genus *Fictor* Paramonov, 1952	1	0	0		
Order ENOPLIDA Filipjev, 1929					
Family Ironidae de Man, 1876	1	0	0	1	0
Genus *Ironus* Bastian, 1865	1	0	0		
Family Tobrilidae Filipjev, 1918	6	0	1		
Genus *Tobrilus* Andrassy, 1959	3	0	1	7	7
Genus *Eutobrilus* Tsalolichin, 1981	1	0	1	9	9
Genus *Brevitobrilus* Tsalolichin, 1981	1	0	0	0	0

Taxon	Total no. of species in Khubsugul	No. of species found only in Khubsugul ('endemics' of Khubsugul)	No. of species found only in Khubsugul and Baikal (and in Angara)	Total no. of species in this genus in Baikal	No. of endemic species in this genus in Baikal
Genus *Neotobrilus* Tsalolichin, 1981	1	0	0	0	0
Family Tripylidae de Man, 1876	2	0	0		
Genus *Tripyla* Bastian, 1865	1	0	0	2	1
Genus *Trischistoma* Cobb, 1913	1	0	0	0	0
Order MONONCHIDA Jairajpuri, 1969					
Family Mononchidae Filipjev, 1934	1	0	0		
Genus *Mononchus* Bastian, 1865	1	0	0	1	0
Order DORYLAIMIDA Pearse, 1942					
Family Dorylaimidae de Man, 1876	2	0	0		
Subfamily Dorylaiminae de Man, 1876					
Genus *Dorylaimus* Dujardin, 1845	1	0	0	1	0
Subfamily Mesodorylaiminae Andrassy, 1969					
Genus *Mesodorylaimus* Andrassy, 1959	1	?	?	2	1
Family Qudsianematidae Jairajpuri, 1965	3	1	0		
Genus *Eudorylaimus* Andrassy, 1959	2	1	0	3	1
Genus *Labronema* Thorne, 1939	1	0	0	0	0
Family Aporcelaimidae Heyns, 1965	2	0	0		
Genus *Aporcelaimellus* Heyns, 1965	2	0	0	0	0
Order MERMITHIDA					
Family Mermithidae Braun	10	10	0		
Genus *Abathymermis* Rubzov, 1971	1	1	0	2	2
Genus *Baicalomermis* Rubzov, 1976	3	3	0	3	3
Genus *Gastromermis* Micoletzky, 1923	3	3	0	6	6
Genus *Lanceimermis* Artyukhovsky, 1969	1	1	0	1	1
Genus *Spiculimermis* Artyukhovsky, 1963	2	2	0	2	2
Class CLITELLATA					
Subclass APHANONEURA Timm, 1987	1	0	0		
Order AEOLOSOMATIDA Timm, 1987					
Family Aeolosomatidae Beddard, 1895	1	0	0		

Taxon	Total no. of species in Khubsugul	No. of species found only in Khubsugul ('endemics' of Khubsugul)	No. of species found only in Khubsugul and Baikal (and in Angara)	Total no. of species in this genus in Baikal	No. of endemic species in this genus in Baikal
Genus *Aeolosoma* Ehrenberg, 1828	1	0	0	2	2
Subclass OLIGOCHAETA	29	2	2		
Order TUBIFICIDA Timm, 1987	14	0	1		
Family Naididae Benham, 1890					
Subfamily Naidinae Lastočkin, 1921					
Genus *Stylaria* Lamarck, 1816	1	0	1	2	0
Genus *Nais* Müller, 1773	9	0	0	16	6
Genus *Specaria* Sperber, 1939	1	0	0	1	0
Genus *Uncinais* Levinsen, 1884	1	0	0	3	2
Subfamily Chaetogastrinae Lastočkin, 1921					
Genus *Chaetogaster* Baer, 1827	2	0	0	11	10
Family Tubificidae Vejdovsky, 1884	8	1	1	11	
Subfamily Rhyacodrilinae Hrabě, 1963					
Genus *Rhyacodrilus* Bretscher, 1901	1	?	?	10	6
Subfamily Tubificinae Eisen, 1879					
Genus *Tubifex* Lamarck, 1816	1	0	0	8	5
Genus *Teneridrilus* Holmquist, 1985	1	0	1	3	2
Genus *Spriosperma* Eisen, 1879	1	0	0	1	0
Genus *Limnodrilus* Claparède, 1862	1	0	0	2	0
Genus *Isochaetides* Hrab@uhat;, 1965	1	1	0	12	11
Genus *Monopylephorus* Levinsen, 1883	1	?	0	0	0
Genus *Potamothrix* Vejdovsky et Mrázek, 1902	1	0	0	1	0
Family Propappidae Coates, 1982	1	0	0		
Genus *Propappus* Michaelsen, 1905	1	0	0	2	1
Family Enchytraeidae Vejdovsky, 1879	5	1	0		
Genus *Enchytraeus* Henle, 1837	3	1	0		0

Taxon	Total no. of species in Khubsugul	No. of species found only in Khubsugul ('endemics' of Khubsugul)	No. of species found only in Khubsugul and Baikal (and in Angara)	Total no. of species in this genus in Baikal	No. of endemic species in this genus in Baikal
Genus *Henlea* Michaelsen, 1889	1	0	0	0	0
Genus *Fridericia* Michaelsen, 1899	1	?	0	0	0
Order LUMBRICULIDA Timm, 1987					
Family Lumbriculidae Vejdovsky, 1884	1	0	0		
Genus *Lumbriculus* Grube, 1884	1	0	0	2	0
Subclass HIRUDINIONES Epstein, 1987	8	0	0		
Family Glossiphoniidae Vaillant, 1890	5	0	0		
Genus *Protoclepsis* Livanov, 1902	1	0	0	2	0
Genus *Glossiphonia* Johnson, 1816	2	0	0	2	0
Genus *Helobdella* Blanchard, 1896	1	0	0	1	0
Genus *Haementeria* de Filippi, 1849	1	?	0	0	0
Family Piscicolidae Johston, 1865	1	0	0		
Genus *Piscicola* Blainville, 1818	1	0	0	1	0
Family Erpobdellidae	2	0	0		
Genus *Erpobdella* Agassiz, 1846	2	0	0	0	0
Class CRUSTACEA					
Order AMPHIPODA Latreille, 1816	3	2	0		
Family Gammaridae Leach, 1813	3	2	0		
Genus *Gammarus* Fabricius, 1775	3	2	0	1	0
Class ARACHNIDA					
Order ACARIFORMES	27	0	1		
HYDRACHNIDIA					
Family Hydrachnidae Leach, 1815	1	0	0		
Genus *Hydrachna* Müller, 1776	1	0	0	0	0
Family Sperchonidae Thor, 1900	3	0	0	0	0
Genus *Sperchon* Kramer, 1877	3	0	0	0	0
Family Lebertiidae Thor, 1900	9	0	0		
Genus *Lebertia* Neuman, 1880	9	0	0	0	0

Taxon	Total no. of species in Khubsugul	No. of species found only in Khubsugul ('endemics' of Khubsugul)	No. of species found only in Khubsugul and Baikal (and in Angara)	Total no. of species in this genus in Baikal	No. of endemic species in this genus in Baikal
Family Oxidae Viets, 1926	2	0	0		
Genus *Oxus* Kramer, 1877	1	0	0	0	0
Genus *Gnaphiscus* Koenike, 1898	1	0	0	0	0
Family Limnesiidae Thor, 1900	2	0	0		
Genus *Limnesia* Koch, 1836	2	0	0	0	0
Family Hygrobatidae Koch, 1842	2	0	0		
Genus *Hygrobates* Koch, 1837	1	0	0	0	0
Family Unionicolidae Oudemans, 1909	1	0	0		
Genus *Unionicola* Haldeman, 1842	1	0	0	0	0
Family Feltridae Viets, 1926	1	0	0		
Genus *Feltria* Koenike, 1892	1	0	0	1	0
Family Pionidae Thor, 1900	2	0	0		
Genus *Tiphys* Koch, 1836	2	0	0	0	0
Family Mideopsidae Koenike, 1910	1	0	0		
Genus *Mideopsis* Neuman, 1880	1	0	0	0	0
Family Arrenuridae Thor, 1900	1	0	0		
Genus *Arrenurus* Duges, 1834	1	0	0	0	0
EUPODINA					
Family Halacaridae Murray, 1876	2	0	1		
Subfamily Limnohalacarinae Viets, 1927	2	0	1		
Genus *Parasoldanellonyx* Viets, 1929				3	1
Class INSECTA					
Order EPHEMEROPTERA	13	1	0		
Family Siphlonuridae	2	0	0		
Genus *Siphlonurus* Eaton, 1868	2	0	0		
Family Ameletidae	1	1	0		
Genus *Ameletus* Eaton, 1865	1	1	0		
Family Metretopodidae	1	0	0		

Taxon	Total no. of species in Khubsugul	No. of species found only in Khubsugul ('endemics' of Khubsugul)	No. of species found only in Khubsugul and Baikal (and in Angara)	Total no. of species in this genus in Baikal	No. of endemic species in this genus in Baikal
Genus *Metretopus* Eaton, 1891	1	0	0		
Family Baetidae	5	0	0		
Genus *Cloeon* Leach, 1815	3	0	0		
Genus *Baetis* Leach, 1815	2	0	0		
Family Heptageniidae	2	0	0		
Genus *Rhithrogena* Eaton, 1881	2	0	0		
Family Ephemerellidae	1	0	0		
Genus *Ephemerella* Walsh, 1862	1	0	0		
Family Caenidae	1	0	0		
Genus *Caenis* Stephens, 1835	1	0	0		
Order PLECOPTERA	17	1	0		
Family Perlodidae	6	0	0		
Genus *Arcynopteryx* Klapálek, 1904	3	0	0		
Genus *Diura* Billberg, 1820	2	0	0		
Genus *Isoperla* Banks, 1906	1	?	?		
Family Perlidae	1	0	0		
Genus *Paragnetina* Klapálek, 1907	1	0	0		
Family Chloroperlidae	3	0	0		
Genus *Alloperla* Banks, 1906	1	?	?		
Genus *Suwallia* Ricker, 1943	2	0	0		
Family Taeiopterygidae	1	0	0		
Genus *Taeniopteryx* Pictet, 1841	1	0	0		
Family Nemouridae	3	0	0		
Genus *Nemoura* Latreille, 1796	3	0	0		
Family Capniidae	3	1	0		
Genus *Capnia* Pictet, 1841	2	1	0		
Genus *Mesocapnia* Rauser, 1968	1	0	0		
Order TRICHOPTERA	18	0	0		
Family Glossosomatidae Wallengren	1	0	0		

Taxon	Total no. of species in Khubsugul	No. of species found only in Khubsugul ('endemics' of Khubsugul)	No. of species found only in Khubsugul and Baikal (and in Angara)	Total no. of species in this genus in Baikal	No. of endemic species in this genus in Baikal
Genus *Eomystra* Martynov, 1934	1	0	0		
Family Hydroptilidae Stephens	1	0	0		
Genus *Hydroptila* Dalman, 1819	1	0	0		
Superfamily Phryganeoidea					
Family Phryganeidae Burmester	3	0	0		
Genus *Agrypnia* Curtis, 1835	2	0	0		
Genus *Phryganea* Linnaeus, 1758	1	0	0		
Family Leptoceridae	1	0	0		
Genus *Oecetis* MacLachlan, 1877	1	0	0		
Family Limnephilidae Kolenati	12	0	0		
Genus *Anabolia* Stephens, 1837	1	0	0		
Genus *Apatania* Kolenati, 1848	3	0	0		
Genus *Asynarchus* MacLachlan, 1880	2	0	0		
Genus *Chaetopteryx* Stephens, 1829	1	0	0		
Genus *Hydatophylax* Wallengren, 1891	1	0	0		
Genus *Limnephilus* Leach, 1815	3	0	0		
Genus *Philarctus* MacLachlan, 1880	1	0	0		
Order DIPTERA					
Family Chironomidae Macquart, 1838	106	1	3		
Subfamily Tanypodinae	11	0	0		
Genus *Macropelopia* Thienemann, 1916	1	0	0	1	0
Genus *Psectrotanypus* Kieffer, 1909	1	0	0		0
Genus *Procladius* Skuse, 1889	3	0	0	1	0
Genus *Psilotanypus* Kieffer, 1906	2	0	0		0
Genus *Thienemannimyia* Fittkau, 1957	1	0	0	2	0
Genus *Larsia* Fittkau, 1962	1	0	0	0	0
Genus *Ablabesmyia* Johannsen, 1905	2	1	0	1	0
Subfamily Diamesinae	8	0	0		
Genus *Protanypus* (Kieffer, 1916) Edwards, 1929	1	0	0	1	0

Taxon	Total no. of species in Khubsugul	No. of species found only in Khubsugul ('endemics' of Khubsugul)	No. of species found only in Khubsugul and Baikal (and in Angara)	Total no. of species in this genus in Baikal	No. of endemic species in this genus in Baikal
Genus *Psedodiamesa* Goetghebuer, 1939	3	1	0	2	0
Genus *Diamesa* Meigen, 1838	2	0	0	3	1
Genus *Potthastia* Keiffer, 1922	1	0	0	2	0
Genus *Monodiamesa* Kieffer, 1921	1	0	0	1	0
Subfamily Orthocladiinae	45	0	2		
Genus *Brillia* Kieffer, 1913	1	0	0	0	0
Genus *Trissocladius* Kieffer, 1908	4	0	0	3	0
Genus *Heterotrissocladius* Spärck, 1922	1	0	0	2	0
Genus *Eukiefferiella* Thienemann, 1920	4	0	0	8	0
Genus *Orthocladius* (Van der Wulper, 1874) Brundin, 1956	8	0	2	12	3
Genus *Cricotopus* (Van der Wulper, 1874) Edwards, 1929	6	0	0	7	0
Genus *Paratrichocladius* Thienemann, 1918	3	0	0	2	0
Genus *Acricotopus* Thienemann, 1935	1	0	0	0	0
Genus *Psectrocladius* Kieffer, 1906	4	0	0	4	0
Genus *Limnophyes* Eaton, 1875	2	0	0	3	0
Genus *Pseudosmittia* (Goetghebuer, 1932) Brundin, 1956	2	0	0	1	0
Genus *Parakiefferiella* (Thienemann, 1936) Brundin, 1956	2	0	0	0	0
Genus *Lapposmittia* Thienemann, 1939	1	0	0	0	0
Genus *Smittia* (Holmgren, 1869) Brundin, 1956	2	0	0	1	?
Genus *Corynoneura* (Winner, 1846) Edwards, 1929	2	0	0	2	0
Genus *Thienemanniella* Kieffer, 1911	2	0	0	2	0
Subfamily Chironominae	42	0	1		
Genus *Constempellina* Brundin, 1947	1	0	0	1	0
Genus *Tanytarsus* Van der Wulp, 1874	4	0	0	3	0
Genus *Paratanytarsus* Bause, 1913	2	0	0	2	1
Genus *Cladotanytarsus* Kieffer, 1922	1	0	0	1	0
Genus *Rheotanytarsus* Bause, 1913	1	0	0	1	0
Genus *Micropsectra* Kieffer, 1909	1	0	0	3	0
Genus *Lauterbornia* Kieffer, 1911	1	0	0	2	0
Genus *Chironomus* Meigen, 1803	5	0	0	6	0

Taxon	Total no. of species in Khubsugul	No. of species found only in Khubsugul ('endemics' of Khubsugul)	No. of species found only in Khubsugul and Baikal (and in Angara)	Total no. of species in this genus in Baikal	No. of endemic species in this genus in Baikal
Genus *Einfeldia* Kieffer, 1924	1	0	0	4	0
Genus *Cryptochironomus* Kieffer, 1918	2	0	0	2	0
Genus *Cryptocladopelma* Lenz, 1941	1	0	0	1	0
Genus *Demicryptochironomus* Lenz, 1941	1	0	0	1	0
Genus *Leptochironomus* Pagast, 1931	1	0	0	1	0
Genus *Paracladopelma* Harnisch, 1923	1	0	0	2	0
Genus *Parachironomus* Lenz, 1921	1	0	0	2	0
Genus *Limnochironomus* Kieffer, 1920	1	0	0	3	0
Genus *Endochironomus* Kieffer, 1918	3	0	0	2	0
Genus *Glyptotendipes* Kieffer, 1913	3	0	0	8	7
Genus *Sergentia* Kieffer, 1921	1	0	0	2	0
Genus *Pentapedilum* Kieffer, 1913	3	0	0	5	0
Genus *Polypedilum* Kieffer, 1913	1	0	0	1	0
Genus *Microtendipes* Kieffer, 1921	1	0	0	2	0
Genus *Paratendipes* Kieffer, 1911	1	0	0	2	0
Genus *Stictochironomus* Kieffer, 1919	4	0	0	2	0
Class BIVALVIA					
Order LUCINIFORMES Stoliczka, 1871	3	0	0		
Family Sphaeriidae Jeffreys, 1862	2	0	0		
Subfamily Sphaeriinae Jeffreys, 1862					
Genus *Sphaerium* Scopoli, 1777	2	0	0	8	4
Family Pisidiidae Gray, 1857	1	?	?		
Subfamily Pisidiinae Gray, 1857					
Genus *Pisidium* Pfeiffer, 1821	1	?	?	3	1
Class GASTROPODA	9	1	0		
Order VIVIPARIFORMES Sitnikova et Starobogatov, 1982					
Family Valvatidae Gray, 1840	1	0	0		

Taxon	Total no. of species in Khubsugul	No. of species found only in Khubsugul ('endemics' of Khubsugul)	No. of species found only in Khubsugul and in Baikal (and in Angara)	Total no. of species in this genus in Baikal	No. of endemic species in this genus in Baikal
Subfamily Valvatinae Gray, 1840					
Genus *Cincinna* Hubner, 1810	1	0	0	9	5
Order LYMNAEIFORMES Ferussac, 1822					
Family Lymnaeidae Rafinesque, 1815	5	0	0		
Genus *Lymnaea* Lamarck, 1799	5	0	0	12	0
Family Physidae Fitzinger, 1883	1	0	0		
Subfamily Physinae Fitzinger, 1883					
Genus *Physa* Draparnaud, 1801	1	0	0	0	0
Family Planorbidae Rafinesque, 1815	2	1	0		
Subfamily Planorbinae Rafinesque, 1815					
Genus *Anisus* Studer, 1820	1	0	0	6	2
Genus *Choanomphalus* Gerstfeldt, 1859	1	1	0	22	22

Biogeography and Species Diversity of Diatoms in the Northern Basin of Lake Tanganyika

C. COCQUYT

I. SUMMARY

The geographical distribution and diversity of the diatoms in the northern basin of Lake Tanganyika (historically by far the most intensively studied area of the African Great Lakes concerning this group) are examined and compared with the data available for Lake Tanganyika as a whole. Most of the diatom taxa recorded in Lake Tanganyika are cosmopolitan in distribution, but *c.* 21% have a distribution restricted to the African continent and *c.* 8% are endemic to Lake Tanganyika. The lowest species diversity seen among the diatom communities examined was at the northernmost part of the northern basin. Effects of the high sedimentation rate of the River Ruzizi, and of pollution originating from urban and industrial sources, were evident in the diatom communities. A clear north–south trend of increased diatom diversity and a change in species composition was evident. These trends were particularly evident among endemic taxa, notably in members of the genus *Surirella*. Diatom communities, and especially their endemic components, may thus be valuable bioindicators of environmental perturbations within this and other ancient lakes.

ADVANCES IN ECOLOGICAL RESEARCH VOL. 31
ISBN 0-12-013931-6

II. INTRODUCTION

The past few decades have seen a renewed interest in the algal flora of the African Great Lakes, especially the diatoms. These investigations culminated in the first comprehensive biogeography of the diatoms from these lakes (Ross, 1983), since when several more studies of diatoms have been conducted and a large number of diatom taxa has been newly reported from this region. A comparative survey of the composition and diversity of the algal flora of Lakes Tanganyika, Malawi and Victoria was provided by Cocquyt and Vyverman (1994) and included these new data.

The present chapter documents and examines the diversity and the biogeographical distribution of the diatom species in the northern basin of Lake Tanganyika, historically the most intensely studied area within the African Great Lakes. This chapter is based on results from several projects, in which a large number of samples of planktonic organisms were taken in the littoral zone, and also of surface sediments at several sites along the Burundian shore. Some of these projects are ongoing.

For many diatom taxa the accurate taxonomic status and phylogenetic relationships, intraspecific variability and ecological characteristics are poorly known, and the true distribution of several taxa is still problematic. Studies are currently underway to address these problems, and have already yielded much new information and improved our knowledge of this group of organisms. As such, this chapter describes the findings from a study which is very much still in progress. However, the preliminary findings presented here indicate clearly the effects of perturbation on diatom communities, and hint at the usefulness of such studies in an applied context.

III. METHODS

Quantitative plankton and qualitative diatom sampling were carried out in the north basin of Lake Tanganyika. This chapter utilizes findings from two separate sampling regimes. In the first sample set, five sites along the Burundian shore of the lake (from north to south: Bujumbura, Gatororongo, Resha, Kibwe and Nyanza-Lac) were sampled at monthly intervals (Figure 1). On each occasion, samples for plankton investigation were taken with a 28 μm mesh plankton net (Cocquyt, 1999a) and epilithon was sampled by scraping material off rocks and stones (Cocquyt, 1999b). Epiphyton, removed from submerged aquatic plant material (e.g. *Vossia cuspidata* Griff), was sampled at three stations. At Kibwe, the epipelon was sampled by collecting the surface layer of several millimetres thickness of the bottom mud.

The second study sampled surface sediments at four transects within 10 km of the Ruzizi delta during the GEORIFT project (ELF-Aquitaine) (Caljon, 1991; Caljon and Cocquyt, 1992). Many of these samples were composed of

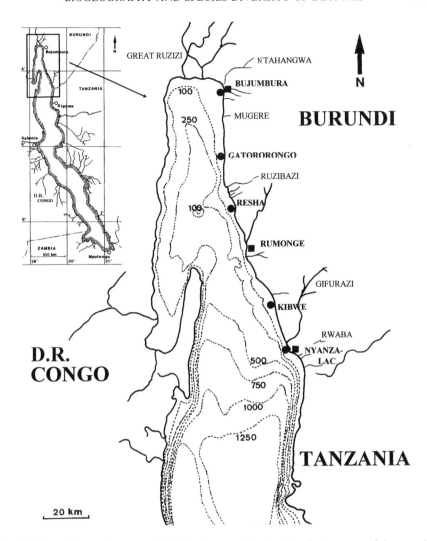

Fig. 1. Map of the northern part of Lake Tanganyika showing the location of the sample stations in the littoral zone along the Burundian coast (Bujumbura, Gatororongo, Resha, Kibwe, Nyanza-Lac), the Ruzizi Delta and some smaller rivers (Mugere, Ruzibazi, Gifurazi/Rwaba).

deposited diatoms, reflecting the pelagic nature of the diatom flora. However, some of the observed taxa were allochthonous in origin and originated from the Ruzizi, the most important inflowing river in this area. Supplementary samples of surface sediments, taken with a ponar grab, were collected from near the mouths of three small rivers, the Mugere, Ruzibazi and Gifurazi/ Rwabe rivers, along the Burundian coast at Ramba, Magara and Mvugo, respectively (Cocquyt, submitted) (Figure 1). When assessing the geographical distribution of diatoms within the northern basin, data on species occurrence in plankton and benthos samples taken in the bay of Bujumbura (Caljon, 1992), and epilithic samples taken on thrombolitic reefs near the Burundian coast (Cocquyt, 1991), were also used.

Permanent slide preparations were made by embedding diatom samples in Naphrax. Taxonomy was based mainly on Krammer and Lange-Bertalot (1986, 1988, 1991a,b), with the exception of the genera *Fragilaria* and *Synedra*, which followed Hustedt (1931–1959, 1961–1966). To make this chapter more accessible to the general reader, the recent revised taxonomy of the diatoms (Round *et al.*, 1990), with its splitting of "classic" and familiar genera such as *Navicula*, is not used here.

IV. GEOGRAPHICAL DISTRIBUTION

In contrast to many other groups of aquatic organisms, most of the diatoms have a cosmopolitan distribution and occur throughout the world. One objective of the present study was to identify taxa with a restricted geographical distribution pattern, and here, five categories – cosmopolitan, pantropical, African, tropical African, and endemic – are used to describe the distributional status of diatoms. The definition of these terms and the geographical distribution patterns of the diatoms from the northern basin of Lake Tanganyika are here presented separately for each of these categories.

A. Cosmopolitan Taxa

The term "cosmopolitan" is used here to describe taxa occurring throughout the world; taxa which are known from the African Great Lakes and also from other specific but widely dispersed habitats, such as alpine regions and from Europe (e.g. *Cyclotella meduanae* Germain, *Cymbella helmckei* Krammer and *Neidium minutissimum* Krasske) are also included.

To date, a total of 358 diatom taxa has been recorded from the northern basin of Lake Tanganyika (Appendix I). Four other taxa identified from the lake, *Amphora tanganyikae* var. 1, *Nitzschia* cf. *epiphyticoides, Surirella* cf. *acuminata* and *S.* cf. *heidenii*, are not included in this number. Further investigation is necessary to establish the precise taxonomic status of these forms, but they probably fall within the intraspecies variability of taxa already described from the lake, and are thus of dubious status.

Of these 358 diatom taxa, 257 (71.8%) have a cosmopolitan distribution, a proportion corresponding well with the 69.8% of cosmopolitan taxa recorded from the whole lake (Cocquyt *et al.*, 1993; Cocquyt and Vyverman, 1994). Results from ongoing studies in the southern basin of the lake will establish whether the proportion of cosmopolitan diatoms in the northern basin is truly representative of the lake as a whole. However, examination of preliminary results of the present study has already revealed small differences in the composition and species richness of the ambient diatom assemblages, between 66.3% and 76%. The largest number of cosmopolitan taxa has been observed in the surface sediment samples taken along four transects in the northernmost part of the lake (173 taxa, 75.9% cosmopolitan) and those taken near Ramba (117 taxa, 76% cosmopolitan). Epiphytic and epipelic taxa were very few (between 34 and 38 taxa) and were mainly composed of cosmopolitan taxa (between 73.5% and 75.6%) (Table 1). As such, the composition of the diatom communities may vary according to locality, and northern communities might not be representative of those found in the extreme south.

B. Pantropical Taxa

A small number of diatoms is known from the tropical areas of different continents, from tropical Asia and Africa, or from tropical America and Africa; here, these are termed "pantropical taxa". Taxa restricted to the African continent are not included in this group, but are treated as separate category. In addition to pantropical taxa, three species reported only from paleotropical regions, *Cyclotella woltereckii* Hust., *Navicula perparva* Hust. and *N. platycephala* O. Müll., were recorded in the present study. Twenty-five (7%) of the 358 taxa found in the northern basin are known to have a tropical distribution, a proportion slightly greater than the 4.8% reported for the total lake (Cocquyt et al., 1993; Cocquyt and Vyverman, 1994). Most of these taxa belong to the genera *Aulacoseira* (three taxa), *Eunotia* (three taxa), *Gomphonema* (two taxa), *Navicula* (seven taxa), *Pinnularia* (two taxa) and *Surirella* (three taxa).

The relative proportion of the tropical taxa differed according to the type of sample. Within the surface sediments taken along four transects off the Ruzizi delta, up to 7% of the taxa had a tropical distribution; for the surface sediments taken off the smaller rivers these numbers were 5%, 5% and 4.3%, respectively. In these samples the tropical taxa were characterized by low species diversity, with only five tropical taxa recorded near Mvugo, and a maximum of 16 taxa in samples taken off the Ruzizi delta (Table 1). These differences in the proportion and number of taxa are due to the influence of the Ruzizi River (Caljon, 1991), but also partly reflect the greater number of samples (n = 49 samples, 262 taxa recorded) investigated near the Ruzizi delta. Near the mouths of the rivers Mugere, Ruzibazi and Gifurazi, fewer samples were studied (n = 6, n = 7 and n = 7, respectively) and the number of taxa observed was lower (173, 187 and 138, respectively). Moreover, these smaller rivers had less influence on the composition of the surface sediments, a feature that has implications for the supply of allochthonous taxa (Cocquyt, unpubl.).

In the littoral samples, the lowest species diversity was recorded in the epiphytic samples (three taxa) and the highest in the plankton samples (nine taxa) (Table 1). The highest proportion of tropical taxa in the littoral samples

Table 1
Synopsis of the number of cosmopolitan, tropical, African, tropical African and endemic diatom taxa, and their proportion in different sample types from the northern basin of Lake Tanganyika: plankton, epilithon, epiphyton and epipelon of five littoral sample stations along the Burundian coast, surface sediments of the northernmost part (Ruzizi), and surface sediments from near Ramba, Magara and Cap de Mvugo (Rivers Mugere, Ruzibazi and Gifurazi, respectively)

	Cosmopolitan taxa	Tropical taxa	African taxa	Tropical African taxa	Endemic taxa
LITTORAL					
Plankton					
Bujumbura	134 (70.9%)	9 (4.8%)	43 (22.8%)	25 (13.2%)	8 (4.2%)
Gatororongo	112 (67.8%)	8 (4.8%)	43 (26.0%)	27 (16.4%)	10 (6.1%)
Resha	101 (66.9%)	6 (4.0%)	41 (27.0%)	26 (17.1%)	11 (7.2%)
Kibwe	116 (69.5%)	9 (5.4%)	41 (24.7%)	26 (15.1%)	9 (5.4%)
Nyanza-Lac	113 (70.2%)	7 (4.3%)	39 (24.2%)	26 (16.2%)	11 (6.8%)
Epilithon					
Bujumbura	45 (66.3%)	5 (7.0%)	19 (26.8%)	10 (14.1%)	1 (1.4%)
Gatororongo	39 (68.4%)	3 (5.3%)	14 (24.6%)	6 (10.5%)	1 (1.8%)
Resha	36 (62.1%)	4 (6.9%)	17 (29.3%)	8 (13.8%)	1 (1.7%)
Kibwe	81 (71.7%)	5 (4.4%)	27 (24.1%)	14 (12.5%)	4 (3.6%)
Nyanza-Lac	77 (70.6%)	4 (3.7%)	27 (24.8%)	14 (12.8%)	4 (3.7%)
Epiphyton					
Bujumbura	34 (75.6%)	3 (6.7%)	8 (18.2%)	3 (6.8%)	0 (0%)
Gatororongo	41 (74.5%)	3 (5.5%)	11 (23.4%)	4 (8.5%)	1 (2.1%)
Kibwe	50 (74.6%)	3 (4.5%)	11 (16.7%)	4 (6.1%)	1 (1.5%)
Epipelon					
Kibwe	58 (73.5%)	4 (5.1%)	18 (22.2%)	8 (9.9%)	2 (2.5%)
SURFACE SEDIMENTS					
Rusizi	173 (75.9%)	16 (7.0%)	39 (17.1%)	25 (11.0%)	9 (3.9%)
Ramba	117 (76.0%)	8 (5.2%)	29 (18.8%)	15 (9.7%)	4 (2.6%)
Magara	113 (72.0%)	8 (5.1%)	36 (22.9%)	22 (14.0%)	8 (5.1%)
Mvugo	81 (70.5%)	5 (4.3%)	29 (25.2%)	17 (14.8%)	6 (5.2%)
Total	257 (71.8%)	25 (7.0%)	76 (21.2%)	46 (12.8%)	17 (4.8%)

was observed in the epilithon (7%) and the epipelon (6.7%) near Bujumbura; the lowest proportions (3.7%) were seen in epilithon samples taken at the most southern sample station, near Nyanza-Lac (Table 1).

C. African Taxa

Some diatoms are known only from the African continent and are here called "African taxa". This group was mainly composed of tropical African taxa; of the 76 taxa in this group, only 29 (38%) have also been reported from non-tropical regions. Most of these species belong to the genera *Navicula* (five taxa), *Nitzschia* (eight taxa), *Rhopalodia* (two taxa) and *Surirella* (three taxa). *Amphora strigosa* Hust., a species previously known only from Israel, is included in this group. Most of the African taxa have been reported from south of the tropic of Capricorn, particularly from South Africa. This reflects not only this region having been subjected to more intense studies (e.g. Archibald, 1983, 1984; Archibald and Schoeman, 1987; Cholnoky, 1953, 1955, 1956, 1957, 1958, 1959, 1960, 1966a, b, 1970; Schoeman, 1973; Schoeman and Archibald, 1976), but also the closer geographical relationship between South Africa and tropical Africa. The part of Africa north of the tropic of Cancer can be regarded as separated from the rest of the continent by the Sahara and Sahel deserts. The largest number of African taxa in northern Lake Tanganyika was the 43 taxa observed in the plankton samples from the littoral zone near Bujumbura (Table 1).

D. Tropical African Taxa

Several of these African taxa are restricted to the tropical part of the continent, and are here termed "tropical African taxa". Up to 13.1% (29 taxa) of all diatoms recorded in the northern basin of Lake Tanganyika are restricted to tropical Africa. These taxa belong mostly to the genera *Thalassiosira* (two taxa), *Capartogramma* (three taxa), *Gomphonema* (two taxa), *Rhopalodia* (three taxa) and *Surirella* (11 taxa). A large number of these tropical African taxa has a distribution restricted to the region of the African Great Lakes (Cocquyt, 1999). In the present study, the greatest percentage of tropical African taxa was seen in the plankton of the littoral near Resha (17.1%, 26 taxa) (Table 1).

E. Endemic Taxa

A number of the tropical African diatoms is known only from Lake Tanganyika, and are here referred to as "endemic taxa". It is this group which is of greatest interest to students of evolution. A total of 38 endemic taxa has been described from Lake Tanganyika, comprising 8% of the total diatom flora. However, although a significant number of Tanganyika's diatom species is endemic, this is a relatively low proportion when compared with many other groups of aquatic organisms of this lake (Coulter, 1991, 1994). Despite this, in terms of its high absolute number of endemic diatom taxa, Lake Tanganyika is unique among the African Great Lakes. The number and proportion of endemic diatoms of Lakes Malawi and Victoria are very low: Lake Victoria contains only three endemic diatoms (*Rhizosolenia curviseta* Hustedt, *R. victoriae* B. Schröder and *Fragilaria longissima* Hustedt), as does Lake Malawi (*Nitzschia pelagica, Cyclostephanos malawiensis* Caspar and Klee and *Stephanodiscus muelleri* Klee and Caspar).

The number of endemic taxa described from the African Great Lakes will certainly increase in the future owing to more widespread and thorough sampling, more detailed studies and an improved taxonomic knowledge. For example, from recent studies of centric diatoms from Lake Malawi, two new taxa have been described (Klee and Casper, 1992). However, it should also be recognized that detailed studies of the other African Great Lakes might also result in a decrease in the number of taxa presently described as endemic to Lake Tanganyika. A case in point is a recent study of the surface sediment flora of northern Lake Malawi (Vyverman and Cocquyt, 1993), which identified two taxa (*Capartogramma amphoroides* Ross and *Cymbellonitzschia minima* Hustedt) hitherto known only as Lake Tanganyika endemics.

Of the diatoms observed in the northern basin of Lake Tanganyika, 17 taxa (4.8%) are endemic (Table 1, Appendix I). However, this is probably an underestimate, as several other taxa (e.g. *S. heidenii* var. 1, *S. nyassae* var. 1, *Navicula* cf. *rhynchocephala, Rhizosolenia* sp., *Gomphocymbella* sp. and *Rhopalodia* sp.) (Cocquyt, 1998) are very probably also endemic. Investigations of these taxa are currently underway. Endemic taxa in the epiphyton and epilithon of the littoral were very rare; the largest number was observed in the plankton of the littoral near Nyanza-Lac (11 taxa, 6.8%) and near Resha (11 taxa, 7.2%) (Table 1).

In this study in the northern basin, 12 taxa previously reported as being endemic to Lake Tanganyika (Hustedt, in Huber-Pestallozi, 1942; Kufferath, 1956; Van Meel, 1954) were not observed. These taxa are *Aulacoseira pyxis* (O. Müll.) Simonsen, *Campylodiscus tanganyicae* Hust., *Fragilaria aethiopica* G.S. West, *Gomphonema apicetrapezon* Kufferath, *Navicula vanmeelii* Kufferath, *Neidium tanganyikae* (G.S. West) Mills, *Nitzschia kampimbiense* Kufferath, *N. profunda* Kufferath, *N. pseudosubrostrata* Kufferath, *Surirella effusa* Hust., *S. margaritifera* Hust., *S. subrobusta* Hust. and *S. vasta* Hust. At present it is not known whether their absence reflects the patchy distribution pattern and sporadic occurrence of these species, or their extirpation from this part of the lake subsequent to the earlier investigations. Three other endemic species, *Gomphocymbella muelleri* Kociolek and Stoermer, *G. rossii* Kociolek and Stoermer and *G. tanganyikae* Kociolek and Stoermer, have been recorded as fossilized forms in material from substrate cores of the lake (Kociolek and Stoermer, 1993) but have not been seen as living organisms.

V. DIVERSITY

Overall, the highest species diversity of diatoms in the littoral zone was observed in the plankton (280 taxa), and the species diversity in the epilithon (154 taxa), and especially in the epipelon (84 taxa) and the epiphyton (81 taxa), was much lower (Table 2).

The diversity of diatoms in the plankton differed along the Burundian coast. The highest diversity (189 species) was observed at Bujumbura, the northern-most sampling station; the number of diatom taxa in the other four stations varied between 152 and 166 (Table 2). However, it is noteworthy that endemic species were less frequent at the Bujumbura site. The greater dominance of benthic taxa in the plankton samples reflects the resuspension of the diatom flora, due to the effects of strong monsoon winds in the dry season (Cocquyt, 1999), when sampling was conducted.

Simpson's and Shannon's diversity indices were calculated for the phytoplankton samples of the five littoral sample stations along the Burundian coast (Table 3). Both for diatoms alone and for all algae (including diatoms and cyanobacteria), the lowest diversity values were obtained at the northern-most sampling station (Bujumbura) and the highest values in the southernmost station (Nyanza-Lac). Diversity values for the stations in between show a southwards increase, except for one station near Kibwe, 108 km south of Bujumbura, where the values are again low. The low diversity at this station is believed to reflect local influences, such as heavy rainfall, inputs from small

Table 2

Total number of diatom taxa observed in the littoral samples (plankton, epilithon, periphyton and epipelon) along the Burundian coast

	Littoral			
	Plankton	Epilithon	Epiphyton	Epipelon
Bujumbura	189	71	44	0
Gatororongo	166	58	45	0
Resha	152	58	0	0
Kibwe	166	112	66	84
Nyanza-Lac	160	109	0	0
Total	280	154	81	84

rivers and local mixing of shallow water, which have a strong impact on the phytoplankton communities in the littoral zone (Cocquyt et al., 1991). Given that such local influences can have such strong effects, the general trend of a northwards decrease in species diversity is even more noteworthy.

Given our present level of knowledge, the most parsimonious explanation for the lower species diversity in the north is in terms of the negative effects of the sediment-loaded Ruzizi River (Cohen et al., 1993; Vandelannoote et al., 1999) and the eutrophication near Bujumbura (Caljon, 1992). Almost all industrial activity of Burundi takes place near Bujumbura, and the small River Ntahangwa, which runs through the industrial area before emptying into the lake, is the most polluted river in that country (Vandelannoote et al., 1999). It is also probable that physiological and ecological adaptations to the very special habitats of the lake have been of great importance to the distribution of the endemic species. Further investigations, over longer periods and including the entire lake area, will give a clearer picture of this problem.

That the effects of perturbations originating at Bujumbura are evident along the eastern shore of the lake, and gradually decrease moving southwards, is in accordance with the presence of a steady clockwise coastal circulation within the lake (Coulter and Spigel, 1991). The north to south current in the upper 5 m along that coast averages 11 cm s^{-1}, and 16 cm s^{-1} in the rainy season.

The negative effects of the increased sedimentation from deforestation-initiated erosion in Central Africa, especially that in the Ruzizi basin during the last decades, have been documented for certain aquatic organisms (Cohen et al., 1993). However, although that study detected a significant decrease in biodiversity in the northern part of the lake within the ostracods and fishes, no significant decrease was detected for the diatoms (Cohen et al., 1993). The present study has shown that the effects of perturbation are especially evident in the endemic taxa, and the final section of this chapter therefore considers the distribution and abundance of endemic taxa only.

VI.   DISTRIBUTION OF ENDEMIC TAXA

The largest number of endemic Tanganyikan diatom taxa belongs to the genus Surirella, and the southwards increase in the number of endemic taxa reflected primarily an increase in the number of Surirella taxa (Figure 2, 3). In contrast, no southwards increase was evident in the endemic taxa belonging to the genera Amphora, Cymatopleura, Gomphonema and Navicula. Some endemic taxa belonging to genera other than Surirella were found throughout the study area (e.g. the very common Navicula amplectens, Amphora tanganyikae and Cymatopleura calcarata), but several other species were observed only sporadically, and in localized populations. For example, Gomphonema paddockii was recorded only near Bujumbura, while G. kilhamii was recorded from only two sites along the Burundian coast. Further studies are necessary to

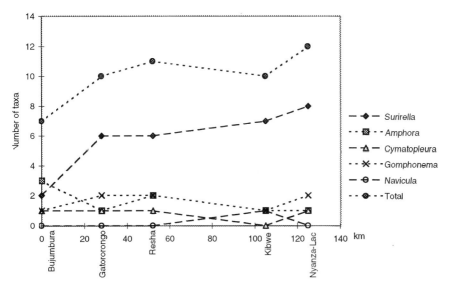

Fig. 2. Number of endemic taxa observed in the littoral zone of the northern basin of Lake Tanganyika. The *x*-axis represents the distance (km) along the Burundian coast, starting at Bujumbura (0 km) going south to Nyanza-Lac in southern Burundi (122 km).

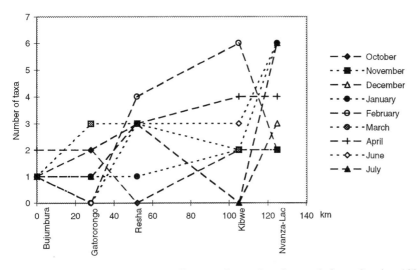

Fig. 3. Number of endemic *Surirella* taxa observed each month from October 1985 to July 1986 in the littoral zone of the northern basin of Lake Tanganyika. The *x*-axis represents the distance (km) along the Burundian coast, starting at Bujumbura (0 km) going south to Nyanza-Lac in southern Burundi (122 km).

describe in detail the microdistribution patterns of such species, and to assess which biotic and physical factors affect the observed distribution patterns.

Table 3
Simpson's (D) and Shannon's (H) diversity indices and their equitability indices (E and J, respectively) for the diatoms and for all algae (including diatoms and cyanobacteria) recorded in plankton samples taken monthly (October 1985–July 1986) at five littoral sampling stations along the Burundian coast (Bujumbura, Gatororongo, Resha, Kibwe and Nyanza Lac)

	Diatoms				All algae			
	D	E	H	J	D	E	H	J
Bujumbura	4.794	0.0799	2.502	0.6011	6.011	0.0661	2.848	0.6314
Gatororongo	10.168	0.2825	2.716	0.7579	17.772	0.3012	3.262	0.8000
Resha	12.836	0.3056	2.994	0.8010	20.307	0.2943	3.388	0.8001
Kibwe	5.150	0.0888	2.342	0.5767	6.536	0.0688	2.746	0.6031
Nyanza-Lac	15.416	0.3032	3.158	0.8032	22.924	0.2439	3.599	0.7921

A. The Endemic *Surirella*

The endemic *Surirella* species of Lake Tanganyika were described by Hustedt (in Huber-Pestalozzi, 1942), and for most of these the type locality was in the vicinity of Rumonge, 73 km south of Bujumbura (Hustedt, in Schmidt, 1874–1959; Van Meel, 1954). Yet one of these taxa, *Surirella margaritifera* Hust., was not observed during the present investigation, despite Rumonge being located between two of the sampling stations (Resha and Kibwe) (Figure 1). However, Hustedt (in Huber-Pestalozzi, 1942) noted the occurrence of this taxon as being sporadic, and perhaps its absence from samples taken near the type locality is therefore not so remarkable. Several other rare endemic *Surirella* taxa were observed only once during the present study (e.g. *S. fuellebornii* var. *tumida* and *S. latecostata*). The low probability of observing such rare and sporadically occurring taxa makes it premature to describe this taxon as being extinct, yet it is surely highly vulnerable.

Two other endemic *Surirella* taxa, *S. effusa* and *S. vasta*, were also not found in the northern part of the lake in the present study, despite their being described by Hustedt (in Huber-Pestalozzi, 1942) as not uncommon and rather frequent, respectively. Unfortunately, the location where these organisms were originally observed (Hustedt, in Schmidt, 1874–1859; Van Meel, 1954) is unknown, and their absence from the present samples might therefore reflect site-specific differences in their distribution patterns. It is noteworthy that some of the endemic *Surirella* taxa described by Hustedt (in Huber-Pestalozzi, 1942) as frequent or rather frequent (e.g. *S. aculeata* and *S. subcontorta*) were observed only once or very sporadically, as were several other species described as sporadic or rare (e.g. *S. gradifera, S. latecostata, S. fuellebornii* var. *tumida* and *S. striolata*).

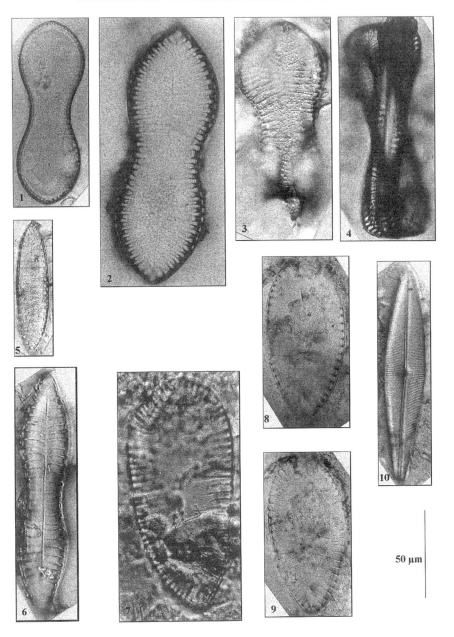

Fig. 4. Examples of several endemic diatom taxa of Lake Tanganyika. 1, *Cymatopleura calcarata*; 2, *Surirella heidenii*; 3, 4, *S. spiraloides*; 5, *S. aculeata*; 6, *S. acuminata*; 7, *S. striolata*; 8, 9, *S. subcontorta*; 10, *Gomphonema paddockii*. Scale bar (1–10): 50 μm.

In this study, endemic *Surirella* taxa could be categorized into two groups according to their distribution pattern through the northern lake basin. One group consisted of taxa observed at all sample stations along the Burundian coast (e.g. *S. acuminata*, *S. heidenii*, *S. lancettula* and *S. sparsipunctata*). The second group consisted of taxa not observed in the northernmost part of the lake (e.g. *S. aculeata*, *S. fuellebornii* var. *tumida*, *S. gradifera*, *S. heidenii* var. 1, *S. latecostata*, *S. spiraloides* and *S. subcontorta*). The genus *Surirella* belongs to the most highly developed diatom family, the Surirellaceae (Round *et al.*, 1990), and all of the endemic forms have probably evolved within the lake. These endemic taxa have become adapted to stenobiontic conditions and have lost their tolerance to changes in environmental conditions, such as those caused by pollution and sedimentation. However, the existence of these two groups suggests the presence within this genus of species-specific differences in tolerance to environmental perturbation.

VII. APPLICATION OF DIATOM STUDIES TO ENVIRONMENTAL MONITORING

At sites where environmental changes have taken place, the proportion of tropical African taxa, and especially the number of endemic *Surirella* taxa, decrease. This was especially evident at the Bujumbura site. These perturbations are usually of anthropogenic origin, and are most often due to eutrophication and to the impact of higher sedimentation rates following deforestation.

Their relatively large size and the ease of identification of these endemic taxa, even under low magnification by light microscope (Figure 4), contrast with many other diatom taxa and are extremely useful characters for biologists of the riparian countries of Lake Tanganyika, where scientific resources and training are at a premium. The present study in the northern basin of Lake Tanganyika clearly shows that the endemic diatom taxa are stenobiontic and extremely sensitive to environmental changes. Some are even endangered with extinction. As such, diatoms, and especially the easily viewed and identified endemic species, have a useful role as bioindicator organisms to monitor anthropogenic perturbations in Lake Tanganyika, and perhaps also in other ancient lakes.

ACKNOWLEDGEMENTS

The author is currently working as senior researcher on a project (GC/DD/10) financed by the Prime Minister's Office – Federal Office for Scientific, Technical and Cultural Affairs, Belgium. Thanks are due to Dr A. Rossiter for his assistance and advice regarding this manuscript.

REFERENCES

Archibald, R.E.M. (1983). The diatoms of the Sundays and the Great Fish Rivers in the Eastern Cape Province of South Africa. *Biblio. Diatomol.* **1,** 1–362.

Archibald, R.E.M. (1984). Diatom illustrations – an appeal. *Bacillaria* **7**, 173–178.
Archibald, R.E.M. and Schoeman, F.R. (1987). Taxonomic notes on diatoms (Bacillariophyceae) from the Great Usutu River in Swaziland. *S. Afr. J. Bot.* **51**, 75–92.
Caljon, A. (1991). Sedimentary diatom assemblages in the northern basin of Lake Tanganyika. *Hydrobiologia* **226**, 179–191.
Caljon, A. (1992). Water quality in the Bay of Bujumbura (Lake Tanganyika) and its influence on phytoplankton composition. *Mitt. Int. Verein. Limnol.* **23**, 55–65.
Caljon, A. and Cocquyt, C. (1992). Diatoms from surface sediments of the northern basin of Lake Tanganyika. *Hydrobiologia* **230**, 135–156.
Cholnoky, B.J. (1953). Diatomeenassoziationen aus dem Hennops-rivier bei Pretoria. *Verh. Zool. Bot. Gesellsch. Wien* **93**, 134–149.
Cholnoky, B.J. (1955). Hydrobiologische Untersuchungen in Transvaal. I. Vergleichung der herbstlichen Algengemeinschaften in Raytonvlei und Leeufontein. *Hydrobiologia* **7**, 137–209.
Cholnoky, B.J. (1956). Neue und seltene Diatomeen aus Africa. II. Diatomeen aus dem Tugela-Gebiet in Natal. *Österreich. Bot. Z.* **103**, 53–97.
Cholnoky, B.J. (1957). Beiträge zur Kenntnis der Südafrikanischer Diatomeenflora. *Port. Acta Biol.* **6**, 53–93.
Cholnoky, B.J. (1958). Beiträge zur Kenntnis der Südafrikanischer Diatomeenflora. II. Einige Gewässer in Waterburg-Gebiet Transvaal. *Port. Acta Biol.* **6**, 99–160.
Cholnoky, B.J. (1959). Neue und seltene Diatomeen aus Africa. IV. Diatomeen aus des Kaap-Provinz. *Österreich. Bot. Z.* **106**, 1–69.
Cholnoky, B.J. (1960). Beiträge zur Kenntnis der Diatomeenflora von Natal. *Nova Hedwig. Beiheft.* **2**, 1–133.
Cholnoky, B.J. (1966a). Die Diatomeen im Unterlauf des Okawango-Flusses. *Nova Hedwig. Beiheft.* **21**, 10–102.
Cholnoky, B.J. (1966b). Diatomeen-Assoziationen aus einigen Quellen in Südwestafrika und Bechuanaland. *Nova Hedwig.* **21**, 163–244.
Cholnoky, B.J. (1970). Die Diatomeenassoziationen in Nonoti-Bach in Natal (Südafrika). *Nova Hedwig. Beiheft.* **31**, 313–329.
Cocquyt, C. (1991). Epilithic diatoms from thrombolitic reefs of Lake Tanganyika. *Belg. J. Bot.* **124**, 102–108.
Cocquyt, C. (1998). Diatoms from the northern basin of Lake Tanganyika. *Biblio. Diatomol.* **39**, 1–276.
Cocquyt, C. (1999a). Seasonal dynamics of diatoms in the littoral zone of Lake Tanganyika, Northern Basin. *Algol. Stud.* **92**, 73–85.
Cocquyt, C. (1999b). Seasonal variations of epithilic diatom communitires in the northern basin of Lake Tanganyika. *Syst. Geogr. Pl.* **69**, 265–273.
Cocquyt, C. and Vyverman, W. (1994). Composition and diversity of the algal flora in the East African Great Lakes: a comparative survey of lakes Tanganyika, Malawi (Nyasa), and Victoria. *Arch. Hydrobiol. Beiheft. Ergebnisse Limnol.* **44**, 161–172.
Cocquyt, C., Caljon, A. and Vyverman, W. (1991). Seasonal and spatial aspects of phytoplankton along the north-eastern coast of Lake Tanganyika. *Ann. Limnol.* **27**, 215–225.
Cocquyt, C., Vyverman, W. and Compère, P. (1993). A check-list of the algal flora of the East African Great Lakes (Malawi, Tanganyika and Victoria). *Scripta Bot. Belg.* **8**, 1–55.
Cohen, A., Bills, R., Cocquyt, C. and Caljon, A. (1993). The impact of sediment pollution on biodiversity in Lake Tanganyika. *Conserv. Biol.* **7**, 667–677.
Coulter, G.W. (Ed.) (1991). *Lake Tanganyika and its Life.* Oxford University Press, Oxford.

Coulter, G.W. (1994). Lake Tanganyika. *Arch. Hydrobiol. Beiheft. Ergebrisse Limnol.* **44**, 13–18.

Coulter, G.W. and Spigel, R.H. (1991). Hydrodynamics. In: *Lake Tanganyika and its Life* (Ed. by G.W. Coulter), pp. 49–75. Oxford University Press, Oxford.

Huber-Pestalozzi, G. (1942). Das Phytoplankton des Süsswassers: Diatomeen. *Binnengewasser* **16**, 367–549.

Hustedt, F. (1931–1959). Die Kieselalgen. In: *Rabenhorst, Kryptogamen-Flora von Deutschland, Österreich und der Schweiz*. Vol. 7, Part 2. Academischer, Leipzig.

Hustedt, F. (1961–1966). Die Kieselalgen. In: *Rabenhorst, Kryptogamen-Flora von Deutschland, Österreich und der Schweiz*. Vol. 7, Part 3. Academischer, Leipzig.

Klee, R. and Casper, J. (1992). New centric diatoms (Thalassiosirales) of Lake Malawi (formerly Lake Nyassa; Malawi, East Africa). *Arch. Protistenkunde* **142**, 179–192.

Kociolek, J.P. and Stoermer, E.F. (1993). The diatom genus *Gomphocymbella* O. Müller: taxonomy, ultrastructure and phylogenetic relationships. *Nova Hedwig. Beiheft.* **106**, 71–91.

Krammer, K. and Lange-Bertalot, H. (1986). Bacillariophyceae: Naviculaceae. *Süsswasserflora von Mitteleuropa*, Vol. 2, Part 1 (Ed. by H. Ettl, J. Gerloff, H. Heybig and D. Mollenhauer). Gustav Fischer, Stuttgart.

Krammer, K. and Lange-Bertalot, H. (1988). Bacillariophyceae: Bacillariaceae, Epithemiaceae, Surirellaceae. *Süsswasserflora von Mitteleuropa*, Vol. 2, Part 2 (Ed. by H. Ettl, J. Gerloff, H. Heybig and D. Mollenhauer). Gustav Fischer, Stuttgart.

Krammer, K. and Lange-Bertalot, H. (1991a). Bacillariophyceae: Centrales, Fragilariaceae, Eunotiaceae. *Süsswasserflora von Mitteleuropa*, Vol. 2, Part 3 (Ed. by H. Ettl, J. Gerloff, H. Heybig and D. Mollenhauer). Gustav Fischer, Stuttgart.

Krammer, K. and Lange-Bertalot, H. (1991b). Bacillariophyceae: Achnanthaceae, Kritische Ergänzungen *Navicula* (Lineolatae) und *Gomphonema*. In: *Süsswasserflora von Mitteleuropa*, Vol. 2, Part 4 (Ed. by H. Ettl, J. Gerloff, H. Heybig and D. Mollenhauer). Gustav Fischer, Stuttgart.

Kufferath, H. (1956). Organismes trouvés dans les carottes de sondages et les vases prélevées aufond du Lac Tanganika. Résultats scientifiques de l'exploration hydrobiologique du Lac Tanganika (1946–1947). *Inst. R. Sci. Nat. Belg.* **4**, 1–74.

Ross, R. (1983). Endemism and cosmopolitanism in the diatom flora of the East African Great Lakes. In: *Evolution, Time and Space: The Emergence of the Biosphere* (Ed. by R.W. Sims, J.H. Price and P.E.S. Whalley), pp. 157–177. Academic Press, London.

Round, F.E., Crawford, R.M. and Mann, D.G. (1990). *The Diatoms. Biology and Morphology of the Genera*. Cambridge University Press, Cambridge.

Schmidt, A. (ed.) (1874–1959). *Atlas der Diatomaceenkunde*. O.R. Reisland, Leipzig.

Schoeman, F.R. (1973). *A Systematical and Ecological Study of the Diatom Flora of Lesotho with Special Reference to Water Quality*. V. & R. Printers, Pretoria.

Schoeman, F.R. and Archibald, R.E.M. (1976). *The Diatom Flora of Southern Africa*. CRIS, Pretoria.

Van Meel, L. (1954). Le phytoplankton. Résultats scientifiques de l'exploration hydrobiologique du Lac Tanganika (1946–1947). *Inst. R. Sci. Nat. Belg.* **4**, 1–681.

Vandelannoote, A., Robbrecht, H., Deelstra, H., Vyumvuhore, F., Bitetra, L. and Ollevier, F. (1996). The impact of Lake Tanganyika, on the water quality of the Lake. *Hydrobiologica* **328**, 161–171.

Vandelannoote, A., Deelstra, H. and Ollevier, F. (1999). The inflow of the Rusizi River to LAke Tanganyika. *Hydrobiologica* **407**, 65–73.

Vyverman, W. and Cocquyt, C. (1994). Depth distribution of living and non-living diatoms in the northern part of Lake Nyasa (Tanzania). *Arch. Hydrobiol. Beiheft. Ergebnisse Limnol.* **44**, 161–172.

Appendix overleaf
List of the diatoms observed in the northern basin of Lake Tanganyika, with their geographical distribution, location where collected and distribution pattern (indicated as widespread or localized, abundant, common, not uncommon, scarce or rare). The probable geographical distribution of each taxon, with reference to an existent taxon (cf. or aff.), is given between parentheses. The status of several taxa previously described as endemic is doubtful, and it has not proven possible to verify identification by examining type material, much of which is lost or of extremely poor quality. Therefore, endemic taxa which were not observed during the present study are not included in this list.

Taxon	Geographic distribution category					Location in Northern Basin								Distribution in Northern Basin						
	Cosmopolitan	Tropical	African	Tropical African	Endemic	surface (off Rusizi)	off small rivers	plankton	epilithon	epiphyton	epipelon	epilithon (others)	Bay Buijumbura	widespread	localised	abundant	common	not uncommon	scarce	rare
Aulacoseira distans (Ehrenb.) Sim.	●					●	●	●						●						
Aulacoseira goetzeana (O. Müll.) Sim.		●				●	●												●	
Aulacoseira granulata (Ehrenb.) Sim.	●					●											●			
Aulacoseira granulata var. *angustissima* (O. Müll.) Sim.	●	●				●												●		
Aulacoseira ikapoënsis (O. Müll.) Sim. var. *minor* (O. Müller) Sim.		●				●														●
Aulacoseira ikapoënsis var. *procera* (O. Müller) Sim.		●				●									●					
Aulacoseira italica (Ehrenb.) Sim.	●					●									●					
Aulacoseira pfaffiana (Reinsch) Krammer	●					●		●					●							●
Cyclostephanos damasii (Hust.) Stoermer & Hakansson				●		●	●						●	●						
Cyclotella krammeri Hakansson	●							●												
Cyclotella meduanae Germain	●					●	●	●					●						●	
Cyclotella meneghiniana Kütz.	●							●	●	●		●	●						●	
Cyclotella ocellata Pant.	●					●		●	●											●
Cyclotella wollereckii Hust.		●													●					
Orthoseira roeseana (Rabenh.) O'Meara	●												●		●					
"*Stephanodiscus astraea*" (Ehrenb.) Grun. complex	●							●					●							
Stephanodiscus carconensis Grun.	●					●							●					●		
Stephanodiscus minutulus (Kütz.) Cl. & Möller	●																●			
Thalassiosira faurii (Gasse) Hasle				●		●		●	●				●							
Thalassiosira rudolfi (Bachm.) Hasle				●				●	●				●							
Rhizosolenia eriensis H.L. Smith var. *pusilla* Wolosz.	(●)							●					●		●					
Rhizosolenia aff. *eriensis* H.L. Smith	(●)														●				●	
Rhizosolenia aff. *longiseta* Zach.															●					
Rhizosolenia spec.					(●)															
Fragilaria africana Hust.				●		●	●	●	●	●	●	●	●							●
Fragilaria brevistriata Grun.	●					●	●	●	●				●	●						
Fragilaria brevistriata var. *trigibba* (Pant.) Hust.	●					●	●	●			●			●						
Fragilaria construens (Ehrenb.) Grun.	●					●	●	●	●		●		●						●	
Fragilaria construens var. *subsalina* Hust.	●					●	●	●	●	●			●	●						
Fragilaria construens var. *venter* (Ehrenb.) Grun.	●					●	●	●	●	●						●				●
Fragilaria heidenii Østrup	●					●	●	●	●						●			●		●
Fragilaria leptostauron (Ehrenb.) Hust. var. *dubia* (Grun.) Hust.	●					●	●		●							●		●		●
Fragilaria pinnata Ehrenb.	●					●	●	●	●					●		●		●		●

Fragilaria pinnata var. lanceolata (Schumann) Hust.
Fragilaria vaucheriae (Kütz.) Boye-Pet. var. capitellata (Grun.) Ross
Meridion cf. circulare (Grev.) Ag.
Opephora martyi Héib.
Synedra acus Kütz.
Synedra goulardii Bréb.
Synedra nyansae G.S. West
Synedra rumpens Kütz.
Synedra rumpens var. fragilarioides Grun.
Synedra ulna (Nitzsch) Ehrenb.
Synedra ulna var. aequalis (Kütz.) Hust.
Synedra ulna var. amphirhynchus (Ehrenb.) Grun.
Synedra ulna var. danica (Kütz.) Grun.
Eunotia arcus Ehrenb.
Eunotia cf. camelus Ehrenb.
Eunotia curvata (Kütz.) Lagerstedt
Eunotia didyma Grun. var. tuberosa Hust.
Eunotia flexuosa Bréb. ex Kütz.
Eunotia flexuosa var. eurycephala Grun.
Eunotia incisa W. Smith ex Greg.
Eunotia microcephala Krasske
Eunotia monodon Ehrenb.
Eunotia pectinalis (O.F. Müll.) Rabenh.
Eunotia pectinalis var. minor (Kütz.) Rabenh.
Eunotia pectinalis var. rostrata Germain
Eunotia pectinalis var. undulata Ralfs
Eunotia pectinalis var. ventricosa Grun.
Eunotia porcellus Choln.
Eunotia rabenhorstii Cl. et Grun. var. africana Hust.
Eunotia rabenhorstii var. monodon Grun.
Achnanthes buccula Choln.
Achnanthes exigua Grun.
Achnanthes inflata (Kütz.) Grun.
Achnanthes inflata var. elata (Leuduger-Fortmüller) Hust.
Achnanthes lanceolata (Bréb.) Grun.
Achnanthes lanceolata var. rostrata Hust.
Achnanthes linearis (W. Smith) Grun.
Achnanthes microcephala (Kütz.) Grun.
Achnanthes minutissima Kütz.
Achnanthes minutissima var. affinis (Grun.) L.-B.
Achnanthes oblongella Østrup
Achnanthes rupestoides Hohn
Achnanthes subhudsonis Hust.

Cocconeis disculus (Schumann) Cl.
Cocconeis placentula Ehrenb.
Cocconeis placentula var. euglypta (Ehrenb.) Cl.
Cocconeis placentula var. rouxii (Brun ex Héribaud) Cl.
Rhoicosphenia abbreviata (Ag.) L.-B.
Amphora coffeiformis (Ag.) Kütz.
Amphora coffeiformis var. acutiuscula (Kütz.) Rabenh.
Amphora copulata (Kütz.) Schoeman & Archibald
Amphora inariensis Krammer
Amphora montana Krasske
Amphora ovalis (Kütz.) Kütz.
Amphora pediculus (Kütz.) Grun.
Amphora strigosa Hust.
Amphora tanganyikae Caljon
Amphora tanganyikae var. 1
Anomoeoneis sphaerophora (Ehrenb.) Pfitzer
Anomoeoneis sphaerophora f. sculpta (Ehrenb.) Krammer
Brachysira brebissonii Ross
Brachysira microcephala (Grun.) Compère
Brachysira serians (Bréb.) Round & Mann
Caloneis aerophila Bock
Caloneis bacillum (Grun.) Cl.
Caloneis hyalina Hust.
Caloneis limosa (Kütz.) Patrick
Caloneis molaris (Grun.) Krammer
Caloneis permagna (Bailey) Cl.
Caloneis ventricosa (Ehrenb.) Meister
Capartogramma amphoroides Ross
Capartogramma crucicula (Grun. ex Cl.) Ross
Capartogramma karstenii (Zanon) Ross
Capartogramma rhombicum Ross
Cymatopleura calcarata Hust.
Cymatopleura solea (Bréb.) W. Smith
Cymatopleura solea var. apiculata (W. Smith) Ralfs
Cymbella affinis Kütz.
Cymbella amphicephala Nägeli
Cymbella caespitosa (Kütz.) Brun
Cymbella cucumis A. Schmidt
Cymbella cuspidata Kütz.
Cymbella grossestriata O. Müll.
Cymbella helmckei Krammer
Cymbella leptoceros (Ehrenb.) Kütz.

Cymbella cf. mendosa VanLandingham
Cymbella microcephala Grun.
Cymbella minuta Hilse ex Rabenh.
Cymbella muelleri Hust.
Cymbellonitzschia minima Hust.
Diploneis elliptica (Kütz.) Cl.
Diploneis finnica (Ehrenb.) Cl.
Diploneis ovalis (Hilse) Cl.
Epithemia adnata (Kütz.) Bréb.
Epithemia argus (Ehrenb.) Kütz.
Epithemia sorex Kütz.
Frustulia rhomboides (Ehrenb.) De Toni
Frustulia rhomboides var. crassinervia (Bréb.) Ross
Frustulia rhomboides var. saxonica (Rabenh.) De Toni
Frustulia vulgaris (Thwaites) De Toni
Gomphocymbella beccarii (Grun.) Forti
Gomphocymbella gracilis Hust.
Gomphocymbella spec.
Gomphonema acuminatum Ehrenb.
Gomphonema aequatoriale Hustedt
Gomphonema affine Kütz.
Gomphonema africanum G.S. West
Gomphonema augur Ehrenb. var. turris (Ehrenb.) L.-B.
Gomphonema clevei Fricke
Gomphonema dichotomum Kütz.
Gomphonema gracile Ehrenb.
Gomphonema kilhamii Kociolek & Stoermer
Gomphonema olivaceum (Hornemann) Bréb.
Gomphonema paddockii Kociolek & Stoermer
Gomphonema parvulum (Kütz.) Kütz.
Gomphonema truncatum Ehrenb.
Gomphonitzschia ungeri Grun.
Gyrosigma acuminatum (Kütz.) Rabenh.
Gyrosigma attenuatum (Kütz.) Rabenh.
Gyrosigma balticum (Ehrenb.) Rabenh.
Gyrosigma nodiferum (Grun.) Reimer
Gyrosigma scalproides (Rabenh.) Cl.
Gyrosigma spenceri (Quekett) Griffith & Henfrey
Gyrosigma wormleyi (Sullivan) Boyer
Hantzschia amphioxys (Ehrenb.) Grun.
Hantzschia amphioxys var. africana Hust.
Hantzschia virgata (Roper) Grun.
Mastogloia elliptica (Agardh) Cl.

Mastogloia elliptica var. dansei (Thwaites) Cl.
Mastogloia smithii Thwaites
Navicula absoluta Hust.
Navicula agrestis Hust.
Navicula amplectens Hust.
Navicula arvensis Hust.
Navicula bacillum Ehrenb.
Navicula barbarica Hust.
Navicula brevissima Hust.
Navicula capitata Ehrenb.
Navicula capitata var. hungarica (Grun.) Ross
Navicula capitatoradiata Germain
Navicula cincta (Ehrenb.) Ralfs
Navicula confervacea (Kütz.) Grun.
Navicula contenta Grun.
Navicula costulata Grun.
Navicula cryptocephala Kütz.
Navicula cryptotenella L.-B.
Navicula cuspidata (Kütz.) Kütz.
Navicula damasii Hust.
Navicula decussis Østrup
Navicula digitoradiata (Greg.) Ralfs
Navicula elaborata Hust.
Navicula elginensis (Greg.) Ralfs
Navicula erifuga L.-B.
Navicula exigua (Greg.) Grun.
Navicula fossalis Krasske var. obsidialis (Hust.) L.-B.
Navicula gallica (W. Smith) V.H.
Navicula gallica var. perpusilla (Grun.) L.-B.
Navicula gastrum (Ehrenb.) Kütz.
Navicula gastrum var. signata Hust.
Navicula goeppertiana (Bleisch) H.L. Smith
Navicula hambergii Hust.
Navicula hasta Pant.
Navicula insociabilis Krasske
Navicula lapidosa Krasske
Navicula marginestriata Hust.
Navicula miramaris Frenguelli
Navicula modica Hust.
Navicula monoculata Hust. var. omissa (Hust.) L.-B.
Navicula mutica Kütz.
Navicula mutica f. intermedia (Hust.) Hust.
Navicula muticoides Hust.

Species distribution matrix (row labels, read bottom to top):

- *Navicula nivaloides* Bock
- *Navicula nyassensis* O. Müll.
- *Navicula oblonga* (Kütz.) Kütz.
- *Navicula paramutica* Bock
- *Navicula perlatoides* (O. Müll.) Hust.
- *Navicula perparva* Hust.
- *Navicula placentula* (Ehrenb.) Kütz.
- *Navicula platycephala* O. Müll.
- *Navicula cf. praeterita* Hust.
- *Navicula pupula* Kütz.
- *Navicula pygmaea* Kütz.
- *Navicula radiosa* Kütz.
- *Navicula cf. rhynchocephala* Kütz.
- *Navicula rotunda* Hust.
- *Navicula schroeteri* Meister
- *Navicula scutelloides* W. Smith ex Greg.
- *Navicula seminuloides* Hust.
- *Navicula subcostulata* Hust.
- *Navicula subrhynchocephala* Hust.
- *Navicula trivialis* L.-B.
- *Navicula tuscula* Ehrenb.
- *Navicula vanheurckii* Patrick
- *Navicula viridula* (Kütz.) Ehrenb.
- *Navicula viridula var. linearis* Hust.
- *Navicula viridula var. rostellata* (Kütz.) Cl.
- *Navicula zanonii* Hust.
- *Neidium affine* (Ehrenb.) Pfitzer
- *Neidium ampliatum* (Ehrenb.) Krammer
- *Neidium dubium* (Ehrenb.) Cl.
- *Neidium iridis* (Ehrenb.) Cl.
- *Neidium minutissimum* Krasske
- *Nitzschia acicularis* (Kütz.) W. Smith
- *Nitzschia adapta* Hust.
- *Nitzschia aequalis* Hust.
- *Nitzschia amphibia* Grun.
- *Nitzschia amphioxoides* Hust.
- *Nitzschia archibaldii* L.-B.
- *Nitzschia asterionelloides* O. Müll.
- *Nitzschia bacillum* Hust.
- *Nitzschia calida* Grun.
- *Nitzschia capitellata* Hust.
- *Nitzschia clausii* Hantzsch
- *Nitzschia communis* Rabenh.

Species list (table rows):

- Nitzschia compressa (Bailey) Boyer
- Nitzschia compressa var. vexans (Grun.) L.-B.
- Nitzschia constricta (Kütz.) Ralfs
- Nitzschia dissipata (Kütz.) Grun.
- Nitzschia epiphyticoides Hust.
- Nitzschia cf. epiphyticoides Hust.
- Nitzschia fonticola Grun.
- Nitzschia frustulum (Kütz.) Grun.
- Nitzschia gracilis Hantzsch
- Nitzschia granulata Grun.
- Nitzschia inconspicua Grun.
- Nitzschia intermedia Hantzsch
- Nitzschia lacustris Hust.
- Nitzschia lancettula O. Müll.
- Nitzschia levidensis (W. Smith) Grun. var. victoriae (Grun.) Choln.
- Nitzschia liebetruthii Rabenh.
- Nitzschia linearis W. Smith
- Nitzschia linearis var. subtilis (Grun.) Hust.
- Nitzschia linearis var. tenuis (W. Smith) Grun.
- Nitzschia microcephala Grun.
- Nitzschia nana Grun.
- Nitzschia nyassensis O. Müll.
- Nitzschia obsoleta Hust.
- Nitzschia obtusa W. Smith
- Nitzschia obtusa var. schweinfurthii Grun.
- Nitzschia palea (Kütz.) W. Smith
- Nitzschia palea var. tenuirostris L.-B.
- Nitzschia paleacea (Grun.) Grun.
- Nitzschia recta Hantzsch
- Nitzschia reversa W. Smith
- Nitzschia sigma W. Smith
- Nitzschia aff. sigmaformis Hust.
- Nitzchia siliqua Archibald
- Nitzschia solita Hust.
- Nitzschia spiculum Hust.
- Nitzschia subacicularis Hust.
- Nitzschia tubicola Grun.
- Nitzschia umbonata (Ehrenb.) L.-B.
- Nitzschia vitrea Norman
- Pinnularia acrosphaeria Rabenh.
- Pinnularia amaniensis Hust.
- Pinnularia appendiculata (Ag.) Cl.
- Pinnularia bogotensis (Grun.) Cl.

Pinnularia borealis Ehrenb.
Pinnularia borealis f. *rectangularis* Carlson
Pinnularia brevicostata Cl.
Pinnularia divergens W. Smith var. *elliptica* (Grun.) Cl.
Pinnularia gibba Ehrenb.
Pinnularia gibba var. *linearis* Hust.
Pinnularia gibba var. *sancta* Grun.
Pinnularia legumen (Ehrenb.) Ehrenb.
Pinnularia mesolepta (Ehrenb.) W. Smith
Pinnularia microstauron (Ehrenb.) Cl.
Pinnularia cf. nobilis (Ehrenb.) Ehrenb.
Pinnularia obscura Krasske
Pinnularia scaettae Zanon
Pinnularia stomatophora (Grun.) Cl.
Pinnularia subcapitata Greg.
Pinnularia subcapitata var. *elongata* Krammer
Pinnularia tropica Hust.
Pinnularia viridis (Nitzsch) Ehrenb.
Rhopalodia gibba (Ehrenb.) O. Müll.
Rhopalodia gibba var. *paralella* (Grun.) H. & M. Peragallo
Rhopalodia gibberula (Ehrenb.) O. Müll.
Rhopalodia gracilis O. Müll.
Rhopalodia hirundiniformis O. Müll.
Rhopalodia hirundiniformis var. *parva* O. Müll.
Rhopalodia hirundiniformis var. *turgida* Fricke
Rhopalodia operculata (Ag.) Hakansson
Rhopalodia rhopala (Ehrenb.) Hust.
Rhopalodia spec.
Stauroneis agrestis Petersen
Stauroneis anceps Ehrenb.
Stauroneis manguinii Guermeur
Stauroneis nana Hust.
Stauroneis obtusa Lgst.
Stauroneis phoenicenteron (Nitzsch) Ehrenb.
Surirella aculeata Hust.
Surirella acuminata Hust.
Surirella cf. acuminata Hust.
Surirella angusta Kütz.
Surirella astridae Hust.
Surirella bifrons Ehrenb.
Surirella biseriata Bréb.
Surirella brebissonii Krammer & L.-B
Surirella brevicostata O. Müll.

	1	2	3	4	5	6	7	8	9	10	11	12	13	14	15	16	17	18	19
Surirella brevicostata var. *constricta* Hust.				•			•	•							•			•	
Surirella debesii Hust.			•					•						•					•
Surirella decipiens Hust.		•						•			•			•					•
Surirella elegantula Hust.		•						•						•					•
Surirella engleri O. Müll.				•		•	•	•	•			•	•			•			
Surirella fasciculata O. Müll.				•				•										•	
Surirella fuellebornii O. Müll.			•			•	•	•				•			•				
Surirella fuellebornii var. *elliptica* O. Müll.				•				•										•	
Surirella fuellebornii var. *tumida* Hust.					•			•						•					•
Surirella gradifera Hust.					•			•						•					•
Surirella heidenii Hust.					•		•	•	•			•			•				
Surirella cf. *heidenii* Hust.					(•)			•						•					•
Surirella heidenii var. 1					(•)			•						•					•
Surirella lancettula Hust.					•	•		•			•			•					
Surirella latecostata Hust.					•			•			•			•					•
Surirella linearis Hust.	•						•	•			•			•					
Surirella margaritacea O. Müll.			•					•											•
"*Surirella muelleri*" Hust. (non Forti)				•				•			•			•					
Surirella nervosa (A. Schmidt) Mayer	•							•						•					
Surirella nyassae O. Müll. var. 1				(•)		•	•	•	•	•	•			•					
Surirella obtusiuscula G.S. West				•		•	•	•			•		•			•			
Surirella plana G.S. West				•				•						•					•
Surirella reicheltii Hust.				•				•						•					•
Surirella sparsipunctata Hust.				•		•	•	•			•	•	•			•			
Surirella spiraloides Hust.					•	•		•			•					•			
Surirella splendida (Ehrenb.) Kütz.	•					•		•			•					•			
Surirella striolata Hust.					•	•	•	•			•							•	
Surirella subcontorta Hust.					•			•						•					•
Surirella tenera Greg.	•					•	•		•		•	•	•			•		•	
Surirella terryi Ward ex Terry		•						•						•					•

Evolution and Endemism in Lake Biwa, with Special Reference to its Gastropod Mollusc Fauna

M. NISHINO and N.C. WATANABE

I. SUMMARY

A detailed bibliographic survey revealed 57 endemic taxa of Lake Biwa. These taxa comprise 50 species, five subspecies and two varieties, among which 11 species were newly described after 1990. They include two species of pleurocerid gastropod belonging to the subgenus *Semisulcospira* (*Biwamelania*) which, except in the lake itself, have been recorded only in its natural outlet, the River Seta. Several species described recently (such as eight species of chironomid; Sasa and Kawai, 1987; Sasa and Nishino, 1995, 1996), but which have an unknown geographical distribution, will probably be added to the inventory in the future.

Approximately 70% of the endemic species are benthic invertebrates and 20% are fishes. Although insects are numerically dominant among the benthic invertebrates, they contain only three endemic species. In contrast, molluscs comprise almost half (27 species and two subspecies) of the endemic species of Lake Biwa. However, with the exception of the pleurocerid gastropod

ADVANCES IN ECOLOGICAL RESEARCH VOL. 31
ISBN 0-12-013931-6

Semisulcospira (*Biwamelania*), only one or two endemic species and subspecies have been reported for each respective genus or subgenus from the lake. The subgenus *Biwamelania* is composed of 15 endemic species, and shows the greatest variations in the number and morphology of chromosomes among the species. Members of this subgenus are widely distributed in the littoral areas; some species, such as *S.* (*B.*) *niponica* and *S.* (*B.*) *morii*, are restricted to rocky and/or gravel bottoms and others, such as *S.* (*B.*) *reticulata* and *S.* (*B.*) *arenicola*, to sandy and/or muddy bottoms, and yet others, such as *S.* (*B.*) *habei* and *S.* (*B.*) *decipiens*, are widely distributed over various kinds of lake bottom. The *Biwamelania* group comprises a species flock. However, only one recent species, *S.* (*B.*) *habei*, occurs as fossils from the Palaeo-Lake Biwa formation, and it seems probable that a *S.* (*B.*) *habei*-like ancestor might have evolved into the *Biwamelania* group within the lake.

II. INTRODUCTION

Lake Biwa is located in the temperate zone of East Asia. The lake consists of two basins, the large and deep northern basin and the smaller shallow southern basin, and is categorized as a subtropical lake, having a circulation period in winter. Formerly, both basins were oligotrophic, but have changed into mesotrophic and eutrophic, respectively, owing to eutrophication since the late 1960s. More than 400 rivers flow into the lake, but there is only one natural outlet, the Seta River. Around the present Lake Biwa, several ancient lake beds of Palaeo-Lake Biwa are distributed, which are rich in lacustrine fossils (Research Group for Natural History of Lake Biwa, 1994).

   The unusual nature and high level of endemicity seen in the Lake Biwa fauna was first noted by Uéno (1943a), who posited that some species were northern relicts, and also indicated the corophiid amphipod *Kamaka biwae* as a northern marine relict species (Uéno, 1943b). A combined glacial and marine relict origin for many Biwa endemics was also the conclusion arrived at by Horie (1961), based on glacial evidence. Successive biologists have discussed the origin and evolution of Lake Biwa biota, and have proposed intralacustrine evolution for some species groups, and a glacial or marine relict origin for others (Tomoda, 1967; Kawanabe, 1975a,b, 1978; Uéno, 1975, 1984; Tokui and Kawanabe, 1985; Takahashi, 1989). However, these hypotheses have yet to be critically evaluated.

   The first attempt to inventory all endemic species of Lake Biwa was that of Mori (1970), who listed about 1000 species of fauna and flora in the lake, 42 of which were endemic. The list was revised twice (Mori and Miura, 1980, 1990), with a total of 54 endemic taxa being recognized in the last revision. However, it ignored six species already recorded as endemic in Lake Biwa (Davis, 1969; Miyadi *et al.*, 1976; Nakamura, 1984; Morino, 1985; Kuroda and Habe, 1987), and also recorded three species as endemic which had already been described or recorded as distributed in localities other than Lake Biwa (Okugawa, 1930; Mori, 1933; Miyadi et al., 1976).

Recently, more than 30 species have been newly described or redescribed, some of which have been reported only from the lake (e.g. Sasa and Kawai, 1987; Okubo, 1990; Ohtaka, 1994; Nishimoto, 1994; Sasa and Nishino, 1995, 1996; Watanabe and Nishino, 1995; Ishiwata, 1996; Malzacher, 1996; Watanabe, 1996). Among them, two species have subsequently been described as endemic to Lake Biwa (Ohtaka and Nishino, 1995; Tanida et al., 1999), whereas endemicity was denied for six species, based on the taxonomic revision of oligochaetes (Ohtaka and Nishino, 1995) and ecological studies of Trichoptera and Odonata (Uenishi, 1993; Ishida, 1996). Clearly, the species lists of the biota of the lake should be re-examined based on a comprehensive bibliographic survey, supplemented by taxonomic re-evaluation of certain groups.

The present report lists those species thus far reported as endemic to Lake Biwa and examines the hypotheses on the origin of endemicity based on recent findings. The existence of a species flock of pleurocerid gastropods belonging to the subgenus *Biwamelania* is described for the first time, and the origins of this flock are discussed. Prospects for future studies on endemism in Lake Biwa are also presented.

III. REVISED LIST OF ENDEMIC SPECIES IN LAKE BIWA

In the present study, an endemic species is defined as one described from specimens collected only from Lake Biwa, or its effluent and affluent rivers, and which has never been reported from other localities. However, it is difficult to assess accurately the endemic status of a species, especially for recently described species. This difficulty is due mainly to the poor level of taxonomic and biogeographical data for several Japanese taxa. For example, it is not yet possible to consider the endemicity of chironomids: among the chironomid species described recently, *Hydrobaenus biwaquartus*, which Mori and Miura (1990) regarded as endemic to Lake Biwa, was later found to be distributed in the Kiso River, the Jinzu River and Lake Kojima (Sasa and Kikuchi, 1995). The biogeographical distribution of these taxa should be examined in detail to confirm their endemicity. A further difficulty stems from the repeated introductions of exotic animals and plants into Lake Biwa, along with the introduction of a variety of Japanese and foreign fishes, shrimps, clams and toads, intending to improve fishery resources (e.g. Furukawa and Awano, 1969; Maehata, 1969; Maehata, 1987; Rossiter, this volume). Thus, in this study, endemicity was assigned only to those taxa with a comprehensive history of taxonomic, biogeographical and ecological studies.

A detailed bibliographic survey resulted in 57 taxa, including 50 species, five subspecies and two varieties being regarded as endemic to Lake Biwa (Table 1).

Table 1
Endemic species reported thus far from Lake Biwa and its drainage system

Taxonomic group	Scientific name
Protozoa	*Difflugia biwae* Kawamura
Desmidia	*Pediastrum biwae* Negoro, *P. biwae* var. *triangulatum* Negoro, *P. biwae* var. *ovatum* Negoro
Plantae	*Vallisneria biwaensis* Ohwi, *Potamogeton biwaensis* Miki
Animalia	
Platyhelminthes	*Bdellocephala annandalei* Ijima et Kaburaki
Gastropoda	*Heterogen longispira* (Smith), *Valvata biwaensis* Preston, *Semisulcospira* (*Biwamelania*) *arenicola* Watanabe et Nishino[a], *S.* (*B.*) *dilatata* Watanabe et Nishino[a], *S.* (*B.*) *fuscata* Watanabe et Nishino[a], *S.* (*B.*) *decipiens* (Westerlund), *S.* (*B.*) *fluvialis* Watanabe et Nishino[a], *S.* (*B.*) *habei* (Davis), *S.* (*B.*) *morii* (Watanabe), *S.* (*B.*) *multigranosa* (Boeetger), *S.* (*B.*) *nakasekoae* (Kuroda), *S.* (*B.*) *niponica* (Brot), *S.* (*B.*) *ourense* Watanabe et Nishino[a], *S.* (*B.*) *reticulata* (Kajiyama et Habe), *S.* (*B.*) *rugosa* Watanabe et Nishino[a], *S.* (*B.*) *shiraishiensis* Watanabe et Nishino[a], *S.* (*B.*) *takeshimensis* Watanabe et Nishino[a], *Radix onychia* (Westerlund), *Gyraurus biwaensis* (Preston), *G. amplificatus* (Mori)
Bivalvia	*Hyriopsis schlegeri* (v. Martens), *Unio douglasiae biwae* (Kobelt), *Inversiunio reinianus* (Kobelt), *Lanceolaria oxyrhyncha* (v. Martens), *Cristaria plicata clessini* (Kobelt), *Synanodonta calipygos* (Kobelt), *Oguranodonta ogurae* Kuroda et Habe, *Corbicula* (*Corbicula*) *sandai* Reinhardt, *Pisidium* (*Eupisidium*) *kawamurai*
Oligochaeta	*Embolocephalus yamaguchii* (Brinkhurst)[a]
Hirudinea	*Ancyrobdella biwae* Oka
Crustacea	*Daphnia biwaensis* (Uéno), *Jesogammarus* (*Annanogammarus*) *annandalei* (Tattersal), *J.* (*A.*) *naritai* Morino, *Kamaka biwae* Uéno
Insecta	*Ephoron limnobium* Ishiwata[a], *Aphelocheirus kawamurae* Matsumura, *Apatania biwaensis* Nishimoto[a]
Pisces	*Salmo* (*Oncorhynchus*) *masou rhodurus* Jordan et McGregor, *Ischikauia steenackeri* (Sauvage), *Gnathopogon caerulescens* (Sauvage), *Sarcocheilichthys variegatus microoculus* Mori, *S. biwaensis* Hosoya, *Squalidus chankaensis biwae* (Jordan et Snyder), *Carassius cuvieri* Temminck et Schlegeri, *Carassius carassius grandoculis* Temminck et Schlegeri, *Silurus biwaensis* (Tomoda), *Silurus lithophilus* (Tomoda), *Chaenogobius isaza* Tanaka, *Cottus reinii* Hilgendorf

[a]Species described after 1990.

In this list, two species of pleurocerid gastropod are recorded only from the Seta River and its downstream, the River Uji. From the list of Mori and Miura (1990), 10 endemic species are eliminated and 40 endemic species are included in both lists. Eighteen species are added to the list, among which 11 species were newly described after 1990 (Table 1). As the geographical distributions were relatively well studied for these 11 species, they can safely be included in the list of endemic species.

However, the present list still includes some dubious taxa, such as the desmid *Pediastrum biwae* and its two varieties, the species validity of which was denied by Tanaka (1994), although he did not present any evidence for his decision. As such, the present list is still tentative.

IV. THE PECULIARITY OF ENDEMISM IN LAKE BIWA

As shown in Table 1, one species of Protozoa, three taxa of desmids, two of vascular plants and 51 animal taxa (46 species and five subspecies) are presently regarded as endemic to Lake Biwa. The endemic animals comprise one species of flatworm, 29 molluscs, two annelids, seven arthropods (four crustacean species and three insect species) and 12 fish taxa.

A preliminary estimate of the biodiversity and degree of endemicity of Lake Biwa biota was made by a bibliographic survey. This was not an exhaustively comprehensive survey, but was sufficient for the present purposes. A total of about 1000 taxa was recorded, which consisted of some 480 plankton species (*c.* 240 species each of phytoplankton and zooplankton, respectively), 60 nectonic species (fishes), 60 species of aquatic plants, 400 species of benthic invertebrates, 24 species of parasitic animal and fewer than 10 pleustonic species.

More than 90% of the endemic species are either benthic invertebrates (*c.* 70%, 39 spp.) or fishes (*c.* 20%, 12 taxa) (Figure 1). Insects (211 taxa) and rotifers (192 taxa) are highly diverse groups in the lake, but despite their speciosity only two endemic insect taxa and no endemic rotifers were recorded. In contrast to this pattern, the relatively species-poor molluscs and fishes contain 29 and 12 endemic taxa, respectively. The proportion of endemic taxa within the total benthic invertebrate taxa is 10% (39 spp.) and that of the endemic fishes is 25% (12 taxa). Endemic species comprise about 6% (57 taxa) of the total recorded taxa. This value is markedly lower than that seen in other ancient lakes, e.g. Lake Baikal (52%; Kozhova and Izmest'eva, 1998) and Lake Tanganyika (38%; Coulter, 1991).

There were formerly four endemic genera described from Lake Biwa, i.e. the viviparid gastropod *Heterogen*, the tubificid oligochaete *Kawamuria*, the glossiphoniid leech *Ancyrobdella* and the unionid bivalve *Oguranodonta ogurae*. However, Brinkhurst (1971) denied the generic and specific validity of *Kawamuria japonica*, and regarded it as a synonym of the cosmopolitan species, *Branchiura sowerbyi*. Recently, Nesemann (1997) assigned the widely

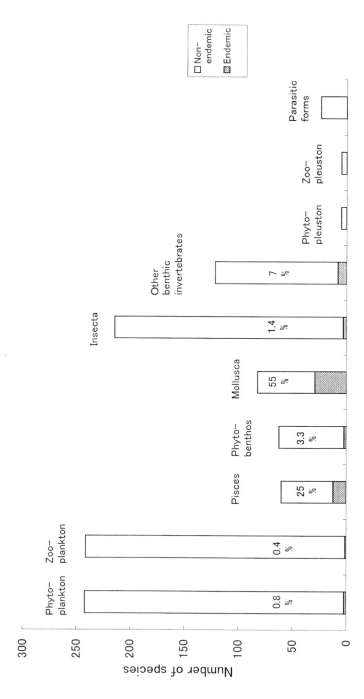

Fig. 1. Number of species of various lifeforms in Lake Biwa. The shaded area indicates the number of endemic species. Data are derived from Kawamura (1915), Fujita (1927a,b, 1928), Hiura (1967), Tsuda (1971), Negoro (1971), Ohwi (1972), Tanaka (1979), Wakabayashi and Ichise (1982), Harada (1984), Sasa and Kawai (1987), Kawanabe and Mizuno (1989), Ishida (1993, 1997), Nishino (1991, 1992, 1993), Shimazu (1993, 1994), Kawakatsu and Nishino (1994), Ohtaka and Nishino (1995, 1999), Sasa and Nishino (1995, 1996), Research Group for Microorganisms (1996), and Tanida *et al.* (1999). Exotic species are not included.

distributed Japanese glossiphonid *Batracobdella smaragdina* to *Ancylobdella*, and hence this genus is no longer endemic to Lake Biwa. Matsuoka's (1987) study resulted in *Oguranodonta* being assigned to *Synanodonta*, a genus widely distributed in Japan and China. However, some malacologists still recognize *Oguranodonta* (Fukuhara *et al.*, 1997) and the status of this genus requires re-examination. Thus, there seems to be only one unequivocally endemic genus, *Heterogen*, in Lake Biwa. In other genera or subgenera, with the exception of the subgenus *Semisulcospira* (*Biwamelania*), only one or two endemic species have been reported from the lake, even in the molluscs. As a whole, endemics of Lake Biwa exhibit low evolutionary diversity.

V. INTRALACUSTRINE VARIATIONS IN SOME ENDEMIC SPECIES

Several Lake Biwa endemic species show intralacustrine variation.

The limnaeid gastropod *Radix onychia* is widely distributed in the littoral areas of Lake Biwa. However, the local population from Chikubu Islet in the northern basin differs from its conspecifics in the southern basin in such diverse characters as shell morphology, buccal mass, salivary gland, radula, caecum, hermaphrodite gland and duct, egg membrane gland, prostate gland and male copulatory gland (Itagaki, 1959). The morphology of the southern basin population closely resembles the congeneric species *R. japonica*, which is widely distributed in the freshwaters of Japan. Itagaki (1959) considered these differences as being insufficient to separate them into two distinct species, and instead posited that they might constitute two ecological forms.

For insects, the two local populations of the caddisfly *Apatania biwaensis*, distributed geographically apart in the northern basin, exhibit morphological variations in male genitalia (Nishimoto, 1994): the male dorsal process in tergum IX varies in length, which is the easiest visible character to distinguish this species from the congeneric *A. aberrans*. The southern population includes various kinds of males, some with and others without fully developed dorsal processes. Most males of the northern population lack the dorsal processes. Nishimoto (1994) classified five types according to the length of the dorsal process and showed that the relative abundance of these types differs significantly between the southern and northern populations.

For fishes, Hosoya (1989) reported three morphological types of the endemic cyprinid fish, *Sarcocheilichthys variegatus microoculus*: the long-head, medium-head and short-head types. The long-head type is distributed in the northern rocky and cobble shore, whereas the short-head type lives in the outlet at the south end of Lake Biwa and the medium-head type lives throughout the entire littoral zone. He thought that the congeneric endemic species *S. biwaensis*, distributed in the northern rocky and cobble habitat, had evolved from the long-head type, but presented no evidence to support this hypothesis.

The congeneric planorbid gastropods, *Gyraurus biwaensis* and *G. amplifica-tus*, both found widely in the littoral, provide another example. In the original description of *G. amplificatus*, Mori (1938) distinguished this species as having a large shell, a remarkably large and inflated body whorl, and an enlarged aperture. However, he admitted that this "species" might be only a full-grown form of *G. biwaensis*, since in some specimens the shell morphology was very similar to the latter. There also exist forms intermediate between the two species (Nishino, 1991).

A similar case is seen in the congeneric unionid bivalves, *Inversiunio reinianus* and *I. hiraseus*. Kondo (1982) regarded *I. hiraseus* as a variety of *I. reinianus*, because their differences were observed only in the angle between the lateral cardinal teeth on the right valve of the shell. However, Matsuoka (1987, 1994) recorded *Inversiunio* fossils from the two formations of Palaeo-Lake Biwa (Palaeo-Lake Katata) and insisted that only the fossils of *I. hiraseus* were found from the older formation, whereas those of both species were reported from the youngest formation. The above two cases must be re-examined to assess whether they constitute distinct species or intralacustrine variations of a single species.

Some endemic species exhibit considerable morphological intraspecific variations, both sympatrically and allopatrically. In order to clarify the variations within the lake, detailed surveys will be needed of the morphology, karyotype, molecular taxonomy and ecological characteristics of these species.

VI. INTRASPECIFIC VARIATIONS IN THE WIDELY DISTRIBUTED SPECIES

There are some species widely distributed in Japan or Asia in which the population from Lake Biwa shows quite different features in morphology or ecological characteristics from those in other localities.

Ohtaka and Nishino (1999) reported the morphological variations of the tubificid oligochaete *Branchiura sowerbyi* both between the populations in Lake Biwa and other localities, and within the lake. This species is widely distributed in Asia, but the gill-absent form has been recorded only from Lake Biwa. In the profundal region of the northern basin, almost all individuals lacked gill filaments and only a few individuals had vestigial gills. In the littoral region, however, worms with gills were abundant, yet more than half were still devoid of gills in the northern basin, whereas gilled worms were more abundant in the shallow southern basin. In the lagoons adjacent to Lake Biwa, all worms possessed developed gills.

Further examples are presented by the caridean shrimp *Palaemon paucidens* and the Ayu fish *Plecoglossus altivelis altivelis*. Both species are widely distributed in Japan and its adjacent areas, and the population from Lake Biwa exhibits different features from those in other localities: the eggs are much smaller and more numerous in the Lake Biwa population for both species, compared with those borne by same-sized females from other populations (Azuma, 1973; Nishino, 1980).

For *P. paucidens*, the reproductive effort, i.e. the ratio of total egg weight vs female body weight, was almost identical among the geographical populations, and the single egg weight was inversely correlated to the relative brood size, i.e. the ratio of egg number to respective female body weight (Nishino, 1980). The Biwa population exhibits the smallest size of eggs and largest relative brood size known in this species. Under identical rearing conditions, newly hatched larvae of *P. paucidens* from Lake Biwa were the smallest in size and took two more moults in their larval development, which in turn led to a much longer larval period than those in Lakes Towada and Ikeda, where females bear much larger eggs (Nishino, 1984, unpubl.).

Through electrophoretic analysis of the 20 geographical populations of *P. paucidens* in Japan, Chow *et al.* (1988) differentiated between two types: the A-type included Lake Biwa and 11 populations from all over Japan inhabiting lakes, ponds and rivers, whereas the B-type was found only in rivers in East Honshu. The A-type populations exhibit great variation both genetically and in egg size, whereas little variation was evident among the B-type populations. Mating experiments demonstrated a low attraction between adult individuals of A- and B-types, and if mating did occur, embryos failed to develop. This suggests that premating and postmating isolation mechanisms are present, and that the A- and B-types are different biological species. It is noteworthy that the Lake Biwa population differs greatly both genetically and in egg size from other geographical populations of A-type distributed throughout Japan.

The Ayu is an osmerid fish found throughout Japan. The territorial behaviour of the land-locked Lake Biwa population is much stronger and more stable than that seen in amphidromous populations, and the breeding period is about 1 month earlier (Azuma, 1973). Kawanabe (1976) considered this strong territoriality as being a "social relict behaviour" remaining from the glacial age. The Biwa population also differs morphologically from amphidromous populations in such features as head and gill raker lengths (Azuma, 1981). Genetic differences among the Biwa and the amphidromous populations of the Amori River in southern Kyushu, and those of the Ryukyu Islands, were studied using isozyme data, and revealed large genetic differences between the Ryukyu population and the Biwa and Amori populations (Nishida, 1985). The results indicated that the Ryukyu population had evolved as a distinct genetic unit since the middle to late Quaternary, and that the Biwa population diverged from the Amori population *c.* 100 000 years ago. Genetic investigation of Ayu populations in Lake Biwa and 10 amphidromous populations in Japan and Korea also indicated that the Biwa population differs genetically from the amphidromous populations, except for those in the Ryukyu Islands (Seki *et al.*, 1988). The Ryukyu population was subsequently redescribed as a new subspecies, *P. altivelis ryukyuensis* (Nishida, 1988). Based on genetic data and a knowledge of the geographical history of Japan and its adjacent areas, the most likely scenario is that the ancestor of the Japanese and Korean amphidromous populations diverged from the Ryukyu population *c.* 1.37 Mya, and the Biwa population then diverged from the amphidromous populations during the Würm Glacial age (100 000 years ago) (Seki *et al.*, 1988).

A more complicated scenario is presented by the striated spined loach *Cobitis taenia* complex. In the large Japanese islands of Honshu and Shikoku three races and three local forms of *C. taenia* were known, the large and middle races, and the small race, which was subdivided into the small, spotted-small and Biwa-small forms. Among these, the large race and the Biwa-small form occur sympatrically in Lake Biwa, and differ in maturation size and colour spot patterns. Mating experiments among the races showed that these two races in Lake Biwa were almost completely isolated from each other through hybrid sterility (Minamori, 1956). It was also shown that considerable isolation exists between the Biwa-small and the middle races, in contrast to the small degree of isolation evident between the Biwa-small and small forms.

Later, the small race was further divided into six forms, i.e. the small, spotted-small, Biwa-small, Yodo-small, Tokai-small and Kyushu forms (Saitoh, 1996). As not all have been formally described, they will be referred to here as "races" or "forms". These races and forms differ not only in maturation size and colour spot patterns but also in the number and morphology of chromosomes. Two races occur exclusively in Lake Biwa, the large race and one form of the small race (Biwa-small form). The number of chromosomes in the former and latter are tetraploid ($2n = 98$) and diploid ($2n = 49$–50), respectively (Saitoh *et al.*, 1984). From colour spot patterns and DNA similarities, Saitoh (1996) proposed that the small race was an ancestor of the *C. taenia* complex, and that the large race was derived from hybridization between females of the congeneric diploid species *Cobitis biwae* ($2n = 50$), which is widely distributed in Honshu and Shikoku and exhibits a different chromosome morphology from the *C. taenia* complex, and males of the Biwa-small form.

All of the above-mentioned species exhibit geographical variations morphologically, genetically or in their karyotypes, and the respective populations from Lake Biwa represent markedly different features from other populations. The evidence to date suggests that, for each of these widely distributed species, incipient speciation may be present in their Lake Biwa populations.

VII. HYPOTHESES ON THE ORIGIN OF ENDEMISM IN LAKE BIWA

A. Glacial, Marine and Continental Relicts

There has long been an interest in the origin and biogeographical attributes of the fauna of Lake Biwa. As early as 1922, Annandale compared the origin of the fauna of Lake Biwa with that of lakes in Asia and Europe, and concluded that Lake Biwa fauna was derived from both the northern and southern regions. Uéno (1937, 1943b) also noted the biogeographical peculiarity of its fauna and regarded some organisms, such as the triclad *Bdellocephala annandalei*, the valvatid gastropod *Valvata* and the amphipod *Jesogammarus* (*Annanogammarus*) *annandalei*, as northern relicts. Later, Horie (1961) pointed out the presence of marine relict species, such as the corbiculid bivalve *Corbicula sandai*. He also commented on the presence of glacial relicts such as *B. annandalei*, *J.* (*A.*) *annandalei*, the sphaeriid bivalve *Pisidium*, the salmonid fish *Salmo* (*O.*) *masou rhodurus* and the gobiid fish *Chaenogobius isaza* (see Table 2).

Uéno (1984) considered the glacial relicts as comparable to the plant relicts at the end of the Pleistocene, regarding them as organisms transported from the northern areas during the several glacial periods in the Pleistocene. He also noted the presence of glacial marine relicts, such as the amphipod *Kamaka biwae*, whose relatives were distributed in brackish habitats in Hokkaido, Sakhalin and the Kuril Islands. Takaya (1963) categorized three kinds of origin for endemic species, i.e. continental relicts, marine relicts and those evolved within the lake. He regarded the unionid bivalve *Hyriopsis schlegeri* and the pleurocerid gastropod *Semisulcospira libertina* [possibly not *S. libertina* but one of the *S.* (*Biwamelania*) species] as continental elements and the terrestrial plant *Pinus thumbergii* as a marine element, although the latter was not a freshwater inhabitant. He also insisted that *Corbicula sandai* had diverged from the brackish species *C. japonica* about 1 Mya, when Palaeo-Lake Biwa was located near Osaka Bay. However, it is now known that Palaeo-Lake Biwa was never located near Osaka Bay (Research Group for Natural History, 1994).

Kawanabe (1975a, 1978) classified the origin of the endemic organisms into two categories; those species which differentiated in Lake Biwa and relict species. He assigned the five endemic species of fishes, *Gnathopogon caerulescens*, *Carassius cuvieri*, *Silurus biwaensis*, *S. lithophilus* and *Chaenogobius isaza* and the planktonic cladoceran *Daphnia biwaensis* to the former group. He insisted that all of these species were pelagic or open-water forms, unlike their assumed ancestor species, which were littoral inhabitants and showed a benthic mode of life.

Kawanabe (1978) subdivided the relict species into four subcategories: freshwater relicts of warm water, those of cold water, marine relicts of warm water and marine glacial relicts (Table 2). The first subcategory was represented by four kinds of cyprinid fishes, *Ischikauia steenackeri*, *Squalidus*

Table 2

Hypotheses presented by various authors concerning the origin of some Lake Biwa endemic species and their allied taxa

Uéno (1937, 1943b)	Horie (1961)	Kawanabe (1975a, 1978)	Morino (1994)
1. Northern relict	1. Glacial relict	1. Differentiated in Lake Biwa	1. Different ancestors
Bdellocephala annandalei	*Bdellocephala annandalei*	*Daphnia biwaensis*	*Jesogammarus (A.) annandalei*
Valvata	*Valvata*	*Gnathopogon caerulescens*	*J. (A.) naritai*
Anisogammarus annandalei	*Pisidium*	*Carassius cuvieri*	
[= *Jesogammarus (A.) annandalei*]	*Anisogammarus annandalei*	*Silurus biwaensis*	
	[= *Jesogammarus (A.) annandalei*]	*Silurus lithophilus*	
	Oncorhynchus masou	*Chaenogobius isaza*	
	[= *Salmo (O.) masou rhodurus*]		
2. Marine relict	2. Marine relict	2. Relict	
Oncorhynchus masou	*Corbicula sandai*	2.1 Freshwater relict of warm water	
		Ischikauia steenackeri	
[= *Salmo (O.) masou rhodurus*]		*Squalidus chankaensis biwae*	
Plecoglossus altivelis altivelis[a]		*Squalidus japonicus japonicus*	
		Opsariichthys uncirostris uncirostris[b]	

Table 2 *continued*

Uéno (1937, 1943b)	Horie (1961)	Kawanabe (1975a, 1978)	Morino (1994)
3. Marine glacial relict *Kamaka biwae*		2.2 Freshwater relict of cold water (= glacial relicts) *Valvata biwaensis* *Pisidium biwaensis* (= *kawamurae*) 2.3 Marine glacial relict *Kamaka biwae* 2.4 Marine relict of warm water *Corbicula sandai* *Plecoglossus altivelis altivelis*	

[a]Although this species is not endemic to Lake Biwa, the land-locked population there exhibits unique characteristics in egg size and territorial behaviour.
[b]This species is found only in Lake Biwa and Lake Mikata, 22 km away.

chankaensis biwae, S. japonicus japonicus and *Opsariichthys uncirostris uncirostris*. However, as the last species is also distributed in Lake Mikata, a water body 22 km north-west of Lake Biwa and not connected with it by any waterway, it should not be included here. Each of the remaining three taxa has related species or subspecies on the Chinese mainland or on the Korean peninsula, and it is most parsimonious to assume that their ancestors came from the continental mainland through the land bridge which existed during the Pliocene and earlier periods. These species were probably widely dispersed throughout Japan, or at least its western area, for a time, but then became extinct, except for the Lake Biwa population.

Examples of freshwater glacial relict species are provided by the molluscs *Valvata biwaensis* and *Pisidium kawamurae*. Their relatives were distributed in many lakes of the Kamchatka Peninsula and Sakhalin, Hokkaido and Kuril Islands, and in a few high-altitude mountain lakes of central Honshu. Kawanabe (1978) thought that their ancestors might have lived in lakes in Japan during some glacial ages and then disappeared, except for the Biwa population.

Marine glacial relicts were represented by the corophiid amphipod *Kamaka biwae*, other family members of which live in marine habitats. Its nearest relative is *K. kuthae*, which occurs in brackish or freshwater coastal waters around the Okhotsk Sea. However, Kawanabe (1978) denied the possibility of the marine glacial relict hypothesis, because during the glacial ages, when the sea level was quite low and the mouth of the outlet river was located far from the lake, this animal hardly invaded Lake Biwa.

Kawanabe (1978) also regarded the example of the fourth subcategory of marine relicts of warm water as the bivalve *Corbicula sandai* and the lacustrine form of Ayu fish. He assumed that these species invaded the lake during the interglacial age.

B. Intralacustrine Evolution

Kawanabe (1975a) posited that five endemic fish species, which were later assigned to the first category of Kawanabe (1978), had evolved into plankton feeders from their ancestors which lived in the shallow ponds or rivers adjacent to Lake Biwa. He considered that *Gnathopogon caerulescens* was derived from *G. elongatus elongatus, Carassius cuvieri* from *C. carassius, Silurus biwaensis* and *S. lithophilus* from *S. asotus*, and *Chaenogobius isaza* from *C. urotaenia*. Takahashi (1989) accepted Kawanabe's (1978) idea and added the following species to the first category, i.e. species differentiated within the lake: two cyprinids, *Carassius auratus grandoculis* and *Sarcocheilichthys biwaensis*; and seven gastropods, *Semisulcospira decipiens, S. multigranosa, S. reticulata, S. nakasekoae, S. niponica* and *S. habei*, all of which belong to the subgenus *Biwamelania*. However, none of the *Biwamelania* species, nor *Sarcocheilichthys biwaensis*, was pelagic or lived in open water: living in these habitats were features which Kawanabe (1978) indicated as common to organisms in the first category. Takahashi (1989) did not present any comments on this point.

VIII.   RECENT EVIDENCE ON THE ORIGIN OF ENDEMISM IN LAKE BIWA

A. Cladistic Analysis

Based on a cladistic analysis of their morphology, Morino (1994) concluded that two endemic gammarids, *Jesogammarus* (*A.*) *annandalei* and *J.* (*A.*) *naritai*, had evolved from different ancestors: the former was from an ancestor distributed over Korea and Japan and which invaded in the lake about 100 000 years ago, and the latter was from an ancestor distributed in Japan and which invaded afterwards. He added a third category to Kawanabe's (1978) classification of endemism, that of species of relict origin but which differentiated in the lake, and assigned the two gammarid species to this category, since their sister species were distributed not in the lake but in the neighbouring waters.

A similar case of evolution is known for two endemic catfishes, *Silurus biwaensis* and *S. lithophilus* (Kobayakawa, 1994). Morphologically, the former is closer to *S. meridionalis* of the Yangtze River in China, whereas *S. lithophilus* is the sister species of *S. asotus*, a species distributed widely in Japan and China. These facts suggest that the two catfish species were derived from different ancestors.

These two cases can be explained by Mayr's (1963) multiple invasion scenario: both *J.* (*A.*) *naritai* and *S. lithophilus* are regarded as secondary invaded species after *J.* (*A.*) *annandalei* and *Silurus biwaensis* had already colonized and differentiated in Lake Biwa (Table 3).

B. Molecular Analysis

From comparison of the genetic divergence among four species and four subspecies of the triclad *Bdellocephala* from Japan, Kamchatka and Lake Baikal, Kuznedelov *et al.* (1997) divided *Bdellocephala* into two groups: species from Kamchatka and Japan, and those from Lake Baikal. They proposed that these two groups split in recent geological time and showed that *B. annandalei* from Lake Biwa was closely related to the species from Kamchatka (*Bdellocephala* sp.), having an identical sequence within the DNA region studied. However, *B. annandalei* differs in one nucleotide position from the related Japanese species *B. brunnea*. Although the Kamchatka species was not identified, these facts together suggest that *B. annandalei* differentiated from the Kamchatka species after splitting from Japanese *B. brunnea*. It thus seems

Table 3
Assumed origin of the living endemic species of Lake Biwa

Palaeoendemic		Neoendemic
Fossils found from the Palaeo-Lake Biwa group	Glacial relict origin?: *Bdellocephala annandalei*	Species flock: *Semisulcospira (B.) habei*
1. Katata Formation	*Valvata biwaensis*	S. (*B.*) *dilatata*
Heterogen longispira	*Pisidium kawamurae*	S. (*B.*) *rugosa*
Unio douglasiae biwae	*Jesogammarus (A.) annandalei*	S. (*B.*) *arenicola*
		S. (*B.*) *decipiens*
Lanceolaria oxyrhyncha	Marine relict origin?:	S. (*B.*) *multigranosa*
Cristaria plicata clessini	*Kamaka biwae*	S. (*B.*) *fluvialis*
Inversiunio reinianus		S. (*B.*) *fuscata*
Synanodonta calipygos		S. (*B.*) *reticulata*
Oguranodonta ogurae		S. (*B.*) *takeshimensis*
Corbicula sandai		S. (*B.*) *shiraishiensis*
Semisulcospira (B.) habei		S. (*B.*) *ourense*
Carassius cuvieri		S. (*B.*) *morii*
Ischikauia steenackeri		S. (*B.*) *rugosa*
		S. (*B.*) *nakasekoae*
2. Fossils found from both Palaeo-Lake Katata and other localities *Hyriposis schlegeri* *Inversiunio (=hiraseus)*		Differentiated into two species: *Gyraurus biwaensis* *Gyraurus amplificatus* More than two forms: *Radix onychia* *Apatania biwaensis*
3. Fossils found from Palaeo-Lake Ohyamada *Silurus biwaensis*		Second invasion: *Jesogammarus (A.) naritai* *Silurus lithophilus*

Two kinds of origin are discussed in the text: (1) palaeoendemic, i.e. those species whose fossils have been found; and (2) neoendemic, i.e. species of assumed glacial or marine relict origin, species which evolved within the lake and species which represent intraspecific variations.

likely that the Biwa endemic *B. annandalei* is a glacial relict, a status loaned support by the presence of a land bridge which connected Japan with the Kamchatka Peninsula during the Pleistocene (Ichikawa *et al.*, 1970). The authors are unaware of any other studies of endemic Biwa organisms from the viewpoint of molecular evolution, but molecular-level studies are certain to play an increasingly important role in the elucidation of the origins and evolution of the Lake Biwa fauna.

C. Karyotypic and Electrophoretic Analyses

Three species of corbiculid bivalves are endemic to Japan, where *Corbicula leana* and *C. japonica* are widely distributed in fresh and brackish waters, respectively, and *Corbicula sandai* occurs only in Lake Biwa. Analysis of the karyotypes of these three species revealed *C. japonica* and *C. sandai* to be diploid, with 38 and 36 chromosomes, respectively, whereas *C. leana* was triploid with 54 chromosomes (Okamoto and Arimoto, 1986). These authors concluded that *C. sandai* was derived from the ancestral species of *C. japonica*, and that the ancestral species of the hermaphroditic species, including *C. leana*, originated from the ancestral species of *C. sandai* by polyploidy.

Sakai *et al.* (1994) analysed the three species electrophoretically at 12 isozyme-coding loci. *Corbicula leana* had almost monomorphically the same allele as the most (11 loci) or the second most (one locus) frequently observed allele of *C. sandai*. Between *C. japonica* and the other two species, however, allelic displacement was observed at six loci. They hypothesized that the lacustrine *C. sandai* was derived from brackish *C. japonica* and that the fluvial *C. leana* was then derived from *C. sandai* through triploidization. They also assumed that both *C. sandai* and *C. leana* were differentiated in the Asian continent. Based on karyotype differences, Sakai *et al.* (1994) also disproved the marine origin theory of Takaya (1963).

From electrophoretic analysis on 12 isozyme-coding loci, Hatsumi *et al.* (1995) confirmed the findings of Sakai *et al.* (1994) and showed that *C. japonica* had diverged from the ancestral species about 8 Mya, and that *C. sandai* and *C. leana* had diverged from the ancestral species about 900 000 years ago.

D. Fossil Records

The geological history of Lake Biwa is both long and complex, but is well understood. Recently, however, much misunderstanding and incorrect information has been written about the history of Lake Biwa. For example, Martin (1996) stated that the present lake was filled with fluvial sediments twice during its history, while Morino (1994) and Morino and Miyazaki (1994) gave the lake's age as 0.1 My and Noakes (1998) as 2 My. All of these details are erroneous.

During the Pliocene and Pleistocene, there were at least five kinds of old lakes of different ages, collectively known as Palaeo-Lake Biwa, located apart from the present Lake Biwa. The oldest of these, Palaeo-Lake Ohyamada, was formed 4 Mya about 50 km south-east of the present lake, and the youngest, Palaeo-Lake Katata (1–0.4 Mya), appeared in the western portion of the southern basin of the present lake (Yokoyama, 1984). The northern deep basin of Lake Biwa has existed continuously for 430 000 years (Meyers *et al.*, 1993).

Many fossils of desmids, diatoms, protists, sponges, turbellarians, molluscs, crustaceans, chironomids and fishes, including some endemic mollusc and fish species, have been recovered from the Palaeo-Lake Biwa group (Research Group for Natural History of Lake Biwa, 1994). Studies of the mollusc fossils have shown that extinction of molluscs occurred at least four times: the first three extinctions occurred at about 3.1 Mya, 2.4–2.5 Mya and 1.8–1.9 Mya, and the most recent extinction occurred at the end of Palaeo-Lake Katata (0.3–0.4 Mya) (Matsuoka, 1987). No endemic mollusc species survived throughout the whole history of Palaeo-Lake Biwa, i.e. from Palaeo-Lake Ohyamada to the most recent Palaeo-Lake Katata. Only nine endemic and eight non-endemic mollusc species have been continuously present since the formation of Palaeo-Lake Katata.

Matsuoka (1987) also indicated that there were three phases in the origin of the extant endemic molluscs, some of which were found as fossils from the Palaeo-Lake Katata formation. The first period (1–0.6 Mya) was represented by an extremely sparse fauna comprising the viviparid snail *Heterogen longispira*, and the unionid bivalves *Unio douglasiae biwae* and *Inversiunio hiraseus*. The second period (0.6–0.4 Mya) was more species rich, with the pleurocerid gastropod *Semisulcospira* (*Biwamelania*) *habei*, three kinds of unionid bivalve, *Hyriopsis schlegeri, Lanceolaria oxyrhyncha* and *Synadonta calipygos*, and the corbiculid bivalve *Corbicula sandai* being recorded. *Inversiunio reinianus* and *I. hiraseus* first appeared in this period. The third period (0.3 Mya to present) comprised the remaining endemic molluscs, whose fossils remain unknown.

From biogeographical studies of the present distribution of Lake Biwa mollusc genera, Matsuoka (1994) divided the mollusc fossils from the Palaeo-Lake Biwa group into eight groups. Only four groups include extant endemic species: the continental (from the Amur basin to India and Malaysia), the east Asian (from the Amur basin to China), the world-wide (except for polar regions) and the endemic groups. He assigned *Unio douglasiae biwae* and *Synadonta calipygos* to the continental group, *Semisulcospira* (*Biwamelania*) spp., *Lanceolaria oxyrhyncha, Inversiunio hiraseus* and *I. reinianus* to the east Asian group, *Corbicula sandai* to the world-wide group and *Heterogen longispira, H. schlegeri* and *Oguranodonta ogurae* to the endemic group, i.e. distributed exclusively in the present Lake Biwa and its tributaries.

In contrast, catfish fossils, closely resembling the present endemic fish *Silurus biwaensis* in bone morphology, have been recovered from the Palaeo-Lake

Ohyamada formation (Kobayakawa, 1994). This indicates that the ancestor of *S. biwaensis* had already differentiated in Palaeo-Lake Oyamada (3.2 Mya) and had survived throughout the history of Palaeo-Lake Biwa with little morphological change, but with increased body size. Thus, in contrast to Kawanabe's (1978) assertion, this species should not be categorized as having evolved within the lake (Table 3). *Silurus biwaensis* is the only extant endemic species whose fossils have been found in the Palaeo-Lake Biwa group in layers older than the Palaeo-Lake Katata formation. Other kinds of fish fossil, such as two cyprinid taxa, *Ischikauia steenackeri* and *Carassius cuvieri*, were found only from Palaeo-Lake Katata (Table 3).

Several endemic mollusc fossils have been found not only from Palaeo-Lake Biwa but also in other localities. The fossils of *Inversiunio hiraseus* (derived from east Asia) or *Hyriopsis schlegeri* have also been found in Miocene strata in south-western Japan. These endemic species are clearly relicts.

Core samples taken from the deep northern basin of the present Lake Biwa contain many microfossils of diatoms, green algae, protozoans, turbellarians, cladocerans, and chironomids and other insects (Kadota, 1975, 1976; Mori, 1975). Almost 200 taxa of diatoms, two genera of green algae, two genera of Protozoa, 17 species of Cladocera and one genus of chironomids have been identified from upper core samples less than 200 m deep (Kadota, 1976, 1984; Mori, 1975). Among the faunal microfossils recovered, cladocerans were most numerous, accounting for 63% of specimens, followed by turbellaria (25%) (whose cocoons were found as microfossils), Insecta (10.9%) and Protozoa (1.0%). The microfossils of turbellarian cocoons (Kadota, 1975) completely differ from those of *B. annandalei* in both size and morphology (Kawakatsu and Nishino, 1994). No microfossils identified as belonging to endemic species were recovered, although many microfossils are still unidentified.

Mollusc and fish fossils are relatively abundant in the Palaeo-Lake Biwa group, but they have never been found as microfossils in the core samples. This is partly attributable to the extremely small amounts of sediments used for analysis (Kadota, 1975, 1976) and his preparation technique, which entailed using hydrogen fluoride and hydrochloric acid (Kadota, 1976): these chemicals can dissolve carbonates in the sediments. Although molluscs and fishes comprise the most species-rich endemic groups in Lake Biwa, there is little possibility of finding their fossils from the core samples unless many samples are taken and examined using non-destructive techniques, as was done for core samples from the Palaeo-Lake Biwa formation.

IX. EXAMINATION OF THE HYPOTHESES

Endemic species which were classed as glacial relicts are *Bdellocephala annandalei, Kamaka biwae, Jesogammarus (A.) annandalei, Salmo (O.) rhodurus masou, Valvata biwaensis* and *Pisidium kawamurae* (see Table 2) (Uéno, 1937;

Horie, 1961). No fossils of any of these species have been found from the Palaeo-Lake Biwa Group.

Kamaka biwae lives in the littoral area of Lake Biwa, where water temperatures can reach almost 30°C in summer. As this is inconsistent with the glacial relict hypothesis for *K. biwae*, Uéno (1943a) explained this distribution as indicating an adaptation to high temperatures. However, congeneric species also occur in shallow brackish lakes in Japan, such as Lakes Shinji, Kasumiga-ura and Hinuma (Takahashi, 1989), which are much less than 10 000 years old and where summer water temperatures also reach almost 30°C. This may support a marine or brackish origin for *K. biwae*, but there is currently little evidence to prove or disprove the glacial relict origin of this species.

Valvata biwaensis and *P. kawamurae* occur not only in the profundal but also in the littoral and sublittoral zones in the northern and southern basins of Lake Biwa (Nishino, 1991, pers. obs.). Morino (1994) insisted that the seasonal migration within the lake shown by the gammarid *J. (A.) annandalei* (Narita, 1976) indicated it had a different origin from *J. (A.) naritai*, which remains in the littoral zone throughout the year. However, again there is little evidence to support or refute a glacial relict origin of these species.

The fossils of *Corbicula sandai* were found only from Palaeo-Lake Katata. Electrophoretic analysis indicates that *C. sandai* had diverged from the ancestral *C. japonica* long before the Second Period of Palaeo-Lake Katata (Sakai *et al.*, 1994; Hatsumi *et al.*, 1995). These facts indicate that *C. sandai* is not a marine relict but a palaeoendemic species, which was widely distributed in Japan and China, as Sakai *et al.* (1994) and Hatsumi *et al.* (1995) suggested. However, several authors have indicated that *C. sandai* also occurred in Lake Tai-Hu, China (Liu *et al.*, 1980; Matsuoka, 1994), and if true, this would mean that *C. sandai* is a relict species from east Asia. However, as little taxonomic examination of *C. sandai* and congeneric species in China has been carried out, the phylogeny of this group is still unknown and such conclusions are thus premature.

X. THE *BIWAMELANIA* SPECIES FLOCK

Although many scientific studies have been conducted on the biota of Lake Biwa, there is still little information concerning the intralacustrine evolution of the Lake Biwa endemics. In addition, the presence of a species flock in the lake has not been recognized.

From Lake Biwa 18 species and one subspecies of the genus *Semisulcospira* have been reported, among which three species of the subgenera *Semisulcospira* are widely distributed throughout Japan and 15 endemic species of the subgenus *Biwamelania* occur in Lake Biwa, 13 species in the littoral zone of the lake and two species in the lake outlet (Kuroda, 1929; Davis, 1969; Urabe and Yoshida, 1995; Watanabe and Nishino, 1995).

Davis (1969) classified Japanese *Semisulcospira* into nine species and grouped them into two species complexes, the *S. libertina* complex (*S. libertina, S. reiniana* and *S. kurodai*) and the *S. niponica* complex (*S. niponica, S. reticulata, S. multigranosa, S. decipiens, S. habei yamaguchii* and *S. nakasekoae*). These groups differ in the number of chromosomes (*n* = 18–20 in the *S. libertina* complex and *n* = 7–14 in the *S. niponica* complex), of basal cords (> 7 vs 2–6, respectively) and of embryos per female (generally ≥ 100 vs usually ≤ 35, respectively). He also indicated that the genus *Semisulcospira* was a part of the Oriental fauna characteristically found in abundance in the belt of regional overlap throughout Honshu, Kyushu, Korea, the Ryukyu Islands, Formosa and China. He suggested that one or two representatives of different genetic stock, i.e. the *S. libertina* group and the *S. niponica* group, had invaded Japan from east Asia several times, to give rise to the species complexes seen today.

The genus *Biwamelania* (Habe, 1978) was established to distinguish species endemic to Lake Biwa and its drainage system from those of the genus *Semisulcospira*, which is widely distributed in Japan. However, Habe (1978) gave no diagionistic features necessary for the establishment of a new taxon. Later, the genus *Semisulcospira* was divided into two subgenera, *Semisulcospira* and *Biwamelania* (Matsuoka and Nakamura, 1981), corresponding to the *S. libertina* group and the *S. niponica* group of Davis (1969), respectively. Currently, a total of 15 endemic species of *Biwamelania* is recognized from Lake Biwa, including eight newly described species (Watanabe and Nishino, 1995).

Matsuoka (1985) redefined the subgenus *Biwamelania* based on the definition of Davis's (1969) *S. niponica* group, placing six endemic species and one subspecies of the genus *Semisulcospira* in the subgenus *Biwamelania*. Based on the number of basal cords, he also assigned two members of the *Semisulcospira cancellata* group, *S. cancellata* and *S. amurensis*, species widely distributed in the Amur basin, Korean Peninsula and China, to the subgenus *Biwamelania*, and stated that this subgenus is not endemic to Lake Biwa. However, the genus *Semisulcospira* is ovoviviparous (Morrison, 1954) and as *S. cancellata* in China is oviparous (Liu *et al.*, 1979), this species clearly does not belong to this genus. To confirm phylogenic and taxonomic relationships between the subgenera *Semisulcospira* and *Biwamelania*, detailed morphological comparisons of the members of the genus *Semisulcospira* over the whole distributional areas are necessary.

Matsuoka (1987) also insisted that the fossils of *Semisulcospira* found in middle Miocene strata from south-western Japan (Sasebo formation) (e.g. Suzuki, 1941) belong to *Biwamelania*. However, no post-Pliocene fossils of *Biwamelania* have been found, except in Palaeo-Lake Biwa, where fossils of nine *Biwamelania* species have been recovered (Matsuoka, 1987) (Figure 2).

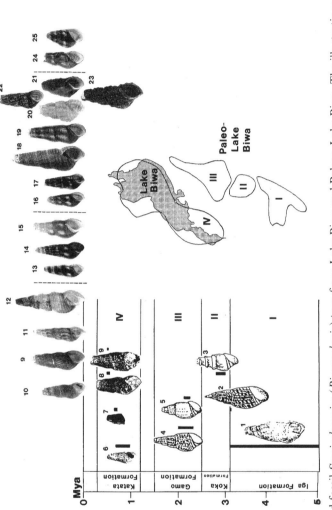

Fig. 2 Extant and recorded fossil *Semisulcospira* (*Biwamelania*) taxa from Lake Biwa and Palaeo-Lake Biwa. The illustrations of the fossil species and the simplified map of Palaeo-Lake Biwa were derived and modified from Matsuoka (1987, 1993). I, Iga formation (including Palaeo-Lake Ohyamada); II, Koka formation; III, Gamo formation; IV, Katata formation (Palaeo-Lake Katata). 1–9, Fossil species: 1, *S.* (*B.*) *praemultigranosa*; 2, *S.* (*B.*) sp. A; 3, *S.* (*B.*) sp. B; 4, *S.* (*B.*) sp. C; 5, *S.* (*B.*) sp. D; 6, *S.* (*B.*) *spinifera* MS; 7, *S.* (*B.*) *pusilla* MS; 8, *S.* (*B.*) *multigranosa nakamurai*; 9, *S.* (*B.*) *habei* of Matsuoka (1987). Bars indicate the periods of occurrence of the respective fossil species. 9–15, Living species: 9, *S.* (*B.*) *habei*; 10, *S.* (*B.*) *multigranosa*; 11, *S.* (*B.*) *decipiens*; 12, *S.* (*B.*) *reticulata*; 13, *S.* (*B.*) *arenicola*; 14, *S.* (*B.*) *rugosa*; 15, *S.* (*B.*) *dilatata*; 16, *S.* (*B.*) *fuscata*; 17, *S.* (*B.*) *ourense*; 18, *S.* (*B.*) *shiraishiensis*; 19, *S.* (*B.*) *takeshimensis*; 20–23, local races of *S.* (*B.*) *niponica*; 24, *S.* (*B.*) *fluvialis*; 25, *S.* (*B.*) *nakasekoae*.

Most of them were extinct at the end of Palaeo-Lake Katata, except for only one recent species, *S.* (*B.*) *habei*, whose fossils were found from the most recent Palaeo-Lake Katata formation (1–0.4 Mya) (Matsuoka, 1987). This means that the ancestors of the endemic *Biwamelania* species of Lake Biwa had invaded Japan at least in the Miocene, and then restricted their distribution to both the Palaeo- and present Lake Biwa.

The three species of the subgenus *Semisulcospira* have a limited distribution within the lake, whereas *Biwamelania* is represented throughout the littoral areas (Watanabe and Nishino, 1995). Some *Biwamelania* species, such as *S.* (*B.*) *niponica* and *S.* (*B.*) *morii*, are restricted to rocky and/or gravel substrates, while others, such as *S.* (*B.*) *reticulata* and *S.* (*B.*) *arenicola*, occur only on the sandy and/or muddy areas. Some species, e.g. *S.* (*B.*) *habei* and *S.* (*B.*) *decipiens*, are widely distributed throughout the littoral zone of Lake Biwa irrespective of substratum type (Watanabe and Nishino, 1995). Had most of the extant endemic *Biwamelania* evolved during the Palaeo-Lake Biwa ages, some species widely distributed in the present lake, such as *S.* (*B.*) *decipiens* and *S.* (*S.*) *reiniana*, or those distributed on the muddy or sandy bottoms, such as *S.* (*B.*) *reticulata* and *S.* (*B.*) *multigranosa*, should have good fossil records. However, since no fossils other than *S.* (*B.*) *habei* have been recovered from the Palaeo-Lake Biwa group, it can be assumed that species other than *S.* (*B.*) *habei* were not present during Palaeo-Lake Biwa periods and must have evolved later. It is most parsimonious to propose that the *S.* (*B.*) *habei*-like ancestor, widely distributed in the littoral of Palaeo-Lake Biwa, evolved into the endemic *Biwamelania* group. Some species may have adapted to rocky and gravel bottoms, others to sandy and/or muddy bottoms, while other species have retained the ancestral eurysubstratic habit.

Biwamelania species display large variations in chromosome morphology and numbers (Burch, 1968; Davis, 1969; Kobayashi, 1986; T. Nakamura, pers. comm.), ranging from $2n = 14$ or 18 for *S.* (*B.*) *habei* to 32 for *S.* (*B.*) *morii*, whereas the chromosome number of *Semisulcospira* (*S.*) *reiniana* is 36 or 40, despite close similarities in the total amount of DNA of these two subgenera. This suggests that the evolution of *Biwamelania* in Lake Biwa might be associated with changes in chromosome number and morphology. If *S.* (*B.*) *habei* was the ancestor of the *Biwamelania* group in Lake Biwa, the number of chromosomes would have increased from $n = 14$ or 18 during the intralacustrine evolution of this group.

The term "species flock" was first defined and then standardized by Mayr (1942, 1963); Greenwood (1984) and Ribbink (1984) also contributed to the clarification of this concept. Under the definition of Ribbink (1984), the 15 endemic *Biwamelania* species constitute a species flock, which was defined by monophyly, endemism and speciosity. They comprise 82% of the species of the genus *Semisulcospira* and 100% of the subgenus *Biwamelania* in Lake Biwa. As already noted, fossil records support the hypothesis that the *Biwamelania* group in Lake Biwa evolved from a *habei*-like ancestor, suggesting their monophyly. The endemic status of this subgenus is also clear. It is noticeable that no closely related *Biwamelania* species occur in the vicinity of Lake Biwa or in Japan, except for fossils in Palaeo-Lake Biwa and in south-western Japan in the middle Miocene. This combination of distributional and fossil evidence suggests strongly that they have evolved within the lake.

Speciosity is defined as a disproportionate abundance of closely related species within a geographically circumscribed area (Ribbink, 1984). With the exception of the subgenus *Biwamelania*, endemic species in Lake Biwa are represented by only one or two species within their respective genera or subgenera. Thus, only *Biwamelania* can be regarded as a species flock.

To elucidate the origin and evolutionary processes which gave rise to the endemic *Biwamelania* species flock of Lake Biwa, it is essential that studies also include *Biwamelania* species from other localities in east Asia. To date, the only studies of such fundamental features as the sizes and numbers of embryos, and karyological studies, have been made in Lake Biwa.

XI. PROSPECTS

As indicated in the previous section, the biota of Lake Biwa have different evolutionary histories from those of other ancient lakes, such as Lakes Baikal and Tanganyika. Lake Biwa is unique among ancient lakes in being located in east Asia and in the temperate zone. The present Lake Biwa is much younger (*c.* 0.4 My old; Meyers *et al.*, 1993) than other ancient lakes such as Lakes Baikal (*c.* 28 My old; Mats, 1993) and Tanganyika (9–12 My old; Cohen *et al.*, 1993). The origin and the mode of evolution of each lake's endemic species must differ according to the lake's geological location and history.

This paper has examined the many hypotheses presented to explain the evolution of endemic forms in Lake Biwa. However, despite the availability of many data the origin and evolutionary process of the endemic species of Lake Biwa remain unknown. To address this deficiency, future research should focus on two kinds of study. The first is on the origin and evolutionary mechanisms of the *Biwamelania* species flock, with special reference to the phylogenic relationships with east Asian group members. The hypothesis concerning a *habei*-like ancestor should also be examined by karyotypic, electrophoretic and DNA- or RNA-based studies, both within and among the species.

The second focus of research should be an examination of intraspecific variation both among widely distributed species and endemic species. Some species, such as *Branchiura sowerbyi*, *Palaemon paucidens*, *Plecoglossus altivelis altivelis* and the *Cobitis taenia* complex, are widely distributed in Japan and its adjacent areas, and often exhibit great geographical variation in morphology, isozymes, allozymes or karyotypes. The Biwa populations of respective species show unique characteristics among their intraspecific variants and may represent incipient evolution. Some endemic species also represent intraspecific variations, such as *Radix onychia* and *Apatania biwaensis*. Together with studies on the taxonomic status and phyogenetic relations of certain "problem" taxa, such as *Gyraurus biwaensis* and *G. amplificatus*, such investigations will provide a firm foundation for future studies. Coupled with the low degree of evolution of respective endemic species, these intraspecific variations may be related to the history of Lake Biwa, which is but an adolescent among the ancient lakes of the world.

REFERENCES

Annandale, T.N. (1922). The macroscopic fauna of Lake Biwa. *Annotat. Zool. Jpn.* **10**, 127–153.

Azuma, M. (1973). Studies on the variability of the landlocked Ayu-fish *Plecoglossus altivelis* T. et S., in Lake Biwa. IV. Considerations on the grouping and features of variability. *Jpn. J. Ecol.* **23**, 255–265 (in Japanese).

Azuma, M. (1981). On the origin of Koayu, a landlocked form of amphidromous Ayu-fish, *Plecoglossus altivelis*. *Verh. Int. Verein. Limnol.* **21**, 1291–1296.

Brinkhurst, R.O. (1971). Part 8. Family Tubificidae. In: *Aquatic Oligochaeta of the World* (Ed. by R.O. Brinkhurst and B.G.M. Jamieson), pp. 444–625. Oliver and Boyd, Edinburgh.

Burch, J.B. (1968). Cytotaxonomy of some Japanese *Semisulcospira* (Streptoneura: Pleuroceridae). *J. Conchyliol.* **107**, 3–51.

Chow, S., Fujino, Y. and Nomura, T. (1988). Reproductive isolation and distinct population structures in two types of the freshwater shrimp *Palaemon paucidens*. *Evolution* **42**, 804–813.

Cohen, A.S., Soreghan, M.J. and Scholz, C.A. (1993). Estimating the age of formation of lakes: an example from Lake Tanganyika, East African Rift system. *Geology* **21**, 511–514.

Coulter, G.W. (ed.) (1991). *Lake Tanganyika and Its Life*. Oxford University Press, Oxford.

Davis, G.M. (1969). A taxonomic study of some species of *Semisulcospira* in Japan (Mesogastropoda:Pleuroceridae). *Malacologia* **7**, 211–294.

Fujita, T. (1927a). Some nematodes parasitic to the fishes of Lake Biwa. *Zool. Mag.* **39**, 39–45 (in Japanese).

Fujita, T. (1927b). Some nematodes parasitic to the fishes of Lake Biwa (2). *Zool. Mag.* **39**, 157–161 (in Japanese).

Fujita, T. (1928). Further studies on nematodes from fishes of Lake Biwa. *Zool. Mag.* **40**, 303–314 (in Japanese).

Fukuhara, S., Kihira, H., Matsuda, M., Tabe, M. and Kondo, T. (1997). Breeding season of *Oguranodonta ogurae* (Bivalvia:Unionidae) *Venus, Jpn. J. Malacol.* **4**, 299–304.

Furukawa, M. and Awano, K. (1969). List of transplanted plant and animals. *Res. Rep. Shiga Prefect. Fish. Exp.* (in Japanese). **22**, 245–250.

Greenwood, P.H. (1984). What is a species flock? In: *Evolution of Fish Species Flocks* (Ed. by A.A. Echelle and I. Kornfield), pp. 13–19. University of Maine at Orono Press, Orono, ME.

Habe, T. (1978). Freshwater molluscs in Korea and Japan as the intermediate hosts of trematodes. *Yonsei Rep. Trop. Med.* **9**, 91–96.

Harada, E. (1984). Benthos. In: *Lake Biwa* (Ed. by S. Horie), pp. 324–330. Junk, Dordrecht.

Hatsumi, M., Nakamura M., Hosokawa, M. and Nakao, S. (1995). Phylogeny of three *Corbicula* species and isozyme polymorphism in the *Corbicula japonica* populations. *Venus, Jpn. J. Malacol.* **54**, 185–193.

Hiura, I. (1967). Studies on the distribution of Japanese freshwater and semi-freshwater Hemiptera. 1. Examination of the specimens in Osaka Museum of Natural History. *Bull. Osaka Mus. Nat. Hist.* **20**, 65–81 (in Japanese).

Horie, S. (1961). Paleolimnological problems of Lake Biwa. *Mem. Fac. Sci. Univ. Kyoto, Ser. B* **28**, 53–71.

Hosoya, K. (1989). Genus *Sarcocheilichthys*. In: *Freshwater Fishes of Japan* (Ed. by H. Kawanabe and N. Mizuno), pp. 310–313. Yama-Kei, Tokyo (in Japanese).

Ichikawa, K., Fujita, Y. and Shimadu, M. (1970). *The Developmental History of Geological Structure in Japan.* Tsukiji, Tokyo (in Japanese).

Ishida, K. (1996). *Monograph of Odonata Larvae in Japan.* Hokkaido University Press, Sapporo (in Japanese).

Ishida, T. (1993). Rare copepods from fresh and brackish waters in Japan. *Jpn. J. Limnol.* **54**, 163–169.

Ishida, T. (1997). *Eucyclops roseus*, a new Eurasian copepod, and the *E. serrulatus-speratus* problem in Japan. *Jpn. J. Limnol.* **58**, 349–358.

Ishiwata, S. (1996). A study of the genus *Ephoron* from Japan (Ephemeroptera, Polymitarcyidae). *Can. Entomol.* **128**, 551–572.

Itagaki, H. (1959). Anatomy of two forms of *Lymnaea onychia* Westerlund and their ecological notes. *Venus, Jpn. J. Malacol.* **20**, 274–288.

Kadota, S. (1975). A quantitative study of the microfossils in a 200-meter-long core sample from Lake Biwa. *Paleolimnol. Lake Biwa Jpn. Pleistocene* **3**, 354–367.

Kadota, S. (1976). A quantitative study of the microfossils in a 200-meter-long core sample from Lake Biwa. *Paleolimnol. Lake Biwa Jpn. Pleistocene* **4**, 297–307.

Kadota, S. (1984). Animal microfossils. In: *Lake Biwa* (Ed. by S. Horrie), pp. 545–555. Junk, Dordrecht.

Kawakatsu, M. and Nishino, M. (1994). A list of publications on turbellarians recorded from Lake Biwa-ko, Honshu, Japan. Addendum 1. A supplemental list of publications and a revision of the section Platyhelminthes in the papers by Mori (1970) and Mori & Miura (1980, 1990). *Bull. Fuji Women's Coll., Ser. II* **32**, 87–103.

Kawamura, T. (1915). On the two kinds of order Aspidocotylea. *Zool. Mag.* **27**, 475–480 (in Japanese).

Kawanabe, H. (1975a). Freshwater fishes in Japan and in Lake Biwa. *Tansuigyo* **1**, 4–6 (in Japanese).

Kawanabe, H. (1975b). On the origin of Ayu-fish (Pisces, Osmeridae) in Lake Biwa (Preliminary Notes). *Paleolimnol. Lake Biwa Jpn. Pleistocene* **3**, 317–320.

Kawanabe, H. (1976). A note on the territoriality of Ayu, *Plecoglossus altivelis* Temmink et Schlegel (Pisces:Osmeridae) in the Lake Biwa stock, based on the "relic structure" hypothesis. *Physiol. Ecol. Jpn* **17**, 395–399.

Kawanabe, H. (1978). Some biological problems. *Int. Assoc. Theor. Appl. Limnol.* **20**, 2674–2677.

Kawanabe, H. and Mizuno, N. (1989). *Freshwater Fishes of Japan.* Yama-Kei Publishers, Tokyo (in Japanese).

Kobayakawa, M. (1994). Catfish fossils from the sediments of ancient Lake Biwa. *Arch. Hydrobiol. Beiheft. Ergebnisse Limnol.* **44**, 425–531.

Kobayashi, T. (1986). Karyotypes of four species of the genus *Semisulcospira* in Japan. *Venus, Jpn. J. Malacol.* **45**, 127–137 (in Japanese with English summary).

Kondo, T. (1982). Taxonomic revision of *Inversidens* (Bivalvia:Unionidae). *Venus, Jpn. J. Malacol.* **41**, 181–198.

Kozhova, O.M. and Izmest'eva, L.R. (eds) (1998). *Lake Baikal. Evolution and Biodiversity.* Backhuys, Leiden.

Kuroda, T. (1929). On the species of Japanese *Kawanina* (*Semisulcospira*). *Venus, Jpn. J. Malacol.* **1**, 179–193 (in Japanese).

Kuroda, T. and Habe, T. (1987). Description of *Oguranodonta ogurae* gen. et sp. nov. *Venus, Jpn. J. Malacol.* **45**, 215–218 (in Japanese).

Kuznedelov, K.D., Timoshkin, O.A. and Goldman, E. (1997). Genetic divergence of Asiatic *Bdellocephala* (Turbellaria, Tricladida, Paludicola) as revealed by partial 18S rRNA gene sequence comparisons. *Zh. Obshchei Biogii* **58**, 85–93.

Liu, Y., Zhang, W., Wang, Y. and Wang, E. (1979). *Monograph of Economical Animals of China. Freshwater Mollusca.* Academia Sinica, Peking (in Chinese).

Liu, Y., Zhang, W. and Wang, Y. (1980). Bivalves (Mollusca) of the Tai Hu and its surrounding waters, Jianzu Province, China. *Acta Zool. Sin.* **26**, 365–369.

Maehata, M. (1987). The status of the large mouth black bass in Lake Biwa. *Tansuigyo* **13**, 44–49 (in Japanese).

Malzacher, P. (1996). *Caenis nishinoae*, a new species of the Family Caenidae from Japan (Insecta:Ephemeroptera). *Stuttgart. Baitrage Naturkunde, Ser.* A (*Biol.*) **547**, 1–5.

Martin, P. (1996). Oligochaeta and Aphanoneura in ancient lakes: a review. *Hydrobiogia* **334**, 63–72.

Mats, V.D. (1993). The structure and development of the Baikal rift depression. *Earth Sci. Rev.* **34**, 81–118.

Matsuoka, K. (1985). Pliocene freshwater gastropods from the Iga Formation of the Kobiwako Group, Mie Prefecture, central Japan. *Proc. Palaeontol. Soc. Jpn* **139**, 180–195.

Matsuoka, K. (1987). Malacofaunal succession in Pliocene to Pleistocene non-marine sediments in the Omi and Ueno Basins, central Japan. *J. Earth Sci.* **35**, 23–115.

Matsuoka, K. (1993). Evolution of molluscs in Lake Biwa from the viewpoints of molluscan fossils. *Oumia* **45**, 4–6 (in Japanese).

Matsuoka, K. (1994). Mollusca. In: *The Natural History of Lake Biwa.* (Ed. by Research Group for Natural History of Lake Biwa), pp. 116–168. Yasaka-shobo, Tokyo (in Japanese).

Matsuoka, K. and Nakamura, T. (1981). Preliminary report of the freshwater molluscs from the Pleistocene Katata Formation of the Kobiwako Group in Shiga Prefecture, Japan. *Bull. Mizunami Fossil Mus.* **8**, 105–126.

Mayr, E. (1942). *Systematics and the Origin of Species.* Columbia University Press, New York.

Mayr, E. (1963). *Animal Species and Evolution.* Belknap Press, London.

Meyers, P.A., Takemura, K. and Horie, S. (1993). Reinterpretation of late Quaternary sediment chronology of Lake Biwa, Japan, from correlation with marine glacial–interglacial cycles. *Q. Res.* **39**, 154–162.

Minamori, S. (1956). Physiological isolation in Cobitidae. IV. Speciation of two sympatric races of Lake Biwa of the striated spinous loach. *Jpn. J. Zool.* **12**, 89–104.

Miyadi, D., Kawanabe, H. and Mizuno, N. (1976). *Coloured Illustrations of the Fresh Water Fishes of Japan.* Hoikusha, Osaka.

Mori, S. (1933). On the classification of Japanese *Sphaerium. Venus, Jpn. J. Malacol.* **IV**, 149–158.

Mori, S. (1970). List of plant and animal species living in Lake Biwa. *Mem. Fac. Sci. Kyoto Univ. Ser. B* **3**, 22–46.

Mori, S. (1975). Vertical distribution of diatoms in core sample from Lake Biwa. *Paleolimnol. Lake Biwa Jpn. Pleistocene* **3**, 368–392.

Mori, S. and Miura, T. (1980). List of plant and animal species living in Lake Biwa, Rev. edn. *Mem. Fac. Sci. Kyoto Univ. Ser. B* **8**, 1–33.

Mori, S. and Miura, T. (1990). List of plant and animal species living in Lake Biwa, corrected 3rd edn. *Mem. Fac. Sci. Kyoto Univ. Ser. B* **14**, 13–32.

Morino, H. (1985). Revisional studies on *Jesogammarus-Annanogammarus* group (Amphipoda: Gammaroidea) with descriptions of four new species from Japan. *Publ. Itako Hydrobiol. Stn* **2**, 9–55.

Morino, H. (1994). The phylogeny of *Jesogammarus* species (Amphipoda: Anisogammaridae) and life history features of two species endemic to Lake Biwa, Japan. *Arch. Hydrobiol. Beiheft. Ergebnisse Limnol.* **44**, 257–266.

Morino, H. and Miyazaki, N. (1994). *Lake Baikal–Field Science of an Ancient Lake.* University of Tokyo Press, Tokyo (in Japanese).

Morrison, J.P.E. (1954). The relationships of old and new world melanians. *Proc. US Nat. Mus.* **103**, 357–394.

Nakamura, M. (1984). *Keys to the Freshwater Fishes of Japan Fully Illustrated in Colors.* Hokuryukan, Tokyo (in Japanese).

Negoro, K. (1971). Plankton in Lake Biwa. In: *Biwa-ko Kokutei Koen Gakujyutsu Chosa Houkokusho*, pp. 245–274. Shiga Prefectural Government, Otsu (in Japanese).

Nesemann, H. (1997). Rediscovery of the leech genus *Ancyrobdella* (Hirudinea, Glossiphoniidae). *Misc. Zool. Hung.* **11**, 5–10.

Nishida, M. (1985). Substantial genetic differentiation in Ayu *Plecoglossus altivelis* of the Japan and Ryukyu Islands. *Bull. Jpn. Soc. Scient. Fish.* **51**, 1269–1274.

Nishida, M. (1988). A new subspecies of the ayu, *Plecoglossus altivelis*, (Plecoglossidae) from the Ryukyu Islands. *Jpn. J. Ichthyol.* **35**, 236–242.

Nishimoto, H. (1994). A new species of *Apatania* (Trichoptera, Limnephilidae) from Lake Biwa, with notes on its morphological variation within the lake. *Jpn. J. Entomol.* **62**, 775–785.

Nishino, M. (1980). Geographical variations in body size, brood size, and egg size of a freshwater shrimp, *Palaemon paucidens* de Haan, with some discussions on brood habit. *Jpn. J. Limnol.* **41**, 185–202.

Nishino, M. (1984). Developmental variation in larval morphology among three populations of the freshwater shrimp, *Palaemon paucidens* de Haan. *Lake Biwa Study Monogr.* **1**, 1–118.

Nishino, M. (ed.) (1991). *Handbook of Zoobenthos in Lake Biwa. I. Mollusca.* Lake Biwa Research Institute, Otsu (in Japanese).

Nishino, M. (ed.) (1992). *Handbook of Zoobenthos in Lake Biwa. II. Aquatic Insects.* Lake Biwa Research Institute, Otsu (in Japanese).

Nishino, M. (ed.) (1993). *Handbook of Zoobenthos in Lake Biwa. III. Porifera, Platyhelminthes, Annelida, Tentaculata and Crustacea.* Lake Biwa Research Institute, Otsu (in Japanese).

Noakes, D.L.G. (1998). A new perspective on lakes: Kawanabe's latest achievements. *Envir. Biol. Fishes* **52**, 391–394.

Ohtaka, A. (1994). Redescription of *Embolocephalus yamaguchii* (Brinkhurst, 1971) comb. nov. (Oligochaeta, Tubificidae). *Proc. Jpn. Soc. Syst. Zool.* **52**, 34–42.

Ohtaka, A. and Nishino, M. (1995). Studies on the aquatic oligochaete fauna in Lake Biwa, central Japan. I. Checklist with taxonomic remarks. *Jpn. J. Limnol.* **56**, 167–182.

Ohtaka, A. and Nishino, M. (1999). Studies on the aquatic oligochaete fauna in Lake Biwa, central Japan. III. Records and taxonomic remarks of nine species. *Hydrobiologia* **406**, 33–47.

Ohwi, J. (1972). *New Flora of Japan* (Rev.). Shibundo, Tokyo (in Japanese).

Okamoto, A. and Arimoto, B. (1986). Chromosomes of *Corbicula japonica, C. sandai* and *C. (Corbiculina) leana* (Bivalvia: Corbiculinae). *Venus, Jpn. J. Malacol.* **45**, 203–209.

Okubo, I. (1990). Seven new species of freshwater Ostracoda from Japan. *Res. Crustac.* **19**, 1–12.

Okugawa, K. (1930). A list of fresh-water rhabdocoelids found in Middle Japan, with preliminary descriptions of new species. *Mem. Coll. Sci. Kyoto Imp. Univ., Ser. B* **V**, 75–88.

Research Group for Microorganisms (1996). Rotifers in Lake Biwa. *Res. Rep. Lake Biwa Mus. Proj. Office* **5**, 1–134 (in Japanese).

Research Group for Natural History of Lake Biwa (1994). *The Natural History of Lake Biwa.* Yasaka-shobo, Tokyo (in Japanese).

Ribbink, A.J. (1984). Is the species flock concept tenable? In: *Evolution of Fish Species Flocks* (Ed. by A.A. Echelle and I. Kornfield), pp. 21–25. University of Maine at Orono Press, Orono, ME.

Saitoh, K. (1996). *Cobitis taenia.* In: *Freshwater Fishes of Japan* (Ed. by H. Kawanabe and N. Mizuno), pp. 386–391. Yama-Kei Publishers, Tokyo (in Japanese).

Saitoh, K., Takai, A. and Ojima, Y. (1984). Chromosomal study on the three local races of the striated spined loach (*Cobitis taenia striata*). *Proc. Jpn. Acad., Ser. B* **60**, 187–190.

Sakai, H., Kamiyama, K., Jeon, S. and Amio, M. (1994). Genetic relationships among three species of freshwater bivalves genus *Corbicula* (Corbiculiidae) in Japan. *Bull. Jpn. Soc. Scient. Fish.* **60**, 605–610 (in Japanese with English summary).

Sasa, M. and Kawai, K. (1987). Studies on the chironomid midges of lake Biwa (Diptera, Chironomidae). *Lake Biwa Study Monogr.* **3**, 1–119.

Sasa, M. and Kikuchi, M. (1995). *Chironomidae [Diptera] of Japan.* University of Tokyo Press, Tokyo.

Sasa, M. and Nishino, M. (1995). Notes on the chironomid species collected in winter on the shore of Lake Biwa. *Jpn. J. Sanit. Zool.* **46**, 1–8.

Sasa, M. and Nishino, M. (1996). Two new species of Chironomidae collected in winter on the shore of Lake Biwa, Honshu, Japan. *Jpn. J. Sanit. Zool.* **47**, 317–322.

Seki, S., Taniguchi, N. and Jeon, S. (1988). Genetic divergence among natural populations of Ayu from Japan and Korea. *Nippon Suisan Gakkaishi* **54**, 559–568 (in Japanese with English summary).

Shimazu, T. (1993). Redescription of *Paraproteophalus parasiluri* (Yamaguti, 1934) n. comb. (Cestoda:Proteocephalidae), with notes on four species of the genus *Proteocephalus*, from Japanese freshwater fishes. *J. Nagano Prefect. Coll.* **48**, 1–9.

Shimazu, T. (1994). A new species of the genus *Gangesia* (Cestoidea:Proteocephalidae) from the Biwa catfish of Japan. *Proc. Jpn. Soc. Syst. Zool.* **51**, 3–7.

Suzuki, K. (1941). The Palaeogene corbiculids of northwestern Kyushu. *J. Fac. Sci. Imp. Univ. Tokyo, Sect. II* **6**, 39–62.

Takahashi, S. (1989). A review of the origins of endemic species in Lake Biwa with special reference to the goby fish, *Chaenogobius isaza. J. Paleolimnol.* **1**, 279–292.

Takaya, Y. (1963). Stratigraphy of the Paleo-Biwa group and the paleogeography of Lake Biwa with special reference to the origin of the endemic species in Lake Biwa. *Mem. Coll. Sci. Univ. Kyoto, Ser. B* **30**, 81–119.

Tanaka, M. (1979). The trophic status of Japanese lakes from the viewpoint of plankton. 23. Lakes in Kinki District. *Mizu* **21**, 34–39 (in Japanese).

Tanaka, M. (1994). Plankton. In: *The Natural History of Lake Biwa* (Ed. by Research Group for Natural History of Lake Biwa, pp. 73–98. Yasaka-shobo, Tokyo (in Japanese).

Tanida, K., Nishino, M. and Uenishi, M. (1999). Trichoptera of Lake Biwa: a check-list and the zoogeographical prospect. *Proceedings of the 9th International Symposium on Trichoptera 1998*, pp. 389–410. University of Chiang Mai, Chiang Mai.

Tokui, T. and Kawanabe, H. (1984). Fishes. In: *Lake Biwa* (Ed. by S. Horie) pp. 339–360. Junk, Dordrecht.

Tomoda, Y. (1967). Evolutional studies on the fishes of Lake Biwa. *J. Michurin Biol.* **3**, 150–162 (in Japanese).

Tsuda, M. (1971). Aquatic insects in Lake Biwa. In: *Scientific Report of the National Park of Lake Biwa*, pp. 285–299. Shiga Prefectural Government, Otsu (in Japanese).

Uenishi, M. (1993). Genera and species of leptocerid caddisflies in Japan. In: *Proceedings of the 7th International Symposium on Trichoptera* (Ed. by C. Otto), pp. 79–84. Backhuys, Leiden.

Uéno, M. (1937). *Fauna Nipponica. Order Branchiopoda (Class Crustacea) 9*. Sanseido, Tokyo (in Japanese).

Uéno M. (1943a). *Kamaka biwae*, a new amphipod of marine derivative found in Lake Biwa. *Bull. Biogeog. Soc. Jpn* **13**, 139–143.

Uéno, M. (1943b). Freshwater biota in the West Pacific. In: *Taiheiyou no Kaiyou to Rikusui* (Ocean and Freshwater of the Pacific). (Ed. by Taiheiyou-Kyoukai), pp. 817–884. Iwanami, Tokyo (in Japanese).

Uéno, M. (1975). Evolution of life in Lake Biwa–a biogeographical observation. *Paleolimnol. Lake Biwa Jpn. Pleistocene* **3**, 5–13.

Uéno, M. (1984). Biogeography of Lake Biwa. In: *Lake Biwa* (Ed. by S. Horie), pp. 625–633. Junk, Dordrecht.

Urabe, M. and Yoshida, K. (1995). Distribution of lotic snails (genus *Semisulcospira*: *S. libertina* species group) in Lake Biwa water system. *Nara Sci. Res. Soc. Inland Wat. Biol.* **10**, 7–17 (in Japanese).

Wakabayashi, T. and Ichise, S. (1982). *The Plankton of Lake Biwa*. Shiga Prefectural Institute of Public Health and Environmental Science, Otsu (in Japanese).

Watanabe, N.C. and Nishino, M. (1995). A study on taxonomy and distribution of the freshwater snail, genus *Semisulcospira* in Lake Biwa, with descriptions of eight new species. *Lake Biwa Study Monogr.* **6**, 1–36.

Watanabe, M. (1996). Studies on planktonic blue–green algae 6. Bloom forming species in Lake Biwa (Japan) in the summer of 1994. *Bull. Nat Sci. Mus. Tokyo, Ser. B* **22**, 1–10.

Yokoyama, T. (1984). Stratigraphy of the Quaternary system around Lake Biwa and geohistory of the ancient Lake Biwa. In: *Lake Biwa* (Ed. by S. Horie), pp. 43–138. Junk, Dordrecht.

Endemism in the Ponto-Caspian Fauna, with Special Emphasis on the Onychopoda (Crustacea)

H.J. DUMONT

I. SUMMARY

The Caspian Lake originated in the Miocene as the eastern half of the Ponto-Caspian basin. During the Pliocene, this water body sometimes merged with the Black Sea, also engulfing the Aral Lake and during its > 5 My existence has experienced strong and repeated fluctuations in salinity. These salinity changes have had important repercussions for the colonization of the lake and the evolution of its fauna.

Speciation in the lake has produced an overall level of endemism of c. 42%, somewhat lower than that of Lake Baikal (c. 54%). In both lakes endemism has reached the genus level, and in some faunal groups subject to adaptive radiation, levels of endemism close to 100% are seen. The main difference between these lakes is in absolute species richness and in the nature of the groups that radiated. In the Caspian Lake, selection systematically favoured groups with wide salinity tolerances. Among the Crustacea, this produced species flocks in the Mysidacea, Cumacea, the corophiid amphipods and the cytherid ostracods, while in the Onychopoda a unique pelagic speciation event occurred. The ancestors of all of these taxa were probably either estuarine (living in estuaries that flowed into the Tethys or Paratethys Seas) or freshwater animals. It is postulated that the Onychopoda are monophyletic and evolved *in situ* from the Sarmatian onwards.

Of the pelagic onychopods, seven species expanded into the oceans, and only two (possibly three) species are almost exclusively freshwater forms. In recent decades, however, the two freshwater species, and Caspian species tolerant of freshwater, have expanded their ranges to outside the Caspian. Artificial constructions (reservoirs, canals) as well as transport facilities (ships: ballast water) and incidental introduction with other, intentionally displaced Caspian species are believed to have facilitated these range expansions. Caspian onychopods are now found in the Laurentian Great Lakes of North America, the Baltic Gulf, lakes in Siberia and Central Asia, and drinking-water reservoirs in Western Europe. At least one "marine" species reinvaded the Caspian Lake.

It is known that speciation of the Onychopoda took place within the Ponto-Caspian, yet more "species" co-occur in the open water at any one time than ecological theory can explain. This unusual situation is loaned added interest by the fact that these excellent swimmers do not utilize the water column beyond 50 m depth (maximum recorded depth = 100 m), a consequence of the risk of anoxic conditions that may occur at greater depths about twice per century. In addition to detailed ecological investigations, a taxonomic revision of the speciose genera *Corniger* and *Cercopagis* is recommended.

Adaptations to the risk of anoxia in deeper waters seem to be of two kinds: living in littoral bays, such that resting eggs do not sink too deep and are able to reach an oxygenated bottom or, alternatively, abandoning sexual reproduction altogether. The latter adaptation is most unusual and evolutionarily hazardous, and merits further study.

II. INTRODUCTION

The origin and geological history of the Ponto-Caspian basin is both long and involved. During the Miocene, the Arabian subcontinent collided with western Asia, pushing up the Caucasus and Elburz mountains and severing the pre-existing marine waterway between the Atlantic and Indian Oceans. At the same time, the Tethys Sea gave way to the Sarmatian Lake, filling up the Ponto-Caspian basin, including most of the present Black and Caspian Sea areas.

Although it is possible that there was an occasional connection between this giant brackish water body and the ocean (Zaitsev and Mamaev, 1997), the Ponto-Caspian basin has existed as a lacustrine body for at least the past 5 My. This explains why true marine groups like the Echinodermata and the Chaetognatha were unable to colonize the Pontocaspian until the Holocene, when the connection between the Black Sea and the Mediterranean Sea transformed the former from a lake to a true hybrid between lake and sea. Following this event, salinity in the Black Sea rose to a level sufficient to permit invasion by exclusively marine faunas, although even to the present day, this invasion has remained quite selective. At the same time, the increasing salinity levels forced most of the exclusively freshwater species among the Onychopoda (Crustacea) out of the pelagic and into estuarine and marginal freshened zones, such as the north-west area of the lake and the Sea of Azov.

By the early Pliocene, a deep regression, coupled to continued crustal movements, had separated the western (Black) and the eastern (Caspian) sub-basins. This long regressive phase, corresponding to an increase in salinity from an almost fresh to a moderately saline water body, only reversed at the Pliocene–Pleistocene transition, when two high lake stands of long duration (the Akchagylian and the Apsheronian), the second lasting until c. 700, 000 years ago, reunited the Black, Caspian and Aral Lakes into a single gigantic brackish-water lake (Figure 1). During the intervening regressions, separate saline-water lakes were formed again (Dumont, 1998a).

Fluctuations in the level of the Caspian have continued to the present (for details, see Kaplin, 1997), with the lake surface first sinking below world ocean level possibly not later than c. 40 000 years ago. After that, it continued sinking to an absolute low of c. −40 m. In historical times, the lake level has never exceeded the −20 m mark, and has averaged c. −27 m. During the twentieth century, the amplitude of variation has been about 4 m, with minimum levels reached in 1977 and a rapid rise thereafter. As noted later, lake levels exceeding

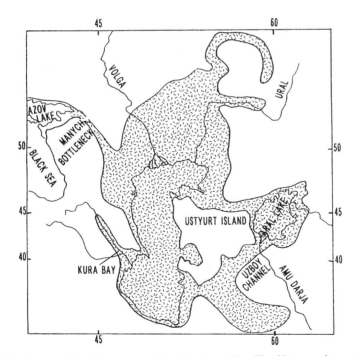

Fig. 1. The Akchagylian high stand of the Caspian Lake. The Ustyurt plateau became an island, the Aral Lake was engulfed and a temporary connection with the Black Sea became established through the Manyach depression. Similarly high lake levels were also present in the Caspian during the Apsheronian and early Khvalyn stages. From Dumont, in Rivier (1998), with permission of Backhuys Publishers.

the -26 m mark are likely to be of considerable significance for the fauna of Caspian Lake.

III. CASPIAN SPECIES RICHNESS AND LEVEL OF ENDEMISM

Ignoring Protista, the total species richness of the Caspian is currently estimated at c. 950 species (Kasymov, 1987), with about 400 species endemic to the lake. This is less than the species richness of Lake Baikal, presently standing at 1874 species (Kozhova and Izmest'eva, 1998). However, these recent estimates for Lake Baikal are themselves considerably higher than former estimates. Around 1960, only 902 species, of which 618 were endemic, had been counted in Baikal (Kozhov, 1963). Further studies of the Caspian fauna might also result in a large increase in the number of species described from this lake. As noted below, for certain faunal groups in the Caspian, estimates of species diversity differ greatly.

Although the overall level of endemism in the Caspian appears to be lower than that of Lake Baikal, there are several reasons why such a direct comparison might be misleading and of little value.

First, it may be argued that all faunal elements shared between the Caspian, the Black Sea and Lake Aral should be classed as endemics of a single water body: the Ponto-Caspian.

Secondly, the Caspian has a steep north–south gradient in salinity, from the shallow fresh waters of the Volga and Ural in the northern third of the lake, to the deep brackish basins of the center and the south, where salinity may sometimes exceed 13%. In the northern third, a freshwater fauna of wide biogeographic distribution is found, together with only few true Caspian elements. Including this fauna as Caspian increases total species richness, but decreases percentage endemism. For example, the delta region of the Volga River, which enters the Caspian, harbours 226 species of rotifers, 110 species of cladocerans *s.l.* (including onychopods) and 75 species of copepods (Chuikov, 1994). In comparison, the Caspian is home to 78 species of rotifers, 55 species of cladocerans and 41 species of copepods (Kasymov, 1987). These differences are large, and reflect the wide variety of wetlands and shallow, enclosed lakes of the deltaic zone, which contain a rich but wide-ranging fauna. However, exactly the same is true of the marginal bays (sors) and shallows of Lake Baikal (see Sideleva; Kozhova *et al.*; Takhteev, this volume). Much of the discrepancy between the 1960s and the 1990s faunal estimates for Baikal is due to the inclusion in the most recent list of biota from the sors. After including the sors biota, apparent endemism in Lake Baikal decreased from over 80% to 54%, a figure comparable with that for the Caspian (42%, using Kasymov's 1987 figures).

Thirdly, it should also be stressed that, as the Ponto-Caspian graben formed, mountain ranges were pushed up, creating rivers that have the same age as the lake. These rivers (e.g. the western Rivers Kura, Samur, Sulak and Terek, all draining the Caucasus mountains) all have an endemic species component in their mollusc, insect and fish faunas. It is debatable whether these should be included with the fauna of the Caspian proper or not, but they are indisputably part of the fauna of the basin.

An extreme example is that of the Amu and Syr Darja Rivers, in the eastern Caspian basin. Both currently drain the Tiyan Shan and Pamir mountains towards Lake Aral in the north. However, the Amu Darja is noted for changing the direction of its flow: even in historical times, it has drained alternatively to the Caspian, either directly or via the Sarykamish depression, or to Lake Aral (Aladin, 1996). Both rivers are home to unusual (presently endangered) sturgeon fish of the genus *Pseudoscaphyrhynchus* (Acipenseridae, Subfamily Scaphyrhynchinae). The three species in this genus are restricted to the Lake Aral basin. The closest relatives to these sturgeons are the three species of the genus *Scaphyrhynchus* (also subfamily Scaphyrhynchinae), which live in the Mississippi basin in North America. Evidence suggests that ancestors of these currently land-locked species lived in rivers that drained central Asia to the Indian Ocean before the uprise of the Himalayas (Bemish and Kynard, 1997). These ancestors may have been anadromous (as are many acipenserids today) and inhabited the Tethys Sea, extending as far west as the North American shores of the Atlantic.

In conclusion, if one were to compare basins and not strictly lakes, endemism within the Ponto-Caspian fauna would perhaps turn out to be even higher than that of the (comparatively small) Baikal basin. Such comparisons are best made on taxa which are well represented in both lakes, such as molluscs, crustaceans and fishes (Table 1). In details, these figures may be subject to future change, yet they eloquently show that in several families of gastropods and fish, and in several crustacean orders, in both lakes the level of endemism may approach 100%. Endemism in these lakes reaches at least the genus level, and sometimes higher. From Table 1 it is also evident that the differences between these lakes resides not in the degree of endemism, but in the nature of the taxa that underwent adaptive radiation.

IV.   THE NATURE OF SPECIATION IN THE CASPIAN

Taking, henceforth, the Crustacea as an example (but the same trends apply also to other groups), it is clear that speciation in the Ponto-Caspian proceeded along a different path from that in Lake Baikal. All Caspian groups that radiated are typical of estuarine environments. Some of these are marine groups with few (Mysidacea) or no (Cumacea) representatives in limnic ecosystems. Others, such as the Ostracoda and the Amphipoda, have a more

Table 1
Comparison of the level of endemism between the Caspian Lake and Lake Baikal

	Caspian (~ 1978)		Baikal (~ 1963)	
	Total spp.	Endemic	Total spp.	Endemic
Pisces	126	63	50	23
Acipenseridae	7	7	1	1
Clupeidae	18	18	–	–
Gobiidae	35	35	–	–
Comephoridae	–	–	25	23
Mollusca				
Bivalvia	25	21	12	3
Gastropoda	82	82	72	53
Crustacea				
Cladocera	35	–	14	4
Onychopoda	20	20[a]	–	–
Mysidacea	20	16	–	–
Cumacea	20	20	–	–
Amphipoda	74	74[b]	240	239[c]
Isopoda	2	1	5	5
Decapoda	5	3	–	–
Copepoda (parasites excluded)	41	16	73	55
Ostracoda	46	24	33	31

[a]23, and [b]69, according to Mordukhai-Boltovskoy (1979).
[c]Based on data in Kohzov (1963).

balanced distribution between marine and continental waters, yet their Caspian representatives (the Cytheridae and Corophiidae, respectively) are, at least in part, related to marine or brackish-water families. Even the sole Caspian cnidarian possessing a medusa phase, *Moerisia pallasi*, belongs to a family and genus whose species occur mainly in estuarine environments (Dumont, 1994a).

This intriguing phenomenon has not been well analysed in the literature and calls for more investigation. It reflects a history of repeated changes in salinity of the lake(s), whereby natural selection was invariably picking out taxa for their euryhalinity and broad capacity for osmoregulation (Aladin, 1996).

However, whether the ancestors of the current Caspian endemics were of marine or freshwater origin is a different matter. There are probably no true marine-derived species in the Caspian, and thus the term "Tethyan relicts" strictly does not apply (*contra* Dumont, 1998b). The ancestral forms of the Moerisiidae, Cumacea, Corophiidae, Gobiidae, Clupeidae, etc., of the modern

Caspian Lake probably lived in the estuaries and tidal zones of rivers that emptied into the Tethys before the closing of the Sarmatian basin. Their osmoregulatory abilities meant they were preadapted for life in the brackish Sarmatian lake. All other Caspian biota must be considered to have been derived from freshwater stock.

V.   THE CASE OF THE ONYCHOPOD CRUSTACEA

The order Onychopoda represents the only case of adaptive radiation among the branchiopod crustaceans. Formerly classified with the Cladocera, and still often indicated by that name, they were separated in a specific order by Fryer (1987). Ongoing molecular work, using the full sequence of the 18S rDNA nuclear gene, has not only confirmed this position, but also suggests that the Onychopoda might even be more closely related to the Anostraca than with the cladoceran orders Ctenopoda and Anomopoda (Weekers *et al.*, unpublished).

That the Onychopoda and Anostraca might share a common ancestry is indeed revealing, because they are clearly primarily freshwater animals. Yet some extant genera, and even families, of the Anostraca (the Artemiidae and Parartemiidae, as well as some *Branchinecta* species) have a capacity for osmoregulation that permits them to inhabit saline waters, from slightly brackish to almost saturated brines. If such a capacity was already present in the ancestor, which appears a reasonable supposition, this ancestor was preadapted to the nascent Ponto-Caspian. This hypothesis assumes a monophyletic origin of the Onychopoda, in the Ponto-Caspian basin itself. Their radiation must have started as soon as marine conditions had been replaced by a lacustrine environment, i.e. in the early Sarmatian, although the ancestral form itself might be significantly older.

This evolutionary scenario runs counter to the opinion of Rivier (1998), who postulates a far more complex, possibly diphyletic, origin for the Onychopoda, with the podonids (one of the three onychopod families) derived from a pure marine ancestry. The podonids have seven marine species, in addition to nine Ponto-Caspian species (Table 2); a situation which lends some support to the diphyletic origin hypothesis. However, several other points argue strongly against Rivier's position.

First, even the so-called marine species often thrive best at below-seawater salinities and several reach maximum abundance in the lower zones of estuaries, or in marine lagoons diluted with inflowing freshwater, or in brackish-water seas such as the Baltic. Moreover, except for *Evadne spinifera* and the tropical *Pleopis schmackeri*, all of the marine species are restricted to coastal or near-coastal zones.

Secondly, of the seven marine onychopods, five approach a world-wide distribution or are widely distributed in the northern hemisphere. All but *P. schmackeri* are present in the Black Sea, and none of the other six species shows

H.J. DUMONT

Table 2
Distribution of the Onychopoda

	Marine (and Black Sea)	Caspian Lake	Freshwater	Comments
Pleopsis schmackeri	X			Tropical seas only
Pleopsis polyphemoides	X			Reinvaded Caspian in 1957
Pseudevadne tergestina	X			
Podon leuckarti	X			
Podon intermedius				
Evadne anonyx	X			
Evadne prolongata				
Evadne spinifera				
Evadne nordmanni				
Podonevadne trigona		X		Found in Black Sea lagoons and rivers, and in Lake Chalkar
Podonevadne angusta		X		
Podonevadne camptonyx		X		
Cornigerius maeoticus		X		Lagoons of Black Sea, Azov Sea, up rivers to fresh water
Cornigerius lacustris			X	Lake Hazar, Anatolia
Cornigerius bicornis		X		Black Sea, up to fresh water
Cornigerius arvidi		X		
Caspievadne maximovitchi		X		
Polyphemus pediculus			X	In brackish Baltic gulfs
Polyphemus exiguus		X		
Bythotrephes longimanus			X	Tolerates brackish water

Table 2 *Continued*

	Marine (and Black Sea)	Caspian Lake	Freshwater	Comments
Cercopagis neonilae		X		Black Sea, Azov
Cercopagis socialis		X		
Cercopagis pengoi		X		Black Sea, Azov, Volga reservoirs, Black Sea lakes and reservoirs, Baltic Sea
Cercopagis prolongata		X		
Cercopagis spinicaudata		X		
Cercopagis longiventris		X		
Cercopagis robusta		X		
Cercopagis micronyx		X		
Cercopagis anonyx		X		
(Apagis) ossiani		X		Lake Chany (W. Novosibirsk)
(Apagis) cylindrata		X		
(Apagis) beklemishevi		X		Lake Chany (W. Novosibirsk)
(Apagis) longicaudata		X		

signs of speciation or subspeciation in restricted marine zones. This absence of incipient speciation is a feature strongly suggestive of a recent conquest of the ocean. Their entry to the ocean may be as recent as the Holocene, when the Black Sea became connected to the Mediterranean. This event also provided an entry route for true marine elements to the Black Sea, as an exit for those Ponto-Caspian species with a high salinity tolerance. Some of these species may have gone extinct in the Caspian itself, at a time when local salinity was too low for them to survive. That this is currently no longer the case was demonstrated by the successful reinvasion of the Caspian by *Pleopis polyphemoides* after 1957.

Evidence of the inability of some onychopods to tolerate high salinity levels is provided by the rapid salinization of the drying out Aral Lake, where at 24‰ salinity *Podonevadne camptonyx* was the only one of four onychopod species to survive into the 1990s (Aladin, 1995). In the Black Sea also, and for the same reason (salinity of the open water becoming too high after the inflow of Mediterranean water), several species survive under precarious conditions by colonizing marginal habitats, in coastal wetlands and limans (coastal lagoons) (Wilson and Moser, 1994), estuaries and the fresher Azov Sea (Table 2).

From these varied sources of evidence, the marine onychopods may be best considered as salt-tolerant Ponto-Caspian animals that recently expanded from the Black Sea to the ocean, via the Mediterranean Sea. More examples of recent range expansions are given later.

Studies of the marine onychopods in Chesapeake Bay demonstrated that they partition ecological space by living at specific salinity–temperature combinations, and thereby reduce or avoid niche overlap (Bryan and Grant, 1974). A similar strategy appears to be present in Caspian species (Rivier, 1998), with the sole difference that each occupies a specific range of below-marine salinities. Several of these taxa (*Podonevadne trigona, Cornigerius maeoticus, Cercopagis pengoi*) are perfectly capable of surviving in true freshwater, although tolerant of salinities of up to 13%, the current maximum in the south-east Caspian. An isolated population of *P. trigona* in the freshwater Lake Chalkar (Behning, 1928; Rivier, 1998) is believed to be a relict of a former lake transgression into the valley of the Ural River. Both other species migrate up rivers, either actively (which seems improbable), or as resting eggs transported in the intestine of fish, with birds or by other means. If stagnant water bodies are encountered *en route*, this upstream migration can be effective and rapid. This is evinced by the upstream expansion of the three above-cited species in several rivers feeding the Black and Caspian Lakes (the Danube, Don, Dniepr and Ural Rivers) after reservoirs had been constructed across their valleys. The most spectacular case is that of the Volga River, dammed in cascade since the 1930s, and where at least three former exclusively Caspian onychopods have now colonized all these artificial reservoirs.

There are two (or three, if *Cornigerius lacustris* is indeed a valid species; see below) true freshwater species. *Bythotrephes longimanus* is capable of surviving in slightly brackish water. The haloxenic Caspian species *Polyphemus exiguus* is not universally accepted as specifically distinct from the widespread, freshwater *Polyphemus pediculus*. This, the only onychopod known to live in temporary waters and to have a resting egg that is drought resistant does, however, also occur in the brackish Baltic Sea. Both are thus so closely related to typical Caspian species that a recent common ancestor is obvious, and both have also maintained a capacity for euryhalinity.

In their expansion outside the lake, these two species have managed to cover a Holarctic or north Eurasian range and, more recently, both have begun expanding their ranges even further. It has become clear that the ability of large *B. longimanus* to colonize new biotopes is restricted mainly by the nature and depth of the new habitat: it requires stagnant waters with a minimum depth of *c*. 5–10 m. Consequently, while it may occur as isolated specimens in river water, in the newly created reservoirs of the Volga it has flourished, and now coexists there with the two or three aforementioned Caspian species. In the course of the twentieth century, no doubt aided by human activity (ships carrying ballast water), it reached North America, where its populations increased greatly during the late 1980s (Bur *et al.*, 1986; Lehman, 1987; Sprules *et al.*, 1990). Recently, it expanded from the north of Europe to The Netherlands and Belgium (Ketelaars and van Breemen, 1993). Ketelaars and Gille (1994) state that this invasion was facilitated by the construction of large, deep (up to 15 m) drinking-water reservoirs in an area that used to consist of swamps and shallow lakes, but also note the close proximity of several major ports, suggesting that shipping might also have played a role. Possibly, the invasion proceeded in a sequence of small steps, with Antwerp currently its southern border of distribution in Atlantic western Europe.

In 1992, *Cercopagis pengoi* was first recorded from the Gulf of Riga, but soon was found across the whole brackish-water Baltic Sea (Krylov and Parrot, 1998). Gorokhova *et al.* (2000) believe that it may have been present here much longer, but either was ignored or misidentified as *Bythotrephes*, and that the morphological variation in the Baltic populations overlaps with that of at least two other *Cercopagis* species, and even with that of *Apagis* (see below). Intentional introductions of Caspian animals to other parts of the Soviet Union were common, especially in the 1960s and 1970s (Mordukhai-Boltovskoy, 1979), and the earlier, undocumented presence of this species at low densities is thus a strong possibility. Mysids were a popular object for such "acclimatizations", as were amphipods, polychaetes, molluscs and fish. In the 1960s, large water containers with live mysids (and probably also onychopods) were transported to some reservoirs on the Baltic rivers Daugava and Neman. Zjuravel (1965) explicitly called these nuclei ("otchagi") of further dispersal. Over 30 years later, the facts show him to have been correct.

The introduction of two species of the cercopagid subgenus *Apagis* (*A. cylindrata* and *A. longicaudata*; Table 2) into the large, fluctuating, saline Lake Chany, west of Novosibirsk in Siberia, may have taken place by the same means (large water containers) (Vizen, 1986). In *Apagis*, no resting stages are known (see below), and provided the Lake Chany populations indeed belong to this subgenus, live transport is here an even more probable scenario.

In conclusion, given a suitable means of transportation, the considerable osmotic tolerance of the Ponto-Caspian onychopods makes them ideally preadapted to invade new environments, whether the ocean (for the most salt-tolerant species) or fresh water (for the least salt-tolerant species). Their range expansion has been slowed down more by a scarcity of easily accessible deep, pelagic biotopes than by salinity. That until recently so few of them had been able to expand beyond the Ponto-Caspian also testifies to the relative isolation of that region and to a rather recent origin of the group. Current range expansions seem almost exclusively due to humans, either intentional, or facilitated by constructions (reservoirs, canals) or vehicles (ships).

VII. WHY SO MANY PELAGIC ONYCHOPODS IN THE CASPIAN?

In an earlier paper on the zooplankton of ancient (pre-Pleistocene) lakes, the present author argued that, with time, a simplification of the pelagic foodweb structure should occur (Dumont, 1994b). This process would eventually result in a single efficient filter-feeding algivore, usually a calanoid copepod, becoming dominant, while Cladocera, in spite of possible attempts at reinvasion, would be consistently barred from success in the limnetic environment (Dumont, 1994b).

The Caspian was not included in the list of lakes examined at that time. Indeed, relative to other ancient lakes, its pelagic presents several complications: the shallow, freshwater north area of the lake is very different from the saline centre and south. The north freezes in winter, the centre partly freezes and turns over, yet the surface water temperature in the south never drops below 10°C. Circulation here is driven by salinity increases at the surface as water evaporates, and is extremely slow (Kosarev and Yablonskaya, 1994; Dumont, 1998b). This physical heterogeneity, and immigration waves of northern animals that arrived via the Volga following deglaciation in Russia and through the artificial Volga–Don canal in the twentieth century, have made it possible for five or six species of pelagic calanoid to become established in the lake, partitioning its ecological space in several salinity–temperature compartments.

Cladocerans (anomopods and ctenopods) seldom occur in the centre and south, but at least 23 species of Onychopoda have been claimed to inhabit the lake. They are partly or entirely predacious and therefore may not compete with the calanoids. As stated earlier, they, too, partition lake space in temperature–salinity zones. In addition, some are more littoral in habitat choice and others more strictly pelagic. In spite of this, four or more congeners may co-occur at any one time. All of these fall into two groups, similar in shape and size, and show no evidence for partitioning of food items (Dumont, 1998a, b).

This assemblage of closely related, similar species seems to defy one of ecology's basic tenets: Gause's rule. Indeed, the coexistence of so many species, with little evidence of niche partitioning, leads one to question whether the number of "species" has not been grossly exaggerated. For example, the four small-sized species of the genus *Cornigerius* (including *C. lacustris*, the "endemic" of the East Anatolian freshwater Lake Hazar) only differ in details of body form and in the shape of the horn(s) on their heads. The latter, perhaps antipredator devices, may be transient (predator triggered) morphologies, or represent cyclomorphotic traits such that, in reality, there is perhaps only a single species (cf. Mourguiart, this volume). How the Anatolian "species" reached Lake Hazar is unknown. This elongated lake lies isolated between mountain ranges and therefore phoretic transport by birds seems the most plausible mechanism.

Similar concerns apply to the large-sized species of the genus *Cercopagis (s.l.)* (13 named species). The pointed brood pouch in the *C. pengoi* group, for example, may be an environmentally introduced antipredator morphology without taxonomical meaning. The same is true of the presence or absence of spinules on the cauda, the length of the cauda and even the presence or absence of a loop in the cauda (the main distinctive character between *Cercopagis s.s.* and *Apagis*). Although the distinction of two subgenera on morphological grounds is supported by ecological differences (see below), a taxonomic revision, based on a two-track approach utilizing morphology and molecular information, is needed to resolve this question.

One obvious mechanism for reducing interspecific competition in situations where congeners co-occur is via a different vertical distribution within the water column, but this is not present in the onychopods. All onychopods are eyed animals that hunt visually, and had they coexisted for a long time, specializations would inevitably have been selected for, one of which could involve a vertical segregation. Yet these excellent swimmers and vertical migrators do not extend beyond a zone reaching from the surface to a depth of 50 m, with only the occasional solitary specimen venturing down to a maximum depth of 100 m (Mordukhai-Boltovskoy and Rivier, 1987). As such, *c.* 800 m of the 900 m water column in the lake is not being used, except by the occasional specimen of *Apagis* (Rivier, 1998).

The reason why these animals utilize only a limited depth range is perhaps the same as for the absence of a true deep-water benthos (Dumont, 1998b): at high stands of the Caspian, when more fresh water flows in than evaporates, the density-driven circulation of the deep basins breaks down and meromictic conditions develop. Anoxia in the deeper waters was observed during the high lake levels of the 1930s, with meromixis setting in at near the -26 m mark. Since that mark may be reached or exceeded about twice per century, it is clear that the low oxygen content of the deep waters of the Caspian, and possibly the entire water mass below 200 m (an analogy with Lake Tanganyika is obvious here), with unpredictable anoxic pulses, has been impossible for animals to adapt to. In all probability, mass mortalities in the hypolimnion have been a recurrent phenomenon and plankton has been selected not to migrate beyond a critical depth. This extreme compression of the pelagic space to the upper 50 m has important ecological consequences in that competition here is exacerbated. This, in turn, makes the indefinite coexistence of so many similar species all the more improbable (see Simm and Ojaveer, 1999).

VIII.   WHY RESTING EGGS?

The capacity to produce eggs (more correctly, encysted gastrulas) resistant to adverse conditions and with a long dormant period is an asset in circumstances where the environment may present sudden changes of undetermined duration. Thus, as an adaptation to life in temporary waters anostracans routinely produce cysts that are drought resistant (*Polyphemus pediculus* being the only onychopod known to be capable of the same). Many anomopods and *Artemia*, living in (sometimes deep) lakes, produce resting stages that float to the surface. As a rule, cysts are produced shortly before the temporary disappearance of active populations. In contrast, all onychopods with cyclical parthenogenesis produce resting eggs that sink to the bottom (Rivier, 1998).

What might be the adaptive significance of this life tactic in a lake with an episodic anoxic hypolimnion? Even if the hypolimnion was well oxygenated, it seems unlikely that environmental cues to trigger hatching would be perceived by sedimented cysts, this environment being too cold and too dark. It is known (pers. obs.) that in low-salinity environments *Artemia* cysts sink, and will become suspended and float only after evaporation results in an increased salinity (and density) of the water. Perhaps, cysts in a monimolimnion becoming more saline might be floated to the top of the anoxic layer (if sufficiently long lived and still viable after immersion in anoxic waters), but this certainly does not appear to be an efficient mechanism for survival.

Avoidance of the deepest basins, and selecting bays or shallows in which to live would seem a more rewarding strategy, and this is exactly what seems to be happening in several species. However, in a significant number of other species (seven in all, the marine *Pleopis schmackeri*, the Caspian *Evadne prolongata, Caspievadne maximowitchi*, and all four *Apagis* species), a different avenue has been explored. All of these species seem to have completely abandoned sexual reproduction, thereby eliminating the risk of losing an investment in a sexually produced cyst. The cost of this solution is considerable: risk of accumulating a genetic load, as well as the need for an active population to be present throughout the year. Yet, in light of the history and conditions in the Caspian, this solution is more appropriate and seems to have arisen three times independently. In view of the rarity of animals completely abandoning sex, even under severe selective pressure, further research on the crustacean fauna of the Caspian Lake is clearly warranted.

REFERENCES

Aladin, N.V. (1995). The conservation ecology of the Podonidae from the Caspian and Aral Seas. *Hydrobiologia* **307**, 85–97.
Aladin, N.V. (1996). Salienostnye adaptatsii Ostracoda i Branchiopoda. (Saline adaptations of ostracods and branchiopods). *Trudy Zool. Inst. Akad. Nauk St Petersburg* **265**, 1–206 pp.
Behning, A.L. (1928). On plankton of Lake Chalkar. *Russ. Gidrobiol. Zh.* **7**, 220–228 (in Russian).
Bemish, W.E. and Kynard, B. (1997). Sturgeon rivers: an introduction to acipenserid biogeography and life history. *Envir. Biol. Fishes* **48**, 167–183.
Bryan, B.B. and Grant, G.C. (1974). The occurrence of *Podon intermedius* (Crustacea, Cladocera) in Chesapeake Bay, a new distributional record. *Chesapeake Sci.* **15**, 120–121.
Bur, M.T., Klarer, D.M. and Krieger, K.A. (1986). First records of the European cladoceran, *Bythotrephes cederstroemi*, in Lakes Erie and Huron. *J. Great Lakes Res.* **12**, 144–146.
Chuikov, Y.S. (1994). *Zooplankton of the Northern Fore-Caspian and of the Northern Caspian.* Committee for Ecology and Nature Resources of the Astrakhan Region, Astrakhan.
Dumont, H.J. (1994a). The distribution and ecology of the fresh- and brackish-water medusae of the world. *Hydrobiologia* **272**, 1–12.
Dumont, H.J. (1994b). Ancient lakes have simplified pelagic food webs. *Arch. Hydrobiol. Beiheft. Ergebnisse Limnol.* **44**, 223–234.
Dumont, H.J. (1998a). The Caspian cradle. In: *The Predatory Cladocera (Onychopoda: Podonidae, Polyphemidae, Cercopagidae) and Leptodorida of the World. Guides to the Identification of the Microinvertebrates of the Continental Waters of the World*, Vol. 13, pp 9–15. Backhuys, Leiden.
Dumont, H.J. (1998b). The Caspian Lake: history, biota, structure and function. *Limnol. Oceanogr.* **43**, 44–52.
Fryer, G. (1987). Morphology and classification of the so-called Cladocera. *Hydrobiologia* **145**, 19–28.
Gorokhova, E., Aladin, N. and Dumont, H.J. (2000). Further expansion of the genus *Cercopagis* in the Baltic Sea, with notes on the taxa present and their ecology. *Hydrobiologia* (in press).
Kaplin, P.A. (1997). Implications of climate change and water-level rise in the Caspian Sea region. *Regional Review 1995*. UNEP, Geneva. Kasymov, A.G. (1987). *The Animal Kingdom of the Caspian.* Elm Publishers, Baku.
Ketelaars, H.A.M. and Breemen, L.W.C.A. van (1993). The invasion of the predatory cladoceran *Bythotrephes longimanus* Leydig and its influence on the plankton communities in the Biesbosch reservoirs. *Verh. Int. Verein. Limnol.* **25**, 1168–1175.
Ketelaars, H.A.M. and Gille, L. (1994). Range extension of the predatory cladoceran *Bythotrephes longimanus* Leydig 1860 (Crustacea, Onychopoda) in Western Europe. *Neth. J. Aquat. Ecol.* **28**, 175–180.
Kosarev, A.N. and Yablonskaya, E.A. (1994). *The Caspian Sea.* SPB, The Hague.
Kozhov, M. (1963). *Lake Baikal and its Life.* Dr W. Junk, The Hague.

Kozhova, O.M. and Izmest'eva, L.R. (1998). *Lake Baikal. Evolution and Biodiversity.* Backhuys, Leiden.

Krylov, P. and Panov, V.E. (1998). Resting eggs in the life cycle of *Cercopagis pengoi*, a recent invader of the Baltic Sea. *Arch. Hydrobiol. Beiheft.* **52**, 383–392.

Lehman, J.T. (1987). Palearctic predator invades North American Great Lakes. *Oecologia* **74**, 478–480.

Mordukhai-Boltovskoy, P.D. (1979). Composition and distribution of Caspian fauna in the light of modern data. *Int. Rev. Gesamten Hydrobiol.* **64**, 1–38.

Mordukhai-Boltovskoy, P.D. and Rivier, I.K. (1987). *Chitchnie vetvistousnie Podonidae, Polyphemidae, Cercopagidae i Leptodoridae fauni mira.* Nauka, Leningrad.

Rivier, I.K. (1998). *The Predatory Cladocera (Onychopoda: Podonidae, Polyphemidae, Cercopagidae) and Leptodorida of the World. Guides to the Identification of the Microinvertebrates of the Continental Waters of the World*, Vol. 13. Backhuys, Leiden.

Simm, M. and Ojaveer, H. (1999). Occurrence of different morphological forms of *Cercopagis* in the Baltic Sea. *Proc. Estonian Acad. Sci. Biol. Ecol.* **48**, 169–172.

Sprules, G.W., Riessen, H.P. and Jin, E.H. (1990). Dynamics of the *Bythotrephes* invasion of the St Lawrence Great Lakes. *J. Great Lakes Res.* **16**, 346–351.

Vizer, L.S. (1986). Zooplankton of Lake Chany. In: *Ekologyia Ozera Chany*, pp. 105–115. Nauka, Novosibirsk.

Wilson, A.M. and Moser, M.E. (1994). Conservation of Black Sea wetlands. A review and preliminary action Plan. *IWRB Publ.* **33**, 1–76.

Zaitsev, Y. and Mamaev, V. (1997). *Biological Diversity in the Black Sea. A Study of Change and Decline.* Black Sea Environmental Series, United Nations Publications, New York.

Zjuravel, P.A. (1965). O Protsessie obrazovania novogo moschnogo otsaga fauni limanno–kaspiiskogo kompleksa v vodoemach bas. Baltiiskogo moria. (The process of forming a new powerful focus of fauna of the Liman–Caspian complex in waters of the Baltic Sea basin). *Trudy Akad. Nauk Litovisk SSR, Biol.* **2**((37), 77–83 (in Russian).

Trends in the Evolution of Baikal Amphipods and Evolutionary Parallels with some Marine Malacostracan Faunas

V.V. TAKHTEEV

I. SUMMARY

The taxonomically rich amphipod fauna of Lake Baikal (at present 257 described species and 74 subspecies) shows great ecological diversity and a variety of evolutionary directions. The main trends in the intralacustrine evolution of this group are: habitat partitioning by depth, substrate type or layer; trophic differentiation; differentiation by season of reproduction; transition to marsupial parasitism; the appearance of giant and dwarf forms; colonization of Baikal's gulfs; and geographical (allopatric) differentiation. Each of these ecological/evolutionary mechanisms operates at a different level in the taxonomic hierarchy. The prevalence of sympatric speciation in these amphipods is examined, and the occurrence of numerous cases of morphological convergence (parallelism) between Baikal and other, especially marine, amphipod faunas noted. Several examples of these parallelisms are given and are considered within the framework of nomogenetic evolution.

ADVANCES IN ECOLOGICAL RESEARCH VOL. 31
ISBN 0-12-013931-6

II. INTRODUCTION

The amphipods are among the most speciose groups within the macrofauna of Lake Baikal. To date, 257 species and 74 subspecies of amphipod have been described from this lake (Takhteev, 1997), only one of which is non-endemic or subendemic: the Palaearctic distributed *Gammarus lacustris* Sars, 1863, and this number constitutes almost 5% of the entire species diversity of the known freshwater amphipod fauna. Different groups within the amphipod fauna of Lake Baikal are at different evolutionary stages, some primitive, some advanced, and through studies of this fauna many evolutionary patterns can be identified.

However, until very recently, discussion of the mechanisms of speciation in Baikalian amphipods was conducted only in hypothetical terms, and while the diversity itself was enthusiastically catalogued by generations of taxonomists, the mechanisms whereby these forms arose remained poorly understood. This situation was due to two factors. First, evolutionary thought in Russia was long dominated by a dogmatic stereotype, namely, that speciation occurs only (or mainly) allopatrically. As a result, there was no tradition of diverse research on alternative speciation mechanisms. Secondly, modern approaches and technologies have been slow to become accepted and available to Russian biologists working in Baikal, and accordingly, very few karyological, genetic and molecular biological studies have been conducted on the amphipods of this lake (e.g. Salemaa and Kamaltynov, 1994; Mashiko *et al.*, 1997a, b; Ogarkov *et al.*, 1997). A non-speculative description of the mechanisms of speciation is impossible without such research.

Studies of the taxonomy of Baikal amphipods, their gross distribution patterns and ecological characteristics have been carried out since the 1940s (e.g. Bazikalova, 1945, 1948 a, b 1962, 1971, 1975; Bekman, 1959, 1983a, b, 1984, 1986; Mekhanikova and Takhteev, 1991; Tachteew, 1995; Tachteev, 1997; Takhteev and Mekhanikova, 1996; Weinberg and Kamaltynov, 1998a, b). For several species, biological parameters such as period of reproduction, growth rate, fertility, feeding, oxygen absorption ability, osmotic characteristics of the haemolymph, reaction to changes in hydrostatic pressure and thermopreferences have been investigated (Bazikalova, 1941, 1951, 1954; Bazikalova *et al.*, 1946a, b; Gavrilov, 1949; Bekman and Bazikalova, 1951; Bekman, 1962; Vilisova, 1962; Nikolayeva, 1964, 1967; Brauer *et al.*, 1980a, b; Rudstam *et al.*, 1992; Tachteew and Mekhanikova, 1993; Melnik *et al.*, 1993, 1995; Timofeev *et al.*, 1997; Zerbst-Boroffka *et al.*, 1998, 1999). Studies have also begun on the daily vertical migrations of various benthic species and the ecological role of these migrations (Bessolitsyna, 1999; Takhteev and Bessolitsyna, 1999). By identifying and defining trends in the intralacustrine evolution of this group of animals in Baikal, several striking parallels can be revealed between the development of Baikal's amphipod fauna and those of other water bodies, especially the sea.

One aspect that will be emphasized throughout this chapter is that the various evolutionary trends which are discussed below are often manifest at discrete levels within the taxonomic hierarchy: one trend may be expressed among species, another among genera, and yet another at the subspecies level. Similarly, the evolutionary trends seen among Baikal's amphipods have not proceeded uniformly: some pertain to a large set of species and genera, whereas others are observed in only a few taxa. However, it is also important to pay attention to the latter, as in other freshwater bodies these trends are sometimes much more evident. Some processes, for example the formation of a parasitic malacostracan fauna, are represented in Lake Baikal by earlier stages in comparison with the seas. Nonetheless, it is clear that in both cases this process has proceeded in a very similar direction.

This chapter utilizes information from literary sources and original data to present, for the first time, a descriptive synthesis of the basic trends of the intralacustrine evolution of the amphipods of Lake Baikal. Much of this information is dispersed among (sometimes obscure) publications in Russian, and is here gathered together and presented in English for the first time. First, several examples of evolutionary trends displayed by Baikal amphipods are presented, with emphasis on the, in many respects, parallel character of development of this fauna and amphipod faunas of the seas. Then, the evolutionary trends seen among the amphipods of Lake Baikal are interpreted within a nomogenetical evolutionary framework. It is hoped that this chapter will draw the attention of biologists to the unsolved problems and the rich potential of studies on the amphipod crustaceans of Lake Baikal.

III. BASIC TENDENCIES IN THE EVOLUTION OF BAIKALIAN AMPHIPODS

A. Bathymetric Segregation

The great depth of the lake, which is oxygenated throughout, offers many opportunities for bathymetric segregation, and this is shown by several polytypic and oligotypic genera of Baikal amphipods, e.g. the polytypic genus *Poekilogammarus* (Tachteew, 1995) and the genera *Acanthogammarus, Echiuropus, Eulimnogammarus* (*sensu* Bazikalova, 1945) and *Odontogammarus*, and also to a lesser degree in other genera. However, it is appropriate to mention here that the number of abyssal forms, those occurring at depths greater than 500 m, is insignificant. Although Bekman (1984) listed 18 such forms, ongoing research has reduced this list; with few exceptions Bekman's data included only species occurring rarely, and even those described from single occurrences, i.e. species whose true depth range is unknown. While the abyssal zone (> 500 m depth) might not be as well populated as previously believed, the shallower waters, especially those of the littoral and sublittoral zones, harbour an amphipod fauna which is abundant and taxonomically diverse.

Most species and subspecies of amphipod can be placed in one of three groups: (i) littoral (coastal, strongly stenobenthic, inhabiting only the first few metres of depth); (ii) sublittoral (found in the tens of metres of depth, and which do not occur in shallow waters or extreme depths; and (iii) those taxa that are eurybenthic and inhabit a wide range of depths, often hundreds of metres, but may sometimes be encountered even in the sublittoral zone.

Although quantitative ecological studies are few, several examples of bathymetric segregation, where different species inhabit different depth ranges, have been documented. The depth ranges of these species often overlap, but usually one species, or subset of a species, prevails at shallower depths and a different species, or subset thereof, prevails at greater depths. For example, two species of the genus *Garjajewia*, *G. cabanisii* (Dybowsky, 1874) and *G. sarsi* Sowinky, 1915 (Figure 1), co-occur over a significant range of depths, but the relative abundance of each varies with depth (Tachteew and Mekhanikova, 1996).

The disparity seen between the amphipod faunas of the littoral and abyssal zones is largely due to such depth partitioning. How this is achieved is

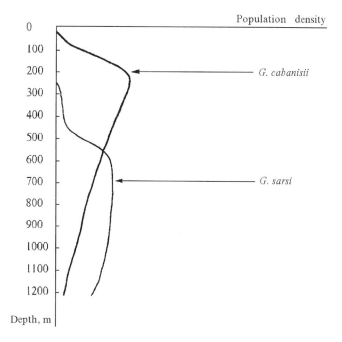

Fig. 1. Bathymetric partitioning shown by two allied species, *Garjajewia cabanisii* and *G. sarsii.*

unknown. Experimental studies have shown that in several littoral species of the genera *Pallasea* and *Eulimnogammarus*, the first changes in behaviour (increased locomotory activity) were evident only at pressures of 15 atmospheres or more, which corresponds to the pressure at depths in excess of 150 m (Brauer *et al.*, 1980b). As such, at the shallower depths inhabited by most amphipod species, differences in hydrostatic pressure play only a minor, if any, role in effecting isolation of species. Similarly, the increase of water depth during the formation of the Baikal basin seems not to have played a major role in the origin of the new forms of amphipods. In this context, it should be noted that Lake Baikal has become an ultra-deep-water lake (with depths more than 500 m) only since the Pleistocene (Popova *et al.*, 1989).

B. Differentiation by Season of Reproduction

This discussion is restricted to closely related forms, living at shallow depths, and which are characterized by temporal segregation of their periods of reproduction and brooding of eggs (Figure 2). For example, species such as *Gmelinoides fasciatus* (Stebbing, 1899), *Micruropus wohlii* (Dyb., 1874) and *Eulimnogammarus cyaneus* (Dyb., 1874) reproduce in the period from summer to autumn, whereas during the winter to spring period, other species, such as *Eulimnogammarus verrucosus* (Gerstfeldt, 1858), *E. viridis* (Dyb., 1874), *E. maackii* (Dyb., 1874), *E. marituji* (Bazikalova, 1945), *Pallasea cancellus* (Pallas,

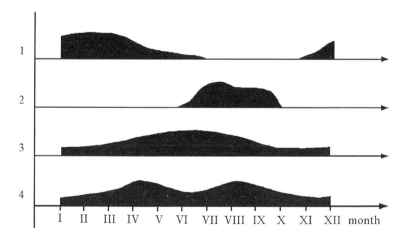

Fig. 2. Examples of differentiation by the season of reproduction. 1, *Eulimnogammarus viridis* (Dyb., 1874), a shallow-water inhabitant; 2, *E. cyaneus* (Dyb., 1874), a shallow-water inhabitant; 3, *Ommatogammarus albinus* (Dyb., 1874), a deep-water inhabitant; 4, *Macrohectopus branickii* (Dyb., 1874), a pelagic species. Based on data given in Bazikalova (1941) and Melnik *et al.* (1995).

1776) and *P. cancelloides* (Gerstf., 1858) (Bazikalova, 1941; Gavrilov, 1949), reproduce. In contrast to the shallow-water inhabitants, many deep-water, benthic amphipods, such as several species of *Garjajewia, Paragarjajewia, Ommatogammarus* and *Acanthogammarus*, and also the pelagic amphipod *Macrohectopus*, reproduce continuously during the year. However, even in these taxa one or two seasonal maxima in reproductive output are usually evident (Figure 2).

What cues determine the different seasonality of reproduction in these species, and what is the adaptive advantage of this behaviour? Bazikalova (1941) proposed that the temperature conditions of the reproductive period are encoded in the hereditary "memory" of a species. Bazikalova posited a relationship between a species' reproductive period and its phylogenetic history and, based on this, identified three discrete groups of amphipods: (i) species with a summer–autumn period of reproduction. These are phylogenetically more ancient, having originated in the warmer climatic conditions of the Tertiary Period; (ii) species with a winter–spring period of reproduction. These are younger, having originated under the conditions of decreasing temperature regime which took place during the Pleistocene; and (iii) species with year-round reproduction, primarily abyssal species. These taxa are phylogenetically the youngest, their evolutionary origin being connected to the rather recent transformation of Baikal into an ultra-deep-water lake.

Several major objections can be raised against this hypothesis. First, the representatives of the summer-reproducing genera *Hyalellopsis, Micruropus* and *Gmelinoides* are not primitive, but are morphologically advanced forms. Bazikalova has subsequently acknowledged this fact. Secondly, the reproduction of many shallow-water species of the winter complex begins in September–October, at which time water temperatures exceed 6°C. This is warmer than the *c.* 3–4°C seen during May to June, when reproduction of species in the summer complex takes place. Thirdly, it would seem logical to expect that species with winter reproduction would be more cold-loving than those with summer reproduction. However, a study of thermal preferences of littoral Baikalian amphipods (Timofeev, unpublished) has shown that the winter-reproducing *Eulimnogammarus verrucosus* (preferred temperature 5–13°C) and *E. vittatus* (Dyb., 1874) (preferred temperature 9–13°C) are no more cold-loving than the summer-reproducing *E. cyaneus* (preferred temperature 5–13°C) and *Micruropus wohlii* (Dyb., 1874) (preferred temperature *c.* 11°C). Among the investigated species having a summer reproduction period, only for *Gmelinoides fasciatus* is the preferred range of temperature rather high (10–19°C).

Thus, the factors determining the timing of amphipod reproduction remain unknown. It is likely that not only temperature but also photoperiod is involved. However, the ecological implications of differentiation by season of reproduction are clear: it prevents the simultaneous output and synchronous development of the young of different species, and thus acts to reduce potential interspecific competition, which might be expected to be especially intense among the similar-sized early life stages.

C. Segregation in Terms of Substrate Layer

Studies have revealed several examples of closely related forms, such as congeneric species, inhabiting different "floors", such as benthic, suprabenthic and various layers within the bottom sediment. The various species occupying these different layers have diverged in body shape to different ecomorphs, as is seen in members of the genus *Plesiogammarus* (Figure 3). The initial members of this lineage, *P. zeinkowiczii* (Dyb., 1874) and *P. longicornis* Sow., 1915, are nectobenthic, while the final member, *P. brevis* Bazikalova, 1975, with two subspecies, is a burrower, tunnelling in the silt to a depth of 3 cm (Takhteev, 1997). The morphological changes associated with these different habitats include (Figure 3, left to right) a trend of reduction in the length of antenna I, the substitution of the plumose setae on uropods III by simple ones, and then the shortening of this pair of limbs and the gradual reduction of the eyes from large, dark structures of *P. zeinkowiczii* to the white and dotted structures seen in *P. brevis*.

Among the nectobenthic amphipods three ecomorphs can be distinguished:

(i) permanent inhabitants of the near-bottom water layer. All representatives of the genera *Garjajewia, Paragarjajewia* and *Leptostenus*, most members of the genus *Poekilogammarus*, and also probably *Abyssogammarus*, are in this category. Characteristic features of this group include a weak, elongated body, long, thin and fragile extremities, and uropods with plentiful plumose setae. The shape of these animals most closely resembles that of the marine planktobionts;

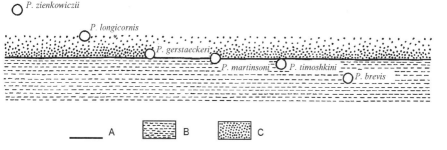

Fig. 3. Schematic respresentation of the partitioning of benthopelagic and benthic strata by several species of *Plesiogammarus*. A, Substrate surface; B, overlying sediment layer; C, pelogenic layer. Based on Takhteev (1997).

(ii) species which are capable of walking on the bottom and of swimming significant distances. This group includes *Plesiogammarus zeinkowiczii* and species of the genus *Acanthogammarus*. Characteristics features are a large body size, strong cuticle and widely outstretched extremities;

(iii) "plaice" species, which rest on their side but are capable of swimming significant distances. Representatives include members of the genera *Ceratogammarus* and *Parapallasea*. The body of these species exhibits a strong lateral flattening and their legs are not elongated.

Another version of layer differentiation is the transition of benthic amphipods to a nectobenthic lifestyle and, as a final stage of this process, to a pelagic lifestyle. Within Lake Baikal, the pelagic environment has been mastered by only one species, *Macrohectopus branickii* (Dyb., 1874), which belongs to the endemic family Macrohectopodidae Sow., 1915. Vinogradov (1988) assigns this species to the lifeform of "gnat-like, poorly mobile" crustaceans. It is probable that the transition to a pelagic lifestyle was made through the first group of nectobenthic animals mentioned above. The most recent view is that *Macrohectopus* originated from a form close to the common stem of the benthic genus *Pallasea* and the nectobenthic genus *Poekilogammarus* (Tachteew, 1995). Molecular investigations are necessary to confirm or refute this scenario.

D. Differentiation by Substrate

Partitioning of substrate has led to the evolution of several stenotopic groups of benthic amphipods: psammophiles (members of the genus *Crypturopus* and many species of *Micruropus*), pelophiles (series of species from the genera *Plesiogammarus, Macropereiopus* and *Homocerisca*); lithophiles (*Hyalellopsis* and many *Eulimnogammarus* species) and phytophiles [e.g. *Micruropus vortex vortex* (Dyb., 1874), which lives on the alga *Ulothrix zonata*].

One distinct group of amphipods is spongiophiles, and are found only on Baikal sponges of the family Lubomirskiidae. For the sponge *Lubomirskia baikalensis* (Pallas), the following species are characteristic: *Brandtia* (*Spinacanthus*) *parasitica* (Dyb., 1874), *Eulimnogammarus* (*Eurybiogammarus*) *violaceus* (Dyb., 1874) and *Poekilogammarus* (*Onychogammarus*) *erinaceus* Tachteew, 1992 (Kamaltynov *et al.*, 1993).

In the phytophile and spongiophile taxa, differentiation by substrate has been combined with trophic differentiation. Underwater observations have revealed that representatives of the genus *Pallasea* (*P. cancellus* and others) feed on the algae that they inhabit, and that the spongiophile *Brandtia parasitica* eats the tissue of the sponge on which it lives. It is also believed that the phytophilic amphipods eat aquatic vegetation, although detailed studies are needed to confirm this. It has been suggested that the relationship between the sponge *Lubomirskia* and members of its consortium (which include the above-named amphipods) is primarily trophic (stenophagic) and that stenotopy is incidental (Kamaltynov *et al.*, 1993). The same may also be true of certain phytophilic species.

In nectobenthic forms, differentiation by substrate is expressed poorly or is absent. For example, within the genus *Poekilogammarus*, the majority of species is characterized by a nectobenthic lifestyle, and only *P. crassimanus* (stones, frequently with sponges), *P. erinaceus* (only sponges) and *P. longipes* (black sand-silt substrates with detritus) are benthic forms (Tachteew, 1995).

E. Trophic Differentiation

Many species of amphipods are euryphagic and opportunistic generalists (see Morino *et al.*, this volume). However, one group has become specialized necrophages and has evolved morphological adaptations to search actively for the bodies of dead animals. This group includes species of the genera *Ommatogammarus* and *Polyacanthisca* (Tachteev, 1995). Scavengers are usually characterized by a compact, streamlined body, biting mouthparts, moderate to short pereopods with tenacious claws, and well-developed pleopods and uropods III bearing plumose setae.

In other species, notably members of the genus *Odontogammarus*, these adaptations are still at an early stage (Takhteev, 1999). In such taxa the pleopods and uropods are not well developed, but the mandible possesses a strongly developed molar edge and lacinia, and the maxilliped has a long palp. However, the body is not streamlined, but is slightly elongated, and whereas the claws of pereopods III–V are rather strong, those of pereopods VI and VII (or occasionally only pair VII) are greatly reduced and are unable to serve as organs of attachment. Based on these anatomical features, the *Odontogammarus* species can be placed in the trophic category of facultative scavengers.

Some parasitic species are highly specialized. For example, the members of the parasitic genus *Pachyschesis* (see below) are specialized for devouring the ova of their hosts (Tachteew and Mekhanikova, 1993).

F. Transition to Marsupial Parasitism

The genus *Pachyschesis* originated through a transition to marsupial parasitism and is presently placed in the same family as the Caspian genus *Iphigenella* (Bousfield, 1977). However, more recent studies have questioned this classification (*Pachyschesis* has evolved within Lake Baikal completely independently of the Caspian genera). It is therefore proposed to place this group in a newly erected endemic and monotypic family, the Pachyschesiidae (Takhteev, unpubl.). This grouping unites 17 supposed species, only four of which have been described.

These animals live in the gill cavity and marsupium of large nectobenthic amphipods. Many of the parasite species are host specific (Takhteev, unpubl.), an adaptation which may be viewed as an extreme version of differentiation by substrate. The basic morphological features of *Pachyschesis* include dwarf males, a thick spherical body in females, with an abundance of lipid inclusions, short legs with strong and tenacious claws, and poorly developed mouthparts.

G. Appearance of Giant Forms

The phenomenon of gigantism is present in many groups of Baikalian fauna and flora. Within the amphipods of the lake gigantism is displayed by many species of the genera *Acanthogammarus, Garjajewia, Abyssogammarus, Para-pallasea* and *Ceratogammarus*, and also by *Plesiogammarus zienkowiczii* and several other amphipod species. Perhaps the most striking example is that of *Acanthogammarus grewingkii* (Dyb., 1874), in which body length can reach 6 cm in females and 9 cm in males.

Several theories have been proposed to explain the occurrence of gigantism in Baikal amphipods. However, until recently, few of these explanations were convincing or were based on actual material. Vereschagin (1940) believed that the gigantism of certain Baikal fauna and the great differences between the lake's fauna and that of other lakes in East Siberia was due to the presence of heavy isotopes of oxygen and some "unique dissolved substance" (which he does not name) in the abyssal waters of the lake. These assumptions have proved to be completely unfounded. Vereschagin (1940) also dismissed the possibility of dissolved gasses being involved in the presence of giant forms, and considered the composition and concentration of dissolved gasses in Baikal to be unremarkable, and typical of lakes in general.

An alternative explanation was provided by Bazikalova (1948a), who posited that the intralacustrine origin of giant (and dwarf) amphipods took place as an evolutionary response to predation pressure, especially that from fishes of the family Cottidae. By this logic, giant forms were difficult for the cottids to handle, whereas the small size of the dwarf forms enabled them to hide easily from predators within interstices in the rocky substrate of the Baikal littoral. This explanation is also unsatisfactory. Sideleva and Mechanikova (1990) investigated food specialization in six abyssal species of cottoid fishes. From stomach content analyses of 175 fish specimens they identified 18 amphipod species. Examination of this list reveals that 15 of the 18 species exceed 20 mm body length when mature, and that 12 of them exceed 30 mm. It is thus apparent that larger sized amphipods are favoured food items for cottoid fishes, and are actively selected over smaller individuals and species.

The most recent explanation for gigantism in Baikal invertebrates is that of Chapelle and Peck (1999), who compared Baikal and oceanic waters from polar to tropical zones, and obtained a positive correlation between the number of giant species present and the degree of oxygen saturation in the water. The authors consider that in the highly oxygenated waters of Baikal and the polar seas, the haemolymph of large species is well able to supply sufficient oxygen to all sites of the body. Conversely, in waters with a low dissolved oxygen concentration, such as the tropical seas, normal interchange of gases is possible only for small species. Here, natural selection exerts pressure on the large forms and does not affect small ones. While this finding is undoubtedly important, in my opinion, gigantism among the amphipods of Lake Baikal cannot be explained exclusively in terms of oxygen availability. This question demands further study.

H. Appearance of Dwarf Forms (Hypomorphosis)

Dwarf forms have also appeared in various phyletic lineages. Examples include *Micruropus pusillus* Baz, 1962 (body length 1.5–2 mm), some other species of *Micruropus*, species of *Homocerisca*, and *Plesiogammarus martinsoni impransus* Takhteev, 1997 and *P. timoshkini* Takhteev, 1997. Both of the latter forms have evolved in the abyssal zone of northern Baikal, which is characterized by the most oligotrophic conditions throughout the lake (Takhteev, 1997).

Dwarf males are known in several ecologically diverse species, all of which exhibit some degree of hypomorphosis, e.g. the pelagic *Macrohectopus branickii*, some burrowing species of *Micruropus* and the parasitic genus *Pachyschesis* (Figure 4). Bekman (1958) proposed that the development of dwarf males facilitates an economy of resources, and thus a higher output of progeny. Studies of marsupial parasites of the genus *Pachyschesis* (Tachteew and Mekhanikova, 1993) lend strong support to this hypothesis: for this group, the habitat in the marsupium of the host means that not only food resources (eggs of the host) but also space are limiting factors.

I. Occupation of the Shallow Gulf Habitats of the Lake

The temperature and chemical conditions in the shallow-water gulfs (sors) differ greatly from those in the open waters of the lake. The sor habitats are protected from the influence of the open lake and are inhabited mainly by Euro-Siberian fauna (Kozhov, 1972; Sideleva; Kozhova *et al.*, this volume), including the Palaeoarctic gammarid *Gammarus lacustris*. Few Baikal species live in the sors, and the amphipod fauna of these habitats comprise the most eurytopic species and those with high intraspecific variability (e.g. *Gmelinoides fasciatus*). It has been shown that species with high variability and intraspecific diversity are evolutionary the most stable (Severtsov, 1990), and such polymorphic Baikal species which show these characters are capable of extending their range beyond Baikal proper. For example, the Yenisey River is home to more than 10 species of Baikal amphipods (Bazikalova, 1945),

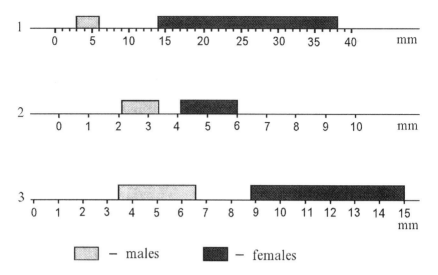

Fig. 4. Adult body size (length) in several species of Baikal amphipods in which dwarf males are present. 1, *Macrohectopus branickii*, a pelagic species; 2, *Micropurus ciliodorsalis* Sow., 1915, a burrowing psammophile; 3, *Pachyschesis acanthogammarii* Takhteev, sp. n., a parasitic species. Based on data given in Bekman (1958), Timoshkin *et al.* (1995) and Tachteew and Mekhanikova (1993).

including *E. cyaneus, E. viridis, G. fasciatus* and *M. wohlii. Eulimnogammarus viridis* has also been reported from Lakes Taymyr, Khantayskoye and Norilsk in the north of Siberia (Grese, 1957; Vershinin, 1960; Vershinin *et al.*, 1967). These examples provide the very few exceptions to the general rule of discrete distributions of Baikalian and Siberian faunas.

However, in those localities where the habitat is intermediate between sor-like and typically Baikalian, several Baikal species can be found. Nevertheless, they are here most often represented by derived types, such as subspecies and varieties. For example, the nominotypical form of *Eulimnogammarus verrucosus verrucosus* inhabits the stony littoral zone of the open coast of Baikal, whereas in the intermediate habitat of the Chivyrkuy Gulf an unusual subspecies, *E. verrucosus olicacanthus* Baz., 1945, occurs. Similarly, *Poekilogammarus megonychus perpolitus* subsp. n. (Takhteev, 2000) inhabits the open waters of Baikal, but in Chivyrkuy Gulf and in the warmer and shallow waters of the southern half of Maloe More Strait is replaced by *P. megonychus megonychus* Sow., 1915.

The tendency for specially adapted forms to inhabit the sor zone is especially evident in the middle part of Chivyrkuy Gulf, where there are shallow-water bays that warm up during the summer (e.g. Zmeinaya Bay and Ongokonskaya Bay). The shallow waters of Zmeinaya Bay (< 4 m depth) reach 20°C in August, and here Baikalian amphipod taxa are represented mostly by known subspecies (*E. verrucosus oligacanthus, P. megonychus megonychus*) and by (presumed) new subspecies such as *E. lividus* subsp. n. and *Micruropus wohlii* subsp. n. (Tachteev, unpubl.). The high incidence of unique subspecies within the sor habitats can best be understood in terms of ongoing "classical" Darwinian microevolution taking place in representatives of diverse genera, presumably as an adaptation to the unique conditions of the sor habitat.

J. Geographical Differentiation

Several examples of geographical differentiation (allopatric speciation) are known among the Baikalian amphipods. The clearest examples are those at the shoal near the Ushkany Islands, where several local endemic forms have arisen under conditions of insular isolation. Typical examples include *Brandtia* (*Spinacanthus*) *insularis* (Dorogostajsky, 1930), *Hyalellopsis insularis* (Baz., 1936), *Micruropus macroconus tenuis* Baz., 1962, *M. pupilla* Baz., 1962 and *M. ushkani* Baz., 1945.

As noted earlier, allopatric speciation was for a long time the only evolutionary mechanism considered possible by Russian biologists, and this mechanism still dominates evolutionary thought world-wide. However, the author believes that allopatric speciation has not played a major role in the diversification and speciation of the Baikalian amphipod fauna. Instead, the major factor influencing the formation of the faunal differences among the separate regions of Baikal (Selenga region, Chivyrkuy Gulf, Olkhonskiye Vorota Strait, etc.) is not spatial isolation, but the ecological conditions unique to each of these regions.

Allopatric speciation requires that populations be physically separated, but analyses of the distribution of pelophylic species have concluded that within Lake Baikal there are no impassable barriers preventing species dispersal (Mekhanikova and Takhteev, 1991). In particular, studies in the northern basin of the lake found seven pelophylic species and one subspecies restricted to that basin. Throughout the lake, at depths greater than 25 m the substrate is dominated by silts, yet the thresholds between separate lake basins (northern, middle and southern) are located much deeper, at 250–500 m depth. The silt layer is thus continuous through all three basins and there are no barriers to the dispersal of pelophytic species. The restriction of these taxa to the northern basin cannot, therefore, be due to geographical isolation.

Mashiko *et al.* (1997a, b) found population-specific genetic differences within the littoral amphipod *Eulimnogammarus cyaneus* and concluded that "intralacustrine speciation by localised topographic changes played a substantial role in the steady increase of indigenous species in ancient lakes." It should be noted that, despite the genetic differences between populations, morphologically they all belong to a single species. At the same time, in the Angara River, which drains from Lake Baikal, there are two morphologically

distinct subspecies of *E. cyaneus*. The upper section of the Angara River is late
Pleistocene in age, i.e. less than 15 000 years old (Popova *et al.*, 1989). Hence,
the habitat conditions in the Angara River, even during its geologically short
existence, have exerted a more substantial influence on the morphology of *E.
cyaneus* than the numerous geological events connected to the formation of
modern Baikal and the genetic divergence in *E. cyaneus* populations around
the shoreline of the lake proper (see Mashiko, this volume). This supports the
thesis that spatial isolation has played an insignificant role in the speciation of
amphipods in Lake Baikal.

Speciation related to bathymetric and substrate differentiation is most often
interpreted as being allopatric, but in relation to the amphipods, this view is
incorrect. The majority of amphipod species is highly mobile and capable of
extensive migrations. Furthermore, many benthic species (at least 25) make
daily nocturnal migrations into the water column (Bessolitsyna, 1999;
Takhteev and Bessolitsyna, 1999). These vertical migrations are made by both
adult animals and young, and during the summer period the density of
amphipod aggregations in the water column can reach 250 individuals m^{-3}.
(To emphasize the point made in the paragraph above, *E. cyaneus* is one of the
most common species in the nocturnal migratory complex of amphipods.)
Carried by currents, migrants can come to rest a significant distance away from
the place where they left the substrate. For example, consider that average
speeds of horizontal currents in open Baikal during the ice-free period are *c.*
1.5–6 cm s^{-1} and during storms they can reach 10–18 cm s^{-1} (Shimaraev *et al.*,
1994). Simple calculations reveal that amphipods remaining in the water
column at a current speed of 6 cm s^{-1} for only 5 h in a single night will travel in
excess of 1 km. That such transportation of amphipods occurs is evinced by
individuals of the benthic littoral-inhabiting *Micruropus wohlii* being found in
the surface layers of open Baikal, where depths exceeded 400 m. Therefore, the
complete isolation of populations in separate habitats or depth zones is
exceedingly unlikely.

Bathymetric and habitat segregation are sometimes included under the
umbrella of parapatric speciation (Martens, 1997). In the present author's
opinion, this term is applicable only to poorly mobile aquatic animals, such as
molluscs, where movement to an adjacent habitat patch is difficult. However,
for mobile organisms, such as the amphipods, the patchy distribution of
suitable habitat presents no barrier to the mixing of different populations. For
such groups speciation can only be considered meaningfully when viewed as an
allopatric or a sympatric mechanism. It would be more informative and
interesting to examine the genetic mechanisms that result in speciation under
these conditions – a metapopulation scenario, with essentially sympatric
populations.

IV. PARALLEL (NOMOGENETIC) DEVELOPMENT OF THE BAIKALIAN AND MARINE MALOCOSTRACAN FAUNAS

The term "nomogenesis" was coined by the Russian scientist L.S. Berg to define an evolutionary process which proceeded on the basis of laws, as opposed to the widely accepted view of evolution as a causal process (tychogenesis), as characterized by the theory of Darwinian evolution (Berg, 1922; there are three English editions, including Berg, 1969).

The key theses of Berg's theory can be expressed as follows:

(i)   Evolution is not a casual, but a law-governed process whose direction is as predetermined as the direction of individual development of each organism (ontogenesis).

(ii) Natural selection is not the main evolutionary factor, but acts only to eliminate those forms unsuited to the specific conditions of a given environment.

(iii) The diversity of organisms arises under the influence of internal laws, which are inherent in the given group of a fauna or a flora (i.e. phylogenetic development is predetermined, just as the individual development of an organism is predetermined by its genetic programme).

(iv) Many characters of an organism will be formed independently of natural selection, but can be supported or eliminated by it.

(v)   Parallelisms (evolutionary development in the same directions) are not a casual phenomenon in nature, but are completely normal and governed by laws.

(vi) If parallelisms can be demonstrated within a certain group of animals or plants, it is possible to discern an evolutionary law for this group and to predict its future course.

The theory of nomogenesis certainly contains several incorrect hypotheses. However, its summary dismissal, most often by supporters of Darwinism and a synthetic theory of evolution, was premature and entailed the rejection of several reasonable ideas. For example, in Russia the ideas of nomogenesis have been applied only in the works of a rather small group of scientists, yet these studies have resulted in major advances in the development of a general theory of systems (Urmantsev, 1988).

Unfortunately, page limitations prevent deeper discussion here on the philosophy and logic of the theory of nomogenesis. One should at least note that it offers a seemingly high potential for developing a general evolutionary law for specific animal groups, such as the amphipods. Therefore, attention should be paid to any cases of parallel occurrence of the same morphological attributes seen in organisms inhabiting different water bodies, irrespective of whether or not these are environmentally similar.

Evolutionary parallels can be divided into three categories: ecological, phylogenetic and nomogenetic. However, the first and second categories are but special cases of the third; they can be explained either by similar living conditions (first category) or by the close phyletic relationship of the organisms (second category). In contrast, nomogenetic parallels *per se* cannot be explained in terms of either reason. They reflect system laws ("refrains" *sensu* Meyen, 1978), according to which there is ordering of the diversity in nature. Thus, similar morphological characters can arise in amphipods which are related only distantly, or which live in completely different conditions.

How can nomogenetic similarities be distinguished from ecological ones? In the case of an ecological parallelism (convergence), correlation of an aggregate of morphological attributes is observed. Similar living conditions stimulate morphological changes in a given direction and this process is reflected in the concept of the life form. Some examples of ecologically caused evolutionary parallels between the faunas of amphipods of Lake Baikal and the sea are given in Table 1.

Table 1
Ecological parallels seen between Baikal amphipods and marine crustaceans

Ecological group	Lake Baikal	Marine habitats
Nectobenthic scavengers; compact active swimmers with a streamlined body	Genera *Ommatogammarus, Polyacanthisca*, partly *Odontogammarus*	*Hirondella gigas* Birst. et Vinogr., many species of the family Lysianassidae, etc.
Gnat-like pelagobionts	*Macrohectopus branickii*	Families Vitjazianidae, Hyperiopsidae (partly); genera *Eusirogenes, Pareusirogenes, Eurisella*, etc.
Burrowing psammophiles with a thick compact body	Genus *Crypturopus*	Genus *Niphargoides*
Burrowing pelophyles with a compact body	Genera *Micruropus* and *Macropereiopus*	Genus *Pontoporeia*
Burrowing amphipods with an elongate body and reduced uropods III	*Plesiogammaris brevis* (with two subspecies) and *P. timoshkinii*	Family Corophiidae; suborder Ingolfiellidae (extreme case)
Parasites of gill cavities and marsupial chambers of Malacostraca; poorly mobile crustaceans with tenacious claws	Genus *Pachyschesis*	Many Isopoda (e.g. Bopyridae); to a lesser extent the Amphipoda (e.g. the Caspian *Cardiophilus baeri* Sars, 1896

Data are taken partly from Birstein and Romanova (1968), Osadchikh (1977), Ginetsinskaya and Dobrovol'skiy (1978), Müller (1989), Belyajev (1989), Vinogradov (1992), Tachteew (1995), Tachteev (1995) and Takhteev (1997).

In purely nomogenetic parallelisms, similar morphological features arise in animals of different lifeforms. These similar features are most often unique or are correlated with only a few other features in the morphology of the animal; the probability of the independent concurrence of a large set of attributes (not correlated with each other) is extremely insignificant. The nomogentic similarity of organisms may be viewed as nature arriving at identical constructive "decisions" as the solution to different adaptive problems. In amphipods, examples of nomogenetic similarities can be found in the cuticular armature of the body surface, and the remainder of this chapter focuses on these examples.

The similarity in gross morphology between the Baikalian and Caspian amphipod faunas has long been recognized (Bazikalova, 1940) and has led to some Baikalian genera being placed within Caspian families (e.g. Bousfield, 1977). However, if in this case one assumes a common genesis of the fauna in both basins, then must one not similarly explain the equally striking similarity of armature of representatives of the Baikalian gammarid genus *Acanthogammarus* (family Acanthogammaridae; Bousfield, 1977), Titicacan amphipods of the genus *Hyalella* (family Orchestriidae) and even those members of the family Epimeriidae from the waters of the Antarctic Ocean (see Martens, 1997, Figure 3)?

An even more interesting case of nomogenetic similarity in the structure of the cuticular armature of the body is shown by the two pairs of species illustrated in Figure 5. In the Atlantic species *Laetmatophilus armatus* the metasomal segments are supplied by precisely the same vertically positioned teeth as seen in representatives of the Baikalian subgenus *Propachygammarus*, and also *Acanthogammarus maximus* (Garjajew, 1901) (the present author's unpublished revision of *A. maximus* places it within *Propachygammarus*, which is elevated to the rank of genus). The *Propachygammarus* taxa are ecologically diverse; *P. bicornis* is an endemic of the Olkhonskiye Vorota Strait and lives at shallow depths (35–40 cm) among clumps of aquatic vegetation, whereas *P. dryshenkoi* (Garj., 1901) and *P. lamellispinus* Baz., 1945 are mostly deep-water forms found on silt substrates. In the two species depicted in Figure 5a and b the antennae are arranged differently, and one can thus infer that they also function differently. These two species clearly belong to different life forms and no single adaptive interpretation can explain the formation of similar armature of segments of their bodies.

In the marine species *Pereionotus testudo*, the dorsal and marginal rows of eminences on the body are remarkably similar to those seen in Baikalian amphipods of the genus *Brandtia* (*sensu* Bazikalova, 1948) (Figure 5c, d). Species of this genus are usually found on stones and on sponges and are weakly mobile. However, they also show some similar characters of armature to some Baikalian species of the genus *Hyalellopsis*, which inhabit a wide spectrum of substrate types, from stones to silts (Bazikalova, 1945). In contrast, the marine amphipod species which occur on sponges sometimes have a completely smooth body [e.g. *Perrierella audouiniana* (Bate, 1857) or *Aristias*

214 V.V. TAKHTEEV

neglectus Hansen, 1887; Lincoln, 1979, pp. 46–47, 60–61]. These examples comprise species with different lifestyles but similar armatures, and species with similar lifestyles but different armatures, and here also, an adaptive interpretation for the similarities seen in the body armature of these species is not obvious.

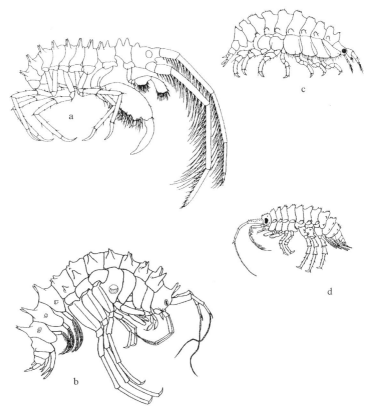

Fig. 5. Nomogenetical similarities in the character of the body armature of Baikal amphipods and marine malacostracan taxa. Similarities are evident between the dorsal row of eminences (a, b), and in the dorsal and lateral rows (c, d), despite the vast taxonomic and ecological differences between these component species pairs. (a) *Laetmatophilus armatus* (Norman, 1869) (family Podoceridae): a marine species, found in the north-east Atlantic, north and west Norway, the Bay of Biscay and the Mediterranean Sea, at depths of 35–900 m. Figure and data adapted from Lincoln (1979). (b) *Pallasea (Propachygammarus) bicornis* Dorogostajsky, 1930: a Baikal endemic, found only in the Olkhonskiye Vorota Strait at depths of 35–40 m; probably a phytophile. (c) *Pereionotus testudo* (Montague, 1808) (family Phliantidae): a marine species, found in the Mediterranean and Red Seas, and along the Atlantic coast of Europe, in the intertidal–shallow subtidal zone. Figure and data adapted from Lincoln (1979). (d) *Brandtia (Spinacanthus) parasitica* (Dybowsky, 1874): a Baikal endemic; a spongiophage, found only on sponges, at depths of 1–60 m. Figure adapted from Kamaltynov *et al.* (1993).

Another example of parallelism is seen in the sharp, rear-directed teeth on the back edges of metasomal segments, which have arisen independently in the Baikalian genus *Pallasea* (nominative subgenus), the North Atlantic species *Gitanopsis bispinosa* (Boeck, 1871) (family Amphilochidae; Lincoln, 1979, Figure 73), the boreal-Atlantic *Melita gladiosa* Bate, 1862 (family Melitidae; Lincoln, 1979, Figure 141) and the circumpolar Antarctic *Gnathiphimedia sexdentata* (Schellenberg, 1926) (family Acanthozomatidae; Sieg and Wegele, 1990, Figure 277). It should be noted that the latter species inhabits depths greater than 700 m (Sieg and Wegele, 1990, p. 138), whereas the Baikalian *Pallasea* are phytophilic species inhabiting shallow waters.

Many other examples could be given, such as that of amphipods with a variety of lifestyles, but in which only one dorsal row of eminences is formed as flat keels, cut off in front and pointed behind. Among the Baikalian forms this is seen in *Carinogammarus wagii* (Dyb., 1874) (Bazikalova, 1945; Figure 2) and among marine forms in representatives of the families Atylidae, Epimeriidae (= Paramphithoidae), Gammarellidae (= Calliopiidae), Eusiridae, Gammaracanthidae, etc. (Lincoln, 1979; Bousfield, 1989; Sieg and Wegele, 1990; Bousfield and Kendall, 1994).

These examples of similarity may be considered as demonstrations of nomogenesis (directed development) in the evolution of amphipods. It is hoped that detailed karyological, genetic and physiological research will to clarify the taxomony of the Baikalian amphipods and help to reveal the evolutionary mechanisms, and perhaps even allow an understanding of the parallelisms seen between this and other faunas.

ACKNOWLEDGEMENTS

The author thanks all colleagues from whose discussions and correspondence the ideas of this article were born: Prof. Ya.I. Starobogatov, Drs O.A. Timoshkin, I.V. Mekhanikova, V.R. Alekseev, V.N. Ryabova, E.L. Bousfield and H.-G. Andres, Prof. I. Zerbst-Boroffka, employees of Laboratory of the Marine Research of the Zoological Institute RAS (St Petersburg) and the Chair of Invertebrate Zoology, Irkutsk State University. Thanks also to Dr A. Rossiter for his kindness and for improving the style and content of this article. This work was supported by the Russian Foundation for Basic Research (RFFI), grant 96-04-49766.

REFERENCES

Bazikalova, A.J. (1940). Caspian elements in the fauna of Baikal. *Trudy Baikal Limnol. Inst. SO AN SSSR* **10**, 357–367 (in Russian with French summary).

Bazikalova, A.J. (1941). Materials on the study of the amphipods of Baikal. II. Reproduction. *Izvestya Acad. Sci. USSR, Biol.* **3**, 407–426 (in Russian with French summary).

Bazikalova, A.J. (1945). The amphipods of Lake Baikal. *Trudy Baikal Limnol. Inst. SO AN SSSR* **11**, 1–440 (in Russian with French summary).

Bazikalova, A.J. (1948a). Adaptive importance of the sizes of Baikalian amphipods. *Doklady Akad. Nauk SSSR* **61**, 569–572 (in Russian).

Bazikalova, A.J. (1948b). Notes on amphipods of Baikal. *Trudy Baikal Limnol. Inst. SO AN SSSR* **12**, 20–32 (in Russian).

Bazikalova, A.J. (1951). On the growth of some amphipods from Baikal and Angara. *Trudy Baikal Limnol. Inst. SO AN SSSR* **13**, 206–216 (in Russian).

Bazikalova, A.J. (1954). Some data on the biology of *Acanthogammarus* (*Brachyuropus*) *grewingki* (Dyb.). *Trudy Baikal Limnol. Inst. SO AN SSSR* **14**, 312–326 (in Russian).

Bazikalova, A.J. (1962). The systematics, ecology and distribution of the genera *Micruropus* Stebbing and *Pseudomicruropus* nov. gen. (Amphipoda, Gammaridae). *Trudy Limnol. Inst. SO AN SSSR* **22**, 3–140 (in Russian).

Bazikalova, A.J. (1971). Benthic fauna. In: *Limnologiya prideltovykh prostranstv Baikala. Selenginskiy rayon*, pp. 95–114. Nauka, Leningrad (in Russian).

Bazikalova, A.J. (1975). On the systematics of Baikal amphipods (the genera *Carinogammarus* Stebbing, *Eucarinogammarus* Sowinsky, *Echiuropus* (Sow.) and *Asprogammarus* gen. n.). In: *Novoye o faune Baikala*, Part 1, pp. 31–81. Nauka, Novosibirsk (in Russian).

Bazikalova, A.J., Birstein, Ya. A. and Taliev, D.N. (1946a). Osmotic pressure of the cavity liquid of sideswimmers of Lake Baikal. *Doklady Akad. Nauk SSSR* **53**, 293–295 (in Russian).

Bazikalova, A.J., Birstein, Ya.A. and Taliev, D.N. (1946b). The osmoregulatory abilities of sideswimmers from Lake Baikal. *Doklady Akad. Nauk SSSR* **53**, 381–384 (in Russian).

Bekman, M.Y. (1958). Dwarf males in Baikal endemics. *Doklady Akad. Nauk SSSR* **120**, 208–211 (in Russian).

Bekman, M.Y. (1959). Some patterns in the distribution and productivity of mass species of the zoobenthos in the Maloye More Strait. *Trudy Baikal Limnol. Inst. SO AN SSSR* **17**, 342–381 (in Russian).

Bekman, M.Y. (1962). Ecology and production of *Micruropus possolskii* Sow. and *Gmelinoides fasciatus* Stebb. *Trudy Limnol. Inst. SO AN SSSR* **22**, 141–155 (in Russian).

Bekman, M.Y. (1983a). Amphipods. In: *Ecology of Southern Baikal* (Ed. by G.I. Galuzii), pp. 128–143. Nauka, Irkutsk (in Russian).

Bekman, M.Y. (1983b). Benthos of sites near river mouths. In: *Limnologiya Severnogo Baikala* (Ed. by G.I. Galuzii and V.T. Bogdanov), pp. 103–108. Nauka, Novosibirsk (in Russian).

Bekman, M.Y. (1984). The deep-living amphipod fauna. In: *Sistematika i evolutsiya bespozvonochnykh Baikala* (Ed. by A.A. Linevich), pp. 114–123. Nauka, Novosibirsk (in Russian).

Bekman, M.Y. and Bazikalova, A.J. (1951). Biology and production potentials of some Baikalian and Siberian sideswimmers. In: *Problemy gidrobiologii vnutrennikh vod*, Part 1, pp. 61–67. Nauka, Leningrad (in Russian).

Belyajev, G.M. (1989). *The Deep-water Oceanic Troughs and Their Fauna.* Nauka, Moscow (in Russian).

Berg, L.S. (1922). *Nomogenesis, or Evolution Determined by Laws.* State Publishing House, St Petersburg (in Russian).

Berg, L.S. (1969). *Nomogenesis, or Evolution Determined by Laws.* MIT Press, Massachusetts.

Bessolitsyna, I.A. (1999). The change of structure of gammarid migratory complexes under the influence of anthropogenic factors. In: *Problemy ekologii i prirodopolizovaniya Baikalískogo regiona* (Ed. by S.B. Kuzmin), pp. 36–39. Irkutsk State Technical University Publishers, Irkutsk (in Russian).

Birstein, Y.A. and Romanova, N.N. (1968). Order sideswimmers: Amphipoda. In: *Atlas bespozvonochnykh Kaspiyskogo morya,* pp. 241–289. Pischevaya Promyshlennost Publishers, Moscow (in Russian).

Bousfield, E.L. (1977). A new look at the systematics of gammaroidean amphipods of the world. *Crustaceana* **4**, Suppl., 282–316.

Bousfield, E.L. (1989). Revised morphological relationships within the amphipod genera *Pontoporeia* and *Gammaracanthus* and the "glacial relict" significance of their postglacial distributions. *Can. J. Fish. Aquat. Sci.* **46**, 1714–1725.

Bousfield, E.L. and Kendall, J.A. (1994). The amphipod superfamily Dexaminoidea on the North American Pacific coast; families Atylidae and Dexaminidae: systematics and distributional ecology. *Amphipacifica* **3**, 3–66.

Brauer, R.W., Bekman, M. Yu., Keyser, D.L., Nesbitt, S.G., Sidelew, G.N. and Wright, S.L. (1980a). Adaptation to high hydrostatic pressures of abyssal gammarids from Lake Baikal in eastern Siberia. *Comp. Biochem. Physiol.* **65A**, 109–128.

Brauer, R.W., Keyser, D., Nesbitt, D.L., Wright, S.L., Bekman, M.Y. and Sidelyov, G.N. (1980b). Reaction of gammarids from Baikal to hydrostatic pressure. *J. Evol. Biochem. Physiol.* **16**, 545–550 (in Russian with English summary).

Chapelle, G. and Peck, L.S. (1999). Polar gigantism dictated by oxygen availability. *Nature* **399**, 114–115.

Gavrilov, G.B. (1949). On the problem of the time of reproduction in amphipods and isopods in Lake Baikal. *Doklady Akad. Nauk SSSR* **64**, 739–742 (in Russian).

Ginetsinskaya, T.A. and Dobrovol'skiy, A.A. (1978). *Special Parasitology. Parasitic Worms, Molluscs and Arthropods.* Vysshaya Schkola Publishers, Moscow (in Russian).

Grese, V.N. (1957). General features of the hydrobiology of Taymyr Lake. *Trudy Vsesoyusnogo Gidrobiol. Obshestva* **8**, 183–218 (in Russian).

Kamaltynov, R.M., Chernykh, V.I., Slugina, Z.V. and Karabanov, E.B. (1993). The consortium of the sponge *Lubomirskia baicalensis* in Lake Baikal, East Siberia. *Hydrobiologia* **271**, 179–189.

Kozhov, M.M. (1972). *Characteristic Features of the Natural History of Baikal.* Vostochno-Sibirskoye knizhn. Izd-voi Publishers, Irkutsk (in Russian).

Lincoln, R.J. (1979). *British Marine Amphipoda: Gammaridea.* British Museum (Natural History), London.

Martens, K. (1997). Speciation in ancient lakes. *Trends Ecol. Evol.* **12**, 177–182.

Mashiko, K., Kamaltynov, R.M., Sherbakov, D.Y. and Morino H. (1997a). Genetic separation of gammarid (*Eulimnogammarus cyaneus*) populations by localized topographic changes in ancient Lake Baikal. *Arch. Hydrobiol.* **139**, 379–387.

Mashiko, K., Kamaltynov, R.M., Sherbakov, D.Y. and Morino H. (1997b). Speciation of gammarids in ancient Lake Baikal. In: *Animal Community, Environment and Phylogeny in Lake Baikal* (Ed. by N. Miyazaki), pp. 51–56. Otsuchi Marine Research Center, Ocean Research Institute, University of Tokyo, Tokyo.

Mekhanikova, I.V. and Takhteev, V.V. (1991). The amphipod fauna of Northern Baikal. In: *Morfologiya i evolutsiya bespozvonochnykh* (Ed. by A.A. Linevich and E.L. Afanasyeva), pp. 199–210. Nauka, Novosibirsk (in Russian).

Melnik, N.G., Timoshkin, O.A., Sideleva, V.G., Pushkin, S.V. and Mamylov, V.S. (1993). Hydroaccoustic measurement of the density of the Baikal macrozooplankter *Macrohectopus branickii*. *Limnol. Oceanogr.* **38**, 425–434.

Melnik, N.G., Timoshkin, O.A. and Sideleva, V.G. (1995). Distribution of *M. branickii* and some characteristics of its ecology. In: *Guide and Key to Pelagic Animals of Baikal (with Ecological Notes)* (Ed. by O.A. Timoshkin), pp. 511–522. Nauka, Novosibirsk (in Russian).

Meyen, S.V. (1978). The basic aspects of the typology of organisms. *J. Gen. Biol.* **39**, 495–508 (in Russian with English summary).

Müller, H.-G. (1989). Asseln als Parasiten mariner Krebse und Fische. *Microcosmos* **78**, 42–45.

Nikolayeva, E.P. (1964). Materials on the feeding of a pelagic Baikalian sideswimmer. In: *Sbornik kratkikh soobscheniy i dokladov o nauchnoy rabote po biologii I pochvovedeniyu (Irkutskiy Universitet)*, pp. 31–35. Vostochno-Sibirskoye knizhn. Izd-voi Publishers, Irkutsk (in Russian).

Nikolayeva, E.P. (1967). Some data on the reproduction biology of pelagic Baikalian sideswimmer *Macrohectopus branickii* (Dyb.). *Izvestiya Biologo-geograficheskogo nauchno-issledovatelískogo instituta pri Irkutskom gosudarstvennom universitete* **20**, 28–33 (in Russian).

Ogarkov, O.B., Kamaltynov, R.M., Belikov, S.I. and Sherbakov, D.Y. (1997). The analysis of phylogenetical relations of Baikalian endemic amphipods (Crustacea, Amphipoda) on the basis of comparison of nuclear sequences of a mitochondrial gene CO III fragment. *Molek. Biol.* **31**, 32–37 (in Russian).

Osadchikh, V.F. (1977). Finding of the crustacean *Cardiophilus baeri* in a marsupium of corophiids (Amphipoda, Gammaridae). *Zool. Zh.* **56**, 156–158 (in Russian with English summary).

Popova, S.M., Mats, V.D. and Chernyaeva, G.P. (1989). *Paleolimnological Reconstructions: The Baikal Rift Zone* (Ed. by N.A. Logachev). Nauka, Novosibirsk (in Russian).

Rudstam, L.G., Melnik, N.G., Timoshkin, O.A., Hansson, S., Pushkin, S.V. and Nemov, V. (1992). Diel dynamics of an aggregation of *Macrohectopus branickii* (Dyb.) (Amphipoda, Gammaridae) in the Barguzin Bay, Lake Baikal, Russia. *J. Great Lakes Res.* **18**, 286–297.

Salemaa, H. and Kamaltynov, R. (1994). Chromosomal relationships of the endemic Amphipoda (Crustacea) in the ancient lakes Ohrid and Baikal. In: *Genetics and Evolution of Aquatic Organisms* (Ed. by A.R. Beaumont), pp. 405–414. Chapman and Hall, London.

Severtsov, A.S. (1990). Intraspecific variety as a cause of evolutionary stability. *J. Gen. Biol.* **51**, 579–589 (in Russian with English summary).

Shimaraev, M.N., Verbolov, V.I., Granin, N.G. and Sherstyankin, P.P. (1994). *Physical Limnology of Lake Baikal: A Review*. Baikal International Center for Ecological Research, Irkutsk.

Sideleva, V.G. and Mechanikova, I.V. (1990). Feeding preference and evolution of the cottoid fishes of Lake Baikal. *Proc. Zool. Inst. Leningrad* **222**, 144–161 (in Russian with English summary).

Sieg, J. and Wegele J.W. (eds) (1990). *Fauna der Antarktis*. Paul Parey Verlag, Berlin.

Tachteev, V.V. (1995). On the ecology of a rare species of Baikal amphipods *Polyacanthisca calceolata* in the context of the problem of parallel geneses of the abyssal Baikal and oceanic faunae. *Zool. Zh.* **74**, 141–143 (in Russian with English summary).

Tachteew, V.V. (1995). The gammarid genus *Poekilogammarus* Stebbing, 1899, in Lake Baikal, Siberia (Crustacea Amphipoda Gammaridea). *Arthropoda Selecta* **4**, 7–64.

Tachteew, V.V. and Mekhanikova, I.V. (1993). Some data on life history of the Baikal parasitic gammarids of the genus *Pachyschesis* (Amphipoda, Gammaridae). *Zool. Zh.* **72**, 18–28 (in Russian with English summary).

Takhteev, V.V. (1999). Revision of the genus *Odontogammarus* (Crustacea, Amphipoda, Gammaridae) from Lake Baikal. *Zool. Zh.* **78**, 796–810 (in Russian with English summary).

Takhteev, V.V. (2000). On the revision of a genus *Poekilogammarus* Stebbing, 1899. (Crustacea, Amphipoda, Gammaridae) from Lake Baikal. *Zool. Zh.* **79**, in press (in Russian with English summary).

Takhteev, V.V. (1997). The gammarid genus *Plesiogammarus* Stebbing, 1899, in Lake Baikal, Siberia (Crustacea Amphipoda Gammaridea). *Arthropoda Selecta* **6**, 31–54.

Takhteev, V.V. and Bessolitsyna, I.A. (1999). Some scientific problems of baikalogia and their importance for ecological education. In: *Ecology, Education, Health*, Part 1, pp. 20–26. Irkutsk State University Publishers, Irkutsk (in Russian).

Takhteev, V.V. and Mekhanikova, I.V. (1996). Distribution of the endemic nectobenthic amphipods in the Lake Baikal. *Bull. Moscow Soc. Nat. Biol. Ser.* **101**, 39–48 (in Russian with English summary).

Timofeev, M.A., Stom, D.I., Gil, T.A. and Naumova, E.Y. (1997). Ascertainment of the temperature preference of amphipods and their reaction to high temperatures. *Chelovek. Sreda. Vselennaya* **1**, 75–77. Irkutsk State Technical University Publishers, Irkutsk (in Russian).

Timoshkin, O.A., Mekhanikova, I.V. and Shubenkov, S.G. (1995). Morphological peculiarities of *M. branickii*. In: *Guide and Key to Pelagic Animals of Baikal (with Ecological Notes)* (Ed. by O.A. Timoshkin), pp. 485–511. Nauka, Novosibirsk (in Russian).

Urmantsev, Y.A. (1988). The general theory of systems: a condition, application and prospects of the development. In: *Systema. Simmetriya. Garmoniya*, pp. 38–130. Mysl Publishers, Moscow (in Russian).

Vereschagin, G.J. (1940). The origin and history of Baikal and its flora and fauna. *Trudy Baikal Limnol. Inst. SO AN SSSR* **10**, 73–239 (in Russian with French summary).

Vershinin, N.V. (1960). On the question of the genesis of relict fauna in the Norilsk lakes group. *Doklady Akad. Nauk SSSR* **64**, 739–742 (in Russian).

Vershinin, N.V., Sychyova, A.V. and Syrygina, F.F. (1967). On the invertebrate fauna of Khantayskoye Lake. *Trudy Krasnoyargskogo otdeleniya Sibirskogo NII Rybnogo Khosyajstva* **9**, 214–230 (in Russian).

Vilisova, I.K. (1962). On the ecology of a Baikalian pelagic sideswimmer *Macrohectopus branickii* (Dyb.). *Trudy Limnol. Inst. SO AN SSSR* **22**, 156–171 (in Russian).

Vinogradov, G.M. (1988). System of life forms of pelagic sideswimmers. *Doklady Akad. Nauk SSSR* **298**, 1509–1512 (in Russian).

Vinogradov, G.M. (1992). The probable ways of the gammarid (Amphipoda: Crustacea) invasion in the pelagic zone: analysis of the life forms. *J. Gen. Biol.* **53**, 328–339 (in Russian with English summary).

Weinberg, I.V. and Kamaltynov, R.M. (1998a). Zoobenthos communities at a stony beach of Lake Baikal. 1. *Zool. Zh.* **77**, 158–165 (in Russian with English summary).

Weinberg, I.V. and Kamaltynov, R.M. (1998b). Zoobenthos communities at a stony beach of Lake Baikal. 2. *Zool. Zh.* **77,** 259–265 (in Russian with English summary).

Zerbst-Boroffka, I., Grospietsch, T., Mekhanikova, I. and Knopse, M. (1998). Osmotic and ionic adjustment in abyssal gammarids after acclimation to different water depths in Lake Baikal. *Zoology* **101,** 71.

Zerbst-Boroffka, I., Grospietsch, T. and Mekhanikova, I. (1999). Osmotische und ionale Anpassung von abyssalen Amphipoden an verschiedene Wassertiefen im Baikal. In: *Crustaceologen Tagung*, p. 58. Humbolt Universität Publishers, Berlin.

Insights into the Mechanism of Speciation in Gammarid Crustaceans of Lake Baikal using a Population-genetic Approach

K. MASHIKO

I. SUMMARY

Through use of a schematic multilocus gene model, it is postulated that for reproductive isolation to occur between populations independent of ecological (resource utilization) or morphological divergence of organisms, the physical (geographic) isolation of conspecific individuals is a prerequisite. A key factor affecting the species diversity in ancient lakes is thus intralacustrine allopatric speciation. In Lake Baikal, neighbouring populations of the endemic gammarid amphipod crustacean *Eulimnogammarus cyaneus* were genetically differentiated at the outlet of the major effluent, the Angara River. Speciation through microgeographical environmental changes which have not catastrophically impacted upon the whole biotic community is of great significance. Such types of mechanism are likely to have substantially enriched the abundance of species in certain groups of organisms in ancient lakes. A high genetic variation (allozymic average heterozygosity, H) was detected within populations of some species of Baikalian gammarids. This variation could be an intrinsic factor acting to accelerate speciation of this group of animals in this lake.

II. INTRODUCTION

Lake Baikal, located in East Siberia, is the deepest and oldest lake in the world, and contains a unique and peculiar biota (Kozhov, 1963; Martin, 1994;

ADVANCES IN ECOLOGICAL RESEARCH VOL. 31
ISBN 0-12-013931-6

222   K. MASHIKO

Kozhova and Izmest'eva, 1998). The most diverse group of metazoan animals in Baikal is the gammarid amphipod crustaceans (Figure 1). Around 260 species and 80 subspecies, allocated to 45 genera mostly within two major families, the Acanthogammaridae and the Gammaridae, are known from the lake (Kamaltynov, 1992). Almost all of these taxa (98%) are endemic to Lake Baikal, and together comprise c. 20% of all freshwater gammarids hitherto recorded world-wide (Barnard and Barnard, 1983).

One of the diversifications seen in this fauna is in morphology, especially in the well-developed cuticular ornamentation with processes, which is character-istic to Baikalian gammarids (Barnard and Barnard, 1983). Recent studies of molecular phylogeny (Sherbakov et al., 1998) suggest strongly that these ornamentations have arisen independently in several taxa. As a group, the amphipods of Lake Baikal also show high ecological diversity, inhabiting a wide range of biotopes on the lake bottom, from the surf zone to depths beyond 400 m. They utilize a variety of food resources, such as encrusting algae, leaf litter, plant material, sponges, detritus and, occasionally, decayed fish corpses (Kozhov, 1963; Kozhova and Izmest'eva, 1998; Morino et al., this

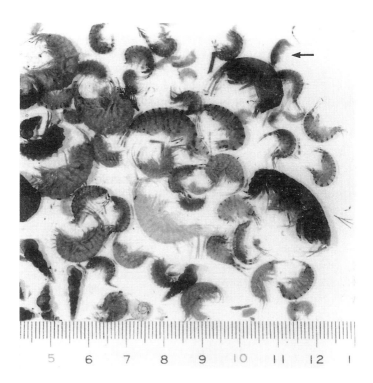

Fig. 1. Gammarid crustaceans in littoral zones of Lake Baikal. The arrow indicates *E. cyaneus*, a species which is discussed in the later part of this chapter.

volume; Tachteev, this volume). Only one species, *Macrohectopus branickii*, is pelagic, and shows a large-scale nocturnal vertical migration (Melnik *et al.*, 1993).

Salemaa and Kamaltynov (1994) examined 33 species and two subspecies within 18 endemic genera of Baikal amphipod, and found all to have the same genome size, $2n = 52$, a standard genome size of the non-endemic widespread genus *Gammarus*. This observation, together with recent molecular phylogenetic studies (Ogarkov *et al.*, 1997; Sherbakov *et al.*, 1998) supports the view that, despite spectacular morphological and ecological divergence, gammarids in Lake Baikal have descended from a few common ancestral immigrants and have diversified in this water body of geologically long duration (Bazikalova, 1945). However, the precise mechanism of their diversification is still controversial (Brooks, 1950; Kozhov, 1963; Martens *et al.*, 1994; Tachteev and Morino *et al.*, this volume).

This chapter begins by examining potential speciation mechanisms with a schematic model and considering how these might apply to the Baikal amphipods. Then, the significance of intralacustrine speciation in the high species richness of amphipod crustaceans of Lake Baikal is discussed.

III. INSIGHTS INTO SPECIES DIVERSITY AND SPECIATION

Ancient lakes are famed for their usually high biodiversity and the coexistence of many species within diverse communities. From a community perspective, biological diversity is usually addressed from two main themes: ecological divergence and species abundance or richness. Ecological divergence has resulted from adaptation to various biotic and abiotic environments, in which natural selection has played an important role in the context of resource utilization (e.g. Greenwood, 1984) and prey–predator relationships (e.g. Neill, 1992). More recently, however, special attention has been paid to the question of species abundance. It was MacArthur and Wilson (1967) who first tried to explain the abundance of species in an area (island) by two ecological processes, the increase and decrease of species by immigration and extinction, respectively. However, more recent studies have concluded that the factor that ultimately acts to increase the number of species is not immigration, but speciation (Cracraft, 1985). For ancient lakes, with many endemic species, this perspective is particularly important.

A. Reproductive Isolation and Speciation

According to the modern biological species concept (Mayr, 1957, 1963, 1992), speciation is the process whereby reproductive isolation is established in a group of panmictic individuals. Reproductive isolation in populations is most effectively achieved through changes in the premating system, especially through change in the specific mate-recognition systems (Paterson, 1980, 1981; Kaneshiro, 1980; Sved, 1981; Ritchie and Gleason, 1995; Buckley *et al.*, 1997). In contrast, unless there further emerges some special mechanism to prevent interbreeding between populations, postmating reproductive isolation (through gametic incompatibility, infertility and inviability of hybrids, hybrid break-down, etc.) in itself does not result in populations speciating. This is because when two contiguous populations are reproductively isolated only by postmating mechanisms, one of them (usually the smaller one) will eventually go extinct as a result of unsustainable wastage of gametes (Paterson, 1978; Chandler and Gromko, 1985). To overcome this theoretical difficulty, Dobzhansky (1937) first proposed the idea of reinforcement, whereby if two divergent populations produce hybrids of low fitness where they come into contact, natural selection will favour an increase in assortative mating, and thus progress towards speciation (Butlin, 1995). However, whether reinforce-ment actually operates in nature is controversial (Paterson, 1978; Butlin, 1989; Coyne, 1992).

Mate recognition in animals occurs mainly through visual, acoustic, chemical and other species-specific signal clues (e.g. Robertson and Paterson, 1982; Henry, 1994; Buckley *et al.*, 1997). In gammarids, water-borne pheromones serve as a key signal in all stages of mating activities: location, pairing induction, exact pairing and copulation (Dunham, 1978; Borowsky and Borowsky, 1987; Borowsky, 1991). Both theoretical (Sved, 1981; Nei, 1987; Coyne, 1992) and empirical (Prakash, 1972; Kawanishi and Watanabe, 1981; Henry, 1994) studies indicate that animal mating systems are controlled by a set of (usually additive in function) multilocus genes (i.e. polygene) in which each gene does not play an individually decisive role.

B. Changes in Mate Recognition: A Genetic Model

The origin of premating reproductive isolation through a change in the mate-recognition system can be readily understood when interpreted using multi-locus genetic changes (Mashiko, in press). Imagine one Mendelian population in which mate recognition is controlled by three gene loci, *A, B* and *C* (Figure 2). Initially, the genotype of all individuals was [*aabbcc*] (stage 1). This population was then physically (geographically) separated into two popula-tions (stage 2). In one of these populations, mutation takes place in an individual at locus A, to give genotype [*axbbcc*]. Owing to its additive polygenic nature (Sved, 1981), this mutation does not severely degrade the mating activity of this individual: supporting evidence is provided by studies of lepidoptera, where sex pheromone-mediated recognition was maintained despite experimentally manipulated high variations in pheromone components (e.g. Buckley *et al.*, 1997). As the gene locus concerned is due to the nature of the signal system *per se*, adaptively neutral to environmental conditions, no

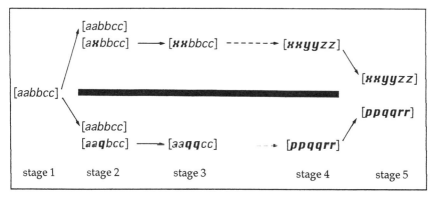

Fig. 2. A schema indicating the rise of premating reproductive isolation in which three additive gene loci are involved. A bar indicates a physical barrier which is acting to separate populations.

reduction in survivorship of [axbcc] occurs, and the mutant is able to mate with surrounding individuals [aabbcc]. In due course, through a stochastic process of random genetic drift the mutation gene *x* is fixed with some probability (Buri, 1956) and all individuals therefore eventually change into genotype [xxbcc] (stage 3). The smaller the population, the more rapidly this stochastic genetic change takes place (Kimura, 1983). Repeating this process at all three loci, the genotype of all individuals finally becomes [xxyyzz]. Similarly, all individuals in another population change to genotype [ppqqrr] (stage 4). When the two populations become sympatric at that stage (stage 5), their entirely different genotypes mean that they no longer interbreed, i.e. reproductive isolation (speciation) is accomplished. Nei *et al.* (1983) used a similar line of argument to formulate a mathematical model explaining the formation of reproductive isolation.

C. Physical Separation: Allopatric Speciation

Genetic differentiation of populations by random genetic drift is readily contradicted by gene flow (migration of individuals). It is predicted that, independent of population size, the migration between populations of even one individual per generation is enough to prevent genetic differentiation (Slatkin, 1985). As such, physical (geographical) separation of conspecific individuals has long been acknowledged as a key factor to initiate speciation (e.g. Mayr, 1959, 1963, 1992; Prakash, 1972; Ayala, 1975; Futuyma and Mayer, 1980). Moreover, the above scheme predicts that reproductive isolation occurs only with physical separation of populations, and irrespective of any morphological and ecological changes of organisms, i.e. that reproductive isolation arises as incidental genetic changes (e.g. Lessios and Cunningham, 1990). In *Drosophila*, the degree of reproductive isolation is correlated with allozymically detected genetic distances (Coyne and Orr, 1989). This finding supports the above line of reasoning, since allozymic protein polymorphisms are mostly, if not entirely, attributable to stochastic genetic changes which occur independently of environmental adaptations (Mukai, 1978; Kimura, 1983; Nei, 1987).

Ecological divergence of resource utilization (niche shift) in organisms has long been thought indispensable for, or inseparably combined with, speciation (Mayr, 1963) and this scenario is one template for sympatric speciation (e.g. Schliewen et al., 1994). In fact, repeated speciation accompanied by large-scale ecological divergence can result in spectacular adaptive radiation, as is seen in *Drosophila* of the Hawaiian Islands (Zimmerman, 1958), the cichlid fishes of the African Great Lakes (Fryer and Iles, 1972) and the amphipods of Lake Baikal addressed in this chapter. From the perspective of genetic mechanics, however, the above thinking appears inadequate because genetic linkage (which may integrate ecological divergence and reproductive isolation inot a single supergene) is rarely seen (Mukai, 1978). Under the conditions of chromosomal and genetic recombination which occur every generation, an individual's genes to control matings cannot be invariably combined with genes to control that individual's resource-utilization characteristic (Slatkin, 1987).

The irrelevance of reproductive isolation to ecological divergence could be easily examined through reference to sibling (or cryptic) species, which often occur in the same habitat and show no morphological, ecological or life-historical difference (e.g. Marsh et al., 1981). In Lake Malawi, for instance, species flocks of cichlid fishes which differ markedly in specific mate-recognition traits (body colours and patterns; see Turner and Seehausen, this volume) coexist in the same biotope and utilize identical food material (Ribbink et al., 1983), often without any mutually exclusive interaction (Marsh et al., 1981). Thus, ecological divergence cannot be a prerequisite of speciation, even though it may result from speciation.

Speciation is principally a fission of the system whereby genes are transmitted over generations. It is therefore important to differentiate between the species as a unit of reproduction (deme) and as a unit of economy in nature (ecological niche) (Eldredge and Cracraft, 1980). It is also important to realize that morphological divergence in itself does not always result in reproductive isolation: in closely related species, the degree of morphological difference is correlated with neither reproductive isolation (Rubinoff and Rubinoff, 1971; Turner, 1974) nor molecular genetic changes (Scheepmaker, 1987; Turner, 1974).

IV. CONSEQUENCES OF ALLOPATRIC SPECIATION THROUGH MICROGEOGRAPHICAL CHANGES

How might conspecific individuals be physically separated within a single lacustrine system? This is a fundamental problem when considering speciation in ancient lakes. In heterogeneous limnetic environments, many aquatic organisms are patchily distributed within fragmented habitats, and this observation has supported the idea of intralacustrine speciation by ecological segregation (Owen et al. 1990; Fryer, 1991). However, it remains questionable whether, in the absence of any extrinsic barriers, the gene flow among populations is sufficiently impeded to allow speciation to occur (Slatkin, 1987).

The basic idea that organisms speciate through geographic (spatial) isolation was applied to organisms in ancient lakes by Brooks (1950). Conspecific individuals would be completely isolated by the splitting of a whole lake basin into sub-basins owing to large-scale lake-level fluctuations caused by tectonic or climatic changes. Such large-scale fluctuations have been recognized for Lakes Malawi (Scholz and Rosendahl, 1988; Owen et al., 1990), Titicaca (Mourguiart, this volume), Baikal (Mats, 1993), Tanganyika (Scholz and Rosendahl, 1988; Tiercelin and Mondeguer, 1991) and Victoria (Johnson et al., 1996). However, it should be noted that such severe environmental changes can act to exterminate numerous species and thereby result in a decreased species abundance (Allomon, 1992; Sturmbauer and Meyer, 1992). Consequently, the problem is to assess what environmental changes induce intralacustrine speciation without impacting catastrophically upon the whole biotic community. Studies of the amphipod fauna of Lake Baikal shed some light on this question.

Eulimnogammarus (= *Phylolimnogammarus*) *cyaneus* is a gammarid crustacean endemic to Baikal, where it inhabits the stony shoreline of the littoral zone mainly at depths less than 15–20 cm (Weinberg and Kamaltynov, 1998). Within and among location(s) in the lake individuals vary markedly in body colour, from red–orange to dark blue, although they consist of a randomly mating population in each location (Mashiko et al., in press). Allozyme analyses of 11 local populations along the south-west coast of the lake (Figure 3) revealed that genetic composition significantly differed between the two groups of northern and southern populations divided by the Angara River outlet (Mashiko et al., 1997). The distinction was especially evident in the *ARK* (arginine kinase) locus (Figure 3): all northern populations had ARK^{82} at high frequencies, but this allele was not present in any of the southern populations. Thus, the gene flow between populations is completely obstructed by the river outlet. Yet, despite distinct genetic differences, no significant morphological differences have been recognized between these northern and southern groups of populations and, as noted by Tachteev (this volume), they are recognized as the same species.

The Angara River is currently the only drainage of Lake Baikal, but according to recent geological observations (Kononov and Mats, 1986; Ryazanov, 1993; Mats, 1993), Lake Baikal previously discharged through an ancient Kultuk–Irkut River system by a more southerly exit (see Figure 3) and, prior to that, from a northern opening of the pre-Manzurka River in the middle Pleistocene. The present Angara outlet first opened in the second half of

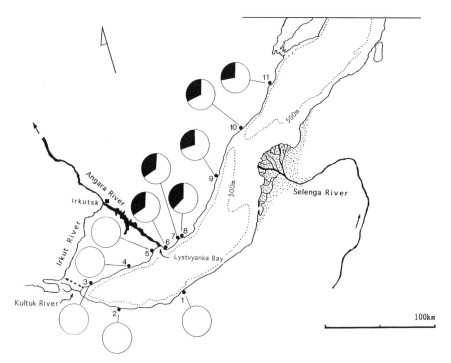

Fig. 3. Frequencies of ARK^{100} (white portion of circle) and ARK^{82} (black portion) in 11 populations of *E. cyaneus* in southern Baikal (Mashiko *et al.*, 1997). The dotted line at the southernmost part of the lake represents an inferred previous drainage (Kultuk–Irkut River system) from the lake.

the Late Pleistocene (approximately 20 000–80 000 years bp) as a result of extensive earth-crust movements and an overthrust fault around Lystvyanka.

Nei's genetic distance (Nei, 1987) between the two groups of northern and southern populations was 0.0126 (Figure 4). If one assumes one unit of Nei's genetic distance to be equivalent to 5 My (Nei, 1987), this value means that an absolute time of 60 000 years has elapsed since the two groups began to differentiate genetically. This result accords well with the geological evidence on the timing of formation of the Angara River drainage, the geological event which separated *E. cyaneus* into two groups.

What is maintaining allopatry and preventing these two groups from mixing? *Eulimnogammarus cyaneus* does not occur in the Angara River, but instead two closely related species (designated as subspecies of *E. cyaneus*) are found there (Bazikalova, 1957). *Eulimnogammarus cyaneus* is a shallow-water lentic species and it is likely that the rapid water flow of the Angara River (c. 1.1 m s^{-1} in average, estimated from hydrological data), combined with the depth at the outlet (4 m maximum), presents an impassable obstacle to this

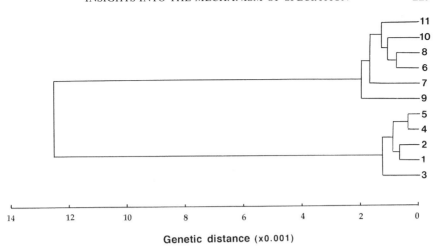

Fig. 4. An allozymic UPGMA dendrogram based on the Nei's genetic distance among 11 local populations, where 21 gene loci encoding 19 enzymes were investigated (Mashiko *et al.*, 1997). Numerals indicate the locality number shown in Figure 3.

species. Interestingly, Michel *et al.* (1992) observed that gastropods of the genus *Lavigeria* which occur along the shoreline of Lake Tanganyika are markedly differentiated between the two areas divided by a large inflow, the Ruzizi River. This is a phenomenon comparable to *E. cyaneus* in Lake Baikal.

Another feature that might restrict the potential for mixing of these populations is their reproductive strategy. In *E. cyaneus*, as in other gammarids, newborn young remain for a while within the brooding pouch of the female and then settle into benthic life as juveniles. This life history without a pelagic larval stage restricts the possibility of dispersal via the open waters of lake, and has further facilitated the local differentiation of populations in the presence of a physical barrier. Brooding of young has been suggested as promoting speciation in a variety of lacustrine taxa (Cohen and Johnson, 1987).

Such genetic fragmentation by microgeographic barriers is of particular significance when considering the mechanism of speciation in some ancient lake organisms. While regionally restricted topographic changes have a lesser impact on the whole biotic community, they can nevertheless induce speciation. With expected low extinction rates of species in ancient lakes under conditions of long geological stability, splitting of species in this fashion will steadily increase the number of indigenous species. When conspecific individuals are separated (or reproductively isolated) in this way at qualitatively different habitats within an area of lake basin, they will adaptively vary their life history traits as a secondary phase of species diversification, as observed, for instance, in atyid shrimps of Lake Tanganyika (Mashiko *et al.*, 1991). Sympatric speciation of gammarids in Lake Baikal has been interpreted in terms of adaptation to various depths (Kozhov, 1963), but it seems more likely that bathymetric divergence took place secondarily after the horizontal isolation of populations (and resultant speciation) in the lake (e.g. Mashiko *et al.*, in press).

V.   WITHIN-POPULATION GENETIC VARIATION

In a long-isolated biotope, high species richness in comparison to other biotopes is often attributed to either higher rates of speciation or lower rates of extinction, or both. It is likely that in ancient lakes of high environmental stability extinction rates of species are generally low, although this has not been satisfactorily demonstrated. The rate of speciation (i.e. rate of multiplication of species per unit time) is influenced by many biotic and abiotic factors (e.g. Cracraft, 1985). Among the more important of these is the genetic variation contained within a population: this determines the tempo of evolution, i.e. of both speciation and phyletic evolution (e.g. Fisher, 1958; Nei, 1987).

Can the high diversity of Baikal gammarids be attributed to a high genetic variation? An insight into this question is obtained by assessing whether the genetic variation contained within populations of gammarids in Lake Baikal is significantly higher than those in other regions.

Although data of use in addressing this problem are limited, the genetic variation within populations (average heterozygosity, H) of Baikal species and that of non-Baikal species are compared in Table 1. The *Gammarus* species listed in this table are found in rivers or marine coastal waters of Europe. *Gammarus lacustris* was collected from a small pond adjacent to Lake Baikal (Mashiko, unpublished). In the group of *Gammarus*, average heterozygosities of all marine species (0.08–0.10, except for one species with an extremely low value of 0.02) are greater than those of riverine species (0.02–0.07). Here, the equation $H = 4Nv/(4Nv + 1)$ holds between the average heterozygosity (H) and the effective population size (N): v represents mutation rate (Kimura, 1983). Since average mutation rates within a closely related taxonomic group are almost identical (Nei, 1987), the large average heterozygosity values shown by the marine species are explicable in terms of each species having a large effective population size in a vast and rather homogeneous habitat. For the same reasons, genetic variation in marine fishes is also higher than that seen in freshwater (riverine and lacustrine) taxa (Gyllensten, 1985).

Interestingly, the average heterozygosities in Baikal gammarids of the genera *Eulimnogammarus, Spinacanthus* and *Brandtia* (0.05–0.11, average = 0.08) are as high as those of marine *Gammarus* species. This may mean that effective population sizes in gammarids of Lake Baikal are as great as those in the sea; indeed, in terms of its vast size and volume, Baikal resembles more a sea than a lake.

Table 1

Comparison of the average heterozygosity (*H*) in populations of Baikalian and non-Baikalian gammarids. Only the results obtained for more than 18 gene loci are cited

Species	H	Habitat	Reference
Gammarus ibericus	0.03	River	1
Gammarus gauthieri	0.04	River	1
Gammarus pulex pulex	0.03	River	2
Gammarus pulex gallicus	0.05	River	3
Gammarus fossarum	0.05	River	3
Gammarus wautieri	0.02	River	3
Gammarus stupendus	0.07	River	3
Gammarus orinos	0.04	River	3
Gammarus lacustris	0.03	River	4
Mean	0.04		
Gammarus duebeni	0.08	Seashore	5
Gammarus zaddachi	0.10	Seashore	5
Gammarus salinus	0.10	Seashore	5
Gammarus oceanicus	0.09	Seashore	5
Gammarus locusta	0.02	Seashore	5
Mean	0.08		
Eulimogammarus cyaneus	0.08	Baikal	4, 6
Eulimogammarus vittatus	0.10	Baikal	4
Eulimnogammrus sp.	0.08	Baikal	4
Spinacanthus parasiticus	0.08	Baikal	7
Spinacanthus insualis	0.05	Baikal	7
Spinacanthus armatus	0.08	Baikal	7
Brandia lata	0.11	Baikal	7
Mean	0.08		

References: 1, Scheepmaker *et al.* (1988); 2, Scheepmaker and van Dalfsen (1989); 3, Scheepmaker (1990); 4, Mashiko, unpublished data; 5, Siegismund *et al.* (1985); 6, Mashiko *et al.* (1997); 7, Yampolsky *et al.* (1994).

Following conditions where populations have been spatially separated within the lake, such as the opening of the Angara River, it is probable that the large genetic variation in gammarids of Lake Baikal contributed to accelerate the tempo of speciation. Further investigations are underway to examine the role of intraspecific genetic variation in the speciation of gammarids in Lake Baikal.

REFERENCES

Allomon, W.D. (1992). A causal analysis of stages in allopatric speciation. *Oxford Surv. Evol. Biol.* **8,** 219–257.

Ayala, F. (1975). Genetic differentiation during the speciation process. *Evol. Biol.* **8,** 1–78.

Barnard, J.L. and Barnard, C.M. (1983). *Freshwater Amphipoda in the World. I. Evolutionary Patterns, and II. Handbook and Bibliography.* Hayfield Associates, Mt. Vernon, VA.

Bazikalova, A.Y. (1945). Amphipods of Lake Baikal. *Trudy Baikal Limnol. Inst. SO AN SSSR* **11,** 5–440 (in Russian).

Bazikalova, A.Y. (1957). The amphipods of the River Angara. *Trudy Baikal Limnol. Inst. SO AN SSSR* **15,** 377–387 (in Russian).

Borowsky, B. (1991). Patterns of reproduction of some amphipod crustaceans and insight into the nature of their stimuli. In: *Crustacean Sexual Biology* (Ed. by R.T. Bauer and J.W. Martin), pp. 33–49. Columbia University Press, New York.

Borowsky, B. and Borowsky, R. (1987). The reproductive behavior of the amphipod crustacean *Gammarus palustris* (Bousfield) and some insights into the nature of their stimuli. *J. Exp. Mar. Biol. Ecol.* **107,** 131–144.

Brooks, J.L. (1950). Speciation in ancient lakes. *Q. Rev. Biol.* **25,** 30–60, 131–176.

Buckley, S.H., Tregenza, T. and Butlin, R.K. (1997). Speciation and signal trait genetics. *Trends Ecol. Evol.* **12,** 299–301.

Buri, P. (1956). Gene frequency in small populations of mutant *Drosophila. Evolution* **10,** 367–402.

Butlin, R.K. (1989). Reinforcement of premating isolation. In: *Speciation and Its Consequences* (Ed. by D. Otte and J.A. Endler), pp. 158–179. Sinauer Associates, Sunderland, MA.

Butlin, R.K. (1995). Reinforcement: an idea evolving. *Trends Ecol. Evol.* **10,** 432–434.

Chandler, C.R. and Gromko, M.H. (1985). On the relationship between species concepts and speciation process. *Sys. Zool.* **38,** 116–125.

Cohen, A.S. and Johnston, M.R. (1987). Speciation in brooding and poorly dispersing lacustrine organisms. *Palaios* **2,** 426–435.

Coyne, J.A. (1992). Genetics and speciation. *Nature* **355,** 511–515.

Coyne, J.A. and Orr, H.A. (1989). Two rules of speciation. In: *Speciation and Its Consequences* (Ed. by D. Otte and J.A. Endler), pp. 180–207. Sinauer, Sunderland, MA.

Cracraft, J. (1985). Biological diversification and its cause. *Ann. Missouri Bot. Gardens* **72,** 794–822.

Dobzhansky, T. (1937). Speciation as a stage in evolutionary divergence. *Am. Nat.* **74,** 312–321.

Dunham, P.J. (1978). Sex pheromones in crustacea. *Biol. Rev.* **53,** 555–583.

Eldredge, N. and Cracraft, J. (1980). *Phylogenetic Patterns and the Evolutionary Process.* Columbia University Press, New York.

Fisher, R.A. (1958). *The Genetical Theory of Natural Selection.* Dover, New York.

Fryer, G. (1991). Comparative aspects of adaptive radiation and speciation in Lake Baikal and the great rift lakes of Africa. *Hydrobiologia* **211,** 137–146.

Fryer, G. and Iles, T.D. (1972). *The Cichlid Fishes of the Great Lakes of Africa. Their Biology and Evolution.* Oliver and Boyd, Edinburgh.

Futuyma, D.J. and Mayer, G.C. (1980). Non-allopatric speciation in animals. *Syst. Zool.* **29,** 254–271.

Greenwood, P.H. (1984). African cichlids and evolutionary theories. In: *Evolution of Fish Species Flocks* (Ed. by A.A. Echelle and I. Kornfield), pp. 141–154. University of Maine at Orono Press, Orono, ME.

Gyllensten, U. (1985). The genetic structure of fish: differences in the intraspecific distribution of biochemical genetic variation between marine, anadromous, and freshwater species. *J. Fish Biol.* **26**, 691–699.

Henry, C.S. (1994). Singing and cryptic speciation in insects. *Trends Ecol. Evol.* **9**, 388–392.

Johnson, T.C., Scholz, C.A., Talbot, M.R., Kelts, K., Ricketts, R.D., Ngobi, G., Beuning, K., Ssemmanda, I. and McGill, J.W. (1996). Late Pleistocene desiccation of Lake Victoria and rapid evolution of cichlid fishes. *Science* **273**, 1091–1093.

Kamaltynov, R.M. (1992). On the present state of systematics of the Lake Baikal amphipods (Crustacea, Amphipoda). *Zool. Zh.* **71**, 24–31 (in Russian).

Kaneshiro, K.Y. (1980). Sexual isolation, speciation and the direction of evolution. *Evolution* **34**, 437–444.

Kawanishi, M. and Watanabe, T.K. (1981). Genes affecting courtship song and mating preference in *Drosophila melanogaster, Drosophila simulans* and their hybrids. *Evolution* **35**, 1128–1133.

Kimura, M. (1983). *The Neutral Theory of Molecular Evolution.* Cambridge University Press, Cambridge.

Kononov, E.E. and Mats, V.D. (1986). History of Baikal water runoff. *Izv. Vyss. Ucheb. Zaved. Geol. Raved.* **1986**, 91–98 (in Russian).

Kozhov, M. (1963). *Lake Baikal and Its Life.* Junk, The Hague.

Kozhova, O.M. and Izmest'eva, L.R. (1998). *Lake Baikal. Evolution and Biodiversity.* Backhuys, Leiden.

Lessios, H.A. and Cunningham, C.W. (1990). Gametic incompatibility between species of the sea urchin *Echinometra* on the two sides of the Isthmus of Panama. *Evolution* **44**, 933–941.

MacArthur, R.M. and Wilson, E.O. (1967). *The Theory of Island Biogeography.* Princeton University Press, Princeton, NJ.

Marsh, A.C., Ribbink, A.J. and Marsh, B.A. (1981). Sibling species complex in sympatric populations of *Petrotilapia* Trewavas (Cichlidae, Lake Malawi). *Zool. J. Linn. Soc. Lond.* **71**, 253–264.

Martens, K., Coulter, G. and Goddeeris, B. (1994). Speciation in ancient lakes–40 years after Brooks. *Arch. Hydrobiol. Beiheft. Ergebnisse Limnol.* **44**, 75–96.

Martin, P. (1994). Lake Baikal. *Arch. Hydrobiol. Beiheft. Ergebnisse Limnol.* **44**, 3–11.

Mashiko, K. (in press). Researching the origin of species diversity in freshwater prawns. In: *Egg Sizes in Aquatic Organisms–Biology of Species Diversity and Speciation* (Ed. by A. Goto, and K. Eguchi). Kaiyusya, Tokyo (in Japanese).

Mashiko, K., Kawabata, S. and Okino, T. (1991). Reproductive and populational characteristics of a few caridean shrimps collected from Lake Tanganyika, East Africa. *Arch. Hydrobiol.* **122**, 69–78.

Mashiko, K., Kamaltynov, R., Sherbakov, D. and Morino, H. (1997). Genetic separation of gammarid (*Eulimnogammarus cyaneus*) populations by localized topographic changes in ancient Lake Baikal. *Arch. Hydrobiol.* **139**, 379–387.

Mashiko, K., Kamaltynov, R., Morino, H. and Sherbakov, D.Yu. (in press). Genetic differentiation among gammarid (*Eulimnogammarus cyaneus*) populations in Lake Baikal, East Siberia. *Arch. Hydrobiol.*

Mats, V.D. (1993). The structure and development of the Baikal rift depression. *Earth Sci. Rev.* **34**, 81–118.

Mayr, E. (1957). Species concepts and definitions. *Am. Assoc. Advance. Sci.* **50**, 1–22.

Mayr, E. (1959). Isolation as an evolutionary factor. *Proc. Am. Phil. Soc.* **103**, 221–230.

Mayr, E. (1963). *Animal Species and Evolution*. Harvard University Press, Cambridge, MA.

Mayr, E. (1992). Controversies in retrospect. *Oxford Surv. Evol. Biol.* **8**, 1–34.

Melnik, N.G., Timoshkin, O.A., Sideleva, V.G., Pushkin, S.V. and Mamylov, V.S. (1993). Hydroacoustic measurement of the density of the Baikal macrozooplankter *Macrohectopus branickii*. *Limnol. Oceanogr.* **38**, 425–434.

Michel, A.E., Cohen, A.S., West, K., Johnston, M.R. and Kat, P.W. (1992). Large African lakes as natural laboratories for evolution: examples from the endemic gastropod fauna of Lake Tanganyika. *Mitt. Int. Verein. Limnol.* **23**, 85–99.

Mukai, T. (1978). *Population Genetics*. Kodansha, Tokyo (in Japanese).

Nei, M. (1987). *Molecular Evolutionary Genetics*. Columbia University Press, New York.

Nei, M., Maruyama, T. and Wu, C.-I. (1983). Model of evolution of reproductive isolation. *Genetics* **103**, 553–579.

Neill, W. (1992). Population variation in the ontogeny of predator-induced vertical migration of copepods. *Nature* **356**, 54–57.

Ogarkov, O.B., Kamaltynov, R.M., Belikov, S.I. and Sherbakov, D.Yu. (1997). Phylogenetic relatedness of the Baikal Lake endeminal amphipods (Crustacea, Amphipoda) deduced from partial nucleotide sequences of the cytochrome oxidase subunit III genes. *Mol. Biol.* **31**, 32–37.

Owen, R.B., Crossley, R., Johnson, T.C., Tweddle, D., Kornfield, I., Davison, S., Eccles, D.H. and Engstrom, D.E. (1990). Major low levels of Lake Malawi and their implications for speciation rates in cichlid fishes. *Proc. R. Soc. Lond. B* **240**, 519–553.

Paterson, H.E.M. (1978). More evidence against speciation by reinforcement. *S. Afr. J. Sci.* **74**, 369–371.

Paterson, H.E.M. (1980). A comment on "mate recognition systems". *Evolution* **34**, 330–331.

Paterson, H.E.M. (1981). The continuing research for the unknown and unknowable: a critique of contemporary ideas on speciation. *S. Afr. J. Sci.* **77**, 113–119.

Prakash, S. (1972). Origin of reproductive isolation in the absence of apparent genetic differentiation in a geographic isolate of *Drosophila pseudoobscura*. *Genetics* **72**, 143–155.

Ribbink, A.J., Marsh, B.A., Marsh, A.C., Ribbink, A.C. and Sharp, B. J. (1983). A preliminary survey of the cichlid fishes of rocky habitats in Lake Malawi. *S. Afr. J. Zool.* **18**, 149–310.

Ritchie, M.G. and Gleason, J.M. (1995). Rapid evolution of courtship song pattern in *Drosophila willistoni* sibling species. *J. Evol. Biol.* **8**, 463–479.

Robertson, H.M. and Paterson, H.E.H. (1982). Mate recognition and mechanical isolation in *Enallagma* damselflies (Odanta:Coenagrionidae). *Evolution* **36**, 243–250.

Rubinoff, R.W. and Rubinoff, I. (1971). Geographic and reproductive isolation in Atlantic and Pacific populations of Panamanian *Bathygobius*. *Evolution* **25**, 88–97.

Ryazanov, G.V. (1993). *The Unique Localities Around Lake Baikal*. Nauka, Novosibirsk (in Russian).

Salemaa, L. and Kamaltynov, R. (1994). The chromosome numbers of endemic Amphipoda and Isopoda–an evolutionary paradox in the ancient lakes Ohrid and Baikal. *Arch. Hydrobiol. Beiheft. Ergebnisse Limnol.* **44**, 247–256.

Scheepmaker, M. (1987). Morphological and genetic differentiation of *Gammarus stupendus* Pinkster. *Bijdragen Dierkunde* **57**, 1–18.

Scheepmaker, M. (1990). Genetic differentiation and estimated levels of gene flow in members of the *Gammarus pulex*-group (Crustacea, Amphipoda) in western Europe. *Bijdragen Dierkunde* **60**, 3–30.

Scheepmaker, M. and Dalfsen, J. van (1989). Genetic differentiation in *Gammarus fossarum* and *G. capari* (Crustacea, Amphipoda) with reference to *G. pulex pulex* in northwestern Europe. *Bijdragen Dierkunde* **59**, 127–139.

Scheepmaker, M., Meer, F. van der and Pinkster, S. (1988). Genetic differentiation of the Iberian amphipods *Gammarus ibericus* Margalef, 1951 and *G. gautheiri* S. Karaman, 1935, with reference to some related species in France. *Bijdragen Dierkunde* **58**, 205–226.

Schliewen, U.K., Tautz, D. and Paabo, S. (1994). Sympatric speciation suggested by monophyly of crater lake cichlids. *Nature* **368**, 629–632.

Scholz, C.A. and Rosendahl, B.R. (1988). Low lake stands in Lakes Malawi and Tanganyika, East Africa, delineated with multifold seismic data. *Science* **240**, 1645–1648.

Sherbakov, D.Yu., Kamaltynov, R.M., Ogarkov, O.B. and Verheyen, E. (1998). Patterns of evolutionary change in Baikalian gammarids inferred from DNA sequences (Crustacea, Amphipoda). *Mol. Phylogenet. Evol.* **10**, 160–167.

Siegismund, H.R., Simonsen, V. and Kolding, S. (1985). Genetic studies of *Gammarus* I. Genetic differentiation of local populations. *Hereditas* **102**, 1–13.

Slatkin, M. (1985). Rare alleles as indicators of gene flow. *Evolution* **39**, 53–65.

Slatkin, M. (1987). Gene flow and the geographic structure of natural populations. *Science* **236**, 787–792.

Sturmbauer, C. and Meyer, A. (1992). Genetic divergence, speciation and morphological stasis in a lineage of African cichlid fishes. *Nature* **358**, 578–581.

Sved, J.A. (1981). A two-sex polygenic model for the evolution of premating isolation. I. Deterministic theory for natural populations. *Genetics* **97**, 197–215.

Tiercelin, J.-J. and Mondeguer, A. (1991). The geology of the Tanganyika trough. In: *Lake Tanganyika and Its Life* (Ed. by G.W. Coulter), pp. 7–47. Oxford University Press, London.

Turner, B.J. (1974). Genetic divergence of Death Valley pupfish species: biochemical versus morphological evidence. *Evolution* **28**, 281–294.

Weinberg, I.W. and Kamaltynov, R. (1998). Zoobenthos communities at stony beach of Lake Baikal. *Zool. Zh.* **77**, 158–165 (in Russian).

Yampolsky, L.Yu., Kamaltynov, R.M., Ebert, D., Filatov, D.A. and Chernykh, V.I. (1994). Variation of allozyme loci in endemic gammarids of Lake Baikal. *Biol. J. Linn. Soc.* **53**, 309–323.

Zimmerman, E.C. (1958). 300 species of *Drosophila* in Hawaii? – a challenge to geneticists and evolutionists. *Evolution* **12**, 557–558.

Explosive Speciation Rates and Unusual Species Richness in Haplochromine Cichlid Fishes: Effects of Sexual Selection

O. SEEHAUSEN

I. SUMMARY

Ancient lakes are often unusually species rich, mostly as a result of radiation and species-flock formation having taken place in only one or a few of many taxa present. Understanding why some taxa radiate and others do not is at the heart of understanding biodiversity. In this chapter I discuss possible explanations for disproportionally large species numbers in some cichlid fish lineages in East African Great Lakes: the haplochromine cichlid fishes in Lakes Victoria and Malawi. I show that speciation rates in this group are higher than in any other lacustrine fish radiation.

Against this background, I review hypotheses put forward to explain diversity in cichlid species flocks. The evolution of species diversity requires three processes: speciation, ecological radiation and anatomical diversification, and it is wrong to consider hypotheses that are relevant to different processes as alternatives to each other. The African cichlid species flocks show unusually high ecological species packing in several phylogenetic groups and unusually high speciation rates in haplochromines. Therefore, it maybe concluded that at least two evolutionary models are required to explain the difference between cichlid diversity and other fish diversity in East African Lakes: one for speciation in haplochromines and one for coexistence.

Subsequently I review work on speciation in haplochromines, and in particular studies aimed at testing the hypothesis of speciation by sexual selection. Haplochromines have a polygynous mating system, conducive to sexual selection, but other polygynous cichlids are not particularly species rich. This suggests that more than just strong sexual selection is required to explain haplochromine species richness. Recent palaeoecological evidence undermines the previously popular hypotheses that explained the species richness of Lake Victoria in terms of speciation under varying natural or sexual selection regimes in satellite lakes or in isolated lake basins. I summarize experimental and comparative studies, which provide evidence for two mechanisms of sympatric speciation by disruptive sexual selection on polymorphic coloration. Such modes of speciation may explain (i) the high speciation rates in colour polymorphic lineages of haplochromine cichlids under conditions where colour variation is visible in clear water, and (ii) in combination with factors that affect population survival, the unusual species richness in haplochromine species flocks.

I argue that sexual selection, if disruptive, can accelerate the pace of adaptive radiation because the resultant genetic population fragmentation allows a much increased rate of differential response to disruptive natural selection. Hence, the ecological pattern of diversity resembles that produced by disruptive natural selection, with the difference that disruptive sexual selection continues to cause (gross) speciation even after niche space is saturated. This may explain the unusually high numbers of very closely related and ecologically similar species in haplochromine species flocks. The role of disruptive sexual selection is twofold: it not only causes speciation, but also maintains reproductive isolation in sympatry between species that have evolved in sympatry or allopatry. Therefore, the maintenance of diversity in species flocks that originated through sexual selection depends on the persistence of the selection regime within the environmental signal space under which that diversity evolved.

II. INTRODUCTION

Ancient lakes are known for their unusually high species diversity, which is usually caused by adaptive radiation and formation of species flocks in only one or a few out of many taxa that inhabit a lake (Brooks, 1950; Martens *et al.*, 1994). This phenomenon is known from groups as distantly related as fishes, crustaceans, molluscs, turbellarians and sponges. Within each group only a handful of taxa has formed species flocks, and some have done so repeatedly, for example, African cichlid (e.g. Snoeks; Turner; Kuusipalo, this volume), cyprinodontiform (e.g. Northcote, this volume) and cyprinid fishes; amphipod (e.g. Takhteev; Mashiko; Morino *et al.*, this volume), ostracod and copepod crustaceans (e.g. Boxshall and Jaume; Park and Downing; Mourguiart, this volume); and prosobranch molluscs (e.g. West and Michel, this volume). Recent reviews that tried to extract commonalities have demonstrated that too little is known about the evolutionary processes associated with individual cases to allow meaningful generalizations (Fryer, 1996; Martens, 1997). It has often been presumed that the strongest commonality is the great age of the lakes, supposedly providing a long, relatively undisturbed history. Indeed, Boxshall and Jaume (this volume) have noted that the only factor common to ancient lakes is their great age. However, recent geological and biochemical results cast doubts on the presumption that a strong relationship exists between the age of a species flock and its species richness (e.g. Johnson *et al.*, 1996a; A. Parker, in McCune, 1997).

In this chapter, I discuss possible causes of the particularly high species richness of some cichlid lineages in the East African Great Lakes and I suggest an explanation. I first review the evidence, then the hypotheses of sexual selection and sympatric speciation and their relationship with other hypotheses that seek to explain cichlid diversity. I then summarize recent research which aims at testing the hypotheses that (i) speciation is caused by mate-choice-mediated sexual selection and (ii) speciation can occur in sympatry if sexual selection is disruptive. Finally, I discuss the results in the context of adaptive radiation, and show how an improved understanding of evolutionary processes may help to explain recent changes in species richness.

A. Cichlids in the Great Lakes of Eastern Africa

The biotas of the Great Lakes of Eastern Africa are outstanding in terms of the richness and composition of their fish faunas. Several of these lakes contain as many or more fish species than all the rivers and lakes of Europe together (Lowe-McConnell, 1987; Kottelat, 1997). The species richness and composition of a biota are determined by a combination of speciation, dispersal and biotic dispersion as diversity-enriching processes, and extinction as a diversity-limiting process (Cracraft, 1994). About 90% of the fish species in the three largest East African Great Lakes, Malawi, Tanganyika and Victoria, belong to a single family, the Cichlidae (Teleostei, Perciformes), and for each lake most, if not all, taxa present are endemic to that lake. Estimates of their phylogenies suggest that the endemic species have evolved within their lakes, and that those of Lakes Victoria and Malawi each are derived from one or a few closely related ancestral species (Lippitsch, 1993; Meyer, 1993; Nishida, 1997). Because extralacustrine dispersal of these fishes is negligible, explanations for the outstanding species richness, as well as for variation in richness among lakes, must be sought in the balance between speciation and extinction within each lake. Either the rate of extinction in relation to the number of species already present in a community must be lower than in other fish groups, allowing species to accumulate longer before competition-driven extinction rates match speciation rates (Figure 1), or gross speciation rates must be higher, and possibly so high that they outrun competitive exclusion over a part of the curve (Dominey, 1984).

B. Geological Evidence for High Speciation Rates in Cichlid Species Flocks

Geological evidence leaves no doubt that gross and net rates of speciation are extraordinarily high in some of the Great Lakes. The three Great Lake basins differ considerably in age. Lake Tanganyika is the oldest, estimated at 9–12 My old (Cohen *et al.*, 1993), and Malawi is estimated at 0.6–2 My old (Fryer and Iles, 1972; Turner, 1999). Victoria is estimated at 0.25–0.75 My old (Fryer, 1996), but had probably dried up 200 000 years ago (Martens, 1997), and seems to have dried up again in the late Pleistocene and filled up again only 13 200 (Beuning *et al.*, 1997) to 12 400 years ago (Johnson *et al.*, 1996a).

Contradicting conventional wisdom, the young lakes contain considerably more endemic cichlid species than the older Lake Tanganyika. Recent figures of known species stand at about 200 for Lake Tanganyika (Snoeks *et al.*, 1994),

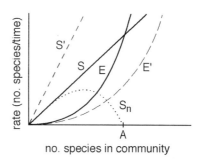

Fig. 1. Schematic representation of the regulation of species diversity in a purely speciation/extinction-regulated community. Solid lines represent relationships between gross speciation rate (*S*, measured in units of time between subsequent speciation events) and the number of species in the community, and between extinction rate (*E*) and the number of species in the community. The dotted line represents the resulting net speciation rate (S_n). The net speciation rate initially rises with the number of species that can speciate, until competitive exclusion begins to cause extinctions. The extinction rate rises with increasing saturation of the community until net speciation is zero. "A" represents the equilibrium species number. Variation in species richness among speciation/extinction-regulated communities must be explained by variation in either speciation rates (variation in susceptibility of populations to break-up of genetical coherence) or extinction rates (variation in susceptibility to competitive exclusion and stochastic population-size-related effects) or both (*S'*, *E'*, indicated by dashed lines).

compared with more than 500 for Lake Malawi (Konings, 1995) and about 500
for Lake Victoria (Seehausen, 1996). Given the limitations of the sampling
programmes, it is certain that large numbers of species are still unknown in
both lakes. Notwithstanding that some biologists have questioned whether
Lake Victoria was indeed entirely dry in the late Pleistocene (Fryer, 1997), it is
beyond doubt that, if not entirely dried up, only a shallow, although extensive,
swamp could have persisted, possibly with seasonal pools (Figure 2). Very few
haplochromine species can survive in such habitats, as the current swamp fish
fauna around Lake Victoria demonstrates (Chapman *et al.*, 1995; Rosenberger

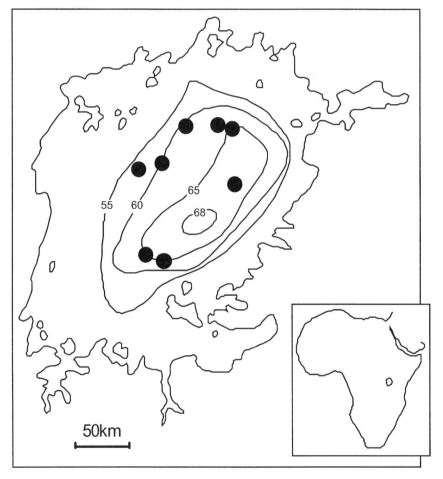

Fig. 2. Map of Lake Victoria showing the bathymetry where the water is deeper than 55
m and core locations as given by Johnson *et al.* (1996a). Water depths are in metres.
Modified after Johnson *et al.* (1996a).

and Chapman, 1999). Significantly, the two haplochromines commonly found in swamps, *Pseudocrenilabrus multicolor* and *Astatoreochromis alluaudi*, belong not to the Victorian species flock but to older lineages that have not speciated in the Lake Victoria basin.

There is a possibility that individual lineages survived the drought in Lakes Edward and Kivu west of Lake Victoria, or in the Malagarasi system to the south, and recolonized the lake after the drought. These regions are now inhabited by cichlids that are obviously closely related to those of Lake Victoria. With depths of 117 m and 485 m, respectively, these lakes are deeper than Lake Victoria (*c.* 70 m) and may not have entirely dried up as recently as that lake. Lake Edward is connected with Lake Victoria via the 4 m deep Lake George and the Katonga River. Faunal exchange between these systems is currently constrained by a swamp that separates the westward flowing from the eastward flowing part of the Katonga River and is apparently impenetrable to all but air-breathing fishes (Greenwood, 1973). This barrier to fish migration may have been less severe 8000 years ago, when the water level of Lake Victoria was 18 m higher than at present.

A Holocene migration route from Lake Kivu to Victoria is more difficult to envisage. However, Lake Kivu is connected to Lake Tanganika via the Rusizi River, and a connection to the Malagarasi and the Lake Tanganyika basin might have existed if Lake Victoria had a second, southern overflow during the Holocene lake-level high stand (Stager *et al.*, 1986). If surviving lineages recolonized Lake Victoria from Lake Edward, Lake Kivu or the Malagarasi during the Holocene, they must still have undergone explosive speciation and radiation in Lake Victoria, because Victoria shares no endemic species and very few, if any, of its 22 described genera with any of these systems (Lippitsch, 1997). Therefore, irrespective of whether or not some endemic cichlid species survived the drought inside or outside the lake basin, cichlid speciation in Lake Victoria must have been truly explosive.

III. REVIEW OF HYPOTHESES

A. Can Unusual Mate Choice Behaviour Explain Unusual Diversity?

1. Sexual Selection Through Mate Choice

An appreciation of sexual selection theory may be very relevant to explaining speciation and the origin of species flocks (Dominey, 1984). Because traits under intersexual selection are important in mate choice, their diversification might be particularly effective at facilitating reproductive isolation independently of other factors (but see Price, 1998). The first to suggest explicitly the importance of mate choice in cichlid speciation was probably Kosswig (1947). Assuming monogamy as the dominant mating system in African lacustrine cichlids, he suggested that mating preferences may lead to the behavioural isolation of family groups. Modifying Kosswig's idea, Fryer and Iles (1972) and Dominey (1984) suggested that microgeographical divergence in courtship may have been important in speciation of those taxa that are monogamous. However, speciation rates in monogamous lacustrine cichlids are no higher than in lacustrine fish that lack pairbonds and complex courtship (Table 1).

The strongly asymmetric investment in parental care in polygynous cichlids (females invest much more than males) is conducive to sexual selection upon male characters. As predicted by theories of sexual selection through female choice, many polygynous cichlids have strongly sexually dimorphic breeding coloration and courtship behaviour. Males are colourful and conspicuous, whereas females in most such species are cryptically coloured. The most species-rich lineages of cichlids are polygynous (although not all polygynous lineages are species rich). However, studies on the function of the sexual dimorphisms that characterize most polygynous cichlids have been relatively few, and studies related to speciation even fewer. McElroy and Kornfield (1990) studied male courtship behaviour among closely related rock-dwelling Lake Malawi cichlids and found that it was highly conserved. They concluded that selection on courtship behaviour was unlikely to have played an important role in speciation. Others have studied female mate choice among lekking sand-dwelling cichlids in Lakes Malawi and Tanganyika, where males construct sand bowers in which spawning takes place. Because bower shape varies among species, McKaye (1991) hypothesized that sexual selection on this extended male phenotypic trait could lead to speciation. While female preference for higher bowers was found in Lake Malawi cichlids (McKaye *et al.*, 1990; McKaye, 1991), no such preference was found in Lake Tanganyika species (Rossiter, 1994; Karino, 1997; Rossiter and Yamagishi, 1997) and preference for specific bower shapes has not yet been tested. In any case, selection on bower shape or size cannot explain the high speciation rates in the majority of the haplochromines that do not make bowers.

The egg dummies on the anal fins of males in many Great Lake cichlid taxa are a conspicuous sexually dimorphic trait, and in rock-dwelling species from Lake Malawi, females indeed preferred males with egg dummies on their anal fins over males with egg dummies removed (Hert, 1989). Goldschmidt and de Visser (1990) demonstrated correlations between egg dummy morphology and light conditions among Lake Victoria cichlids, and suggested that sexual selection on egg dummy morphology could lead to speciation. However, studies to produce evidence for microevolutionary processes that might support the hypothesis are lacking. Although at least two studies have analysed patterns in male cichlid nuptial coloration (McElroy *et al.*, 1991; Deutsch, 1997) and several studies have found circumstantial evidence in support of speciation via selection on nuptial coloration in haplochromine cichlids (Marsh *et al.*, 1981; McKaye *et al.*, 1982, 1984), experimental tests of components of this hypothesis had not been undertaken until very recently (Seehausen *et al.*, 1997; Seehausen and van Alphen, 1998; Knight *et al.*, 1998).

Table 1

Species richness in relation to age of adaptive radiations in monogamous and polygynous cichlids compared with other freshwater fish

Taxon	Habitat	Age (My)	Number of extant species	Number of founders	Speciation interval[a]	Reference
Monogamous cichlids						
Tilapia spp.	L. Bemin	~1	9	1	320	Schliewen *et al.* (1994)
Lamprologines	L. Tanganyika	~10	~85	1	1500	Sturmbauer *et al.* (1994), Stiassny (1997)
Mesoamerican heroines	Mesoamerican land bridge	50	~100	1	7526	Roe *et al.* (1997)
Polygynous cichlids						
Haplochromines	L. Victoria proper	0.012	500–1000	1?	1.2–1.3	Johnson *et al.* (1996a), Seehausen (1996)
	L. Victoria region	0.2	750–1000 +	1	2.0–2.1	Meyer *et al.* (1990); L. Kaufman (pers. comm.)
Oreochromis	L. Malawi	1–2	3	1	630–1261	Sodsuk *et al.* (1995)
Sarotherodon-derived	L. Barombi Mbo	1–2	11	1	289–578	Schliewen *et al.* (1994)
Haplochromines	L. Malawi	0.6–2	600–1000	1	60–217	Meyer *et al.* (1990), Konings (1995), Turner (1999)
Ectodines	L. Tanganyika	3.7	~30	1	754	Sturmbauer and Meyer (1993)
Non-cichlids						
Pupfish (*Cyprinodon*)	L. Chichancanab	0.008	5	1	3	Humphries (1984), Strecker *et al.* (1996)
Arctic charr	Thingvallavatn	0.015	4	1	7.5	Skulasson and Smith (1995)

Table 1 *Continued*

Taxon	Habitat	Age (My)	Number of extant species	Number of founders	Speciation interval[a]	Reference
Orestias	L. Titicaca	0.06–0.15	~22	1	13–34	A. Parker in McCune (1997)
Barbs (*Puntius* spp.)	L. Lanao	0.09–2	~17	1	22–490	Kornfield and Carpenter (1984)
Bathyclarias	L. Malawi	1–2	10	1	301–602	Lowe-McConnell (1987)
Barbs (*Barbus* spp.)	L. Tana	~2	~13	1	540	Nagelkerke *et al.* (1995)
Sculpins	L. Baikal	2–3	29	1 or 2	412–778	Slobodyanyuk *et al.* (1995)
Afromastacembelus	L. Tanganyika	~10	11	1	2891	Coulter (1991)
Characiformes	Neotropics/Africa	100	~1500	1	9478	Orti and Meyer (1997)

[a]In 1000 years units; calculated as mean doubling time of species number, L (Turner, 1999):

$$L = \frac{t \cdot \log_e(2)}{\log_e N_t - \log_e N_0}$$

where N_t = number of extant species, N_0 = number of founding species, and t = age of radiation.

2. Sympatric Speciation Through Mate Choice?

In classical allopatric models, speciation occurs if pleiotropic effects of genetic drift, adaptation to different selection regimens, or founder effects, either directly affect the mating system or cause reduced hybrid fitness which, upon secondary contact, selects for prezygotic reproductive isolation by reinforcement (Mayr, 1942, 1963). In contrast, in sympatric speciation the mating system must be directly under disruptive selection, which is why, theoretically, sympatric speciation should proceed more rapidly than allopatric speciation (Maynard Smith, 1966; Kondrashov and Mina, 1986; Dieckmann and Doebeli, 1999; Kondrashov and Kondrashov, 1999). If cichlids do speciate in sympatry, this may explain their propensity to rapid radiation.

Some of the early writers on fish speciation, facing the difficulty of explaining the diversity of endemic lacustrine cichlids, speculated that cichlids might speciate in unusual ways. Woltereck (1931) coined the term "schizotypic speciation", by which he implied rapid splitting of an ancestral gene pool into a multitude of sexually isolated gene pools in sympatry after invasion of a new adaptive zone. Mayr (1942, 1963) rejected Woltereck's hypothesis and argued that cichlid species flocks could be explained by multiple invasions. Conequently, cichlid research between the 1960s and the 1980s subscribed almost entirely to allopatric models. The most influential propositions were that cichlid species richness in Lake Victoria could be accounted for by speciation in separate protolake basins and/or in marginal lagoons (satellite lakes) (Fryer and Iles, 1972; Greenwood, 1974; Kaufman and Ochumba, 1993; Meyer, 1993; Fryer, 1996; but see Hoogerhoud et al., 1983), whereas the evolution of the cichlid species flock in Lake Malawi, which has probably never been split into separate lake basins, could be explained by microallopatric intralacustrine speciation in habitat patches (Fryer and Iles, 1972; Ribbink et al., 1983).

In reality, however, the extent to which habitat stenotopy limits dispersal in cichlids, a prerequisite for microallopatric speciation, has rarely been critically tested (but see Hert, 1992) and supporting evidence is equivocal. On one hand, population genetical studies using molecular markers seem to suggest very low dispersal between habitat patches (van Oppen et al., 1997; Arnegard et al., 1999). On the other hand, in a study of colonization of artificial reefs by rock-dwelling cichlids in Lake Malawi, the first immigrants, implicitly the best dispersers, belonged to the most species-rich taxa (McKaye and Gray, 1984). Moreover, in Lakes Victoria and Malawi even species living in continuous habitats have speciated very rapidly (Turner, 1996), while in Lake Tanganyika, the stenotopic petrophilous genus *Tropheus* has become an oft-cited example for evolutionary stasis and lack of speciation (Sturmbauer and Meyer, 1992).

Probably the strongest evidence for speciation in protolakes comes from Lake Tanganyika, where speciation in historically separate lake basins has been inferred from molecular phylogeographical patterns (Sturmbauer and

Meyer, 1992; Verheyen *et al.*, 1996; Rüber *et al.*, 1999) and from species distribution patterns (Coulter, 1991; Rossiter, 1995; Snoeks, this volume).

The strongest evidence for speciation in marginal lagoons comes from Lake Nabugabo, a satellite lake of Victoria that contains five endemic haplochromine species (Greenwood, 1965). The case of Nabugabo was long believed to exemplify the mechanism by which the Victorian species flock evolved (Futuyma, 1986). The hypothesized scenario was that when Lake Victoria was formed, it began to fill from several sides such that several isolated lake basins persisted over a considerable period (Fryer and Iles, 1972). Smaller peripheral water bodies were created and reunited repeatedly owing to fluctuations in lake levels during the unstable history of the lake, a process which generated new taxa and then added them to the species flock of the main lake (Greenwood, 1981; Kaufman and Ochumba, 1993).

Elsewhere, I have argued that distribution patterns and ecological specializations among the recent fauna provide biological arguments for an intralacustrine origin of species diversity (Seehausen, 1996). This view received support (Kaufman *et al.*, 1997) from new geological evidence on the shape of the lake basin and for a very recent desiccation (complete or incomplete) (Johnson *et al.*, 1996a; Figure 2), evidence which strongly supported the scenario of speciation within a single water body. Palaeoclimatic evidence suggests that it is most unlikely that satellite lakes persisted during the drought (Johnson *et al.*, 1996a) and basin morphology rules out the possibility that more than one basin existed when the lake filled up again (Johnson *et al.*, 1996a; Figure 2). Palaeohydrological evidence proves that since the refill, lake level fluctuations, which might have provided a situation for allopatric speciation in satellite lakes, could not account for any significant part of the 500-plus speciation events. The new data suggest that after the drought, the lake level rose rapidly until it peaked at 18 m above the present level about 7200 years ago. This was followed by several thousand years of lake-level decline after the incision and subsequent opening of the outlet to the Nile. Around 5000 years ago the lake level was still about 12 m higher than at present, but by 4000 years ago the lake had fallen to a level only 3 m higher than it is at present (Beuning *et al.*, 1997). It was probably around this time that the peripheral lagoon Lake Nabugabo was finally isolated from the main lake (Greenwood, 1965).

Against this revised time-frame for evolution in Lake Victoria, the lagoon's stability has turned the Nabugabo case into evidence that speciation in satellite lakes alone is very unlikely to explain Lake Victoria's species richness and diversity. Separated from the main lake only by a sand bar, Lake Nabugabo has persisted stably for one-third of the time that has passed since the reflooding of the main lake basin, but it contains only five endemic species.

Although microallopatric population differentiation is known to occur within Lake Victoria (Seehausen, 1996; Witte *et al.*, 1997; Seehausen *et al.*, 1998a; Bouton *et al.*, 1999), and it probably results in speciation among littoral cichlids (Seehausen and van Alphen, 1999), it is difficult to explain the explosive rate of speciation and ecological diversification in Lake Victoria in this way alone. The case of the Lake Victoria cichlids challenges the conventional view that all speciation occurs allopatrically, as a byproduct of population differentiation in geographical isolation (Mayr, 1963; Paterson, 1985; Fryer, 1996).

The main theoretical argument against sympatric speciation is the difficulty for disruptive selection to overcome the genetical homogenizing effect of recombination in a contiguous population (Rice, 1987). In the absence of associated differences in habitat preference, a build-up of genetic covariance between fitness and mate-choice traits, leading to sympatric speciation, would require that disruptive natural selection be unrealistically intense or assortative mating be nearly perfect (Felsenstein, 1981). Several classes of sympatric speciation models suggest that this difficulty can be overcome (reviewed in Bush, 1994).

One class of model suggests that the recombination effect can be overcome in a population that is polymorphic for habitat preference and habitat-associated fitness, if mating takes place strictly in the preferred habitat resulting in positive assortative mating (Diehl and Bush, 1989; Kawecki, 1996). In other types of sympatric speciation models, active mate choice is required to overcome the recombination effect. Maynard Smith's (1966) original model and related subsequent models (e.g. Rosenzweig, 1978; Kondrashov and Mina, 1986; Johnson et al., 1996b) require polymorphisms in mate preference, habitat preference and habitat-based fitness. More recent models suggest that variation in any trait related to resource utilization can cause sympatric speciation if there is heritable variation in a mate-choice trait and in the preference for the latter. In such situations, density-dependent (Kodrashov and Kondrashov, 1999) and frequency-dependent (Dieckmann and Doebeli, 1999) natural selection can recruit disruptive mate preferences.

Other models suggest that a build-up of covariance between mate preference in the choosy sex and the chosen trait in the other sex by divergent sexual selection, can be sufficient for speciation in clines (Lande, 1982) and for fully sympatric speciation, independent of possible covariance between fitness and mate-choice traits (Wu, 1985; Turner and Burrows, 1995; Noest, 1997; Payne and Krakauer, 1997; Van Doorn et al., 1998). Covariance between fitness and mate-choice traits may evolve either simultaneously with, or subsequent to, speciation.

The possibility of sympatric speciation under disruptive selection in cichlids was stimulated by the discovery of what appeared to be balanced polymorphisms in some Central American cichlids (Sage and Selander, 1975; Barlow, 1976; Kornfield et al., 1982; Liem and Kaufman, 1984; Meyer, 1990), some of which tended to display assortative mate choice (reviewed in Barlow, 1983), and by the realization that what were believed to be colour morphs of Lake Malawi haplochromines were actually reproductively isolated species (Holzberg, 1978; Marsh et al., 1981; McKaye et al., 1982, 1984). However, the same discoveries were also used as evidence against sympatric speciation: why would the polymorphic Central American species maintain balanced polymorphisms and not speciate (see e.g. Barlow, 1998)? Polymorphic species with restricted gene flow between morphs are not known from Lake Malawi, and had until recently not been found in Lake Victoria (Seehausen and Bouton, 1996; Witte et al., 1997). The first strong evidence for sympatric speciation in cichlids was produced by Schliewen et al. (1994) who, by reconstructing a molecular phylogeny of species flocks in Cameroonian crater lakes from mitochondrial DNA sequences, demonstrated that each flock probably evolved within the confines of its lake. These lakes are much smaller and more homogeneous than the East African Great Lakes, allowing Schliewen et al. to infer sympatric origin from sympatric occurrence.

I believe the two main reasons for the reluctance of many biologists to seriously consider sympatric modes of speciation in the evolution of cichlid species flocks are: (i) the absence of empirical demonstration of the mechanics of sympatric speciation in cichlid fishes; and (ii) the use of species concepts that accommodate sympatric speciation only with great difficulty, if at all (but see Turner, 1994, and this volume).

3. Other Hypotheses

Selective mate-choice hypotheses are highly relevant to explain the rapid speciation but, alone, cannot explain the adaptive radiation in cichlid fish species flocks. Other hypotheses are more relevant in this regard, and I outline the most influential of these as a background within which studies of speciation should be considered.

a. The New Adaptive Zone Hypothesis. It has been argued that cichlids entering the developing African Great Lakes experienced ecological character release in then depauperate fish communities. However, the non-cichlid fish faunas of the African Great Lakes are not depauperate (Greenwood, 1981; Fryer, 1996). Moreover, if the lacustrine fish assemblages had been depauperate, the radiation of the cichlids would, as Fryer (1996) pointed out, have quickly rendered the situation otherwise and speciation should have stopped at a level of species richness comparable to that in other fish communities. These two lines of argument suggest that certain biological properties of cichlids must be involved in their prolific speciation.

b. Physiological Preadaptation* Hypotheses. Cichlids are secondary fresh-water fishes, and their euryhalinity and their closed (physoclist) swimbladder have been suggested as two physiological properties that enabled them to

---

*The term preadaptation is considered inappropriate by many authors because it implies goal-orientated evolution (Williams, 1966; Ribbink, 1994).

tolerate the challenging physicochemical conditions in large lakes, and to adapt to deep water (Poll, 1980).

The intensive parental care of cichlids has probably been of even greater importance. This behaviour, which entails fanning of eggs and larvae or oral incubation, has released the cichlids from the constraints to intralacustrine reproduction posed by insufficient oxygen supply for eggs and larvae, to which other fish groups are vulnerable. Furthermore, the mouthbrooding habits of some cichlid lineages allow them to breed and rear young in habitats that do not provide shelter (Fryer and Iles, 1972). However, the large numbers of silurids, cyprinids and clupeids in the African Great Lakes leave little doubt that physostomes and species without parental care can cope very well with lacustrine conditions and thrive at great depths. Moreover, clupeids are, like cichlids, secondary freshwater fish, and are thus also "preadapted". Hence, being physiologically and ecologically successful in lakes does not necessarily lead to a taxon becoming species rich there. Physiological and behavioural properties, and in particular their highly developed parental care, enable cichlids successfully to inhabit lakes, but on their own, do not explain their tendency to diversify.

   c. Cichlid Design and Versatility of the Jaws as a Key Evolutionary Innovation. The mouthparts of cichlids are a versatile apparatus for capturing and processing food. In most non-marine fishes the lower and upper pharyngeal bones each form a pair of bones, but in cichlids and their marine relatives (Suborder Labroidei) these have become fused and bonded, respectively, and function as a second pair of jaws (Liem, 1973; Kaufman and Liem, 1982; Stiassny and Jensen, 1987). This fusion of bones doubles the force that can be applied onto the jaws (Galis and Drucker, 1996). Probably even more importantly, the functional flexibility of the labroid pharyngeal jaws is enhanced by a series of changes that probably preceded their fusion, including two decouplings that allow independent movement of the upper and lower jaw and that are considered key evolutionary innovations (Liem, 1973; Galis and Drucker, 1996). This increase in versatility of the pharyngeal jaws allows efficient handling of many different prey items (Liem, 1973; Stiassny and Jensen, 1987; Galis and Drucker, 1996) and the resulting flexibility allows improved phenotypic response to changing requirements, which may subsequently be genetically assimilated (Galis and Metz, 1998). Phenotypic plasticity (reviewed by Witte *et al.*, 1997) is predicted to have similar effects. Both increase the likelihood of longer term population survival. However, it remains unknown whether phenotypic plasticity is more apparent in cichlids than in other taxa.

   A related hypothesis is that the cichlid design allows for ecologically efficient changes to be effected through overall or relative growth (Greenwood, 1974, 1981, 1994). Strauss (1984) showed that ecologically very diverse cichlids of Lake Victoria differ little in anatomy, except for size-related effects on shape. Barel (1984) and Wilhelm (1984) demonstrated that another part of the ecological differentiation is achieved through differences in allometric growth coefficients of structures in the head. Gould (1977) suggested that these kinds of changes could be induced by changes in timing during development, coded by just a few genes. It remains unknown whether the phenotypic control of body size and that of relative growth rates in cichlids differ from those in other fish.

4. Are These Really Alternative Hypotheses?

These hypotheses, and sexual selection versus anatomical versatility in particular, are sometimes treated implicitly or explicitly as alternatives, but they are not!

The evolution of species diversity requires three conceptually independent, although often interrelated, processes: speciation, ecological radiation and anatomical diversification. That ecological and anatomical radiation are not necessarily linked to speciation has been shown in cichlids and other fishes (Turner and Grosse, 1980; Liem and Kaufman, 1984). That the extent of ecological radiation is not necessarily reflected in the extent of anatomical diversification has been shown in cichlids too (Strauss, 1984). If an objective assessment of hypotheses is to be achieved, then explicit acknowledgement of this conceptual independence is necessary. Hypotheses that differ in relevance to these processes are not alternatives to each other.

A reappraisal of what is unusual in cichlid species flocks should help to focus attention on which components of their striking diversity might require unusual explanations. Compelling evidence exists for unusual ecological radiation and unusually high ecological species packing in several lineages of African cichlids. Most lacustrine species flocks of other fish taxa (Table 1) are either ecologically less diverse or, if a similar number of feeding niches is occupied, fewer species share each niche (cf. Greenwood, 1981, and papers on non-cichlids in Echelle and Kornfield, 1984; Nagelkerke, 1997). Despite frequent claims to the opposite, the evidence that the anatomical radiation of cichlids is unusual is less compelling when compared with other ecologically radiated fish groups of comparable geological age. Greenwood (1974, 1981) and Strauss (1984) observed that the anatomical diversity in the haplochromine species flock of Lake Victoria is muted and may result from only very limited genetic differentiation. Dominey (1984) pointed out that the anatomical radiation of cichlids, even in the oldest lake, Lake Tanganyika, is only moderate in relation to the age of the species flock.

Speciation rates of cichlids during species-flock formation do not generally differ from those in other fish taxa either (McCune, 1997; Table 1), with the notable exception of the haplochromines in Lakes Victoria and Malawi. As the data in Table 1 show, the latter speciated by up to an order of magnitude more rapidly than most other fish in radiations of similar age (Lake Victoria < 1 My age and Lake Malawi 1–2 My age), and is the only group that speciates rapidly in already species-rich communities. Two-thirds of the more than 2000 known cichlid species belong to the haplochromines, which form just one of 15 major lineages (tribes) within the family. Species numbers in the other lineages range from about 15 (Etroplini) to about 150 (Heroini). More than 1200 haplochromine species live in the African Great Lakes, but the riverine sister taxa to the lacustrine species flocks have undergone no more diversification than have other riverine cichlid lineages. This can be firmly concluded despite considerable uncertainty regarding the exact sister group relations because there is no species-rich or anatomically diverse haplochromine group known from any river system. African rivers between Israel and South Africa harbour about 50 haplochromine species, assigned to nine genera (Greenwood, 1979), and similar species numbers are obtained for tilapiine (Trewavas, 1983), chromidotilapiine (Greenwood, 1987) and tylochromine (Stiassny, 1990) cichlids.

Three conclusions can be drawn from this review. (i) The true uniqueness of the African lacustrine cichlid species flocks lies in ecological radiation and unusually high species packing (α diversity) in several lineages, and in the unusually rapid rate of speciation in the haplochromine lineage. (ii) Because of this, an explanation of cichlid diversity might require more than one evolutionary model: one for speciation in haplochromines and one for ecological coexistence. (iii) General cichlid properties alone cannot explain the species richness of cichlids, nor can intrinsic haplochromine properties alone explain their species richness. Explanations for the species richness of haplochromines have to be sought in properties that have their impact on lineage diversification under some extrinsic conditions, but not under others.

In the next section I discuss ongoing studies which are designed to identify these properties in haplochromine cichlids. In particular, I explore the hypothesis of speciation caused by disruptive sexual selection upon coloration, a mechanism that may not require geographical isolation.

IV.   THE ROLE OF MATE CHOICE IN HAPLOCHROMINE SPECIATION

A. Patterns

To identify elements of cichlid colour pattern that evolve under sexual selection, we conducted a phylogenetic comparative analysis of the evolution of colour patterns across the larger East African cichlid radiation (Lakes

---

Fig. 3. Tracing gain and loss of a polygynous mating system, and gain and loss of male nuptial coloration on a phylogenetic tree of the East African cichlids shows that the origin of polygynous mating systems generally precedes the origin of male nuptial coloration. A polygynous mating system is more conducive to sexual selection than other mating systems. The tree is a strict supertree. 1, monogamy; 2, harem polygyny; 3, open polygyny. Revised after Seehausen *et al.* (1999a).

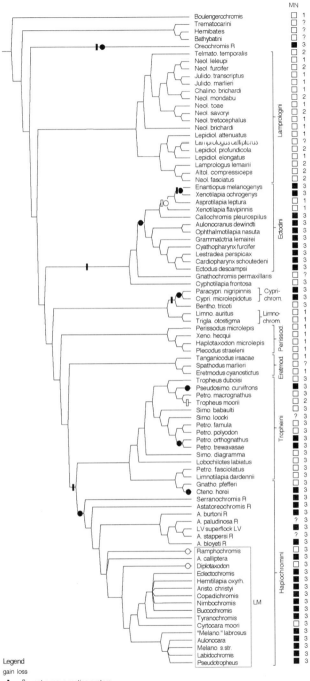

MN

Boulengerochromis □ 1
Trematocarini □ ?
Hemibates □ ?
Bathybatini □ ?
Oreochromis R ■ 3
Telmato. temporalis □ 2
Neol. leleupi □ 1
Neol. furcifer □ 2
Julido. transcriptus □ 1
Julido. marlieri □ 1
Chalino. brichardi □ 1
Neol. mondabu □ 2
Neol. toae □ 1
Neol. savoryi □ 2
Neol. tretocephalus □ 1
Neol. brichardi □ 1
Lepidiol. attenuatus □ 1
Lamprologus callipterus □ ?
Lepidiol. profundicola □ 2
Lepidiol. elongatus □ 1
Lamprologus lemairii □ 2
Altol. compressiceps □ 2
Neol. fasciatus □ 2
Enantiopus melanogenys ■ 3
Xenotilapia ochrogenys ■ 3
Asprotilapia leptura □ 1
Xenotilapia flavipinnis □ 1
Callochromis pleurospilus ■ 3
Aulonocranus dewindti ■ 3
Ophthalmotilapia nasuta ■ 3
Grammatotria lemairei ■ 3
Cyathopharynx furcifer ■ 3
Lestradea perspicax ■ 3
Cardiopharynx schoutedeni ■ 3
Ectodus descampsi ■ 3
Gnathochromis permaxillaris □ ?
Cyphotilapia frontosa □ 3
Paracypri. nigripinnis □ 3
Cypri. microlepidotus ■ 3
Bentho. tricoti □ 3
Limno. auritus □ 1
Trigla. otostigma □ 1
Perissodus microlepis □ 1
Xeno. hecqui □ 1
Haplotaxodon microlepis □ 1
Plecodus straeleni □ 1
Tanganicodus irsacae □ ?
Spathodus marlieri □ 1
Eretmodus cyanostictus □ 1
Tropheus duboisi □ 3
Pseudosimo. curvifrons ■ 3
Petro. macrognathus ■ 3
Tropheus moorii □ 2
Simo. babaulti □ 3
Simo. loocki ? 3
Petro. famula □ 3
Petro. polyodon □ 3
Petro. orthognathus ■ 3
Petro. trewavasae ■ 3
Simo. diagramma □ 3
Lobochilotes labiatus □ 3
Petro. fasciolatus □ 3
Limnotilapia dardennii □ 3
Gnatho. pfefferi □ 3
Cteno. horei ■ 3
Serranochromis R ■ 3
Astatoreochromis R ■ 3
A. burtoni R ■ 3
A. paludinosa R ? 3
LV superflock LV ■ 3
A. stappersi R ? 3
A. bloyeti R ■ 3
Ramphochromis □ 3
A. calliptera ■ 3
Diplotaxodon □ 3
Eclectochromis ■ 3
Hemitilapia oxyrh. ■ 3
Aristo. christyi ■ 3
Copadichromis ■ 3
Nimbochromis ■ 3
Buccochromis ■ 3
Tyranochromis ■ 3
Cyrtocara moori □ 3
"Melano." labrosus ■ 3
Aulonocara ■ 3
Melano. s.str. ■ 3
Labidochromis ■ 3
Pseudotropheus ■ 3

Lamprologini

Ectodini

Cypri-chrom.
Limno-chrom.
Perissod.
Eretmod.
Tropheini
Haplochromini
LM

Legend

gain loss

■ □ polygynous mating system

● ○ male nuptial hue (MN)

Tanganyika, Malawi, Victoria and connected river systems; Seehausen *et al.*, 1999a). Performing comparative tests, using a phylogenetic supertree of the East African radiation and information on coloration, ecology and mating system (as a measure for sexual selection) from a large number of species, we found that the appearance of male nuptial coloration is generally preceded by the evolution of a polygynous mating system (Figure 3) and that in clades with this mating system changes in hue occurred in frequent association with speciation events. The evolution of the diverse melanic stripe patterns of cichlids also, although only slightly phylogenetically conserved, proved to be ecologically constrained. The frequency of change in stripe patterns, unlike the frequency of change in nuptial hues, is not associated with the strength of sexual selection as measured by mating system (Seehausen *et al.*, 1999a). In fact, very few speciation events in the fast-evolving haplochromines are associated with changes in stripe pattern (Seehausen *et al.*, 1998a).

As a model system for further study we chose the haplochromines of the rocky shores in Lake Victoria. Since 99% of the species and most of the genera at rocky shores were undescribed (during our study we discovered more than 100 new species), their systematics had to be unravelled and a subset of taxa described (three genera, 15 species) to make the system accessible (Seehausen *et al.*, 1998a). Morphological evidence suggests that many of the rock-dwelling Lake Victoria cichlids form a species-rich monophyletic lineage, remarkably similar to the rock-dwelling cichlids (mbuna) of Lake Malawi. The implication is that shifts from other habitats to rocks occurred in both lakes only a few times, followed by much speciation within the rocky habitat.

A systematic survey of rock-dwelling cichlids in south-eastern Lake Victoria was conducted (Seehausen, 1996) to study distribution patterns and to identify sibling species pairs and intraspecific polymorphisms. We found intraspecific polymorphism in five sets of characters: male nuptial body coloration, male fin coloration, female (X-linked) body coloration, dentition and dorsal head profile. All of these characters also differentiate closely related species, though some more often than others. It was found that closely related species with fully overlapping geographical distribution almost always differ in body coloration, where often one has blue and the other one yellow–red males. However, in certain species groups, males of sympatric species can have the same body coloration, and one species has only plain females whereas the other has many blotched females. This cannot be explained in terms of reinforcement of species recognition, because closely related species with different distribution ranges but some geographical overlap often have identical body coloration in both sexes and differ only in male fin coloration. In several cases that we studied, such species spawned at the same time and the same place, yet coexist without hybridization (Seehausen *et al.*, 1998b). Different fin coloration alone is, hence, sufficient to maintain reproductive isolation. Species of different genera, whether sympatric or not, may not differ at all in coloration, which suggests that reproductive isolation among distantly related species involves more cues than only colour (Seehausen et al., 1998a).

In a series of studies of the ecological structure of island assemblages dense ecological species packing was found very much like in similar habitats in Lake Malawi (Ribbink et al., 1983). Many sympatric species, and in particular those that are closely related, overlap broadly in microdistribution (Seehausen and Bouton, 1997), feeding behaviour (Seehausen and Bouton, 1998) and diet (Bouton et al., 1997). Statistically significant differences could be shown between any two species, but not between colour morphs of one species (Neochromis omnicaeruleus Seehausen and Bouton, 1998). Correlations between the anatomy of the feeding apparatus and microdistribution suggest that interspecific differences are functional rather than stochastic (Seehausen and Bouton, 1997).

These patterns are consistent with the hypothesis that change in coloration is the first step in haplochromine speciation, followed by divergence in other characters. Specifically: (i) there are conspecific colour morphs that do not differ ecologically; (ii) closely related species with identical geographical distribution are conspicuously different in coloration but usually differ little in ecology; (iii) closely related species with different geographical distributions but considerable overlap often differ less conspicuously in coloration and also differ little in ecology; and (iv) sympatric and allopatric species that are not closely related may not differ in coloration at all, but usually differ distinctly in ecology.

From these findings we derived three predictions regarding mate choice and reproductive isolation: (i) coloration is important in interspecific mate choice among closely related species; (ii) closely related species are reproductively isolated by nothing else than active and direct mate choice; and (iii) coloration affects intraspecific mate choice, such that colour polymorphism is associated with mate preference polymorphism.

B. Mechanisms

Testing females of sympatric populations of the sibling species *Pundamilia pundamilia* Seehausen & Bouton 1998 (blue males) and *P. nyererei* (Witte-Maas and Witte 1985) (males with red dorsum) for preferences among conspecific and heterospecific males in a two-way choice design demonstrated that females preferentially respond to courtship by conspecific males, but that this preference is lost when the differences in coloration between the two kinds of male are masked by light conditions (Seehausen and van Alphen, 1998). In the latter case, females of both species may exhibit the same preferences for size and display rates of males, but these are normally overruled by the effects of coloration when colours are visible. Hence, coloration is important in interspecific mate choice among closely related sympatric species. If colour effects also dominate other effects in intraspecific mate choice, colour polymorphism might be associated with mate preference polymorphism. To evaluate this possibility two tests were carried out.

First, females of one population of *P. nyererei* were tested for intrapopulation variation in preferences for male nuptial colouration. Male nuptial coloration in this population varies between blue and dull red in the dorsal fin, and between blue–grey and dull red on the dorsum (Seehausen, 1997: population of Luanso Island). The response of each female towards males that differed in the relative amounts of red and blue was measured in a four-way choice design. Testing each female four times with several days between tests revealed significant between-female variation of preference for male colour types (Seehausen, unpubl.). Where-as most females consistently and significantly preferred blue over red, some were unselective and others tended to prefer red males. Hence, male colour poly-morphism is associated in this population with female preference polymorphism.

Secondly, mate preferences were studied in a species that is polymorphic for plain versus orange-blotched and black-and-white blotched coloration, an X-linked polymorphism that affects particularly females. Strong mate preferences were found in males of both morphs for the coloration of their mother. There were also strong preferences in the ancestral (plain) morph females for plain males, whereas blotched females were not choosy (blotched males are very rare in nature). Hence, mainly female colour polymorphism was associated with mainly male mate-preference polymorphism. Cross-breeding experiments revealed that the genetics of the triple female colour polymorphism (one plain and two blotched morphs of different hue) can be explained by a minimum of three colour-determining loci, two X-linked and at least one autosomal. These experiments further revealed strong sex-ratio distortion in certain matings between original and incipient species. The X-linked blotch genes are tightly linked to a dominant female determiner (w) and the autosomal gene is tightly linked to a male determining (sex-ratio rescue) gene. Mate preferences avoid sex-ratio distortion and may have evolved under sex-ratio selection (Lande and Wilkinson, 1999). By using knowledge of the Mendelian genetics of the polymorphism to interpret phenotype frequencies in nature, it could be demonstrated that the mate preferences measured in these experiments translate into selective non-random mating in nature which exerts disruptive sexual selection. Intermediate phenotypes were distinctly less abundant than expected under random mating (Seehausen *et al.*, 1999b).

This colour polymorphism has properties of an incipient stage in sympatric speciation. The morphs were anatomically indistinguishable, and diversifica-tion in coloration and mate preferences has clearly evolved prior to ecological diversification. We concluded that to terminate gene flow, a female preference gene is required to spread through the derived morph, which may currently be selected against by rarity of males of the derived morph (Seehausen *et al.*, 1999b). Comparing the evidence from taxonomy, female colour polymorph-isms appear to be less frequently associated with speciation than male colour polymorphisms. Nevertheless, there are sibling species pairs in Lake Victoria, and also in Lake Malawi, that differ greatly in female coloration but little if at all in male coloration.

To test the prediction that closely related species are reproductively isolated only by mate choice we studied the timing, location and behaviour of reproduction in one island community. Closely related species (members of the same genus) are isolated neither by timing nor by location of reproductive activities, but display highly assortative mate choice. Distantly related species are often isolated by depth of spawning sites (Figure 4). This makes it most likely that direct and active mate choice is the factor that keeps closely related species genetically isolated and maintains species diversity in the absence of postzygotic barriers to hybridization (Seehausen et al., 1997, 1998b). Similar conclusions can be drawn from experimental work on sympatric closely related Lake Malawi cichlids (Knight et al., 1998).

Combining the evidence from these studies, it may be concluded that speciation among haplochromine cichlids can be initiated by diversification of coloration and mate preferences, which may be caused by sexual selection. If sexual selection is disruptive, it appears to be able to break up population

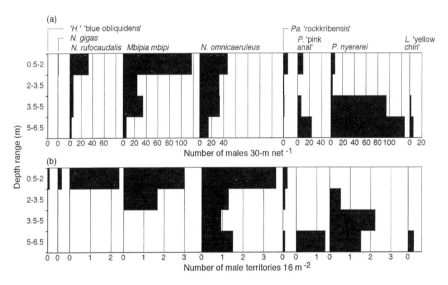

Fig. 4. Depth distribution of populations and of spawning sites of haplochromine cichlid species at the rocky shore of Makobe Island (Speke Gulf, Lake Victoria). (a) Depth distribution of populations based on gillnet catches. (b) Spawning site distribution, measured as mean number of male territories per unit area. Male territories were counted along transects by a SCUBA diver. Note that spawning sites of closely related (congeneric) species overlap broadly while those of distantly related species can be isolated by depth. 'H'., "Haplochromis"; N., Neochromis; Pa., Paralabidochromis; P., Pundamilia; L., Lithochromis. From Seehausen et al. (1998b).

coherence in sympatry. The slight but significant ecological differences that are often observed between closely related species (but see also Genner *et al.*, 1999) may evolve under disruptive natural selection, the response to which probably increases enormously as a consequence of reduction or termination of gene flow. To test further the hypothesis of speciation by disruptive sexual selection, we investigated patterns of species diversity to determine whether the signature of the process by which it arose could be traced.

C. Patterns Revisited

Two predictions yielded by the hypothesis that sympatric speciation caused by disruptive mate choice for yellow–red versus blue male coloration and plain versus blotched female coloration contributed to the origin of haplochromine species diversity are: (i) The proportions of geographical overlap between closely related species have a bimodal frequency distribution with peaks at fully allopatric and fully sympatric; and (ii) the proportion of pairs of closely related species that are heteromorphic with regard to either one of the two kinds of colour polymorphism is higher among sympatric pairs than among allopatric pairs.

Results of analyses of patterns of geographical distribution of 41 species in relation to phylogeny and colour variation supported both predictions, consistent with the hypothesis that disruptive sexual selection accounts for speciation events (Seehausen and van Alphen, 1999, 2000). Particularly strong

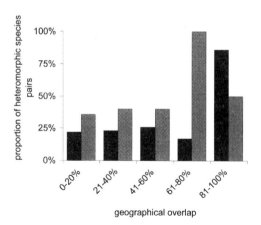

Fig. 5. Proportion of species pairs that are heteromorphic with respect to body coloration, plotted against geographical distribution overlap. The two polymorphisms are shown separately: dark bars stand for the proportions of species pairs among all species pairs in the genera *Mbipia, Pundamilia* and *Lithochromis*, in which one has mainly blue and the other mainly yellow–red males (based on, from left to right, 126, 14, 15, 6 and 22 species comparisons); light bars indicate the proportion of species pairs among all *Neochromis* species pairs, in which one has exclusively plain females and the other many blotched females (based on, from left to right, 14, 5, 5, 1 and 10 species comparisons).

support for the second prediction came from yellow–red versus blue species pairs (Figure 5). The estimate of the proportion of sympatric species pairs at 17.5% (Seehausen and van Alphen, 1999) may be considered conservative for Lake Victoria cichlids because it is based on the species group that is most stenotopic among all Lake Victoria haplochromines, and is found over an extremely patchily distributed type of habitat. None of the species has ever been caught over any substrate other than rocks. Hence, rock-dwelling haplochromines should speciate in intralacustrine allopatry more often than species from other habitats.

Because the perception of colour depends on light, the effects of sexual selection on coloration and, in turn, those of colour variation on mate choice, are influenced by ambient light conditions. Variation in coloration and/or mate preferences can be picked up by sexual selection only where light conditions allow perception of the variation. This yielded several predictions that we tested in the field: (i) the distinctiveness of interspecific differences in coloration (hue) is a function of the width of the ambient light spectrum. (ii) Intraspecific variation in coloration (polymorphism) is more developed, and (iii) the number of closely related species coexisting at any place is higher, the broader the ambient light spectrum.

Among a large number of ecological variables that we measured at two series of rocky islands, the ambient light regime (Figure 6A) explained better than any other variable the variation between islands in (i) the distinctiveness of nuptial hue difference between sympatric species (Figure 6C), (ii) the number of sympatric colour morphs within a species, and (iii) the number of coexisting species within genera (Figure 6B). The frequency of phenotypes that are intermediate between sympatric forms is negatively correlated with water transparency. Finally, the brightness and distinctiveness of the hue of male nuptial coloration in a population have decreased within the past decade following a decrease in water transparency (Seehausen et al., 1997).

V. DISCUSSION

A. The Role of Sexual Selection in Speciation and Adaptive Radiation

I conclude that the high speciation rates in some lineages of haplochromine cichlids under certain external conditions can probably be explained by disruptive sexual selection on coloration. In combination with factors that affect population survival (see hypothesis on cichlid design), this can adequately explain the unusual species richness of haplochromine species flocks. Such selection can be exerted by female preference polymorphism with bimodal preferences for male nuptial hues, but also, although perhaps less often, by what is mainly male preference polymorphism with bimodal preferences for female coloration. Judging from male colours, the same female-choice polymorphism (for

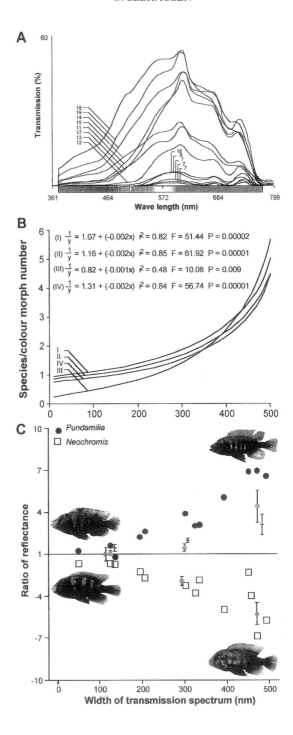

A

Transmission (%)

60

18
19
14
15
11
17
13
12

16
9
7
7
7

361   464   572   684   799

Wave length (nm)

B

Species/colour morph number

6

(I) $\frac{1}{y}$ = 1.07 + (-0.002x) r^2 = 0.82 F = 51.44 P = 0.00002

(II) $\frac{1}{y}$ = 1.16 + (-0.002x) r^2 = 0.85 F = 61.92 P = 0.00001

(III) $\frac{1}{y}$ = 0.82 + (-0.001x) r^2 = 0.48 F = 10.08 P = 0.009

(IV) $\frac{1}{y}$ = 1.31 + (-0.002x) r^2 = 0.84 F = 56.74 P = 0.00001

5

4

3

2

I
II
IV
III

1

0

0   100   200   300   400   500

C

10

● *Pundamilia*
□ *Neochromis*

Ratio of reflectance

7

4

1

-4

-7

-10

0   100   200   300   400   500

Width of transmission spectrum (nm)

blue versus yellow–red males) is found in all genera of Lake Victoria haplo-chromines studied (*Neochromis, Mbipia, Pundamilia, Lithochromis*) and in many others, as well as in species of Lake Edward and many genera of Lake Malawi haplochromines. Judging from female colours, the male-choice polymorphism (for plain versus blotched females or females of different hue) also occurs in many genera in Lake Victoria (*Neochromis, Paralabidochromis, Ptyochromis, Macro-pleurodus, Hoplotilapia* and *Lipochromis*), in species of Lake Kivu and in several genera of Lake Malawi (mbuna) cichlids (Seehausen *et al.*, 1999c).

Fig. 6. Illustration of the influence of aquatic light conditions on coloration, intraspecific colour polymorphism and species richness, through sexual selection. (A) Transmission light spectra at a water depth of 2 m at 13 research stations along a 70 km long south–north transect in southern Lake Victoria. Numerals indicate the number of coexisting rock-restricted haplochromine species for each station. Each curve is based on the mean from 10 spectral scans measured in 1996. The width of transmission spectra (nm) is related to turbidity (cm Secchi disc) as: $y = 59.73 + 1.91x$, $r^2 = 0.88$, $F = 77.1$, $p < 0.00001$. (B) Relationship between width of the transmission spectrum at 2 m water depth and the number of coexisting species in the genera *Neochromis* (I), *Pundamilia* (II) and *Paralabidochromis* (III), and the number of colour morphs of the *Neochromis greenwoodi/omnicaeruleus* superspecies (IV). Curves are best fits of 13 data points representing the 13 stations on the research transect. (C) Hue of male nuptial body coloration as a function of light transmission. The redness of *Pundamilia nyererei* and the blueness of *Neochromis greenwoodi/omnicaeruleus* in 13 populations along the research transect are quantified as the ratio of reflectance 610 nm/515 nm and 515 nm/610 nm, respectively, at each island (*P. nyererei* was absent from one). The horizontal line stands for equal reflectance of red (610 nm) and blue (515 nm) light. Above the line, more red than blue light is reflected, while below it, more blue than red light is reflected (negative ratio on the *y*-axis). Three kinds of data are presented: (i) reflection ratios from the brightest males (filled circles and open squares); brightness (relative to population average) indicates sexual activity; (ii) population means ± SE for two extreme and one intermediate population of each species. They were obtained by measuring (from left to right) five, five and seven individuals of *P. nyererei* (red points) and six, three and five individuals of *Neochromis* (blue points), covering the range of variation in each population; and (iii) means ± SE for laboratory-bred individuals of the same three populations of *P. nyererei* (horizontal bars). Populations were spawned and raised separately under standardized conditions. After the fishes had attained maturity, they were kept in three groups of 10 individuals. From each group, the five most intensively coloured males that became sexually active were eventually photographed for measurements. The difference (*y*) in hue between wild males of the red and the blue species is related to spectral band-width (*x*) as: $y = -1.83 + 0.02x$, $r^2 = 0.88$, $F = 71.54$, $p < 0.00001$. The difference between the mean values for wild males from dull and colourful populations is significant in both species (*P. nyererei*: $Z = 1.95$, $p = 0.05$; *Neochromis*: $Z = 2.65$, $p < 0.01$) and is also significant between the captive-bred fishes ($Z = 2.10$, $p < 0.05$), demonstrating that the colour differences are heritable. The larger standard errors in more colourful populations reflect the larger colour diversity found in clear-water populations. (Reprinted with permission from Seehausen, O., van Alphen, J.J.M. and Witte, F. (1997). Cichlid fish diversity threatened by entrophication that curbs sexual selection. *Science* **277**, 1808–1811. Copyright 1997 American Association for the Advancement of Science.)

   The role of disruptive sexual selection is twofold: it can cause speciation in sympatry and it maintains reproductive isolation in sympatry between species that have evolved in sympatry or allopatry. Intraspecific sympatric poly-morphisms in Lake Victoria haplochromines were found also in three other character sets – hue of male fin coloration, oral dentition and head shape – but there is no compelling evidence that these have led to sympatric speciation (Seehausen et al., 1998a). Variation in sensory bias (Ryan, 1998) may be responsible for disruptive selection upon male coloration (yellow–red versus blue), while disruptive selection on female coloration can be explained by effects of sex ratio selection, as a form of "good genes" selection (Seehausen et al., 1999b).

   Speciation during adaptive radiations of fishes in lacustrine environments tends to be faster than in rivers (McCune, 1997). This is not unique to cichlids but holds for diverse taxa (Table 1; compare continental, largely riverine faunas with lacustrine faunas). This difference may reflect a difference in complexity and spatial dimensionality of the environment. The higher dimensionality of lake environments may support sympatric ecological speciation, which is predicted to proceed more rapidly than allopatric speciation because of selection on the mating system (Kondrashov and Mina, 1986; Dieckmann and Doebeli, 1999; Kondrashov and Kondrashov, 1999). Studies of fish speciation in several small, young and ichthyologically depauperate lakes suggest that sympatric speciation there follows the ecological model: disruptive natural selection causes the evolution of resource polymorphism with pleiotropic or selective effects on mate choice (e.g. assortative mating by size or shape; Skulason and Smith, 1995). Examples include sticklebacks (Schluter, 1996) and Arctic charr (Skulason et al., 1989) in postglacial lakes, and adaptive radiations of Mexican pupfish (Humphries, 1984). Ecological and phylogenetic patterns of diversity observed in larger and older species flocks suggest that ecological speciation was involved in their origin too. Examples are Lake Tana barbs (Nagelkerke et al., 1995; Berrebi and Valiushok, 1998) and African cichlid species flocks. The deepest branches in the phylogeny of the cichlid species flock in the Cameroonian crater lake Barombi Mbo separate feeding guilds (Schliewen et al., 1994), and the deepest branches in the phylogenies of the Lake Tanganyika and the Lake Malawi cichlid flocks separate pelagic from littoral cichlids, followed by one major and several smaller splits of the littoral clades into rock and sand dwellers (Sturmbauer and Meyer, 1993; Sturmbauer et al., 1994; Kocher et al., 1995; Nishida, 1997).

   Under conditions where sexual selection is strong and disruptive, sympatric speciation by sexual selection is predicted to occur even more rapidly than ecological speciation because the disruptive selection on the mating system is present initially rather than being recruited by natural selection against intermediate phenotypes (Wu, 1985; Turner and Burrows, 1995; Payne and Krakauer, 1997; Van Doorn et al., 1998). When ecological space is available, speciation by disruptive sexual selection will interact with and accelerate the adaptive radiation process by rapidly generating sympatric genetically isolated incipient species that can respond more quickly than interbreeding resource utilization morphs to the disruptive natural selection that seems to be characteristic for unsaturated lake communities. The phylogenetic pattern produced in such situations with rapid ecological character displacement is very similar to that produced in situations with real ecological speciation. The deepest splits in a phylogeny will, in both cases, separate ecological guilds. The difference is that taxa that speciate by disruptive sexual selection continue to speciate as niche space fills up. Species numbers will increase until the extinction rate (competitive exclusions and stochastic extinctions) equals the speciation rate. The species number at which equilibrium is reached depends on the relative importance of density-dependent (resource competition) and density-independent (e.g. predation) mortality, the species-carrying capacity for that habitat and the survivability of new species. Because ecological differentiation that is sufficient to facilitate coexistence may be less evident than ecological differentiation that would cause assortative mate choice (which is required for ecological speciation), species packing is expected to be higher in those taxa that speciate by disruptive sexual selection.

The available information on the phylogeny of the rock-dwelling haplochromines in Lake Victoria (Seehausen et al., 1998a; E. Lippitsch and O. Seehausen, unpubl.) supports predictions from a speciation by disruptive sexual selection model that incorporates the effects of disruptive natural selection. In the monophyletic rock-dwelling lineage that comprises the genera *Neochromis, Mbipia, Pundamilia* and *Lithochromis*, at least some of the first splits separate distinct feeding guilds. *Neochromis* and *Mbipia* feed on firmly attached and loose algae, respectively, while most species of *Pundamilia* and *Lithochromis* eat insect larvae and zooplankton. Most later splits are associated only with relatively minor ecological differentiation (Figure 7). A very similar picture emerges from studies on rock-dwelling Lake Malawi cichlids (Albertson et al., 1999).

Even though the ecological and anatomical differences between sympatric (and allopatric) sister species are small, the existence of such differences (Seehausen et al., 1998a) indicates that species are not mere short-lived mate-choice morphs, but persist over evolutionary time. That ecological and anatomical differences tend to correlate (Seehausen and Bouton, 1997; Bouton et al., 1999) suggests that species affect each other in terms of resource utilization and do not evolve in an entirely competition-free space with continuously superfluous resources. Survival of a new species will depend on its ability to respond to ecological demands, and it is here where versatile anatomy, phenotypic plasticity and learning capability, that together give rise to ecological and behavioural flexibility, would become critical (Galis and Metz, 1998). Such flexibility probably contributes greatly to the unusually high species packing in cichlid communities.

Because it simultaneously fulfils the two critical requirements of disruptive selection and assortative mating, speciation by disruptive sexual selection overcomes theoretical difficulties inherent with sympatric speciation (Felsenstein, 1981; Rice, 1987). It offers an explanation for the rapidity of cladogenesis in Lake Victoria and for the absence of much cladogenesis in riverine environments. Light conditions in most rivers are either permanently or temporarily unsuitable to allow disruptive sexual selection on coloration to

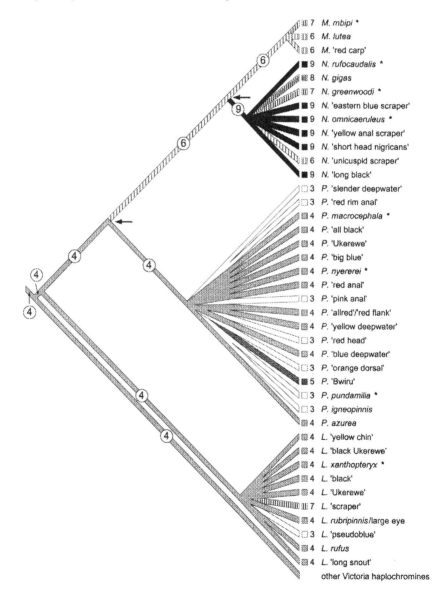

cause evolution of lasting mate-choice polymorphism. The proposed speciation mechanisms, moreover, can explain the rarity of interspecific haplochromine hybrids in Lakes Victoria and Malawi (Dominey, 1984). If most speciation events occurred in accordance with the traditional model of allopatric speciation, interspecific hybrids would be expected to be common in contact zones (Barton and Hewitt, 1981; Shaw, 1981). However, observations suggestive of such situations are restricted to cases where males of sister species belong to the same body coloration morph (blue, red dorsum or red ventrum) and have the same fin coloration: in Lake Victoria, the ring species pair *Pundamilia pundamilia* and *P.* 'Big blue' (Seehausen, 1996, p. 275) and the allopatric pair *P. nyererei* and *P. igneopinnis* Seehausen & Lippitsch 1998 (Seehausen, 1996, p. 114), and in Lake Malawi, translocated populations of *Cynotilapia afra* (Günther 1894) and *Maylandia zebra* (Boulenger 1899) (Stauffer *et al.*, 1996).

Placing these results in the larger context of explaining adaptive radiation suggests that sexual selection, if disruptive, can considerably increase the speed of adaptive radiation because the genetic fragmentation of populations allows an increased rate of differential response to disruptive natural selection. Disruptive sexual selection is, therefore, a likely explanation for dispropor- tionally fast adaptive radiations such as those of haplochromine cichlids in several African Great Lakes. That only some haplochromine lineages have undergone explosive cladogenesis and radiation might be explained by

Fig. 7. Morphology-derived estimate of diet traced on the phylogenetic tree of the rock- dwelling Vertical Bar Mbipi lineage (Seehausen *et al.*, 1998a). The tree is based on the cladistic analysis of scale and squamation characters. Diet is estimated from morphology along a continuum from items obtained predominantly by suction feeding (plankton and macrobenthos), via items that require a combination of suction and biting (loose algae cover), to items that require forceful biting (firmly attached algae; Bouton *et al.*, 1999). The estimates are obtained from a combination of three morphological characters that are highly correlated with diet: the lower jaw length/ width ratio, the crown shape of the outer row teeth, and the number and arrangement of inner tooth rows. A species is given between 1 and 3 points for each character, where the biting component increases from 1 to 3 (lower jaw length/width: 1, > 1.4; 2, 1.4–1.2; 3, < 1.2; crown shape of outer teeth: 1, unicuspid/weakly bicuspid; 2, unequally bicuspid; 3, subequally bicuspid; inner tooth rows: 1, ≤ 3 not closely set; 2, > 3 not closely set; 3, > 3 closely set). A typical plankton/macrobenthos feeder obtains a sum of 3 or 4, a typical scraper of firmly attached algae obtains a sum of 8 or 9, and loose algae feeders obtain intermediate sum values, represented by the shading of the tree branches (light shading indicates feeding on loose items, and dark shading feeding on firmly attached items). Morphological characters are taken from Table 3 in Seehausen (1996). The diet of eight of the species has been studied in detail (Bouton *et al.*, 1997) and matches closely the morphology-derived predictions. Bold arrows point to nodes that separate feeding guilds.

differences in the genetics of coloration and/or the mate-choice mechanism (Seehausen, 1999).

Models that seek to predict species diversity, or the effects of environmental change on species diversity in assemblages regulated by speciation and extinction, might be improved if sexual niche space (signal space) is considered as a limiting "resource" axis, in addition to the traditional resource axes. This is because the maintenance of diversity in such communities depends not only upon ecological resources but also on the persistence of the selection regime and the environmental signal space under which diversity evolved. Changes in the environment that relax selection or its disruptive character, or reduce signal space, are predicted to lead to an erosion of diversity by the amalgamation of gene pools (Seehausen *et al.*, 1997; Van Doorn *et al.*, 1998).

ACKNOWLEDGEMENTS

I thank the Tanzanian Commission for Science and Technology for research permission, the Tanzanian Fisheries Research Institute (Prof. P.O.J. Bwathondi and E.F.B. Katunzi) for hospitality and support, the Tanzanian Fisheries Division for permits to export live fishes, J.J.M. van Alphen, F. Witte, N. Bouton, F. Galis, R. Lande and P.J. Mayhew for collaboration, M.K. Kayeba, R. Enoka, A. Samwel-Terry, M. Haluna, P. Snelderwaard, L. Assembe, S. Ober and A. Rozier for technical assistance, J.J.M. van Alphen, G.F. Turner, F. Witte, F. Galis and A. Rossiter for comments on the manuscript, J. Endler and M.E. Knight for very helpful discussions, R. Kitery, H. Gonza Mbilinyi, H. Nielsen, A. Samwel-Terry, L. Schadhauser and C. Schaefer for logistical support, M. Brittijn for making or improving the figures, and WOTRO (grant no. 84-282) and NWO (grant no. 805-36.025) for financial support.

REFERENCES

Albertson, R.C., Markert, J.A., Danley, P.D. and Kocher, T.D. (1999). Phylogeny of a rapidly evolving clade: the cichlid fishes of Lake Malawi, East Africa. *Proc. Nat Acad. Sciences USA* **96,** 5107–5110.

Arnegard, M.E., Markert, J.A., Danley, P.D., Stauffer, J.R. Jr, Ambali, A.J. and Kocher, T.D. (1999). Population structure and colour variation of the cichlid fish *Labeotropheus fuelleborni* Ahl along a recently formed archipelago of rocky habitat patches in Southern Malawi. *Proc. R. Soc. Lond. B* **266,** 119–130.

Barel, C.D.N. (1984). Form-relations in the context of constructional morphology: the eye and suspensorium of lacustrine Cichlidae (Pisces, Teleostei). *Neth. J. Zool.* **34,** 439–502.

Barlow, G. (1976). The midas cichlid in Nicaragua. In: *Investigations of the Ichthyofauna of Nicaraguan Lakes* (Ed. by T.B. Thorson), pp. 333–358. School of Life Sciences, University of Nebraska-Lincoln, Lincoln, NE.

Barlow, G.W. (1983). The benefits of being gold: behavioral consequences of polychromatism in the midas cichlid, *Cichlasoma citrinellum. Envir. Biol. Fishes* **8**, 235–247.

Barlow, G.W. (1998). Sexual-selection models for exaggerated traits are useful but constraining. *Am. Zool.* **38**, 59–69.

Barton, N.H. and Hewitt, G.M. (1981). Hybrid zones and speciation. In: *Evolution and Speciation: Essays in Honour of M.J.D. White* (Ed. by W.R. Atchley and D.S. Woodruff), pp. 109–145. Cambridge University Press, Cambridge.

Berrebi, P. and Valiushok, D. (1998). Genetic divergence among morphotypes of Lake Tana barbs. *Biol. J. Linn. Soc. Lond.* **64**, 369–384.

Beuning, K.R.M., Kelts, K. and Ito, E. (1997). Paleohydrology of Lake Victoria, East Africa, inferred from $^{18}O/^{16}O$ ratios in sediment cellulose. *Geology* **25**, 1083–1086.

Bouton, N., Seehausen, O. and Alphen, J.J.M. van (1997). Resource partitioning among rock-dwelling haplochromines (Pisces:Cichlidae) from Lake Victoria. *Ecol. Freshwat. Fish* **6**, 225–240.

Bouton, N., Witte, F., Alphen, J.J.M. van, Schenk, A. and Seehausen, O. (1999). Local adaptations in populations of rock-dwelling haplochromines (Pisces:Cichlidae) from southern Lake Victoria. *Proc. R. Soc. Lond. B* **266**, 355–360.

Brooks, J.L. (1950). Speciation in ancient lakes. *Q. Rev. Biol.* **25**, 30–60, 131–176.

Bush, G.L. (1994). Sympatric speciation in animals: new wine in old bottles. *Trends Ecol. Evol.* **9**, 285–288.

Chapman, L.J., Chapman, C.A., Ogutu-Ohwayo, R., Chandler, M., Kaufman, L. and Keiter, A.E. (1995). Refugia for endangered fishes from an introduced predator in Lake Nabugabo, Uganda. *Conserv. Biol.* **10**, 554–561.

Cohen, A.S., Soreghan, M.J. and Scholz, C.A. (1993). Estimating the age of formation of lakes: an example from Lake Tanganyika, East African Rift system. *Geology* **21**, 511–514.

Coulter, G.W. (ed.) (1991). *Lake Tanganyika and Its Life*. Oxford University Press, Oxford.

Cracraft, J. (1994). Species diversity, biogeography, and the evolution of biotas. *Am. Zool.* **34**, 33–47.

Deutsch, J.C. (1997). Colour diversification in Malawi cichlids: evidence for adaptation, reinforcement or sexual selection? *Biol. J. Linn. Soc. Lond.* **62**, 1–14.

Dieckmann, U. and Doebeli, M. (1999). On the origin of species by sympatric speciation. *Nature* **400**, 354–357.

Diehl, S.R. and Bush, G.L. (1989). The role of habitat preference in adaptation and speciation. In: *Speciation and its Consequences* (Ed. by D. Otte and J.A. Endler), pp. 345–365. Sinauer, Sunderland, MA.

Dominey, W. (1984). Effects of sexual selection and life history on speciation: species flocks in African cichlids and Hawaiian *Drosophila*. In: *Evolution of Fish Species Flocks* (Ed. by A.A. Echelle and I. Kornfield), pp. 231–249. University of Maine at Orono Press, Orono.

Echelle, A.A. and Kornfield, I. (eds.) (1984). *Evolution of Fish Species Flocks*. University of Maine at Orono Press, Orono, Maine.

Felsenstein, J. (1981). Scepticism towards Santa Rosalia, or why are there so few kinds of animals. *Evolution* **35**, 124–138.

Fryer, G. (1996). Endemism, speciation and adaptive radiation in great lakes. *Envir. Biol. Fishes* **45**, 109–131.

Fryer, G. (1997). Biological implications of a suggested Late Pleistocene desiccation of Lake Victoria. *Hydrobiologia* **354**, 177–182.

Fryer, G. and Iles, T.D. (1972). *The Cichlid Fishes of the Great Lakes of Africa: Their Biology and Evolution*. Oliver and Boyd, London.

Futuyma, D.J. (1986). *Evolutionary Biology*. Sinauer, Sunderland, MA.

Galis, F. and Drucker, E.G. (1996). Pharyngeal biting mechanics in centrarchids and cichlids: insights into a key evolutionary innovation. *J. Evol. Biol.* **9**, 641–670.

Galis, F. and Metz, J.A.J. (1998). Why are there so many cichlid species? *Trends Ecol. Evol.* **13**, 1–2.

Genner, M.J., Turner, G.F., Barker, S. and Hawkins, S.J. (1999). Niche segregation among Lake Malawi cichlid fishes? Evidence from stable isotope signatures. *Ecol. Lett.* **2**, 185–190.

Goldschmidt, T. and Visser, J. de (1990). On the possible role of egg mimics in speciation. *Acta Biotheor.* **38**, 125–134.

Gould, S.J. (1977). *Ontogeny and Phylogeny*. Cambridge and Belknap Press, London.

Greenwood, P.H. (1965). The cichlid fishes of Lake Nabugabo, Uganda. *Bull. Br. Mus. Nat. Hist. (Zool.)* **12**, 315–357.

Greenwood, P.H. (1973). A revision of the *Haplochromis* and related species (Pisces, Cichlidae) from Lake George, Uganda. *Bull. Br. Mus. Nat. Hist. (Zool.)* **25**, 139–242.

Greenwood, P.H. (1974). The cichlid fishes of Lake Victoria, East Africa: the biology and evolution of a species flock. *Bull. Br. Mus. Nat. Hist. (Zool.)* **6**, Suppl., 1–134.

Greenwood, P.H. (1979). Towards a phyletic classification of the "genus" Haplochromis (Pisces, Cichlidae) and related taxa. Part I. *Bull. Br. Mus. Nat. Hist. (Zool.)* **35**, 265–322.

Greenwood, P.H. (1981). Species flocks and explosive speciation. In: *Chance, Change and Challenge – The Evolving Biosphere* (Ed. by P.H. Greenwood and P.L. Forey), pp. 61–74. Cambridge University Press and British Museum (Natural History), London.

Greenwood, P.H. (1987). The genera of pelmatochromine fishes (Teleostei, Cichlidae). A phylogenetic review. *Bull. Br. Mus. Nat. Hist. (Zool.)* **53**, 139–203.

Greenwood, P.H. (1994). The species flock of cichlid fishes in Lake Victoria – and those of other African Great Lakes. *Arch. Hydrobiol. Beiheft. Ergebnisse Limnol.* **44**, 347–354.

Hert, E. (1989). The function of the egg-spots in an African mouth-brooding cichlid fish. *Anim. Behav.* **37**, 726–732.

Hert, E. (1992). Homing and home-site fidelity in rock-dwelling cichlids (Pisces:Teleostei) of Lake Malawi, Africa. *Envir. Biol. Fishes* **33**, 229–237.

Holzberg, S. (1978). A field and laboratory study of the behaviour and ecology of *Pseudotropheus zebra* (Boulenger) an endemic cichlid of Lake Malawi (Pisces, Cichlidae). *Z. Zool. Syst. Evol. Forschell.* **16**, 171–187.

Hoogerhoud, R.J.C., Witte, F. and Barel, C.D.N. (1983). The ecological differentiation of two closely resembling *Haplochromis* species from Lake Victoria (*H. iris* and *H. hiatus*; Pisces, Cichlidae). *Neth. J. Zool.* **33**, 283–305.

Humphries, J.M. (1984). Genetics of speciation in pupfishes from Laguna Chichancanab, Mexico. In: *Evolution of Fish Species Flocks* (Ed. by A.A. Echelle and I. Kornfield), pp. 129–139. University of Maine at Orono Press, Orono, ME.

Johnson, T.C., Scholz, C.A., Talbot, M.R., Kelts, K., Ricketts, R.D., Ngobi, G., Beuning, K., Ssemmanda, I. and McGill, J.W. (1996a). Late Pleistocene desiccation of Lake Victoria and rapid evolution of cichlid fishes. *Science* **273**, 1091–1093.

Johnson, P.A., Hoppenstaedt, F.C., Smith, J.J. and Bush, G.L. (1996b). Conditions for sympatric speciation: a diploid model incorporating habitat fidelity and non-habitat assortative mating. *Evol. Ecol.* **10**, 187–205.

Karino, K. (1997). Female mate preference for males having long and symmetric fins in the bower-holding cichlid *Cyathopharynx furcifer*. *Ethology* **103**, 883–892.

Kaufman, L.S. and Liem, K.F. (1982). Fishes of the Suborder Labroidei (Pisces:Perciformes): phylogeny, ecology and evolutionary significance. *Breviora* **472**, 1–19.

Kaufman, L.S. and Ochumba, P. (1993). Evolutionary and conservation biology of cichlid fishes as revealed by faunal remnants in northern Lake Victoria. *Conserv. Biol.* **7**, 719–730.

Kaufman, L.S., Chapman, L.J. and Chapman, C.A. (1997). Evolution in fast forward: haplochromine fishes of the Lake Victoria region. *Endeavour* **21**, 23–30.

Kawecki, T.J. (1996). Sympatric speciation driven by beneficial mutations. *Proc. R. Soc. Lond. B* **263**, 1515–1520.

Knight, M.E., Turner, G.F., Rico, C., Oppen, M.J.H. van and Hewitt, G.M. (1998). Microsatellite paternity analysis on captive Lake Malawi cichlids supports reproductive isolation by direct mate choice. *Mol. Ecol.* **7**, 1605–1610.

Kocher, T.D., Conroy, J.A., McKaye, K.R., Stauffer, J.R. and Lockwood, S.F. (1995). Evolution of NADH dehydrogenase subunit 2 in East African cichlid fish. *Mol. Phylogenet. Evol.* **4**, 420–432.

Kondrashov, A.S. and Kondrashov, F.A. (1999). Interactions among quantitative traits in the course of sympatric speciation. *Nature* **400**, 351–354.

Kondrashov, A.S. and Mina, M.V. (1986). Sympatric speciation: when is it possible? *Biol. J. Linn. Soc. Lond.* **27**, 201–223.

Konings, A. (1995). *Malawi Cichlids in Their Natural Habitat*, 2nd ed. Cichlid Press, St Leon-Rot, Germany.

Kornfield, I.I. and Carpenter, K.E. (1984). Cyprinids of Lake Lanao, Phillippines: taxonomic validity, evolutionary rates and speciation scenarios. In: *Evolution of Fish Species Flocks* (Ed. by A.A. Echelle and I. Kornfield), pp. 69–84. University of Maine at Orono Press, Orono.

Kornfield, I.I., Smith, D.C., Gagnon, P.S. and Taylor, J.N. (1982). The cichlid fish of Cuatro Cienegas, Mexico: direct evidence of conspecificity among distinct morphs. *Evolution* **36**, 658–664.

Kosswig, C. (1947). Selective mating as a factor for speciation in cichlid fish of East African lakes. *Nature* **159**, 604–605.

Kottelat, M. (1997). European freshwater fishes. *Biologia* **52**, Suppl. 5, 1–271.

Lande, R. (1982). Rapid origin of sexual isolation and character divergence in a cline. *Evolution* **36**, 1–12.

Lande, R. and Wilkinson, G.S. (1999). Models of sex-ratio meiotic drive and sexual selection in stalk-eyed flies. *Genet. Res.* **74**, 245–253.

Liem, K.F. (1973). Evolutionary strategies and morphological innovations: cichlid pharyngeal jaws. *Syst. Zool.* **22**, 425–441.

Liem, K.F. and Kaufman, L.S. (1984). Intraspecific macroevolution: Functional biology of the polymorphic cichlid species *Cichlasoma minckleyi*. In: *Evolution of Fish Species Flocks* (Ed. by A.E. Echelle and I. Kornfield), pp. 203–215. University of Maine at Orono Press, Orono, Maine.

Lippitsch, E. (1993). A phyletic study on lacustrine haplochromine fishes (Perciformes, Cichlidae) of East Africa, based on scale and squamation characters. *J. Fish Biol.* **42**, 903–946.

Lippitsch, E. (1997). Phylogenetic investigations on the haplochromine Cichlidae of Lake Kivu (East Africa), based on lepidological characters. *J. Fish Biol.* **51**, 284–299.

Lowe-McConnell, R.H. (1987). *Ecological Studies in Tropical Fish Communities*. Cambridge University Press, Cambridge.

McCune, A. (1997). How fast is speciation? Molecular, geological, and phylogenetic evidence from adaptive radiations of fishes. In: *Molecular Evolution and Adaptive*

Radiation (Ed. by T.J. Givnish and K.J. Sytsma), pp. 585–610. Cambridge University Press, Cambridge.

McElroy, D.M. and Kornfield, I. (1990). Sexual selection, reproductive behaviour, and speciation in the mbuna species flock of Lake Malawi (Pisces:Cichlidae). *Envir. Biol. Fishes* **28**, 273–284.

McElroy, D.M., Kornfield, I. and Everett, J. (1991). Colouration in African cichlids: diversity and constraints in Lake Malawi endemics. *Neth. J. Zool.* **41**, 250–268.

McKaye, K.R. (1991). Sexual selection and the evolution of the cichlid fishes of Lake Malawi, Africa. In: *Cichlid Fishes. Behaviour, Ecology and Evolution* (Ed. by M.H.A. Keenleyside), pp. 241–257. Chapman and Hall, London.

McKaye, K.R. and Gray, W.N. (1984). Extrinsic barriers to gene flow in rock-dwelling cichlids of Lake Malawi: macrohabitat heterogeneity and reef colonization. In: *Evolution of Fish Species Flocks* (Ed. by A.A. Echelle and I. Kornfield), pp. 169–183. University of Maine at Orono Press, Orono, Maine.

McKaye, K.R., Kocher, T., Reinthal, P. and Kornfield, I. (1982). A sympatric sibling species complex of *Petrotilapia* Trewavas from Lake Malawi analysed by enzyme electrophoresis (Pisces:Cichlidae). *Zool. J. Linn. Soc. Lond.* **76**, 91–96.

McKaye, K.R., Kocher, T., Reinthal, P., Harrison, R. and Kornfield, I. (1984). Genetic evidence for allopatric and sympatric differentiation among color morphs of a Lake Malawi cichlid fish. *Evolution* **38**, 215–219.

McKaye, K.R., Louda, S.M. and Stauffer, J.R., Jr (1990). Bower size and male reproductive success in a cichlid fish lek. *Am. Nat.* **135**, 597–613.

Marsh, A.C., Ribbink, A.J. and Marsh, B.A. (1981). Sibling species complexes in sympatric populations of *Petrotilapia* Trewavas (Cichlidae, Lake Malawi). *Zool. J. Linn. Soc. Lond.* **71**, 253–264.

Martens, K. (1997). Speciation in ancient lakes. *Trends Ecol. Evol.* **12**, 177–181.

Martens, K., Goddeeris, B. and Coulter, G. (eds) (1994). Speciation in ancient lakes. *Arch. Hydrobiol. Beiheft. Ergebnisse Limnol.* **44.**

Maynard Smith, J. (1966). Sympatric speciation. *Am. Nat.* **100**, 637–650.

Mayr, E. (1942). *Systematics and the Origin of Species.* Columbia University Press, New York.

Mayr, E. (1963). *Animal Species and Evolution.* Harvard University Press, Cambridge, MA.

Meyer, A. (1990). Ecological and evolutionary aspects of the trophic polymorphism in *Cichlasoma citrinellum* (Pisces:Cichlidae). *Biol. J. Linn. Soc. Lond.* **39**, 279–299.

Meyer, A. (1993). Phylogenetic relationships and evolutionary processes in East African cichlid fishes. *Trends Ecol. Evol.* **8**, 279–284.

Nagelkerke, L.A.J. (1997). The barbs of Lake Tana, Ethiopia. Unpublished PhD Thesis, University of Wageningen, The Netherlands.

Nagelkerke, L.A.J., Mina, M.V., Wudneh, T., Sibbing, F.A. and Osse, J.W.M. (1995). In Lake Tana, a unique fish fauna needs protection. *BioScience* **45**, 772–775.

Nishida, M. (1997). Phylogenetic relationships and evolution of Tanganyikan cichlids: a molecular perspective. In: *Fish Communities in Lake Tanganyika* (Ed. by H. Kawanabe, M. Hori and M. Nagoshi), pp. 1–23. Kyoto University Press, Kyoto.

Noest, A.J. (1997). Instability of the sexual continuum. *Proc. R. Soc. Lond. B* **264**, 1389–1393.

Orti, G. and Meyer, A. (1997). The radiation of characiform fishes and the limits of resolution of mitochondrial ribosomal DNA sequences. *Syst. Biol.* **46**, 75–100.

Paterson, H.E.H. (1985). The recognition concept of species. In: *Species and Speciation* (Ed. by E.S. Vrba), pp. 21–29. Transvaal Museum Monograph No. 4, Pretoria.

Payne, R.J.H. and Krakauer, D.C. (1997). Sexual selection, space, and speciation. *Evolution* **51**, 1–9.

Poll, M. (1980). Ethologie comparee des poissons fluviatiles et lacustres africains. *Acad. R. Belg. Mem. Classe Sci. 5e Ser.* **66**, 78–97.

Price, T. (1998). Sexual selection and natural selection in bird speciation. *Phil. Trans. R. Soc. Lond. B* **353**, 251–260.

Ribbink, A.J. (1994). Alternative perspectives on some controversial aspects of cichlid speciation. *Arch. Hydrobiol. Beiheft. Ergebnisse Limnol.* **44**, 101–125.

Ribbink, A.J., Marsh, B.A., Marsh, A.C., Ribbink, A.C. and Sharp, B.J. (1983). A preliminary survey of the cichlid fishes of rocky habitats in Lake Malawi. *S. Afr. J. Zool.* **18**, 149–309.

Rice, W.R. (1987). Selection via habitat specialization: the evolution of reproductive isolation as a correlated character. *Evol. Ecol.* **1**, 301–314.

Roe, K.J., Conkel, D. and Lydeard, C. (1997). Molecular systematics of Middle American cichlid fishes and the evolution of trophic-types in "*Cichlasoma (Amphilophus)*" and "*C. (Thorichthys)*". *Mol. Phylogenet. Evol.* **3**, 366–376.

Rosenberger, A.E. and Chapman, L.J. (1999). Hypoxic wetland tributaries as faunal refugia from an introduced predator. *Ecol. Freshwat. Fish* **8**, 22–34.

Rosenzweig, M.L. (1978). Competitive speciation. *Biol. J. Linn. Soc. Lond.* **10**, 275–289.

Rossiter, A. (1994). Territory, mating success, and the individual male in a lekking cichlid fish. In: *Animal Societies: Individuals, Interactions and Organisation* (Ed. by P.J. Jarman and A. Rossiter), pp. 43–55. Kyoto University Press, Kyoto.

Rossiter, A. (1995). The cichlid fish assemblages of Lake Tanganyika, ecology, behaviour and evolution of its species flock. *Adv. Ecol. Res.* **26**, 187–252.

Rossiter, A. and Yamagishi, S. (1997). Intraspecific plasticity in the social system and mating behaviour of a lek-breeding cichlid fish. In: *Fish Communities in Lake Tanganyika* (Ed. by H. Kawanabe, M. Hori and M. Nagoshi), pp. 193–218. Kyoto University Press. Kyoto.

Rüber, L., Verheyen, E. and Meyer, A. (1999). Replicated evolution of trophic specializations in an endemic cichlid fish lineage from Lake Tanganyika. *Proc. Nat. Acad. Sci. USA* **96**, 10230–10235.

Ryan, M.J. (1998). Sexual selection, receiver biases, and the evolution of sex differences. *Science* **281**, 1999–2003.

Sage, R.D. and Selander, R.K. (1975). Trophic radiation through polymorphism in cichlid fishes. *Proc. Nat Acad. Sci. USA* **72**, 4669–4673.

Schliewen, U.K., Tautz, D. and Paabo, S. (1994). Sympatric speciation suggested by monophyly of crater lake cichlids. *Nature* **368**, 629–633.

Schluter, D. (1996). Ecological speciation in postglacial fishes. *Phil. Trans. R. Soc. Lond. B* **351**, 807–814.

Seehausen, O. (1996). *Lake Victoria Rock Cichlids–Taxonomy, Ecology and Distribution.* Verduijn Cichlids, Zevenhuizen, The Netherlands.

Seehausen, O. (1997). Distribution of and reproductive isolation among colour morphs of a rock-dwelling Lake Victoria cichlid (*Haplochromis nyererei*). *Ecol. Freshwat. Fish* **6**, 59–66.

Seehausen, O. (1999). Speciation and species richness in African cichlids: effects of sexual selection by mate choice. PhD Thesis, University of Leiden, The Netherlands. Published as *Speciation and Species Richness in African Cichlids.* Thela Thesis Publishers, Amsterdam.

Seehausen, O. and Alphen, J.J.M. van (1998). The effect of male coloration on female mate choice in closely related Lake Victoria cichlids (*Haplochromis nyererei* complex). *Behav. Ecol. Sociobiol.* **42**, 1–8.

Seehausen, O. and Alphen, J.J.M. van (1999). Can sympatric speciation by disruptive sexual selection explain rapid evolution of cichlid diversity in Lake Victoria? *Ecol. Lett.* **2**, 262–271.

Seehausen, O. and Alphen, J.J.M. van (2000). Reply: Inferring modes of speciation from distribution patterns. *Ecol. Lett.* **3**, 169–171.

Seehausen, O. and Bouton, N. (1996). Polychromatism in rock dwelling Lake Victoria cichlids: types, distribution, and observations on their genetics. *Cichlids Yearbook* **6**, 36–45.

Seehausen, O. and Bouton, N. (1997). Microdistribution and fluctuations in niche overlap in a rocky shore cichlid community in Lake Victoria. *Ecol. Freshwat. Fish* **6**, 161–173.

Seehausen, O. and Bouton, N. (1998). The community of rock-dwelling cichlids in Lake Victoria. *Bonner Zool. Beitraege* **47**, 301–312.

Seehausen, O., Alphen, J.J.M. van and Witte, F. (1997). Cichlid fish diversity threatened by eutrophication that curbs sexual selection. *Science* **277**, 1808–1811.

Seehausen, O., Lippitsch, E., Bouton, N. and Zwennes, H. (1998a). Mbipi, the rock-dwelling cichlids of Lake Victoria: description of three new genera and fifteen new species (Teleostei). *Ichthyol. Explor. Freshwat.* **9**, 129–228.

Seehausen, O., Witte, F., Alphen, J.J.M. van and Bouton, N. (1998b). Direct mate choice maintains diversity among sympatric cichlids in Lake Victoria. *J. Fish Biol.* **53**, Suppl. A, 37–55.

Seehausen, O., Mayhew, P.J. and Alphen, J.J.M. van (1999a). Evolution of colour patterns in East African cichlid fish. *J. Evol. Biol.* **12**, 514–534.

Seehausen, O., Alphen, J.J.M. van and Lande, R. (1999b). Color polymorphism and sex ratio distortion in a cichlid fish as an incipient stage in sympatric speciation by sexual selection. *Ecol. Lett.* **2**, 367–378.

Seehausen, O., Alphen, J.J.M. van and Witte, F. (1999c). Can ancient colour polymorphisms explain why some cichlid lineages speciate rapidly under disruptive sexual selection? *Belg. J. Zool.* **129**, 279–294.

Shaw, D.D. (1981). Chromosomal hybrid zones in orthopteroid insects. In: *Evolution and Speciation: Essays in Honour of M.J.D. White* (Ed. by W.R. Atchley and D.S. Woodruff), pp. 146–170. Cambridge University Press, Cambridge.

Skulason, S. and Smith, T.B. (1995). Resource polymorphisms in vertebrates. *Trends Ecol. Evol.* **10**, 366–370.

Skulason, S., Noakes, D.L.G. and Snorrason, S.S. (1989). Ontogeny of trophic morphology in four sympatric morphs of arctic char *Salvelinus alpinus* in Thingvallavatn, Iceland. *Biol. J. Linn. Soc. Lond.* **38**, 281–301.

Slobodyanyuk, S.J., Kirilchik, S.V., Pavlova, M.E., Belikov, S.I. and Novitsky, A.L. (1995). The evolutionary relationships of two families of cottoid fishes of Lake Baikal (East Siberia) as suggested by analysis of mitochrondrial DNA. *J. Mol. Evol.* **40**, 392–399.

Snoeks, J., Rüber, L. and Verheyen, E. (1994). The Tanganyika problem: comments on the taxonomy and distribution patterns of its cichlid fauna. *Arch. Hydrobiol. Beiheft. Ergebnisse Limnol.* **44**, 355–372.

Sodsuk, P.K., MacAndrew, B.A. and Turner, G.F. (1995). Evolutionary relationships of the Lake Malawi *Oreochromis* species: evidence from allozymes. *J. Fish Biol.* **47**, 321–333.

Stager, J.C., Reinthal, P.N. and Livingstone, D.A. (1986). A 25,000-year history for Lake Victoria, East Africa, and some comments on its significance for the evolution of cichlid fishes. *Freshwat. Biol.* **16**, 15–19.

Stauffer, J.R., Jr, Bowers, N.J., Kocher, T.D. and McKaye, K.R. (1996). Evidence of hybridization between *Cynotilapia afra* and *Pseudotropheus zebra* (Teleostei:Cichlidae) following an intralacustrine translocation in Lake Malawi. *Copeia* **1996**, 203–208.

Stiassny, M.L.J. (1990). *Tylochromis*, relationships and the phylogenetic status of the African Cichlidae. *Am. Mus. Novitates* **2993**, 1–14.

Stiassny, M.L.J. (1997). A phylogenetic overview of the lamprologine cichlids of Africa (Teleostei, Cichlidae): a morphological perspective. *S. Afr. J. Sci.* **93**, 513–523.

Stiassny, M.L.J. and Jensen, J.S. (1987). Labroid intrarelationships revisited: morphological complexity, key innovations, and the study of comparative diversity. *Bull. Mus. Comp. Zool.* **151**, 269–319.

Strauss, R.E. (1984). Allometry and functional feeding morphology in haplochromine cichlids. In: *Evolution of Fish Species Flocks* (Ed. by A.A. Echelle and I. Kornfield), pp. 217–229. University of Maine at Orono Press, Orono, Maine.

Strecker, U., Meyer, C.G., Sturmbauer, C. and Wilkens, H. (1996). Genetic divergence and speciation in an extremely young species flock in Mexico formed by the genus *Cyprinodon* (Cyprinodontidae, Teleostei). *Mol. Phylogenet. Evol.* **6**, 143–149.

Sturmbauer, C. and Meyer, A. (1992). Genetic divergence, speciation and morphological stasis in a lineage of African cichlid fishes. *Nature* **358**, 578–581.

Sturmbauer, C. and Meyer, A. (1993). Mitochondrial phylogeny of the endemic mouthbrooding lineages of cichlid fishes from Lake Tanganyika in Eastern Africa. *Mol. Biol. Evol.* **10**, 751–768.

Sturmbauer, C., Verheyen, E. and A. Meyer. (1994). Mitochondrial phylogeny of the Lamprologini, the major substrate spawning lineage of cichlid fishes from Lake Tanganyika in eastern Africa. *Mol. Biol. Evol.* **11**, 691–703.

Trewavas, E. (1983). *Tilapiine Fishes of the Genera Sarotherodon, Oreochromis and Danakilia*. British Museum (Natural History), London.

Turner, G.F. (1994). Speciation mechanisms in Lake Malawi cichlids: a critical review. *Arch. Hydrobiol. Beiheft. Ergebnisse Limnol.* **44**, 139–160.

Turner, G.F. (1996). *Offshore Cichlids of Lake Malawi*. Cichlid Press, Lauenau, Germany.

Turner, G.F. (1999). Explosive speciation of African cichlid fishes. In: *The Evolution of Biological Diversity* (Ed. by A.E. Magurran and R.M. May), pp. 217–229. Oxford University Press, Oxford.

Turner, G.F. and Burrows, M.T. (1995). A model of sympatric speciation by sexual selection. *Proc. R. Soc. Lond. B* **260**, 287–292.

Turner, B.J. and Grosse, D.J. (1980). Trophic differentiation in *Ilyodon*, a genus of stream dwelling goodeid fishes: speciation versus ecological polymorphism. *Evolution* **34**, 259–270.

Van Doorn, G.S., Noest, A.J. and Hogeweg, P. (1998). Sympatric speciation and extinction driven by environment dependent sexual selection. *Proc. R. Soc. Lond. B* **265**, 1915–1919.

van Oppen, M.J.H., Turner, G.F. and Rico, C. (1997). Unusually fine-scale genetic structuring found in rapidly speciating Malawi cichlid fishes. *Proc. R. Soc. Lond. B* **264**, 1803–1812.

Verheyen, E., Rüber, L., Snoeks, J. and Meyer, A. (1996). Mitochondrial phylogeography of rock-dwelling cichlid fishes reveals evolutionary influence of historical lake level fluctuations of Lake Tanganyika, Africa. *Phil. Trans. R. Soc. Lond. B* **351**, 797–805.

Wilhelm, W. (1984). Interspecific allometric growth differences in the head of three haplochromine species (Pisces, Cichlidae). *Neth. J. Zool.* **34**, 622–628.

Williams, G.C. (1966). *Adaptation and Natural Selection*. Princeton University Press, Princeton, NJ.

Witte, F., Barel, K.D.N. and Oijen, M.J.P. van (1997). Intraspecific variation of haplochromine cichlids from Lake Victoria and its taxonomic implications. *S. Afr. J. Sci.* **93**, 585–594.

Woltereck, R. (1931). Wie entsteht eine endemische Rasse oder Art? *Biol. Zentrabl.* **51**, 231–253.

Wu, C. (1985). A stochastic simulation study on speciation by sexual selection. *Evolution* **39**, 66–82.

Phylogeny of a Gastropod Species Flock: Exploring Speciation in Lake Tanganyika in a Molecular Framework

E. MICHEL

I. SUMMARY

A phylogenetic tree based on mitochondrial DNA (mtDNA) sequence data is presented that provides a basic framework for elucidating relationships of the endemic Tanganyikan freshwater gastropod genera *Lavigeria* and "Nov. gen." (yet to be formally described). Analyses indicate that these two genera are sister taxa, yet their diversification patterns contrast markedly. Each contains definable species-level clades. The two species in the genus Nov. gen. are morphologically conservative and genetically divergent from each other and from *Lavigeria*. The five species of *Lavigeria* delineated in this analysis form a species flock. In comparison with Nov. gen., the *Lavigeria* flock is morphologically more variable both within and among species, but shows less genetic differentiation. Relationships within *Lavigeria* are resolved among the largest species, which form a single clade, but the basal taxon is still equivocal.

The phylogeny presented allows for examination of characters such as habitat specificity, reproductive anatomy, and shell and radula morphology. These genera show clear differences in depth preference, with Nov. gen. species dominating deep-water habitats and *Lavigeria* species diversifying in shallow to mid-depths. According to the current tree, hard-substrate specificity is basal for both genera, but in the absence of an unambiguous closest outgroup for these gastropods, this must remain questionable. Perhaps the most significant character difference between Nov. gen. and *Lavigeria* is in reproductive mode: Nov. gen. species lay eggs whereas *Lavigeria* species brood their young in the pallial oviduct. Trophic differences, as implied by morphological differences in the radular teeth, and verified by differences in stable isotopes in the body tissues, are also present.

The sister clades of *Lavigeria* and Nov. gen. contrast in morphological, genetic and life history characters, but have diversified as highly derived endemics in the same ancient lake. This means that they have experienced the same geological history, and because their distributions are largely sympatric, they provide ideal material for studies of speciation and for comparative phylogenetic studies. Clearly, the time is ripe for detailed investigations of the gastropod faunas of Lake Tanganyika and of other ancient lakes (e.g. see Geary *et al.*, this volume).

II. INTRODUCTION

Hotspots of biodiversity, such as the Great Lakes of Africa, are often home to large numbers of endemic species. Some of these groups of organisms form species flocks, which by definition include clades of closely related species endemic to a geographically circumscribed area (Greenwood, 1984). The existence of species flocks raises questions about standard allopatric models for divergence and maintenance of species boundaries, and alerts us to the potential influence of other biological characters that may lead to divergence in sympatry. Lake Tanganyika, the oldest and deepest of the African Great Lakes, harbours species flocks in numerous higher taxa, in contrast to the other African Great Lakes, where species flocks occur in far fewer taxa (predominantly cichlid fishes). Of the approximately 1400 described animal and plant species in Lake Tanganyika, approximately half are endemic, with endemism approaching 100% in some groups (Coulter, 1991).

Among the endemic taxa of Lake Tanganyika the gastropods are a particularly striking group, as they exhibit high endemism, high taxomomic diversity and great morphological disparity. Their shell morphologies converge on those of marine gastropods and this similarity stimulated the first biological expeditions to the lake (see Fryer, this volume). For several reasons they constitute important material for comparative studies of speciation with other diverse taxa, such as the cichlid fishes: (i) gastropods are the major benthic grazers in the rocky littoral zone, and thus occupy an ecologically central role; (ii) they occur sympatrically with other diverse taxa, which allows for quantitative comparative evaluation of hypotheses of extrinsic speciation mechanisms, such as palaeoallopatry; (iii) they exhibit a species-flock like variation similar to that seen in other groups, with some taxa being hypervariable and showing indistinct species boundaries (implying retained ancestral polymorphism or hybridization), and other taxa being clearly definable morphologically; (iv) gastropod shells are common components within the fossil record, permitting an increased understanding of evolutionary histories. Shells also retain a record of ontogeny from the juvenile to the adult as a result of accretionary growth, enabling ontogenetic studies to be performed within an evolutionary perspective.

Gastropod speciation in the African lakes has played a central role in one of the twentieth century's most heated evolutionary debates, that concerning macroevolution versus microevolution. The championing of the Plio-Pliocene fossils of the endemic gastropod fauna of Lake Turkana as the prime example of punctuated equilibrium (Williamson, 1981a, b) led to intense disagreement over the role and importance of shell variation in these animals (Fryer *et al.*, 1983, 1985; Williamson, 1985). In essence, the argument revolved around answering one fundamental question: if we are to understand the world's living diversity, should not we be studying the deep time scales available only through fossil samples, with attendant focus on morphology? Or should we instead adopt a uniformitarian perspective, studying characters such as genetics, behaviour and ecology, visible only from observations of living animals, and therefore at geologically brief time scales?

Although this debate has mellowed in recent years, the potential of gastropod radiations to illuminate both sides of the issue has not decreased. Other than molluscs, few taxa have a sufficiently complete fossil record with which to address this question, yet surprisingly few evolutionary studies have addressed molluscan taxa in general. In this chapter both molecular and morphological data on the Tanganyikan gastropod species flock are utilized to understand speciation in this ancient lake and to provide the first synthetic species-level phylogeny of any African Great Lake gastropods.

The thiarid gastropods of Lake Tanganyika are notable for the diversity of taxa that seem to have evolved *in situ*, yet coexist in sympatry (Brown, 1980, 1994; Brown and Mandel-Barth, 1987; Coulter, 1991; West and Michel, this volume). In addition to addressing fundamental questions of ecology and coexistence mechanisms, studies of the Tanganyikan gastropods can provide a model system for understanding evolutionary processes on a number of different levels. The genus *Lavigeria* and that referred to here as "Nov. gen." (it is newly recognized and lacks an available name; see discussion below) are sister clades that contrast in patterns of radiation, levels of variation and several life history characters, and are the most logical group for testing models of species diversification.

The *Lavigeria* group forms a species flock of 20 or more taxa, and is similar to species flocks of some endemic Tanganyikan cichlid fishes and ostracods in having relatively high numbers of sympatric, yet closely related species, some of whose species boundaries are fuzzy with respect to geographical differences in morphology and genetics. In some localities, suites of *Lavigeria* taxa are easily differentiated, yet in other regions these morphological boundaries become unclear. These features imply that (i) this group may be in the process of speciating, with instances of convergence and cryptic species, or (ii) speciation happened rapidly, allowing multiple states of retained ancestral polymorphisms, or (iii) hybridization occurs among the taxa. In contrast, the two taxa that comprise the Nov. gen. subclade are well differentiated in all localities where they co-occur, and provide a meaningful comparison with their species flock-forming sister clade.

To construct a phylogenetic framework for determining species relationships and testing speciation hypotheses, the mitochondrial genes for cytochrome oxidase subunit I (*COI*) and *16s-ND1* in *Lavigeria* and Nov. gen. were investigated. Results are presented from initial analyses of these genetic data that demonstrate a sister clade relationship between the genera *Lavigeria* and Nov. gen. and monophyly in morphologically recognizable species-level taxa. Using this initial tree, patterns of character evolution in the morphology of shells and radulae, habitat specificity and reproductive strategy are then explored.

III. MATERIALS AND METHODS

A. Collection and Preservation

Specimens of *Lavigeria* and Nov. gen. were collected at sites along the shorelines of Burundi, Zaire (now Democratic Republic of the Congo, DRC), Zambia and Tanzania during 1989–1995. Sites for samples in this study are shown in Figure 1 and Table 1. Collections of animals were made with the aid of SCUBA or snorkel in the rocky and sandy littoral zone to 40 m water depth. After collection, mollusc shells were broken and the soft tissue was preserved in 95% ethanol. Voucher specimens of shell material were retained for all sites. Gastropod diversity in Lake Tanganyika is highest in the littoral zone (Coulter, 1991; Michel, 1994; Alin *et al.*, 1999), and at 40 m depth both species diversity

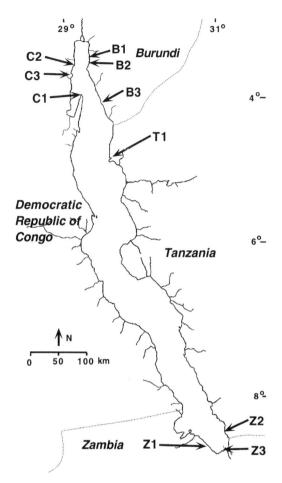

Fig. 1. Map of Lake Tanganyika showing collection sites of samples used for sequencing, anatomy/radulae and isotope studies. Site descriptions are given in Table 1. Although site Z2 is in Tanzania, it is given as a Z site so that it groups with the other far southern sites in Zambia.

and abundance decline rapidly. It is therefore probable that maximum diversity was sampled in this study. Depth and substrate type at sampling points was also recorded.

Sampling was designed to test (i) whether sympatric, but morphologically distinct individuals are truly representatives of different clades; and (ii) whether individuals that look similar yet are from widely separated populations are members of the same clade. Therefore, collections of representative specimens were made at a small number of sites located at the southern or northern ends of the lake. Specifically, the study concentrated on representatives from the two

Table 1
Site, ecological and morphological information for sequenced samples in this study

Site code	Site name	Site coordinates	Species collected	Substrate	Depth (m)	Morphological variation within clade
C1	DR Congo, Cape Banza (Ubwari Peninsula)	4°02'S, 29°15'E	Nov. gen. n. sp.[a]	Rocks and stromatolites	Dominantly deep, k:20[b], m:40[b]	Moderate
			Nov. gen. *guillemei*[a]	Rocks and stromatolites	Dominantly deep, l:20[b], n:40[b]	Moderate
			*L.*cf. *giraudi* "FR"[a]	Rocks	Shallow–middle, e:2[b]	High
			L. cf. *paucicostata*[a]	Sand and cobbles	Shallow–middle, f:2[b], i:20[b]	High
			L. n. sp.[a]	Rocks	Shallow–middle, g:2[b], j:20[b]	Moderate
			L. grandis[a]	Boulders	Shallow, h:2[b]	Moderate
C2	DR Congo, Luhanga	3°38'S, 29°09'E	*L.* n. sp.			
C3	DR Congo, Bemba	3°49'S, 29°07'E	Nov. gen. n. sp.			
B1	Burundi, km 19	3°30'S, 29°20'E	*L.* cf. *paucicostata*	Sand	Shallow	
B2	Burundi, km 29	3°38'S, 29°20'E	*L.* cf. *paucicostata*[a]	Sand	d:1[b]	
			L. cf. *giraudi* "FR"[a]	Rocks and cobbles	a:1[b]	
			Nov. gen. n. sp.[a]	Stromatolites	b:20[b]	

Table 1 *continued*

Site code	Site name	Site coordinates	Species collected	Substrate	Depth (m)	Morphological variation within clade
B3	Burundi, km 114	4°17'S, 29°34'E	*L.* cf. *paucicostata*[a]	Sand	c:20[b]	
T1	Kigoma, Tanzania	4°50'S, 29°39'E	*L. grandis,* *L. coronata*	Boulders and rocks	Shallow–middle	Moderate
Z1	Zambia, Nakaku	8°40'S, 30°54'E	Nov. gen. n. sp., Nov. gen. *guillemei,* *L.* cf. *paucicostata,* *L. grandis*			
Z2	S. Tanzania, Kasanga	8°26'S, 31°09'E	*L.* cf. *giraudi* "TM", *L.* cf. *giraudi* "TF"			
Z3	Zambia, Kasenga Pt.	8°43'S, 31°09'E	*L.* cf. *giraudi* "TM"			

[a] Samples also used for isotope study.

[b] Samples used for isotope study, with specific letter from Figure 6 shown with depth.

Depth: shallow, to wave base; medium, littoral (to c. 20 m); deep, sublittoral (> 20 m). For the two groups with high morphological variation, there are probably many more species nested within clade boundaries than are indicated by the sequence phylogeny. Generalized substrate, depth and morphological information is given only on the first mention of a species in this table.

most widely separated sites, Cape Banza, DRC (site C1), and Nakaku, Zambia (site Z1), which are 500 km apart (Figure 1), but samples from other sites were added to expand the comparison at the species level. The distribution of *L. coronata* is very limited, and at only one site (Kigoma, Tanzania, site T1), were *L. grandis* and *L. coronata* found together. Therefore, samples of these two species from this site were used. As these two species share some unique morphological characters, it was important to test for species distinction that was independent of geographical distance.

Two outgroups were selected, one a common but highly endemic thiarid from Lake Tanganyika, *Paramelania damoni* (Smith, 1881b), and the other the thiarid *Melanoides admirabilis* (Smith, 1880), which is found in the Malagarasi River flowing into Lake Tanganyika. Shell morphologies of some *Paramelania* species closely resemble those of some *Lavigeria* species, to the extent that the generic names have been confused at certain points in the taxonomic history of these groups (e.g. Leloup, 1953), and they have occasionally been suggested as being sister taxa. *Melanoides* is a thiarid genus occurring throughout much of the palaeotropics, and includes both ubiquitous species found in many habitats (e.g. *M. tuberculata*) and endemic species, such as a potential species flock in Lake Malawi (Brown, 1994). *Paramelania damoni* is oviparous and occurs on soft substrate in deep water, whereas *M. admirabilis* is probably ovoviviparous (although reproduction has not been studied in this species; Brown, 1980) and lives in muddy bottoms along river banks in the Tanganyika drainage.

B. Taxonomy and Terminology

A taxonomic revision of the species comprising *Lavigeria* (Bourguignat, 1888) and those of the undescribed genus referred to here as Nov. gen. is in progress (Michel and Todd, unpubl. data). The complexity of this group requires not only analysis of independent data sets such as morphology, anatomy and genetics, but also broad geographical sampling. Species within *Lavigeria* are sometimes widely distributed and sometimes narrowly endemic, and may also be morphologically highly variable or sometimes morphologically clearly delimited. To distinguish taxa within this group it is necessary to examine variation across the whole range of their distribution within the lake, and to make comparisons among conspecifics and congenerics from distant sites.

Because the revision is still in progress, it was thought prudent to use open nomenclature, indicated by use of "cf.", to provide some indication of the taxonomic affinity of species and species groups recognized herein to those available and more frequently employed species names in the literature. The use of "cf." indicates that the species is similar, although not necessarily identical, to the name-bearing taxon. For example, when referring to *Lavigeria* cf. *giraudi*, this indicates that a distinct taxon is recognized which will be redescribed as *L. giraudi*, or will be described as a closely related species within an informal "*L. giraudi* group" when the latter is probably subdivided upon revision.*

The earliest taxonomic work to document the diversity within what is now considered to be one genus, *Lavigeria*, recognized 46 species based on shell material alone (Bourguignat, 1885, 1888, 1890). The specimens were often in poor condition, most having been collected as shell drift from beaches at three sites in the southern part of the lake. This number is a gross overestimate of what today corresponds to about 10 species, with perhaps an additional four more of dubious identity, from that region.

A serious attempt at revising this group was made by Leloup (1953), who sampled sites all around the lake during the Belgian Hydrobiological Expedition of 1946–1947. He, however, was apparently overwhelmed by the diversity and apparent lack of discontinuous morphological characters by which to define species. As a result he ended up synonymizing all *Lavigeria* into one species, with a series of "formes" or morphs that, judging from his collections, he left with fluid taxonomic boundaries. Other taxonomic treatments (Pilsbry and Bequaert, 1927; Brown, 1980, 1994) reworked earlier data, tidying up taxonomic loose ends from earlier revisions, but without adding new data to solve the persistent problems. A recent attempted revision (Bandel, 1998) was based on samples from essentially a single site (Kigoma, Tanzania), thereby repeating the limitations of the geographically inadequate sampling of Bourguignat. The same study also employs species names plucked from Bourguignat without further discussion, for the Kigoma samples. However, by ignoring the geographical variation in these taxa, Bandel (1998) has largely misidentified and confused the Kigoma species with Bourguignat's taxa, and has used species names which, pending revision, can be applied properly only to the type series from the extreme south of the lake (Michel and Todd, unpubl. data.).

There are, however, some species for which we currently have reliable names. Specifically, *L. coronata* (Bourguignat, 1888), *L. grandis* (Smith, 1881a) and Nov. gen [*Nassopsis*] *guillemei* Martel and Dautzenberg, 1899 are morphologically clearly defined and taxonomically stable. *Lavigeria* n. sp. and Nov. gen. n. sp. are clearly defined by their morphology, but their nomenclature is still under examination. Morphological variation is high within the clades labelled here as *Lavigeria* cf. *giraudi* (Bouguignat, 1888) and *Lavigeria* cf. *paucicostata* (Smith, 1881b), and it is likely that, following detailed investigations, several

---

*In the course of intensive study of museum types and earlier publications the author was surprised to discover that common species that can be recognized easily in nature may have no clear nomenclatural heritage, and may even end up, temporarily, with the least enlightening names. For instance, Nov. gen. n. sp. (which means "new genus new species") may be easily and consistently distinguished, even by non-biologists, from among a large mixed group of Tanganyikan gastropod shells, yet still has no available name in the taxonomic literature.

taxa included here in these groups will be divided into more species (Table 1). This notwithstanding, here the species name in open nomenclature for the most well-defined member of the group will be used, with the caveat that this nomenclature is preliminary. In addition, an abbreviation is included for morphologically distinguishable forms that other data indicate are likely to be separate species (Michel, 1995).

C. Molecular Methods

DNA was obtained from soft-tissue samples using a CTAB extraction technique (T. Collins, pers. comm.; Palumbi, 1996) and a 642 base pair (bp) region from the mitochondrial *COI* gene was amplified using universal primers (Folmer *et al.*, 1994). Other gene fragments *12s*, cytochrome *b* and *ITS 2* were assessed; however, there were problems with phylogenetic information content, alignability or amplifiability. In contrast, the *COI* fragment amplified easily from *Lavigeria*, Nov. gen. and other taxa of interest (West and Michel, this volume) and gave no spurious amplifications in the negative control.

The 1400 bp fragment that spans the *16s-ND1* gene region is useful for elucidating relationships within *Lavigeria*, and sequences were obtained from 22 individuals in this clade. However, this fragment did not amplify well for other many other Tanganyikan genera and it proved possible to obtain amplifications from only four individuals in the Nov. gen. clade and none from any outgroups. The tree derived from the *16s-ND1* data will briefly be considered in light of that resulting from the *COI* sequence, but molecular variation in this fragment will be discussed elsewhere.

The *COI* data presented here are a subset of approximately 100 individuals sequenced for that mitochondrial gene fragment. Analyses of the larger *COI* data set used at least two individuals from each representative clade. These individuals were chosen so as to maximize geographical separation between sites but, where possible, were picked randomly from within clades (i.e. not biasing phylogenetic separation). Sequencing of gene-cleaned *COI* polymerase chain reaction (PCR) products was done on an ABI 377 automated sequencer, whereas most of the *16s-ND1* sequence was obtained manually. In all cases sequences were obtained in both directions and repeat sequenced if nucleotide identities were in doubt. Alignment by eye using SeqApp (Gilbert, 1994) was unambiguous. Translation to proteins was done with MEGA (invertebrate mtDNA code) and unusual amino acid changes were double checked in the sequence chromatograms.

D. Phylogenetic Analysis

To construct sequence-based trees, the maximum parsimony criterion in a branch and bound search in PAUP 4.0b1 (Swofford, 1998) was applied to unweighted sequence data. Pairwise sequence divergence was assessed between species using the LogDet/paralinear distance model, which corrects for multiple substitutions (Lake, 1994; Lockhart *et al.*, 1994). Relative information content for the molecular characters was assessed in plots of transition (TS) or transversion (TV) distances versus uncorrected genetic p-distances between pairs of samples.

Exploratory analyses of weighting transversions to transitions revealed that the parsimony results were robust to weighting (TV:TS 2:1, 3:1, 10:1, 14:1 and TV only). The weighting of 14:1 was calculated based on TV:TS values from within the *Lavigeria* clade, as that is where the relationships were least robustly resolved. Branch support was calculated with decay indices [or Bremer Support (Bremer, 1988) performed with heuristic searches for trees successively one step longer than the minimum] and with bootstrap analysis (1000 replicates each with 10 random addition sequences, saving two trees at each step). Alternative trees were compared using a maximum likelihood criterion (Goldman, 1993) based on both the *COI* and *16s-ND1* data sets, using the same weightings as in the parsimony trials. MacClade 3.06 (Maddison and Maddison, 1996) was used to manipulate trees and explore character evolution. Because individuals in which both genes were successfully sequenced were fewer and more unevenly distributed across clades than were individuals in which only one gene was successfully sequenced, phylogenies from the *COI* and *16s-ND1* sequences are here considered separately.

E. Anatomical Dissections

Soft-part anatomy can provide a wealth of useful phylogenetic characters for differentiation of species, yet earlier researchers have paid very little attention to anatomical variation in Tanganyikan gastropods. The only detailed dissections on this group were conducted on *L. grandis* by Moore (1899), who showed that despite some apparent conchological similarities, *Lavigeria* was not closely related to *Paramelania*. Moore enthused over the taxonomic value of *Lavigeria* anatomy when he wrote "... it will be readily accepted that in [*Lavigeria*] we are presented with an animal which in the future will probably constitute one of the most important prosobranchiate archetypes of which we are in search" (Moore, 1899, p. 188). However, despite this accolade and Moore's dismissive comments on the utility of shell characters for gastropod phylogenetics, subsequent work on the *Lavigeria* group has been primarily conchological (Pilsbry and Bequaert, 1927; Leloup, 1953).*

---

*This enthusiasm was probably based on Moore's mistaken idea that *Lavigeria* was a relict representative of Jurassic marine fauna that had been stranded in Lake Tanganyika as it became isolated from the sea. Thus, the anatomy of *Lavigeria* would reveal, he hoped, details of a basal marine taxon. Moore's phylogenetic and biogeographic hypotheses were debunked by Pelseneer (1906), who showed freshwater origins for the gastropods based on anatomical data (see Fryer, this volume). Anatomical data remain central in gastropod systematics, although difficult to collect.

The utility of soft-part anatomy in differentiating among members of the studied clades was assessed by dissecting 14–21 freshly killed individuals from each of six morphologically identified *Lavigeria* and Nov. gen. species collected from sites at the northern end of the lake (Figure 1: C3, Bemba and C2, Luhanga, DRC; B1, km 19 and B2, km 29, Burundi; and three sites around T1, Kigoma). Here, an overview is presented of major differences between the genera based on those dissections; the detailed morphometric comparisons are presented and discussed elsewhere (Michel, 1995).

F. Radula Samples

The radula, the chitinous ribbon of teeth with which gastropods feed, was examined to determine its utility as a taxonomic indicator and to assess whether trophic partitioning has been important in diversification of these clades. Radulae were extracted from the animals used previously in the anatomy study and preserved in alcohol. Surrounding tissue was cleaned off with 5% NaClO solution and the radula was rinsed in distilled water with a sonicator. Teeth were mounted on a glass coverslip coated with a 50% Elmer's glue mixture and coated with three or four layers of 30 nm gold sputter. Scanning electron micrographs of the three central teeth were taken at magnifications of 250–700× from a standardized viewing angle of 90°. This method facilitated the direct comparison of the teeth that are most important in food gathering and most variable in form (Michel, 1995).

G. Isotope Samples

Variation in stable isotopes of carbon and nitrogen can be used to detect differences in trophic position and energy sources among animals (Hecky and Hesslein, 1995). The ratio of $^{13}C/^{12}C$ ($\delta^{13}C$) varies with a plant's photosynthetic pathway and this characteristic ratio is then passed on to the consuming herbivore. This leaves a $\delta^{13}C$ signature in the tissue of the consumer, which indicates not only what the animal has ingested (as do gut content analyses) but, more importantly, what it has assimilated into its body tissues. Furthermore, $^{15}N/^{14}N$ ($\delta^{15}N$) increases by 3–4% with each trophic transfer and the ratio can thus be used to assess the trophic level of the consumer. The soft tissue from 17 gastropods was assessed for stable isotope variation using methods given in Hesslein *et al.* (1991) and Ramlal *et al.* (1994), using representative samples across all taxa collected from site B1, km 29, and BS, km 114, in Burundi and from site C1, Cape Banza, DRC (Table 1).

IV. RESULTS

A. Patterns of Variation in Molecular Characters

Pairwise sequence divergences in this clade were much higher than those normally seen in animal taxa. There was *c*. 30% divergence between ingroup and outgroup taxa, 17–20% divergence for comparisons between *Lavigeria* and Nov. gen. species and between the two Nov. gen. species, and 8–15% divergence for comparisons between *Lavigeria* species. Species-level differences among metazoans are usually in the order of 3%, humans and chimps differ in only 1.5% of their DNA (Li, 1997), and many cichlid species in Lakes Malawi and Victoria have almost no consistent genetic differences (Parker and Kornfield, 1997). These high sequence divergences indicate that the Tanganyikan gastropods either have high rates of molecular evolution, or diverged a long time ago, and/or have retained variation that was present in their ancestral species.

The *COI* sequences showed some A–T bias in nucleotide frequencies along the coding strand (23.3% adenine, 19.3% cytosine, 20.4% guanine and 37.1% thymine, giving 60.4% A–T), but with values consistent within clades of this tree and with patterns among genera (see West and Michel, this volume). Out of 642 total base pairs there were 255 variable sites, among which 200 were phylogenetically informative. Eighty-four sites were two-fold degenerate (one of three possible changes was synonymous) and 97 were four-fold degenerate (all possible changes were synonymous). The variation in first, second and third codon positions was 14.7%, 1.4% and 83.9%, respectively (variable site ratio of 10.5:1:59.9). Most variation was in third positions, as would be expected in a species-level analysis of a protein-coding gene. Over the whole tree (including outgroups), out of 214 coded amino acids there were 23 variable amino acids, six of which were phylogenetically informative. No obvious hotspots or conserved regions in overall variation, transitions or transversions were evident. Sequences for *COI* and the *16s-ND1* fragment are available from the author on request.

Plots of pairwise distances in transitions or transversions versus overall genetic distance revealed several interesting patterns (Figure 2). Both transitions and transversions are clumped in groups that correspond to major clade boundaries, indicating strong phylogenetic structure to the data, and genetic discontinuities at the clade boundaries. The number of transitions increased linearly in comparisons both within and among species in the same genus. However, in comparisons between genera at a genetic distance of 15%, transitions show saturation (indicated by an asymptote in the curve), even in the comparison between *Lavigeria* and Nov. gen. species. As expected, at low levels of comparison, such as within and among *Lavigeria* species, transversions are not informative, but increase linearly in comparisons among genera. Transversion distances between Nov. gen. species lie on the linear part of the

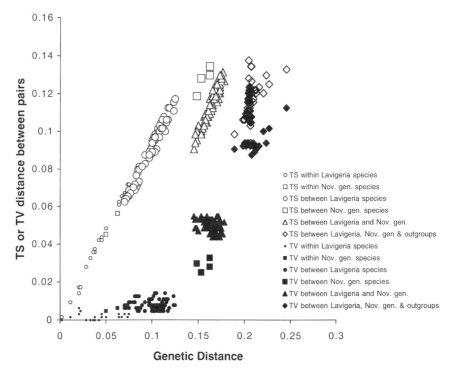

Fig. 2. Pairwise transition or transversion distances versus overall genetic distance. Note that both transitions and transversions clump in groups that correspond to major clade boundaries, indicating phylogenetically structured data. Transitions increase linearly within and among species in the same genus, but reach an asymptote (which indicates saturation) in comparisons between genera at a genetic distance of 15%, which includes the comparison of *Lavigeria* with Nov. gen. species. Transversions increase linearly in comparisons among genera, but are generally not informative in relationships among species within a genus. The transversion distances between Nov. gen. species are on the linear part of the transversion curve, which reflects the generally longer branch lengths between these species than among *Lavigeria* species.

transversion curve, which reflects the generally longer branch lengths between these species than those seen between *Lavigeria* species (see below).

B. Phylogenetic Analysis

An unweighted branch and bound search of 642 bp of the *COI* sequence from 22 individuals (two outgroups, 20 *Lavigeria* and Nov. gen.) produced a single most parsimonious tree of 701 steps (Figure 3). This tree length indicates the data to be highly phylogenetically structured: a plot of 10 000 randomized trees based on the same data set produced a range of much longer tree lengths (10 000 trees, lengths between 990 and 1230 steps, mean at 1140 steps).

Measures of clade robustness showed very strong support for species-level clades of Nov. gen. n. sp., Nov. gen. *guillemei*, *Lavigeria coronata*, *L. grandis*, *L. n. sp.* and *L.* cf. *paucicostata* (bootstrap values 99–100% and decay indices of > 10 steps) and weaker support for the clade that includes many variable forms in the *L.* cf. *giraudi* group (bootstrap values 56%, but 94% for an interior clade in this group; decay indices of 1 and 7, respectively). Relationships among these species-level clades were also indicated in the most parsimonious tree, with three clades that consistently recurred in all analyses: (i) the two Nov. gen. species, which together form a genus level clade; (ii) "large snails" of *L. coronata*, *L. grandis* and *L. n. sp.*; and (iii) all of the *Lavigeria* species together. Furthermore, the sister relationship between Nov. gen. and *Lavigeria* was also strongly supported in this analysis (bootstrap value of 98% and decay index of > 10) and using a wide range of Tanganyikan genera as outgroups (West and Michel, this volume).

Changing the relative importance of some kinds of characters over others in an analysis (weighting) can reveal patterns previously obscured by noise in the data. In analyses that weighted transversions greater than transitions, basic tree structure was consistent with the results from the unweighted analysis (Figure 3), with the following exceptions. In weighting TV:TS 2:1, the three lower-level *Lavigeria* nodes collapsed in a trichotomy with the relationships between *L.* cf. *giraudi*, *L.* cf. *paucicostata* and the "large" clade unresolved. Weighting TV:TS as 3:1, 10:1 or 14:1 resolved structure within lower levels of the *Lavigeria* clade, however, the *L.* cf. *paucicostata* group became basal and the "large" clade and *L.* cf. *giraudi* became sister groups. The tree length with this topology was 702 steps, or one step longer than that derived from the unweighted data (both lengths measured on unweighted data).

Likelihood is another statistically powerful way of comparing alternative tree topologies. In comparing likelihood values of alternative trees, the tree with *L.* cf. *giraudi* as basal was significantly more likely if transitions and transversions were not assigned different probabilities of change analogous to TS:TV of 1:1; $2\Delta\ln(\text{Lik})L = -8.58$, $p < 0.05$ based on a χ^2 approximation (Edwards, 1972; Goldman, 1993). As the TS:TV ratio increased, the differences between the likelihoods decreased for the two trees, becoming non-significantly different as TS:TV exceeded 3:1. The *COI* data are therefore equivocal in resolving this part of the topology of the *Lavigeria* radiation, as branch support is weak in both a parsimony and a likelihood context.

When two different data sets present the same solution to a phylogenetic question, one can be more confident that the resulting tree approximates evolutionary history. It is therefore meaningful that the *16s-ND1* consensus of 12 equally parsimonious trees is consistent with the topology of the *COI* tree (Figure 3) as is the combination of *COI* and *16s ND1* sequence data (not shown). Similarly, in addressing the uncertain topology at the base of *Lavigeria*, manipulation of branches in the *16s-ND1* data set revealed that, when tree length was increased by one step, from 866 steps, in the most

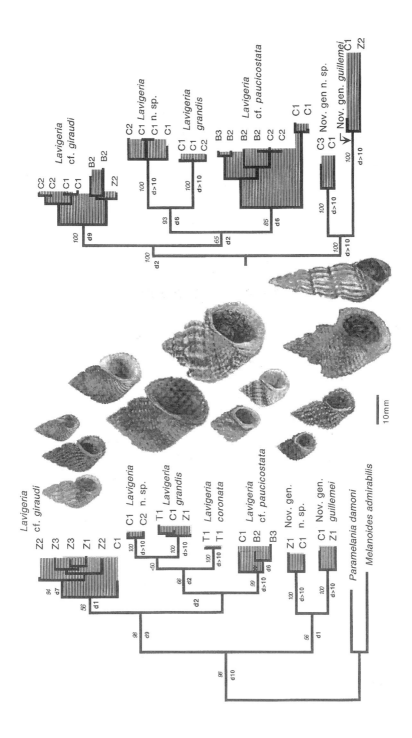

parsimonious tree (with *L.* cf. *giraudi* basal), changed to a tree with *L.* cf. *paucicostata* as basal. As in the *COI* analysis, the alternative tree likelihoods were significantly different for unweighted data, $2\Delta\ln(\text{Lik})L = -14.44$, $p < 0.05$), with the alternatives becoming non-significantly different as TS:TV exceeded 4:1.

C. Anatomy

Dissections of specimens revealed the surprising result that both species in the genus Nov. gen. are oviparous (egg layers), whereas all *Lavigeria* species are ovoviviparous, brooding their young in an anterior enlargement of the pallial oviduct (Figure 4). In *Lavigeria* females the interior surfaces of the pallial oviducts are equipped with a suite of lamellae for protecting the embryos in the brood. The character and quantity of lamellar tissue vary among species, as do brood and embryo sizes. *Lavigeria grandis* exhibits the extreme of this morphology, possessing curtains of tissue that may separate each of the embryos, which are significantly sculptured and quite large just before birth. Females of both Nov. gen. species have a ciliated groove extending from the base of the pallial oviduct to a small pit in the foot. This pit is likely to be an oviposition gland. The groove lacks pigment, is visible on the extended animal and is a reliable diagnostic character of Nov. gen. species, as it is absent in *Lavigeria*. However, it is of little use in sexing live animals, as males of both Nov. gen. taxa have an unpigmented stripe along this area which, although not as distinct as in females, may make them appear to be females.

There is no clearly distinguishable sexual dimorphism in either clade, and at the current state of knowledge animals can be reliably sexed only by killing and examining the gonads (Figure 4). Trematode parasites are occasionally found in the gonad and reproductive tract and no broods have been found in parasitized animals; thus, parasites may exert a significant selective pressure on Tanganyikan gastropods (Michel, pers. obs.). Sex ratios generally average to 1:1, which indicates that these animals are likely to be obligately gonochoristic.

Fig. 3. (a) Parsimony tree based on 642 bp of the *COI* sequence, TS:TV 1:1. A single tree of length 701 steps, CI = 0.54, resulted from this analysis. Sites are shown at the end of each branch. Shading indicates nominal species. Examples of individuals from each clade, as currently construed, are shown next to the tree. *Lavigeria* cf. *giraudi* and *L.* cf. *paucicostata* are likely to encompass many more species; several specimens of each (three and two, respectively) are shown, to illustrate the variation present within these species. *Lavigeria* n. sp. and *L. coronata* are to the right of *L. grandis*: their body whorls align with the species name on the left. *Melanoides admirabilis* photo from Brown (1980), with permission. (b) Consensus of 12 equally parsimonous trees based on parsimony analysis of 1400 bp of the *16s-ND1* sequence, tree length 866 steps, CI = 0.97. Bootstrap support values are above the branches, decay indices are below the branches. Branch lengths are proportional to amount of sequence divergence.

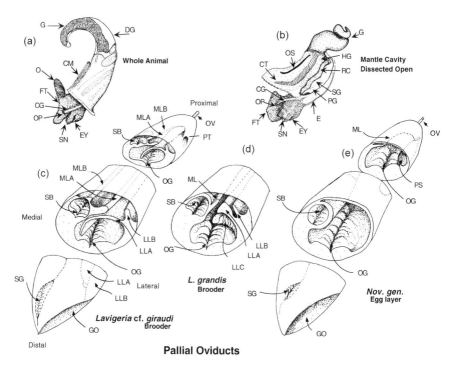

Fig. 4. Reproductive anatomy of brooding *Lavigeria* and egg-laying Nov. gen. (a) The oviposition groove and gland are present and visible on whole animal only in Nov. gen. females. (b) The dissected, opened animal is representative of the morphology of both genera. (c–d) Cross-sections of pallial oviduct of two representative *Lavigeria* species, *L. cf. giraudi* and *L. grandis*, show extensive lateral and medial lamellae. These structures are used to hold the brood securely in the oviduct. Species-specific differences in lamellae correlate with differences in brood and embryo sizes. The thicker lateral lamellae of *L. grandis*, with its pendant curtains of tissue, is shown here. (e) The pallial oviduct of a representative Nov. gen. species (Nov. gen. n. sp.) exhibits few lamellae and a well-developed proximal sac. Lateral lamellae are thicker and have pendant curtains of tissue. CG, ciliated groove; CM, columellar muscle; CT, ctenidium; DG, digestive gland; E, oesophagus; EY, eye; FT, foot; G, gonad; GO, gonoduct opening; SG, sperm gutter; SB, spermatophore bursa; LL (A and B), lateral lamella (A and B); LLC, lateral lamella curtain; ML (A and B), medial lamella (A and B); O, operculum; OV, oviduct; PT, proximal tuck; OG, oviducal groove; OP, oviposition pore; OS, osphradium; PS, proximal sac.

Anatomical differences are apparent between *Lavigeria* species, but because individual character differences are often continuous, these can be demonstrated quantitatively only through multivariate morphometric analyses. However, although these anatomical differences are subtle, their multivariate clustering correlates with species boundaries based on shell morphology and

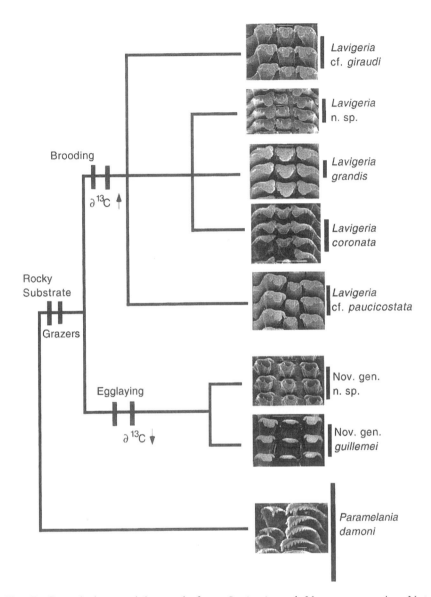

Fig. 5. Central three radular teeth from *Lavigeria* and Nov. gen. species. Note differences in the shape of the central, or rachidian tooth, and the number of cusps on the lateral teeth. Scale bars = 100 μm; note that there are significant size differences. The tree is a schematic representation of the results from phylogenetic analysis of sequence data in this study; branch lengths here are not proportional to amount of change, weakly resolved nodes are shown as polytomies, and character state changes of reproductive mode (Figure 4) and stable isotope differences (Figure 6) are indicated at the base of the nodes.

genetics, lending support to taxonomic units identified with the latter data (Michel, pers. obs.).

D. Radulae

The variation in radulae among species of *Lavigeria* and Nov. gen. is phylogenetically informative. Representative examples of the three central teeth (one rachidian and two lateral teeth, excluding the two marginal teeth on each side) from each species are illustrated in Figure 5. A more detailed treatment, including a multivariate morphometric approach to intra- and interspecific variation, is presented in Michel (1995). The rachidian (central) tooth in Nov. gen. species has a pair of small wings that sweep laterally outwards at the base of the cusps, pointing towards the lateral teeth. *Lavigeria* species lack these wings, although they may have small cusps in this inflection point of the rachidian tooth. Rachidian teeth of Nov. gen. species point nearly straight out (90°) from the radular ribbon and appear to be considerably less robust than *Lavigeria* teeth, with a greater gap between the rachidian and the lateral teeth. The rachidian tooth in *Lavigeria* species points towards the posterior of the radula, giving the tooth a strongly recurved profile, and is relatively wide, filling the space between the lateral teeth. *Lavigeria* cf. *giraudi* and *L.* cf. *paucicostata* have three dominant cusps on the rachidian tooth, with a large, rounded rectangular central cusp that is several times larger than the two flanking cusps. The lateral teeth in both species have three dominant cusps (but not bilaterally symmetric), with other cusps very reduced or non-existent. *Lavigeria* n. sp. has a squared rachidian tooth shape similar to *L.* cf. *giraudi* and *L.* cf. *paucicostata*; however, there are more cusps evident on the lateral teeth. The rachidian teeth of *L. coronata* and *L. grandis* are distinctive, with scalloped edges from multiple, rounded cusps along their margins. The strongly recurved and thickened teeth of *L. coronata* and *L. grandis* are the most robust within the *Lavigeria*–Nov. gen. clade, although when animals or radulae of similar size are compared *L. grandis* teeth are approximately twice as large as those of *L. coronata*.

E. Isotopes

Plots of stable isotope values (Figure 6) indicate that *Lavigeria* and Nov. gen. have distinct food sources. There is a strong separation between these genera in $\delta^{13}C$ values, which indicates differences in the photosynthetic pathway of the algae grazed by the component species. This is a conservative indicator that these gastropods are trophically differentiated, as they may also be feeding on different algae species that have the same $\delta^{13}C$ values. *Lavigeria* found in deep water have $\delta^{13}C$ values that are distinct from the values for Nov. gen. specimens. $\delta^{15}N$ values of *Lavigeria* and Nov. gen. samples overlap, which is

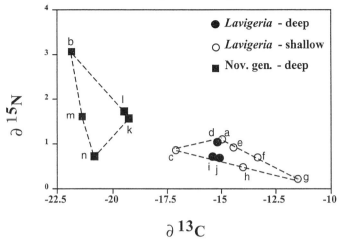

Fig. 6. Plots of nitrogen and carbon stable isotopes for representative *Lavigeria* and Nov. gen. samples. The water depth at which samples were collected is categorized as shallow or deep (see Table 1 for specific collection information). Strong separation in $\delta^{13}C$ values indicates differences in the photosynthetic pathway of the algae grazed by these genera. Polygons encompass points for each genus, showing lack of overlap between samples. This is a conservative indicator that these gastropods are trophically differentiated, as they may also feed on distinct algae species that do not have different $\delta^{13}C$ values. $\delta^{15}N$ values overlap among *Lavigeria* and Nov. gen. samples, which is consistent with our understanding that these animals are primary herbivores and do not differ in their trophic level.

consistent with our understanding that both animals are primary herbivores and occupy the same trophic level.

V. DISCUSSION

A. Clade Structure

The structure of the phylogeny for a species flock provides insights into the mechanisms behind past speciation and offers prospects for future determination of species boundaries. The trees resulting from analyses of *COI* and *16s-ND1* sequences indicate a consistency in the phylogeny of mtDNA for the *Lavigeria* and Nov. gen. clades. Based on these and other data (Michel, 1995; West, 1997; West and Michel, this volume), *Lavigeria* and Nov. gen. are sister clades, making them ideal for comparisons of evolutionary processes of benthic organisms in Lake Tanganyika. Within each of these genus-level clades, there are consistent species-level clades. Nov. gen. consists of two genetically delimited species, Nov. gen. *guillemei* and Nov. gen. n. sp., which correspond closely with morphological differences (Michel, 1995). *Lavigeria* is the larger genus, with five species-level clades revealed in this analysis. However, further sampling in other localities and investigation of alternative data sets will result in the identification of additional species both by discovery of new groups and splitting of current groups.

Although *Lavigeria* forms a robust clade, the relationships among its component species are still uncertain. Two alternative topologies can be supported, depending on the relative weighting given to transitions and transversions. If transitions and transversions are given equal phylogenetic weight, *L.* cf. *giraudi* is a sister clade to the other *Lavigeria* species. Alternatively, if transitions are downweighted relative to transversions, *L.* cf. *paucicostata* becomes a sister clade to the other species. The lack of resolution of basal relationships in *Lavigeria* could be a result of either (i) a lack of resolution of the mtDNA sequence for recording this evolutionary event, or (ii) rapid radiation in this genus, with incomplete lineage sorting and retained ancestral polymorphisms. If the first hypothesis is correct, it should be possible to utilize other data, such as nuclear DNA sequences, allozymes or morphology, to resolve *Lavigeria* relationships. However, if the second hypothesis is supported, it is unlikely that any data set which can adequately resolve the relationships at this node will be found.

Preliminary examination of data from an ongoing study (Michel, unpubl.) suggests that the two species-level clades, *L.* cf. *giraudi* and *L.* cf. *paucicostata*, are likely to contain definable lineages, yet are also likely to have a looser phylogenetic structure, with less distinct species boundaries within them than their sister taxa. This may be the result of either retained ancestral polymorphisms or present hybridization, the type of phylogenetic structure present within many species flocks (Greenwood, 1984). The suggestion that more lineages exist than are recognized in this analysis is supported by the often discrete morphological and life-history variation that is evident between populations in these groups. The strong node support for branches within the *COI*-defined clades, which indicates that there is further phylogenetic structuring, is also suggestive.

Further evidence that the number of species in the *Lavigeria* clade is probably much higher than indicated in the current tree structure is provided by geographical variation. Examination of shell collections from several hundred sites has uncovered examples of morphologically different *Lavigeria* populations that are likely to be distinct species, but that have only limited distributions. One such example is included in this analysis; *L. coronata* had not been recorded from the lake since 1985, until a thriving, although geographically limited, population was discovered in Kigoma in 1998 (Michel, pers. obs.). Although it was not initially clear from gross morphology whether *L. coronata* was merely a distinct morph of *L. grandis*, with which it is sympatric, the genetic data indicate this to not be a single polymorphic species, but instead to comprise two sister species which share some apomorphies and differ in others.*

B. Shell Characters

Despite its inclusion in numerous evolutionary debates, the phylogenetic usefulness of shell differences for freshwater thiarid gastropods remains largely untested. This situation is improved by the phylogeny presented here, which can be used to begin to explore the evolution of shell morphology in the *Lavigeria*–Nov. gen. clade. The two Nov. gen. species share a regular, tuberculate sculpture that lacks any rib-like axial elements, and both have a tendency to have dark shells, often with a purple apertural lining. Nov. gen. *guillemei* is tall-spired, while Nov. gen. n. sp. is squat. In comparison, as a group and also, in some cases, within species, *Lavigeria* taxa are more highly variable, but most have moderate to strongly raised ribbing, that may be very fine or very coarse. *Lavigeria* species span a range of sizes from < 10 mm to > 30 mm shell height at adulthood, and also exhibit differences in shell shape, from tall-spired to squat. Their colour and patterns range from uniform browns and beiges to striped. Pattern characters appear to correlate with some species distinctions.

It appears that large, massive shells evolved once within *Lavigeria*, forming the "large clade", *L. grandis*, *L. coronata* and *L.* n. sp. Additional undescribed and poorly known species probably belong here too. That this occured only once is somewhat surprising, as functional explanations have been presented as to how the evolution of massive shells could occur readily (West and Cohen, 1996). Change in shape, however, does appear to be more labile, as tall-spired *Lavigeria* and Nov. gen. are seen in several independent groups, and even among populations of some species (Michel *et al.*, 1992). Parsimony analysis indicates that a well-developed columellar tooth or bump evolved once in the clade comprising *L. grandis*, *L. coronata* and *L.* n. sp.

---

*Specifically, both species are very large, with a columellar tooth, have robust scalloped radular rachidian teeth and are found on large boulders in shallow water. They differ in that *L. grandis* has a wide geographical distribution around most of the central and southern lake, with populations that vary somewhat in shell morphology (especially colour and sculpture), with a much weaker axial sculpture and occuring in shallower waters than the geographically restricted *L. coronata*. Furthermore, the larval shell and early teleoconch markedly differ between these species and allow for their easy discrimination (Michel and Todd, unpubl. data).

C. Habitat Specificity

For most organisms habitat specificity is an important factor in speciation mechanisms, because it determines susceptibility to allopatric separation and probability of population fragmentation. The Nov. gen. clade has a habitat synapomorphy in that both component species are found in increasing numbers in deep water (> 20 m depth), where they are frequently the only common gastropod macrofauna on hard substrates. At a few sites, both Nov. gen. *guillemei* and Nov. gen. n. sp. may be found in water as shallow as 3 m depth; however, they are both typically deep-water species. Because branch lengths tend to be long both within and between species in the *COI, 16s-ND1* and allozyme trees (Figure 3 and Michel, 1995), the phylogeny suggests that these species both have strong population subdivision. It is intriguing that these species appear to have remained morphologically conservative despite strong genetic differences between sites. Future studies should investigate whether this genetic structuring is a result of habitat or life-history characters.

Within the genus *Lavigeria* the majority of the species is hard-substrate specialists, occurring across a range of depths and with differing degrees of geographic subdivision. The present analysis indicates that only one clade, *L.* cf. *paucicostata*, is found on soft substrates. If *L.* cf. *paucicostata* is basal in the *Lavigeria* clade, this might indicate that the ancestor of the *Lavigeria* radiation was a substrate generalist, with diversification of the many crown species preceded by a single event of hard substrate specialization. If, however, *L.* cf. *giraudi* is basal and *L.* cf. *paucicostata* arose within another *Lavigeria* clade, then because the sister group Nov. gen. is hard substrate limited, the ancestor is also likely to have been a substrate specialist. This would suggest that evolution of substrate specificity occurs at the species level. The most parsimonious reconstructions with the mtDNA data and the likelihood analyses of unweighted data both support the latter scenario, suggesting that substrate specificity did not change when the *Lavigeria*–Nov. gen. clade bifurcated.

D. Anatomy

Ecological and population genetic interactions are strongly influenced by life-history variables, and the discovery that *Lavigeria* and Nov. gen. have contrasting reproductive modes thus opens avenues for multiple comparisons of extrinsic and intrinsic effects on diversification. In general, brooders have more strongly subdivided populations and thus a higher diversification potential than do egg layers (Cohen and Johnston, 1987). However, these Tanganyikan gastropod taxa may provide some test cases of the limits of this generality. *Lavigeria* species brood their young and comprise a more species-rich clade than Nov. gen., and so follow this generalization at the generic level. However, the evidence that Nov. gen. populations and species are genetically more divergent than are *Lavigeria* species suggests that egg-laying may, in this case, also permit significant population subdivision. As Nov. gen. eggs have not been seen (with one exception, K. West, pers. comm.), the method of encapsulation, deposition, developmental time, etc., all of which affect divergence potential, are not known (Michel, 1994). Furthermore, there are large differences in brood and embryo sizes among *Lavigeria* species (Michel, unpubl.), which permit comparisons of differences within the brooding habit on divergence potential.

E. Radulae and Isotopes

A perspective on feeding relationships is central to understanding the ecology of any organism. Nov. gen. and *Lavigeria* species differ greatly in their radulae, with the most dramatic contrast being that between the fine, presumably weak teeth of Nov. gen. n. sp. and the robust teeth of *L. grandis*. The different radula types correlate with differences in stable isotopes in the body tissues and indicate that these animals are consuming different food resources. It is therefore possible that, together with reproductive mode, trophic differences may comprise one of the major divergences between the sister clades. A more exhaustive study is required to address this hypothesis, incorporating larger sample sizes for stable isotope analyses, complete descriptions of the available algal foods and collection of microsympatric individuals of different gastropod species.

Variation in radular tooth morphology within the *Lavigeria* clade exhibits subtle but consistent differences. These differences are useful in determining species differences, but their effects on trophic separation are not known. A preliminary investigation of character displacement in radular morphology found tentative evidence for increased divergence between potentially competing species when they were sympatric relative to when they were allopatric (Michel, 1995). Further studies of the fine-scale differences among populations and taxa are necessary to test fully the role of trophic partitioning in the divergence and coexistence of this gastropod species flock.

ACKNOWLEDGEMENTS

This work would not have been possible without the support of the University of Burundi (notably G. Ntakimazi), the Centre de Recherche en Hydrobiologie in Congo (especially M.-M. Gashagaza), TAFIRI in Tanzania (particularly D. Chitamwebwa) and the Department of Fisheries, Mpulungu, Zambia (with thanks to L. Mwape) in facilitating collections and fieldwork. Zambian collections were made as part of a research cruise sponsored by the Royal Belgian Institute of Natural Sciences in 1995, at the kind invitation of E. Verheyen and K. Martens. T. Collins, J. Boore and other colleagues at the

Brown Laboratory at the University of Michigan provided unflagging support during the sequencing and analysis, and A. Cohen and K. West were prime motivators for pursuing evolutionary questions in Lake Tanganyika. R. Hecky, P. Ramlal and R. Hesselein provided isotope values and the stimulation to investigate trophic differences with this technique. Thanks to A. Cohen, M. Medina, S. Menken, A. Mooers, J. Todd, K. West, and two anonymous reviewers for critical reading of the manuscript. This work was supported by an NSF Post-Doctoral Fellowship No. DEB 9303281 and NSF No. BIR-9113362 (RTG in Biological Diversification) to the author.

REFERENCES

Alin, S., Cohen, A., Bills, R., Gashagaza, M.-M., Michel, E., Tiercelin, J., Martens, K., Coveliers, P., Mboko, S., West, K., Soreghan, M., Kimbadi, S. and Ntakimazi, G. (1999). Effects of landscape disturbance on animal communities in Lake Tanganyika, East Africa. *Conserv. Biol.* **13**, 1071–1033.

Bandel, K. (1998). Evolutionary history of East African fresh water gastropods interpreted from the fauna of Lake Tanganyika and Lake Malawi. *Zbl. Geol. Palaeont. Teil I. H.* **1/2**, 233–292.

Bourguignat, J.R. (1885). *Notice Prodromique sur les Mollusques Terrestres et Fluviatiles recueilles par M. Victor Giraud dans la Région Méridionale du Lac Tanganika*, 110 pp. V. Tremblay, Paris.

Bourguignat, J.R. (1888). *Iconographie malocologique des animaux mollusques fluvatiles du Lac Tanganyika.* Corbeil, Paris.

Bourguignat, J.R. (1890). Histoire Malacologique du Lac Tanganika (Afrique Équitoriale). *Ann. Sci. Nat. Paris, Zool. Paléontol. VII Ser.* **10**, 1–258.

Bremer, K. (1988). The limits of amino acid sequence data in angiosperm phylogenetic reconstruction. *Evolution* **42**, 795–803.

Brown, D.S. (1980). *Freshwater Snails of Africa and their Medical Importance.* Taylor and Francis, London.

Brown, D.S. (1994). *Freshwater Snails of Africa and Their Medical Importance.* Revised 2nd ed. Taylor and Francis, London.

Brown, D.S. and Mandahl-Barth, G. (1987). Living molluscs of Lake Tanganyika: a revised and annotated list. *J. Conchol.* **32**, 305–327.

Cohen, A.S. and Johnston, M.R. (1987). Speciation in brooding and poorly dispersing lacustrine organisms. *Palaios* **2**, 426–435.

Coulter, G.W. (ed.) (1991). *Lake Tanganyika and its Life.* Oxford University Press, Oxford.

Edwards, A.W.F. (1972). *Likelihood.* Cambridge University Press, Cambridge.

Folmer, O., Black, M., Hoeh, W., Lutz, R. and Vrijenhoek, R. (1994). DNA primers for amplification of mitochondrial cytochrome *c* oxidase subunit I from diverse metazoan invertebrates. *Mol. Mar. Biol. Biotechnol.* **3**, 294–299.

Fryer, G., Greenwood, P.H. and Peake, J.F. (1983). Punctuated equilibria, morphological stasis and the palaeontological documentation of speciation: a biological appraisal of a case history in an African lake. *Biol. J. Linn. Soc. Lond.* **20**, 195–205.

Fryer, G., Greenwood, P.H. and Peake, J.F. (1985). The demonstration of speciation in molluscs and living fishes. *Biol. J. Linn. Soc. Lond.* **26**, 325–336.

Gilbert, D.G. (1994). *SeqApp: A Biosequence Editor Application, Version 1.9a169.* Biology Department, Indiana University, Bloomington, IN.

Goldman, N. (1993). Statistical tests of models of DNA substitution. *J. Mol. Evol.* **36,** 182–198.

Greenwood, P.H. (1984). What is a species flock? In: *Evolution of Fish Species Flocks* (Ed. by A. Echelle and I. Kornfield), pp. 13–20. University of Maine at Orono Press, Orono, ME.

Hecky, R.E. and Hesslein, R.H. (1995). Contributions of benthic algae to lake food webs as revealed by stable isotope analysis. *J. N. Am. Benthol. Soc.* **14,** 631–653.

Hesslein, R.H., Capel, M.J., Fox, D.E. and Hallard, K.A. (1991). Stable isotopes of sulfur, carbon, and nitrogen as indicators of trophic level and fish migration in the lower Mackenzie River Basin, Canada. *Can. J. Fish. Aquat. Sci.* **48,** 2258–2265.

Lake, J.A. (1994). Reconstructing evolutionary trees from DNA and protein sequences: paralinear distances. *Proc. Nat Acad. Sci. USA* **91,** 1455–1459.

Leloup, E. (1953). *Gastéropodes. Résultats scientifiques de l'exploration hydrobiologique du Lac Tanganyika (1946–1947), Vol. 3, part 4.* Institut Royal des Sciences Naturelles de Belgique, Brussels.

Li, W. (1997). *Molecular Evolution.* Sinauer, Sunderland, MA.

Lockhart, P.J., Steel, M.A., Hendy, M.D. and Penny, D. (1994). Recovering evolutionary trees under a more realistic model of sequence evolution. *Mol. Biol. Evol.* **11,** 605–612.

Maddison, W. and Maddison, D. (1996). *MacClade 3.06.* Sinauer, Sunderland, MA.

Martel, H. and Dautzenberg, P. (1899). Observations sur quelques mollusques du Lac Tanganyika recueilles par le R.P. Guillemé et descriptions de formes nouvelles. *J. Conchyliol.* **4,** 163–181.

Michel, E. (1994). Why snails radiate: a review of gastropod evolution in long-lived lakes, both recent and fossil. *Arch. Hydrobiol. Beiheft. Ergebnisse Limnol.* **44,** 285–317.

Michel, E. (1995). Evolutionary diversification of Rift Lake gastropods: morphology, anatomy, genetics, and biogeography of *Lavigeria* (Mollusca:Thiaridae) in Lake Tanganyika. Unpublished PhD Thesis, University of Arizona, Tucson, AZ.

Michel, A.E., Cohen, A.S., West, K.A., Johnston, M.R. and Kat, P.W. (1992). Large African lakes as natural laboratories for evolution: examples from the endemic gastropod fauna of Lake Tanganyika. *Mitt. Int. Verein. Limnol.* **23,** 85–99.

Moore, J.E.S. (1899). The molluscs of the Great African Lakes. IV. *Nassopsis* and *Bythoceras. Q. J. Microsc. Sci.* **42,** 187–201.

Palumbi, S.R. (1996). Nucleic acids II: The polymerase chain reaction. In: *Molecular Systematics* (Ed. by D.M. Hillis, C. Moritz and B.K. Mable). Sinauer, Sunderland, MA.

Parker, A. and Kornfield, I. (1997). Evolution of the mitochondrial DNA control region in the mbuna (Cichlidae) species flock of Lake Malawi, East Africa. *J. Mol. Evol.* **45,** 70–83.

Pilsbry, H. and Bequaert, J. (1927). The aquatic mollusks of the Belgian Congo, with a geographical and ecological account on Congo malacology. *Bull. Am. Mus. Nat. Hist.* **53,** 69–602.

Ramlal, P.S., Hesslein, R.H., Hecky, R.E., Fee, E.J., Rudd, J.M.W. and Guilford, S.J. (1994). The organic carbon budget of a shallow Arctic tundra lake on the Tuktoyaktuk Peninsula, N.W.T., Canada. *Biogeochemistry* **24,** 1–28.

Smith, E.A. (1880). On the shells of Lake Tanganyika and of the neighbourhood of Ujiji, Central Africa. *Proc. Zool. Soc. Lond.* 1880, 344–352.

Smith, E.A. (1881a). On a collection of shells from Lakes Tanganyika and Nyassa and other localities in East Africa. *Proc. Zool. Soc. Lond.* 1881, 276–300.

Smith, E.A. (1881b). Descriptions of two new species of shells from Lake Tanganyika. *Proc. Zool. Soc. Lond.* 1881, 558–561.

Swofford, D. (1998). *PAUPv.4.0b1: Phylogenetic Analysis Using Parsimony (and other methods)*. Sinauer, Sunderland, MA.

West, K. (1997). Perspectives on the diversification of species flocks: systematics and evolutionary mechanisms of the gastropods (Prosobranchia:Thiaridae) of Lake Tanganyika, East Africa. Unpublished PhD Thesis, University of California, Los Angeles, CA.

West, K. and Cohen, A. (1996). Shell microstructure of gastropods from Lake Tanganyika, Africa: adaptation, convergent evolution and escalation. *Evolution* **50**, 672–681.

Williamson, P.G. (1981a). Palaeontological documentation of speciation in Cenozoic molluscs from the Turkana Basin. *Nature* **293**, 437–443.

Williamson, P.G. (1981b). Morphological stasis and developmental constraint: real problems for neo-Darwinism. *Nature* **294**, 214–215.

Williamson, P.G. (1985). Punctuated equilibrium, morphological stasis and the palaeontological documentation of speciation: a reply to Fryer, Greenwood and Peake's critique of the Turkana Basin mollusc sequence. *Biol. J. Linn. Soc. Lond.* **26**, 307–324.

Implications of Phylogeny Reconstruction for Ostracod Speciation Modes in Lake Tanganyika

L.E. PARK and K.F. DOWNING

I. SUMMARY

Speciation in the East African lakes has been remarkable and has resulted in many endemic species flocks, consisting of several to hundreds of species. Lake Tanganyika, in particular, supports one of the most diverse faunas of any lake system. The origin of the Tanganyikan fish and gastropod species flocks has been attributed to divergence of populations through the development of barriers to interbreeding, such as the formation of separate lakes or habitat fragmentation within subregions of a single lake during major falls in lake level. Similar speciation mechanisms may have been important in the history of the diverse ostracod species flocks of Lake Tanganyika.

ADVANCES IN ECOLOGICAL RESEARCH VOL. 31
ISBN 0-12-013931-6

In order to test hypotheses concerning character evolution and speciation patterns in this group, a phylogenetic analysis was conducted of the ostracod clade *Gomphocythere*. This analysis yielded high levels of homoplasy (CI = 0.56), permitting the development of a qualitative model to explain the observed phylogenetic patterns.

This model considers two major speciation scenarios. The first involves a hypothetical radiation of a "parent" species and "daughter" species that evolve in a single lake system, taking advantage of available niche space. The resulting pattern is one with each daughter species evolving into its own niche space. In contrast, a scenario of multiple radiations from two lakes or lake sub-basins, with species brought together via a serendipitous invasion, or from water-level fluctuations having caused lakes to join together, suggests a complex phylogenetic reconstruction that could account for the high level of homoplasy seen in the analysis of *Gomphocythere* in Lake Tanganyika. Phylogeny reconstruction of the *Gomphocythere* species flock of Lake Tanganyika and the high homoplasy frequency revealed in this study support the importance of multiple radiations as a primary speciation mechanism in this ancient lake system.

II. INTRODUCTION

A. Biological Diversification in Lake Tanganyika

The East African lake system has long been known as a site of megadiversity (*sensu* Mittermeier, 1988; Mittermeier and Werner, 1990), particularly with respect to the large, endemic species flocks that originated within several lakes in this geographical area (Coulter, 1991, 1994; Sturmbauer and Meyer, 1992; Snoeks *et al.*, 1994; Rossiter, 1995). For example, almost 190 cichlid fish species (over 180 of which are endemic), 68 gastropod species (45 of which are endemic) and 80 ostracod species (almost all of which are endemic) have been described from Lake Tanganyika alone. For all of these groups, the given numbers are underestimates and many newly discovered taxa await formal description (e.g. see Snoeks; Michel; West and Michel, this volume). These speciose faunas provide excellent material for studies of diversification processes and diversity changes of endemic organisms over time. Analyses of cichlid fishes (Fryer and Iles, 1972; Sturmbauer and Meyer, 1992; Lowe-McConnell, 1993; Sultmann *et al.*, 1995; Verheyen *et al.*, 1996; Rossiter and Yamagishi, 1997; Mayer *et al.*, 1998) and thiarid molluscs (Cohen and Johnston, 1987; Michel *et al.*, 1992) show that diversification patterns are often linked to environmental differences, and to incidences of multiple invasions and subsequent radiations in the lake. Whether or not similar patterns of evolution have produced the diversity observed in the lesser known groups found in these lakes, such as the ostracods, can be tested through analyses of their phylogeny.

B. Ostracods in Lake Tanganyika

Eighty described ostracod species, assignable to 25 genera, have been found in Lake Tanganyika. Almost all of these species are endemic to the lake (Rome, 1962) and provide an excellent means for studying diversification processes and diversity changes of endemic organisms. Most of them belong to three families, the Candonidae, the Cytherideidae and the Limnocytheridae. The genus *Gomphocythere* is a member of the Limnocytheridae. Very little is known about speciation patterns of ostracods in the African lakes, although it has been widely speculated that allopatric speciation in isolated basins during low lake levels may contribute to their radiation (e.g. Martens, 1994). However, little work has been done on documenting and developing phylogenetically based speciation models for any ostracod clade in any African lake system.

C. Speciation Models in Ancient Lakes

Species flocks in ancient lakes in general, and Tanganyika in particular, have originated in various ways that may be taxon dependent. However, two basic patterns have emerged from studies of fishes and invertebrates. In some cases, species closely resembling the ancestral taxa are extant in the rivers, ponds and lakes surrounding the lake basins, and the invading species appear to have given rise to a monophyletic flock within the lake. In other cases, there is evidence of multiple radiations, with subsequent radiations either establishing themselves along with older ones or replacing them. In the latter case, some of the older radiations have suffered catastrophic extinctions either by repeated desiccation (e.g. Lakes Turkana and Victoria) or through volcanic eruptions (e.g. Lake Kivu) (Eccles, 1984; Coulter, 1991; Kolding, 1992; Bootsma and Hecky, 1993; Coulter, 1994).

Two types of barrier to gene flow have been proposed as possible mechanisms which act to promote allopatric speciation within lakes. Extrinsic barriers involve the physical features of the lake environment, including lake-level changes that can result in the development of isolated basins, and therefore include geographical barriers, as well as intrabasinal differences in temperature, light, pressure, density, water chemistry and substrate (Smith and Todd, 1984; Mayr, 1984; Coulter, 1994). Lake Tanganyika has been separated into isolated sub-basins throughout its history during times of low lake-level stand (*c.* 35 000–15 000 years bp, 16 000–14 000 years bp and 3500–1400 years bp) (Haberyan and Hecky, 1987; Scholz and Rosendahl, 1988; Tiercelin and Mondeguer, 1991) (Figure 1) and it has been speculated that these level fluctuations have had a profound effect on speciation of littoral cichlid fish

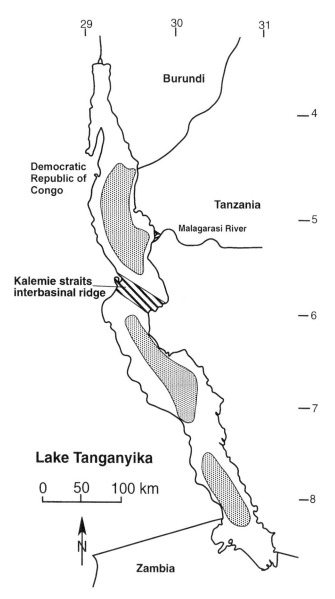

Fig. 1. Lake Tanganyika with its three major sub-basins, Kigoma, Kalemie and East Marungu, shown at lake levels 600 m lower than at present, as estimated to have occurred 35 000–15 000 years ago (Haberyan and Hecky, 1987; Scholz and Rosendahl, 1988; Tiercelin and Mondeguer, 1991). Sub-basins have been successively connected and isolated throughout Lake Tanganyika's history.

(Coulter, 1994). Whether these same conditions also had a major effect on ostracod speciation, and were a more important factor than disjunct shorelines, substrate barriers to dispersal, or differences in water quality or chemistry, remains unknown.

Intrinsic or biotic barriers are those that include characteristics of the organisms themselves, such as errant homing behaviour in fish (Horrall, 1981), food preference (Smith, 1987), colour preference (McKaye *et al.*, 1984) and brooding mechanisms (Cohen and Johnston, 1987). Although it is possible to observe potential barriers to gene flow between extant populations in lakes, trying to establish the historical development of a barrier or sequence of barriers that actually resulted in speciation events is much more difficult.

D. The Ostracod Genus *Gomphocythere*

The genus *Gomphocythere* is a diverse taxon whose individual members show specificity to various environmental conditions, such as substrate and depth, making it an ideal group in which to document and understand diversification processes of ostracods in Lake Tanganyika. Fourteen described species of *Gomphocythere* can be found today in Africa and parts of the Middle East (e.g. Israel) (Figure 2). The sister or subgroup taxon, *Cytheridella*, is found in South America, North America and Australia, whereas *Gomphodella* is restricted to Australia.

The distribution of *Gomphocythere* in East African lakes is poorly understood. Several species appear to be endemic to individual, large, inland lakes; however, there are also widespread species that can be found in other lakes throughout Africa (Sars, 1924; Rome, 1962; Martens, 1990).

In Lake Tanganyika *Gomphocythere* is represented by five described endemic species: *G. alata, G. cristata, G. curta, G. lenis* and *G. simplex* (Rome, 1962), and four additional, as yet formally undescribed species: *G. "coheni"* n. sp., *G. "downingi"* n. sp., *G. "wilsoni"* n. sp. and *G. "woutersi"* n. sp. (Park and Martens, unpubl.) (Table 1). Despite their diversity, Rome (1962) is the only previous author to address specifically the taxonomy of *Gomphocythere* species in Lake Tanganyika.

Gomphocythere species occur throughout the Tanganyikan basin, with certain species being more widely distributed and locally abundant than others. While certain species may be concentrated in a single sub-basin, no species is strictly limited to any sub-basin (Park, 1995). All members of the genus *Gomphocythere* are brooders, but the dispersal ability of this group is not well known; indeed, whether brooding organisms are better or poorer dispersers is still a point of contention (Fryer, 1996; Horne *et al.*, 1998). For Tanganyikan *Gomphocythere* species, the action of water currents or fish migration may be a major factor influencing dispersal. The ostracods' small size may also contribute to their dispersal ability, as they can be carried throughout the

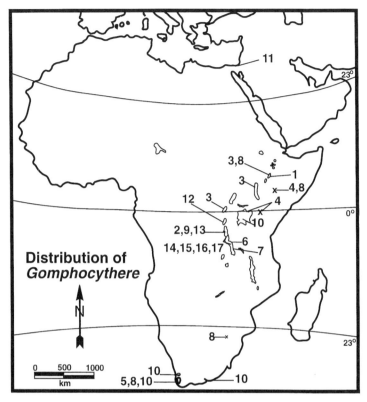

Fig. 2. Distribution map of *Gomphocythere* species in Africa. The numbers indicated on the diagram represent the following species: 1, *Gomphocythere aethiopis*; 2, *G. alata*; 3, *G. angulata*; 4, *G. angusta*; 5, *G. capensis*; 6, *G. cristata*; 7, *G. curta*; 8, *G. expansa*; 9, *G. lenis*; 10, *G. obtusata*; 11, *G. ortali*; 12, *G. parcedilatata*, 13, *G. simplex*; 14, *G. "coheni"* n. sp.; 15, *G. "downingi"* n. sp.; 16, *G. "wilsoni"* n. sp.; 17, *G. "woutersi"* n. sp.

lake by many different agents. It has been speculated that the distribution of the various species is related more to niche specifications and population dynamics than to radiations in different sub-basins, such as would occur through isolation owing to lake-level fluctuations (Cohen, 1995). Only a phylogenetic analysis, such as the one used here, can elucidate such a pattern.

III.   STUDY OBJECTIVES

The primary purpose of this study was to document the phylogenetic pattern of one group of ostracods, *Gomphocythere*, in Lake Tanganyika, and to use that reconstruction to delimit the speciation patterns of the clade, thus providing a model for speciation of ostracods in large rift lakes. Despite the ostracods

Table 1
Taxonomy of species used in this analysis, including the four newly designated species
endemic to Lake Tanganyika

Subclass Ostracoda, Latreille, 1806
  Order Podocopida, Müller, 1894
    Suborder Podocopa, Sars, 1866
      Superfamily Cytheracea, Baird, 1850
        Family Limnocytheridae, Klie, 1938
          Subfamily Timiriaseviinae, Martens, 1995
            Tribe Cytheridelli, Danielopol and Martens, 1990
              Genus *Gomphocythere*, Sars, 1924
                Gomphocythere aethiopis, Rome, 1970
                Gomphocythere alata, Rome, 1962
                Gomphocythere angulata, Lowndes, 1932
                Gomphocythere angusta, Klie, 1939
                Gomphocythere capensis, Müller, 1914
                Gomphocythere n. sp. "*coheni*", Park and Martens, unpubl.
                Gomphocythere cristata, Rome, 1962
                Gomphocythere curta, Rome, 1962
                Gomphocythere n. sp. "*downingi*", Park and Martens, unpubl.
                Gomphocythere expansa, Sars, 1924
                Gomphocythere lenis, Rome, 1962
                Gomphocythere obtusata, Rome, 1962
                Gomphocythere ortali, Rome, 1962
                Gomphocythere parcedilatata, Rome, 1977
                Gomphocythere simplex, Rome, 1962
                Gomphocythere n. sp. "*wilsoni*", Park and Martens, unpubl.
                Gomphocythere n. sp. "*woutersi*", Park and Martens, unpubl.
              Genus *Cytheridella* Daday, 1905
                Cytheridella chariessa, Rome, 1977
           Subfamily Limnocytherinae, Klie, 1938
          Tribe Limnocytherini, Klie, 1938
             Genus *Limnocythere*, Brady, 1867
               Limnocythere thomasi, Martens, 1990
             Genus *Leucocythere*, Kaufmann, 1892

being one of the better studied and most diverse taxonomic groups within the lake, only one group, the Megalocypridinaea, has had a phylogenetically derived hypothesis presented (Martens and Coomans, 1990). However, their study was not a quantitative cladistic analysis using Hennigian principles, and the present chapter is the first such cladistic approach towards investigating any ostracod group in the East African rift lakes.

IV. METHODS

A. Taxa and Characters

This analysis includes 16 *Gomphocythere* species from Lake Tanganyika and elsewhere in Africa, including the West Transvaal region (Republic of South Africa), the Ethiopian rift lakes, and Lakes Albert, Kivu and Turkana. The study used all known *Gomphocythere* available from the authors' own collections, collections of the Royal Belgian Institute of Natural Sciences and those documented in the literature. The species from Lake Tanganyika are *Gomphocythere alata*, *G. curta*, *G. cristata*, *G. lenis* and *G. simplex*, and the new species *Gomphocythere* n. sp. "*coheni*", *G.* n. sp. "*downingi*", *G.* n. sp. "*wilsoni*" and *G.* n. sp. "*woutersi*". Species from outside Lake Tanganyika are *G. aethiopis*, *G. angulata*, *G. angusta*, *G. capensis*, *G. obtusata*, *G. ortali* and *G. parcedilatata*. Outgroups, *Cytheridella chariessa*, *Limnocythere dadayi* and *Leucocythere* sp., were chosen from the family Limnocytheridae (Table 1).

Characters for phylogenetic analysis were based on homologous structures and are coded as a numeric or alphabetic symbol that represents a particular character state (Wagner, 1989; Pogue and Mickevich, 1990). In total, 44 characters were defined and coded for all 19 taxa in the analysis (Appendix I). Similar numbers of hard-part characters (21) as well as soft-part characters (23) were identified for the composite analysis, using male and female data sets separately to avoid the problems caused by sexual dimorphism (Figure 3, 4). The complete matrix showing the presence and absence of characters and multistate values is given in Appendix I. Nine of 15 multistate characters were then coded, ordered as if they had successively additive states. Separate analyses were carried out for data sets with all unordered characters, and with the multistate characters that were coded as ordered.

B. Phylogeny Reconstruction

The computer program PAUP (Swofford, 1998) was used to compute the most parsimonious tree from a character set of the 44 characters. Two types of run were undertaken. In the first runs, any change from one state to another was counted as one step. In subsequent analyses, nine characters were redefined as

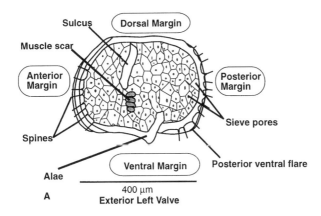

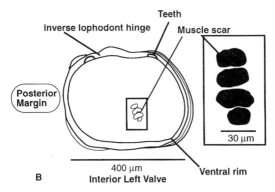

Fig. 3. Schematic drawings of hard-part morphologies of *Gomphocythere* ostracods. Anterior, posterior, dorsal and ventral margins are labelled on the upper diagram.

ordered, based on their successive, additive states. Separate analyses were then performed using the dataset with the redefined characters.

Because the data matrix was too large for the algorithm and computer capabilities, exact methods, which guarantee optimal reconstructions, could not be used. Therefore, the data matrix was analysed using the heuristic searching option. The tree bisection and reconstruction (TBR) search option was used on trees reconstructed with the random addition sequence. To increase the likelihood of finding all islands of equally parsimonious trees (*sensu* Maddison, 1991), 100 random replications were included in each analysis. An island of equally parsimonious trees is a set of trees in which each tree in an island is connected to every other tree through a series of trees, each member of the series differing from the next by a single, minor rearrangement of branches.

All characters in the initial analysis were both unpolarized and undirected, following Hauser and Presch (1991) and Swofford and Maddison (1987),

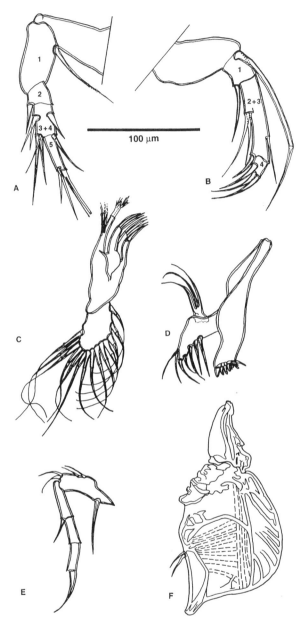

Fig. 4. Drawings of soft-part morphologies of *Gomphocythere "downingi"* n. sp. (A) Antennula, (B) antenna, (C) maxillula, (D) mandibula, (E) P3-walking limb, (F) hemipenis of male. Limb terminology follows Broodbakker and Danielopol (1982).

making the initial trees produced by PAUP unrooted. Since the outgroup taxa are included in the PAUP parsimony analysis, the assumption of monophyly of the ingroup is tested in the analysis. The trees were then rooted using outgroup analysis (Maddison, 1991). The computer program MacClade (Maddison and Maddison, 1992) was used to explore equally parsimonious character distributions within the minimal-length topology discovered by PAUP. Trees were compared with respect to their phylogenetic structure and then compared with trees produced with characters that were randomized over the taxa (*sensu* Archie, 1989).

V.   RESULTS

Using heuristic and branch and bound searches conducted with the phylogenetic reconstruction program, five trees of 99 steps were found (CI = 0.56) (Figure 5). The skewness of tree-length distribution and permutation tests revealed significant phylogenetic structure in the data. Nodes were supported by one to 10 character state changes and these character changes were sometimes reversed or paralleled elsewhere, accounting for much of the homoplasy in the reconstructions. Additional analyses removing the more homoplastic characters (12 characters, defined as CI < 0.5) failed to improve the resolution markedly. Therefore, it was determined that the favoured tree structure can be supported even without the homoplastic characters. *Gomphocythere* was monophyletically distributed on the tree, with two major subclades being supported in each reconstruction: subclade A (*G. aethiopis, G. obtusata, G. angulata, G.* n. sp. "*wilsoni*" and *G.* n. sp. "*downingi*") and subclade B (*G. alata, G. cristata, G.* n. sp. "*coheni*", *G.* n. sp. "*woutersi*", *G. angusta, G. simplex, G. curta* and *G. lenis*) (Figure 5). A majority rule consensus tree yielded a single phylogeny in which the monophyly of *Gomphocythere* is supported, but the monophyly of a Tanganyikan endemic *Gomphocythere* clade is not (Figure 5).

A. Phylogenetic Patterns

1.   Interlake Distribution

A plot of occurrence data for *Gomphocythere* in Africa showed no congruence between species distribution and phylogeny (Figure 5), and many geographically distant lakes share similar *Gomphocythere* species. This may be an artefact of the inadequate sampling of the ostracod faunas of many of the East African lakes but, even so, those species known do not support the hypothesis that there is a congruence between *Gomphocythere* phylogeny and species distribution in adjacent African lakes. For example, Lakes Kivu, Tanganyika

Fig. 5. Majority-rule tree of the five most parsimonious reconstructions using unordered characters. The five initial trees have 99 steps, CI = 0.56, HI = 0.44, RI = 0.63 and three islands. Endemic Tanganyikan *Gomphocythere* species have been mapped onto the majority consensus tree. Note that endemics do not occur within a monophyletic, Tanganyikan clade.

and Victoria, which are adjacent to one another, do not share any of the same *Gomphocythere* species (Figures 2, 5).

2.   Endemic Tanganyikan Gomphocythere

Gomphocythere species endemic to Lake Tanganyika include *G. alata, G. cristata, G. curta, G. lenis, G. simplex, G.* n. sp. "*coheni*", *G.* n. sp. "*downingi*", *G.* n. sp. "*wilsoni*" and *G.* n. sp. "*woutersi*". Endemic species, as traced over the phylogenetic tree, do not cluster together in a single clade, but instead are interspersed with non-endemics (Figures 5, 6). This indicates that either there were multiple radiations of *Gomphocythere* species in Lake Tanganyika, or Tanganyikan endemics subsequently dispersed from the lake. The latter possibility is unlikely and would be contrary to the pattern shown by other organisms. Furthermore, *G. angulata* is known in Late Miocene sediments from northern Kenya. These sediments are only slightly younger than the maximum age for Lake Tanganyika itself (Cohen, 1982) and a Tanganyikan origin for this genus thus appears most improbable. The fact that an endemically constrained tree is not the most parsimonious tree, but has a higher tree length than an unconstrained tree, further supports this conclusion.

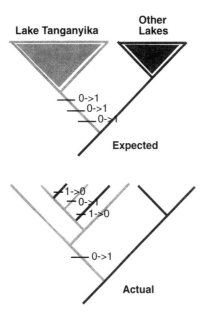

Fig. 6. Expected pattern resulting from a single invasion and subsequent radiation of a single species in Lake Tanganyika. $0 \rightarrow 1$ indicates character state changes.

3. Intralake Distribution

Very little information is available concerning the ostracod faunas of the
western (Democratic Republic of Congo) shores of Lake Tanganyika, and
most collected material was from the eastern (Tanzania) shore of the lake.
However, despite this sampling bias, no species of *Gomphocythere* was
recorded from only one sub-basin or area of the lake.

VI. DISCUSSION

A. Speciation Model

Our speciation model considers two major evolutionary scenarios. The first
involves a hypothetical radiation of a "parent" species (p) and five descendant,
or "daughter" species (d1–d5) that evolve in a single lake system, taking
advantage of the available niche space (ns1–ns6) (Figure 7). Two patterns
could result, one with relatively low homoplasy, since each species would
evolve to occupy its own niche space, and the species from Lake Tanganyika

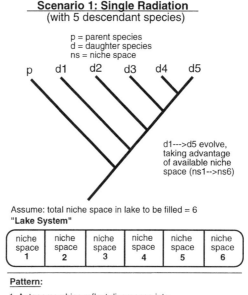

Fig. 7. Speciation scenario of a single radiation with five descendant species in a large
lake; p, parent species; d, daughter species; ns, niche space. Following such a speciation
scenario one would expect to find low levels of homoplasy and autopomorphies that
reflect the divergence of the daughter species into well-spaced niches.

would nest within their own subclade; or, alternatively, one with higher homoplasy as the populations are isolated and acquire parallel characters independently. This divergence into well-spaced niches would be reflected by the presence of autapomorphies.

In contrast, a second scenario involving multiple radiations from two lakes or different sub-basins would reveal a pattern different from that of a single parent and subsequent radiation. Consider the case of six hypothetical species found in one lake, but derived from two different lakes or from different sub-basins within a single lake (Figure 8). Lake or lake sub-basin "A" supports, as before, six individual niche spaces. The daughter species in this lake or sub-basin evolve, taking advantage of the most suitable niche spaces available (ns2–ns4). Similarly, in the second lake or lake sub-basin "B", the daughter species also evolve, again taking advantage of the most suitable niche spaces available in that lake or sub-basin (ns2 and ns3, where ns2 and ns3 are ecologically equivalent in both lakes or sub-basins) (Figure 8).

When, for whatever reason, the species of these two lakes or lake sub-basins are brought together, several patterns could emerge, complicating parts of the phylogeny reconstruction, or even making them spurious (Figure 9). One potential pattern is where the dichotomy represents the actual evolution of the sister groups within each lake or lake sub-basin, and reflects actual homoplasy. A second possibility is a pattern resulting from a situation where the number of homoplastic characters overwhelms the phylogenetic sorting role of the true synapomorphies (i.e. autapomorphies playing a role of synapomorphies). This could yield spurious associations of sister groups or, alternatively, imply a type of pseudoreticulation. Because d4 and d5 would evolve in the same approximate niche space as d1 and d2, demonstrating parallelism of adaptively derived characters, the patterns emerging from this type of scenario can potentially include high levels of homoplasy (Figure 9). In such a case the autapomorphies would reflect the divergence into well-spaced niches, but similar autapomorphies would be perceived as synapomorphies. Paradoxically, such an example could therefore actually lower homoplasy values for the tree.

Which of these two scenarios, single or multiple radiation, is more likely? Lake Tanganyika has been separated into sub-basins during intervals of low lake-level stand (Tiercelin and Mondeguer, 1991), and this would be consistent with the model if the species within these sub-basins acquired similar character states due to similar environmental parameters. Furthermore, comparison of the favoured phylogeny with the currently known distribution of *Gomphocythere* species suggests that the diversity of *Gomphocythere* in Lake Tanganyika cannot be explained solely in terms of a simple radiation from adjacent lakes. Taken together, these various sources of evidence lead the authors to favour the latter of these two scenarios, one of multiple radiation from adjacent sub-basins of the Tanganyikan trough, as an explanation for the high level of homoplasy seen in the analysis of *Gomphocythere* in Lake Tanganyika.

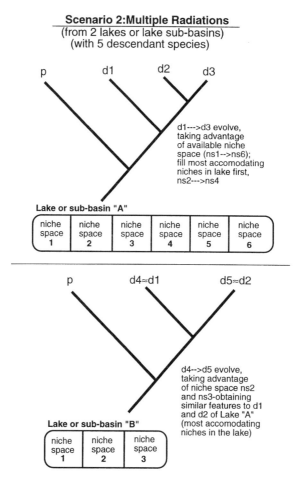

Fig. 8. Speciation scenario of two adjacent lakes or lake sub-basins, A and B, that will subsequently be joined. Lake A has six niche spaces, while B has three. Daughter species d1–d3 evolve to occupy niche spaces ns2–ns4 in lake basin A, while in lake B daughter species d4 and d5 evolve to occupy niche spaces ns2 and ns3.

It is also possible that the *Gomphocythere* lineage was derived from Lake Tanganyika, and then spread outwards via widely dispersed species. This scenario can be tested using fossil evidence to assess the minimum age of any species within this subclade. *Gomphocythere angulata* has been found in the Late Miocene Lothagam III Formation from Lake Turkana and the Middle to Late Pliocene Koobi Fora Formation from east Lake Turkana, showing that its associated branching event must pre-date 7 Mya. *Gomphocythere obtusata* is also known from the Koobi Fora Formation of Middle to Late Pliocene age (approximately 2.5 Mya). Neither of these species is present in Lake

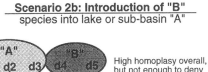

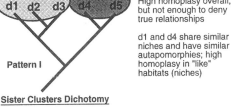

High homoplasy overall, but not enough to deny true relationships

d1 and d4 share similar niches and have similar autapomorphies; high homoplasy in "like" habitats (niches)

Pattern I

Sister Clusters Dichotomy

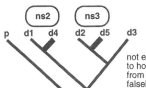

not enough synapomorphies to hold d4 and d5 separate from d1-->d3; d1 and d4 falsely represented as sistergroups and d2 and d5 falsely represented as sistergroups; d1 and d4 share same niche space, as does d2 and d5

Pattern II

Pattern:

1. Autapomorphies can either be reversed (gained or lost), as in Pattern I; or could become false synapomorphies, forcing the construction of false sister groups.

2. Homoplasy relatively high because d4 and d5 evolve in the same approximate niche space as d1 and d2-----> parallelism results

Fig. 9. Speciation scenario of two joined adjacent lakes or lake sub-basins, A and B, and the patterns produced by the overlapping distribution of two similar clades. In the first pattern (upper diagram), the true relationships would be evident, as the tree produced is an unresolved dichotomy. In pattern II, however (lower diagram), d1 and d4, and d2 and d5, would be erroneously reconstructed as sister taxa, having similar autapomorphies that are falsely recognized as synapomorphies. This latter pattern could be misconstrued as reticulation (i.e. pseudoreticulation). Thick or normal lines indicate similar autapomorphies; daughter species d1–d3 are from lake basin A, and d4 and d5 are from lake basin B, as indicated in Figure 8.

Tanganyika, but both are present in other African lakes, except for Lake Turkana. Lake Tanganyika is estimated to be 9–12 My old (Cohen *et al.*, 1993). The Koobi Fora Formation (3.3–2.6 Mya) and Lothagam III Formation (4.5–3.5 Mya) are much younger than Tanganyika, and the occurrence of *G. angulata* and *G. obtusata* in these formations suggests that these two *Gomphocythere* species are younger than the formation of Lake Tanganyika. This complicates testing of the Lake Tanganyikan origin hypothesis using these two fossil species. Unfortunately, other than the Lothagam Hill sites, there are few deposits known to be older than Early Pliocene, making further testing of the "Out-of-Tanganyika hypothesis" untenable. The minimum age estimate, based on palaeontological evidence for *G. angulata* and *G. obtusata*, also suggests that at least parts of the clade could be as old as or older than Lake Tanganyika itself.

B. Speciation in the Tanganyikan Basin

Many questions arise in the study of large numbers of closely related taxa occurring within a single lake basin. What is the origin of these faunas? What is their age? When did the ancestors invade the basin? When did the radiation occur? How did these intralacustrine radiations come into existence?

Many of the speciation models proposed for species flocks in East African lakes are specific to individual lakes and their basin histories. As such, speciation scenarios in Tanganyika are not necessarily the same as in other ancient lakes. Despite this, the speciation patterns of *Gomphocythere* (i.e. high homoplasy) in Tanganyika can be compared with other groups that are characterized by morphological divergence and parallel character state acquisitions resulting from multiple radiations.

The species assemblages of cichlid fishes in Lakes Victoria, Malawi and Tanganyika have been well studied (Sturmbauer and Meyer, 1992; Meyer, 1993; Lowe-McConnell, 1993; Sturmbauer *et al.*, 1994; Rossiter, 1995; Sultmann *et al.*, 1995; Fryer, 1996; Verheyen *et al.*, 1996; Mayer *et al.*, 1998). The genetic variation within the Tanganyikan cichlid flocks reveals a high degree of within-lake endemism among genetically well-separated lineages, distributed along the inferred shorelines of three historically intermittent lake basins. The three-clade–three-basin phylogeographical pattern demonstrated by Verheyen *et al.* (1996) is found twice within the Eretmodini tribe of cichlids. This phylogeographical pattern suggests that major fluctuations in the lake level have been important in shaping the adaptive radiation and speciation within this group. The mitochondrially defined clades are in conflict with the current taxonomy of the group (see Verheyen and Rüber, this volume) and suggest that there has been convergent evolution in trophic morphology, particularly in the shapes of teeth, taxonomically the most diagnostic character of the three genera. This evolutionary scenario, of multiple radiations from adjacent sub-basins, is similar to that of scenario II in the present model. Evidence presented by the phylogenetic pattern for *Gomphocythere* suggests it to be the most likely scenario for ostracods in the Tanganyikan trough also.

Other authors have also attributed the high amount of genetic diversity within the cichlid fish flocks to geographical isolation due to lake-level fluctuations. For example, in their study of the Tanganyikan genus *Tropheus*, Sturmbauer and Meyer (1992) propose that, after an initial invasion and radiation, secondary radiations occurred, triggered by fluctuations in lake level. These abiotic factors may have strongly affected the distribution of many species, and probably led to widespread extinctions and fusions of isolated populations. This, too, is supported by the present phylogenetic analysis of *Gomphocythere*. Such lake-level fluctuations are likely to have been of great importance, since they have occurred many times in the Tanganyikan trough, and the resultant geographical isolation would potentially have promoted speciation events on each occasion.

C. The Single Invasion Scenario

If the diversity of *Gomphocythere* species observed in Lake Tanganyika were the result of a single radiation, then one would expect a single endemic, monophyletic clade within the phylogenetic tree. However, the distribution of non-endemic species within endemic subclades throughout the phylogenetic tree (Figure 5) suggests strongly that there was more than one radiation in the lake and that the *Gomphocythere* species in Lake Tanganyika are not the result of a single radiation. It is probable that some species which arose during these evolutionary bursts became extinct. Evidence for their existence might be uncovered in a longer fossil record, if and when it becomes available.

D. The Multiple Invasion Scenario

An immigration scenario explaining the origin of species flocks or clusters requires that dispersal of a sister population from a lake with subsequent reproductive isolation occurs, as well as reintroduction to the lake through immigration. Were this the case, then many of the *Gomphocythere* species within Lake Tanganyika should have a sister group outside the lake (Smith and Todd, 1984). The presence of some *Gomphocythere* species throughout Africa, and the restriction of others to specific lake systems, could then be explained in terms of multiple invasions. Dispersal of *Gomphocythere* species inhabiting shallow lakes probably takes place via the feet of wading birds.

It might also be argued that the wide distribution of *Gomphocythere* species can be explained by their age alone. If they originated long ago in a single lake or in a series of connected drainages, the presence of *Gomphocythere* species in these various lakes today could be the result of geologically distant isolation and not immigration. However, the presence of *Gomphocythere* conspecifics in lakes that were never connected in the geological past makes this scenario improbable (Figure 1).

E. Fluctuation in Lake Levels

Another hypothesis for species flock formation involves lake-level fluctuations and multiple radiations. Low lake levels would isolate populations into separate water bodies. The different evolutionary pressures (or sufficient time for random mutations) in each water body would mean that each population would be reproductively isolated and allopatric speciation would take place. When the lake rose again, the two (or more) newly evolved species would become sympatric. Repeated connection and isolation of basins would provide opportunities for invasions and also for extinctions of species, which would open up possible new niche space for these invading species to occupy. The repeated separation and mixing of species which resulted from lake-level changes in Lake Tanganyika may have promoted species diversity in certain taxa, and has been described as a "species pump" (Rossiter, 1995).

There is abundant evidence of lake-level fluctuation in Tanganyika's late Pleistocene history. For example, 50 000–45 000 years ago and 25 000–15 000 years ago there were low stands of at least 200 m below present water levels. High lake-level stands, 20 m above present levels, occurred between 15 000 and 5000 years ago, when the Kivu basin was open to the Tanganyikan trough (Haberyan and Hecky, 1987; Scholz and Rosendahl, 1988; Tiercelin and Mondeguer, 1991). Lake levels are presently rising, but are 20 m below the maximum lake-level estimates (Jolly *et al.*, 1994). Although a fluctuation in lake levels is an intuitively appealing hypothesis, there is no evidence to support the importance of this mechanism in the evolutionary history of *Gomphocythere*. For example, one might consider the possibility of vestigial populations, with newly evolved species, separated from each other in sub-basins of Lake Tanganyika (i.e. not yet having become sympatric) as circumstantial evidence supporting the fluctuation hypothesis. However, for *Gomphocythere*, species distributions in Lake Tanganyika are well mixed between sub-basins.

F. Tempo and Mode of *Gomphocythere* Speciation

There is no unequivocal evidence that supports gradualistic or explosive speciation for *Gomphocythere* in particular, or for ostracods in general, in ancient lakes. Less than 1000 years are represented in the record of cores currently available for Lake Tanganyika and therefore rates of evolution cannot be evaluated by this means. This has led to general evolutionary models of ostracod evolution in these lakes being hypothetical only. For example, Martens (1990) interpreted the contast between the great number of limnocytherid species now present in African lakes and their potential origins within the Holocene as evidence of rapid speciation. He hypothesized that there has been a number of discrete bursts of speciation, interspersed by relatively long periods of stasis. However, there are no geological data to support or refute this idea, and it remains wholly speculative.

In fact, evidence suggests that, unlike the haplochromine cichlid fish radiations of Lake Victoria and Lake Malawi (Sturmbauer and Meyer, 1992), the radiation of *Gomphocythere* occurred early in the lake's history, with little evidence for cladogenesis since the Miocene. Based on current knowledge, it is therefore unrealistic to assume that the diversification of *Gomphocythere* and other ostracod groups is strictly a Late Pleistocene or Holocene phenomenon.

The presence of non-endemics within Lake Tanganyikan subclades revealed in the present study suggests that the tempo of speciation of *Gomphocythere* within Lake Tanganyika may have been slow. The data do not support a recent and rapid radiation from a single ancestor. Instead, the topology of the tree suggests that many species are old (pre-Pliocene) and that there have been multiple radiations in the lake. The rate of speciation of the *Gomphocythere* clade prior to *G. angulata* remains unknown.

Whether the ancestors and their descendants spread over an increasingly large geographical area, with descendant species filling the same or very similar ecological niches in different locations, is not known. However, the distribution pattern resulting from such a situation would be one of a widely distributed ancestor with overlapping ecological distributions, with younger species occupying different ecological niches. It is just such a pattern that the *Gomphocythere* species show in Lake Tanganyika, where younger taxa show species-specific affinities for different substrates (Park, 1995).

VII. CONCLUSIONS

The analysis suggests that *Gomphocythere* diversified many times. This interpretation is supported, in part, by the occurrence of non-endemics and endemics in the same subclades. Biogeographical distributions of all *Gomphocythere* species in Africa indicate that there are no systematic corridors or connected pathways between *Gomphocythere* faunas in different lakes. In addition, the position of many of the species that occur widely in Africa is at the base of various subclades, suggesting that they might be older than the subclades above them on the tree. Their position, interspersed within, on the *Gomphocythere* phylogenetic tree also suggests a mosaic pattern of multiple speciation events.

ACKNOWLEDGEMENTS

The authors thank A. Cohen, K. Martens, K. Wouters, W. Maddison and D. Maddison for their help in this project. They are also grateful to the reviewers of this manuscript, and especially A. Rossiter for his support and helpfulness. This work was part of L.E. Park's dissertation research at the University of Arizona, and was generously supported by funding from the Geological Society of America, Sigma Xi, Chevron, University of Arizona Analysis of Biological Diversification Research Training Grant and the Sulzer Fund.

REFERENCES

Archie, J.W. (1989). A randomization test for phylogenetic information in systemic data. *Syst. Zool.* **38**, 239–252.

Bootsma, H.A. and Hecky, R.E. (1993). Conservation of the African Great Lakes: a limnological perspective. *Conserv. Biol.* **7**, 644–656.

Broodbakker, N. and Danielopol, D.L. (1982). The chaetotaxy of the Cypridaea (Crustacea, Ostracoda) limbs; proposals for a descriptive model. *Bijdragen Dierkunde* **52**, 103–120.

Cohen, A.S. (1982). Ecological and paleoecological aspects of the rift valley lakes of East Africa. Unpublished PhD Thesis, University of California, Davis, CA.

Cohen, A.S. (1995). Paleoecological approaches to the conservation biology of benthos in ancient lakes: a case study from Lake Tanganyika. *J. N. Am. Benthol. Soc.* **14**, 654–668.

Cohen, A.S. and Johnston, M. (1987). Speciation in brooding and poorly dispersing lacustrine organisms. *Palaios* **2**, 426–435.

Cohen, A.S., Soreghan, M.S. and Scholz, C.A. (1993). Estimating the age of formation of lakes: an example from Lake Tanganyika, East African Rift system. *Geology* **21**, 511–514.

Coulter, G.W. (1991). *Lake Tanganyika and its Life*. Oxford University Press, London.

Coulter, G.W. (1994). Lake Tanganyika. *Arch. Hydrobiol. Beiheft. Ergebnisse Limnol.* **44**, 13–18.

Eccles, D.H. (1984). On the recent high levels of Lake Malawi. *S. Afr. J. Sci.* **80**, 461–468.

Fryer, G. (1996). Endemism, speciation and adaptive radiation in great lakes. *Envir. Biol. Fishes* **45**, 109–131.

Fryer, G. and Iles, T.D. (1972). *The Cichlid Fishes of the Great Lakes of Africa: Their Biology and Evolution*. Oliver and Boyd, Edinburgh.

Haberyan, K.A. and Hecky, R.E. (1987). The late Pleistocene and Holocene stratigraphy and paleolimnology of Lakes Kivu and Tanganyika. *Palaeogeogr. Palaeoclimatol. Palaeoecol.* **61**, 169–197.

Hauser, D.L. and Presch, W. (1991). The effect of ordered characters on phylogenetic reconstruction. *Cladistics* **7**, 243–265.

Horne, D.J., Baltanas, A. and Paris, G. (1998). Geographical distribution of reproductive modes in living non-marine ostracods. In: *Sex and Parthenogenesis: Evolutionary Ecology of Reproductive Modes in Non-marine Ostracods* (Ed. by K. Martens), pp. 77–100. Backhuys, Leiden.

Horrall, R.M. (1981). Behavioral stock-isolating mechanisms in Great Lakes fishes with special reference to homing and site imprinting. *Can. J. Fish. Aquat. Sci.* **38**, 1481–1496.

Jolly, D., Bonnefille, R. and Roux, M. (1994). Numerical interpretation of a high resolution Holocene pollen record from Burundi. *Palaeogeogr. Palaeoclimatol. Palaeoecol.* **109**, 357–370.

Kolding, J. (1992). A summary of Lake Turkana: an ever-changing mixed environment. *Mitt. Int. Verein. Limnol.* **23**, 25–35.

Lowe-McConnell, R. (1993). Fish faunas of the African Great Lakes–Origins, diversity and vulnerability. *Conserv. Biol.* **7**, 634–643.

McKaye, K.R., Kocher, T., Reinthal, P., Harrison, R. and Kornfield, I. (1984). Genetic evidence for allopatric and sympatric differentiation among color morphs of a Lake Malawi cichlid fish. *Evolution* **38**, 215–219.

Maddison, D.R. (1991). The discovery and importance of multiple islands of most-parsimonious trees. *Syst. Zool.* **40**, 315–328.

Maddison, W.P. and Maddison, D.R. (1992). *MacClade: Analysis of Phylogeny and Character Evolution.* Sinauer, Sunderland, MA.

Martens, K. (1990). Revision of African *Limnocythere* s.s. BRADY, 1867 (Crustacea, Ostracoda) with special reference to the Eastern Rift Valley Lakes: morphology, taxonomy, evolution and (palaeo)ecology. *Arch. Hydrobiol.* **83,** Suppl., 453–524.

Martens, K. (1994). Ostracod speciation in ancient lakes: a review. *Arch. Hydrobiol. Beiheft. Ergebnisse Limnol.* **44,** 203–222.

Martens, K. and Coomans, A. (1990). Phylogeny and historical biogeography of the Megalocypridinae ROME, 1965 with an updated checklist of this subfamily. In: *Ostracoda and Global Events, Proceedings of the 10th International Symposium on Ostracoda* (Ed. by R.C. Whatley and C. Maybury), pp. 545–556. Chapman and Hall, London.

Mayer, W.E., Tichy, H. and Klein, J. (1998). Phylogeny of African cichlid fishes as revealed by molecular markers. *Heredity* **80,** 702–714.

Mayr, E. (1984). Evolution of fish species flocks: a commentary. In: *Evolution of Fish Species Flocks* (Ed. by A.A. Echelle and I. Kornfield), pp. 3–11. University of Maine Press, Orono, ME.

Meyer, A. (1993). Phylogenetic relationships and evolutionary processes in East African cichlid fishes. *Trends Ecol. Evol.* **8,** 279–284.

Michel, E., Cohen, A.S., West, K., Johnston, M.R. and Kat, P.W. (1992). Large African lakes as natural laboratories for evolution: examples from the endemic gastropod fauna of Lake Tanganyika. *Mitt. Int. Verein. Limnol.* **23,** 85–99.

Mittermeier, R.A. (1988). Primate diversity and the tropical forest: case studies from Brazil and Madagascar and the importance of the megadiversity countries. In: *Biodiversity* (Ed. by E.O. Wilson and F.M. Peter), pp. 145–154. National Academic Press, Washington, DC.

Mittermeier, R.A. and Werner, T.B. (1990). Wealth of plants and animals unites "megadiversity" countries. *Tropicus* **4,** 4–5.

Park, L.E. (1995). Assessing diversification patterns in an ancient tropical lake: *Gomphocythere* (Ostracoda) in Lake Tanganyika. Unpublished PhD Thesis, University of Arizona, Tucson, AZ.

Pogue, M.G. and Mickevich, M.F. (1990). Character definitions and character state delineation: the bete noire of phylogenetic inference. *Cladistics* **6,** 319–361.

Rome, D.R. (1962). *Exploration hydrobiologique du Lac Tanganyika (1946–1947): Ostracods.* Royal Belgian Institute of Natural Sciences, Brussels.

Rossiter, A. (1995). The cichlid fish assemblages of Lake Tanganyika: ecology, behaviour and evolution of its species flocks. *Adv. Ecol. Res.* **26,** 187–252.

Rossiter, A. and Yamagishi, S. (1997). Intraspecific plasticity in the social system and mating behaviour of a lek-breeding cichlid fish. In: *Fish Communities in Lake Tanganyika* (Ed. by H. Kawanabe, M. Hori and M. Nagoshi), pp. 193–218. Kyoto University Press, Kyoto.

Sars, G.O. (1924). The freshwater Entomostraca of the Cape Province (Union of South Africa): Ostracoda. *Ann. S. Afr. Mus.* **20,** 2–20.

Scholz, C.A. and Rosendahl, B.R. (1988). Low lake stands in Lake Malawi and Tanganyika, East Africa, delineated with multi-fold seismic data. *Science* **240,** 1645–1648.

Smith, G.R. (1987). Fish speciation in a Western North American Pliocene Rift Lake. *Palaios* **2,** 436–446.

Smith, G.R. and Todd, T.N. (1984). Evolution of species flocks in north temperate lakes. In: *Evolution of Fish Species Flocks* (Ed. by A.A. Echelle and I. Kornfield), pp. 45–68. University of Maine Press, Orono, ME.

Snoeks, J., Rüber, L. and Verheyen, E. (1994). The Tanganyika problem: comments on the taxonomy and distribution patterns of its cichlid fauna. *Arch. Hydrobiol. Beiheft. Ergebnisse Limnol.* **44**, 355–372.

Sturmbauer, C. and Meyer, A. (1992). Genetic divergence, speciation and morphological stasis in a lineage of African cichlid fishes. *Nature* **358**, 578–581.

Sturmbauer, C., Verheyen, E. and Meyer, A. (1994). Mitochondrial phylogeny of the Lamprologini, the major substrate spawning lineage of cichlid fishes from Lake Tanganyika in eastern Africa. *Mol. Biol. Evol.* **11**, 691–703.

Sultmann, H., Mayer, W.E., Figueroa, F., Tichy, H. and Klein, J. (1995). Phylogenetic analysis of cichlid fishes using nuclear-DNA markers. *Mol. Biol. Evol.* **12**, 1033–1047.

Swofford, D.L. (1998). *PAUP: Phylogenetic Analysis Using Parsimony, Beta Version 4.0.* Computer Program, Smithsonian Institution. Sinauer, Sunderland, MA.

Swofford, D.L. and Maddison, W.P. (1987). Reconstructing ancestral character states under Wagner parsimony. *Math. Biosci.* **87**, 199–229.

Tiercelin, J.-J. and Mondeguer, A. (1991). The geology of the Tanganyika trough. In: *Lake Tanganyika and its Life* (Ed. by G.W. Coulter), pp. 7–48. Oxford University Press, London.

Verheyen, E., Rüber, L., Snoeks, J. and Meyer, A. (1996). Mitochondrial phylogeography of rock-dwelling cichlid fishes reveals evolutionary influence of historical lake level fluctuations of Lake Tanganyika, Africa. *Phil. Trans. R. Soc. Lond. B* **351**, 797–805.

Wagner, G.P. (1989). The origin of morphological characters and the biological basis of homology. *Evolution* **43**, 1157–1171.

Appendix I:
Characters used in cladistic analysis

1. Tubercle
   0: absent
   1: present
2. Venteroposterior flare
   0: absent
   1: present
3. Surface reticulation
   0: reduced
   1: robust
4. Brood pouch on female
   0: absent
   1: present
5. Valve shape (dorsal view: anterior–posterior)
   0: round
   1: oval
   2: square
6. Dorsal view of valve shape
   0: convex
   1: heart
   2: triangular
7. Reticulation density of carapace
   0: absent
   1: < 11 (per 5 μm^2)
   2: 12–17 (per 5 μm^2)
8. Sieve pore
   0: normal
   1: radial
9. Central muscle scar
   0: straight
   1: posterior
10. Node position
    0: absent
    1: anteroventral
    2: posteroventral
    3: mediodorsal
11. Node number
    0: absent
    1: 1
    2: 3

12. Maximum dorsal node size
    0: absent
    1: < 0.1 mm
    2: > 0.1 mm
13. Maximum ventral node size
    0: absent
    1: < 0.1 mm
    2: > 0.1 mm
14. Sulcus
    0: absent
    1: present
15. Alae
    0: absent
    1: present
16. Ventrolateral expansion; alar
    prolongation
    0: absent
    1: present
17. Ornamental medial ridge
    0: absent
    1: present
18. Hinge angle
    0: parallel
    1: acute
19. Shell thickness
    0: thin
    1: thick
20. Marginal pore canals
    0: < 5/0.1 mm
    1: > 5/0.1 mm
21. Hingement
    0: lophodont
    1: inverse lophodont
    2: merodont
22. Total number of furca setae
    (female)
    0: 2 setae each
    1: 2 setae and 3 lobes
    2: more than 2 setae
23. Distal lobe on hemipenis
    0: fixed
    1: movable

24. Position of furca (male)
    0:   above copulatory
    1:   below copulatory
25. Distal lobe apex
    0:   ridged
    1:   smooth
26. A1 number of podomeres on endopodite
    0:   4
    1:   5
    2:   6
27. A1 character of 3rd and 4th podomeres
    0:   separated
    1:   fused
28. A1 2nd endopodite podomere dorsal apical setae
    0:   absent
    1:   present
29. A1 number of claws on the last podomere of endopodite
    0:   2 + 2
    2:   2 + 1
30. A1 number of mediodorsal setae on 3rd and 4th podomeres
    0:   1
    1:   2
    2:   3
31. A1 number of medioventral spines on 3rd and 4th podomeres
    0:   0
    1:   1
32. A1 number of dorsal apical setae on 3rd and 4th podomeres
    0:   1
    1:   2
    2:   3
33. A1 position of ventral setae on 1st endopodite
    0:   absent
    1:   apically inserted
    2:   medially inserted

34. A1 number of ventral apical setae on 3rd and 4th podomeres
    0:   0
    1:   1
    2:   2
35. A2 number of podomeres on endopodite
    0:   3
    1:   4
36. A2 1st endopodite podomere apical–ventral setae
    0:   absent
    1:   present
37. A2 1st endopodite podomere shape
    0:   rectangular
    1:   square
38. A2 number of ventral apical setae on 2nd and 3rd podomeres of endopodite
    0:   2
    1:   1
39. A2 number of mediodorsal setae on 2nd and 3rd podomeres
    0:   1
    1:   2
40. A2 number of medioventral setae on 2nd and 3rd podomeres
    0:   3
    1:   2
    2:   1
41. Mandibular palp
    0:   bent knee
    1:   normal
42. Mandibular palp setae
    0:   bifurcated
    1:   straight
43. Maxillula palp
    0:   reduced
    1:   normal
44. P3 size and shape of terminal claw
    0:   very elongated
    1:   slightly elongated and curved
    2:   short and slightly curved

Appendix II:
Data matrix for *Gomphocythere* species and outgroups

Characters/Character States

Taxa	1	2	3	4	5	6	7	8	9	10	11	12	13	14	15	16	17	18	19	20	21	22	23	24	25
G. aethiopis	1	1	1	1	0	0	1	1	1	0	0	0	0	1	0	0	0	0	0	1	1	1	1	1	0
G. alata	0	0	1	1	1	2	1	1	1	0	0	0	0	1	1	0	0	0	1	0	1	1	1	1	0
G. angulata	0	1	1	1	0	0	1	1	?	0	0	0	0	1	0	1	0	0	0	1	1	1	1	1	0
G. angusta	0	0	0	1	0	2	0	1	1	0	0	0	0	1	0	0	0	0	0	0	1	1	1	1	1
G. capensis	1	1	0	1	0	0	1	1	0	0	0	0	0	1	0	0	0	0	0	0	1	1	1	1	1
G. coheni	?	0	1	1	1	2	1	1	1	1	2	2	2	1	1	0	0	0	1	1	1	1	1	1	0
G. cristata	0	0	1	1	0	2	1	1	1	0	0	0	0	1	1	0	0	0	1	1	1	1	1	1	0
G. curta	1	1	0	1	0	0	0	1	1	0	0	0	0	1	0	0	1	0	1	1	1	1	1	1	1
G. downingi	1	1	1	1	0	0	0	1	1	0	0	0	0	1	0	1	0	0	0	0	1	1	1	1	0
G. lenis	0	1	0	1	0	0	0	1	1	0	0	0	0	1	0	0	0	0	0	0	1	1	1	1	1
G. obtusata	1	1	0	1	0	0	1	1	1	0	0	0	0	1	0	0	0	0	0	1	1	1	1	1	0
G. ortali	0	1	0	1	0	0	1	1	1	0	0	0	0	1	0	0	0	1	0	0	1	1	1	1	0
G. parcedilatata	0	0	0	1	0	0	1	1	1	0	0	0	0	1	0	0	0	0	0	1	1	1	1	1	1
G. simplex	0	0	0	1	0	0	0	1	1	0	0	0	0	1	0	0	0	0	0	0	1	1	1	1	1
G. wilsoni	1	?	0	1	0	2	1	1	1	1	1	0	2	1	0	1	0	0	1	1	?	?	1	1	1
G. woutersi	0	0	1	1	1	0	0	1	1	1	0	0	0	1	0	0	0	0	1	0	?	0	0	1	0
C. chariessa	0	0	0	1	0	0	0	1	1	0	0	0	0	0	0	0	0	1	0	0	2	2	1	?	1
Leucocythere	0	0	0	0	0	0	1	0	0	0	0	0	0	1	0	0	0	0	0	0	2	2	0	0	1
L. dadayi	0	0	0	0	0	0	1	0	0	1	2	1	1	0	0	0	0	0	0	0	0	0	0	0	1

Taxa	26	27	28	29	30	31	32	33	34	35	36	37	38	39	40	41	42	43	44
G. aethiopis	1	0	?	?	?	?	?	2	?	1	0	0	?	?	?	1	1	1	0
G. alata	0	1	1	1	0	0	?	2	0	0	0	1	0	0	2	1	1	1	2
G. angulata	0	1	1	0	1	1	?	2	0	0	1	1	0	1	1	1	1	1	1
G. angusta	?	?	?	?	?	?	?	?	?	?	?	?	?	?	?	1	1	1	?
G. capensis	0	1	1	0	1	0	2	2	0	0	1	1	0	1	0	1	1	1	0
G. coheni	0	1	1	1	1	0	0	1	0	0	1	1	0	0	2	1	1	1	0
G. cristata	0	1	1	1	0	0	2	2	0	0	1	1	0	1	0	1	1	1	2
G. curta	0	1	1	1	1	0	1	1	0	0	1	1	0	0	1	1	1	1	2
G. downingi	0	1	1	1	1	0	2	1	0	0	1	1	0	0	0	1	1	1	1
G. lenis	0	1	1	1	1	0	2	1	0	0	1	1	0	0	0	1	1	1	2
G. obtusata	1	0	1	0	2	0	2	2	0	0	1	1	0	0	1	1	1	1	0
G. ortali	0	1	1	0	1	0	2	1	0	0	1	1	0	0	0	1	1	1	0
G. parcedilatata	0	1	1	0	1	0	2	2	0	0	1	1	0	0	0	1	1	1	0
G. simplex	0	1	1	1	0	0	1	1	1	0	1	1	0	0	0	1	1	1	2
G. wilsoni	0	1	1	0	0	0	2	1	0	0	1	1	0	0	1	1	1	1	1
G. woutersi	0	1	1	1	1	0	0	1	0	0	0	1	0	0	2	1	1	1	0
C. chariessa	0	1	1	0	1	1	2	2	0	0	1	0	1	1	1	1	1	1	0
Leucocythere	0	1	1	0	0	0	2	0	0	0	1	1	0	1	0	0	0	0	0
L. dadayi	0	0	0	0	0	0	1	0	1	0	1	1	0	1	0	0	0	0	0

The Dynamics of Endemic Diversification: Molecular Phylogeny Suggests an Explosive Origin of the Thiarid Gastropods of Lake Tanganyika

K. WEST and E. MICHEL

I. SUMMARY

The endemic gastropod fauna of Lake Tanganyika is remarkable not only for its great species richness, but also for its unusually ornate and heavily calcified shell morphologies that are convergent with diverse marine forms. The origin and intralacustrine radiation of these thiarid gastropods have been debated since the late nineteenth century, as they are perhaps the most dramatic lacustrine radiation of gastropods in the world. They parallel the endemic cichlid fish fauna of the African Great Lakes in their potential for providing information about the mechanisms of evolution.

This chapter presents the first molecular phylogenetic treatment of 12 of the 18 endemic gastropod genera of Lake Tanganyika, based on a mitochondrial gene fragment of cytochrome oxidase I (*COI*). The endemic thiarid fauna of Lake Tanganyika was found to be paraphyletic, but a larger clade including *Cleopatra*, a thiarid genus widely distributed throughout East Africa, is monophyletic. The data reveal five robust clades within this larger monophyletic group: (1) ((*Reymondia, Cleopatra*) *Spekia*), (2) (*Stanleya, Tanganyicia*) as sister group to group 1, (3) the trochiform genera ((*Bathanalia, Chytra*) *Limnotrochus*) as a clade and (4) sister-taxon pairings for (*Lavigeria*, Nov. gen.) and (5) (*Anceya, Paramelania*).

Analyses using parsimony, neighbour-joining and maximum likelihood analyses agreed on sister-taxon relationships at terminal nodes, but were unable to resolve relationships among these Tanganyikan clades. This may be interpreted as an indication of rapid, burst-like radiation at the time of origin of this fauna. The term "superflock" (*sensu* Ribbink) may be used to describe the generic level radiation of Tanganyikan gastropods, as it preserves the information that this is a group of closely related endemics that have radiated *in situ*, but does not imply complete monophyly.

II. INTRODUCTION

Among biologists, the African Great Lakes have long been renowned for their species flocks of cichlid fishes. Lakes Malawi, Tanganyika and Victoria each host endemic flocks of several hundred morphologically, genetically and ecologically distinct cichlid species (e.g. Fryer and Iles, 1972; Greenwood, 1981; Seehausen, 1996; Turner, 1996; Kawanabe *et al.*, 1997). The advent of molecular systematics has provided a host of new characters from which to address questions of phylogenetic affinities among cichlids. These characters have shed light on several major questions. First, it is now known that the rates of molecular and morphological evolution in cichlids may be almost wholly decoupled (Meyer *et al.*, 1990; Sturmbauer and Meyer, 1992). Secondly, molecular studies have loaned support to the anatomically based findings of Stiassny (1981) that the striking morphological similarities between some Tanganyikan and Malawian cichlids is an example of convergence, and is not indicative of the cichlids of the African Great Lakes having once constituted a "superflock" in which sister taxa had dispersed to different lakes. Thirdly, evidence now exists that the Malawi and Victoria cichlid flocks are monophyletic and that both were probably derived from single lineages in the Tanganyikan flock (Kocher *et al.*, 1993; Meyer, 1993), a finding which again supports results from earlier, traditionally based taxonomic studies (Fryer and Iles, 1972). In short, the advent of molecular systematics has revolutionized how cichlid phylogenies are reconstructed and strengthened our understanding of evolutionary history and diversification in this group.

While cichlid species flocks have been the focus of intense study, the other species flocks in the African Rift Lakes have received comparatively little attention from modern molecular techniques. The thiarid gastropods of Lake Tanganyika comprise a spectacular endemic radiation in their own right. Rich in species and diverse in form, the Tanganyikan thiarids (Figure 1) have unusually ornate and heavily calcified shells which look compellingly like marine shells (Moore, 1903), a similarity which stimulated the early scientific explorations of the lake (Fryer, this volume). Anatomical (Smith, 1904; Pelseneer, 1906) and molecular (West, 1997) studies show, however, that this resemblance to marine shells is the result of convergent evolution and not phylogenetic affinities. Several different lines of evidence support the hypotheses that the ornate and heavily calcified Tanganyikan shells are an adaptive response to predation pressure (West et al., 1991; West and Cohen, 1994, 1996). However, much work remains in analysing the patterns of diversification, phylogenetic relationships and mechanisms driving diversification in this group. This study of the molecular systematics of the thiarid gastropods of Lake Tanganyika explores some of these issues.

The thiarid gastropods of Lake Tanganyika include 18 endemic genera encompassing approximately 70 species. The high levels of endemicity in this group suggest that they may have radiated within the basin, some time after the lake's origin, 9–12 Mya (Ebinger, 1989; Cohen et al., 1993). It is perhaps the high levels of thiarid endemism in Lake Tanganyika that have led many people to assume that these gastropods constitute a species flock (Boss, 1978; Brown, 1994; Coulter, 1991; Michel, 1995; West, 1997). However, monophyly of this group has not been rigorously investigated to date, and in reality the taxonomy of Lake Tanganyika's thiarid gastropods is muddled and confused, the result of a century-long battle between taxonomic "lumpers" and "splitters". A recent taxonomic treatment of the Tanganyikan gastropods recognizes a total of 60 prosobranch and pulmonate species (Brown and Mandahl-Barth, 1987), but molecular studies indicate that, at least for the thiarids, these taxonomies significantly underestimate species-level diversity (E. Michel and K. West, unpubl.). Accordingly, these taxonomies are presently under revision.

It is unusual that in most taxonomies of the Tanganyikan thiarids, with the exception of Bourguignat (1885, 1890), the genera are either monotypic (e.g. *Chytra, Stormsia*) or highly speciose (e.g. *Lavigeria, Paramelania, Reymondia*). Perhaps the greatest difficulty encountered when studying this group is in defining and delineating species in these speciose clades. Morphological or anatomical studies alone have their limitations, and a more diverse approach, preferably one that incorporates modern biochemical tools, is necessary to address this problem. For example, through a combination of allozyme, conchological, soft-part anatomical, biogeographical and DNA sequence data, Michel (1995; this volume) was able to designate nine species of *Lavigeria*. Similar studies should be undertaken for the genera *Paramelania* and *Reymondia*.

Higher level systematic relationships among the Tanganyikan thiarid genera are also not clear. A recent contribution (Bandel, 1998) reassigned the Tanganyikan thiarids into three different families (the majority being placed

into the Pleuroceridae) and seven subfamilies therein. Because some of these groupings are contradicted by the present study and others (West *et al.*, in press; Michel and Todd, unpubl. data) and because the hypotheses of these relationships have not yet been tested through character analysis, this taxonomy (Bandel, 1998) has not been adopted in this study.

The present phylogenetic investigation focuses on generic-level relationships and aims to establish the coarse structure of the radiation of thiarid gastropods in Lake Tanganyika. While considerable work remains in delimiting taxonomic boundaries within the speciose genera, the taxonomic boundaries between genera, monotypic and speciose alike, are well established and provide confidence in the genus-level taxonomy (Brown, 1994) examined in this study. Of particular interest are the following questions. Was there a single thiarid invasion into Lake Tanganyika, with subsequent radiation forming the diversity seen today? Are the Tanganyikan thiarids monophyletic, or have they also given rise to non-Tanganyikan species? Within the radiation, do speciose clades give rise to other speciose clades, or do they arise independently from non-speciose clades? Are there any key morphological features associated with the speciose clades?

In order to establish a phylogeny with which to explore these questions, we analysed 640 base pairs (bp) of the cytochrome *c* oxidase subunit I (*COI*) region of the mitochondrial genome. This gene is associated with electron

Fig. 1. Examples of the endemic thiarid gastropods of Lake Tanganyika. Note the heavily calcified and ornate shells. A, *Paramelania damoni*, form: typica; B, *Paramelania iridescens*; C, *Reymondia horei*; D, *Lavigeria grandis*; E, *Bathanalia howesii*; F, *Limnotrochus thomsoni*; G, *Spekia zonata*; H, *Paramelania iridescens*; I, *Mysorelloides multisulcata*; J, *Paramelania damoni*, form: *imperialis*; K, *Tiphobia horei*; L, *Lavigeria nassa*, form: typica; M, *Chytra kirki*; N, *Paramelania damoni*, form: *mpalaensis*; O, *Lavigeria nassa*, form: *paucicostata*; P, *Hirthia globosa*; Q, *Paramelania damoni*, form: *crassigranulata*; R, Nov. gen. n. sp.; S, *Paramelania damoni*, form: *imperialis*; T, Nov. gen. *guillemei*; U, *Tanganyica rufofilosa*. Scale: the maximum length of shell N, *P. damoni* form: *mpalaensis*, is 3.95 cm. Species identifications were based on Leloup (1953) and Brown (1994). However, the taxonomies of *Lavigeria* and *Paramelania* are currently under revision.

transport in cells (Frohlich *et al.*, 1996) and is one of the more conservative protein-coding genes of the mitochondrial genome (Brown, 1985). Many studies attest to the phylogenetic utility of *COI* at the genus, species and population levels across a variety of taxa (Crozier *et al.*, 1989; Folmer *et al.*, 1994; Frohlich *et al.*, 1996; Pederson, 1996; Michel, this volume). The differential rates of evolution of first, second and third position nucleotides also offer multiple levels of resolution within a study.

III. MATERIALS AND METHODS

A. Sample Collection and Preservation

Thiarid gastropods were collected from the Burundi and the Democratic Republic of Congo (formerly Zaïre) coastlines of Lake Tanganyika in 1992–1993 and the Tanzania and Zambia coastlines in 1995. Thiarid and non-thiarid outgroups were sampled from other East African lakes and rivers during this same time. Live specimens of the ingroup (Tanganyikan) and outgroup (non-Tanganyikan) taxa were obtained by snorkelling, SCUBA diving and/or dredging in 0.25–60 m of water. Sample lots (populations of a single species from the same locality) were randomly split into three sub-lots to form reference collections for conchological, anatomical and molecular studies, respectively. The shells of the gastropods reserved for molecular studies were cracked and peeled completely and the soft tissues were placed directly in 95% ethanol.

B. Molecular Methods

Except for one taxon for which only a single individual was collected, two or more individuals from a representative species of each thiarid genus in Lake Tanganyika, and all of the outgroups, were prepared for molecular analyses (see Table 1 for taxonomic information and sampling locales). Following the molecular methods of T. Collins (pers. comm.) and Palumbi (1996), DNA was extracted from alcohol-preserved tissues using a CTAB protocol. In the larger specimens (body > 0.5 cm), the heart and kidney and portions of the gonad were dissected out and used to prepare the DNA extract. In the smaller specimens (body < 0.5 cm), the whole soft body was used. The mucopolysaccharides secreted by molluscs sometimes inhibit DNA amplification. Several problematic DNA templates were successfully amplified after being treated with one-half volume 8 M lithium chloride for 1 h at 65°C (Palumbi, 1996).

  Exploratory analyses were conducted to identify an appropriate mitochondrial gene region for this study. In initial studies, the 16S to ND1, 16S RNA, 12S RNA and cytochrome *b* regions proved to be either less informative and/or more difficult to amplify or sequence than cytochrome *c* oxidase I. This gene fragment was readily amplified and sequenced from the several taxa used in the preliminary analyses.

Table 1

Gastropods (family Thiaridae) collected from Lake Tanganyika and its environs

Genus	Species	Author	Sample no.	Locality	Depth, substrate
Anceya	giraudi	Bourguignat 1885	95 KW 45	Gitaza, BR	8 m, rocks
Bathanalia	howesi	Moore 1898	95 KW 17	Cameron Bay, ZМ	57 m, mud
Bridouxia	giraudi	Bourguignat 1885	95 KW 10	Wonzye Pt., ZM	2 m, under rocks
Chytra	kirki	Smith (1880b)	93 KW 01	Gitaza, BR	12 m, sand or silt
Cleopatra	ferruginea	Lea and Lea 1850	93 KW 57	Kinango Dam, KN	0.25 m, silt or mud
Hirthia	globosa	Ancey 1898	95 KW 28	Mtossi, TZ	5 m, rocks
Lavigeria	cf. nassa	Woodward 1859	95 KW 01	Kasenga Pt. ZM	5 m, rocks
Limnotrochus	thomsoni	Smith (1880a)	95 KW 12	Myiamba, ZM	4-13 m, sand
Martelia	tanganyicensis	Dautzenberg 1907	93 KW 31	Rwaba, BR	8 m, rocks
Melanoides	admirablis	Smith (1880b)	95 KW 42	Malagarasi R., TZ	1 m, mud
	tuberculata	Müller 1774	93 KW 55	Mazeras, KN	0.5 m, macrophytes
Mysorelloides	multisulcata	Bourguignat 1888	95 KW 31	Kala Bay, TZ	30 m, dredge
Nov. gen.	n. sp.	Michel (unpubl.)	93 KW 06	Gitaza, BR	12 m, rocks
Paramelania	damoni: imperialis	Giraud 1885	93 KW 36	Nyanza-Lac, BR	30 m, sand and silt
Reymondia	horei	Smith (1880a)	95 KW 04	Kasenga Pt., ZM	5-9 m, under rocks
Spekia	zonata	Woodward 1859	93 KW 04	Gitaza, BR	<1m, rocks
Stanleya	neritinoides	Smith (1880b)	95 KW 29	Msamba, TZ	5 m, silt and sand
Stormsia	minima	Smith 1907	95 KW 08	Wonzye Pt. ZM	<1m, rocks
Syrnolopsis	minuta	Bourguignat 1885	93 KW 15	Muguruka, BU	8 m, sand
Tanganyicia	rufofilosa	Smith (1880a)	95 KW 13	Myiamba, ZM	4-13 m, sand
Tiphobia	horei	Smith (1880a)	95 KW 21	Nkamba, Bay, ZM	30 m, mud

All samples were collected alive. Species author, sample identification number, locality (BR, Burundi; KN, Kenya; TN, Tanzania; ZR, Democratic Republic of Congo (formerly Zaïre); ZM, Zambia), depth and substrate data are included.

Using the polymerase chain reaction (PCR) (Saiki *et al.*, 1988), an approximately 640 bp region of the mitochondrial *COI* gene region was amplified with universal primers developed by Folmer *et al.* (1994). Amplification reactions, including a negative control, were carried out following typical protocols (Palumbi, 1996). Amplified DNA products were separated in a 2% agarose gel in TAE buffer (0.04 M Tris acetate, 0.001 M EDTA) and visualized with ethidium bromide. Bands, approximately 640 bp in length, were excised and purified using the UltraClean Kit (Mo Bio Laboratories) following the manufacturer's directions. Both strands of the resultant double-stranded DNA product were sequenced directly using slightly modified sequencing primers (J. Staton, pers. comm.). Sequencing reactions were carried out following the manufacturer's instructions for the ABI PRISM Cycle Sequencing Kit. Excess dye terminators were removed from cycle-sequenced product by purifying the product in Centri-Sep spin columns (Princeton Separations). The cycle-sequenced products were visualized with an Applied Biosystems 377 automated sequencer and supporting ABI Prism software.

Spectrogram sequences were proof-read by eye. Nucleotide sequences were translated into amino acid sequences using DNA Strider 1.2 (Marck, 1995) with the *Drosophila* mitochondrial genome genetic code and three-phase translation option, so as to identify any reading-frame violations. Sequences were readily aligned by eye in SeqApp· (version 1.9a169; Gilbert, 1994).

C. Phylogenetic Analyses

As a means of measuring hierarchical signal in the data, the skewness of the distribution (i.e. the g_1 statistic) of 100 000 randomly generated trees was calculated (Hillis, 1991; Hillis and Huelsenbeck, 1992). Because the inclusion of even one set of closely related taxa can significantly skew a frequency distribution, thereby giving a false sense of confidence in the signal of the data, the skewness of distributions of randomly generated trees was explored iteratively, at first with the entire data set and subsequently with subsets of the data (Lara *et al.*, 1996).

Because the efficiency and fidelity of different methods may vary depending on intrinsic characteristics of the data set (Kim, 1993), the data were analyzed using a suite of phylogenetic methods. Phylogenetic reconstruction using methods of maximum parsimony, neighbour-joining (Saitou and Nei, 1987) and maximum likelihood (Felsenstein, 1981) were employed in phylogenetic analyses using the Phylogenetic Analysis Using Parsimony (PAUP) computer program, version 4.0 (Swofford, 1997). Support for specific nodes was assessed using both bootstrap analysis (Felsenstein, 1985) (with 1000 replicates each with 10 random addition sequences per replicate and two trees saved at each step) and decay indices (Bremer, 1988) calculated by Autodecay version 2.9.6

(Eriksson, 1997). Character evolution at the amino acid level was explored using the MacClade computer program (Maddison and Maddison, 1996).

IV. RESULTS

A. PCR Reaction Amplification and DNA Sequencing of Preserved Tissues

Despite successful extractions of high molecular weight DNA from several individuals, some taxa (including *Bridouxia, Hirthia, Martelia, Mysorelloides, Stormsia* and *Syrnolopsis*) consistently failed to amplify. Even after applying a wide variety of trouble-shooting measures,* no amplified product could be obtained. With the exception of *Hirthia*, taxa which failed to amplify were of very small body size (body < 0.5 cm). It is possible that digestive enzyme activity during preservation and the age of the tissues used in the analyses (2–5 years in preservative) are especially detrimental in small taxa because of high surface to volume ratios, and are perhaps ultimately responsible for PCR failure. Alternatively, these taxa may share a biochemical characteristic which inhibited DNA amplification.

In contrast, the *COI* region of many other taxa consistently amplified. In each of the following taxa, genomic DNA were successfully amplified, and both strands of the *COI* fragment were successfully sequenced from at least two individuals: ingroup taxa: *Anceya giraudi* (95KW45), *Bathanalia howesi* (95KW17), *Chytra kirki* (93KW01), *Lavigeria* (95KW01), *Limnotrochus thomsoni* (95KW12), Nov. gen. n. sp. (genus and species yet to be formally described; see Michel, this volume, 93KW06), *Paramelania damoni* form: *imperialis* (93KW16), *Reymondia horei* (95KW04), *Spekia zonata* (93KW04), *Stanleya neritinoides* (95KW29), *Tanganyicia rufofilosa* (95KW13), *Tiphobia horei* (95KW21); outgroup taxa: *Cleopatra ferruginea* (93KW57), *Melanoides admirabilis* (95KW42), and *Bellamya* sp., *Melanoides nodicincta, Melanoides tuberculata* and *Melanoides turritispira* (collected from Lake Malawi by S. Drill). Hereafter, these taxa will be referred to by their generic names. Sequences are available from the authors upon request.

B. Sequence Analyses

COI sequences from individuals of the same population did not vary. Because genetic haplotypes were identical for the two individuals examined from each

---

*Trouble-shooting measures included adjusting DNA template and/or magnesium chloride concentrations, adjusting thermal cycling parameters, reprecipating total DNA in lithium chloride, gene-cleaning total DNA, extracting total DNA from different individuals and/or different populations, and extracting total DNA using the Chelex method.

lot, in subsequent analyses a single sequence was retained to represent each taxon. Two taxa were deleted from subsequent analyses. *Melanoides tuberculata* and *Melanoides turritispira* shared identical sequences, perhaps owing to an error in species identification. Resequencing produced a single sequence which was confirmed to be *M. tuberculata* and consequently *M. turritispira* was excluded from further analysis. The *M. nodocincta* sequence was also excluded because it was a distant outlier and behaved strangely in analyses, features which suggest contamination of the sample.

Aligned sequences of 640 bp of the *COI* gene, representing 12 Tanganyikan thiarid and four outgroup taxa, yielded an average nucleotide composition (along the coding strand) of 24.3% adenine, 18.5% cytosine, 20.5% guanine and 36.7% thymine. These taxa show an A–T bias, with 61% of the sites being adenine or thymine. Inspection revealed 337 invariant sites (48%) and 303 (52%) variable sites. Of the variable sites, 234 (39% of the total sites) were phylogenetically informative and 69 (13%) were phylogenetically uninformative (autapomorphic or symplesiomorphic for the ingroup). Variable sites were distributed across codons as follows: 69 first positions, 27 second positions and 207 third positions, yielding a ratio of 2.6:1:7.7 for variable sites across codon positions. This value differs slightly from the 2:1:5 ratio commonly reported for variable sites in mitochondrial protein coding genes.

Transition and transversion ratios for all pairwise comparisons of taxa were averaged. In these analyses using PAUP, transitions outnumbered transversions 1.43 to 1. This lower than expected ratio suggests that some kinds of substitution may be saturated (C. Marshall, pers. comm.). To explore further the degrees of transition and transversion saturation in the data set, the relationship between transitions and transversions at each codon position was plotted relative to maximum likelihood distances among taxa (Figure 2). For first and second positions, the relationships between either transitions or transversions and maximum likelihood distance are linear. At third positions, however, transitions and transversions both increase rapidly, with transitions reaching an asymptote, and probably saturation, at a maximum likelihood distance of approximately 0.2. Transversions, however, maintain a relatively linear increase (Figure 2).

Pairwise sequence divergence, estimated using the LogDet/paralinear distance model (Lockhart *et al.*, 1994; Lake, 1994) to correct for multiple substitutions, was 13–31% among ingroup taxa, except for the *Stanleya–Tanganyicia* comparison (3%) (Table 2).* Divergences between ingroup and outgroup taxa ranged from 18 to 30% (Table 2).

---

*West will argue elsewhere that the genus *Stanleya* should be synonymized with *Tanganyicia* based on a shared unique brooding system.

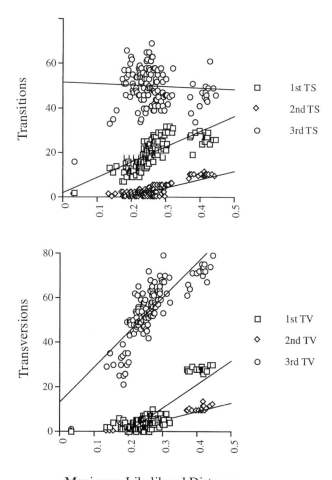

Fig. 2. Comparisons of cytochrome oxidase I sequences, showing relationships between the number of transitions (upper plot) and transversions (lower plot) at each codon position, and the maximum likelihood distance between all pairs. These plots serve as a visual estimate of nucleotide saturation levels.

C. Assessment of Phylogenetic Signal

The strength of phylogenetic signal in the data set was assessed by quantifying the skewness (g_1) of a distribution of the tree lengths of 100 000 randomly generated trees. When all taxa were considered, there was a significant negative skew in the tree length distribution ($g_1 = -0.9437$, $p < 0.01$), indicating the information content of the data to be significantly more structured than random (Hillis and Huelsenbeck, 1992). In addition, the shortest parsimony tree (1086 steps) was 371 steps shorter than the shortest random tree (1457

Table 2

Estimated pairwise sequence divergence between taxa, calculated using the LogDet/paralinear distance model (Lockhart et al., 1994; Lake, 1994) to correct for multiple substitutions

	1	2	3	4	5	6	7	8	9	10	11	12	13	14	15	16
1. Anceya	–															
2. Bathanalia	0.261	–														
3. Chytra	0.267	0.135	–													
4. Lavigeria	0.305	0.263	0.230	–												
5. Limnotrochus	0.265	0.156	0.146	0.215	–											
6. Nov. gen.	0.266	0.250	0.226	0.254	0.223	–										
7. Paramelania	0.232	0.249	0.252	0.274	0.214	0.267	–									
8. Reymondia	0.253	0.221	0.222	0.259	0.215	0.258	0.232	–								
9. Spekia	0.262	0.217	0.193	0.246	0.210	0.252	0.227	0.185	–							
10. Stanleya	0.260	0.227	0.213	0.260	0.200	0.272	0.230	0.212	0.199	–						
11. Tanganyicia	0.264	0.226	0.204	0.246	0.197	0.266	0.233	0.206	0.190	0.033	–					
12. Tiphobia	0.234	0.215	0.207	0.239	0.182	0.240	0.205	0.240	0.211	0.229	0.229	–				
13. Cleopatra	0.271	0.248	0.243	0.273	0.216	0.272	0.242	0.182	0.199	0.203	0.195	0.246	–			
14. M. admirabilis	0.289	0.282	0.287	0.261	0.253	0.293	0.264	0.279	0.266	0.252	0.259	0.272	0.283	–		
15. M. tuberculata	0.293	0.220	0.247	0.246	0.231	0.245	0.255	0.246	0.232	0.241	0.242	0.211	0.265	0.185	–	
16. Bellamya	0.285	0.307	0.295	0.302	0.257	0.327	0.284	0.281	0.290	0.264	0.270	0.284	0.280	0.283	0.260	–

Taxa number 1–12 are genera endemic to Lake Tanganyika, 13–16 are outgroup taxa collected in East Africa, and 14 and 15 are from the genus *Melanoides*.

steps). However, the inclusion of a pair of very closely related taxa in the data sets can significantly skew the distribution of random tree lengths, and thereby produce an unrealistically high confidence level in the estimated phylogenetic signal (Lara *et al.*, 1996). To minimize this possibility, the skewness was measured repeatedly, after iteratively collapsing nodes with the highest bootstrap support. When this was done, in all permutations the distribution of random tree lengths remained significantly negatively skewed.

D. Phylogenetic Analyses

Phylogenetic analyses using maximum parsimony (specifically, a heuristic search with 1000 random addition replicates, 10 random addition sequences per replicate and two trees held at each step) produced a single tree 1086 steps in length (Figure 3a). Bootstrap and Bremer support values (Bremer, 1988) estimating support for various nodes were mapped on to the shortest tree (Figure 3a). Multiple heuristic searches with identical parameters except for the order of taxa in the data set yielded topologies identical to the shortest tree. Because of the high saturation rates in third position transitions, which accounted for much of the phylogenetic signal in the data set, phylogenetic analyses were conducted using first positions only, second positions only, first and second positions combined, and amino acids. These analyses produced topologies broadly similar to those obtained from analyses of the entire data set, but with considerably less resolution among the terminal nodes, which are less likely to be affected by saturation than deeper nodes. Consequently, all sites were retained in the analyses.

Neighbour joining (Saitou and Nei, 1987) using the LogDet/paralinear distance model (Lockhart *et al.*, 1994; Lake, 1994) produced the tree shown in Figure 3b, on to which bootstrap values were mapped. Maximum likelihood (Felsenstein, 1981) analyses, using the transition:transversion ratio calculated from the data, 1.43:1, produced the tree shown in Figure 3c. Bootstrap values were also mapped on to this tree.

E. Phylogenetic Relationships of the Tanganyikan Thiarid Gastropods

The parsimony, neighbour-joining and maximum likelihood trees all divide the Tanganyikan thiarids into four clades that are identical, except for the placement of *Tiphobia*. All topologies support (i) the monophyly of ((*Reymondia, Cleopatra*) *Spekia*) with the (*Stanleya, Tanganyicia*) sister clade to this grouping, (ii) the monophyly of the trochiform taxa ((*Bathanalia, Chytra*) *Limnotrochus*), and (iii) the sister-taxon relationship between (*Lavigeria*, Nov. gen.) and (*Anceya, Paramelania*). Both neighbour-joining and maximum likelihood analyses place *Tiphobia* as the sister taxon to the (*Anceya, Paramelania*) clade.

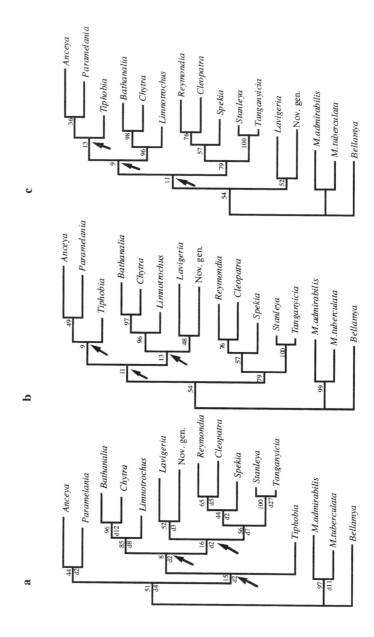

Fig. 3. (a) Parsimony, (b) neighbour-joining, and (c) maximum likelihood trees depicting relationships of Tanganyika thiarid gastropods. Bootstrap and Bremner support values, quantifying support for various relationships, are mapped on to phylogram nodes where appropriate. Arrows indicate nodes that are in discordance among the three topologies.

Although results from parsimony, neighbour-joining and maximum like-lihood analyses show considerable agreement in generic sister-taxon relation-ships at the terminal nodes, and all support the monophyly of the Tanganyikan thiarids + *Cleopatra* clade (bootstrap values ranging from 54 to 100 and decay indices ranging from 4 to 27 for these relationships), they do not resolve the deeper branching order of these four clades concordantly (Figure 3). Parsimony analyses (Figure 3a) place the (*Lavigeria*, Nov. gen.) clade as the sister group to the ((*Reymondia, Cleopatra*) *Spekia*) (*Stanleya, Tanganyicia*) clade, with the trochiform taxa as the sister group to this larger clade, and the (*Anceya, Paramelania*) grouping basal among the Tanganyikan thiarids. Neighbour-joining analyses (Figure 3b), however, place the (*Lavigeria*, Nov. gen.) clade as sister group to the trochiform genera, with the (*Anceya, Paramelania*) *Tiphobia*) clade sister to this grouping, and the ((*Reymondia, Cleopatra*) *Spekia*) (*Stanleya, Tanganyicia*) clade basal among the Tanganyi-kan thiarids. Maximum likelihood analyses (Figure 3c) produced a sister grouping of two larger clades, the trochiform + ((*Anceya, Paramelania*) *Tiphobia*) group, and the ((*Reymondia, Cleopatra*) *Spekia*) (*Stanleya, Tanga-nyicia*)) clade, with the (*Lavigeria*, Nov. gen.) clade basal among the Tanganyikan thiarids. While these trees do differ, it is noteworthy that for all trees the internode branch lengths between these four clades are very short, and bootstrap and Bremer support for these deeper nodes is weak (bootstrap values ranging from 8 to 15 and decay indices of 2).

V. DISCUSSION

A. Rationale for Selecting Outgroups

Three thiarid outgroups, *Melanoides admirabilis, M. tuberculata* and *Cleopatra ferruginea*, were used in this study. The former species is endemic to the Malagarasi River, which presently drains into Lake Tanganyika and pre-dates the formation of the lake, whereas the latter two are distributed throughout southern Asia, the Middle East and eastern Africa (Brown, 1994). *Melanoides tuberculata* is known from lower Miocene (pre-rift) fossils in East Africa and from lake deposits (Schouteden, 1933; Gautier, 1970; Van Damme, 1984). *Cleopatra ferruginea* is known from lower Miocene deposits in East Africa and probably evolved prior to the splitting of Madagascar from the African continent (Fuchs, 1936; Gautier, 1970; Van Damme, 1984; Brown, 1994). Clearly, these three outgroup species have a long history in East Africa, and although none of these taxa is currently found within the lake proper, they are common in rivers and swamps fringing the lake. This, and the fact that they are the only cosmopolitan thiarid gastropods currently found within the Tanganyikan drainage basin, make them excellent candidates for outgroups to the endemic Tanganyikan thiarids.

There is no *a priori* evidence to favour one of these outgroups over the other. In fact, both seem equally appropriate: *Melanoides* broods its young, whereas *Cleopatra* lays eggs, and both reproductive strategies are found among the Tanganyikan thiarids. In addition, *Bellamya*, a cosmopolitan viviparid gastropod, was added to the analyses to determine which of the more proximal outgroups, *Cleopatra* or *Melanoides*, was more distant. There are other potential outgroups with far more limited distributions which may be appropriate for inclusion in this investigation, notably *Potadamoides* and *Potadoma*, and future fieldwork will endeavour to sample these and other outgroup taxa.

B. Simultaneous Diversification of Major Tanganyikan Thiarid Lineages

There is relatively strong support for four distinct monophyletic clades of Tanganyikan thiarid gastropods with *Cleopatra* nested within (bootstrap values ranging from 54 to 100 and decay indices ranging from 4 to 27 for these relationships). There is little support, however, for resolved relationships among these four clades (bootstrap values ranging from 8 to 15 and decay indices of 2). The poor resolution of relationships among the four clades is partly due to the very high and similar levels of sequence divergence in the data set (Table 2). These equally high levels of sequence divergence across taxa can be explained in two ways.

The high levels of sequence divergence and weak support for deeper nodes may reflect the poor resolving power of *COI* for this systematic problem. Because third-position transitions are unconstrained at the amino acid level and occur rapidly, they would be the first type of substitution to record initial divergences. In the present data set, however, third-position transitions are saturated and therefore unable to contribute much meaningful phylogenetic information for the deeper nodes of the tree. Sequences from a more conservative gene, such as 16S or 12S RNA, and/or additional taxa could be used to assess whether *COI* has indeed evolved too quickly to be of use in resolving this systematic problem. However, it should be noted that *COI* has provided phylogenetically informative data in other studies at similar taxonomic levels with similar or considerably older divergence times (Crozier *et al.*, 1989; Folmer *et al.*, 1994; Frolich *et al.*, 1996; Pedersen, 1996), and therefore this explanation is considered unlikely.

An alternative explanation for the similar levels of sequence divergence seen in the data is that the four major Tanganyikan thiarid gastropod clades evolved simultaneously, or nearly so, such that *COI*, and probably any other molecule, could not capture the event(s). Rapid simultaneous or nearly simultaneous divergence would result in a true ("hard") polytomous relationship or star phylogeny. The following lines of evidence are consistent with a star radiation of the deeper nodes of the Tanganyikan thiarid gastropods.

First, while estimated pairwise sequence divergences are rather hetero-geneous among terminal nodes on the tree (ranging from 3 to 18%), distances for each of the four major clades (averaged among clade members) to the *Cleopatra* outgroup were relatively uniform (ranging from 24 to 27%). Similar levels of sequence divergence are consistent with an evolutionary scenario of simultaneous divergence times from a common ancestor.

Secondly, poor resolution is concentrated in only one region of all of the trees, namely, in the relationships of the four major Tanganyikan thiarid clades. Nodes both terminal and basal to this region show greater resolution. If *COI* was completely saturated, one would expect to see progressively less support for the more basal nodes and the Tanganyikan thiarid clades perhaps pairing off with the *Melanoides* outgroup clade. In fact, low bootstrap or Bremer support values are concentrated where the four clades converge, but the monophyly of Tanganyikan thiarid + *Cleopatra* clade is comparatively well supported.

Finally, phylogenetic analyses of anatomical and allozyme data sets for many of these same taxa (West, 1991) produced strong support for species-level relationships at terminal nodes, which were united by long branches into unresolvable polytomies. Analyses of the *COI* sequences and congruent topologies from multiple independent data sets (West, 1991) strongly support the near-simultaneous diversification of several Tanganyikan gastropod clades.

C. Paraphyly of the Tanganyikan Thiarid Gastropods

Although strongly supported, it was none the less surprising to find *Cleopatra* positioned in the terminal branches of the phylogenetic trees. This positioning renders the endemic Tanganyikan thiarids paraphyletic. There is some support for the monophyly of *Cleopatra* + the endemic Tanganyikan thiarids. Bootstrap values of 51–54 for this clade jumped to 79 after deletion of the very distant outgroup *Bellamya*. This node is also supported by an amino acid substitution, which is an otherwise rare event. Bootstrap values supporting relationships between the *Melanoides* taxa and the Tanganyikan thiarids were considerably weaker (10–20) and there are no amino acid substitutions supporting such pairings.

The non-monophyly of the Tanganyikan thiarid gastropods is also supported by other studies. West (1991) collected electrophoretic data across 30 presumptive genetic loci and examined 25 conchological characters for *Cleopatra, Melanoides* and 16 Tanganyikan species. Although the topologies differ somewhat from trees produced in the present study, these allozyme and conchological data sets, whether treated independently or combined, could not be resolved to yield a single monophyletic Tanganyikan thiarid clade (West, 1991).

In converting these cladograms (Figure 3a–c) into phylogenies (Figure 4a, b), there are two ways to interpret the position of *Cleopatra* and the paraphyly of the endemic Tanganyikan gastropods.

a

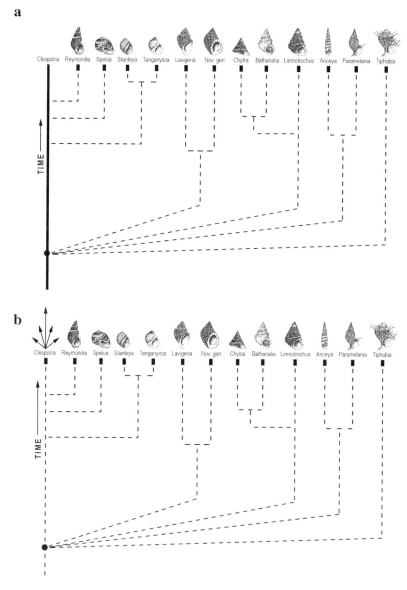

Fig. 4. Two different phylogenies consistent with a consensus of the parsimony, neighbour-joining and maximum likelihood cladograms derived from *COI* sequences. (a) A *Cleopatra*-like ancestor gives rise to the Tanganyika thiarid gastropods; (b) the cosmopolitan outgroup *Cleopatra* is the product of the Tanganyika thiarid radiation.

First, perhaps *Cleopatra* was the source from which all the Tanganyikan taxa were diverging (Figure 4a). In this model, a *Cleopatra*-like ancestor lived in or near to the proto-Lake Tanganyika and diverged four to seven times to produce the thiarid gastropod diversity in the lake. In this case, "*Cleopatra*" is a paraphyletic stem taxon. Early Miocene *Cleopatra* fossils (Van Damme, 1984) indicate that *Cleopatra* was established in the region before the formation of the lake, and thus provide anecdotal support for this scenario.

Alternatively, rather than the source, *Cleopatra* might the product of the Tanganyikan thiarid radiation (Figure 4b). Perhaps, after diverging from a common ancestor with *Reymondia* in the lake, *Cleopatra* dispersed from the Tanganyika basin and colonized other African waters, from the Nile Valley to South Africa, and Senegal to Madagascar. This interpretation assumes that the *Cleopatra* species used in this study is not genetically related to fossils that have been attributed to this genus. Although some studies have demonstrated a decoupling of molecular and morphological evolution in gastropods (e.g. Palmer, 1985), it seems imprudent to rule out the considerable palaeo-record of this species.

To distinguish between these two different interpretations, further sampling of *Cleopatra* taxa throughout Africa is necessary. Monophyly of *Cleopatra* species positioned high in the tree would suggest that *Cleopatra* is the product of the Tanganyikan thiarid radiation, whereas paraphyly of *Cleopatra* species, with some species rooting basal to the Tanganyikan thiarid radiation, would suggest that this genus is the source of Tanganyikan thiarid diversity.

It is also possible that *Cleopatra* fossils are paraphyletic with respect to the Tanganyikan radiation. The earlier *Cleopatra* lineages may have gone extinct and the modern *Cleopatra* may have descended from a Tanganyikan ancestor; in such a case, *Cleopatra* would be monophyletic on the tree and unrelated to the fossils.

D. A Tanganyikan Thiarid Gastropod "Superflock"

Species flocks are monophyletic groups of organisms that are endemic to a geographically circumscribed area and possess unusual species richness or diversity relative to other members of the taxon (Greenwood, 1984). Species flocks can occur hierarchically, across several taxonomic scales (Ribbink, 1984). For example, in Lake Tanganyika one may consider the species flock of lamprologine cichlid fishes (an endemic, monophyletic tribe comprised of nine genera) or the flock of nine species of *Neolamprologus* (an especially rich genus in the lake), nested within the larger lamprologine flock (Sturmbauer *et al.*, 1994).

The thiarid gastropods of Lake Tanganyika, even at the generic level, are endemic to the Tanganyika basin, and are unusually rich and diverse compared to thiarids elsewhere. For these reasons many workers have assumed that they constitute a species flock. A recent study lacking rigorous character analysis and phylogenetic treatment (Bandel, 1998) proposed polyphyletic origins for the Tanganyikan gastropods. Data in the present and other studies (West *et al.*, 2000; Michel and Todd, unpubl. data) support some of these findings, but conflict with others. A polyphyletic origin of the Tanganyikan gastropods remains to be convincingly demonstrated, pending character analysis and phylogenetic treatment with relevant outgroups.

However, the paraphyly of the Tanganyikan thiarid gastropods, established in this study, brings their status as a species flock into question. Because they are not monophyletic, should we refrain from referring to the larger radiation of thiarid gastropods in Lake Tanganyika as a species flock, reserving the term instead for lower level monophyletic components of the radiation, such as the *Lavigeria* species flock or the *Paramelania* species flock? Ribbink has argued that, in cases such as this, the term species flock should be maintained, for "it does not really matter whether the flock is the product of one or several ancestors entering the lake. The collection of closely related species within the confines of the lake has the qualities of a species flock" (Ribbink, 1984). The present authors disagree. Monophyly is a critically important aspect to the species flock concept. It is only after species have been parsed into monophyletic groupings (or true polytomies) that one can begin to explore the biotic and abiotic factors governing the diversification of these groups and thus begin to explore what, exactly, makes them so unique.

Before the striking morphological similarities between cichlids in Malawi and Tanganyika were shown to be due to convergence (Kocher *et al.*, 1993; Meyer, 1993; but see Stiassny, 1981), Greenwood (1983, 1984) used the term "superflock" to describe a larger cichlid species flock in which some sister taxa were dispersed to different lakes, thus circumventing the "endemic to a geographically circumscribed area" aspect of the species flock definition. Studies by the present authors indicate that the thiarid gastropods endemic to Lake Tanganyika diverged rapidly within the lake. The nature of their ancestor(s) remains to be resolved. In addition, the position of *Cleopatra*, nested among the ingroup taxa, renders the Tanganyikan thiarid gastropods paraphyletic, thus violating the monophyly requirement of the species flock definition. The term "superflock" could be resurrected to refer to the rapid paraphyletic radiation of Tanganyika thiarid gastropod genera.

ACKNOWLEDGEMENTS

The authors are grateful to numerous colleagues in Africa and the USA for assistance that enabled them to undertake this study. Thanks are due in particular to G. Ntakimazi and the Université du Burundi for facilitating collections in Burundi, M.-M. Gashagaza and the Centre de Recherche en Hydrobiologie for assistance with the necessary arrangements for fieldwork and collections in Democratic Republic of Congo (former Zaïre), and E. Verheyen and K. Martens for an invitation to participate on a research cruise of Lake Tanganyika sponsored by the L'Institut Royal des Sciences Naturelles de Belgique in 1995. Their support, with facilitation by L. Mwape, G. Milindi and P.-D. Plisinier in Zambia, and A. Kihakwe, D. Chitamwebwa and P. Mannini in Tanzania, enabled the authors to collect samples from the Zambian and Tanzanian coasts of Lake Tanganyika and the Malagarasi River. The authors thank colleagues from the Jacobs Laboratory at University of California Los Angeles and the Brown Laboratory at University of Michigan, notably J. Boore, J. Clabaugh, T. Collins, F. Kraus and J. Staton, for advice, assistance and constructive criticism throughout the laboratory work and analyses. A. Cohen, D. Jacobs, C. Marshall, B. Runnegar and E. Verheyen critically read and greatly improved the manuscript. This work was funded by a Graduate Enhancement Fellowship from the National Security Education Program, a UCLA Dissertation Year Fellowship and research grants from the Western Society of Malacologists and the Conchologists of America to K.W., and a NSF Post-Doctoral Fellowship No. DEB 9303281 to E.M.

REFERENCES

Bandel, K. (1998). Evolutionary history of East African fresh water gastropods interpreted from the fauna of Lake Tanganyika and Lake Malawi. *Z61. Geol. Palaont. Teil. I.H.* **1–2**, 233–292.

Boss, K.J. (1978). On the evolution of gastropods in ancient lakes. In: *Pulmonates*, Vol. 2A, *Systematics, Evolution and Ecology* (Ed. by V. Fretter and J. Peake), pp. 385–428. Academic Press, London.

Bourguignat, J.R. (1885). *Notice prodromique sur les mollusques terrestres et fluviatiles recueillis par M. Victor Giraud dans la region mèridionale du Lac Tanganika.* Tremblay, Paris.

Bourguignat, J.R. (1890). Histoire Malacologique du Lac Tanganika. *Ann. Sci. Nat. Paris, Zool. Ser.* **7**, 1–267.

Bremer, K. (1988). The limits of amino acid sequence data in angiosperm phylogenetic reconstruction. *Evolution* **42**, 795–803.

Brown, D.S. (1994). *Freshwater Snails of Africa and their Medical Importance*, rev. 2nd ed. Taylor & Francis, London.

Brown, D.S. and Mandahl-Barth, G. (1987). Living molluscs of Lake Tanganyika: a revised and annotated list. *J. Conchol.* **32**, 305–327.

Brown, W.M. (1985). Evolution of the animal mitochondrial DNA genome. In: *Molecular Evolutionary Genetics* (Ed. by R.J. MacIntyre), pp. 95–130. Plenum Press, New York.

Cohen, A.S., Soreghan, M.J. and Scholz, C.A. (1993). Estimating the age of formation of lakes: an example from Lake Tanganyika, East African Rift system. *Geology* **21**, 511–514.

Coulter, G.W. (ed.) (1991). *Lake Tanganyika and Its Life*. Oxford University Press, Oxford.

Crozier, R.H., Crozier, Y.C. and Mackinley, A.G. (1989). The CO-I and CO-II region of the honeybee mitochondrial DNA: evidence of variation in insects mitochondrial DNA. *Mol. Biol. Evol.* **6**, 399–411.

Dautzenberg, P. (1907). Description de coquilles nouvelles de diverses provenances et de quelques cas tèratologiques. *J. Conchyliol.* 327–341.

Ebinger, C.J. (1989). Tectonic development of the western branch of the East African rift system. *Geol. Soc. Am. Bull.* **101**, 885–903.

Eriksson, T. (1997). *AutoDecay 2.09.* Botany Department, Stockholm University, Stockholm.

Felsenstein, J. (1981). Evolutionary trees from DNA sequences: a maximum likelihood approach. *J. Mol. Evol.* **17**, 368–376.

Felsenstein, J. (1985). Confidence limits on phylogenies: an approach using the bootstrap. *Evolution* **39**, 783–791.

Folmer, O., Black, M., Hoeh, W., Lutz, R. and Vrijenhoek, R. (1994). DNA primers for amplification of mitochondrial cytochrome *c* oxidase subunit I from diverse metazoan invertebrates. *Mol. Mar. Biol. Biotechnol.* **3**, 294–299.

Frohlich, D.R., Stevenson, B.A., Peterson, A.M. and Wells, M.A. (1996). Mitochondrial cytochrome *c* oxidase subunit I of *Manduca sexta* and a comparison with other invertebrate genes. *Comp. Biochem. Physiol.* **113**, 785–788.

Fryer, G. and Iles, T.D. (1972). *The Cichlid Fishes of the Great Lakes of Africa; Their Biology and Evolution.* Oliver and Boyd, Edinburgh.

Fuchs, V.E. (1936). Extinct Pleistocene Molluscs from Lake Edward, Uganda, and their bearing on the Tanganyika problem. *J. Linn. Soc. Lond.* **40**, 93–105.

Gautier, A. (1970). Fossil freshwater molluscs of the Lake Albert–Lake Edward Rift (Uganda). *Annals du Musée Royal de l'Afrique Centrale,* Vol. 67. Musée Royal de l'Afrique Centrale, Tevuren.

Gilbert, D.G. (1994). *SeqApp: A Biosequence Editor Application. Version 1.9a169.* Biology Department, Indiana University, Bloomington, IN.

Greenwood, P.H. (1981). *The Haplochromine Fishes of the East African Lakes. Collected Papers.* Kraus International, Munich.

Greenwood, P.H. (1983). On *Macropleurodus, Chilotilapia* (Teleostei, Cichlidae), and the interrelationships of African cichlid species flocks. *Bull. Br. Mus. Nat. Hist. (Zool.)* **45**, 209–231.

Greenwood, P.H. (1984). What is a species flock? In: *Evolution of Fish Species Flocks* (Ed. by A. Echelle and I. Kornfield), pp. 13–19. University of Maine at Orono Press, Orono, ME.

Hillis, D.M. (1991). Discriminating between phylogenetic signal and random noise in DNA sequences. In: *Phylogenetic Analysis of DNA Sequences* (Ed. by M.M. Miyamoto and J. Cracraft), pp. 278–294. Oxford University Press, Oxford.

Hillis, D.M. and Huelsenbeck, J.P. (1992). Signal, noise and reliability in molecular phylogenetic analyses. *J. Hered.* **83**, 189–195.

Kawanabe, H., Hori, M. and Nagoshi, M. (eds) (1997). *Fish Communities in Lake Tanganyika.* Kyoto University Press, Kyoto.

Kim, J. (1993). Improving the accuracy of phylogenetic estimation by combining different methods. *Syst. Biol.* **42**, 331–340.

Kocher, T.D., Conroy, J.A., McKaye, K.R. and Stauffer, J.R. (1993). Similar morphologies of cichlid fish in Lakes Tanganyika and Malawi are due to convergence. *Mol. Phylogenet. Evol.* **2**, 158–165.

Lake, J.A. (1994). Reconstructing evolutionary trees from DNA and protein sequences: paralinear distances. *Proc. Nat. Acad. Sci. USA* **91**, 1455–1459.

Lara, M.C., Patton, J.L., Nazareth, M. and Da Silva, F. (1996). The simultaneous diversification of South American echimyid rodents (Hystricognathi) based on complete cytochrome *b* sequences. *Mol. Phylogenet. Evol.* **5**, 403–413.

Leloup, E. (1953). Gastéropodes. Résultats scientifiques de l'exploration hydrobiologique du Lac Tanganika (1946–1947). *Inst. R. Sci. Nat. Belg.* **3**, 1–272.

Lockhart, P.J., Steel, M.A., Hendy, M.D. and Penny, D. (1994). Recovering evolutionary trees under a more realistic model of sequence evolution. *Mol. Biol. Evol.* **11**, 605–612.

Maddison, W. and Maddison, D. (1996). *MacClade 3.06*. Sinauer, Sunderland, MA.

Marck, C. (1995). *DNA Strider 1.2, A C Program for DNA and Protein Sequences Analysis*. Department de Biologie Cellulaire et Moléculaire, Direction des Sciences de la Vie, CEA. Gif-sur-Yvette, France.

Meyer, A. (1993). Phylogenetic relationships and evolutionary processes in East African cichlid fishes. *Trends Ecol. Evol.* **8**, 279–284.

Meyer, A., Kocher, T.D., Basasibwaki, P. and Wilson, A.C. (1990). Monophyletic origin of Lake Victoria cichlid fishes suggested by mitochondrial DNA sequences. *Nature* **347**, 550–553.

Michel, E. (1995). Evolutionary diversification of rift lake gastropods: morphology, anatomy, genetics, and biogeography of *Lavigeria* (Mollusca:Thiaridae) in Lake Tanganyika. Unpublished PhD Thesis, University of Arizona, Tucson, AZ.

Moore, J.E.S. (1903). *The Tanganyika Problem*. Hurst and Blackett, London.

Palmer, R.A. (1985). Quantum changes in gastropod shell morpholoy need not reflect speciation. *Evolution* **39**, 699–705.

Palumbi, S.R. (1996). Nucleic acids II: The polymerase chain reaction. In: *Molecular Systematics* (Ed. by D.M. Hillis, C. Moritz and B.K. Mable), pp. 205–247. Sinauer, Sunderland, MA.

Pederson, B.V. (1996). A phylogenetic analysis of cuckoo bumblebees (*Psithyrus*, Lepeletier) and bumblebees (*Bombus*, Latreille) inferred from sequences of the mitochondrial gene cytochrome oxidase I. *Mol. Phylogenet. Evol.* **5**, 289–297.

Pelseneer, P. (1906). Halolimnic faunas and the Tanganyika problem. *Rep. Br. Assoc. Adv. Sci.* **1906**, 602.

Ribbink, A.J. (1984). Is the species flock concept tenable? In: *Evolution of Fish Species Flocks.* (Ed. by A. Echelle and I. Kornfield), pp. 21–25. University of Maine at Orono Press, Orono, ME.

Saiki, R.K., Gelfand, D.H., Stoffel, S., Scharf, S.J., Higuchi, R.H., Horn, G.T., Mullis, K.B. and Erlich, H.A. (1988). Primer-directed amplification of DNA with a thermostable DNA polymerase. *Science* **239**, 487–491.

Saitou, H. and Nei, M. (1987). The neighbor-joining method: a new method for reconstructing phylogenetic trees. *Mol. Biol. Evol.* **4**, 406–425.

Schouteden, M.H. (1933). Les mollusques aquatiques vivants et subfossiles de la règion du lac Kivu. *Bull. Inst. R. Sci. Nat. Belg.* **4**, 519–527.

Seehausen, O. (1996). *Lake Victoria Rock Cichlids–Taxonomy, Ecology and Distribution.* Verduijn Cichlids, Zevenhuizen, The Netherlands.

Smith, E.A. (1904). Some remarks on the Mollusca of Lake Tanganyika. *Proc. Malacol. Soc. Lond.* **6**, 77–104.

Stiassny, M.L.J. (1981). Phylogenetic versus convergent relationships between piscivorous cichlid fishes from Lakes Malawi and Tanganyika. *Bull. Br. Mus. Nat. Hist. (Zool.)* **40**, 67–101.

Sturmbauer, C. and Meyer, A. (1992). Genetic divergence, speciation and morphological stasis in a lineage of African cichlid fishes. *Nature* **358**, 578–581.

Sturmbauer, C., Verheyen, E. and Meyer, A. (1994). Mitochondrial phylogeny of the Lamprologini, the major substrate spawning lineage of cichlid fishes from Lake Tanganyika in Eastern Africa. *Mol. Biol. Evol.* **11**, 691–703.

Swofford, D. (1997). *PAUP: Phylogenetic Analysis Using Parsimony (and other methods) 4.0.* Sinauer, Sunderland, MA.

Turner, G.F. (1996). *Offshore Fishes of Lake Malawi*. Cichlid Press, Lauenau, Germany.

Van Damme, D. (1984). *The Freshwater Mollusca of Northern Africa, Distribution, Biogeography and Paleoecology* (Ed. by H.J. Dumont). Dr. W. Junk, Dordrecht.

West, K. (1991). Resolution and congruence of paleontological and biological data sets in the reconstruction of Tanganyikan gastropod phylogeny. Unpublished MSc Thesis, University of Arizona, Tucson, AZ.

West, K. (1997). Perspectives on the diversification of species flocks: systematics and evolutionary mechanisms of the gastropods (Prosobranchia:Thiaridae) of Lake Tanganyika, East Africa. Unpublished PhD Thesis, University of California, Los Angeles, CA.

West, K. and Cohen, A. (1994). Predator–prey coevolution as a model for the unusual morphologies of the crabs and gastropods of Lake Tanganyika. *Arch. Hydrobiol. Beiheft. Ergebnisse Limnol.* **44**, 267–283.

West, K. and Cohen, A. (1996). Shell microstructure of gastropods from Lake Tanganyika: adaptation, convergent evolution and escalation. *Evolution* **50**, 672–681.

West, K., Cohen, A. and Baron, M. (1991). Morphology and behavior of crabs and gastropods from Lake Tanganyika, Africa: implications for lacustrine predator–prey coevolution. *Evolution* **45**, 589–607.

West, K., Nakai, K. and Martens, K. (2000). Two new members of Lake Tanganyika's gastropod species flock (Cerithiodea:Thiaridae). *J. Moll. Stud.* (in press).

Phenetic Analysis, Trophic Specialization and Habitat Partitioning in the Baikal Amphipod Genus *Eulimnogammarus* (Crustacea)

H. MORINO, R.M. KAMALTYNOV, K. NAKAI and K. MASHIKO

I. SUMMARY

The pattern of adaptive radiation in the Baikal amphipod genus *Eulimnogammarus* was investigated through phenetic studies, gut contents analysis and a distributional survey. Eight species of *Eulimnogammarus*, together with representatives of related genera from Baikal or other regions, were phenetically analysed. This genus exhibited the highest affinity to Baikal-endemic *Philolimnogammarus* among six genera studied, but not to European "*Eulimnogammarus*".

Of the eight members of the Baikal genus *Eulimnogammarus* investigated, seven species were closely affiliated to one another, but one, *E. grandimanus*, showed a weak affinity not only to other congeners but also to other genera.

Detailed studies of mouthpart morphology revealed interspecific differences in the structure of the mandibles and maxillae I. Among the taxa examined, *Eurybiogammarus violaceus* and *Corophiomorphus kietlinskii* had the most specialized mouthparts. Nine species of *Eulimnogammarus* and *Philolimnogammarus* were found in the littoral zone and were distributed in one of three habitat zones. Gut contents analysis of these taxa demonstrated that *E.*

violaceus is a sponge feeder, but the other species examined were all generalist omnivores. Despite differences in mouthpart morphologies, high trophic overlap was evident between species, and the foods taken reflected the availability of local fauna and flora rather than mouthpart morphology. Possible mechanisms which permit the observed species overlap are considered.

II. INTRODUCTION

Lake Baikal is the oldest lake in the world and is characterized by a rich indigenous fauna, exemplified by the gammaridean amphipods. In the most recent list of Baikal amphipods, 51 genera, 265 species and 81 subspecies are included (Kamaltynov, 1999a; cf. Takhteev, this volume). This number of species, all but one of which are endemic to the lake and connecting rivers, comprises *c.* 20% of the known gammaridean freshwater fauna world-wide.

The first report on the Baikal amphipods was that of Pallas (1772), but it was Dybowsky (1874) who laid the foundation for the taxonomy of the Baikal amphipods, describing 97 species and 22 varieties from south Baikal. These species were reviewed by Stebbing (1899) in the light of generic concepts widely accepted in Europe at that time. Thereafter, taxonomic knowledge on these animals has been constantly expanded by the contributions of, for example, Sowinsky (1915), Dorogostaisky (1930) and Bazikalova (1945), to name but a few. Even today, new taxonomic discoveries of Baikal's amphipod fauna continue to be made (e.g. Takhteev, 1997; this volume).

Yet despite the long history of taxonomic studies of this fauna, the familial allocation of Baikal species is still in a state of flux (Kamaltynov, 1999a). This situation reflects the high degree of morphological specialization exhibited by these taxa, especially in the development of cuticular body processes and armaments. These features present difficulties in taxonomic evaluations, since there are no comparable features in gammarids of other freshwater bodies. One approach towards clarifying this situation would be to perform a phenetic analysis of Baikal amphipods including non-Baikal taxa, in which affinity among species is assessed based on overall similarity, and is not restricted to only a few "important" characters.

It was Bazikalova's (1945) comprehensive volume on Baikal's amphipods which truly attracted the attention and interest of evolutionists world-wide, and the origin and radiation of this group has subsequently presented one of the more challenging problems for both amphipod systematists and evolutionary biologists. Various hypotheses have been advanced to explain the origin and evolution of the high diversity of amphipods and other endemic faunas of this lake (e.g. Brooks, 1950; Kozhov, 1963; Fryer, 1991; Kamaltynov, 1999b). More recently, attempts have been made to shed new light on this subject using genetic and molecular approaches. For example, within one littoral amphipod species, populations separated by a topographic barrier have been shown to be genetically discrete (Mashiko *et al.*, 1997; Mashiko, this volume). Ogarkov *et al.* (1997) and Sherbakov *et al.* (1998) used a molecular approach to analyse the genetic relationships between representative Baikal species and to obtain the general pattern and possible age of the differentiation. However, the lack of substantial supporting evidence has meant that none of these hypotheses has yet received general support (see also Timoshkin, 1999). As has been stressed repeatedly (e.g. Barnard and Barnard, 1983a, b), this situation is partly due to the almost complete absence of ecological information on these animals. Most evidence available for evolutionary consideration has been obtained indirectly, through inference of function based on morphological comparisons and functional analogies. This situation has persisted despite a number of ecological surveys having been conducted on littoral communities in Lake Baikal. Although very few of these studies have analysed the community to the species level, some nevertheless contain much useful information (see Kozhova and Izmest'eva, 1998). To obtain an insight into species diversity and its maintenance in Lake Baikal, a new approach is required.

The presence of species flocks is an important feature common to ancient lakes (Martens *et al.*, 1994) and several genera of Baikal amphipods are believed to represent species flocks. By definition, a species flock comprises "an aggregation of several species whose members are endemic to a geographically circumscribed area and are each others' closest living relatives" (Greenwood, 1984). As such, a study which focuses on a species flock of Baikal amphipods would be especially informative regarding questions about species diversity and its maintenance. As noted earlier, the Baikal amphipods are all (except one) endemic, and have evolved within the lake, and often several species can be found coexisting within the same habitat. By using an approach which combines phylogenetic analysis and ecological surveys, including food habits and habitat utilization, to study the component species of a species flock it should prove possible to reveal the pattern of species diversity and the mechanisms for its maintenance.

The genus *Eulimnogammarus, sensu* Barnard and Barnard (1983a, b), contains 11 species and two subspecies. Most of them co-occur within pebble/rubble substrates in the littoral zone of Lake Baikal (see Bazikalova, 1945) and thus present ideal material for this type of study. Here, a phenetic analysis of the *Eulimnogammarus* species is carried out to infer the phylogenetic interrelationships within the genus. A recent taxonomic examination of this genus found it to be of questionable status (Morino, 1998) and therefore representatives of related genera are also included in the analysis. Effort was made to examine specimens directly, but when this was not possible, reference was made to descriptions and figures. In the character analysis, special attention was paid to the mouthparts: it is here that evolutionary modifications to increase trophic specialization are manifest. Findings from gut contents analysis and an investigation of the distribution of *Eulimnogammarus* species are considered within a framework of potential mechanisms which might facilitate the coexistence of species within the littoral zone area.

III.   MATERIALS AND METHODS

Twenty species of amphipods, eight of which belong to the genus *Eulimnogammarus*, were included in the phenetic analysis. Representatives of the Baikal genera *Heterogammarus sophianosi, Corophiomorphus kietlinskii, C. tenuipes, Eurybiogammarus violaceus, Philolimnogammarus cyaneus, P. viridis* and *P. vittatus*, and the Holarctic and Palaeoarctic genera *Gammarus lacustris, G. salinus, Echinogammarus macrocarpus, E. meridionalis* and *E. obtusatus* were also examined (Table 1). For most species, a single male specimen only was inspected. These Baikal genera, especially *Philolimnogammarus*, are believed to be very closely related to *Eulimnogammarus* (see Morino, 1998, for details). Indeed, some of the *Echinogammarus* species (here *E. macrocarpus* and *E. obtusatus*) have hitherto been placed within the genus *Eulimnogammarus* (Stock, 1969).

Forty-six characters were selected for this study (Table 2). These characters correspond to those used in recent analyses of other freshwater gammaroidean genera (e.g. Bousfield and Morino, 1992; Morino, 1994) and also include characters routinely utilized as standard diagnostic features in Baikal amphipod genera (see Bazikalova, 1945; Barnard and Barnard, 1983a, b). Each character was classified into two to four character states, which were then coded in binary (Table 2). The character states matrix (Appendix I) was converted to a phenogram with the group average clustering method, after calculating similarities based on the simple matching coefficient (Appendix II). However, because the pertinent data for character 8 (Table 2) were not available for some species, this character was omitted from this procedure. In this analysis, *Gammarus lacustris* was regarded as an outgroup, except with respect to the dentition formula of maxilla I (character 8).

For the trophic analysis, 10 taxa, including eight *Eulimnogammarus* species, were examined (Table 1). For each species, two to five adult specimens were collected and placed immediately in 10% buffered formalin. Collections were made at south Baikal during the summer. In the laboratory, these individuals were dissected and gut contents were examined under a microscope.

The distribution of amphipod species in the littoral zone of Lake Baikal was studied in the summer. A transect line 160 m long was set on a gravel–sandy bottom from the shore at Bolshye Koty (south Baikal) (see Figure 4). The water depth at the point furthest offshore was 16 m. An area of gravel substrate was chosen at six depths along the line and at two depths near the shore, and animals on the surface and beneath each gravel patch were carefully collected by a SCUBA diver. Samples were placed separately in plastic bags, to which was added 10% buffered formalin, and then returned to the laboratory for identification and enumeration.

Table 1
List of species included in the phenetic analyses

Genus *Eulimnogammarus* Bazikalova, 1945

1. *burkani* Bazikalova 1945
2. *cruentus* (Dorogostaiski, 1930)[a]
3. *czerskii* (Dybowsky, 1874)[a]
4. *grandimanus* Bazikalova, 1945[a]
5. *heterochirus* Bazikalova, 1945
6. *lividus* (Dybowsky, 1874)[a]
7. *maacki* (Gerstfeldt, 1858)[a]
8. *verrucosus* (Gerstfeldt, 1858)[a]

Genus *Heterogammarus* Stebbing, 1899

9. *sophianosi* (Dybowsky, 1874)[a]

Genus *Corophiomorphus* Barnard and Barnard, 1983

10. *kietlinskii* (Dybowsky, 1874)
11. *tenuipes* (Sowinsky, 1915)

Genus *Eurybiogammarus* Barnard and Barnard, 1983

12. *violaceus* (Dybowsky, 1874)[a]

Genus *Philolimnogammarus* Barnard and Barnard, 1983

13. *cyaneus* (Dybowsky, 1874)[a]
14. *viridis* (Dybowsky, 1874)
15. *vittatus* (Dybowsky, 1874)[a]

Genus *Echinogammarus* Stebbing, 1899, amend. Barnard and Barnard, 1983

16. *macrocarpus* (Stock, 1969)
17. *meridionalis* Pinkster, 1973
18. *obtusatus* (Dahl, 1938)

Genus *Gammarus* Fabricius, 1775

19. *lacustris* Sars, 1863
20. *salinus* Spooner, 1947

---

[a]Taxa also used in the trophic analysis. Details of specimens are available from the senior author.

IV.   RESULTS

A. Description of Mandibles and Maxillae I

Anatomical examinations and consultation of literature revealed interspecific differences in mandibles and maxilla I, as briefly described below.

Table 2

Characters, character states and codings for the species of *Eulimnogammarus* and related genera

1. Eye shape: small reniform (00), reniform (01), elongate reniform (11)

2. Lateral cephalic lobe: straight (0), concave (1)

3. Mandible, palp article I: bare (0), setose (1)

4. Mandible, distal part: normal (0), distal part enlarged as compared with palp (1)

5. Mandible molar: normal size and triturating type (0), small, protruded (1)

6. Maxilla I, no. of distal spines of outer plate: 11 (0), > 12 (1)

7. Maxilla I, distal armaments of left and right mandible palps: dimorphic (0), monomorphic (1)

8. Maxilla I, outer plate distal spines: dentition formula: VI–IX + II–IV (000), X–XIII + IV–XI (001), III + I–II (011), III + IV–VIII (111)

9. Pereonites, dorsal armaments: bare (0), long setose (1)

10. Pereonite VII: non-spinose (0), spinose (1)

11. Pleonites I–III: all non (or weakly) armed (00), PIII armed (01), all armed (11)

12. Pleonites I–III: setose (00), spinose (or spines dominate) (01), spinose and setose (11)

13. Pleonite III: dorsal armament: weak (00), medium (01), strong (11)

14. Antenna I, peduncular setae: shorter than peduncular diameter (0), longer than peduncular diameter (1)

15. Antenna I, peduncular article I: non-spinose (0), distally and/or marginally spinose (1)

16. Antenna I, peduncular articles II and III, dorsal margin: non-spinose (0), spinose (1)

17. Antenna I, peduncular articles II and III, ventral margin: none or setose (0), spinose (1)

18. Antenna I, no. of setal clusters on peduncular articles I and II: ped. II much more than ped. I (0), ped. II subequal to ped. I (1)

19. Antenna I, no. of setal clusters on peduncular article III: < 3 (0), > 4 (1)

20. Antenna I, ratio of peduncular article II to ped. art. I: < 0.90 (0), ≥ 1.00 (1)

21. Antenna I, ratio of peduncular article III to ped. art. I: ≥ 0.50 (0), < 0.45 (1)

22. Antenna I, no. of accessory flagellum articles: < 7 (0), ≥ 7 (1)

23. Antenna I, ratio of width of peduncular article I to width of ped. art. IV of ant. II: ≥ 1 (0), < 1 (1)

24. Antenna II, gland cone: directed anteriorly (0), directed ventrally (1)

25. Antenna II, peduncular article I: setose (0), spinose (1)

26. Antenna II, peduncular articles IV and V: long-setose (0), short-setose (1)

27. Antenna II, dorsal margin of peduncular article IV: setose or bare (0), spinose (1)

28. Antenna II, peduncular article, no. of setal clusters: ≤ 5 (00), 6–11 (01), ≥ 14 (11)

29. Antenna II, ratio of peduncular article V to ped. art. IV: ≥ 1.21 (00), 1.01–1.20 (01), < 1.00 (11)

30. Gnathopod I, coxal plate: as wide as or narrower than deep (0), distinctly wider than deep (1)

31. Gnathopod I, propod palm: not angled with posterior margin (0), angled (1)

32. Gnathopod I, propod palm: straight or convex (0), concave or notched (1)

33. Gnathopod I, propod palm, medial spine: present (0), absent (1)

34. Gnathopod I, ratio of propod to carpus: < 1.9 (0), ≥ 2.0 (1)

35. Gnathopod II, coxal plate shape: normal (0), attenuated (1)

36. Gnathopod II, propod, ratio of length to width: ≤ 1.90 (00), 2.0 2.3 (01), ≥ 2.4

37. Gnathopod II, ratio of propod to carpus: ≥ 1.40 (00), 1.21–1.39 (01), ≤ 1.20 (11)

38. Gnathopods I and II, propod length: gn. II longer than gn. I (0), gn. II equal to or shorter than gn. I (1)

39. Pereopod VII, basis: medium expanded without distinct distal lobe (00), expanded with distinct, more or less sharp lobe (01), slender or very slender without lobe (11)

40. Abdominal side plates II and III: posteroventrally not slanted (0), slanted (1)

41. Uropod I: non or hardly setose (0), setose (1)

42. Uropod I, rami: subequal in length (0), outer ramus shorter than inner (1)

43. Uropod III, ratio of outer ramus to peduncle: ≤ 3.0 (0), > 3.0 (1)

44. Uropod III, ratio of inner ramus to outer ramus: > 0.5 (00), 0.2–0.4 (01), < 0.2 (11)

45. Uropod III, terminal article of outer ramus: present (0), absent (1)

46. Telson, ratio of length to width: ≥ 0.90 (00), 0.89–0.70 (01), < 0.69 (11)

1. Mandibles

Three basic types of mandible were recognized in this study:

(a) *Eurybiogammarus violaceus* (Figure 1-2): the basal article is enlarged compared with the size of the palp, and the distal part of the incisor is strongly developed;

(b) *Corophiomorphus kietlinskii* (Figure 1-3): the basal article is subtriangular in lateral profile, and the molar is small, sitting on the protruded ridge. The lacinia is weak;

(c) in all other species examined, the mandibles were typical of those seen in most Gammaridae (e.g. see Figure 1-1, for *E. verrucosus*).

Thus *E. violaceus* and *C. kietlinskii* possess distinctive mandibles that are quite dissimilar to those of other species.

2. Maxilla I

Maxilla I displayed variation in the number and ornamentation of the distal spines of the outer plate and in the dimorphism of the palp. It was sometimes difficult to distinguish clearly the distal spines at the middle part of the margin, and therefore only the innermost and the second-innermost spines were

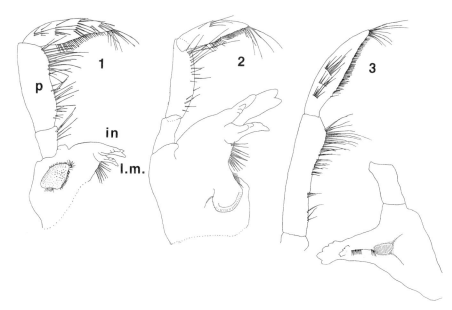

Fig. 1. Left mandibles. 1, *Eulimnogammarus verrucosus*; 2, *Eurybiogammarus violaceus*; 3, *Corophiomorphus kietlinskii*. Mandibular palp of *C. kietlinskii* is of the right mandible. Mandibles are not drawn to the same scale. Abbreviations: in, incisor; l.m., lacinia mobilis; p, mandibular palp.

considered. These are expressed as the following dentition formula: X (the number of dentition of the innermost spine) + X (that of the second-innermost spine). Three *Echinogammarus* species and *Gammarus salinus* were not inspected.

On the basis of these characters, maxillae I of the 16 species examined was classified into four types:

(a) type 1 (Figure 2A-1, B-1): 11 or fewer spines, spines moderately attenuated, dentition formula VI–IX + II–IV; palp dimorphic. Species: *Eulimnogammarus burkani, E. cruentus, E. lividus, E. czerskii, E. maacki, E. verrucosus, Philolimnogammarus vittatus* and *Heterogammarus sophianosi*;

(b) type 2 (Figure 2B-2): 11 or fewer spines, spines moderately attenuated, dentition formula X–XIII + IV–XI, dentitions of one or both spines brush-like; palp dimorphic. Species: *Eulimnogammarus grandimanus, E. heterochirus, Philolimnogammarus cyaneus, P. viridis* and *Gammarus lacustris*;

(c) type 3 (Figure 2B-3): 11 spines, spines moderately attenuate, dentition formula III + I–II; palp dimorphic. Species: *Corophiomorphus kietlinskii* and *C. tenuipes*;

(d) type 4 (Figure 2A-2, B-4): 14 spines, spines slender and elongate, dentition formula III + IV–VIII; palp monomorphic. Type 4 was also distinctive in

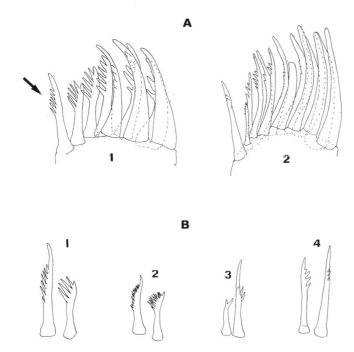

Fig. 2. Distal spines on outer plates of maxilla I. (A) 1, *Eulimnogammarus verrucosus* with nine spines, and 2, *Eurybiogammarus violaceus* with 14 spines. Arrow shows the innermost spine. (B) Four types of distal spine based on the number of dentitions on innermost and second-innermost spines: 1, *Eulimnogammarus verrucosus* (type 1); 2, *Philolimnogammarus cyaneus* (type 2); 3, *Corophiomorphus tenuipes* (type 3); 4, *Eurybiogammarus violaceus* (type 4). For 1 and 2, the spines on the left are the innermost ones, whereas for 3 and 4, the spines to the right are the innermost ones. Figures are not all drawn to the same scale.

that the number of dentitions on the second-innermost spine exceeded that on the innermost spine. Species: *Eurybiogammarus violaceus.*

B. Phenogram

The phenetic relationships and similarity indices (the simple matching coefficient) between 20 species of *Eulimnogammarus* and related genera are shown in Figure 3 and Appendix II, respectively. Among *Eulimnogammarus*, six species (*E. cruentus, E. lividus, E. czerskii, E. heterochirus, E. maacki* and *E. verrucosus*) composed a large cluster, which linked to the *Philolimnogammarus* cluster, which contained three species.

Two species, *Eulimnogammarus burkani* and *E. grandimanus*, formed another cluster, although this cluster was not fully supported by similarity indices (Appendix II). *Eulimnogammarus burkani* showed high similarity (index value

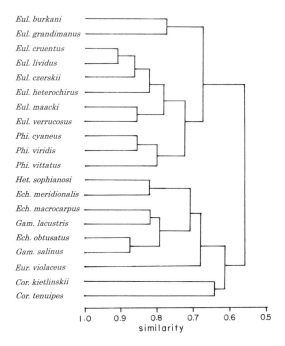

Fig. 3. Phenogram of 20 species of *Eulimnogammarus* and related genera, converted from Appendix with the group average clustering method. Similarities were calculated using the simple matching coefficient.

> 0.73) to *Eulimnogammarus* species, and the similarity to *E. grandimanus* was the second highest behind *E. czerskii*. However, the highest similarity of *E. grandumanus* was to *Echinogammarus obtusatus* and the second highest to *Eulimnogammarus burkani*. Similarity values of *E. grandimanus* to other *Eulimnogammarus* species were lower than those to *Philolimnogammarus cyaneus* and *P. viridis*. Thus, it is difficult to position *E. grandimanus* among the species examined.

Heterogammarus sophianosi presented the highest similarity to *Echinogammarus meridionalis*, although other Baikal species, especially of *Corophiomorphus*, showed rather low similarity indices, which were hard to resolve.

C. Gut Contents Analysis

Table 3 shows the results of gut contents analysis of 10 species which coexist in the littoral zone of Lake Baikal. *Eulimnogammarus grandimanus*, *E. cruentus*, *E. lividus*, *E. czerskii* and *E. maacki* fed mainly on filamentous algae (including blue-green and green algae) and small crustaceans (mainly copepods and cladocerans). The first two species also fed on oligochaetes. *Eulimnogammarus lividus* showed a wider range of food habits than the other species in this group,

Table 3
Gut contents, body size and mouthpart types of littoral Baikal amphipods collected in
the present study

Species	Body size (mm)	Gut contents	Type of mouthparts
Eulimnogammarus grandimanus	10–11	Filamentous algae, sponge spicules, oligochaetes, crustaceans	2
E. cruentus	21–26	Filamentous algae, sponge spicules, oligochaetes, crustaceans, rotatorians, detritus, inorganic matter	1
E. lividus	24	Filamentous algae, sponge spicules (packed), rotatorians, amphipods, other crustaceans, detritus	1
E. czerskii	11.5–15	Filamentous algae, sponge spicules, crustaceans, detritus, inorganic matter	1
E. maacki	17–26	Filamentous algae, sponge spicules, gastropod radulae, crustaceans, detritus, inorganic matter	1
E. verrucosus	17–27	Diatoms, filamentous algae, chironomid larvae	1
Philolimnogammarus cyaneus	10–13	Diatoms, filamentous algae, chironomid larvae, detritus	2
P. vittatus	13–18	Diatoms, filamentous algae, sponge spicules, crustaceans, chironomid larvae, detritus, inorganic matter	1
Heterogammarus sophianosi	17	Sponge spicules, crustaceans, detritus, inorganic matter	1
Eurybiogammarus violaceus	11.5–17	Filamentous algae, sponge spicules (packed)	4

and also included lubomirskiid sponges and amphipods in its diet. *Eulimnogammarus verrucosus, Philolimnogammarus cyaneus* and *P. vittatus* consumed a similar diet, in which diatoms and chironomid larvae were well represented. *Heterogammarus sophianosi* differed from the above species, since no algal material was evident among the gut contents. *Eurybiogammarus violaceus* was unique among the investigated taxa in that the gut contents of this species consisted almost exclusively of spicules of lubomirskiid sponges. Although sponge spicules were also observed in the guts of the other species, their relative scarcity there (except in *E. lividus*) suggests that they had been ingested incidentally with the main food material, as in the case of inorganic matter.

D. The Distribution of Amphipod Species

More than 40 species of amphipods from 13 genera were identified from the gravel. Figure 4 illustrates the distribution of nine species of *Eulimnogammarus* and *Philolimnogammarus*. Three more species of these genera and *Eurybiogammarus violaceus* were also present, but in very low numbers. From the pattern of species distribution of both amphipod genera, three habitat zones could be distinguished: the surfbelt zone (0–0.5 m deep: *E. verrucosus, P. cyaneus* and *P. vittatus*), the littoral zone (1–15 m deep: *E. maacki, E. lividus, E. cruentus, E. grandimanus* and *E. viridis*) and the sublittoral zone (> 15 m deep: *E.* cf. *czerskii*). The surfbelt and littoral zones were inhabited by species from both genera.

V. DISCUSSION

A. Mouthparts and Functional Implications

The mouthparts of Baikal amphipods were first described in detail by Sowinsky (1915), but have since been accorded far less attention. The species treated here showed differences in the structure of maxillae I and the

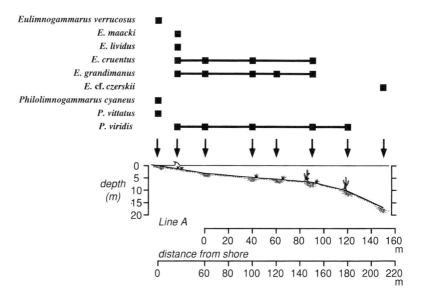

Fig. 4. Profile and distributions of selected amphipod species along a transect line at Bolshye Koty (August, 1997). The bottom substratum consists of gravel and sand; filamentous algae and tree-like lubomirskiid sponges are shown. Arrows indicate sampling points.

mandibles. *Eurybiogammarus violaceus* is peculiar in having 14 slender distal spines on the outer plate and monomorphic palp of maxilla I. This number is exceptional in the superfamily Gammaroidea, members of which have a maximum 11 spines (Bousfield, 1977). Slender spines similar to those of *E. violaceus* are also suggested in the description of *Eurybiogammarus fuscus* (Bazikalova, 1945), although the number is not specified.

The number of dentitions on the distal spines also varied, as noted by Sowinsky (1915). Among the *Eulimnogammarus–Philolimnogammarus* cluster, some correspondence between the types of the dentition and the genus was evident: type 1 dentition was mainly found in *Eulimnogammarus* species and type 2 dentition was mainly found in *Philolimnogammarus* species. Some of the dentitions were frayed apically, or even at the base. This was easily confirmed by comparing with the newly formed complete spines under the tegument. The high incidence of frayed dentitions suggests that the animals actively use maxillae I in feeding.

However, despite these anatomical differences, the high degree of food overlap indicated by the gut contents analysis suggests that no functional differences exist between types 1 and 2. That these different dentition types have no functional implications is strongly suggested when considering *Eulimnogammarus grandimanus* and *Philolimnogammarus cyaneus*. Both species have type 2 maxillae, but eat different foods to each other, yet show similar food preferences to those type 1 species with which they co-occur (Table 3, Figure 4). Quantitative fine-scale analysis of gut contents and ambient food availability (food electivity investigations) should be carried out to assess the functional consequences of these types.

Eurybiogammarus violaceus and *Corophiomorphus kietlinskii* have specialized mandibles. The large cutting edges of the mandible and numerous, smooth distal spines of maxillae I in *E. violaceus* suggest a scraping function. This species frequently occurs on tree-like sponges of the genus *Lubomirskia* (Kamaltynov *et al.*, 1993) and the peculiar mouthpart structure of this species may reflect a specialized food habit related to sponges. This is strongly supported by the finding that the gut contents of this species consist almost exclusively of spicules of lubomirskiid sponges. It should be noted that *Eulimnogammarus lividus*, one of the largest amphipods among those taxa examined, also feeds on the sponges (Table 3). It is likely that large and strong mouthparts are required to graze the spiculate tissues of sponges. Coleman (1990) reported that those Antarctic amphipods which feed on the firm tissues of sponges or holothurians possess mandibles with dentate incisors and a broad lacinia mobilis, which have special cutting mechanisms. By the same logic, the small and protruded molars and weak left lacinia in *C. kietlinskii* suggest that this species may feed on softened material.

This preliminary study revealed extensive trophic overlap among several species showing different mouthpart morphologies. Few species showed any indication of trophic specialization, and most seemed to be generalist omnivores. This situation differs markedly from that seen in another ancient lake species flock, the cichlid fishes of Lake Malawi. Here, speciation has been accompanied by great diversity in trophic apparatus, feeding behaviours and foods eaten, and most species are trophic specialists (see Kuusipalo, this volume).

The presence of several amphipod species with extremely high trophic overlap living within the same habitat is also an apparent contradiction of one of ecology's basic tenets, Gause's rule, which posits that close competitors cannot coexist indefinitely. Further research on the functional morphology and trophic ecology of a variety of amphipod species is needed to clarify phylogenetic implications of mouthpart diversity and its role, if any, in permitting high trophic overlap among coexisting amphipods in Baikal.

B. Phenetic Analysis of *Eulimnogammarus* Species and Related Genera

In the phenetic analysis, a high similarity was seen between all species of *Eulimnogammarus*, with the exception of *E. grandimanus*. It is interesting that a recent mitochondrial DNA (mtDNA) analysis also supports this finding and demonstrates that *E. grandimanus* is not directly related to *Eulimnogammarus verrucosus* or *Philolimnogammarus* (*P. cyaneus* and *P. vittatus*), but to another Baikal genus, *Plesiogammarus* (*P. gerstaeckeri*) (Ogarkov *et al.*, 1997).

The present analysis also suggested that *Eulimnogammarus* and *Philolimnogammarus* might comprise a species flock. However, the similarities of *Eulimnogammarus* to *Eurybiogammarus, Heterogammarus* and *Corophiomorphus* were rather weak, and the placement of *Echinogammarus macrocarpus* and *E. obtusatus* in the genus *Eulimnogammarus*, as proposed by Stock (1969), is not tenable. Stock (1969) claimed that no reliable characters existed by which *Eulimnogammarus* and *Philolimnogammarus* could be separated. Bousfield (1977) followed Stock's claim and also included *Corophiomorphus* species within his *Eulimnogammarus*. In contrast, Barnard and Barnard (1983a, b) later placed *Eulimnogammarus* and *Philolimnogammarus* in different generic groups. The results of the present phenetic study did not support Bousfield's opinion (1977), in that *Corophiomorphus* was shown to stand far apart from *Eulimnogammarus*, nor did the findings support Barnard and Barnard's decision to place *Eulimnogammarus* and *Philolimnogammarus* in different generic groups.

One objective of the present study was to derive a phylogenetic relationship based on the overall similarity among *Eulimnogammarus* species. Thus, the characters used will tend to be those showing high diversity in this genus. In addition, *Corophiomorphus, Philolimnogammarus* and *Eurybiogammarus* are large genera, comprising 11, 13 and 29 species, respectively (Barnard and Barnard, 1983a, b). Further analyses including the remaining species are needed to resolve satisfactorily the relationship between these genera.

C. Habitat Partitioning, Trophic Specialization and Co-occurrence in *Eulimnogammarus* Species

Among the eight species of *Eulimnogammarus* examined, six were collected from subsurface interstices within the gravel bed of the shallow littoral area. On the basis of the distribution of these six species and of *Philolimnogammarus* species, three habitat zones were recognized within the littoral area. Each zone can be clearly characterized by abiotic and biotic environmental factors: (i) the surfbelt zone, with the highest wave action and lowest predation pressure; (ii) the littoral zone, with low wave action, and highest primary production and predation pressure; and (iii) the sublittoral zone, with no wave action, constant low temperature, and high predation pressure. A comparison between the distribution of species (Figure 4) and the phenogram (Figure 3) reveals that sister species occupy different habitat zones, i.e. the same habitat zone is inhabited only by the more distantly related members of the same group, or by species which belong to a species group of a different lineage.

All of these species exhibited a basically omnivorous food habit, with a diet dominated by filamentous algae, with significant amounts also of gastropods, oligochaetes and chironomids. Algae is abundant in the littoral zone of Lake Baikal, especially during the summertime (Kozhova and Izmest'eva, 1998), and gastropods, oligochaetes and chironomids are numerically dominant members of the macrobenthos in the gravel substrates (Weinberg and Kamaltynov, 1998). This highly productive environment thus contains large amounts of a diverse range of food resources. Such trophic conditions are often reflected in a high species richness, and two explanations for this have been offered. One states that a high diversity and abundance of potential food items provides a situation where trophic partitioning among species could evolve through specialization. A second explanation posits that, where food resources are abundant, decreased competition for food can operate to facilitate the coexistence of different species. How do these theories apply to the Baikal amphipods?

While trophic specialization was implied by the different structure of the mouthparts (cf. *Eulimnogammarus verrucosus*, *Philolimnogammarus cyaneus* and *P. vittatus* in Table 3), most co-occurring species showed high overlap in the food items ingested, and in this group of amphipods there was little evidence for trophic segregation. Instead, these amphipods appeared to be non-selective omnivores and their gut contents reflected the trophically available fauna and flora of respective habitat zones. This could be verified by examining the algal foods and benthic algae at a finer taxonomic level, since several taxa of benthic algae show a clear depth-related zonation in the littoral area (Kozhova and Izmest'eva, 1998).

Were trophic competition and overlap high, then other niche dimensions would be expected to become segregated more strongly. For example, a study of North American gammarids inhabiting a cave under conditions of low food availability revealed strong microhabitat segregation among the three species present (Culver, 1970). In Lake Baikal, however, species utilizing the same foodstuffs sometimes showed extensive habitat overlap. As such, neither of the above explanations can satisfactorily account for the coexistence of Baikal amphipods within the littoral zone. Here, *Eulimnogammarus* and *Philolimnogammarus* species utilize a wide variety of food types, and the most parsimonious interpretation based on current evidence might be that their generalist foraging strategy has contributed to the species co-occurrence within this habitat with high food availability. However, theoretical considerations indicate that even where resources are abundant, coexistence between species showing high food overlap cannot be sustained. Whether this trophic overlap is indeed real, or simply an artefact of the gut contents not having been examined in sufficiently fine detail remains unknown, but further studies of this phenomenon are clearly warranted.

Under certain conditions, predation can act to promote species coexistence and maintain a high species diversity. In the littoral of Baikal, abundant sculpin fishes feed almost exclusively on amphipods (Morino, pers. obs.). Such predation may keep some species below their carrying capacities, thereby reducing the intensity and importance of direct competition for resources, and permitting much more niche overlap and a greater richness of species than in a community dominated by competition. The importance of predation in lowering competition within the littoral amphipod community is unknown.

Body-size differences have long been interpreted as indicating segregation of some resource, usually spatial or trophic (e.g. Brown, 1975; Diamond, 1975). Such body-size differences are best understood in terms of selective pressures to maximize the differences in size and resource utilization among coexisting species. Circumstantial evidence suggests the presence of such a partitioning mechanism with the Baikal amphipods. For example, the large amphipod species *Eulimnogammarus lividus* showed a wider food habit than other co-occurring species of the same mouthpart type. In addition, in the three amphipod species inhabiting the surfbelt zone, *Eulimnogammarus verrucosus, Philolimnogammarus cyaneus* and *P. vittatus*, modes of body size distributions in the summer do not overlap, and the breeding season and life history of *E. verrucosus* differ from those of *P. cyaneus* (Morino *et al.*, 1999). However, the presence of species-specific or size-dependent shifts in food preference among the littoral amphipods remains unsubstantiated.

The findings of the present study both emphasize the paucity of knowledge concerning the community ecology of Baikal's amphipods, and suggest several avenues for future investigations. It was revealed that *Eurybiogammarus violaceus* and *Corophiomorphus* species have markedly different mouthparts and belong to discrete species flocks. The component species of these genera are distributed over a wider depth range and the environmental factors in their habitats are very different from those of *Eulimnogammarus*. As such, a study of these genera may well reveal mechanisms different from those which maintain species diversity and coexistence within the littoral *Eulimnogammarus* species flock. A general explanation for the coexistence and high diversity of the amphipod species in Baikal's littoral zone presently remains elusive.

ACKNOWLEDGEMENTS

We thank M. Grachev, K. Numachi and N. Miyazaki for their help during the programme of the Baikal International Center for Ecological Research. We are indebted to O. Coleman and N.L. Tzvetkova for loan of material. Field studies were conducted using the facilities of the Biological Institute, Irkutsk State University, for which thanks are due to O.M. Kozhova and V. Ostroumov, and to T. Narita, A. Goto, V. Sideleva and M. Yamauchi for their assistance in the survey. Thanks go also to the editors, who encouraged this study and refined an earlier draft. This study was financed by the Grants-in-Aid for Overseas Scientific Survey (Nos 04041035 and 07041130) and for the Scientific Research C (No. 05640717), Creative Basic Research Program (09NP1501) from the Ministry of Education, Science, Culture and Sports.

REFERENCES

Barnard, J.L. and Barnard, C.M. (1983a). *Freshwater Amphipoda of the World I. Evolutionary Patterns*. Hayfield Associates, Virginia.
Barnard, J.L. and Barnard, C.M. (1983b). *Freshwater Amphipoda of the World II. Handbook and Bibliography*. Hayfield Associates, Virginia.
Bazikalova, A. (1945). Amphipoda of Baikal. *Acad. Nauk SSSR, Trudy Baikal'skoi Limnol. Inst. SO AN SSSR* **11,** 5–440 (in Russian).
Bousfield, E.L. (1977). A new look at the systematics of gammaroidean amphipods of the world. *Crustaceana* **4,** Suppl., 282–316.
Bousfield, E.L. and Morino, H. (1992). The amphipod genus *Ramellogammarus* in fresh waters of western North America: systematics and distributional ecology. *Contrib. Nat. Sci.* **17,** 1–21.
Brooks, J.L. (1950). Speciation in ancient lakes. *Q. Rev. Biol.* **25,** 30–60, 131–176.
Brown, J.H. (1975). Geographical ecology of desert rodents. In: *Ecology and Evolution of Communities* (Ed. by M.L. Cody and J.H. Diamond), pp. 315–341. Belknap Press of Harvard University Press, Cambridge, MA.
Coleman, C.O. (1990). *Bathyponoploea schellenbergi* Holman & Watling, 1983, an Antarctic amphipod (Crustacea) feeding on Holothuroidea. *Ophelia* **31,** 197–205.
Culver, D.C. (1970). Analysis of simple cave communities: niche separation and species packing. *Ecology* **51,** 949–958.
Diamond, J.M. (1975). Assembly of species communities. In: *Ecology and Evolution of Communities* (Ed. by M.L. Cody and J.H. Diamond), pp. 432–444. Belknap Press of Harvard University Press, Cambridge, MA.
Dorogostaisky, V.C. (1930). New materials on the carcinological fauna of Lake Baikal. *Akad. Nauk SSSR, Komissii Izucheniyo Ozera Baikala* **3,** 49–76 (in Russian).
Dybowsky, B.N. (1874). *Beiträge zur näheren Kenntniss der in dem Baikal-See vorkommenden niederen Krebse aus der Gruppe der Gammariden. Herausgegeben*

von der Russischen Entomologischen Gesellschaft zur St. Petersburg. W. Besobrasoff, St Petersburg.

Fryer, G. (1991). Comparative aspects of adaptive radiation and speciation in Lake Baikal and the great rift lakes of Africa. *Hydrobiologia* **211**, 137–146.

Greenwood, P.H. (1984). What is a species flock? In: *Evolution of Fish Species Flocks* (Ed. by A. Echelle and I. Kornfield), pp. 13–19. University of Maine at Orono Press, Orono, ME.

Kamaltynov, R.M., Chernykh, V.I., Slugina, Z.V. and Karabanov, E.B. (1993). The consortium of the sponge *Lubomirskia baicalensis* in Lake Baikal, East Siberia. *Hydrobiologia* **271**, 179–189.

Kamaltynov, R.M. (1999a). On the higher classification of Lake Baikal amphipods. *Crustaceana* **72**, 921–931.

Kamaltynov, R.M. (1999b). On evolution of Lake Baikal amphipods. *Crustaceana* **72**, 933–944.

Kozhov, M. (1963). *Lake Baikal and its Life.* Dr. W. Junk Publishers, The Hague.

Kozhova, O.M. and Izmest'eva, L.R. (1998). *Lake Baikal: Evolution and Biodiversity.* Backhuys, Leiden.

Martens, K., Coulter, G. and Goddeeris, B. (1994). Speciation in ancient lakes–40 years after Brooks. *Arch. Hydrobiol. Beiheft. Ergebnisse Limnol.* **44**, 75–96.

Mashiko, K., Kamaltynov, R.M., Shervakov, D.Yu. and Morino, H. (1997). Genetic separation of gammarid (*Eulimnogammarus cyaneus*) populations by localized topographic changes in ancient Lake Baikal. *Arch. Hydrobiol.* **139**, 379–387.

Morino, H. (1994). The phylogeny of *Jesogammarus* species (Amphipoda:Anisogammaridae) and life history features of two species endemic to Lake Biwa, Japan. *Arch. Hydrobiol. Beiheft. Ergebnisse Limnol.* **44**, 257–266.

Morino, H. (1998) Specimens of the Baikal amphipods of *Heterogammarus* generic cluster (Crustacea), deposited in the Zoological Institute, St. Petersburg, Museum für Naturkunde, Berlin, and the Zoologishes Institute und Museum der Universität Hamburg. *Nat. Hist. Bull. Ibaraki Univ.* **2**, 1–6.

Morino, H., Yamauchi, M., Kamaltynov, R.M., Nakai, K. and Mashiko, K. (1999). Amphipod association in the surf belt of Lake Baikal. In: *Biodiversity, Phylogeny and Environment in Lake Baikal* (Ed. by N. Miyazaki), pp. 45–60. Otsuchi Marine Research Center, The University of Tokyo, Iwate.

Ogarkov, O.B., Kamaltynov, R.M., Belikov, S.I. and Sherbakov, D.Yu. (1997). Phylogenetic relatedness of the Baikal Lake endemial amphipods (Crustacea, Amphipoda) deduced from partial nucleotide sequences of the cytochrome oxidase subunit III genes. *Mol. Biol.* **31**, 24–29.

Pallas, P.S. (1772). Specilegia zoologica, quibus novae imprimis et obscurae animalium species iconibus, descriptionibus atque commentariis illustrantur cura P. S. Pallas. Fasciculus nonus. Berolini **1772**, 50–80.

Pinkster, S. (1973). The *Echinogammarus berilloni*-group, a number of predominantly Iberian amphipod species (Crustacea). *Bijdr. Dierk.* **43**, 1–39.

Sherbakov, D.Yu., Kamaltynov, R.M., Ogarkov, O.B. and Verheyen, E. (1998). Patterns of evolutionary change in Baikalian gammarids inferred from DNA sequences (Crustacea, Amphipoda). *Mol. Phylogenet. Evol.* **10**, 160–167.

Sowinsky, V.K. (1915). Amphipoda from the Baikal Sea. (Fam. Gammaridae). *Wissensch. Ergebnisse Zool. Exped. Baikal-See* **9**, 1–381 (in Russian and German).

Stebbing, T.R.R. (1899). Amphipoda from the Copenhagen Museum and other sources, Part II. *Trans. Linn. Soc. Lond. (Zool.)* **8**, 395–432.

Stock, J.H. (1969). Members of Baikal amphipod genera in European waters, with description of a new species, *Eulimnogammarus macrocarpus*, from Spain. *Koninklijk Neder. Akad. Wetenshcappen, Amsterdam* **72**, 66–75.

Takhteev, V.V. (1997). The gammarid genus *Plesiogammarus* Stebbing, 1899, in Lake Baikal, Siberia (Crustacea Amphipoda Gammaridea). *Arthropoda Selecta* **6**, 31–54.

Timoshkin, O.A. (1999). Biology of Lake Baikal: "white spots" and progress in research. *Berliner Geowiss. Abh.* **E30**, 333–348.

Weinberg I.W. and Kamaltynov, R.M. (1998). Macrozoobenthos communities at a stone beach of Lake Baikal. *Zool. Zh.* **77**, 158–165 (in Russian).

Appendix I

Character-state matrix for 20 species of amphipods, based on the character states given in Table 2

No. Species	1	2	3	4	5	6	7	8	9	10	11	12	13	14	15	16	17	18	19	20	21	22	23	24	25
1. *Eulimnogammarus burkani*	1,1	0	1	0	0	0	0	0,0,0	0	0	1,1	1,1	1,0,1	0	1	0	1	0	0	1	0	0	0	0	0
2. *E. cruentus*	1,1	0	1	0	0	0	0	0,0,0	1	0	1,1	1,1	1,1	0	1	0	0	0	1	1	0	1	0	0	0
3. *E. czerskii*	1,1	0	1	0	0	0	0	0,0,0	1	0	1,1	0,1	1,1	0	1	0	0	0	1	1	0	0	0	0	0
4. *E. grandimanus*	0,1	0	1	0	0	0	0	0,0,1	0	0	0,1	0,0	0,0	0	1	0	0	0	0	1	0	1	0	0	0
5. *E. heterochirus*	1,1	0	1	0	0	0	0	0,0,0	0	0	1,1	1,1	1,1	1	1	0	0	0	0	1	0	0	0	0	0
6. *E. lividus*	1,1	0	1	0	0	0	0	0,0,0	0	0	1,1	1,1	1,1	1	1	0	0	0	0	1	1	1	0	0	0
7. *E. maacki*	1,1	0	1	0	0	0	0	0,0,0	0	0	1,1	1,1	1,1	1	1	0	0	0	1	1	0	1	0	0	0
8. *E. verrucosus*	1,1	0	1	0	0	0	0	0,0,0	0	1	1,1	1,1	1,1	1	1	1	0	0	0	1	0	0	0	0	1
9. *Heterogammarus sohianosi*	0,1	1	0	0	0	0	0	0,0,0	0	0	1,1	1,1	0,1	0	0	0	0	1	0	0	1	1	0	0	0
10. *Corophiomorphus kietlinskii*	1,1	1	1	0	0	1	0	0,1,1	0	0	0,0	0,0	0,0	0	0	0	0	0	1	1	1	0	1	1	0
11. *C. tenuipes*	0,0	1	0	1	0	0	1	0,1,1	0	0	1,1	0,0	1,1	0	0	0	0	0	1	1	0	1	1	1	0
12. *Eurybiogammarus violaceus*	0,1	0	1	0	0	1	1	1,1,1	0	0	1,1	0,1	0,0	0	0	0	0	1	0	0	1	0	0	0	0
13. *Philolimnogammarus cyaneus*	0,1	1	1	0	0	0	0	0,0,1	0	0	1,1	0,0	1,1	1	0	0	0	0	1	1	0	0	0	0	0
14. *P. viridis*	1,1	1	1	0	0	0	0	0,0,1	0	0	0,1	0,1	0,1	1	0	0	0	0	1	0	0	0	0	0	0
15. *P. vittatus*	1,1	1	1	0	0	0	0	0,0,0	0	0	1,1	0,1	1,1	1	1	0	0	0	1	1	1	0	0	0	0
16. *Echinogammarus macrocarpus*	0,0	0	0	0	0	0	0	ND	0	0	0,0	0,0	0,0	0	0	0	0	1	0	0	0	0	0	0	0
17. *E. meridionalis*	1,1	0	1	0	0	0	0	ND	0	0	1,1	0,1	0,1	1	0	0	0	1	0	0	1	0	0	0	0
18. *E. obtusatus*	1,1	0	1	1	0	0	0	ND	0	0	0,0	0,0	0,0	0	0	0	0	1	0	0	0	0	0	0	0
19. *Gammarus lacustris*	0,0	0	0	0	0	0	0	0,0,1	0	0	0,0	0,0	0,0	0	0	0	0	0	0	0	0	0	0	0	0
20. *G. salinus*	1,1	0	1	0	0	0	0	ND	0	0	0,0	0,0	0,1	0	0	0	0	1	0	0	1	0	0	0	0

Characters	26	27	28	29	30	31	32	33	34	35	36	37	38	39	40	41	42	43	44	45	46
No. Species																					
1. *Eulimnogammarus burkani*	0	1	0, 1	1, 1	1	0	0	0	1	1	1, 1	1, 1	1	0, 0	0	0	0	1	1, 1	1	0, 0
2. *E. cruentus*	1	1	0, 1	0, 1	1	1	0	0	1	0	0, 1	0, 1	1	1, 1	0	0	0	1	1, 1	1	0, 1
3. *E. czerskii*	0	1	1, 1	0, 0	1	0	0	0	1	1	1, 1	1, 1	1	1, 1	0	1	1	1	1, 1	1	0, 1
4. *E. grandimanus*	0	0	0, 0	0, 1	1	0	1	1	1	1	1, 1	1, 1	1	0, 0	0	0	0	0	0, 1	1	0, 0
5. *E. heterochirus*	0	0	0, 1	0, 0	0	0	0	0	1	0	0, 1	1, 1	1	1, 1	0	0	0	1	1, 1	1	0, 1
6. *E. lividus*	0	1	1, 1	0, 1	1	0	0	0	0	0	0, 1	0, 1	1	1, 1	0	0	0	1	1, 1	1	1, 1
7. *E. maacki*	0	1	0, 1	0, 1	1	0	1	0	0	0	0, 0	0, 1	1	1, 1	0	0	0	1	1, 1	1	1, 1
8. *E. verrucosus*	0	1	0, 1	0, 1	1	0	0	0	0	0	0, 0	0, 0	1	0, 1	0	0	0	1	1, 1	1	0, 1
9. *Heterogammarus sohianosi*	0	1	0, 0	1, 1	0	0	0	0	0	0	0, 0	0, 0	0	0, 1	0	1	1	0	0, 0	0	0, 0
10. *Corophiomorphus kietlinskii*	1	0	0, 0	1, 1	1	0	0	0	0	0	0, 0	0, 0	1	1, 1	1	1	1	0	0, 1	1	0, 0
11. *C. tenuipes*	0	0	0, 1	0, 1	1	0	0	0	0	0	0, 1	0, 1	0	1, 1	1	1	0	0	0, 0	0	0, 0
12. *Eurybiogammarus violaceus*	0	1	0, 0	0, 1	0	1	1	0	0	0	1, 1	1, 1	1	0, 1	0	0	0	0	0, 0	0	0, 0
13. *Philolimnogammarus cyaneus*	0	0	0, 1	0, 1	0	0	0	0	0	0	0, 1	1, 1	1	0, 1	0	0	0	0	0, 1	0	0, 1
14. *P. viridis*	0	0	0, 1	0, 1	0	0	0	0	0	0	0, 0	1, 1	1	0, 0	0	0	0	0	0, 1	1	0, 1
15. *P. vittatus*	0	1	0, 1	0, 1	0	1	0	0	0	1	0, 0	0, 1	1	0, 1	0	0	0	0	0, 1	1	0, 1
16. *Echinogammarus macrocarpus*	0	0	0, 0	0, 1	1	1	1	1	0	0	0, 1	1, 1	1	0, 1	0	0	0	0	0, 1	0	0, 0
17. *E. meridionalis*	1	0	0, 0	0, 1	0	0	0	1	0	0	0, 0	0, 0	0	1, 1	0	0	0	0	0, 1	0	0, 1
18. *E. obtusatus*	0	0	0, 0	0, 1	0	0	0	0	0	0	0, 0	1, 1	1	0, 0	0	0	0	0	0, 1	0	0, 0
19. *Gammarus lacustris*	0	0	0, 0	0, 0	0	0	0	0	0	0	0, 0	0, 0	0	0, 0	0	0	0	0	0, 0	0	0, 0
20. *G. salinus*	0	0	0, 0	1, 1	0	0	1	0	0	0	0, 0	0, 0	0	0, 0	0	0	0	0	0, 0	0	0, 0

ND, No data.

Appendix II

Matrix of the simple matching coefficients between species, calculated from the data given in Appendix I

Species	E. bur	E. cru	E. cze	E. gra	E. het	E. liv	E. maa	E. ver	H. soh	C. kie	C. ten	E. vio	P. cya	P. vir	P. vit	E. mac	E. mer	E. obt	G. lac
Eulimnogammarus burkani																			
E. cruentus	0.732																		
E. czerskii	0.786	0.839																	
E. grandimanus	0.768	0.607	0.625																
E. heterochirus	0.750	0.804	0.821	0.661															
E. lividus	0.750	0.911	0.893	0.625	0.857														
E. maacki	0.696	0.786	0.696	0.643	0.804	0.804													
E. verrucosus	0.696	0.786	0.732	0.571	0.839	0.839	0.857												
Heterogammarus sohianosi	0.571	0.518	0.500	0.518	0.500	0.536	0.518	0.518											
Corophiomorphus kietlinskii	0.482	0.536	0.446	0.571	0.482	0.482	0.500	0.500	0.661										
C. tenuipes	0.482	0.571	0.518	0.536	0.554	0.554	0.571	0.536	0.696	0.643									
Eurybiogammarus violaceus	0.607	0.518	0.464	0.661	0.500	0.464	0.518	0.411	0.679	0.482	0.554								
Philolimnogammarus cyaneus	0.643	0.696	0.643	0.732	0.786	0.714	0.696	0.661	0.643	0.482	0.661	0.679							
P. viridis	0.679	0.661	0.607	0.732	0.786	0.679	0.768	0.696	0.607	0.554	0.589	0.607	0.857						
P. vittatus	0.679	0.804	0.714	0.661	0.786	0.786	0.804	0.768	0.607	0.518	0.625	0.571	0.786	0.821					
Echinogammarus macrocarpus	0.518	0.500	0.411	0.679	0.518	0.446	0.464	0.393	0.696	0.571	0.643	0.732	0.732	0.661	0.554				
E. meridionalis	0.571	0.625	0.536	0.554	0.607	0.607	0.661	0.625	0.821	0.589	0.661	0.679	0.714	0.714	0.679	0.768			
E. obtusatus	0.607	0.518	0.464	0.804	0.607	0.536	0.625	0.518	0.643	0.589	0.589	0.679	0.786	0.821	0.643	0.804	0.714		
Gammarus lacustris	0.518	0.464	0.411	0.679	0.518	0.446	0.500	0.464	0.732	0.643	0.679	0.661	0.696	0.696	0.554	0.821	0.732	0.804	
G. salinus	0.518	0.464	0.375	0.679	0.482	0.482	0.536	0.464	0.768	0.679	0.607	0.625	0.661	0.696	0.589	0.750	0.732	0.875	0.821

Evolutionary Inferences from the Scale Morphology of Malawian Cichlid Fishes

L. KUUSIPALO

I. SUMMARY

The evolution of cichlid fishes in Lake Malawi was studied using a small structural character, fine scale morphology. Because features of the scales are presumed to be selectively neutral and to be independent of a species' niche, similarities in scale structure may give an unbiased view of the relationships among extant species.

The scale morphology of 20 endemic cichlid genera was examined. Several structures with possible value for taxonomic and evolutionary studies were identified, including six types of scale shape and three types of interradial denticle. A further character of use was the spiny area in the exposed part of the scale, which varied from non-existent to over 140° wide.

Phylogenetic relationships revealed from scale characteristics were compared with those obtained from other methods which also utilize selectively neutral characters [mitochondrial DNA (mtDNA) and enzyme structure]. The degree of relatedness of taxa was then investigated in relation to the niche occupied. When the studied species were grouped in a dendrogram according to the similarity of their scales, grouping did not reflect the similarity of niches. Furthermore, electrophoretic and mtDNA studies of this species flock suggest

ADVANCES IN ECOLOGICAL RESEARCH VOL. 31
ISBN 0–12–013931–6

that, although niche is a conservative character, fundamentally new niches have been invaded several times during the evolution of the studied species.

The single basin structure of Lake Malawi offers no clear geographical barriers to facilitate allopatric speciation, but lake sediments indicate this seemingly homogeneous environment to have been historically heterogeneous. Geological sediment cores contain evidence of mass mortalities of fishes, and opportunities for niche switches may have resulted from released competition following local fish kills. The relaxation of competition following such events would have provided opportunities for niche switches, which may have driven some of the speciation in this group.

II. INTRODUCTION

The explosively radiated species flock of Malawian cichlids is estimated at 500–1000 endemic species (Owen *et al.*, 1990; Snoeks, this volume) and is a fascinating group in which to investigate speciation. These fishes have diversified to occupy almost all habitats within the lake and they utilize almost all available food resources. Malawian cichlids show many behaviours of interest to the ethologist, including lekking, nest building, territoriality, parental care and mouthbrooding.

The taxonomy of the cichlids has traditionally been based on anatomical differences, mainly characters and features in the jaw apparatus and the head. These features, however, are probably subject to strong selection pressure, and this can act to confound evolutionary analyses and phylogenetic reconstructions based on these characters (see Turner, this volume). Scale morphology provides a multitude of characters of potential value to studies of evolution and speciation. More importantly, in contrast to anatomical features, most of the characters of the scales are stable and independent of environmental factors (Lippitsch, 1989, 1990), and morphology of the scale surface is thus less likely to result from convergent evolution (Lippitsch, 1992).

This study examines the utility of scale characters in taxonomic studies of Lake Malawi cichlid fishes. Using this method, the interrelationships and phylogeny of 20 endemic Malawian genera and species were investigated. These taxa represent *c.* 40% of the cichlid genera endemic to the lake and were chosen to represent a diverse variety of trophic niches. The study also explores the relationship between taxonomic relatedness and similarity of niche, and considers a catastrophic mortality scenario, whereby niche space might have been made available for occupation.

III. MATERIALS AND METHODS

A. Species Examined

Adult individuals of 22 endemic cichlid species were collected from among a commercial trawl catch at Monkey Bay, southern Lake Malawi (Kuusipalo, 1998).

Species were categorized according to trophic niche. The representative species examined and their trophic categories were as follows:

(i)   *Pseudotropheus elegans* (Trewavas): An invertebrate eater;

(ii)   *Otopharynx argyrosoma* (Regan): a deep-water species which picks crustaceans from the surface of sandy substrates;

(iii) *Lethrinops altus* (Trewavas) and *Lethrinops gossei* (Burgess and Axelrod): sifters of invertebrates from sandy substrates;

(iv)   *Alticorpus geoffreyi* and *Aulonocara* "orange": invertebrate hunters with sensory spots in the head;

(v)   *Chilotilapia rhoadesii* (Boulenger): a snail crusher;

(vi)   *Mylochromis anaphyrmus* (Burgess and Axelrod): a deep-water snail crusher;

(vii) *Placidochromis subocularis* (Günther): probably a snail crusher;

(viii) *Platygnathochromis melanonotus* (Regan): an eater of snails and small fishes;

(ix)   *Copadichromis pleurostigmoides* (Iles) and *Copadichromis virginalis* (Iles): "Utaka", pelagic zooplankton feeders, living in large shoals;

(x)   *Corematodus shiranus* (Boulenger): a species which mimics endemic tilapiine fishes and scrapes their scales for food;

(xi)   *Diplotaxodon argenteus* (Trewavas): a pelagic predator of usipa sardines;

(xii) *Buccochromis rhoadesii* (Boulenger): a large semipelagic predator;

(xiii) *Hemitaeniochromis urotaenia* (Regan): a predator of small fish;

(xiv) *Nimbochromis livingstonii* (Günther): an ambush predator which imitates a decaying fish corpse;

(xv)   *Aristochromis christyi* (Trewavas): a predator which stalks eggs and small fish by lying on its side;

(xvi) *Sciaenochromis spilostichus* (Trewavas), *Rhamphochromis* sp. and *Taeniochromis holotaenia* (Regan) (Lewis *et al.*, 1986; Eccles and Trewavas, 1989; Konings, 1989): streamlined predators.

One species of unknown trophic ecology was also included:

(xvii) *Ctenopharynx intermedius* (Günther): a deep-water fish, ecology unknown.

Between three and 11 individuals per species were collected. For two taxa, aquarium-reared juveniles purchased at Salima on the western shore of Lake Malawi, were used:

(xviii) *Cynotilapia axelrodi* (Burgess): a zooplankton feeder;
(xix) *Pseudotropheus zebra* (Boulenger): an aufwuchs scraper.

B. Scale Preparation and Analysis

Several scales were removed from above the upper lateral line of each individual by gentle scraping. Scales were cleaned mechanically, treated briefly with mild sodium hypochlorite solution to remove extraneous tissue, and later examined by both stereomicroscopy and scanning electron microscopy. For scanning electron microscopy, scales were first dried and sputter coated with a 30 nm layer of gold, and electron micrographs were produced for examination. For comparisons, one scale was chosen as a representative for each species.

Scales showed a typical morphology. Each scale had rings of circuli centred around a focus (Figure 1). In the rostral field the circuli support minute denticles that anchor the scale in the tissue: these were visible only under high magnification (> ×2000). The rostral field is partitioned by radii and the

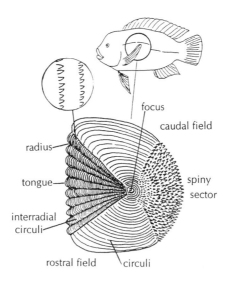

Fig. 1. The scale morphology of Lake Malawi cichlids: each scale has rings of circuli centered around a focus, and in the rostal field the circuli have minute denticles. The rostral field is partitioned by radii, and on the rostral rim the interradial spaces end in smooth, tongue-like projections. In the caudal field a characteristic spiny granulation is present.

interradial spaces of the rostral rim end in smooth tongue-like projections. In the caudal field a characteristic granulation is present, which sometimes obscures the circuli (Lippitsch, 1992).

The microsopic examination revealed interspecific differences in several of these features, namely, scale shape, the characteristics of the denticles, and the type and extent of the granulation. Therefore, to explore the utility of these differences in a taxonomic framework, for each species the following four characters were measured from the representative scale:

(i)   shape: length:width ratio of the scale;
(ii)   focus: distance of focus to the caudal rim in proportion to the total length;
(iii)   angle: angle of the spiny sector within which the spines obscure the circuli, starting from the focus;
(iv)   denticle: the type of interradial denticles in interradial circuli as a nominal variable, where 1 denotes small and sharp, 2, is low and blunt, and 3 is firm and widely spaced denticles.

The first three characteristics were used to construct a phenogram with 10 permutations and random order of adding species, by utilizing CONTML (Continuous Characters Maximum Likelihood Method) of the PHYLIP software package (Felsenstein, 1993). Data on the scales of the cichlid fish *Sarotherodon galilaeus* (Artedi), given in Lippitsch (1992), were used as an outgroup. Interspecific differences in the detected variation were evaluated by first subjecting all four variables to a principal component analysis (PCA) in order to group species by similarity.

IV. RESULTS

Under the microscope, each of the examined scales appeared sectioned with well-developed radii (Lippitsch, 1990). In the interradial space the rostral rim of the scale forms tongue-like projections, which are free of circuli near the rim. The first circuli are straight or convex, the central part of which bulges outwards, and both types were often present in the same scale.

Scales could be categorized according to shape, granulation and denticle characteristics.

A. Shape

Five main shapes of scale were recognized:

(a)   round or hexagonal with focus in the centre: *Chilotilapia rhoadesi* and *Rhamphochromis* sp. (Figure 2a);

(a)   (b)

(c)   (d)

(e)   (f)

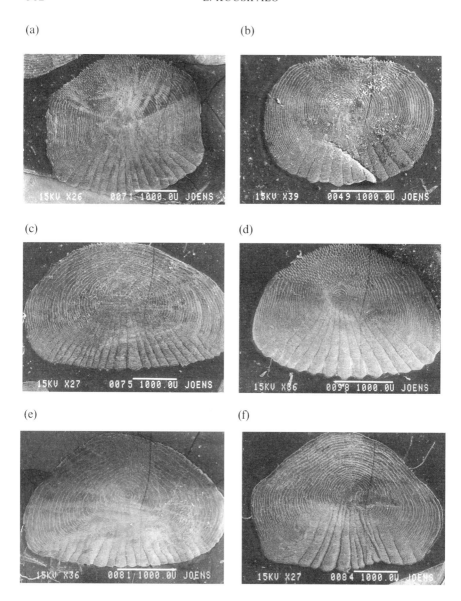

Fig. 2. (a) Round or hexagonal scale with focus in the centre: *Rhamphochromis* sp. (b) Flattened elliptical scale with focus on the caudal side: *Mylochromis anaphyrmus*. Spines cover 90° of the caudal end. (c) Elliptical scale with focus on the caudal side: *Lethrinops altus*. (d) Rounded pentagonal scale with focus on the caudal side: *Pseudotropheus elegans*. Spiny granulation extends 148° and covers almost the entire exposed part of the scale. (e) Spines form a narrow 45° sector on the scale: *Otopharynx argyrosoma*. (f) The caudal field has regular circuli without spines: *Ctenopharynx intermedius*. Reproduced with permission of the Fisheries Society of the British Isles.

(b)   flattened elliptical with focus on the caudal side: *Platygnathochromis melanonotus, Chilotilapia rhoadesii, Nimbochromis livingstonii, Pseudotropheus elegans* and *Mylochromis anaphyrmus* (Figure 2b);

(c)   elliptical with focus on the caudal side: *Lethrinops altus* (Figure 2c), *Hemitaeniochromis urotaenia, Nimbochromis livingstonii, Aristochromis christyi, Aulonocara* "orange" and *Copadichromis virginalis*;

(d)   rounded pentagonal with focus on the caudal side: *Pseudotropheus elegans* (Figure 2d), *Sciaenochromis spilostichus, Alticorpus geoffreyi, Buccochromis rhoadesii, Otopharynx argyrosoma, Placidochromis subocularis, Ctenopharynx intermedius, Hemitaeniochromis urotaenia, Corematodus shiranus, Platygnathochromis melanonotus* and *Copadichromis pleurostigmoides*;

(e)   pentagonal with focus in the centre: *Cynotilapia axelrodi, Aulonocara* "orange" and *Copadichromis virginalis*.

Despite these differences, however, the shape of the scale did not seem to be a taxonomically reliable feature, since in many cases two shapes were present in one species.

B. Granulation

Different amounts of granulation in the form of clear spines were evident. According to the width of the spiny sector, five different types were distinguishable:

(a)   spiny granulation almost totally covers the exposed part of the scale. Typical species: *Pseudotropheus elegans* (Figure 2d), *Corematodus shiranus, Placidochromis subocularis, Sciaenochromis spilostichus* and *Rhamphochromis* sp.;

(b)   spines cover about 90° of the caudal end. Typical species: *Mylochromis anaphyrmus* (Figure 2b), *Copadichromis pleurostigmoides* and *Lethrinops altus* (no sector, spines only on the rim, Figure 2c);

(c)   spines form a 60° sector. Typical species: *Cynotilapia axelrodi*;

(d)   spines form a narrow sector of less than 45°. Typical species: *Otopharynx argyrosoma* (Figure 2e), *Platygnathochromis melanonotus, Chilotilapia rhoadesii, Nimbochromis livingstonii, Aristochromis christyi, Alticorpus geoffreyi* and *Aulonocara* "orange";

(e)   the caudal field bears regular circuli without spines. Typical species: *Ctenopharynx intermedius* (Figure 2f), *Diplotaxodon argenteus, Buccochromis rhoadesii, Hemitaeniochromis urotaenia* and *Copadichromis virginalis*.

C. Denticle Characteristics

Three main types of denticles on the interradial circuli were evident on the examined scales:

(a)   small and sharp: *Pseudotropheus zebra, Lethrinops gossei, Hemitaeniochromis urotaenia, Cynotilapia axelrodi, Pseudotropheus elegans, Copadichromis virginalis* and *Placidochromis subocularis*;

(b)   low and blunt: *Chilotilapia rhoadesi, Copadichromis pleurostigmoides, Lethrinops altus, Aulonocara* "orange", *Alticorpus geoffreyi* and *Platygnathochromis melanotus*;

(c)   firm and widely spaced: *Sciaenochromis spilostichus, Buccochromis rhoadesi, Mylochromis anaphyrmus, Diplotaxodon leucogaster, Ctenopharynx intermedius, Aristochromis christyi, Rhamphochromis* sp., *Nimbochromis livingstonii* and *Corematodus shiranus*.

However, although these three types could be distinguished, the presence of forms intermediate between the main types meant that categorizing of some samples was sometimes ambiguous. A further difficulty was encountered in samples of *Ctenopharynx, Otopharynx* and *Taeniochromis*, in which the denticles were irregularly spaced or missing. Because of these problems, data on denticle characteristics were not included when constructing the dendrogram. The spines on the caudal field showed a growing series from faint to strong: *Copadichromis virginalis*, to *Aristochromis christyi, Corematodus shiranus, Rhamphochromis* sp. and *Mylochromis anaphyrmus*. The width of the spiny sector and the proportional distance of the focus from the caudal rim had the highest eigenvalues in the PCA.

The samples taken from *Lethrinops gossei, Pseudotropheus zebra* and *Taeniochromis holotaenia* contained only scales with a large and unclear focus, indicating regeneration to have taken place, and they were therefore omitted from further studies. The effect of scale regeneration was also detected in the scales of *Pseudotropheus elegans* and *Sciaenochromis spilostichus* (Figure 3f). The data matrix used for the PCA is given in the Appendix and the results are illustrated in Figure 4. The principal component analysis evaluates the detected variation by emphasizing the measured continuous variables – shape of the scale, distance of the focus and width of the spined sector – in factor 1, and brings together the same groups that are found in the phenogram based on the similarity of scale morphology (Figure 5), utilizing the same variables. In factor 2 the nominal variable, type of denticles, has the largest weight (Kuusipalo, 1998).

(a) (b)

(c) (d)

(e) (f)

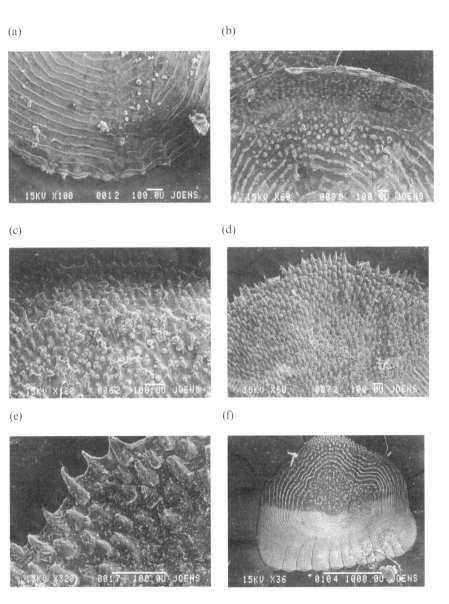

Fig. 3. (a) Faint granules on the caudal field of *Copadichromis virginalis*. (b) Granules and faint spines on the caudal field of *Aristochromis christyi*. (c) Small spines on the caudal field of *Corematodus shiranus*. (d) Spines on the caudal field of *Rhamphochromis* sp. (e) Strong spines on the caudal field of *Mylochromis anaphyrmus*. (f) Regenerated scale of *Sciaenochromis spilostichus* with wide rings, an unclear focus and short radii. Reproduced with permission of the Fisheries Society of the British Isles.

Principal component analysis

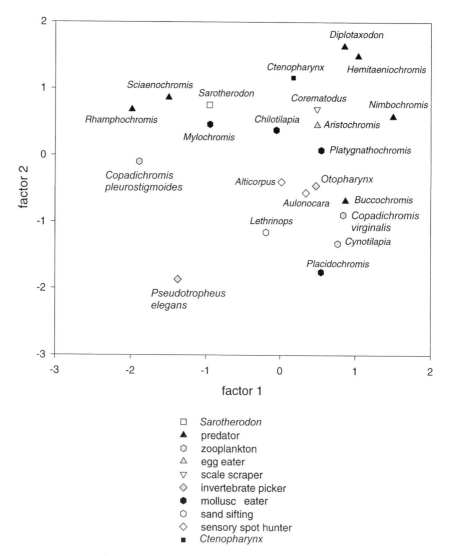

Fig. 4. Result of principal component analysis based on the variables of scale morphology of endemic cichlid genera of Lake Malawi.

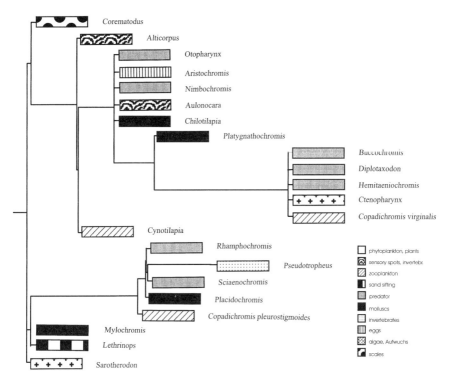

Fig. 5. Dendogram of endemic cichlid genera of Lake Malawi, based on the scale morphology.

V. PHYLOGENETIC UTILITY OF SCALE CHARACTERS

Taxonomic and phyletic studies of fishes have long been based on morphological features, and this approach has been widely used in taxonomical and evolutionary studies of Malawian cichlids (Fryer and Iles, 1972; Ribbink *et al.*, 1983; Greenwood, 1984; Eccles and Trewavas, 1989; Owen *et al.*, 1990) and also for fishery (Allison *et al.*, 1995) and aquarium interests (Konings, 1989). It is only recently that squamation patterns and morphology of scales have been used to study phyletic problems among the lacustrine haplochromines of East Africa (e.g. Lippitsch, 1993; Seehausen *et al.*, 1998).

The present study compared the scale morphology of about 40% of the endemic genera of Lake Malawi, including both benthic and pelagic forms. The surface morphology of sectioned scales is independent of environmental factors and can be distinguished even in aquarium stocks (Sire, 1986; Lippitsch, 1989, 1990). Scale types are usually constant over the whole body, and variation in scale surface morphology is restricted to the number of radii, the closeness of the circuli and the extent of the granulated sector (Lippitsch, 1990, 1992). To avoid the problems associated with ontogenetic-related patterns of juveniles, only adult individuals were sampled: in the cichlid fish *Hemichromis bimaculatus* scale morphology is fully developed by the age of 1 year at a body size of 35 mm (Sire, 1986), and all sampled individuals had reached that size. Lippitsch (1992) reported no clear correlation between the extent of the granular sector and size of the specimen or size of the scale, and also noted that the variation in surface morphology is restricted to the number of radii, closeness of the circuli and extent of the granulated sector. The granulation of exposed scales in Malawian cichlids consists of discrete, well-circumscribed entities and never forms a reticular or crust-like structure. The type of granulation seems to be a phylogenetically useful character.

The pattern described by Lippitsch (1990) for Malawian *Melanochromis auratus* (Boulenger), *Pseudotropheus* and *Aulonocara* was in this study closest to the pattern of *Aristochromis christyi*: granulation is formed by the disintegration of circuli and the entire exposed area is granular; near the centre the granulation consists of round tubercles and near the caudal edge a strip of oblong tubercles is present, with their long axes directed radially and hence at right angles to the circuli. At the rim these tubercles form short, blunt ctenii. In most of the studied Malawian species with an extended spiny sector granulation extends over nearly the whole exposed part and consists entirely of regularly arranged caudally directed ctenii (Kuusipalo, 1998). This resembles the lamprologine species described by Lippitsch (1990), but except for this character, the scales of these groups differ greatly, for example, in the weak ossification of the caudal rim of the scales of lamprologines.

The shape of the first interradial circuli of interradial tongues is a variable character among tilapiine species (Lippitsch, 1990), but its utility for studies on Malawian cichlids remains unclear. The same is true for the different scale shapes and the proportional size of the spines on the caudal field. Minor denticles in the interradial circuli are difficult to classify owing to a gradual shift from one type to another, and are also fairly sensitive to the cleaning methods used on the scale.

Large central and wide rings are features typical of regenerated scales (Sire, 1986), and often occurred in combination with short radii and with spines only on the rim or in a narrow sector. The amount of regenerated scales in wild fish was larger than expected, and in three species the samples taken contained only regenerated scales and they had to be omitted from this study. It is not known whether the original cause of scale loss is related to scale-eating cichlids.

Both ctenoid and derived cycloid scales have evolved several times within the Cichlidae (Lippitsch, 1990). The comparison between species in this study is based on scales removed from a narrow dorsal area surrounding the upper lateral line. For one Malawian cichlid (*Ptyochromis sauvagei*, Pfeffer) Lippitsch (1993) reported that cycloid scales are present above the lateral line and ctenoids below it. Although this was not observed in the present study, the limited sampling regime adopted here may be responsible for only cycloid scales having been recorded in some species.

In the present study, in those species possessing only cycloid scales, all of the scales examined showed no spines, radii were present and the circuli ran without interruption around the caudal end. Interradial denticles were also evident. These species were highly similar in PCA by factor 1 (Figure 4.) and clustered in their own group in the dendrogram (Figure 5).

VI. NICHE OCCUPANCY AND TAXONOMIC RELATEDNESS

The selectively neutral character of variation in scale morphology enables trophic and other niche relationships among species to be explored at a more meaningful level than is possible using most other morphological features. Although this grouping (Figure 5) is based on only several scale characters, the groups of species with cycloid scales accords well with Wright's (1932) concept of speciation. This theory posits that a species is adapted to its environment, and during the process of speciation it is divided so that the new form becomes adapted to somehow different conditions such that neither form faces stronger selection during the process than before separation (Wright, 1932). Such slight differences in resource utilization can lead to divergence under intraspecific competition (Maynard Smith, 1966). Among these species with cycloid scales *Buccochromis, Hemitaeniochromis* and *Copadichromis virginalis* were on the same main branch as *Chilotilapia, Nimbochromis, Aristochromis* and *Otophar-ynx*. Most of these fishes are predators, but each uses a different method to obtain its prey: *Nimbochromis* is an ambush hunter which uses coloration to mimic a rotten fish, *Otopharynx* is a deep-water hunter, *Aristochromis* waits in ambush for young fish, *Hemitaeniochromis* actively hunts them and *Buccochromis* is a large predator in semipelagic waters.

This grouping of species based on presumably neutral characters of scale morphology (Figure 5) can be meaningfully compared with a grouping derived from the similarly neutral structural variation present in metabolic enzymes. Because the genetic control of these enzymes is known, their similarity implies true relationships. The genetic variation among colourful rock-dwelling mbuna is low (Kornfield, 1978, 1984; McKaye *et al.*, 1982, 1984), but the inclusion of benthic and semipelagic species increases genetic variation and facilitates the estimation of the relationships based on moderate variation (Kuusipalo, 1994). The information contained in the phenogram based on enzymatic similarity can be increased by including data on the trophic niches of each species. This is shown in Figure 6, in which it is apparent that taxa which occupy a similar niche are often close relatives: clusters of predators, invertebrate hunters using sensory spots, and algae eaters are evident. However, although the trophic niche is highly similar, component species within each group partition that trophic resource. For example, *Pseudotropheus* algae eaters divide the resource

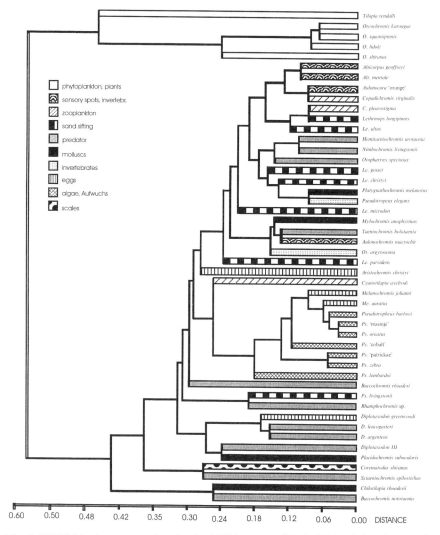

Fig. 6. UPGMA phenogram of endemic cichlid genera of Lake Malawi, based on Nei's genetic distances revealed by enzyme electrophoresis (Kuusipalo, 1994).

by utilizing it in a different way: *P. zebra* scrapes off the algae from the rock, *P. lombardoi* combs the algae cover and *P.* "cobalt" picks up loose algae. *Pseudotropheus* species also display interspecific differences in the size, inclination and depth of the rock substrate that they inhabit.

This fine-scale partitioning of both food and habitat can also be seen in phyletic studies utilizing the neutral variation in mitochondrial DNA (mtDNA) (Moran *et al.*, 1994). In the strict consensus cladogram created using Wagner parsimony and based on the restriction fragment length polymorphism (RFLP) variation in the mtDNA, taxa which utilize the algal aufwuchs on the rocks have divided this niche into smaller fractions (depth, large/small rocks, vertical/horizontal/under rock surface, combining/biting/scraping algae) and speciated accordingly (*Cyathochromis, Labeotropheus, Iodotropheus, Petrotilapia, Pseudotropheus, Labidochromis*) (Figure 7) (Moran *et al.*, 1994). Some mollusc eaters also show bathymetric partitioning and have become specialized to live in deep (*Mylochromis anaphyrmus*) or shallow (*Trematocranus placodon*) water. In the mtDNA tree (Figure 7) the species with

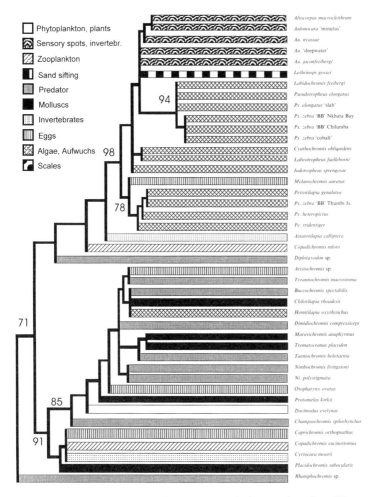

Fig. 7. Niche distribution in a strict consensus cladogram based on Wagner parsimony of the RFLP of mtDNA (modified from Moran *et al.*, 1994).

cycloid scales (*Buccochromis, Hemitaeniochromis* and *Copadichromis*) end up on the same branch together with *Chilotilapia, Nimbochromis, Aristochromis* and *Otopharynx*, as was seen in scale dendrogram (Figure 5.). Although mtDNA classification placed *Diplotaxodon* sp. into another branch (Moran *et al.*, 1994), *Copadichromis* and *Placidochromis* are fairly close in both the scale dendrogram and the mtDNA tree. Overall, there are many strong similarities between these two trees, but differences also exist. Although the number of characters used in the scale study was limited, it seems clear that the evolution of scale morphology and mitochondrial DNA is barely uniform.

The pattern of niche partitioning visible in these three dendrograms (Figures 5–7) has also been documented in other studies (Kornfield and Parker, 1997; Sültmann and Mayer, 1997). In all three dendrograms similar niches occur together much more frequently than expected by chance; however, more than once a contradictory feature is present: species with similar niches are not close relatives. Where distant or unrelated species occupy a similar niche, it suggests that one or other of them has arisen by a switch to a totally new niche, and not as a slightly modified version of a taxon formerly using that niche. The results of the scale morphology, electrophoresis and mtDNA studies suggest strongly that the basic niche types have evolved many times during cichlid evolution in Lake Malawi.

How has evolution proceeded within the cichlids of Lake Malawi? Allopatric scenarios have long been highly influential in interpretations of the speciose cichlid assemblages of the African Great Lakes, and regarding Lake Malawi, microallopatric speciation within discrete habitat patches has been the favoured hypothesis (e.g. Fryer and Iles, 1972; Ribbink *et al.*, 1983). However, more recently, sympatric models have begun to be considered possible (e.g. Turner; Seehausen, this volume). Sympatric speciation involves the evolution of specialization to novel or underused resources, i.e. the invasion of empty niches (Maynard Smith, 1966). The strongest support for such a mechanism has come from studies in the small (4.15 km^2) Cameroonian crater lake Barombi Mbo, which have suggested that the endemic cichlid species found there have evolved sympatrically, first through a basic division of niches, which then, with slight anatomical and habitat modifications in tandem with niche partitioning, led to several endemic species (Schliewen *et al.*, 1994).

When shifting to a fundamentally new niche, such as using a new food resource or habitat, even phenotypes that are poorly adapted might have relatively high fitness if the resource is abundant and underexploited (Maynard Smith, 1966). However, if the community is already diverse and niche space is relatively saturated, poorly adapted phenotypes would be greatly disadvantaged and instead competition would favour species already possessing suitable adaptations.

VII. A SCENARIO FOR SYMPATRIC SPECIATION

The presence of unrelated species within the same niche has already been noted, and interpreted as evidence of one of these taxa having switched to a new niche. If the observed niche switches among the Lake Malawi cichlids are indeed the outcomes of empty niches and released competition, what might have caused these niches to become vacant?

Evidence for mass mortality of fishes has been recovered in three lake-bottom sediment cores dating to 10 740 years ago (Pilskaln and Johnson, 1991). In the central area of the lake, a core with a distinct layer of fish vertebrae dated at 1700 years old was found and a second core 60 km away had a bone layer dated at 1200 years old. A third core with a layer of fish vertebrae dated at 9500 years old was found 150 km away in the middle of the southern end of the lake (Pilskaln and Johnson, 1991). The distinctness of this layer in these three cores indicates that a true mass mortality event occurred at these times. Similarly, that this layer of identical age is present over such a wide area suggests that its effects were truly widespread. What might have caused these?

Although Lake Malawi experiences strong seasonal variations in wind, temperature and rainfall (Eccles, 1974), it is permanently stratified by temperature and salinity, and the waters below c. 200 m depth are anoxic (Eccles, 1974). However, the basin is sensitive to changes in rainfall, since the mean annual evaporation loss is 1.6 m (Patterson and Kachinjika, 1993) and a decrease in lake level of only 5 m cuts off the outflow through the Shire River (Owen et al., 1990). Major drops in water level, to 100 m lower than at present, have occurred 200, 700, 10 740 and 25 000 years ago (Owen et al., 1990). In a closed basin, a period with high evaporation and low rainfall results in increased salinity of the surface waters which, in turn, decreases the differences in density and weight of the layers in the water mass. Under such conditions, the strong seasonal winds might then be capable of inducing an overturn, such as that reported in the permanently stratified Dead Sea in 1979 (Steinhorn, 1985). In Malawi, the rise of the anoxic deeper layers to near the surface would cause large fish kills of sufficient magnitude to account for the widespread distribution of the vertebrae layer.

If this did take place, then local fish kills would have resulted in fairly large areas with abundant and underexploited resources. The removal of many species would have made any niches partially or wholly vacant, and in these circumstances, shifting to a fundamentally new niche, like using a new food resource, would be possible for phenotypes that were poorly adapted for them. Geographically, the areas for this niche availability do not have to be large, as the Barombi Mbo crater lake shows (Schliewen et al., 1994).

Once in a new niche space, cichlids show many characteristics which would facilitate their rapid speciation. The bauplan of cichlid jaws enables the processing of novel food items after only small evolutionary changes (Galis and Drucker, 1996), and such slight differences in resource utilization can lead to divergence under intraspecific competition (Maynard Smith, 1966). Hence, adaptive changes can appear relatively quickly. Such differences require polymorphism and the ancestral cichlid population was indeed genetically variable, at least for their MCH genes (Klein *et al.*, 1993). Cichlids are early maturing (Lowe-McConnell, 1975) and this short generation time would allow rapid spreading of any anatomical adaptations throughout the population. Differences in male coloration can also develop quickly (Owen *et al.*, 1990), allowing for efficient sexual selection (Hert, 1991; McKaye *et al.*, 1984; Turner; Seehausen, this volume) and for new colour forms to establish stable colonies without hybridization (Lewis *et al.*, 1986).

The highly speciose haplochromine cichlid assemblage of Lake Malawi, with its diverse variety of niches scattered among close relatives, both fascinates and challenges ecologists and evolutionists. It has been posited that drought and lowering water levels have decreased the area of Lake Malawi and led to strong selection pressure on cichlid fishes (Owen *et al.*, 1990), but these same conditions may also have resulted in local overturns and fish kills, which make available areas with released competition (Pilskaln and Johnson, 1991). Such disturbance and historical heterogeneity of the environment may act to maintain and promote species richness in Lake Malawi, as it has done in the vegetation of Peruvian Amazon (Salo *et al.*, 1986). Sympatric speciation and interspecific competition in newly inhabited areas after local extinctions following overturns during extreme low lake stands may partly explain the cichlid species richness of Lake Malawi and also account for the niche switches demonstrated in electrophoretic, mtDNA and scale morphology phylogenetic studies of these fishes.

ACKNOWLEDGEMENTS

The author thanks G.F. Turner for identifying the species, H. Hyvärinen for guidance in the techniques, N.F. Broccoli for enthusiastic discussions and G. Delince and A. Rossiter for support. The Government of Malawi is thanked for their permission to publish the results.

REFERENCES

Allison, E.H., Ngatunga, B.P. and Thompson, A.B. (1995). Identification of the pelagic fish. In: *The Fishery Potential and Productivity of the Pelagic Zone of Lake Malawi/Niassa* (Ed. by A. Menz), pp. 159–178. Natural Resources Institute, Chatham, UK.

Eccles, D.H. (1974). An outline of the physical limnology of Lake Malawi (Lake Nyasa). *Limnol. Oceanogr.* **19**, 730–742.

Eccles, D.H. and Trewavas, E. (1989). *Malawian Cichlid Fishes: The Classification of some Haplochromine Genera.* Lake Fish Movies, Herten, Germany.

Felsenstein, J. (1993). *PHYLIP (Phylogeny Inference Package)*, Version 3.5c. University of Washington, USA.

Fryer, G. and Iles, T.D. (1972). *The Cichlid Fishes of the Great Lakes of Africa*. Oliver and Boyd, Edinburgh.

Galis, F. and Drucker, E.G. (1996). Pharyngeal biting mechanics in centrarchid and cichlid fishes: insights into a key evolutionary innovation. *J. Evol. Biol.* **9**, 641–670.

Greenwood, P.H. (1984). African cichlids and evolutionary theories. In: *Evolution of Fish Species Flocks* (Ed. by A.A. Echelle and I. Kornfield), pp. 141–154. University of Maine at Orono Press, Orono, ME.

Hert, E. (1991). Female choice based on egg-spots in *Pseudotropheus aurora* Burgess 1976, a rock-dwelling cichlid of Lake Malawi, Africa. *J. Fish Biol.* **38**, 951–953.

Klein, D., Ono, H., O'hUigin, C., Vincek, V., Goldschmidt, T. and Klein, J. (1993). Extensive MCH variability in cichlid fishes of Lake Malawi. *Nature* **364**, 330–334.

Konings, A. (1989). *Malawian Cichlids in Their Natural Habitat*. Verduijn Cichlids and Lake Fish Movies, Raket B.V., Pijnacker, The Netherlands.

Kornfield, I.L. (1978). Evidence for rapid speciation in African cichlid fishes. *Experientia* **34**, 335–336.

Kornfield, I. (1984). Descriptive genetics of cichlid fishes. In: *Evolutionary Genetics of Fishes* (Ed. by B.J. Turner). Plenum Publishing Corporation, New York.

Kornfield, I. and Parker, A. (1997). Molecular systematics of a rapidly evolving species flock: the mbuna of Lake Malawi and the search for a phylogenetic signal. In: *Molecular Systematics of Fishes* (Ed. by T.D. Kocher and C.A. Stepien), pp. 25–38. Academic Press, London.

Kuusipalo, L. (1994). Speciation of cichlids in Lake Malawi, East-Africa. Unpublished Licentiate Thesis, University of Joensuu, Finland.

Kuusipalo, L. (1998). Scale morphology in Malawian cichlids. *J. Fish Biol.* **52**, 771–781.

Lewis, D., Reinthal, P. and Trendall, J. (1986). *A Guide to the Fishes of Lake Malawi National Park*. WWF, Gland, Switzerland.

Lippitsch, E. (1989). Scale surface morphology in African cichlids (Pisces, Perciformes). *Ann. Mus. R. Afr. Cent. Sci. Zool.* **257**, 105–108.

Lippitsch, E. (1990). Scale morphology and squamation patterns in cichlids (Teleostei, Perciformes): a comparative study. *J. Fish Biol.* **37**, 265–291.

Lippitsch, E. (1992). Squamation and scale character stability in cichlids, examined in *Sarotherodon galilaeus* (Linnaeus, 1758) (Perciformes, Cichlidae). *J. Fish Biol.* **41**, 355–362.

Lippitsch, E. (1993). A phyletic study on lacustrine haplochromine fishes (Perciformes, Cichlidae) of East Africa, based on scale and squamation characters. *J. Fish Biol.* **42**, 903–946.

Lowe-McConnell, R.H. (1975). *Fish Communities in Tropical Freshwaters*. Longman, London.

McKaye, E.R., Kocher, T., Reinthal, P. and Kornfield, I. (1982). A sympatric sibling species complex of *Petrotilapia* (Trewavas) from Lake Malawi analysed by enzyme electrophoresis (Pisces:Cichlidae). *Zool. J. Linn. Soc. Lond.* **76**, 91–96.

McKaye, K.R., Kocher, T., Reinthal, P., Harrison, R. and Kornfield, I. (1984). Genetic evidence for allopatric and sympatric differentiation among color morphs of a Lake Malawi cichlid fish. *Evolution* **38**, 215–219.

Maynard Smith, J. (1966). Sympatric speciation. *Am. Nat.* **100**, 637–650.

Moran, P., Kornfield, I. and Reinthal, P.N. (1994). Molecular systematics and radiation of the haplochromine cichlids (Teleostei:Perciformes) of Lake Malawi. *Copeia* **1994**, 274–288.

Owen, R.B., Crossley, R., Johnson, T.C., Tweddle, D., Kornfield, I., Davidson, S., Eccles, D.H. and Engstrom, D.E. (1990). Major low levels of Lake Malawi and their implications for speciation rates in cichlid fishes. *Proc. R. Soc. Lond. B* **240**, 519–553.

Patterson, G. and Kachinjika, O. (1993). Effect of wind-induced mixing on the vertical distribution of nutrients and phytoplankton in Lake Malawi. *Verh. Int. Verein. Limnol.* **25**, 872–876.

Pilskaln, C.H. and Johnson, T.C. (1991). Seasonal signals in Lake Malawi sediments. *Limnol. Oceanogr.* **36**, 544–557.

Ribbink, A.J., Marsh, B.A., Marsh, A.C., Ribbink, A.C. and Sharp, B.J. (1983). A preliminary survey of the cichlid fishes of rocky habitats in Lake Malawi. *S. Afr. J. Zool.* **18**, 149–310.

Salo, J., Kalliola, R., Häkkinen, I., Mäkinen, Y., Niemelä, P., Puhakka, M. and Coley, P.D. (1986). River dynamics and the diversity of Amazon lowland forest. *Nature* **322**, 254–258.

Schliewen, U.K., Tautz, D. and Pbo, S. (1994). Sympatric speciation suggested by monophyly of crater lake cichlids. *Nature* **368**, 629–632.

Seehausen, O., Lippitsch, E., Bouton, N. and Zwennes, H. (1998). Mpibi, the rock-dwelling cichlids of Lake Victoria. Descriptions of three new genera and fifteen new species (Teleostei). *Ichthyol. Explor. Freshwat.* **9**, 129–228.

Sire, J.-Y. (1986). Ontogenic development of surface ornamentation in the scales of *Hemichromis bimaculatus* (Cichlidae). *J. Fish Biol.* **28**, 713–724.

Steinhorn, I. (1985). The disappearance of the long term meromictic stratification of the Dead Sea. *Limnol. Oceanogr.* **30**, 451–472.

Sültmann, H. and Mayer, W.E. (1997). Reconstruction of cichlid phylogeny using nuclear DNA markers. In: *Molecular Systematics of Fishes* (Ed. by T.D. Kocher and C.A. Stepien), pp. 39–51. Academic Press, London.

Wright, S. (1932). The roles of mutation, inbreeding, crossbreeding and selection in evolution. *Proc. Sixth Int. Congr. Genet. Ithaca, NY* **1**, 356–366.

Appendix
Primary data used for principal component analysis (results are shown in Figure 4)

Taxon	Shape	Focus	Angle	Denticle	FAC1	FAC2
Sarotherodon galilaeus	1.309	0.422	85	3	−0.95	0.75
Rhamphochromis sp.	1.191	0.381	125	3	−1.99	0.68
Copadichromis pleurostigmoides	1.122	0.390	116	2	−1.89	−0.10
Chilotilapia rhoadesii	1.214	0.500	45	2	−0.06	0.38
Nimbochromis livingstonii	1.625	0.575	45	3	1.50	0.58
Copadichromis virginalis	1.425	0.475	0	1	0.84	−0.89
Pseudotropheus elegans	1.357	0.405	148	1	−1.37	−1.87
Diplotaxodon argenteus	1.286	0.548	0	3	0.85	1.63
Mylochromis anaphyramus	1.400	0.400	90	3	−0.95	0.46
Lethrinops altus	1.650	0.400	90	2	−0.19	−1.16
Hemitaeniochromis urotaenia	1.375	0.500	0	1	0.87	−0.68
Aristochromis christyi	1.585	0.463	45	3	0.49	0.45
Aulonocara "orange"	1.561	0.439	45	2	0.34	−0.57
Sciaenochromis spilostichus	1.167	0.452	128	3	−1.50	0.86
Alticorpus "geoffreyi"	1.455	0.455	63	2	0.01	−0.40
Buccochromis rhoadesii	1.350	0.550	0	3	1.03	1.48
Otopharynx argyrosoma	1.537	0.463	45	2	0.47	−0.46
Placidochromis subocularis	1.512	0.561	123	1	0.54	−1.76
Ctenopharynx intermedius	1.386	0.432	0	3	0.17	1.16
Corematodus shiranus	1.452	0.548	80	3	0.49	0.70
Platygnathochromis melanotus	1.386	0.500	29	2	0.54	0.08
Cynotilapia axelrodi	1.477	0.523	60	1	0.77	−1.33

Ecological Interactions Among an Orestiid (Pisces: Cyprinodontidae) Species Flock in the Littoral Zone of Lake Titicaca

T.G. NORTHCOTE

I. SUMMARY

The ecology of seven species of *Orestias* cohabiting the littoral zone of Puno Bay, Lake Titicaca, is reviewed from published works and from unpublished material to elucidate the mechanisms facilitating their millennial coexistence. In addition to marked differences in relative abundance of the species in the littoral zone, there are large temporal (diel, seasonal), spatial (locality to locality, patch to patch) and ontogenetic (size) differences in its use among the species. They are also shown to span differentially a wide range in trophogastric morphology (gill raker number, spacing, length; pharyngeal dentition; alimentary canal length) which in part must relate to demonstrated differences in feeding: one species being exclusively zooplanktivorous, another using a broad coverage of food types and probably feeding processes including phytophagy, zooplanktivory and benthophagy, and still others being largely benthophagous. All four of the mechanisms previously proposed to account for long-term coexistence of closely related species flocks – community/habitat instability, marked environmental patchiness, diel (or other forms of temporal)

ADVANCES IN ECOLOGICAL RESEARCH VOL. 31
ISBN 0–12–013931–6

separation, and trophic/spatial partitioning – are shown to apply in the littoral zone of Puno Bay, and probably also in other parts of Lake Titicaca.

II.   INTRODUCTION

At least since the time of Darwin there has been serious interest by ecologists in competitive interactions among small groups of cohabiting and closely related species (see Hardin, 1960; Northcote, 1995, for reviews). Ancient lakes have been centres of evolution, and most contain highly characteristic faunas with many endemic, and often closely related, species, and thus present ideal "natural laboratories" in which to study this topic. Ecological interactions have been examined in species flocks of several ancient lake fishes, most notably in cichlids of the African Great Lakes (Fryer, 1959; Fryer and Iles, 1972; Rossiter, 1995; Kawanabe *et al.*, 1997). However, surprisingly little work has been specifically directed to this problem for the species flock of cyprinodont fishes in Lake Titicaca, despite almost a century of wide-ranging biological studies there (see Parenti, 1984a, b, for use of the term "species flock" for Lake Titicaca cyprinodonts).

   Through review of some of the Titicaca fish studies in which I was directly involved (Northcote, 1979, 1980; Treviño *et al.*, 1984; Northcote *et al.*, 1989), along with information from those of several other workers (especially Lauzanne, 1982, 1992; Loubens, 1989; Loubens *et al.*, 1984; Vaux *et al.*, 1988), and with the inclusion of some very recently analysed data, this review draws together a focus on orestiid ecological interactions within the littoral zone of this large ancient lake. This chapter will deal mainly with the Puno Bay portion of the lake, but will also refer to littoral studies elsewhere, especially in Lago Pequeño, most of which is in the littoral zone.

III.   LITTORAL LAKE TITICACA

Although Lake Titicaca is steep-sided and deep (maximum depth 284 m) in many parts of its main basin (Lago Grande), even there some large shallow areas occur (Figure 1). In contrast, its north-western and south-eastern basins (Puno Bay and Lago Pequeño, respectively) are highly littoral, if that zone is defined as extending down to the maximum depth of submersed or adnate macrophytes (Wetzel, 1983). Macrophytes (e.g. the bryophyte *Sciaromium*) are found at depths up to 29 m in Lago Grande (Tutin, 1940) but only about 18% of its total surface area has macrophytes (Table 1). Over half of Lago Pequeño is within the littoral zone, as is nearly 80% of Puno Bay, giving an overall littoral surface area approaching 30%. In some regions of the latter basin high turbidity, in part associated with cultural eutrophication, may be restricting the depth to which macrophytes flourish (Levieil *et al.*, 1989).

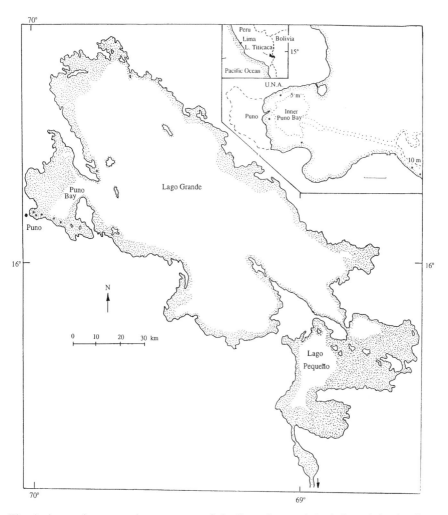

Fig. 1. Approximate maximum extent of the littoral zone (stippled area) in the three main basins of Lake Titicaca (Lago Grande, Lago Pequeño and Puno Bay); see Table 1 for sources. Uppermost inset shows general location of Lake Titicaca; main inset shows inner Puno Bay (scale bar = 1 km). Asterisks show main littoral sampling stations (three in inner Puno Bay, two in outer Puno Bay and one in Chiflon Bay about 5 km north of Puno Bay mouth).

The above view of the depth and areal extent of the Titicaca littoral zone is much broader than the restricted definition adopted by Lauzanne (1992), limiting it to a narrow belt with a maximum depth of only 1 m inside the totora (*Schoenoplectus tatora*) macrophyte zone.

Table 1
Major littoral areas[a] of Lake Titicaca basins and their approximate features

Lake basin	Total surface area (km²)	Surface area with macrophytes		Maximum depth of macrophytes (m)	Turbidity restriction of macrophytes
		km²	%		
Lago Grande	6542	1204	18	30	Slight
Lago Pequeño	1365	758	56	12	Moderate
Puno Bay	602	476	79	9	Severe[b]
Totals	8509	2438	29		

Adapted from data in Northcote (1979), Morales et al. (1984), Wirrman (1992), Iltis and Mourguiart (1992) and Dejoux (1994).
[a]Values given in references vary slightly but do not seriously affect overall percentages.
[b]Especially in inner Puno Bay.

IV. LITTORAL CYPRINODONTS OF LAKE TITICACA

Following the systematic update of the genus *Orestias* in Lake Titicaca (Lauzanne, 1992), there would seem to be some 19 species there now, with another five species not having been found there since the Percy Sladen Expedition of 1937 (but see also Parenti, 1984a, b).

From those specimens available for this study, at least four orestiids commonly found in the littoral zone samples from Puno Bay between 1979 and 1981 were positively identified – *O. ispi, O. pentlandi, O. agassii* and *O. luteus* – using the key of Lauzanne (1982) along with information in Parenti (1984a). Another species also found in most littoral collections examined was almost certainly *O. olivaceus*, although its scalation did not completely fit the descriptions given in Lauzanne (1982). Specimens of a sixth species, tentatively identified as *O. mulleri*, were also found, and are included here even though it is normally considered a deep-water species. Another species, *O. albus*, was only rarely recorded in the littoral sampling of Puno Bay (Treviño et al., 1984). The above six species (not including *O. mulleri*) were also found in macrophyte habitats of Lago Pequeño by Loubens et al. (1984), along with another unidentified species of *Orestias*.

V. RELATIVE ABUNDANCE AND DISTRIBUTION OF LITTORAL ORESTIIDS

A. Overall Orestiid Abundance

One of the first attempts to consolidate the disparate information on the use of the littoral zone habitats in Lake Titicaca by the *Orestias* species flock was that by Lauzanne (1992; see especially his Figure 3). He adopted a restrictive definition of the littoral zone (0–1 m depth) and though citing the work of Treviño *et al.* (1984) did not include any of their relevant information. Nevertheless, Lauzanne (1992) recognized that young and subadults of several species of *Orestias* inhabited the upper littoral (0–1 m) and that juvenile stages of *O. agassii, O. luteus* and *O. olivaceus* occurred in the totora zone (2–3 m depth), and suggested that adult *O. luteus* and *O. olivaceus* were found along the bottom of the *Chara* zone (3–10 m) with *O. agassii* in its overlying waters, as well as *O. ispi* during its September April spawning season (see his Table 1).

The occurrence of several members of the *Orestias* species flock in the inner and outer regions of the littoral zone in Puno and Chiflon bays of Lake Titicaca in 1980 has been summarized from data in Northcote (1980). Four cyprinodont species (mainly *O. agassii*) occurred in littoral waters of Puno Bay close to the bottom (Table 2), with night catches being higher than during the day in both the inner and outer regions of the littoral zone. At Chiflon Bay (Table 3), a strong association of the orestiids with the near-bottom waters was clearly evident, while pejerrey (*Basilichthys bonariensis*) were taken mainly from off-bottom areas.

Subsequent to the above preliminary work, a much more detailed examination was made of orestiid abundance and distribution in littoral waters of Puno Bay during the dry and wet seasons of 1981 and 1982. The primary objective of this work was to compare fish abundance in polluted and unpolluted parts of the bay (Treviño *et al.*, 1984, 1989), but the data set can also be used to examine ecological interactions in the orestiid fishes, a focus not previously attempted. In addition to fish captured in a wide range of gillnet mesh sizes (25–101 mm), depths and locations, replicated catches were made at midday and at night by fine-meshed seining over standardized onshore areas

Table 2
Day and night[a] gillnet catches of *Orestias* spp.[b] in the littoral zone of Puno Bay (Ojerani), Lake Titicaca, 9–11 January 1980

Time	Inner littoral (at 1, 2 and 5 m depths)[c]	Outer littoral/sublittoral (at 10 and 20 m depths)[c]
Day	4 (0)	4 (1)
Night	19 (0)	18 (0)
Total	23 (0)	22 (1)

[a]Six hours of each, before and after midday, midnight; see Northcote (1980) for net mesh and depth details.
[b]Mainly *O. agassii* and a few *O. luteus, O. albus* and *O. ispi*; numbers in parentheses give those > 2 m off bottom.
[c]Monofilament nylon nets with 10 m panels of six different mesh sizes, marked off at 1 m depth intervals and covering surface to bottom at each depth contour.

Table 3

Overnight gillnet catches of *Orestias* spp.[a] and Pejerrey[b] in the littoral zone of Chiflon Bay, Lake Titicaca, 16–17 January, 1980

	Inner littoral (at 2 and 5 m depths)[c]		Outer littoral (at 10 and 20 m depths)[c]	
	Orestias spp.	Pejerrey	*Orestias* spp.	Pejerrey
> 2 m off bottom	2	14	4	40
< 2 m off bottom	23	2	42	2

[a]*Orestias agassii, O. ispi, O. luteus, O. albus.*
[b]*Basilichthys bonariensis.*
[c]See Table 1[c].

(see Northcote, 1980; Treviño *et al.*, 1984, for details). In total, over 74 000 fish were taken by gillnetting and over 10 000 by seining (Table 4), most of which (89.1%) were orestiids. Although overall catch was dominated by the large number of *O. ispi* taken by gillnetting in the wet season (their shore spawning period), many were also taken in the dry season when, according to Lauzanne (1992), there should be no spawning.

Clearly, the littoral zone of Puno Bay is heavily used by at least four species of *Orestias* – *O. ispi, O. agassii, O. olivaceus* and *O. luteus* – with some *O. pentlandi* and *O. albus* also present. *Trichomycterus* sp. is present in low

Table 4

Summary of fish species captured in the littoral zone of Puno Bay by gillnetting and seining in the dry (July–August) and wet seasons (January to February)

Species	Gillnetting		Seining		
	Dry	Wet	Dry	Wet	Totals
O. ispi	1843	68 046	371	454	70 714
O. pentlandi	8	74	–	–	82
O. agassii	444	227	624	819	2114
O. olivaceus	192	727	566	565	2050
O. luteus	496	53	328	196	1073
O. albus	1	3	–	–	4
Trichomycterus sp.	6	4	–	–	10
Pejerrey	1460	1016	2702	4043	9221
Rainbow trout	9	18	–	–	27
Totals	4459	70 168	4591	6077	85 295

Data assembled from appendices in Treviño *et al.* (1984).

numbers, as is the introduced rainbow trout (*Oncorhynchus mykiss*). The most abundant introduced species is the pejerrey, which could be an important predator and/or competitor with orestiids in both the dry and wet seasons.

B. Spatiotemporal Differences in Abundance

Highly significant differences were evident in the relative abundance of four orestiid species, both between locations in Puno Bay and between seasons (Table 5a, b). *Orestias ispi* occurred in very low numbers at inner Puno Bay stations, especially in the dry season, but was one of the most abundant orestiids found at outer Puno Bay stations, not only in the wet season but also in the dry season. Conversely, *O. agassii* was relatively abundant at all stations and seasons, and seemed to be the orestiid least affected in abundance by the severely polluted conditions prevailing in many parts of inner Puno Bay. The other three species commonly found in the littoral waters of Puno Bay were never abundant in the inner bay area (Table 5).

More detailed information on nearshore use of the littoral zone by orestiids can be obtained by comparison of the same mesh-size gillnet catches at 1 m, 2 m and 5 m depth zones (see Treviño *et al.*, 1984, for details). Significantly higher catches per unit effort of four species, *O. ispi, O. luteus, O. agassii* and *O. olivaceus*, were obtained at 1 m and 2 m depth zones, indicating the strong onshore orientation of this group during both dry and wet seasons.

Separation of orestiid gillnet catches between the upper and lower halves of the nets set at 2 m and 5 m depths provides some information on possible vertical differences in their distribution (Table 6). Only for *O. agassii* was there a significant difference in vertical distribution, although for *O. luteus* and *O. olivaceus* mean catches were always much higher in the lower half of the nets.

Seasonal changes in orestiid abundance were most clearly reflected in the gillnet catch data, with significantly higher numbers being recorded in the wet season for all species, especially at inner Puno Bay stations but also in the outer bay (Table 5a, b). For most comparisons between day and night seine catches (Table 7), those at night were significantly higher than during the day (except for *O. olivaceus* in both dry and wet seasons, as well as for *O. luteus* in the wet season), suggesting onshore movement and concentration of orestiids at night, although this difference might reflect seining having been more effective during night-time.

VI. LITTORAL ORESTIID SIZE DIFFERENCES

Size characteristics of four orestiids using the littoral zone of Puno Bay reported by Treviño *et al.* (1984) were largely concerned with examination of gillnet selectivity curves between dry and wet seasons at the severely polluted inner bay region. Further information for some six species is now available

Table 5

(a) Mean catches of orestiids in the littoral zone of Puno Bay, Lake Titicaca, 1981–1982

Method	Season	Area[a]	Species			
			O. ispi	O. luteus	O. agassii	O. olivaceus
Gillnet[b]	Dry	A	0.4	2.5	26.7	0.4
		B	306.5	78.5	29.5	31.3
	Wet	A	3.8	3.3	7.3	9.1
		B	11 334.7	5.0	25.7	113.5
Seine[c]	Dry	A	0	1.7	4.9	0.5
		B	4.6	2.9	4.3	6.6
	Wet	A	0.1	1.9	6.1	2.4
		B	5.1	0.6	4.2	4.4

Data assembled from appendices in Treviño et al. (1984).

[a]A, inner Puno Bay stations; B, outer Puno Bay stations; see Figure 1 for locations.

[b]Catch per standard gillnet set.

[c]Catch per 100 m^2 by standard seine hauls.

(b) Mann–Whitney test of significance matrices (p-values)[d]

Gillnetting

(1) Between areas (A, B):

Season	O. ispi	O. luteus	O. agassii	O. olivaceus
Dry	0.0075**	0.0002***	0.1179	0.0002***
Wet	0.0002***	0.3132	0.0005***	0.0002***

(2) Between seasons (dry, wet):

Area	O. ispi	O. luteus	O. agassii	O. olivaceus
A	0.0115*	0.6305	0.0232*	0.0003***
B	0.0649	0.0022**	0.4848	0.0022**

Seining

(1) Between areas (A, B):

Season	O. ispi	O. luteus	O. agassii	O. olivaceus
Dry	–[e]	0.2921	0.2766	0.0001***
Wet	0.0861	0.0506	0.3818	0.0373*

(2) Between seasons (dry, wet):

Area	O. ispi	O. luteus	O. agassii	O. olivaceus
A	–[e]	0.3743	0.4997	0.0033**
B	1.000	0.3195	0.6276	0.1902

[d]Pooled data (six to 14 variates per test) transformed by square root $x + 0.5$; levels of significance for differences: *significant, **very significant, ***extremely significant.

[e]Tests not possible because of zero standard deviation in area A columns (all zero values), but differences considered highly significant.

Table 6

Comparison of mean orestiid catch in upper and lower halves of gillnets set at 2 m and 5 m depths in the littoral zone of Puno Bay during the dry season (July–August) and wet season (January–February)

	Species							
	O. ispi		*O. luteus*		*O. agassii*		*O. olivaceus*	
Season	Upper	Lower	Upper	Lower	Upper	Lower	Upper	Lower
Dry	34.4	195.0	4.4	51.4	3.0	37.0	2.3	20.8
	(0.4418)		(0.1949)		(0.0002)***		(0.3282)	
Wet	4544.0	3961.5	1.9	4.3	6.8	16.8	18.9	60.3
	(0.9591)		(0.1605)		(0.0650)		(0.2786)	

Assembled from data in Appendix 1, Treviño *et al.* (1984); eight replicates in each cell. Mann–Whitney test of significance (p-values) between upper and lower half catches (data transformed by square root $x + 0.5$). Level of significance: ***extremely significant.

(Figure 2), with standard lengths of those most commonly taken ranging from about 35 to 115 mm. *Orestias agassii* spanned almost this entire range, whereas most *O. ispi* ranged from 45 to 65 mm and *O. pentlandi* from 65 to 85 mm. Most *O. olivaceus* ranged between 35 and 75 mm, with *O. luteus* having a broader size range similar to that of *O. agassii*. The few *O. mulleri* captured were all between 37 and 46 mm.

Table 7

Comparison of day and night seine catches of orestiids (mean number per 100 m^2) in Puno Bay littoral zone during the 1981 dry season (July–August) and the 1982 wet season (January–February)

	Species							
	O. ispi		*O. luteus*		*O. agassii*		*O. olivaceus*	
Season	Day	Night	Day	Night	Day	Night	Day	Night
Dry	0.01	5.92	1.25	3.74	2.93	6.46	4.04	4.21
	(0.0249)*		(0.0249)*		(0.0138) *		(0.6277)	
Wet	0.02	5.56	1.03	1.38	3.88	6.09	2.67	4.28
	(0.0018)**		(0.3969)		(0.0428)*		(0.1439)	

Assembled from data in Appendix 3, Treviño *et al.* (1984); 10–13 replicates in each cell. Mann–Whitney test of significance (p-values) between day and night catches (data transformed by square root $x + 0.5$). Levels of significance for differences: *significant, **very significant.

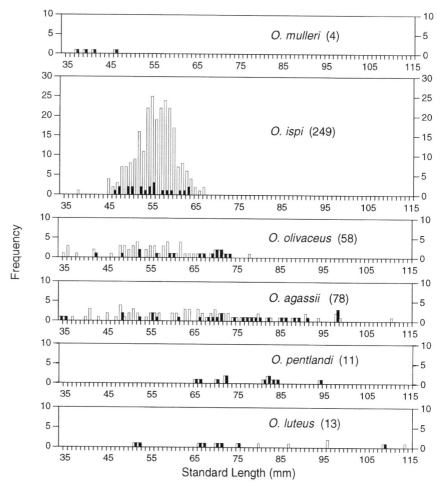

Fig. 2. Length/frequency distributions (open histograms) for the six species of orestiids examined in detail for this study. Black histograms give length frequencies of individuals used for feeding analyses. Numbers in parentheses indicate total number of specimens examined.

Comparison of length/weight regressions for six orestiids (Figure 3) shows at least eight highly significant ($p < 0.01$) differences in regression slope. That for *O. pentlandi* is significantly lower than those for *O. ispi, O. agassii, O. luteus* and *O. olivaceus*, i.e. for a given length, *O. pentlandi* weighs less than the other four species. Likewise, the length/weight regression slope for *O. ispi* is significantly lower than those for *O. agassii* and *O. olivaceus*. However, the slope for *O. olivaceus* is significantly greater than those for *O. agassii* and *O. luteus*.

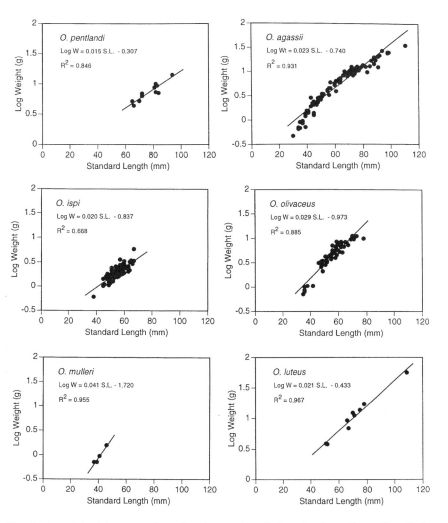

Fig. 3. Length/weight regressions for six species of *Orestias* from Puno Bay, Lake Titicaca. Note the logarithmic (base 10) weight axes.

VII. LITTORAL ORESTIID TROPHOGASTRIC MORPHOLOGY

Although no detailed studies were made of possible mouth size, gape, and outer jaw dentition differences in the littoral orestiids, even superficial examinations revealed that these exist (see also frontal sketches for some species in Parenti 1984a, Figure 8; Muller, 1993, Figure 7). For at least six species there were marked differences in gill raker number (Figure 4), as well as in shape and spacing. *Orestias olivaceus* had the lowest range and mode in gill

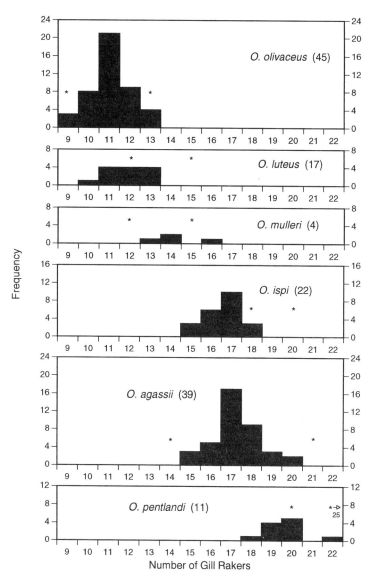

Fig. 4. Gill raker frequency distributions for six species of *Orestias* from Puno Bay, Lake Titicaca. Asterisks show range reported for Lago Pequeño specimens (Lauzanne, 1982).

raker number (9–13), agreeing exactly with that given for this species from Lago Pequeño specimens. *Orestias pentlandi* had by far the highest gill raker number, and also had long, relatively fine gill rakers, some of which were branched near the tip. *Orestias ispi* had shorter, finely pointed gill rakers, with none showing any branching. Those of *O. agassii* and *O. mulleri* were moderately long and pointed, whereas those of *O. olivaceus* and *O. luteus* were much shorter, rounded at the ends and widely spaced.

Length of the alimentary canal also varied considerably between the six Puno Bay orestiids examined (Figure 5). The regressions of gut length against standard length for *O. pentlandi* and *O. ispi* have virtually the same y-axis intercepts but different slopes, whereas the slopes and intercepts of the other four species were virtually the same but very different from the first two. Thus, *O. pentlandi* and *O. ispi* had relatively short alimentary canals compared with those for fish of similar length for the other four species. Zuniga (1941) showed that the intestinal length of *O. pentlandi* (a species that he thought to be phytophagous) was shorter than that of the mollusc-feeding *O. luteus*.

VIII. LITTORAL ORESTIID FEEDING

The six orestiid species taken in the littoral zone of Puno Bay were using at least eight widely differing categories of food (Table 8). Four of the species (*O. ispi, O. pentlandi, O. mulleri* and *O. agassii*) fed heavily on zooplankton, *O. ispi* exclusively so, but with moderately high levels in the other three species also. *Bosmina huaronensis* provided a large part of the zooplankton diet of *O. ispi* and *O. pentlandi*, but *Ceriodaphnia, Boeckella* and cyclopoid copepods were also important components (Table 9). All of these taxa are common zooplankters in the littoral waters of Puno Bay (Muñiz *et al.*, 1989).

Five species (especially *O. agassii* and *O. luteus*) fed heavily on hyalellid amphipods and aquatic insects (mainly chironomid larvae and pupae) (Table 8). Only one species, *O. olivaceus*, used hydrobiid snails (mainly *Littoridina* spp.) to a major extent, along with smaller amounts of sphaerid clams. Fish eggs in various stages of development and similar in size to those found in ripe orestiids were taken by both *O. pentlandi* and *O. luteus*, although they never contributed greatly to mean volumes (Table 8). Plant material (mainly filamentous algae along with occasional *Lemna* and *Chara* fragments) was only found in appreciable amounts in *O. agassii*, especially those fish from the littoral of inner Puno Bay. Oligochaetes and free-living nematodes were found rarely in *O. agassii* alimentary canals.

Only for three species were enough individuals of differing size ranges available to examine the effects of intraspecific body size differences on food type selectivity (Table 10). For *O. pentlandi* there was a shift from heavy zooplanktivory in the smaller size group to more benthic type prey in the larger size group. Such a shift also occurred in *O. olivaceus*, with the larger size group concentrating on snails (*Littoridina* spp.). For *O. agassii* the two larger size groups fed much more on zooplankton than did the small size group, and also broadened their use of different food types.

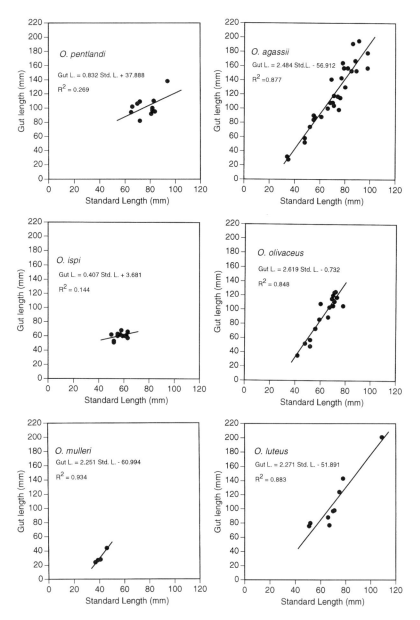

Fig. 5. Regression of alimentary canal length against standard body length for six species of *Orestias* from Puno Bay, Lake Titicaca.

Table 8

Mean per cent occurrence and per cent volume (in parentheses) of various food categories found in the alimentary canal of six species of *Orestias* taken from the littoral zone of Puno Bay, Lake Titicaca

Species	Number	Size range (standard length mm)	Food category							
			1	2	3	4	5	6	7	8
O. ispi	12	50–63	100	—	—	—	—	—	—	—
			(100)							
	10[a]	51–63	100	—	—	—	—	—	—	—
			(100)							
O. pentlandi	11	65–94	54.5	72.7	45.5	—	—	36.4	9.1	—
			(29.8)	(47.8)	(17.3)			(4.7)	(0.4)	
O. mulleri	4	37–46	100	50.0	50.0	—	—	—	—	—
			(48.8)	(17.4)	(33.8)					
O. agassii	30	34–98	46.7	63.3	60.0	13.3	3.3	—	36.7	10.0
			(17.3)	(35.7)	(25.2)	(3.3)	(0.3)		(15.0)	(3.2)
O. olivaceus	16	42–73	12.5	68.8	37.5	68.8	6.3	—	—	—
			(0.2)	(45.9)	(3.7)	(48.3)	(1.9)			
O. luteus	8	51–109	12.5	100	75.0	2.5	2.5	37.5	—	—
			(1.3)	(68.1)	(15.1)	(2.9)	(4.0)	(8.6)		

Food categories: 1, zooplankton; 2, hyalellid amphipods; 3, aquatic insects, mites; 4, hydrobiid snails; 5, sphaeriid clams; 6, fish eggs; 7, algae, macrophytes; 8, oligochaetes, nematodes.

[a] From Uros Islands, outer Puno Bay.

Table 9

Mean per cent occurrence and per cent volume (in parentheses) of major forms in the zooplankton component of prey taken by four *Orestias* species from the littoral zone of Puno Bay

Species	Number	Zooplankton form			
		Bosmina huaronensis	*Ceriodaphnia* spp.	*Boeckella* spp.	Cyclopoid spp.
O. ispi	12	100	33.3	66.6	100
		(71.5)	(7.1)	(8.4)	(12.8)
	10	100	–	60.0	20.0
		(84.1)		(5.3)	(0.6)
O. pentlandi	6	50.0	16.7	33.3	33.3
		(44.2)	(8.3)	(11.7)	(2.5)
O. mulleri	4	50.0	–	50.0	100
		(27.4)		(20.7)	(51.9)
O. agassii	14	14.3	35.7	–	57.1
		(14.3)	(30.2)		(48.4)

Table 10

Mean per cent occurrence and per cent volume (in parentheses) of five major food categories[a] in different size groups of three species of *Orestias* taken from the littoral zone of Puno Bay

Species	Number	Size range (standard length, mm)	Food category				
			1	2	3	4	5
O. pentlandi	5	65–72	60.0	60.0	20.0	–	–
			(40.6)	(53.0)	(2.0)		
	6	81–94	50.0	83.3	83.3	–	–
			(20.8)	(43.3)	(30.0)		
O. olivaceus	7	42–60	14.3	100	42.9	28.6	–
			(0.1)	(85.7)	(7.1)	(7.1)	
	9	66–73	11.1	44.4	33.3	100	11.1
			(0.3)	(14.4)	(1.0)	(80.3)	(3.3)
O. agassii	7	34–55	14.3	100	71.4	–	–
			(2.1)	(79.3)	(9.3)		
	9	56–74	55.6	66.7	33.3	11.1	11.1
			(10.6)	(34.4)	(10.6)	(5.6)	(1.1)
	14	76–98	57.1	42.9	71.4	21.4	–
			(29.3)	(14.3)	(42.1)	(3.6)	

[a]As in Table 8.

IX. DISCUSSION

Ecological interactions among the six to eight species of orestiids found in the littoral zone of Puno Bay are complicated by their marked temporal and spatial differences in habitat use, even at the coarse-grained scale examined in this study. There were large and significant changes in relative abundance between species found in the littoral zone not only between dry and wet seasons, but also between day and night periods (Tables 4–6). Furthermore, considerable and consistent differences in relative abundance of the littoral species were evident between the inner and outer parts of Puno Bay (Treviño *et al.*, 1984, 1989), and the littoral waters close to shore and further offshore at one site (Tables 2, 3). Some similar patterns in the temporal and spatial use of the littoral zone by orestiid species have also been documented in the Lago Pequeño part of Lake Titicaca (Loubens *et al.*, 1984). Preliminary findings from Puno Bay (Treviño *et al.*, 1984, 1989; Northcote, present chapter) indicate the pictorial summary of orestiid and other species use of nearshore waters of Lake Titicaca by Lauzanne (1992) to be far too simplified: seasonal and diel changes in habitat are not considered, and several species such as *O. ispi, O. pentlandi, O. albus* and perhaps *O. mulleri*, some of which may sometimes be present in large numbers in the littoral zone (not necessarily only for spawning-associated activities), are ignored. Certainly, more detailed work is needed to define better the specific macrohabitat and microhabitat use by the orestiid species in Lake Titicaca, and to consider possibilities for short-term and longer term temporal shifts in occupation of the littoral areas as well as fine-grain microhabitat partitioning by the component members of this species flock.

Even from this cursory examination, differences in trophic morphology were apparent between species and among their ontogenetic stages. There is thus also a clear need first to describe accurately the trophic morphology of the various littoral orestiids (e.g. mouth width, gape, dentition; gill raker shape, size and spacing; pharyngeal teeth structure and function) and then to adopt field and laboratory behavioural approaches to examine how these differences might be adaptive in further partitioning competitive interactions in feeding.

The general information already available from previous work on food habits of the littoral orestiids in Lake Titicaca, along with that from the author's Puno Bay studies, suggest that the various species, although overlapping in diet to some degree, exercise considerable selectivity in the categories and size of food. First of all, in the littoral zone there would seem to be several zooplanktivorous orestiids, although it is not entirely clear how much of such prey is actually taken there. Included in this group would be the exclusively zooplanktivorous *O. ispi*, along with *O. pentlandi, O. mulleri* and, to a lesser degree, *O. agassii*. However, these four orestiids show considerable differences in the type of zooplankton selected (Table 9). All 22 *O. ispi* examined had taken large numbers of *Bosmina* (size range 0.31–0.51 mm, $\hat{x} =$

0.41 mm, n = 35) but only a few individuals of the taxa within the three other categories of plankton. In comparison, of the six *O. pentlandi* that had fed on zooplankton, use was spread more broadly among the four types taken, whereas *O. agassii* fed mainly on cyclopoid copepods and *Ceriodaphnia*.

That the *O. ispi* examined had probably been feeding mainly in the littoral waters of Puno Bay is suggested by two lines of evidence:

(i) *Bosmina huaronensis* was the most abundant zooplankter there at that time (Muniz *et al.*, 1989);

(ii) *Daphnia pulex* and *Boeckella titicacae* were the dominant zooplankters taken by *O. ispi* from pelagic waters near the mouth of Puno Bay (Vaux *et al.*, 1988).

Certainly, as Vaux *et al.* (1988) suggest, the pelagic *O. ispi* that they studied (58–82 mm standard length) with a mean gill raker spacing of 0.22 mm should be capable of retaining most size components of pelagic Puno Bay zooplankton, as should littoral *O. ispi* measured by the present author (47–54 mm standard length; mean gill raker spacing of 0.26 mm and raker length of 0.2–0.5 mm) which had fed heavily on 0.3–0.5 mm *Bosmina*. The littoral-feeding *O. agassii* from Puno Bay (71–88 mm standard length) had relatively close gill raker spacing ($\hat{x}$ = 0.35 mm) with fairly long gill raker length ($\hat{x}$ = 0.57 mm) enabling them effectively to take *Ceriodaphnia* (mean body size = 0.69 mm) and even some *Bosmina* in the 0.3–0.5 mm size range.

There are at least five species of heavily benthophagic orestiids (Table 8), although three of them, *O. pentlandi, O. mulleri* and *O. agassii*, can also take appreciable numbers and volumes of zooplankton. The two most clearly benthophagous species, *O. olivaceus* and *O. luteus*, both feed to a large extent on gammarids, but supplemented in the former by considerable use of hydrobiid snails and in the latter by aquatic insects, mainly chironomids. Previous workers have reported similar findings in Lago Pequeño for these species (Lauzanne, 1982; Loubens, 1989).

Orestias agassii had the broadest spectrum of food categories (Table 8), including at times considerable amounts of filamentous algae and macrophyte fragments. These findings are again supported by the comments of workers in Lago Pequeño who considered this species to show high feeding variability in different habitats (Lauzanne, 1982; Loubens and Sarmiento, 1985).

The one littoral orestiid for which no specimens were available for the present analyses was *O. albus*, which occurred only infrequently in the Puno Bay sampling. Five adults from Lago Pequeño examined by Loubens (1989) all contained small orestiids in their stomachs and he considered it to be the only highly piscivorous orestiid now found in the lake, although *O. pentlandi* and *O. luteus* do feed to some extent on fish eggs (Table 8).

Some seven species of *Orestias*, one species of trichomycterid catfish, and two introduced fishes (pejerrey and rainbow trout) are commonly found in the littoral zone of Puno Bay, Lake Titicaca. In comparison with the over 100 species of cichlid fishes alone found in the littoral of Lake Tanganyika (Rossiter, 1995; Kawanabe et al., 1997) and the several hundred found in the littoral of Lake Malawi (Turner; Snoeks, this volume), this number may seem trivial. Yet the low species diversity of the littoral Titicacan orestiids may serve to facilitate insight into ecological mechanisms promoting coexistence, especially in view of the very limited data sets presently available on ecological interactions in that region of the lake. A recent synthesis of limnological information on the Titicaca system (Dejoux and Iltis, 1992) includes a useful review on some of the orestiid work, especially in Lago Pequeño (Lauzanne, 1992), but also serves to emphasize the comparative paucity of information available in relation to that for fish communities in most other ancient lakes. In Lake Tanganyika especially, detailed ecological studies conducted over many years have begun to reveal and unravel the intricate and often subtle mechanisms that facilitate coexistence in the speciose littoral communities of that lake (Kawanabe et al., 1997).

Rossiter (1995) discusses in detail the various mechanisms that may be involved in achievement of coexistence of the high species diversity around the shoreline of Lake Tanganyika, as well as in other tropical systems. It would be of value to consider these now in relation to the Lake Titicaca littoral orestiid assemblage.

The first possibility that he puts forward is that community instability may suppress competitive interactions between closely related cohabiting species, a proposition that he shows clearly cannot apply to Lake Tanganyika. However, most interestingly, it may well do so for Lake Titicaca. Lying at high elevation in the flat Andean Altiplano, where in recent millennia there have been major changes in lake shoreline level (Dejoux and Iltis, 1992; Mourguiart, this volume), even minor shifts in water level of a few metres or less as have occurred in recent decades produce shoreline extensions or recessions of several kilometres along many parts of Puno Bay, Lago Pequeño, and even in some parts of Lago Grande. These must have major effects on the littoral macrophytic and periphytic communities and on their associated planktonic and benthic communities. Human agricultural practices have had to adapt to such lakeshore instability over recent millennia and it would seem reasonable to suggest that so have the orestiids over a much longer time span, with periodic lakeshore shifts and habitat instabilities probably favouring first one species set and then another.

A second way to promote coexistence in a high diversity assemblage of similar species is for it to occur in an arena with high habitat diversity: an environment patchy in habitat type or quality, or both. This mechanism seems to be the key for Lake Tanganyika, where highly complex rocky shoreline diversity prevails in separated patches along an enormous length of stable lake edge (Rossiter, 1995). It may also be a factor in Lake Titicaca, not only in the shorelines which undergo the large-scale horizontal fluctuations noted previously, but also along the steep rocky shorelines of much of its main basin (Lago Grande). Orestiid use of such patchy habitats needs more detailed study along the promising lines of approach outlined by Lauzanne (1992) for macrophyte depth zonation and its orestiid associations.

Diel separation of feeding time, place and prey focus is put forward as a third means of explaining species coexistence via resource partitioning, with the best example being that for the planktivorous cichlids of Tanganyika (Gashagaza, 1988; Rossiter, 1995). Perhaps then, it is not so surprising that it is the highly zooplanktivorous orestiids of Lake Titicaca, especially *O. ispi*, which provide the clearest examples of diel shifts in the use of the shallow littoral zone and its associated zooplanktonic prey. Relaxation of competitive interactions need not occur only by the very short time changes of a diel cycle, but also by somewhat longer periodicities, especially those on a seasonal scale. Again, the Titicaca material presented here shows strong suggestions in this direction for several of the littoral orestiids, specifically in differences in abundance between dry and wet seasons.

The fourth means for promotion of high species diversity and coexistence explored by Rossiter (1995) for the Lake Tanganyika littoral cichlids is that of trophic/spatial partitioning (see especially the several excellent studies on this subject in Kawanabe *et al.*, 1997). Among the Lake Titicaca orestiids also, differences in trophic structure and in fine-scale use of habitat space appear to be important factors involved in reducing competition and permitting coexistence in the littoral zone.

The Titicaca orestiids form a species flock, whose endemism and low species diversity present an ideal system for meaningful studies of mechanisms of speciation and coexistence. There is a clear and loud call for intensive, co-operative approaches of study on Titicacan littoral orestiids following those that have been so successfully used for the cichlid species flocks of Lake Tanganyika.

ACKNOWLEDGEMENTS

First, thanks must go to Andy Rossiter for his not so gentle persuasion that I should attempt this review in such a short time frame, and to my wife Heather for putting up more or less understandingly with the pressures, ill humour and other costs that thereby resulted. J.S. Nelson quickly supplied the few orestiids that the author had donated to the Zoology Museum at the University of Alberta, Edmonton, and J.D. McPhail eventually located the larger collection that I had given to the Fish Museum at the University of British Columbia. Moira S. Greaven provided, as usual, invaluable assistance in locating relevant literature, as did J.D. Green in statistical analyses and computer graphing. Dean E. Moreno, Universidad Nacional del Altiplano at Puno, Peru kindly made available some references that were difficult to obtain, as did L. Lauzanne of ORSTOM, La Paz, Bolivia. Three anonymous external referees made minor suggestions for revision.

REFERENCES

Dejoux, C. (1994). Lake Titicaca. *Arch. Hydrobiol. Beiheft. Ergebnisse Limnol.* **44**, 35–42.

Dejoux, C. and Iltis, A. (eds) (1992). *Lake Titicaca: A Synthesis of Limnological Knowledge*. Kluwer, Dordrecht.

Fryer, G. (1959). The trophic interrelationships and ecology of some littoral communities of L. Nyassa with special reference to the fishes, and a discussion of the evolution of a group of rock-frequenting Cichlidae. *Proc. Zool. Soc. Lond.* **132**, 153–281.

Fryer, G. and Iles, T.D. (1972). *The Cichlid Fishes of the Great Lakes of Africa: Their Biology and Evolution*. Oliver and Boyd, Edinburgh.

Gashagaza, M.M. (1988). Feeding activity of a Tanganyikan cichlid fish *Lamprologus brichardi*. *Afr. Stud. Monogr.* **9**, 1–9.

Hardin, G. (1960). The competitive exclusion principle. *Science* **131**, 1292–1297.

Iltis, A. and Mouguiart, P. (1992). Higher plants: distribution and biomass. In: *Lake Titicaca: A Synthesis of Limnological Knowledge* (Ed. by C. Dejoux and A. Iltis), pp. 241–253. Kluwer, Dordrecht.

Kawanabe, H., Hori, M. and Nagoshi, M. (eds) (1997). *Fish Communities in Lake Tanganyika*. Kyoto University Press, Kyoto.

Lauzanne, L. (1982). Les *Orestias* (Pisces, Cyprinodontidae) du Petit lac Titicaca. *Rev. Hydrobiol. Trop.* **15**, 39–70.

Lauzanne, L. (1992). Native species. The *Orestias*. In: *Lake Titicaca: A Synthesis of Limnological Knowledge* (Ed. by C. Dejoux and A. Iltis), pp. 405–419. Kluwer, Dordrecht.

Levieil, D., Cutipa, Q.C., Goyzueta, C.G. and Paz, F.P. (1989). The socio-economic importance of macrophyte extraction in Puno Bay. In: *Pollution in Lake Titicaca, Peru: Training, Research and Management* (Ed. by T.G. Northcote, S.P. Morales, D.A. Levy and M.S. Greaven), pp. 155–175. University of British Columbia Press, Vancouver.

Loubens, G. (1989). Observations sur les poissons de la partie bolivienne du lac Titicaca. IV. *Orestias* spp., *Salmo gairdneri* et problems d'amenagement. *Rev. Hydrobiol. Trop.* **22**, 157–177.

Loubens, G. and Sarmiento, J. (1985). Observations sur les poissons de la partie bolivienne du lac Titicaca. II. *Orestias agassii*, Valenciennes, 1846 (Pisces, Cyprinidontidae). *Rev. Hydrobiol. Trop.* **18**, 159–171.

Loubens, G., Orsorio, F. and Sarmiento, J. (1984). Observations sur les poissons de la partie bolivienne du lac Titicaca. I. Milieux et peuplements. *Rev. Hydrobiol. Trop.* **17**, 153–161.

Morales, P., Northcote, T.G. and Levy, D.A. (1984). A centre for limnological training and research on Lake Titicaca and the aquatic ecosystems of the Peruvian Altiplano. *Verh. Int. Verein. Limnol.* **22**, 1335–1339.

Muller, R. (1993). Critical comments on the revision of the genus *Orestias* (Pisces: Cyprinodontidae) by Parenti (1984). *Zool. Jahrbuche Anat.* **123**, 31–58.

Muñiz, B.V., Chapman, M.A., Chino, C.B., Azurín, D.E. and Northcote, T.G. (1989). Effects of eutrophication on zooplankton. In: *Pollution in Lake Titicaca, Peru: Training, Research and Management* (Ed. by T.G. Northcote, S.P. Morales, D.A.

Levy, and M.S. Greaven), pp. 81–100. University of British Columbia Press, Vancouver.

Northcote, T.G. (1979). Investigation and recommendations on the hydrobiological resources of the Lake Titicaca system, Peru. *Food and Agriculture Organization of the United Nations*, Report FAO/PER/76/022, 156 pp.

Northcote, T.G. (1980). Methods and recommendations for fish and limnological sampling in the littoral zone of Lake Titicaca. *Food and Agriculture Organization of the United Nations*, Report FAO/PER/76/022, 30 pp.

Northcote, T.G. (1995). Confessions from a four decade affair with Dolly Varden: a synthesis and critique of experimental tests for interactive segregation between Dolly Varden char (*Salvelinus malma*) and cutthroat trout (*Oncorhynchus clarki*) in British Columbia. *Nord. J. Freshwat. Res.* **71**, 49–67.

Northcote, T.G., Morales, S.P., Levy, D.A. and Greaven, M.S. (eds) (1989). *Pollution in Lake Titicaca, Peru: Training, Research and Management*. University of British Columbia Press, Vancouver.

Parenti, L.R. (1984a). A taxonomic revision of the Andean killifish genus *Orestias* (Cyprinodontiformes, Cyprinodontidae). *Bull. Am. Mus. Nat. Hist.* **178**, 110–214.

Parenti, L.R. (1984b). Biogeography of the Andean killifish genus *Orestias* with comments on the species flock concept. In: *Evolution of Fish Species Flocks* (Ed. by A.A. Echelle and I. Kornfield), pp. 85–92. University of Maine at Orono Press, Orono, ME.

Rossiter, A. (1995). The cichlid fish assemblages of Lake Tanganyika: ecology, behaviour and evolution of its species flocks. *Adv. Ecol. Res.* **26**, 187–252.

Treviño, H., Torres, J., Levy, D.A. and Northcote, T.G. (1984). Pesca experimental en aguas negras y limpias del litoral de la Bahia de Puno, Lago Titicaca, Peru. *Inst. Mar. Peru Bol.* **8**, 1–36.

Treviño, H. Torres, J., Choquehuanca, P., Levy, D.A. and Northcote, T.G. (1989). Effects of eutrophication in fish. In: *Pollution in Lake Titicaca, Peru: Training, Research and Management* (Ed. by T.G. Northcote, S.P. Morales, D.A. Levy and M.S. Greaven), pp. 115–128. University of British Columbia Press, Vancouver.

Tutin, T.G. (1940). The macrophytic vegetation of the lake. Report No. 10. The Percy Sladen Trust Expedition to Lake Titicaca in 1937. *Trans. Linn. Soc. Lond. Ser. 3* **1**, 161–189.

Vaux, P., Wurtsbaugh, W.A., Treviño, H., Marino, L., Bustamente, E., Torres, J., Richerson, P.J. and Alfaro, R. (1988). Ecology of the pelagic fishes in Lake Titicaca, Peru–Bolivia. *Biotropica* **20**, 220–229.

Wetzel, R.G. (1983). *Limnology*. Saunders College Publishing, Philadelphia, PA.

Wirrmann, D. (1992). Morphology adn bethymetry. In: *Lake Titicaca: A Synthesis of Limnoloical Knowledge* (Ed. by C. Dejoux and A. Iltis), pp. 16–22. Kluwer, Dordrecht.

Zuniga, E. (1941). Regimen alimentico y longitud del tubo digestivo en los peces del genero *Orestias*. *Mus. Hist. Nat. Javier Prado, Lima* **16**, 79–86.

Effect of Hydrological Cycles on Planktonic Primary Production in Lake Malawi/Niassa

G. PATTERSON, R.E. HECKY and E.J. FEE

I. SUMMARY

A field study measuring depth distribution of chlorophyll and light was carried out in Lake Malawi over 2 years: from these data, modelled estimates of primary production were obtained. It is well known that, in this lake, the rate of nutrient loading to the trophogenic zone from the deeper, nutrient-rich water is dictated by the degree of vertical mixing of water layers. In the present study, estimated values for primary production were strongly correlated with periods when this mixing was assumed to be highest. Primary production in Lake Malawi thus seems driven by physical, climatic factors, notably ambient temperature and wind regimes.

This conclusion was supported by the large differences between estimated production values for 1992 and 1993. The higher production values seen in 1993 were due to the greater degree of seiching activity in that year, which increased the extent of nutrient loading in the upper water layers. The simplicity of the modelling approach used here makes it an effective means of obtaining primary production data, and for Lake Malawi the low spatial variation in algal biomass suggests that whole-lake production could be estimated through extrapolation of data collected from only a small number of sampling locations.

There is a paucity of comparative data on primary productivity in tropical lakes, especially ancient lakes Malawi, Tanganyika and Victoria. Monitoring of whole-lake primary production in these lakes would potentially be a useful management tool where environmental perturbations, first evinced as increased production, need to be monitored and, where possible, controlled.

ADVANCES IN ECOLOGICAL RESEARCH VOL. 31
ISBN 0–12–013931–6

II. INTRODUCTION

Lake Malawi is a meromictic tropical rift lake bordered by Malawi, Tanzania
and Mozambique (Figure 1). The lake is characterized by two barriers to
vertical mixing; a thermocline at 40 m to 100 m depth and a chemocline at
around 230 m depth. These features result in a vertical stratification of
nutrients, with higher concentrations of nitrogen (N), phosphorus (P) and
silicon (Si) occurring in the deeper layers of the lake (Patterson and Kachinjika,
1995). Periods of mixing between the surface waters and the nutrient-rich
deeper waters are most likely in April to September of each year, when cooler
air temperatures and the effect of the strong south-easterly trade winds
combine to reduce the strength of the thermocline. At these times, the
entrainment of nutrient-rich deeper waters into the surface layers increases the

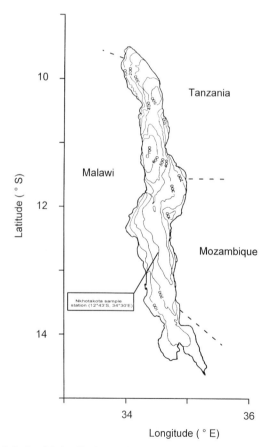

Fig. 1. Map of Lake Malawi/Niassa showing the location of Nkhotakota sample
station.

availability of nutrients to photosynthesizing phytoplankton and results in increased phytoplankton biomass in the trophogenic layer (Patterson and Kachinjika, 1995).

In contrast to the attention paid to its cichlid fish fauna (e.g. Fryer and Iles, 1972; Ribbink et al., 1983; Turner, 1996), investigations of primary production in Lake Malawi have been remarkably few. Degnbol and Mapila (1982) measured planktonic primary production at various depths at a single station off Nkhata Bay throughout one annual cycle (1980–1981), and reported a mean daily carbon (C) fixation value of 740 mg C m^{-2} d^{-1} (range 240–1140 mg C m^{-2} d^{-1}; $n = 21$). This is equivalent to a mean fixation value of 271 mg C m^{-2} y^{-1}. They also showed that, during the year of their study, phytoplankton biomass and primary production rates were maximal during the mixing period of May to September. Subsequently, Bootsma (1993) measured primary production in Lake Malawi during the period 1990–1991 at several locations over the entire length of the lake, and found a similar seasonal pattern of primary production to that reported by Degnbol and Mapila (1982). Bootsma's measurements included some particularly high values from inshore stations, and these contributed to a mean photosynthetic rate of 660 mg C m^{-2} d^{-1}. However, Bootsma's primary production estimates for stations adjacent to the deep location monitored by Degnbol and Mapila in 1980 were less than those recorded by the latter authors. This was especially evident during the 1990 mixing season of May to September, when the rate of photosynthesis was approximately half that recorded in 1980.

This chapter presents estimated rates of primary production based on data from a deep water location in Lake Malawi over a period of 2 years (January 1992 to January 1994). These data were taken concurrently with a range of hydrological, limnological and biological measurements (Patterson and Kachinjika, 1995).

III. MATERIALS AND METHODS

Sampling was conducted at a deep water (310 m water depth) station off Nkhotakota (12°43′ S, 34°30′ E). The station is in a central location and is at least 20 km from the nearest shallow (< 100 m depth) part of the lake (Figure 1).

The station was visited 30 times between January 1992 and January 1994 at 2–5 week intervals (Table 1). Each time, water was sampled from 0, 12.5, 25, 37.5, 50, 75 and 100 m depth using either 3.5 l PWS bottles (Plastic Water Samplers; Hydrobios) or two 2 l Van Dorn bottles. These samples were used to estimate chlorophyll a concentration. First, 750 ml aliquots of lake water were filtered through Whatman GFC filters, which were then stored at −12°C before extraction of pigment in near boiling 90% methanol (Golterman et al., 1978). Fluorescence values for the pigment extracts were obtained using a

Table 1
Sampling dates and vertical light extinction coefficient values (ϵ) for the visits to the sampling station

Date	ϵ (m^{-1})	Date	ϵ (m^{-1})
22 Jan. 1992	0.084	13 Jan. 1993	0.139
29 Feb. 1992	0.091	3 Feb. 1993	0.088
2 Apr. 1992	0.083	26 Feb. 1993	0.081
9 Apr. 1992	0.038	16 Mar. 1993	0.087[a]
23 Apr. 1992	0.088	23 Apr. 1993	0.099
10 May 1992	0.094	18 May 1993	0.094
18 May 1992	0.082	7 Jun. 1993	0.096[a]
14 Jun. 1992	0.156	20 Jul. 1993	0.100
30 Jun. 1992	0.094	4 Aug. 1993	0.115
21 Jul. 1992	0.103	30 Aug. 1993	0.126
26 Aug. 1992	0.106	18 Sep. 1993	0.111
11 Oct. 1992	0.096[a]	15 Oct. 1993	0.110
29 Oct. 1992	0.092	5 Nov. 1993	0.193
10 Nov. 1992	0.078	24 Nov. 1993	0.120
4 Dec. 1992	0.106	21 Jan. 1994	0.092

[a]Estimated value due to insufficient surface irradiance to calculate ϵ.

Turner Series 10 Fluorometer, calibrated against spectrophotometric values made on more concentrated chlorophyll *a* extracts from Lake Malawi. Chlorophyll *a* values were calculated by applying the calibration factor and formula given by Talling and Driver (1963), based on spectrophotometric absorbance at 665 nm and corrected for turbidity by subtracting absorbance at 750 nm.

Concurrent with water sampling, a profile of photosynthetically active radiation (PAR) was made using a custom-designed data-logging instrument (Plymouth Marine Laboratories, UK). The vertical light extinction coefficient (ϵ) between 10 m and 75 m depth was calculated by applying a least-square linear regression to the slope of the natural logarithm of PAR versus the depth curve. On days when there was insufficient surface irradiation to calculate the extinction coefficient, it was estimated assuming a linear change between previous and following values (see Table 1).

Photosynthesis at the sample location for each day of the period 1 January 1992 to 31 December 1993 was estimated using the primary production model of Fee (1990). This model incorporates the light extinction coefficient and the depth profile of chlorophyll *a*. For calculations, the seasonal distribution of photosynthetic efficiency (i.e. the slope of the photosynthesis versus irradiance curve, and the value of photosynthesis at light saturation) was assumed to be the same as that derived from the 1980 data of Degnbol and Mapila (1982) at their nearest sample station to that of this study. The model allows for seasonal variation of PAR at the lake's surface owing to day-length variation and diurnal solar elevation, but assumes only cloudless skies. However, even on cloudy days, this model gives only a slight ($< 15\%$) overestimate of primary productivity (Bootsma, 1993), and one may therefore be confident in the overall accuracy of the interpretation.

IV. RESULTS

During the 2-year sampling period the calculated light extinction coefficient (ϵ) varied from 0.038 m^{-1} (high transparency) to 0.193 m^{-1} (low transparency), with the majority of values around 0.1 m^{-1} (Table 1).

The depth–time distribution of chlorophyll a at the sampling station (Figure 2) shows higher chlorophyll a concentrations to occur during the middle part of each year, and also indicates chlorophyll a abundance during 1993 to have been greater and to have continued longer than in 1992. In 1993 the period of abundance of chlorophyll a extended from late March until almost the end of the year. Throughout most of this period, high concentrations of chlorophyll a occurred to depths of > 50 m and often showed a subsurface maximum. The

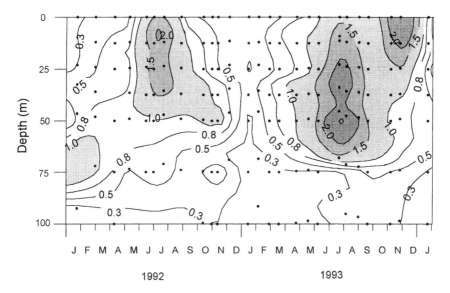

Fig. 2. Depth–time plot of chlorophyll a concentration at the sample station for 1992 and 1993. Asterisks indicate individual samples.

high values of chlorophyll *a* seen in November 1993 were, however, restricted to the upper part of the water column. There was no evidence of elevated levels of chlorophyll *a* during the same period in 1992.

Modelled values of areal rates of primary production during January 1992 to December 1993 indicated periods of relatively high production from May to August in both years (Figure 3). Several smaller production peaks occurred in the latter half of 1992, which was a period of higher productivity than that recorded for January to April of that year. This pattern of low production during the early months followed by relatively high production in the middle of the year was repeated in the early part of 1993. However, during 1993 the peaks for the latter part of the year were considerably higher than those seen during 1992. These peaks during the latter part of 1993 correspond to the high levels of chlorophyll *a* recorded close to the surface during that period, and were principally due to the high ambient population densities of the heterocystous cyanophyte *Anabaena flos-aquae* (Lyngb.) Breb. The estimated mean daily rate of primary production for 1992 and 1993 was 900 mg C m^{-2} d^{-1} and 1420 mg C m^{-2} d^{-1}, respectively. Extrapolation of these values gives total annual production for the sampling location of 329.4 g C m^{-2} y^{-1} for 1992 and 518.3 g C m^{-2} y^{-1} for 1993.

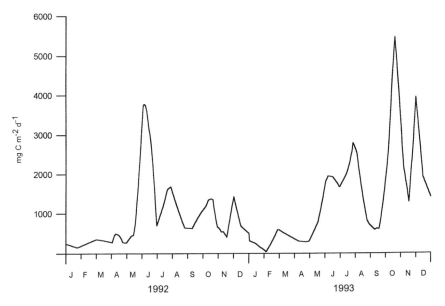

Fig. 3. Estimated rate of primary production for the sample station in Lake Malawi for 1992 and 1993.

V. DISCUSSION

The depth distribution of chlorophyll a concentration is directly related to the availability of light energy for photosynthesis and is therefore a critical factor in the modelling of primary production. The modelled primary production values for the sampling location (Figure 3) reflected conditions where production was nutrient limited, with the peaks in photosynthesis correlating with predicted periods of highest nutrient loading. During the middle period of both years a substantial increase in production was seen over the early parts of the year. In 1992, photosynthesis fell towards the end of the year during August to December, but remained higher than the early part of the year from January to April. This suggests strongly that nutrient loading to the epilimnion extended beyond the wind-driven period of water mixing via the effects of seiching activity, and continued until the lake re-established hydrological equilibrium (Mortimer, 1974). In 1993 there was also a peak of production corresponding to the middle part of the year when the thermocline was weakest. However, in contrast to 1992, when the production levels decreased, the latter part of 1993 (September to December) was characterized by a series of yet higher peaks of production. These peaks coincided with the high biovolume of $A.$ $flos\text{-}aquae$ (Figure 2). Colonies of $A.$ $flos\text{-}aquae$ are able to control their buoyancy and can thus maintain a position close to the water surface, where they receive maximal PAR. The high peaks during the latter part of 1993 were probably due to a greater degree of seiching at that time (Patterson and Kachinjika, 1995) compared with 1992, which resulted in higher rates of nutrient loading from the metalimnion. This high availability of nutrients was utilized by the $A.$ $flos\text{-}aquae$ populations, supplemented by external nitrogen input via N fixation (Patterson and Kachinjika, 1995). High densities of phytoplankton were recorded throughout the lake during the latter part of 1993 compared with 1992 (Patterson and Kachinjika, 1995) and high rates of production were probably common to all parts of the lake at that time.

There are several similarities between in the findings of the present study and those of Degnbol and Mapila in their 1980 study. The total annual primary production estimates were close: 271 g C m^{-2} y^{-1} for 1980, and 329.4 g C m^{-2} y^{-1} and 518.3 g C m^{-2} y^{-1} for 1992 and 1993, respectively. Both studies documented annual cycles with low rates of production at the start of the year, higher values in the middle part and a return to low values towards the end of the year. Finally, both recorded a high degree of variability in production values in the latter part of each year, with overall levels of production higher than those during the first months of the year. The occurrence of higher and variable production values after the windy period is probably due to the effects of seiching.

There is a paucity of quantitative measurements of annual variation of primary production for Lake Malawi: excepting the present study, and those of Degnbol and Mapila (1982) and Bootsma (1993), the authors know of no other data. This lack of information has necessitated the use of inferential evidence when estimating levels of primary production. For example, catches of the small pelagic fish *Engraulicypris sardella* Günther, which is a planktivore, show large interannual variations (Tweddle and Lewis, 1990). These differences may indicate variation in population recruitment, perhaps mediated by annual variation in primary production of their algal food source. However, this interpretation remains unvalidated and there is clearly a strong need for further studies incorporating direct measurements of primary production in Lake Malawi.

During 1990, the rate of photosynthesis close to the region studied by Degnbol and Mapila (1982) during 1980 was significantly less, particularly during the middle part of the year (Bootsma, 1993). This difference was ascribed to less wind in 1990 and therefore reduced mixing and nutrient loading. It is possible that Bootsma's method, which relied mainly on surface samples, underestimated production during the middle part of the year when mixing depths are greatest. However, it seems that the ambient wind and temperature regime, and its effect on mixing, strongly affects rates of nutrient loading to the epilimnion, an interpretation which can also explain the marked difference in the production rates between 1992 and 1993 documented in this paper. The effects of the mixing regime may be subtle and variation in annual rates of production may be highly dependent on the strength of the seiching pattern established after the windy season. More reliable long-term data on the annual rate of primary production of the lake are required to answer these questions. The simplicity of the modelling approach used in this paper may prove an effective means of obtaining such data; the low spatial variation in algal biomass during 1992 and 1993 (Patterson and Kachinjika, 1995) and in 1990 (Bootsma, 1993) suggests that whole-lake production could be estimated through extrapolation of data collected from only a small number of sampling locations.

The entire pelagic system of Lake Malawi appears to respond to fluctuations in primary production, with short time-lags (Allison *et al.*, 1995). This is a characteristic feature of a system in dynamic equilibrium or steady state (DeAngelis and Waterhouse, 1987). Overall rates of primary production in 1992 and 1993 were correlated with the biomass and production values of several organisms at higher trophic levels. These included larvae of the midge *Chaoborus edulis* Verbeke. and those of the fish *E. sardella*, which are principal food items of several predatory fishes. The dependence of biological production of the higher trophic levels on the cycle of internal nutrient loading, and the subsequent increase in primary production, shows many of the characteristics of "bottom-up" control (McQueen *et al.*, 1986). The measurement of primary production is thus likely to assist with predictions of annual fish yield from the lake, and could prove a very powerful management tool for the fishery.

According to the definition of Wetzel (1983; p. 402, Table 15-10), average low chlorophyll a concentrations within the trophogenic zone of Lake Malawi ($c.$ 1 μg l^{-1}) are typical of an oligotrophic lake. However, the production band of 91–365 g C m^{-2} y^{-1} is given by Wetzel as representative of mesotrophy, and the production estimates reported here (329.4–518.3 g C m^{-2} y^{-1}) therefore indicate the lake to be mesotrophic, or even eutrophic. A similar high production:biomass ratio has been described in Lake Tanganyika (Hecky, 1991), which would also be categorized as mesotrophic under the above criterion. This apparent anomaly is probably a reflection of most limnological studies of primary productivity having been made in lakes situated in temperate regions. In tropical lakes, a high production:biomass ratio is probably due to a combination of high ambient water temperatures, which promote relatively high algal growth rates, and rapid carbon transfer to higher trophic levels; the minimal seasonal variation in light input also means that production can continue throughout the year. High production:biomass ratios may well be a feature of large tropical lakes which, as a result, are not strictly comparable with temperate lakes. Our knowledge of the hydrology and production processes in tropical lakes is wanting, and further emphasizes the need for such studies in these lakes. This is especially true for ancient lakes Malawi, Tanganyika and Victoria, where the effects of environmental perturbations, which can deleteriously affect the entire system through cascade effects, need to be monitored.

ACKNOWLEDGEMENTS

The authors thank all agencies and personnel involved with the UK/SADC Pelagic Fish Resource Assessment Project, particularly the Governments of Malawi, Mozambique and Tanzania, the SADC Inland Fisheries Sector Technical Coordination Unit, and the UK's Department for International Development, which funded this study.

REFERENCES

Allison, E.H., Patterson, G., Irvine, K., Thompson, A.B. and Menz, A. (1995). The pelagic ecosystem. In: *The Fishery Potential and Productivity of the Pelagic Zone of Lake Malawi/Niassa* (Ed. by A. Menz), pp. 351–367. Natural Resources Institute, London.

Bootsma, H.A. (1993). Algal dynamics in an African great lake, and their relation to hydrographic and meteorological conditions. Unpublished PhD Thesis. University of Manitoba, Canada.

DeAngelis, D.L. and Waterhouse, J.C. (1987). Equilibrium and non-equilibrium concepts in ecological models. *Ecol. Monogr.* **57**, 1–21.

Degnbol, P. and Mapila, S. (1982). Limnological studies on the pelagic zone of Lake Malawi from 1978–1981. *FAO Tech. Rep. MLW/75/019* **1**, 3–48.

Fee, E.J. (1990). Computer programs for calculating *in situ* phytoplankton photosynthesis. *Can. Tech. Rep. Fish. Aquat. Sci.* **1740**, 1–27.

Fryer, G. and Iles, T.D. (1972). *The Cichlid Fishes of the Great Lakes of Africa; Their Biology and Evolution.* Oliver and Boyd, Edinburgh.

Golterman, H.L., Clymo, R.S. and Ohnstad, M.A.M. (1978). *Methods for Physical and Chemical Analysis of Fresh Waters.* I.B.P. Handbook No. 8. Blackwell Scientific Publications, Oxford.

Hecky, R.E. (1991). The pelagic ecosystem. In: *Lake Tanganyika and its Life* (Ed. by G.W. Coulter), pp. 90–110. Oxford University Press, London.

McQueen, D.J., Post, J.R. and Mills, E.L. (1986). Trophic relationships in freshwater pelagic ecosystems. *Can. J. Fish. Aquat. Sci.* **43**, 1571–1581.

Mortimer, C.H. (1974). Lake hydrodynamics. *Mitt.. Int. Verein. Limnol.* **20**, 124–197.

Patterson, G. and Kachinjika, O. (1995). Limnology and phytoplankton ecology. In: *The Fishery Potential and Productivity of the Pelagic Zone of Lake Malawi/Niassa* (Ed. by A. Menz), pp. 1–67. Natural Resources Institute, London.

Ribbink, A.J., Marsh, B.A., Marsh, A.C., Ribbink, A.C. and Sharp, B.C. (1983). A preliminary survey of the cichlid fishes of rocky habitats in Lake Malawi. *S. Afr. J. Zool.* **18**, 149–310.

Talling, J.F. and Driver, D. (1963). Some problems in the estimation of chlorophyll-a in phytoplankton. *Proceedings: Conference on Primary Productivity Measurements, Marine and Fresh-water,* pp. 117–135. US Atomic Energy Commission TID-7633.

Turner, G.F. (1996). *Offshore Fishes of Lake Malawi.* Cichlid Press, Lauenau, Germany.

Tweddle, D. and Lewis, D.S.C. (1990). The biology of usipa (*Engraulicypris sardella*) in relation to fluctuations in productivity of Lake Malawi and species introductions. *Collected Reports on Fisheries Research in Malawi, Occasional Papers,* Vol. 1. ODA, London.

Wetzel, R.G. (1983). *Limnology.* Saunders College Publishing, Philadelphia, PA.

Macrodistribution, Swarming Behaviour and Production Estimates of the Lakefly *Chaoborus edulis* (Diptera:Chaoboridae) in Lake Malawi

K. IRVINE

I. SUMMARY

Results of a study of the abundance, biomass and production of larvae of the lakefly *Chaoborus edulis* in Lake Malawi are presented. Temporal variations in larval abundance were evident, but no distinct seasonal pattern was discernable. The maximum biomass of larvae occurred in September–October, following increased turbulence of the water column in June–July: this turbulence brings nutrients from the hypolimnion to the epilimnion and promotes enhanced primary and zooplankton production. There was no clear pattern in the macrodistribution of *C. edulis* larvae, other than a general, but inconsistent, tendency for numbers to be lower in the extreme south of the lake.

Swarms of adult flies occurred throughout the year and throughout the lake. In the extreme south, swarms were seen mainly between November and March. In the relatively shallow southern part of the lake, mortality of late instar larvae and pupae from fish predation is likely to be sufficiently severe to prevent numbers of emergent adults from reaching the high densities required for swarming. There was no marked lunar-related periodicity in swarming behaviour, other than an indication that swarms were scarce during about 4–10 days prior to the new moon.

Annual production of *C. edulis* in Lake Malawi was estimated as 2.2 g dry wt m^{-2} in 1992, and 6.4 g dry wt m^{-2} in 1993. Mortalities of second and fourth instar larvae exceeded 75%, probably mainly as a result of food limitation and fish predation, respectively.

II. INTRODUCTION

The occurrence of dense swarms of adult lakeflies, *Chaoborus edulis* (Edwards), is a visually striking feature over Lake Malawi. The swarms often have the structure of a narrow helical column rising from the lake surface and are almost certainly associated with the fly's mating behaviour. Similar swarming behaviour has been observed and documented over several other East African lakes, for example Lake Victoria (MacDonald, 1956; Tjønneland, 1958) and Lake Edward (Verbeke, 1958); however, in Lake George, although *Chaoborus* is abundant, it does not form extensive swarms there (McGowan, 1974). In Lake Victoria, the occurrence of swarms is associated with the period around the new moon (Tjønneland, 1958). The adult phase of *Chaoborus* is brief and the greater part of the life cycle is spent as an aquatic larva, during which it develops through four distinct larval instars. The present chapter examines the spatial and temporal distribution pattern of *C. edulis* larvae and adults in the lake, with special reference to whether any relationship exists between larval distribution and adult emergence, and latitude. Lunar-related synchrony in emergence patterns is also examined. Estimates are also given of overall production rates of *C. edulis* in Lake Malawi.

III. STUDY SITE

Lake Malawi (Figure 1) lies in the East African Rift Valley between latitudes 9°30′ S and 14°30′ S, and has a surface area of 30 800 km^2 and an average depth of 270 m (Gonfiantini *et al.*, 1979; Patterson and Kachinjika, 1993). The lake is meromictic, with a persistent thermocline within the upper 50–100 m of the water column and a permanent zero oxygen boundary at about 230 m (Eccles, 1974). Phytoplankton and crustacean zooplankton are confined to about the upper 70 m, and chlorophyll *a* concentrations rarely exceed 2 μg l^{-1} (Jackson *et al.*, 1963; Patterson *et al.*, this volume). From April to October the lake has a distinct windy season, which induces water turbulence and promotes increased biological production. Between November and March, the lake experiences a calmer, wetter season, during which time vertical stratification is more persistent (Patterson and Kachinjika, 1993; Irvine *et al.*, in press).

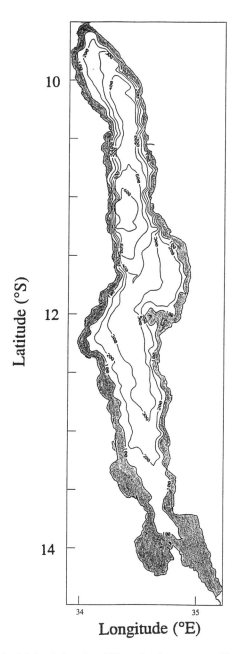

Fig. 1. Map of Lake Malawi showing 100 m depth contours. Shaded area is < 100 m deep.

IV. METHODS

A. Abundance Estimates

Chaoborus edulis larvae were collected with a High Speed Plankton Net
(HSPN) (Milligan and Riches, 1983) during 10 whole-lake surveys between
February 1992 and December 1993. Tows were done at speeds of 3–4 knots
($1.5–2$ m s^{-1}) and sampling profiles extended down to 220 m depth. Details of
sampling appartus and protocols are given in Irvine and Waya (1999).
Restriction of flow owing to net clogging was not a problem in the plankton-
dilute waters of the lake, and filtering efficiency of the main net was assumed to
be 93% (Nichols and Thompson, 1991). Following retrieval, samples were
immediately placed in 4% buffered formaldehyde and were stored until further
analysis. Sampling of adults was not quantitative and emergent flies were
collected opportunistically with a hand-held insect net.

Abundance estimates of larval *C. edulis* were made using subsamples, which
were prepared by pouring the main-net sample through a binary sample
splitter. Each subsample was then poured into a 300 ml measuring cylinder, in
which *C. edulis* larvae floated away from the rest of the sample. Larvae were
then collected by decanting through a 52 μm sieve, transferred to a Petri dish
and counted under a stereomicroscope. Larval instars were distinguished on
the basis of head length. Sampling efficiency for fourth instars was low for
many of the surveys, particularly those in 1992 when routine daytime samples
were collected down to 150 m depth. Even in 1993, when sampling depth was
increased to 220 m, fourth instars were probably underestimated during the
day; it is now known that diel migration may extend to even greater depths
(Irvine, 1997). A series of daytime samples collected during June 1993 to depths
of 260 m and below the zero oxygen boundary suggested that third and fourth
instars have similar abundance (Irvine, 1997), and estimations of the lake-wide
distribution pattern of larvae were therefore based on third instar larvae.

During nine of the 10 whole-lake surveys made during 1992 and 1993, plus
one in January 1994, any sightings of swarms of adult *C. edulis* were recorded
daily. One study objective was to ascertain whether any latitudinal variations
in the timing and frequency of swarming were present. To examine this, the
distribution pattern of fly swarms within boundaries of latitude was
standardized by dividing the number of days during which swarms were seen
by the total number of days spent within each lake sector. The presence of any
lunar synchrony in swarming behaviour was also investigated. Positive records
of swarms of adult flies were standardized to lunar periodicity in the manner
described above. Records were also made of swarms seen from the project base
at Senga Bay ($13°45'$ S $34°35'$ E) between mid-1991 and mid-1994. These
observations were scored on a scale of 1 (minimum) to 10 (maximum),
depending on the density of flies within each swarm and the number of swarms
observed.

B. Biomass and Production Estimates

Dry weight estimates of each life stage were made on freshly collected animals, dried for at least 3 h in an oven at 50–60°C, using an electronic microbalance. First and second instars were weighed in batches of up to 10 individuals, whereas third and fourth instars and adults were weighed individually.

Production of *C. edulis* larvae was estimated by the size-frequency method, originally proposed by Hynes and Coleman (1968) and modified by others (Hamilton, 1969; Benke, 1979; Menzie, 1980; Herman and Heip, 1985). Annual production (*P*) was calculated as

$$P = \sum_{j=1}^{i}(N_j - N_{j+1})(W_j \times W_{j+1})^{0.5} \qquad (1)$$

where

$$N_j = n_j\frac{1}{f_j}\cdot\frac{365}{CPI} \qquad (2)$$

and *i* is the number of size classes, W_j the mean weight of an individual in size class *j* (μg dry wt ind^{-1}), f_j the proportion of the life cycle spent in size class *j*, n_j the annual mean number of individuals observed in size class *j* (m^{-2}), and CPI the cohort production interval (time spent as larvae, days). Multiplying 1/CPI by the constant 365 d y^{-1} gives 365/CPI, which has the dimensions of individuals m^{-2} y^{-1}, and N_j is the estimate of the number of individuals per surface of lake (m^{-2}) that grow into size class *j* during a year. The mean number of individuals was calculated from the mean values estimated from each of the five whole-lake cruises within each year. Based on data obtained from intensive sampling to 260 m depth in 1993 (Irvine, 1995, 1997), abundance estimates of fourth instar larvae and pupae were adjusted assuming that the ratio of third:fourth instars approximated unity and that of pupae:fourth instars was 0.022. W_j was obtained from measured dry weights and f_j was estimated from the ranges of instar development times of *C. edulis* from Lake Victoria given by MacDonald (1956).

The method used calculates production on the basis of production lost between successive life stages. There is, therefore, a need to account for production within the final stage of development which, for estimation of larval production, was considered to be the pupa. This was calculated as the mean biomass multiplied by mean annual recruitment. As the pupal stage is a non-feeding stage, the biomass of pupae was considered to be synonymous with that of adults.

To allow for the apparent rapid development of the first instar of *C. edulis* in Lake Malawi, as evinced by consistently low numbers in the plankton, the development time of the first instar was estimated as 1 day rather than the 4–6 days estimated by MacDonald (1956) for first instar *C. edulis* in Lake Victoria.

Production estimates which incorporated longer development periods for first instars were calculated, but resulted in unrealistic negative production values for this group. Similarly, based on evidence from intensive sampling in June 1993 (Irvine, 1997), 4 days was considered a more reasonable estimate of pupal development period in Lake Malawi than the 2–3 days pupal development noted for *C. edulis* in Lake Victoria (MacDonald, 1956). The estimate of total larval lifespan of 2 lunar months given by MacDonald (1956) is consistent with time-lagged correlations of pupae numbers (Degnbol, 1982) and the present author's unpublished observations.

V.   RESULTS

A. Spatial Pattern of Larvae and Adults

Estimates of mean abundance of *C. edulis* during the 2 year sampling period ranged from 98 to 1651 individuals m^{-2} for first instar larvae, 202 to 6282 individuals m^{-2} for second instars, 95 to 2646 individuals m^{-2} for third instars and 14 to 1865 individuals m^{-2} for fourth instars. The maximum density of pupae recorded was 213 pupae m^{-2} and the mean abundance during June 1993 in samples taken to depths of 260 m was 22 pupae m^{-2}. Abundance estimates of third instar larvae in relation to latitude are shown in Figure 2. Third instar larvae were probably the most representative component of the actual spatial and temporal pattern of *C. edulis* populations within the lake. This is because their occurrence tended to be the least variable of the instars, owing to a longer instar duration than the first and second instars and to their being more efficiently captured during routine sampling than were fourth instars. Mean lake-wide abundance of third instars followed a pattern identical to that estimated for mean standing biomass of all *C. edulis* larvae (Figure 3). The abundance of larvae and pupae in the southernmost part of the lake was usually lower than elsewhere, although higher densities were recorded towards the end of both 1992 and 1993. Elsewhere, there was no consistent pattern of spatial distribution within the lake and seasonal patterns of abundance were more pronounced than spatial patterns.

The temporal pattern of the distribution of *C. edulis* biomass was one of relatively low standing biomass during the first few months of the year, with an increase in October followed by a decline in December. Although this general trend towards an increased biomass during the later part of the year was evident in both sample years, inter-year differences in timing and magnitude were apparent. The total mean [± 95% confidence limits (CL)] biomass of *C. edulis* in 1993 (201 ± 23 mg dry wt m^{-2}, $n = 205$) was greater than that estimated in 1992 (72 ± 10 mg dry wt m^{-2}; $n = 248$; t-test, $p < 0.001$). The highest estimates of standing biomass throughout the lake were seen towards the end of 1993 (Figure 3).

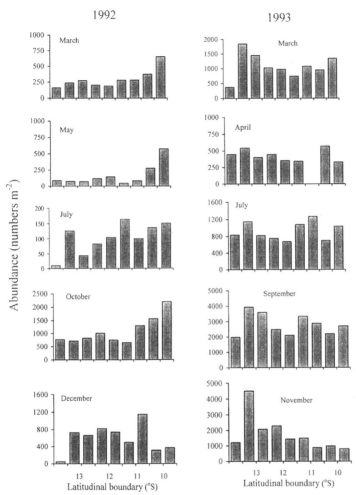

Fig. 2. Mean abundance of third instar larvae of *Chaoborus edulis* found at different latitudes in Lake Malawi during whole-lake surveys in 1992 and 1993. Means were based on *n* values of 1–9; the coefficient of variation (SD/mean) ranged from 7 to 80%. No samples were collected between 10°30' S and 11°S in April 1993.

B. Adult Swarming

Swarms of adults were mostly seen during early morning and seldom after midday, suggesting that swarms develop following the ascent of pupae through the water column at night. Pupae were consistently undersampled during all but the deepest tows with the HSPN, which suggests that they develop either deep in the water column, probably near the permanent oxygen boundary, or in shallow habitats, in the lake's sediment. The typical pattern of swarm development for adult flies was to aggregate near the water surface around

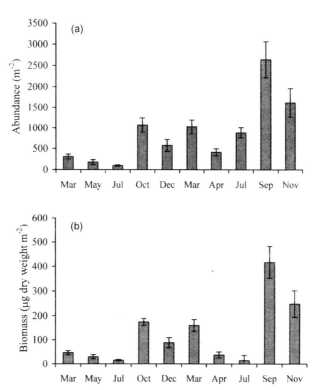

Fig. 3. (a) Mean (±95% CL) abundance of third instar larvae and (b) total standing biomass of *Chaoborus edulis* recorded during each whole-lake survey in 1992 and 1993. Sample size (*n*) was between 33 and 53.

dawn. Over a period of about 1 h these aggregations developed into a dense cloud close to the water surface. Following this, some of the aggregation rose away from the water surface to form the characteristic column, spiralling upwards from the aggregation of flies at the base. Columns often reached a height of about 100 m and maintained their structure even in strong winds. The whole column can move at several knots, depending on wind speed. After a period of time, typically ranging between 30 min and 2 h, the structure of the column began to break down and flies were often carried by the wind to the shore, where they aggregated around trees. In the north of the lake the local people collect these aggregations of flies for food, using them to make *kungu* cake.

Swarms are often observed in clusters over large areas of the lake but, at the same time, are absent from other areas. The frequency of occurrence of swarms of adults in relation to latitude, based on data from standardized observations, is shown in Figure 4. *Chaoborus edulis* swarms were observed most frequently between 11° S and 12° S, where they were present on 22 of the 24 days on which observations were made. Although these preliminary data do not permit

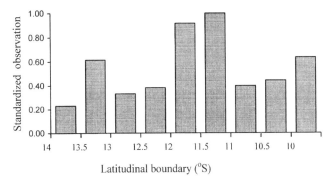

Latitudinal boundary (°S)

Fig. 4. Standardized frequency of swarms of adult *Chaoborus edulis* observed from the survey boat at different latitudes between March 1992 and January 1994.

rigorous testing for any relationship between lake latitude and seasonality, it is noteworthy that swarms were observed in this central part of the lake during every survey and during all seasons. Standardized observations of swarming behaviour in relation to the lunar cycle provided no evidence of any distinct lunar periodicity in this behaviour (Figure 5), but there was a weak bimodality in the adult emergence pattern, with maxima occurring shortly after the new moon and around the period of the full moon. Swarms were least common during the period leading up to a new moon. A total of 46 observations was made within the periods of 7 days before and 7 days after a new moon, and swarms were seen on 26 (57%) occasions. Around the period of the full moon, swarms were observed on 33 of 56 days (59%). Frequencies for the week preceding and that following the new moon were 53% (n = 80 days' observation) and 45% (n = 2 days' observation), respectively. Observations of swarms and sampling of pupae over a 24 day period in June 1993 suggested

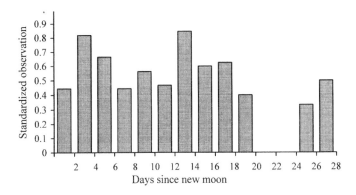

Days since new moon

Fig. 5. Standardized frequency of swarms of adult *Chaoborus edulis* observed from the survey boat at different parts of the lunar cycle between March 1992 and January 1994.

that pupae and emerging adults were absent in the 10 day period following the full moon.

Observations made at Senga Bay revealed swarms in the south of the lake to be prevalent only from November to March (Figure 6). The Senga Bay data indicated that the relative frequency of swarms in the first half of the lunar cycle (77 observations) was greater than that during the second half (24 observations).

C. Lake-wide Production

Dry weight estimates (mean $\pm$ s.e. of n batches or individuals) for first to fourth instar larvae of C. edulis were 1.7 $\pm$ 0.37 mg (n = 5), 6.6 $\pm$ 0.7 mg (n = 10), 41.7 $\pm$ 4.6 mg (n = 10) and 100.3 $\pm$ 15.3 mg (n = 13), respectively. Mean dry weight of adult flies was 300 $\pm$ 27 mg (n = 21). Total annual production during 1992 and 1993 was estimated as 2.2 g dry wt m^{-2} and 6.4 g dry wt m^{-2}, respectively (Table 1).

VI. DISCUSSION

A. Seasonal Variation and Macrodistribution

The pattern of abundance of C. edulis in Lake Malawi was similar in both study years: enhanced production towards the end of the year, with population minima during the middle part of the year. During June and July, strong Mwera winds blowing from the south-east result in enhanced turbulence of the water column that brings nutrients from the hypolimnion into the epilimnion.

Table 1
Relative life-stage duration (F_i), mean annual abundance (m^{-2}, n_j), mean annual recruitment (N_j), mean instar biomass (μg dry weight, W_j), contribution to production of stage-specific mortality (P_j) and overall production (g dry weight m^{-2} y^{-1}) estimated for Chaoborus edulis in 1992 and 1993

Instar	F_i	W_j	n_j 1992	n_j 1993	N_j 1992	N_j 1993	P_j 1992	P_j 1993
I	0.02	1.7	177	659	57 683	214 763	0.40	0.30
II	0.15	6.6	1055	2897	45 842	125 882	0.58	1.55
III	0.26	41.7	446	1300	11 181	32 589	0.35	1.01
IV	0.50	100.3	446	1300	5814	16 946	0.85	2.48
Pupae	0.07	300.0	10	· 29	914	2663	0.27	0.80
Total							2.09	6.13

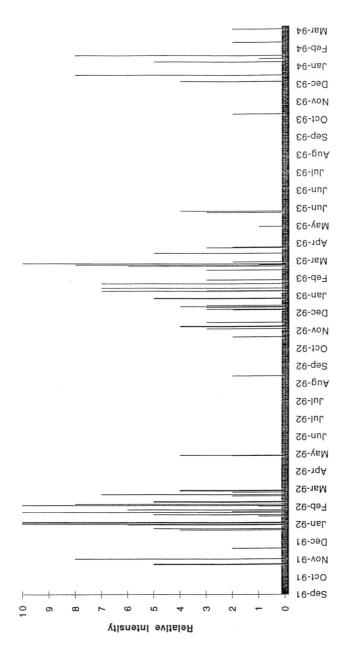

Fig. 6. Observations of swarms of adult *Chaoborus edulis* made from the shore at Senga Bay (13°45′ S 34°35′ E) between September 1991 and March 1994.

These nutrients are utilized by phytoplankton, which show greatly increased productivity at such times (Patterson *et al.*, this volume). The response of phytoplankton populations is closely coupled with that of zooplankton which, in turn, leads to greater food availability for *Chaoborus* larvae and a maxima of standing biomass that manifest itself about 2 months (or one *C. edulis* life cycle) after the annual maxima of zooplankton (Irvine *et al.*, 1999).

There was no clear trend in the macrodistribution pattern of *C. edulis* within Lake Malawi, and larvae were found at all latitudes of the lake. However, in the extreme south of the lake, below 13.5° S, larvae were usually less abundant than at other latitudes, a trend supported by the findings of Twombly (1983a), who reported extremely low densities of *C. edulis* larvae in her zooplankton samples from Cape Maclear (latitude *c.* 14° S). The lowest frequency of adult swarms throughout the lake were also recorded in its southern parts. At Senga Bay, adult swarms were sporadic and seasonal, being more common during the wet and relatively calm period of the year (approximately November–February) than during the windier and drier time (April–September). The periods when adult swarms were observed in the extreme south were generally accompanied by northerly winds, and accordingly the swarms were almost always moving in a southerly direction.

What factors might be responsible for the lower densities of *C. edulis* in the south of the lake? It is possible that populations of late-instar larvae move into shallow southern waters ($> c.$ 200 m) mainly through drift. Late instars have a high-amplitude diel migration distance, which probably functions as a mechanism to reduce fish predation by seeking refuge in deep water at, or close to, the zero oxygen boundary (Irvine, 1997). However, the southern part of the lake does not extend to that boundary, which normally occurs between 180 and 220 m, and at these places fourth instar larvae have been recorded in sediments (Allison *et al.*, 1996b). Here, and presumably also at other parts of the lake that are shallower than that boundary, larvae migrate and pupae develop within the sediment of the lake bed. However, in these situations *C. edulis* larvae and pupae remain vulnerable to predation from demersal and benthic fishes (Allison *et al.*, 1996b), and this is likely to be a less effective predation refuge than the oxycline of the open water. These demersal fish assemblages are often dominated by *Lephrinops* spp., *Synodontis njassae* Keilhack and *Diplotaxodon* spp. (Tomasson and Banda, 1996), for which fourth instar *C. edulis* larvae are an important food (Allison *et al.*, 1996a, b; Irvine *et al.*, in press).

The consequence of this high predation pressure would be to reduce the number of larvae achieving adulthood in those parts of the lake shallower than the zero oxygen boundary. Such large-scale mortality of invertebrate larvae is a common phenomenon in the marine environment (e.g. Rothlisberg *et al.*, 1983; White *et al.*, 1988). High predation, leading to low adult emergence, is consistent with the low densities of *C. edulis* recorded in the shallow southern end of the lake (Twombly, 1983a, b). This is further supported by the observed dominance of pupae and emerging adults of *C. edulis* found in the stomach contents of a variety of pelagic fishes sampled in this area in 1993, during which time no swarms were seen in the vicinity (E.H. Allison, pers. comm.).

B. Lunar Periodicity in Emergence Patterns

The emergence of *C. edulis* was generally associated with the period between the new and full moon, but not ubiquitously so, and emergence was reduced or absent for a short period around 4–10 days prior to the new moon. This finding is consistent with observations on emergence cycles of aquatic insects made in the 1950s in Lake Victoria (MacDonald, 1956; Corbet, 1958; Tjønneland, 1958) and those of Degnbol (1982), who noted that *C. edulis* pupae in Lake Malawi had a 4-weekly cycle of maximum abundance coinciding with the 2 weeks preceding a full moon. An identical trend, of maximum abundance of pupae during the 2 weeks prior to a full moon, was also evident in the present study for samples taken in June 1993 (Irvine, 1997). When sampling commenced on 8 June, the moon had just begun to wane and evenings were well illuminated. No pupae were found in the samples until 16 June and no adult *C. edulis* swarms were observed over the lake until 20 June, coinciding with the new moon. Thereafter, pupae were present in the plankton samples and swarms of adults were seen over the lake nearly every day until sampling ceased on 2 July, 2 days prior to the full moon.

The emergence of *C. edulis* takes place both near the new moon, when moonlight is minimal, and near the full moon, when moonlight is maximal, and thus seems not to be related to ambient lunar illumination. In Lake Malawi, the upward movement of emerging pupae from the lower oxygenated water to the surface is likely to be around 200 m and takes place over a period of several hours (Irvine, 1997). Emergence during the darker nights of the month might be an adaptive behaviour, acting to reduce predation pressure from visually hunting fish on the conspicuous and vulnerable pupae as they ascend through the water column. However, the greatest selective advantage of a lunar cue may be in maintaining a synchrony of larval emergence. The mechanism controlling synchronous emergence of adult *C. edulis* remains unknown, but within temperate chironomid species, which also have synchronous emergence, a seasonal diapause within larval development stages has been postulated as important (Goddeeris, 1987, 1991). A similar mechanism might apply also to *C. edulis*. Lunar illumination may stimulate a short-lived diapause within the fourth instar stage, during which time further growth of well-developed larvae is arrested while other, less mature individuals continue to grow to a similar stage of development. Subsequently, following the breaking of diapause, development to the pupal stage is determined by local feeding conditions. Clearly, much remains to be learned of the cues, mechanisms and adaptiveness of timing of emergence in *C. edulis*.

C. Larval Production

In his studies of *C. edulis* in Lake Victoria, MacDonald (1956) estimated total larval development time to be about 2 lunar months, which is consistent with the findings of the present study on this species in Lake Malawi. A 2-month total development time also agrees well with the time scale of the response of the larval *C. edulis* population to increases in food availability following wind-induced mixing of the water column in June–July of both 1992 and 1993 (Irvine *et al.*, 1999). All larval stages were present throughout the year, but first instar larvae occurred only sporadically, indicating a rapid development time through this stage of probably only 1 or 2 days. Therefore, when estimating larval production, a first instar period of just over 1 day was adopted. This is less than that estimated previously for *C. edulis* from Lake Victoria (MacDonald, 1956), and is also less than the original estimate of 2 days used for the calculation of *C. edulis* larval production in Lake Malawi by Irvine (1995) and Irvine *et al.* (in press), and which gave negative production estimates for first instar larvae. The lower estimation of life-stage duration does not, however, significantly affect previous estimates. In the present study, the difference between first appearance of pupae and adults indicated a pupal development time of around 4 days. Based on the assumption that emergence occurs over a 2-week period, analysis of the cumulative numbers of pupae collected indicated a mortality rate of fourth instars of approximately 75%. Instar-specific annual recruitment, calculated for the estimates of production (Table 1), gave mean mortality estimates of 26%, 75%, 48% and 85% for first to fourth instars, respectively. These high mortality rates, especially those of later instars, are consistent with the intense predation pressure known to occur from fish (Allison *et al.*, 1996a). However, for second instar larvae, which by virtue of their small size and high transparency are less prone to predation by visual fish predators, the high mortality is likely to be mainly a consequence of food limitation. These early instar larvae are restricted in the size range of zooplankton prey that they can ingest (Irvine, 1997).

The higher estimate of *C. edulis* production in 1993 than in 1992 was consistent with differences in the general pattern of primary and secondary productivity observed between the two years (Irvine *et al.*, in press; Patterson *et al.*, this volume). The estimated mean production of 4.3 g dry wt m^{-2} y^{-1} for *C. edulis* is similar to that for predatory crustacean zooplankters in the lake, and about three times greater than that for the zooplantivorous larvae of the endemic cyprinid fish *Engraulicypris sardella* (Günther) (Irvine *et al.*, in press). Collectively, these groups could be considered as tertiary producers and to be potential competitors. However, in the light of new evidence, and in contrast to the assertions of Degnbol (1993), it is now believed that, since energy is subsequently channelled into their fish predators, *C. edulis* larvae are not a significant sink of production. While there is likely to be an associated loss of energy in this trophic exchange it is also true that *C. edulis* larvae concentrate carbon from their zooplankton prey, which tends to be of small body size (Irvine, 1997) and low density (Irvine and Waya, 1999), and thereby provide an energy-rich food source and permit more efficient foraging for a number of fish species. Rather than comprising an energy sink, *C. edulis* larvae probably provide an important trophic link in both the pelagic and benthic foodwebs of Lake Malawi.

D. Ecological Manipulations: A Cautionary Note

Much of our knowledge of *C. edulis* in Lake Malawi was derived from earlier studies of the fisheries potential of the lake, conducted under the auspices of the FAO. These studies concluded that *C. edulis* competed with fish for available zooplankton food and thereby reduced fish production (Turner, 1982). As supporting evidence, envious comparison was drawn with neighbouring Lake Tanganyika, where *Chaoborus* was absent and where planktivorous clupeid fishes *Stolothrissa tanganicae* Regan and *Limnothrissa miodon* (Boulenger) formed the basis of a large and productive fishery. Furthermore, Degnbol (1993), using FAO data with the ECOPATH trophic model (Polovina, 1985), estimated that 98% of primary production of Lake Malawi was transferred through *C. edulis* and concluded that most of that energy was subsequently lost to the system as adult flies leaving the lake. *Chaoborus edulis* was postulated as deleterious to the fisheries potential of Lake Malawi, and led to the recommendation by Turner (1982) that zooplanktivorous clupeid fishes be introduced to Lake Malawi to regulate (through competition and predation) *C. edulis* larvae and thereby simultaneously improve fish yields within the lake and reduce the loss of energy from it.

However, subsequent studies, including the present one, have demonstrated the methodologies and biological assumptions underpinning these recommendations to be seriously wanting. First, the FAO project made no direct estimates of zooplankton standing biomass or production, and obtained their *C. edulis* biomass estimates from only one location, offshore of Nkhata Bay. The inadvisibility of generalizing to a whole-lake scenario from samples taken at a single location is clearly evinced by Twombly (1983a, b), who documented a rarity of *C. edulis* at the southern part of the lake, in stark contrast to the findings of the FAO project, despite their studies running concurrently. Secondly, the fish sampling by FAO was not representative of the whole fish community, and inputs by Degnbol (1990) of fish biomass into the ECOPATH model may therefore have been biased. Thirdly, the absence of *Chaoborus* from Lake Tanganyika may be a consequence of physicochemical and bioenergetic factors, rather than predation pressure from the endemic clupeid fishes. Accordingly, any revival of the proposal by Turner (1982), which was based on the findings of the FAO studies, to introduce zooplanktivorous clupeids into

Lake Malawi in order to remove competition of *Chaoborus* from fish zooplanktivores would be ill advised.

Lastly, it should be recognized that the ecological consequences of introducing an alien species to any lake are uncertain, and therefore always fraught with danger, but especially so in the biologically diverse and poorly understood ecosystems which characterize ancient lakes. The consideration of such introduction is also clearly contrary to the precautionary principle of ecosystem management that is increasingly incorporated into international treaties and guidelines. Any fish introduction into Lake Malawi that effectively redirected production from *C. edulis* into fish could have disastrous consequences on deep-water and demersal stocks of fish that feed on *C. edulis*, and cause serious disruption to ecosystem functioning, biodiversity, fish stocks and, ultimately, protein availability for human consumption.

ACKNOWLEDGEMENTS

The author thanks his colleagues on the UK/SADC Lake Malawi Fisheries Assessment Project for their support and B. Goddeeris for his thoughtful discussions on *Chaoborus* emergence patterns. The author is grateful to the Governments of Malawi, Mozambique and Tanzania for permission to work on the lake, to the British Overseas Development Administration for logistic support and funding, to A. Rossiter for his efforts and assistance and to the important insights provided by N.F. Broccolli.

REFERENCES

Allison, E.H., Irvine, K., Thompson, A.B. and Ngatunga, B. (1996a). Diets and food consumption rates of pelagic fish in Lake Malawi, Africa. *Freshwat. Biol.* **35**, 489–515.

Allison, E.H., Irvine, K. and Thompson, A.B. (1996b). Lakeflies and the deep-water demersal fish community of Lake Malawi. *J. Fish Biol.* **48**, 1006–1010.

Benke, A.C. (1979). A modification of the Hynes method for estimating secondary production with particular significance for multivoltine populations. *Limnol. Oceanogr.* **24**, 168–171.

Corbet, P.S. (1958). Lunar periodicity of aquatic insects in Lake Victoria. *Nature* **182**, 330–331.

Degnbol, P. (1982). Food habits and zooplankton consumption by larval *Chaoborus edulis* in Lake Malawi. FAO FI:DP/MLW/75/019 *Field Document* **24**, 1–19.

Degnbol, P. (1993). The pelagic zone of Lake Malawi – a trophic box model. In: *Trophic Models of Aquatic Ecosystems* (Ed. by V. Christensen and D. Pauly), ICLARM Conference Proceedings **26**, 110–115.

Eccles, D.H. (1974). An outline of the physical limnology of Lake Malawi (Lake Nyassa). *Limnol. Oceanogr.* **19**, 730–742.

Goddeeris, B.R. (1987). The time factor in the niche space of *Tanytarsus* species in two ponds in the Belgian Ardennes (Diptera:Chironomidae). *Entomol. Scand. Suppl.* **29**, 281–288.

Goddeeris, B.R. (1991). Life cycle characteristics in *Tanytarsus debilis* (Meigen, 1830) (Diptera, Chironomidae). *Ann. Limnol.* **27**, 141–156.

Gonfiantini, R., Zuppi, G., Eccles, D.H. and Ferro, W. (1979). Isotope investigation of Lake Malawi. In: *Isotopes in Lake Studies*, pp. 195–217. International Atomic Energy Agency, Vienna.

Halat, K.M. and Lehman, J.T. (1996). Temperature-dependent energetics of *Chaoborus* populations: hypothesis for anomalous distributions in the great lakes of East Africa. *Hydrobiologia* **330**, 31–36.

Hamilton, A.L. (1969). On estimating annual production. *Limnol. Oceanogr.* **14**, 771–782.

Herman, P.M.J. and Heip, C. (1985). Secondary production of the harpacticoid copepod *Paronychocamptus nanus* in a brackish water habitat. *Limnol. Oceanogr.* **30**, 1060–1066.

Hynes, H.B.N. and Coleman, M.J. (1968). A simple method of assessing the annual production of stream benthos. *Limnol. Oceanogr.* **13**, 569–575.

Irvine, K. (1995). Ecology of the Lakefly, *Chaoborus edulis*. In: *Fishery Potential and Productivity of the Pelagic Zone of Lake Malawi/Niassa*. (Ed. by A. Menz), pp. 109–140. Natural Resources Institute, Chatham, UK.

Irvine, K. (1997). Food selectivity and diel vertical distribution of *Chaoborus edulis* (Diptera, Chaoboridae) in Lake Malawi. *Freshwat. Biol.* **37**, 605–620.

Irvine, K. and Waya, R. (1999). Temporal and spatial patterns of zooplankton standing biomass and production in Lake Malawi. *Hydrobiologia* **407**, 191–205.

Irvine, K., Patterson, G., Allison, E.H., Thompson, A.B. and Menz, A. (1999). The pelagic ecosystem of Lake Malawi: trophic structure and current threats. In: *Great Lakes of the World* (Ed. by M. Munawar). Backhuys, Leiden, The Netherlands.

Jackson, P.B.N., Iles, T.D., Harding, D. and Fryer, G. (1963). *Report on the Survey of Northern Lake Nyassa, 1954–1955*. Government Printer, Zomba, Malawi.

MacDonald, W.W. (1956). Observations on the biology of chaoborids and chironomids in Lake Victoria and on the feeding of the "elephant-snout fish" (*Mormyrus kannume* Forsk.). *J. Anim. Ecol.* **25**, 36–53.

McGowan, L.M. (1974). Ecological studies on *Chaoborus* (Diptera, Chaoboridae) in Lake George, Uganda. *Freshwat. Biol.* **4**, 483–505.

Menzie, C.A. (1980). A note on the Hynes method of estimating secondary production. *Limnol. Oceanogr.* **25**, 770–773.

Milligan, S.P. and Riches, B.F. (1983). The new MAFF guideline to high speed plankton samplers. ICES C.M.1987/L7, 1–6 (mimeo).

Nichols, J.H. and Thompson, A.B. (1991). Mesh selection of copepod and nauplius stages of four calanoid copepod species. *J. Plankton Res.* **13**, 661–671.

Patterson, G. and Kachinjika, O. (1993). Effect of wind-induced mixing on the vertical distribution of nutrients and phytoplankton in Lake Malawi. *Verh. Int. Verein. Limnol.* **25**, 872–876.

Polovina, J.J. (1985). An approach to estimating an ecosystem box model. *US Fish. Bull.* **83**, 457–460.

Rothlisberg, P.C., Church, J.A. and Forbes, A.M.G. (1983). Modelling the advection of vertically migrating shrimp larvae. *J. Mar. Res.* **41**, 511–538.

Tjønneland, A. (1953). Observations on *Chaoborus edulis* (Edwards) (Diptera, Culicidae). *Univ. I Bergen Arbok 1958, Naturvitenskapelig rekke* **16**, 1–12.

Tomasson, T. and Banda, M.C. (1996). *Depth Distribution of Fish Species in Southern Lake Malawi. Implications for Fisheries Management*. Fisheries Bulletin No. 34. Fisheries Department, Government of Malawi.

Turner, J.L. (1982). Lake flies, water fleas and sardines. *FAO* FI:DP/MLW/75/019 *Tech. Rep.* **1**, 165–181.

Twombly, S. (1983a). Patterns of abundance and population dynamics of zooplankton in tropical Lake Malawi. Unpublished PhD Thesis, Yale University.

Twombly, S. (1983b). Seasonal and short term fluctuations in zooplankton abundance in tropical Lake Malawi. *Limnol. Oceanogr.* **228**, 1214–1224.

Verbeke, J. (1958). *Exploration du Parc National Albert. Chaoboridae (Diptera Nematocera)*, pp. 1–57. Institute des Parcs Nationaux du Congo, Brussels.

White, R.G., Hill, A.E. and Jones, D.A. (1988). Distribution of *Nephrops norvegicus* (L.) larvae in the western Irish Sea: an example of advective control on recruitment. *J. Plankton Res.* **10**, 735–747.

Age Determination and Growth of Baikal Seals (*Phoca sibirica*)

M. AMANO, N. MIYAZAKI and E.A. PETROV

I. SUMMARY

The ages of 75 Baikal seals (*Phoca sibirica*) were determined in order to study the growth pattern of body length and body weight in this species. Longitudinal decalcified and stained sections of canine teeth were prepared, and growth layer groups (GLGs) in the dentine and cementum were counted. In those specimens with fewer than 10 GLGs, the GLG counts in dentine agreed well with those in cementum. However, in specimens with more than 10 GLGs, the cemental GLGs tended to exceed dentinal ones. Individuals of 4–5 years old were absent from the sample population, suggesting that the recorded mass mortality of Baikal seals in 1987–1988 (4–5 years before sampling) had affected the age composition. Growth of body length, body weight and core weight ceased at around 20 years of age. The growth pattern was similar to that of ringed seals (*Phoca hispida*), but differed from that of other *Phoca* species. Asymptotic body length and core weight were significantly different between sexes, but body weight was not. In the spring sampling period, female body weights tended to be heavier than those of males of the same body length, a feature attributed to the thicker blubber of females. Sexual dimorphism, and the large fluctuation in body weight and presence of secondary sexual characters reported in males, suggested the presence of intermale competition in the Baikal seal, presumably for access to females.

ADVANCES IN ECOLOGICAL RESEARCH VOL. 31
ISBN 0–12-013931–6

II. INTRODUCTION

A. Origins and General Biology

The Baikal seal *Phoca sibirica* is endemic to Lake Baikal, the oldest (*c.* 28 My old; Mats, 1993) and deepest (1632 m; Kozhov, 1963) lake in the world. It is closely related to the Caspian seal *P. caspica* of the Caspian Lake and the ringed seal *P. hispida*, which is widely distributed in Arctic marine waters. These three species share several traits, such as a relatively small body, a slightly built skull and habits adapted to life on the ice, and are sometimes classified together in the subgenus *Pusa* (e.g. Allen, 1880; Scheffer, 1958).

Baikal seals occupy the top position in the foodweb of Lake Baikal, and feed on many kinds of fish, such as the pelagic sculpins known as golomyankas (*Comephorus baicalensis* and *C. dybowskii*), and benthic sculpins, such as *Cottocomephorus grewingki* and *C. comephoroides* (Thomas *et al.*, 1982). However, younger seals also eat amphipods (Pastukhov, 1993), which are numerous in the lake (see Takhteev; Morino *et al.*, this volume). Baikal seals breed in ice lairs from February to March and it is believed that most sexually mature females give birth every year (Thomas *et al.*, 1982).

The evolutionary history and origin of Baikal seals are not fully understood. Two hypotheses, based on fossil records and geological evidence, have been presented. One posits that the recent seals of the subgenus *Pusa* share a common ancestor which inhabited the ancient Paratethys Sea and that the ancestor of the Baikal seal immigrated to Baikal directly from there (Chapskii, 1955). The alternative view is that the common ancestor of ringed and Baikal seals moved into the Arctic from the Parathethys Sea, and the ancestor of Baikal seals then migrated up the Yenisey River to Lake Baikal during the glaciation (Repenning *et al.*, 1979). Speciation then took place following isolation in Lake Baikal. It remains unknown which of these scenarios is the more accurate; however, the close affinity between Baikal and ringed seals revealed by recent and ongoing craniometorical and genetic comparisons lends some support to the latter theory (Koyama, 1996; H. Sasaki and K. Numachi, pers. comm.).

Baikal seals display several characteristic features, which it is believed have evolved as adaptations to life in Lake Baikal. The comparatively large forelimbs and claws of this species are considered as adaptations to excavate holes in the thick ice of the lake (Thomas *et al.*, 1982), and the enlarged eyes have been interpreted as adaptations for feeding in the deep, clear waters of Lake Baikal (Chapskii, 1955; King, 1983). Recently, it has been discovered that the cranial bones and muscles are highly modified to accomodate the enlarged eye (Endo *et al.*, 1998, 1999). Most dives are between 10 and 50 m depth, but studies using satellite-linked depth recorders have revealed that Baikal seals sometimes dive to depths beyond 300 m (Stewart, B.S. *et al.*, 1996).

B. Ageing Studies

Determining the age of these seals provides basic data for documenting their growth, life history and ecology. Furthermore, owing to their high longevity and trophic level, Baikal seals of known age can provide invaluable information on accumulation trends of toxic pollutants, and can thus act as indicator species for monitoring the lake environment. Relatively high accumulations of organochlorine residues and their age-dependent concentration pattern have been reported in Baikal seals (Nakata *et al.*, 1995), and the mass mortality in Baikal seals during 1987–1988 (Grachev *et al.*, 1989; Osterhaus *et al.*, 1989; Nakata *et al.*, 1997) has been attributed to the sublethal toxic effects of these pollutants having lowered the resistance in the seals, and rendered them more vulnerable to the distemper virus which ultimately killed them.

As part of their studies of life-history traits of this species, Russian scientists have made use of age determination of Baikal seals, utilizing growth layers on the claw for younger animals, and dentinal and cemental growth layers in the canine teeth for older animals (e.g. Pastukhov, 1969). However, detailed descriptions of the methods used and information on growth rates of these seals are limited, and growth curves of body length and weight have been reported only for a few captive individuals (Yamagishi *et al.*, 1993).

In an attempt to rectify this paucity of data on wild populations of the Baikal seal, in spring 1992 the biology of this species was investigated in Lake Baikal. As part of this study 101 seals were killed and examined, and the results are reported of age determination of some of these specimens, based on the dentinal and cemental growth layers in their canine teeth. The age composition and growth pattern of Baikal seals are described, and an attempt made to identify life-history adaptations of the Baikal seal to the habitat of Lake Baikal, through comparison of the growth pattern of the Baikal seal with that of other seal species.

III. MATERIALS AND METHODS

Forty-four (27 females and 17 males) Baikal seals collected from Lake Baikal between 14 May and 1 June 1992 were examined. Canine samples and body length and weight data were used from 31 additional specimens collected during the same period. Body weight was measured to the nearest 100 g, and body length along the body axis from the tip of the upper snout to the tip of the tail was measured to the nearest 1 mm. Previous studies have used curvilinear length as an indicator of seal length (e.g. Pastukhov, 1993). However, curvilinear length includes information on the robustness of the body and is not independent of the weight parameters, which vary seasonally. As such, this

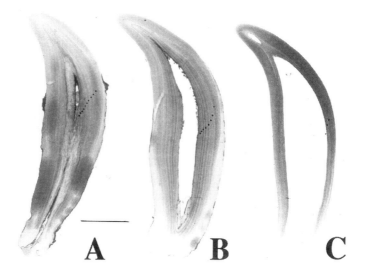

Fig. 1. Dentinal growth layers in decalcified and haematoxylin-stained longitudinal sections of canines of Baikal seals (A, 19.5 + years old; B, 15.5 years old; C, 2.0 years old). Osteodentine was formed in specimen A. Dots indicate dark-stained layers of GLG, excluding very thin layers near the pulp cavity. Scale bar = 5 mm.

length measurement is inappropriate for examining the growth pattern of axial length and length–weight relationships.

The age of each individual was determined using a single canine tooth removed from the right lower jaw of each skull. Longitudinal, decalcified and haematoxylin-stained thin sections (20 μm) of each tooth sampled were prepared following the method of Kasuya (1976) (Figures 1, 2). Although age determination of pinnipeds has traditionally been based on cross-sections of teeth, Stewart, R.E.A. *et al.* (1996), in their ageing studies of ringed seals, found longitudinal sections to be superior, and therefore longitudinal sections were also used in this study. For each canine, all growth layer groups (GLGs; Perrin and Myrick, 1980) were counted in both dentine and cementum, three times independently, and the median value taken as the number of GLGs. For two specimens, poor quality preparations meant that no reliable GLG counts could be obtained, and data from these two individuals were therefore excluded from the analyses. The annual accumulation of a dentinal and cemental GLG is widely accepted for pinniped species (Scheffer, 1950; Laws, 1959; Mansfield and Fisher, 1960) and therefore one GLG was assumed to correspond to 1 year.

Von Bertalanffy's growth equation was fitted by a non-linear least-square method. Sexual differences in the relationship between body length and body and core weights were examined with analyses of covariance (ANCOVA), with body length as a covariate.

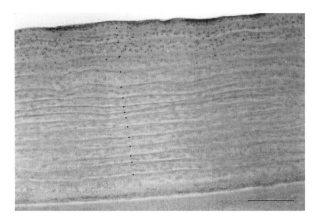

Fig. 2. Cemental growth layers in decalcified and haematoxylin-stained longitudinal section of the canine of a Baikal seal (15.5 years old). Dots indicate dark-stained layers of GLG. Bar = 0.2 mm.

IV. RESULTS AND DISCUSSION

A. Age Determination

In animals with fewer than 10 cemental layers, the number of dentinal GLGs corresponded well to that of cemental GLGs. In the older animals, however, the number of cemental layers tended to exceed that of dentinal ones (Figure 3). A similar tendency has been observed in ringed seals (*Phoca hispida*) (Stewart, R.E.A. *et al.*, 1996), bearded seals (*Erignathus barbatus*) (Benjaminsen, 1973), and Caspian seals (*P. caspica*) (N. Miyazaki, unpubl.). In Baikal seals, pulp cavity closures were observed only in some animals with about 20 or

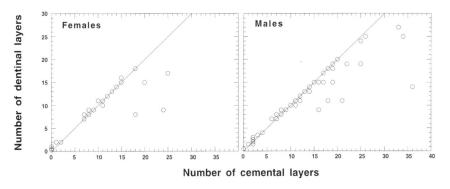

Fig. 3. Comparison between the number of dentinal and cemental growth layers in the canine teeth of Baikal seals. Lines indicate equal accumulation rate.

454   M. AMANO *ET AL.*

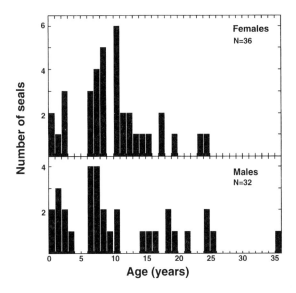

Fig. 4. Age composition of Baikal seals sampled in spring 1992.

more dentinal GLGs, but the irregular or osteodentine began to deposit at around 10 GLGs and seemed to decrease the readability of the sections. This was the main factor causing the discrepancy of readings between dentinal and cemental growth layers. From the above observations, age was estimated as the larger number of either dentinal or cemental GLGs minus 0.5 (Kasuya, 1976).

B. Age Composition

The sample of 73 Baikal seals ranged in age from 0.25 to 24.5 years old for females and 0.5 to 35.5 years old for males (Figure 4). The maximum age in both sexes was markedly younger than the 56 years for females and 52 years for males previously reported for this species (Pastukhov, 1974). The sex ratio was approximately equal, with females comprising 52.9% of the total sample. The mode of the age composition was at 6–7 years for females and 10 years for males. The lower number of young animals less than 3 years old in relation to the number in the 7–11-year-old age group may reflect some sampling bias. It is, however, conspicuous that no 4–5-year-old animals were present in this sample. Their absence can be attributed to the well-documented mass die-off during 1987–1988 of Baikal seals infected with distemper virus (Grachev *et al.*, 1989; Osterhaus *et al.*, 1989). This absence of 4–5-year-old individuals suggests a higher mortality of newborns and/or lower productivity of females during that time, and provides a graphic example of the utility and usefulness of age determination studies.

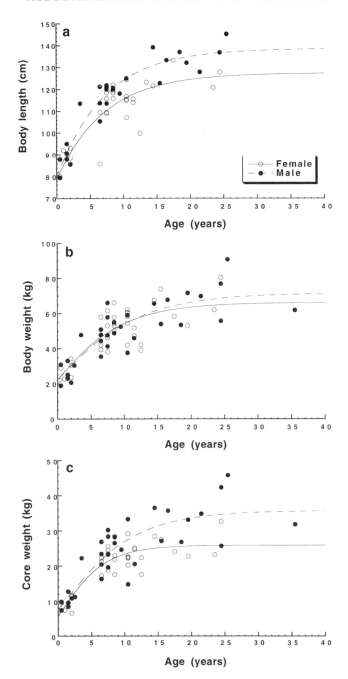

Fig. 5. Relationships of (a) body length, (b) body weight and (c) core weight to age in Baikal seals. Solid and broken curve lines are von Bertalanffy's growth curves for females and males, respectively. For growth equations, see text.

C. Growth

Von Bertalanffy's growth equation was calculated as:

$$\text{Female: BL} = 127.3(1 - 0.381 \exp^{-0.141\,t}) \quad (r = 0.870)$$

$$\text{Male: BL} = 138.5(1 - 0.423 \exp^{-0.140\,t}) \quad (r = 0.953),$$

where BL is body length in cm, and t is age in years. The growth of body length appeared to cease at around the age of 20 years (Figure 5a). Curvilinear length increase ceases at 20–25 years for both sexes (Pastukhov, 1993), and it seems that growth of both linear and curvilinear length stop at a similar age. Baikal seals start to breed by the age of 6 years for females and age 7 years for males (Thomas *et al.*, 1982), and therefore growth continues for at least 10 years after sexual maturity is attained. Asymptotic body length of males was significantly larger than that of females (t-test, $p < 0.001$). Yamagishi *et al.* (1993) documented the growth of Baikal seals in captivity, and obtained asymptotic lengths of 107.8 cm for females and 119.7 cm for males. These values are are much smaller than those of the present study, probably because of the lack of older animals in their sample and to captive conditions being inferior to those in the wild.

Information on the reproductive behaviour of the Baikal seal is scarce, since they are believed to mate underwater (Thomas *et al.*, 1982). However, some inferences concerning the mating system can be drawn from interspecific comparisons of gross morphology. In many mammal groups, a large body size in males relative to females is often indicative of competition between males, whereas an absence of sexual size dimorphism often indicates a monogamous mating system. The large body length difference between male and female Baikal seals is greater than that seen between male and female ringed seals (3.5 cm, eastern Canada, McLaren, 1958; 2–4 cm, Okhotsk, Beiring Sea and Chukchi Sea, Fedoseev, 1975; 4 cm, Baltic Sea, Helle, 1979). The ringed seal is believed to be monogamous (Frost and Lowry, 1981) and the relatively large size of male Baikal seals therefore suggests that intermale competition exists among Baikal seals. This is loaned support by the deeply wrinkled face of the mature male Baikal seal, which is a secondary sexual feature (Pastukhov, 1993).

Population-specific differences in the maximum length of ringed seals (*Phoca hispida*) have been attributed to geographical differences in the stability of the ice on which the pup is nursed (McLaren, 1958; Frost and Lowry, 1981; Fedoseev, 1975). However, despite these size variations, the overall growth pattern of these different populations is similar and individuals reach maximum body length at *c.* 15–20 years old. This is comparable to the growth of Baikal seals, and the close similarity between the growth patterns of these two *Phoca* species suggests a similar evolutionary life-history adaptation to the arctic environment.

Interestingly, the growth rate of Baikal, Caspian and ringed seals (subgenus *Pusa*) is slower than that of other *Phoca* seal species. This might be related to *Pusa* species' breeding in lairs on the fast ice, which is a rather stable habitat (Riedman, 1990; Lydersen, 1995), and in these species the nursing period is relatively long and the growth of the pup is slow. The growth rates (coefficient for age) of von Bertalanffy's growth equation are larger in the harbour seal (0.321 for females, 0.362 for males), (Markussen *et al.*, 1989) and in the harp seal (0.379) (Innes *et al.*, 1981) than in the Baikal seal. However, in *Phoca* species the relatively rapid growth rate is counteracted by a shorter growth period: in harbour (*Phoca vitulina*), largha (*P. largha*) and harp seals (*P. groenlandica*) growth ceases at around 10 years old (Naito and Nishiwaki, 1975; Innes *et al.*, 1981; Markussen *et al.*, 1989). Thus, these two groups differ in their growth strategies, *Pusa* species having a slow growth rate and growth which continues until the age of 15–20 years and *Phoca* species having a relatively high growth rate but one which ceases at around the age of 10 years.

Von Bertalanffy's growth equations for body and core weights of Baikal seals were calculated as follows:

$$\text{Female:} \quad \begin{aligned} \text{BW} &= 66.2(1 - 0.306 \ \exp^{-0.156\,t})^3 \qquad (r = 0.802) \\ \text{CW} &= 25.8(1 - 0.397 \ \exp^{-0.248\,t})^3 \qquad (r = 0.852) \end{aligned}$$

$$\text{Male:} \quad \begin{aligned} \text{BW} &= 71.6(1 - 0.323 \ \exp^{-0.128\,t})^3 \qquad (r = 0.852) \\ \text{CW} &= 35.7(1 - 0.402 \ \exp^{-0.150\,t})^3 \qquad (r = 0.870) \end{aligned}$$

where BW is body weight in kg, CW is core weight in kg, and t is age in years (Figure 5b,c). Growth stops at around the age of 20 years, when body weight reaches about 60 kg (Figure 5b). There was no sex difference in the asymptotic body weight (t-test, $p > 0.05$), unlike in harbour seals, where males are heavier than females by about 10 kg (Markussen *et al.*, 1989). However, male asymptotic core weight was heavier than that of females by about 10 kg (Figure 5c; t-test, $p < 0.001$).

The relationship between body weight and body length in Baikal seals is shown in Figure 6a. No difference was detected between the male and female slopes, but the BL-adjusted mean was significantly larger in females (ANCOVA, $p < 0.05$). However, there was no sex-related difference in the relationship between body length and core weight (ANCOVA, $p > 0.05$; Figure 6b). Equations for these relationships are as follows:

$$\begin{aligned} \text{Female:} &\quad \log(\text{BW}) = 2.370 \ \log(\text{BL}) - 3.149 \qquad (r = 0.943) \\ \text{Male:} &\quad \log(\text{BW}) = 2.370 \ \log(\text{BL}) - 3.186 \qquad (r = 0.944) \\ \text{Female} + \text{male:} &\quad \log(\text{CW}) = 2.860 \ \log(\text{BL}) - 4.527 \qquad (r = 0.942) \end{aligned}$$

where BW is body weight in kg, BL is body length in cm, and CW is core weight in kg. From this information, it is likely that the difference in body weight between the sexes reflects differences in blubber weight, i.e. females are

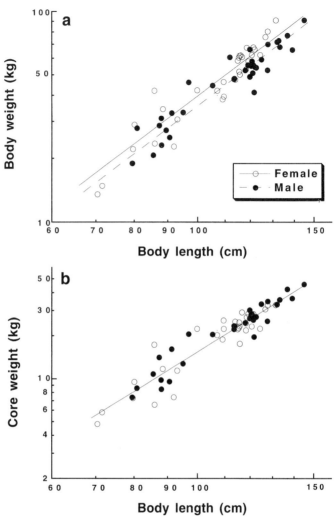

Fig. 6. Relationships of (a) body weight and (b) core weight to body length in Baikal seals. Solid and broken lines in (a) are regression lines for females and males, respectively. Solid line in (b) is the regression line for both sexes. For equations of regression lines, see text.

fatter than males. Blubber weight of phocid seals is known to vary seasonally (Sergeant, 1973). Pastukhov (1993) indicated that the maximum body weight of Baikal seals in June is 64 kg for females and 67 kg for males, values similar to the present study, but increases greatly in November, reaching 84 kg for females and 97 kg for males. It is known that phocid females lose a large portion of their fat through lactating and then retrieve it during the postweaning period (Sergeant, 1973; Fedak and Anderson, 1982), and it is therefore especially noteworthy that the seasonal fluctuation in body weight of Baikal seals was larger in males than in females. Among females, the seasonal fluctuation was reported to be larger in the 1–9-year-old age group, but was larger in males in older individuals (Pastukhov, 1993). However, apart from a report that the condition factor (= girth × 100/length) and blubber thickness of male harp seals decrease during moulting (Sergeant, 1973), there is a paucity of information on seasonal change in the weight of male seals. Accordingly, it is unknown whether the large seasonal weight change in the male Baikal seal is a feature unique to this species. The large weight fluctuation of adult males might reflect their reproductive behaviour, such as territorial defence or intermale competition, as inferred earlier. Future studies will address this question.

Because the present samples were taken in the spring period only, when seals were recovering their body weight and condition, it was not possible to describe seasonal variation in body weight. Core weight, however, is considered a more stable character. In future, by applying the body length and core weight relationship to body length and weight data of live-captured Baikal seals, it should be possible to make non-lethal estimates of blubber and skin weight (sculp weight) which will serve as an indicator of that individual's nutritional condition.

Baikal and ringed seals possibly share a common ancestor and show several similarities. Notable among these is the fact that both breed on the ice and both show a similar slow but prolonged growth pattern. The life history and behaviour of those seal species which breed on ice are considered to have been strongly influenced by the varying ice conditions, including ice break-up and drifting floes, and the presence of predators, such as polar bears and Arctic foxes (Riedman, 1990; McLaren, 1958; Lydersen, 1995). Baikal seals, however, are free from these constraints, since the ice in Baikal is fast and there are no predators other than humans (Thomas *et al.*, 1982). Future studies will utilize modern technologies, such as radio telemetry, hydrophones and underwater video cameras, to investigate free-ranging animals. Detailed comparative studies of this nature, comparing Baikal seals and other *Pusa* species, should reveal ecological traits specific to the Baikal seal and their adaptiveness to the unique habitat of Lake Baikal.

ACKNOWLEDGEMENTS

The authors thank Director M. Grachev, A. Timonin, M. Ivanov and other colleagues of the Limnological Institute, Siberian Division of the Russian Academy of Sciences, and the captain and crew of the R/V Balkhash for their help during the research in Lake Baikal. S. Tanabe, H. Sasaki and M. Ichikawa assisted in the fieldwork. They also thank K. Numachi and T. Kawai for their support and logistic assistance. The manuscript was improved by comments from two anonymous reviewers and the editors. The present study was financially supported by a grant-in-aid from the International Scientific Research (Project nos 04041035, 07041130 and 09041149) and Scientific Research (B)(2) (09460086) of the Ministry of Education, Science, Sports and Culture of Japan.

REFERENCES

Allen, J.A. (1880). History of North American pinnipeds, a monograph of the walruses, sea-lions, sea-bears, and seals of North America. *U.S. Geological and Geographical Survey of the Territories, Miscellaneous Publications* **12**, 1–785.

Benjaminsen, T. (1973). Age determination and the growth and age distribution from cementum growth layers of bearded seals at Svalbard. *Fiskeridirektoratets Skrifter Ser. Havundersokelser* **16**, 159–170.

Chapskii, K.K. (1955). Contribution to the problem of the history of development of Caspian and Baikal seals. *Trudy Zool. Inst. Akad. Nauk SSSR* **17**, 200–216 (in Russian).

Endo, H., Sasaki, H., Hayashi, Y., Petrov, E.A., Amano, M. and Miyazaki, N. (1998). Macroscopic observations of the facial muscles in the Baikal seal (*Phoca sibirica*). *Mar. Mamm. Sci.* **14**, 778–788.

Endo, H., Sasaki, H., Hayashi, Y., Petrov, E.A., Amano, M., Suzuki, N. and Miyazaki, N. (1999). CT examination of the head of the Baikal seal (*Phoca sibirica*). *J. Anat.* **194**, 119–126.

Fedak, M.A. and Anderson, S.S. (1982). The energetics of lactation: accurate measurements from a large wild mammal, the grey seal (*Halichoerus grypus*). *J. Zool.* **198**, 473–479.

Fedoseev, G.A. (1975). Ecotypes of the ringed seal (*Pusa hispida* Schreber, 1777) and their reproductive capabilities. *Rapp. Procès-Verbauz Réunions Conseil Int. Explor. Mer* **169**, 156–160.

Frost, K.J. and Lowry, L.F. (1981). Ringed, Baikal and Caspian seals–*Phoca hispida* Schreber, 1775; *Phoca sibirica* Gmelin, 1788 and *Phoca caspica* Gmelin 1788. In: *Handbook of Marine Mammals*, Vol. 2, *Seals* (Ed. by S.H. Ridgeway and R.J. Harrison), pp. 29–53. Academic Press, San Diego.

Grachev, M.A., Kumarev, V.P., Mammaev, L.V., Zorin, V.L., Baranova, L.V., Denikna, N.N., Belikov, S.L., Petrov, E.A., Kolesnik, V.S., Kolesnik, R.S., Dorofeev, V.M., Beim, A.M., Kudelin, V.N., Nagieva, F.G. and Sidorov, V.N. (1989). Distemper virus in Baikal seals. *Nature* **338**, 209.

Helle, E. (1979). Growth and size of the ringed seal *Phoca* (*Pusa*) *hispida* Schreber in the Bothnian Bay, Baltic Sea. *Z. Säugetierkunde* **44**, 208–220.

Innes, S., Stewart, R.E.A. and Lavigne, D.M. (1981). Growth in northwest Atlantic harp seals *Phoca groenlandica J. Zool.* **194**, 11–24.

Kasuya, T. (1976). Reconsideration of life history parameters of the spotted and striped dolphins based on cemental layers. *Sci. Rep. Whales Res. Inst., Tokyo* **28**, 73–106.

King, J.E. (1983). *Seals of the World.* Oxford University Press, Oxford.

Koyama, Y. (1996). Growth and morphology of three species of the subgenus *Pusa*. Unpublished MSc Thesis, University of Tokyo (in Japanese).

Kozhov, M. (1963). *Lake Baikal and its Life.* Dr. W. Junk, Publishers, The Hague.

Laws, R.M. (1959). Accelerated growth in seals, with special reference to the Phocidae. *Norsk Hvalfangst-Tidende* **9**, 425–452.

Lydersen, C. (1995). Energetics of pregnancy, lactation and neonatal development in ringed seals (*Phoca hispida*). In: *Whales, Seals, Fish and Man* (Ed. by A.S. Blix, L. Walløe and Ø. Ulltang), pp. 319–327. Elsevier, Amsterdam.

McLaren, I.A. (1958). The biology of the ringed seal (*Phoca hispida* Schreber) in the eastern Canadian Arctic. *Bull. Fish. Res. Brd Can.* **118**, 1–97.

Mansfield, A.W. and Fisher, H.D. (1960). Age determination in the harbour seal, *Phoca vitulina* L. *Nature* **186**, 92–93.

Markussen, N.H., Bjorge, A. and Oritsland, N.A. (1989). Growth in harbour seals (*Phoca vitulina*) on the Norwegian coast. *J. Zool.* **219**, 433–440.

Mats, V.D. (1993). The structure and development of the Baikal rift depression. *Earth Sci. Rev.* **34**, 81–118.

Naito, H. and Nishiwaki, M. (1975). Ecology and morphology of *Phoca vitulina largha* and *Phoca kurilensis* in the southern Sea of Okhotsk and northeast of Hokkaido. *Procès-Verbauz Réunions Cons. Int. Explor. Mer* **169**, 379–386.

Nakata, H., Tanabe, S., Tatsukawa, R., Amano, M., Miyazaki, N. and Petrov, E.A. (1995). Persistent organochlorine residues and their accumulation kinetics in Baikal seal (*Phoca sibirica*) from Lake Baikal, Russia. *Envir. Sci. Technol.* **29**, 2877–2885.

Nakata, H., Tanabe, S., Tatsukawa, R., Amano, M., Miyazaki, N. and Petrov, E.A. (1997). Bioaccumulation profiles of polychlorinated biphenyls including coplanar congeners and possible toxicological implications in Baikal seal (*Phoca sibirica*). *Environ. Pollut.* **95**, 57–65.

Osterhaus, A.D., Groen, M.E.J., Visser, I.K.G., Bildt, M.W.G.V.D., Bergman, A. and Klingeborn, B. (1989). Distemper virus in Baikal seals. *Nature* **338**, 209–210.

Pastukhov, V.D. (1969). Onset of sexual maturity in female Baikal seals. In: *Marine Mammals* (Ed. by V.A. Arsen'ev, B.A. Zenokovich and K.K. Chapskii), pp. 127–135. Nauka, Moscow (in Russian).

Pastukhov, V.D. (1974). Some results and problems on population study of Baikal seals. In: *Nature of Baikal*, pp. 235–248. Nauka, Leningrad (in Russian).

Pastukhov, V.D. (1993). *Baikal Seal: Biological Basis of Rational Utilization and Conservation of Resources.* Nauka, Novosibirsk (in Russian).

Perrin, W.F. and Myrick, A.C., Jr (eds) (1980). Report of the workshop. *Report of the International Whaling Commission. Special Issue* **3**, 1–50.

Reppening, C.A., Ray, C.E. and Grigorescu, D. (1979). Pinniped biogeography. In: *Historical Biogeography, Plate Tectonics and the Changing Environment* (Ed. by J. Gray and A.J. Boucot), pp. 357–369. Oregon State University Press, Corvallis.

Riedman, M. (1990). *The Pinnipeds: Seals, Sea Lions, and Walruses.* University of California Press, Berkeley, CA.

Scheffer, V.B. (1950). Growth layers on the teeth of Pinnipedia as an indication of age. *Science* **112**, 309–311.

Scheffer, V.B. (1958). *Seals, Sealions, and Walruses.* Stanford University Press, Stanford.

Sergeant, D.E. (1973). Feeding, growth, and productivity of northwest Atlantic harp seals (*Pagophilus groenlandius*). *J. Fish. Res. Brd Can.* **30**, 17–29.

Stewart, B.S., Petrov, E.A., Baranov, E.A., Timonin, A. and Ivanov, M. (1996). Seasonal movements and dive patterns of juvenile Baikal seals, *Phoca sibirica*. *Mar. Mamm. Sci.* **12**, 528–542.

Stewart, R.E.A., Stewart, B.E., Stirling, I. and Street, E. (1996). Counts of growth layer groups in cementum and dentine in ringed seal (*Phoca hispida*). *Mar. Mamm. Sci.* **12**, 383–401.

Thomas, J., Pastukhov, V., Elsner, R. and Petrov, E. (1982). *Phoca sibirica. Mamm. Species* **188**, 1–6.

Yamagishi, S., Yamamoto, K., Hasagawa, K. and Hayama, S. (1993). On the growth of Baikal seals, *Phoca sibirica*, in captivity at Toba Aquarium. *J. Jpn. Assoc. Zool. Gardens Aquariums* **35**, 1–6.

Ancient Lake Pannon and its Endemic Molluscan Fauna (Central Europe; Mio-Pliocene)

D.H. GEARY, I. MAGYAR and P. MÜLLER

I. SUMMARY

Lake Pannon existed from approximately 12 to 4 Mya, situated in the Pannonian basin of central-eastern Europe. The birth of Lake Pannon is defined by its loss of contact with adjacent marine areas. After these contacts were closed 12 Mya, they were never re-established. The physical history of Lake Pannon, including its palaeogeography, palaeobathymetry, sedimentology and chronostratigraphy, is currently well understood and provides an excellent context in which to investigate a variety of palaeobiological questions.

Lake Pannon harboured a spectacular endemic malacofauna, including over 900 described species and many endemic genera. Among bivalves, the families Cardiidae (> 220 species) and Dreissenidae (> 130 species) predominate. These groups evolved from survivors of the "Sarmatian Sea", the brackish–marine water body that occupied the Pannonian basin prior to the formation of the lake (see Dumont, this volume). Among gastropods, the prosobranch families Hydrobiidae (> 180 species) and Melanopsidae (> 100 species) predominate. Several pulmonate gastropods evolved remarkable endemics adapted to deep basinal habitats.

The fact that Lake Pannon's malacofauna had its ancestry in both the relict marine and surrounding freshwater systems contributed to its overall high diversity. In addition, the long lifetime of the lake (approximately 8 My) probably played a role in the establishment of such high endemism. Although a great many of the hundreds of endemic species apparently evolved geologically rapidly (i.e. $< 10^5$ years), evolution in several lineages appears to have been anagenetic and geologically gradual, with morphological intermediates spanning 1–2 My or more.

II. INTRODUCTION

If Lake Pannon at its maximum existed today, it would be the second largest and third deepest lake in the world. Lake Pannon was also long lived; it existed for about 8 My, from the end of the Middle Miocene to the beginning of the Pliocene (Magyar *et al.*, 1999a,b). Like other long-lived lakes, Lake Pannon maintained a high degree of endemism in its biota. Endemic fishes, gastropods and ostracods of ancient lakes are sometimes called "thalassoid", referring to their morphological similarity to marine organisms (after Bourguignat, 1885; see also Gorthner and Meier-Brook, 1985; West and Cohen, 1994). Whether these freshwater forms did indeed have marine origins is often controversial. In the case of Lake Pannon, it is clear that part of the lake's fauna originated from marine ancestors. The lake formed through isolation from the sea, and maintained a salinity above at least 5‰ through its entire life (Kázmér, 1990). The endemic radiation of molluscs and ostracods (of both marine and freshwater origin), however, was at least as spectacular as those in other freshwater ancient lakes.

In terms of its origin, size, depth, composition of the fauna and water chemistry, the best modern analogue for Lake Pannon is the Caspian Lake. From a biological point of view, however, there is an important difference. The Caspian Lake experienced several episodes of marine migrations through the Black Sea, even after its isolation from the ocean. As far as we know, multiple invasions of marine waters and/or taxa did not affect Lake Pannon; all of its fauna with marine ancestry appear to have evolved from a few Sarmatian survivors.

III. PHYSICAL HISTORY OF LAKE PANNON

Lake Pannon, like the modern Caspian Lake, formed by isolation from the sea. This isolation took place at about 12 Mya, when the newly emerging Carpathian mountain chain cut off the central European Pannonian basin from the south-east European inland sea, the Paratethys (Rögl and Steininger, 1983; Rögl, 1998; Magyar *et al.*, 1999a). With uplift of the Carpathians, the Pannonian basin lost its connection to the sea forever. The lake existed until subsidence of its basin could not keep pace with the enormous amount of clastic material carried in by rivers.

Although Lake Pannon got its name from the ancient Roman province "Pannonia", its geographical extent was much larger than this province. At its maximum, the lake filled the entire intermontane basin system from the eastern Alps in the west to the eastern Carpathians in the east (Figure 1). The extent of Lake Pannon changed dramatically over time, however (Magyar et al., 1999a). Early on, the lake was restricted to relatively shallow and narrow zones mostly near the margins of the Pannonian basin, a reflection of the Middle Miocene continental rift phase of basin evolution (Royden and Horváth, 1988; Kázmér, 1990; Tari et al., 1992). With subsequent subsidence of the central part of the basin due to cooling of the lithosphere (Royden and Horváth, 1988; Teleki et al., 1994; Juhász et al., 1996), the lake filled almost the entire basin, roughly 250 000 km². As rivers brought sediments from the surrounding mountains, Lake Pannon began to shrink and the northern shore of the lake was transformed into alluvial plains (Juhász, 1991). This process lasted for millions of years; eventually, even the most southerly parts of the basin became riverine floodplains and marshes (Figure 2).

Comparison of Lake Pannon molluscs and ostracods with the modern Ponto-Caspian fauna has been used to assess the salinity of the lake (Korpás-Hódi, 1983; Korecz, 1985). These comparisons suggest that the average salinity of Lake Pannon was approximately 10–12‰, although significant freshening

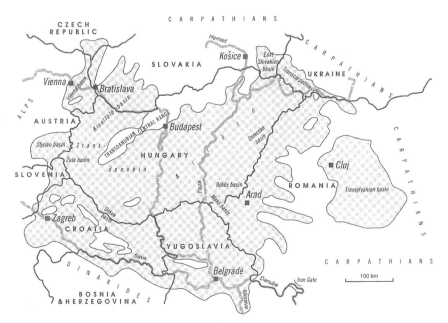

Fig. 1. Map of Pannonian basin region. The modern basin area is shaded; outcrops of the basement (mostly Mesozoic and older) are white.

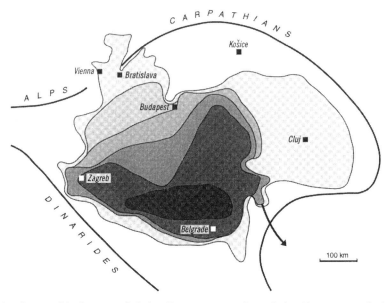

Fig. 2. Geographical area of Lake Pannon over time. Lake Pannon reached its maximum extent about 9.5 Mya. Subsequently, prograding deltas filled the lacustrine basin with sediments. Changes in the position of the shoreline are shown from approximately 9.5 to 4.6 Mya (progressively darker areas); islands have been omitted here. Arrow indicates migration of many Lake Pannon species into the more easterly Dacian basin during the Pontian age. The last vestige of the lake was in the southern part of the Pannonian basin (darkest spot on map) and its water was probably entirely fresh, rather than brackish. After Müller et al. (1999).

occurred locally in shallow parts of the lake, such as the northern delta-dominated "shelf". Analyses of waters contained within the Mio-Pliocene sediments (connate waters) are available from the upper part of the Lake Pannon sequence only (Varsányi et al., 1997). These waters have been diluted with meteoric water. Their salinity (total dissolved solids, TDS) in some cases is higher than 4‰, which thus sets a limit for the freshness of the Lake Pannon waters. Interpretation of the isotopic record from mollusc shells is ambiguous (Geary et al., 1989; Mátyás et al., 1996). Because the salinity of the lake seems to have been rather stable through time (significant perturbations would have caused obvious changes in the fauna), it may be supposed that processes that withdrew salt from the lake, such as temporary overflows, wind dispersion and sinking of brines into the sediments, balanced the salt input by rivers.

The water depth at any part of the basin can be inferred from the geometry of the infilling sediments. Because the sigmoids of seismic reflection profiles correspond to palaeosurfaces, their thicknesses show the vertical distance between the shoreface and the basinal plain, i.e. the palaeodepth. When Lake

Pannon first formed, it was predominantly shallow. Deeper environments appeared only about 1 My later, when general subsidence of the entire basin started (Magyar *et al.*, 1999a). For most of the lake, water depths were 2–400 m. Above the rapidly subsiding sub-basins, such as the Makó or Békés basins in south-eastern Hungary (see Figure 1), however, water depths exceeded 1000 m.

Density stratification and the associated oxygen-depleted conditions of deep waters seem to have been exceptional rather than common. Sediment-loaded density currents close to delta slopes, as well as seasonal overturn, generally supplied oxygenated water to the deeper parts of the lake (Juhász and Magyar, 1992; Magyar, 1995).

Lake-level fluctuations can be studied on seismic reflection profiles using the methods of seismic sequence stratigraphy. Several analyses have been published recently for Lake Pannon, but these often contradict each other (Rögl and Steininger, 1983; Korpás-Hódi *et al.*, 1992; Csató, 1993; Sacchi *et al.*, 1998; Vakarcs *et al.*, 1999). The cause of these contradictions is usually the different chronostratigraphy applied by the authors. Vakarcs *et al.* (1994) and Sacchi *et al.* (1995) both proposed four regional sequence boundaries within Lake Pannon deposits, the first representing the isolation event. These authors estimate that sequence boundaries here were caused by lake level drops of 100–150 m or more. Although such changes would have significantly altered the geography of the basin, there is no evidence for a complex of small, isolated lakes.

IV.   TEMPORAL AND SPATIAL DISTRIBUTION OF HABITATS

Lake Pannon had a range of habitats suitable for molluscs. Some were stable through time, whereas others changed quickly. These habitats are summarized diagrammatically in Figure 3.

Approximately 1 My after the formation of the lake, deep basinal habitats were established, characterized by depths of several hundred metres or more. These habitats were probably stable over time; they were relatively unaffected by water-level fluctuations and they did not shift geographically. These habitats were also persistent; the deep basins of south-eastern Hungary were filled with sediments only by the Pliocene. Although trace fossils and body fossils of benthic organisms such as ostracods and molluscs occur in the deep basinal sediments, it has been argued that fluctuations to dysoxic or anoxic conditions sometimes occured (Kázmér, 1990). The sediments of this environment were prevailingly lacustrine carbonates, with varying amounts of siliciclastics. As the deltas approached, the amount of fine-grained terrestial material increased and distal turbidites were deposited (Juhász, 1991). The deep fauna of the lake was of low diversity, but deposits are sometimes rich in individuals (Juhász and Magyar, 1992). Thin-shelled *Paradacna* (Cardiidae),

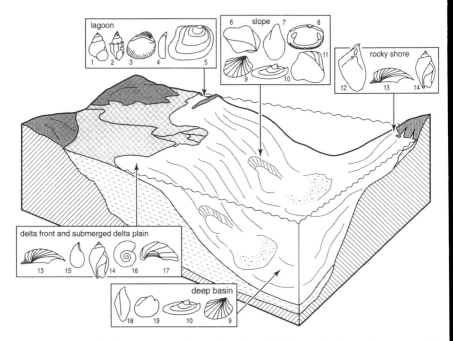

Fig. 3. Generalized representation of various habitats in Lake Pannon, with characteristic mollusc taxa. Vertical relief is exaggerated. Molluscs are: 1, *Viviparus*; 2, small, sculpted *Melanopsis*; 3, *Prosodacnomya*; 4, *Dreissena serbica* Brusina; 5, *Anodonta*; 6, *Congeria rhomboidea* Hörnes; 7, *Congeria zagrabiensis* Brusina; 8, *Lymnocardium majeri* (Hörnes); 9, *Paradacna abichi* (Hörnes); 10, *Valenciennius* (an endemic lymnaeid; 11, *Pteradacna pterophora* (Brusina); 12, *Congeria ungulacaprae* (Münster); 13, large *Lymnocardium*; 14, large *Melanopsis*; 15, *Dreissena auricularis* (Fuchs); 16, *Gyraulus*; 17, *Congeria balatonica* Partsch; 18, *Dreissenomya digitifera* (Andrusov); 19, *"Pontalmyra" otiophora* (Brusina).

Dreissenomya (*Congeria*?; Dreissenidae), *Valenciennius* (Lymnaeidae; Figure 4.3), and planorbid gastropods characterize this environment.

Slope habitats, particularly in their upper parts, were more diverse than those of very deep waters. Characteristic water depths of this environment were several tens of metres and the substratum was muddy. In addition to the deep-water forms mentioned above, certain species of large *Lymnocardium* (Figure 5.3, 5.4) and *Congeria* (Figure 4.7, 4.8) lived on the upper delta slope (Juhász and Magyar, 1992). The delta environment formed soon after the formation of the lake and persisted until its eventual infilling. Spatially, however, these habitats continuously changed position, shifting over time from the north-west and north-east towards the southern part of the basin (Pogácsás *et al.*, 1993).

The massive delta front sands contain almost no fossils. Molluscan shells were probably dissolved from these permeable sand bodies. A diverse fauna (see Figures 3–5) characterizes the overlying sediments, deposited in various

Fig. 4. Representative taxa from Lake Pannon: Gastropoda and Dreissenidae. Unless otherwise noted, scale bar is 1 cm. 1, *Melanopsis caryota* (Brusina); 2, *Melanopsis cylindrica* (Stoliczka); 3, *Valenciennius reussi* Neumayr; 4, 5, *Theodoxus intracarpaticus* Jekelius (scale bar = 5 mm); 6, *Gyraulus* cf. *striatus* (Brusina) (scale bar = 1 mm); 7, 8, *Congeria rhomboidea* Hörnes; 9, *Dreissena auricularis* (Fuchs) (scale bar = 5 mm).

shallow-water habitats, including the delta front, submerged delta plain and interdistributary bays. Various cardiids [including *Lymnocardium* (Figure 5.1, 5.2), *Pseudocatillus, Phyllocardium* (Figure 5.5) and *Parvidacna* (Figure 5.6)], dreissenids [*Congeria, Dreissenomya, Dreissena* (Figure 4.9)], *Unio* (Figure 5.7,

Fig. 5. Representative taxa from Lake Pannon: Cardiidae and Unionidae. Unless otherwise noted, scale bar = 1 cm. 1, 2, *Lymnocardium varicostatum* Vitális (with shells of *Congeria* settled in the valve); 3, 4, *Lymnocardium* (*Budmania*) *semseyi* (Halaváts); 5, *Phyllocardium planum* (Deshayes); 6, *Parvidacna chartacea* (Brusina) (scale bar = 5 mm); 7, 8, *Unio mihanovici* Brusina.

5.8), hydrobiids, planorbids (Figure 4.6), neritids (Figure 4.4, 4.5), lymnaeids and melanopsids (Figure 4.1, 4.2) lived in and on the sandy or muddy substratum and on the aquatic vegetation (Juhász and Magyar, 1992). These environments were changing rapidly in both space and time owing to delta and distributary channel displacements, mouth bar formation, the activity of longshore currents and other nearshore processes (Juhász, 1991).

Lagoonal environments occupied an increasing area during infilling of the lake. Well-aerated lagoons generally had the same fauna as did littoral sands. Restricted lagoons, however, often with lush vegetation and organic-rich muddy substrata, hosted a special lagoonal fauna, consisting of *Prosodacnomya, Dreissena, Anodonta, Viviparus* and certain species of planorbids, neritids and melanopsids. The composition of the fauna suggests that salinity was significantly reduced here compared with non-restricted environments.

The sedimentological record of rocky shores, including breccias and conglomerates, is not ideal for fossil preservation. In some cases, however, identifiable moulds of shells are found in these rocks. They show that rocky shores were populated by a diverse fauna, characterized by thick-shelled *Congeria, Lymnocardium* and *Melanopsis*. It is possible that a strong taphonomic bias exists at these localities and that small or thin shells were preferentially destroyed after death, including during diagenesis. These rocky environments formed around islands and structural highs; they were isolated from each other and were short lived.

V. TEMPORAL RESOLUTION OF LAKE PANNON DEPOSITS

Biostratigraphic study of Lake Pannon deposits started in the nineteenth century and the first comprehensive successions of molluscs were described in the first decade of the twentieth century. These promising attempts went astray, however, when the first deep drillings were performed. The enormous thickness (up to 4000 m) of the Lake Pannon sedimentary sequence surprised geologists, who found an easily recognizable surface in each of the wells. This surface was the base of the delta front sands above the thick, fine-grained delta slope sediments, and it was incorrectly considered an isochronous stratigraphic marker. Thus, time-transgressive lithological units became the basis for "chronostratigraphic" correlation, and the molluscan fauna, which showed significant changes across lithological boundaries, was typically neglected. Biostratigraphic subdivisions were based on vertical facies successions, which seemed to have been universal for the entire basin.

Increased understanding of basin fill geometry through the use of seismic reflection profiles together with the "inexplicable" appearance of certain mollusc species in unorthodox facies successions eventually led to the modern chronostratigraphic framework of Lake Pannon deposits (Magyar, 1991; Korpás-Hódi *et al.*, 1992; Müller and Magyar, 1992; Pogácsás *et al.*, 1993). Integrating K/Ar age determinations from interbedded volcanics, magnetic polarity profiles from reference boreholes, and dinoflagellate and mollusc biostratigraphy permits the establishment of a framework for the history of

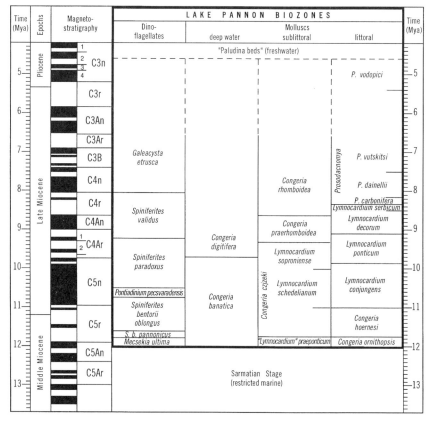

Fig. 6. Basic chronostratigraphic correlation chart for Lake Pannon deposits. For definition of biozones and geochronology of Lake Pannon deposits, see Magyar *et al.* (1999b). Dotted lines indicate uncertainty in chronostratigraphic correlation.

Lake Pannon and its fauna (Magyar *et al.*, 1999b). This work is ongoing and more data is expected to result in a more detailed chronology.

The current chronostratigraphic system is shown in Figure 6. The average resolution of biozones in sublittoral and littoral facies is approximately 1 My. Resolution is considerably worse in the deep basinal environment, where molluscan evolution appears to have been relatively slow.

VI.   MOLLUSCAN EVOLUTION IN LAKE PANNON

A. Taxonomy, Diversity and the Origin of Faunas

Among bivalves, the two families of marine origin produced the greatest diversity by almost an order of magnitude, with more than 220 species of cardiids and more than 130 species of dreissenids (Figure 7). Unionids and sphaeriids, both of freshwater origin, were considerably less diverse (each with fewer than 20 species). The number of species given here is the total number of species described. All groups are in need of taxonomic revision. Many previous authors, including those of the nineteenth century, recognized new taxa for many minor morphological variants. Furthermore, because these deposits lie in many countries, the taxonomic literature is in many different languages and so some redundancy of naming has occurred. For these reasons, revision is highly likely to reduce the number of valid species.

Lake Pannon cardiids descended from a few endemic species that lived in the restricted marine environment that preceded Lake Pannon in the Pannonian basin (the "Sarmatian Sea"; Taktakishvili, 1987). Most authors think that the eventual ancestor of these restricted marine forms was *Cerastoderma*, a highly euryhaline genus (Paramonova, 1977). Unambiguous demonstration of a phylogenetic relationship between the latest Sarmatian Sea cardiids and earliest Lake Pannon cardiids has not yet been established. It is conceivable that morphological evolution was accelerated across the interval of regression and environmental change that came between the Sarmatian Sea and Lake Pannon; however, fossil material here is too scarce and too poorly preserved to be useful in addressing this issue. In any case, the few earliest lacustrine cardiids gave rise to dozens of endemic species belonging to several lineages.

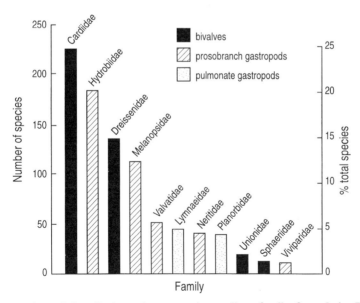

Fig. 7. Number of described species per major mollusc family from Lake Pannon. Percentage of total species number (based on current count of 908) on the left. See Müller *et al.* (1999) for counts by genus and minor taxa.

The systematics of Lake Pannon cardiids requires reconsideration (J.A. Schneider and I. Magyar, unpub.). At the generic level, almost all taxa were originally described from outside the lake (*Paradacna, Prosodacnomya, Prosodacna, Phyllocardium, Pteradacna, Arcicardium, Parvidacna, Didacnomya, Pseudocatillus, Pontalmyra, Caladacna, Plagiodacna*). At the same time, many strikingly distinct species were assigned to the genus "*Lymnocardium*", only because it was the first endemic genus to be described from the sediments of the lake. At the specific level, extreme intraspecific variability and the occurrence of gradual morphological transitions between species sometimes result in considerable taxonomic confusion.

In Lake Pannon, a few marginal marine *Congeria* species survived isolation from the sea and gave rise to a wide variety of epibenthic and infaunal dreissenids (Papp, 1985). Sediments of the lake were long referred to as "Congeria beds" ("Congerienschichten"). Large *Congeria* species (> 10 cm in length) are spectacular and characteristic members of the Lake Pannon fauna. Both *Dreissena* and the deep-burrowing *Dreissenomya* are considered endemic to Lake Pannon, but *Dreissena* is supposed to have originated "repeatedly" from *Congeria* (Papp, 1950; Lueger, 1980; Taylor, in Gray, 1988). Thus, *Dreissena* is likely to be a polyphyletic taxon, requiring revision.

Among gastropods, hydrobiids (>180 species) and melanopsids (>100 species) were most diverse, but neritids, valvatids, viviparids, lymnaeids and planorbids were also significant (Figure 4, 7). In her thorough review of gastropod evolution in long-lived lakes, Michel (1994) noted the predominance of two groups of prosobranch gastropods (the family Hydrobiidae and the superfamily Cerithioidea) and one family (Planorbidae) of pulmonate gastropods. Michel observed a seemingly consistent pattern in long-lived lakes: either the Hydrobiidae or the Thiaridae (from the superfamily Cerithioidea) radiated, but not both. In this regard, the Lake Pannon gastropod fauna is unusually rich, as both hydrobiids and cerithioids (in this case melanopsids) are well represented.

Hydrobiids have been assigned to 26 genera, of which 16 are considered endemic to the lake (*Baglivia, Beogradica, Emmericia, Goniochilus, Gyromelania, Lisinskia, Marticia, Microbeliscus, Micromelania, Odontohydrobia, Pannona, Robicia, Sandria, Scalimelania, Staja* and *Vrazia*). Some of the hydrobiids are of marine origin, whereas others came from terrestrial, freshwater habitats. Their evolution is not well known.

Melanopsid gastropods have been more closely studied (Geary, 1990, 1992; Staley *et al.*, in revision). The two ancestral species of Lake Pannon melanopsids originally inhabited fluvial and marginal marine, estuarine environments. Significant evolutionary changes in this group began only well after the lake formed. Both geologically gradual, anagenetic change and relatively rapid and prolific cladogenesis have been described (Geary, 1990). In the *M. impressa* clade, repeated migration of fluvial forms into the lake proper resulted in iterative changes in shell morphology (Staley *et al.*, in revision).

Most prominent of these changes is an increase in shell shouldering (see Figure 4.1, 4.2), with a concomitant increase in shell thickness and often overall shell size as well. Similar sorts of morphological changes are seen in a variety of other gastropods from other times and places (e.g. Gorthner and Meier-Brook, 1985; Willmann, 1985). The reasons for these iterative changes are uncertain, but adaptive possibilities include a response to predators or differing hydrodynamic conditions.

Some Lake Pannon neritids and valvatids originated from "Sarmatian Sea" forms, whereas viviparids, planorbids and lymnaeids all moved into the lake from freshwater habitats. Adaptation of pulmonates to deep lacustrine waters was a striking yet characteristic feature of Lake Pannon. (Lake Baikal also hosts an endemic, relatively deep-water pulmonate; e.g. Starobogatov, 1989.)

B. The Production and Maintenance of Diversity Within the Lake

One of the focal points of research on ancient lakes is on the mechanisms for the generation of endemic diversity. Two features of Lake Pannon and its history appear to have played a role in the generation of the spectacular molluscan fauna: origins in both the marine and freshwater realms, and the long-term stability of the lake itself.

The Lake Pannon mollusc fauna had multiple sources of origin, which no doubt contributed to its overall taxonomic richness (Müller *et al.*, 1999). Many lineages, including all of the cardiids and dreissenids, and many of the hydrobiids, descended from eurytopic survivors of the marginal marine Sarmatian Sea. The ancestry of other groups is more typical of lacustrine organisms; unionids, viviparids, lymnaeids, planorbids, melanopsids and some neritids immigrated to the lake from surrounding freshwater habitats.

Ancient lakes such as Lake Pannon are rare; only two extant lakes have a history longer than the 8 My lifespan of Lake Pannon (Lakes Baikal and Tanganyika) and few are older than the minimum age of 100 000 years necessary to be termed an ancient/long-lived lake (see Gorthner, 1994). Included in this group are Lakes Biwa, Ohrid, Titicaca, Malawi and Turkana (Martens *et al.*, 1994). Only a very few palaeoancient lakes have been described (e.g. Steinheim in Germany, Kaiso in East Africa; see Gorthner 1994; although many palaeoancient lakes remain to be described from China, A.R. Carroll, pers. comm.; see Chen, this volume). Aside from their great depth, these very ancient lakes are characterized by significant endemic evolution. Thus, although lacustrine species flocks may evolve very rapidly (e.g. Johnson *et al.*, 1996), the long-term stability of an ancient lake system appears fundamentally important to the development of a diverse fauna.

Many authors have suggested that long-term environmental stability is associated with high endemic diversity because it permits the few eurytopic colonizing species to evolve into increasing numbers of stenotopic descendants

(Sanders, 1968; see also Gorther and Meier-Brook, 1985; Coulter, 1994). Among the habitats that existed in Lake Pannon, there does not appear to be a correlation between diversity and stability; the deepest environments of the lake were probably the least subject to physical perturbations, but they harboured a relatively small number of species. Littoral environments must have moved around over time, although assessing the severity of physical or chemical perturbations is difficult. In any case, littoral environments harboured the richest endemic lake fauna. Martens (1994) discusses a similar pattern among the ostracods of Lake Baikal, where despite deep, oxygenated habitats, most speciation has occurred in littoral taxa. The stability–stenotopy hypothesis is not easy to test in the fossil record because both environmental stability and the degree of eurytopy–stenotopy are difficult to establish conclusively and independently. In general, however, Lake Pannon and its fauna appear consistent with the notion that long-term stability permits high endemic diversity.

C. Evolutionary Tempo in Lake Pannon Mollusc Lineages

Many examples of sustained gradual change have been documented among Lake Pannon mollusc lineages (Papp, 1953; Taktakishvili, 1967; Stevanović, 1978; Geary, 1990, 1992; Müller and Magyar, 1992). Given that the lake harboured many hundreds of endemic mollusc species (> 900 before revision), the nature of the events leading to the origin of most of these species will never be resolved in detail. Cladogenesis must have been geologically rapid in a great many cases (i.e. lasting $< 10^5$ years). Even if geologically gradual tempos involve a minority of the many hundreds of endemic species in Lake Pannon, they are relatively common here, especially considering their overall rarity in the fossil record.

Gradual changes have been described in lineages from several families, including cardiids, dreissenids, lymnaeids and melanopsids. They involve a wide range of shell characters, including size, shape and ornamentation. Described trends may be prolonged, lasting for 1–2 My or longer (Geary, 1990, 1992; Müller and Magyar, 1992).

In general, the reasons for the frequency of gradual tempos in the Lake Pannon system are unclear. Sheldon (1987, 1996) has argued that gradualism will be associated with organisms living in (or able to track) narrowly fluctuating, slowly changing environments, whereas stasis will prevail in more widely fluctuating, rapidly changing environments. This generalization is not easily applied to Lake Pannon molluscs. In melanopsid gastropods, for example, prolonged gradual change happens in one lineage living side by side with a related lineage in which speciation appears to have been rapid and cladogenesis frequent (Geary, 1990). Such an association of evolutionary tempos reinforces our confidence in the validity of the patterns found in the fossil record (see Fortey, 1985), but makes generalizations about the causes of such patterns more difficult.

The population structure of various mollusc lineages is poorly understood, but is of obvious relevance to understanding both the tempo and mechanisms of intralacustrine speciation. Given the poor dispersal abilities of most freshwater molluscs, combined with the geographical complexity and overall size of Lake Pannon, it seems likely that most species were subdivided into a great many demes. Differences in the tempo, mode or frequency of speciation might have arisen among even closely related lineages through the evolution of characteristics that influence population structure (brooders versus non-brooders, for example; see Cohen and Johnston, 1987). Paleontologists can gather information relative to population structure through careful examination of intraspecific variability (e.g. Johnston and Cohen, 1987) and such studies are ongoing with several Lake Pannon lineages.

With respect to evolutionary tempo and mode and the production of diversity, it appears to be the case among Lake Pannon molluscs that most lineages exhibiting sustained gradual changes do so anagenetically, whereas the most speciose groups are characterized by apparently rapid cladogenesis. The authors hope to clarify this pattern with additional systematic work.

D. Broad Patterns of Turnover and Species Replacement

Another outstanding issue in palaeontological studies is that of "co-ordinated stasis" (Brett *et al.*, 1996), or the degree to which species (or generic) associations persist or change over time. Major environmental changes marked the end of the Sarmatian Age and the Pliocene turnover. These physical events clearly coincide with major pulses of faunal turnover. These events and their associated faunal turnovers are illustrated in Figure 8, where the overlap of each rectangle with the underlying one represents the percentage of mollusc species that survived the environmental change *in situ*. Whereas many molluscs of the restricted Sarmatian Sea came from the normal marine Badenian Sea, the faunal turnover was almost complete when the brackish Lake Pannon was formed, and significant again when the freshwater Paludina Lake came into existence. Ongoing work is directed towards uncovering evolutionary patterns in the interval between these two turnovers. Available data suggest that considerable evolution, both gradual, within-lineage, and rapid, cladogenic, took place between these turnovers.

VII. STATUS OF RESEARCH

The deposits of Lake Pannon constitute a remarkable opportunity for research into many aspects of ancient lake evolution. While investigation into the physical history of the lake continues, this component is already understood

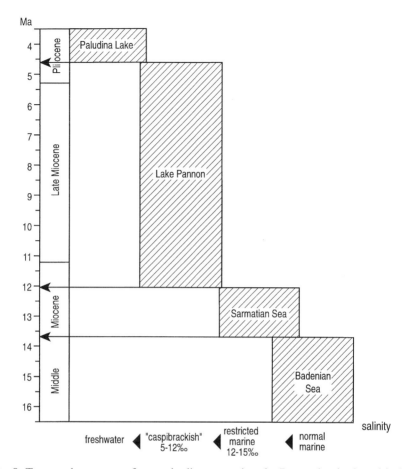

Fig. 8. Temporal sequence of water bodies occupying the Pannonian basin, with their approximate salinities. Salinity axis is not to scale. Overlap of each rectangle with the underlying one represents the percentage of mollusc species that survived the environmental change *in situ*.

well enough to provide an excellent context in which to pursue related palaeobiological studies. Ongoing work on the major mollusc families is directed towards taxonomic revision, phylogenetic analysis, morphometric analysis of temporal and geographical variation, and the history of species-level diversity. These analyses, in which intraspecific variability and diversity can be tracked over long time intervals and placed in a palaeoenvironmental context, provide the natural complement to genetically based studies of living lacustrine organisms.

ACKNOWLEDGEMENTS

This publication was sponsored by the National Science Foundation (EAR-9706230), by the US–Hungarian Science and Technology Joint Fund under project JFNo. 511, by the Hungarian OTKA (project no. T 019679), and by the Albert and Alice Weeks Fund of the Department of Geology and Geophysics, University of Wisconsin, Madison. The authors thank Director István Bérczi, MOL Hungarian Oil and Gas Company, for his support, and Mary Diman for assistance with the figures. Two anonymous reviewers and the editors provided helpful comments on the manuscript.

REFERENCES

Bourguignat, J.R. (1885). *Notice prodromique sur les mollusques terrestres et fluviatiles recueillis par M. Victor Giraud dans la région méridionale du lac Tanganika.* V. Tremblay, Paris.

Brett, C.E., Ivany, L.C. and Schopf, K.M. (1996). Coordinated stasis: an overview. *Palaeogeogr. Palaeoclimatol. Palaeoecol.* **127**, 1–20.

Cohen, A.S. and Johnston, M.R. (1987). Speciation in brooding and poorly dispersing lacustrine organisms. *Palaios* **2**, 426–435.

Coulter, G.W. (1994). Speciation and fluctuating environments, with reference to ancient East African lakes. *Arch. Hydrobiol. Beiheft. Ergebnisse Limnol.* **44**, 127–137.

Csató, I. (1993). Neogene sequences in the Pannonian basin, Hungary. *Tectonophysics* **226**, 377–400.

Fortey, R.A. (1985). Gradualism and punctuated equilibria as competing and complementary theories. *Spec. Pap. Palaeontol.* **33**, 17–28.

Geary, D.H. (1990). Patterns of evolutionary tempo and mode in the radiation of *Melanopsis* (Gastropoda; Melanopsidae). *Paleobiology* **16**, 492–511.

Geary, D.H. (1992). An unusual pattern of divergence between fossil melanopsid gastropods: hybridization, dimorphism, or ecophenotypy? *Paleobiology* **18**, 97–113.

Geary, D.H., Rich, J.A., Valley, J.W. and Baker, K. (1989). Stable isotopic evidence of salinity changes: influence on the evolution of melanopsid gastropods in the Late Miocene Pannonian Basin. *Geology* **17**, 981–985.

Gorthner, A. (1994). What is an ancient lake? *Arch. Hydrobiol. Beiheft. Ergebnisse Limnol.* **44**, 97–100.

Gorthner, A. and Meier-Brook, C. (1985). The Steinheim basin as a paleo-ancient lake. In: *Sedimentary and Evolutionary Cycles* (Ed. by U. Bayer and A. Seilacher), pp. 322–334. Springer, Berlin.

Gray, J. (1988). Evolution of the freshwater ecosystem: the fossil record. *Palaeogeogr. Palaeoclimatol. Palaeoecol.* **62**, 1–214.

Johnson, T.C., Scholz, C.A., Talbot, M.R., Kelts, K., Ricketts, R.D., Ngobi, G., Beuning, K., Ssemmanda, I. and McGill, J.W. (1996). Late Pleistocene desiccation of Lake Victoria and rapid evolution of cichlid fishes. *Science* **273**, 1091–1093.

Johnston, M.R. and Cohen, A.S. (1987). Morphological divergence in endemic gastropods from Lake Tanganyika: implications for models of species flock formation. *Palaios* **2**, 413–425.

Juhász, Gy. (1991). Lithostratigraphical and sedimentological framework of the Pannonian (s.l.) sedimentary sequence in the Hungarian Plain (Alföld), eastern Hungary. *Acta Geol. Hung.* **34**, 53–72.

Juhász, Gy. and Magyar, I. (1992). A pannóniai (s.l.) litofáciesek és molluszka-biofáciesek jellemzése és korrelációja az Alföldön. Review and correlation of the Late Neogene (Pannonian s.l.) lithofacies and mollusc biofacies in the Great Plain, eastern Hungary. *Földtani Közlöny* **122**, 167–194.

Juhász, E., Müller, P., Tóth-Makk, A., Hámor, T., Farkas-Bulla, J., Phillips, R.L., Sütő-Szentai, M. and Ricketts, B. (1996). High-resolution sedimentological and subsidence analysis of the Late Neogene, Pannonian Basin, Hungary. *Acta Geol. Hung.* **39**, 129–152.

Kázmér, M. (1990). Birth, life, and death of the Pannonian Lake. *Palaeogeogr. Palaeoclimatol. Palaeoecol.* **79**, 171–188.

Korecz, A. (1985). Die Ostracodenfauna des Zsámbéker Beckens. In: *Chronostrati-graphie und Neostratotypen. Miozän der Zentralen Paratethys VII, M6, Pannonien* (Ed. by A. Papp, Á. Jámbor and F.F. Steininger), pp. 173–177. Akadémiai Kiadó, Budapest.

Korpás-Hódi, M. (1983). Palaeoecology and biostratigraphy of the Pannonian Mollusca fauna in the northern foreland of the Transdanubian Central Range. *Ann. Hung. Geol. Inst.* **66**, 1–163.

Korpás-Hódi, M., Pogácsás, Gy. and Simon, E. (1992). Paleogeographic outlines of the Pannonian s.l. of the southern Danube-Tisza Interfluve. *Acta Geol. Hung.* **35**, 145–163.

Lueger, J.P. (1980). Die Molluskenfauna aus dem Pannon (Obermiozän) des Fölligberges (Eisenstädter Bucht) im Burgenland (Österreich). *Mitt. Österreich. Geol. Gesellsch.* **73**, 95–134.

Magyar, I. (1991). Biostratigraphic revision of the Middle Pontian (Late Neogene) Battonya Sequence, Pannonian basin (Hungary). *Acta Geol. Hung.* **34**, 73–79.

Magyar, I. (1995). Late Miocene mollusc biostratigraphy in the eastern part of the Pannonian basin (Tiszántúl, Hungary). *Geol. Carpath.* **46**, 29–36.

Magyar, I., Geary, D.H. and Müller, P. (1999a). Paleogeographic evolution of the Late Miocene Lake Pannon in Central Europe. *Palaeogeogr. Palaeoclimatol. Palaeoecol.* **147**, 151–167.

Magyar, I., Geary, D.H., Sütő-Szentai, M., Lantos, M. and Müller, P. (1999b). Integrated bio-, magneto- and chronostratigraphic correlations of the Late Miocene Lake Pannon deposits. *Acta Geol. Hung.* **42**, 5–31.

Martens, K. (1994). Ostracod speciation in ancient lakes: a review. *Arch. Hydrobiol. Beiheft. Ergebnisse Limnol.* **44**, 203–222.

Martens, K., Coulter, G. and Goddeeris, B. (1994). Speciation in Ancient Lakes – 40 years after Brooks. *Arch. Hydrobiol. Beiheft. Ergebnisse Limnol.* **44**, 75–96.

Mátyás, J., Burns, S.J., Müller, P. and Magyar, I. (1996). What can stable isotopes say about salinity? An example from the Late Miocene Pannonian Lake. *Palaios* **11**, 31–39.

Michel, E. (1994). Why snails radiate: a review of gastropod evolution in long-lived lakes, both recent and fossil. *Arch. Hydrobiol. Beiheft. Ergebnisse Limnol.* **44**, 285–317.

Müller, P. and Magyar, I. (1992). Continuous record of the evolution of lacustrine cardiid bivalves in the Late Miocene Pannonian Lake. *Acta Palaeontol. Polon.* **36**, 353–372.

Müller, P., Geary, D.H. and Magyar, I. (1999). The endemic molluscs of the Late Miocene Lake Pannon: their origin, evolution, and family-level taxonomy. *Lethaia* **32**, 47–60.

Papp, A. (1950). Übergangsformen von *Congeria* zu *Dreissena* aus dem Pannon des Wiener Beckens. *Ann. Naturhist. Mus. Wien* **57**, 148–156.

Papp, A. (1953). Die Molluskenfauna des Pannon im Wiener Becken. *Mitt. Geol. Gesellsch. Wien* **44**, 85–222.

Papp, A. (1985). Gastropoda (Neritidae, Viviparidae, Valvatidae, Hydrobiidae, Stenothyridae, Truncatellidae, Bulimidae, Micromelaniidae, Thiaridae) und Bivalvia (Dreissenidae, Limnocardiidae, Unionidae) des Pannonien. In: *Chronostratigraphie und Neostratotypen. Miozän der Zentralen Paratethys VII, M6, Pannonien* (Ed. by A. Papp, À. Jámbor and F.F. Steininger), pp. 276–339. Akadémiai Kiadó, Budapest.

Paramonova, N.P. (1977). On the systematics of Sarmatian and Akchagylian cardiids (Bivalvia). *Paleontol. J.* **11**, 315–324.

Pogácsás, Gy., Müller, P. and Magyar, I. (1993). The role of seismic stratigraphy in understanding biological evolution in the Pannonian lake (SE Europe, Late Miocene). *Geol. Croat.* **46**, 63–69.

Rögl, F. (1998). Paleogeographic considerations for Mediterranean and Paratethys Seaways (Oligocene to Miocene). *Ann. Naturhist. Mus. Wien* **99A**, 279–310.

Rögl, F. and Steininger, F.F. (1983). Vom Zerfall der Tethys zu Mediterran und Paratethys. Die neogene Paläogeographie und Palinspastik des zirkum-mediterranen Raumes. *Ann. Naturhist. Mus. Wien* **85A**, 135–163.

Royden, L.H. and Horváth, F. (1988). The Pannonian Basin. A study in basin evolution. *Am. Assoc. Petrol. Geol. Mem.* **45**, 1–394.

Sacchi, M., Horváth, F. and Magyari, O. (1995). High resolution seismics in the Lake Balaton: insights into the evolution of the western Pannonian Basin. In: *Extensional Collapse of the Alpine Orogene and Hydrocarbon Prospects in the Basement and Basinfill of the Western Pannonian Basin* (Ed. by F. Horváth, G. Tari and Cs. Bokor, pp. 171–185. Guidebook to Field Trip No. 6, Hungary AAPG International Conference and Exhibition, Nice.

Sacchi, M., Tonielli, R., Cserny, T., Dövényi, P., Horváth, F., Magyari, O., McGee, T.M. and Mirabile, L. (1998). Seismic stratigraphy of the Late Miocene sequence beneath Lake Balaton, Pannonian Basin, Hungary. *Acta Geol. Hung.* **41**, 63–88.

Sanders, H.L. (1968). Marine benthic diversity: a comparative study. *Am. Nat.* **102**, 243–282.

Sheldon, P.R. (1987). Parallel gradualistic evolution of Ordovician trilobites. *Nature* **330**, 561–563.

Sheldon, P.R. (1996). Plus ça change–a model for stasis and evolution in different environments. *Palaeogeogr. Palaeoclimatol. Palaeoecol.* **127**, 209–227.

Staley, A.W., Geary, D.H., Müller, P. and Magyar, I. (In revision). Iterative changes in Lake Pannon *Melanopsis* reflect a recurrent theme in gastropod morphological evolution. *Paleobiology*.

Starobogatov, Ya. I. (1989). Molluscs of the family Acroloxidae from Baikal–Chervi, Molluski, Chlenistonogie. In: *Fauna of Baikal*, pp. 42–75. Nauka, Novosibirsk.

Stevanović, P.M. (1978). Neue pannon-pontische Molluskenarten aus Serbien. *Ann. Géol. Péninsule Balkan.* **42**, 315–344.

Taktakishvili, I.G. (1967). *Istoricheskoe Razvitye Semeystva Valenciennid*, 192 pp. Mecniereba, Tbilisi.

Taktakishvili, I.G. (1987). *Sistematika i filogenia pliocenorykh kardiid Paratetiza*, 247 pp. Mecniereba, Tbilisi.

Tari, G., Horváth, F. and Rumpler, J. (1992). Styles of extension in the Pannonian Basin. *Tectonophysics* **208**, 203–219.

Teleki, P.G., Mattick, R.E. and Kókai, J. (Eds.) (1994). *Basin Analysis in Petroleum Exploration. A Case Study from the Békés Basin, Hungary*, 330 pp. Kluwer, Dordrecht.

Vakarcs, G., Vail, P.R., Tari, G., Pogácsás, Gy., Mattick, R.E. and Szabó, A. (1994). Third-order Middle Miocene–Early Pliocene depositional sequences in the prograding delta complex of the Pannonian basin. *Tectonophysics* **240**, 81–106.

Vakarcs, G., Hardenbol, J., Abreau, V.S., Vail, P.R., Tari, G. and Várnai, P. (1999). Correlation of the Oligocene–Middle Miocene regional stages with depositional sequences, a case study from the Pannonian Basin, Hungary. In: *Mesozoic–Cenozoic Sequence Stratigraphy of European Basins* (Ed. by P.-C. De Graciansky, J. Hardenbol, T. Jacquin, P.R. Vail and M.B. Farley), pp. 211–233. SEPM Special Publication No. 60.

Varsányi, I., Matray, J-M. and Kovács, L.O. (1997). Geochemistry of formation waters in the Pannonian Basin (southeast Hungary). *Chem. Geol.* **140,** 89–106.

West, K. and Cohen, A. (1994). Predator–prey coevolution as a model for the unusual morphologies of the crabs and gastropods of Lake Tanganyika. *Arch. Hydrobiol. Beiheft. Ergebnisse Limnol.* **44**, 267–283.

Willmann, R. (1985). Responses of the Plio–Pleistocene freshwater gastropods of Kos (Greece, Aegean Sea) to environmental changes. In: *Sedimentary and Evolutionary Cycles* (Ed. by U. Bayer and A. Seilacher), pp. 295–321. Springer, Berlin.

Using Fish Taphonomy to Reconstruct the Environment of Ancient Shanwang Lake

P.-F. CHEN

I. SUMMARY

Shanwang Lake Basin is a palaeolake formed during the Miocene, and is renowned for its quality and abundance of fossil specimens, including one of the most abundant and diverse Tertiary fish fossil deposits in China. Evidence from studies of these fish fossils has revealed much about the climate and environment of the ancient Shanwang Lake during the Miocene.

Geological studies have identified two kinds of taphonomical facies in the type section of no. 2 locality of the lake basin. One is in the thinly bedded diatom-bearing mudstone and shale typical of the upper part of the section. The poor preservation quality of fossilized fishes in this facies is interpreted as fish having decayed in a hypolimnion warmer than about 15°C, in lake depths less than 8–12 m and under aerobic conditions habitable by scavengers. The second taphonomical facies is in the thinly laminated diatom shale typical of the lower part of the section. That fish fossils here show no examples of scattering, disarticulation or macroscopic scavenger disturbance indicates an environment with low hydraulic energy, anoxic or anaerobic conditions and a temperature lower than 15°C in the hypolimnion.

Seasonal variation is recorded in the thinly laminated diatom shale, which is composed of couplets of light-coloured diatom-rich laminae and dark grey or dark brown organic-rich laminae. The light laminae include a large proportion of deciduous leaves, plant seeds and fishes, and are interpreted as autumn and winter deposits, while the dark laminae, which contain few deciduous leaves and a relatively high percentage of insects, are thought to be spring and summer deposits. The length–frequency histogram of 474 specimens of *Plesioleucicus miocenicus* (Pisces: subfamily Leuciscinae) shows a series of distinct peaks, which are interpreted as averaged year classes produced by repeated, annual winter deaths of fishes. The main cause of fish death is suggested to be overturn-induced anoxia, which occurred in a monomictic lake with annual winter circulation.

II. INTRODUCTION

Studies of fish biology can contribute to an understanding of the environment of ancient palaeolakes, most often through taphonomic investigations, and through a comparative approach which utilizes taxonomic and functional analogues from modern fishes. A major advantage of using a taphonomic approach is its testability. Interpretation of fish fossil skeletal patterns of disarticulation, disorientation and displacement can be made from comparison of such specimens with those of modern fish remains obtained by known processes. This experimental and comparative approach allows the possibility of determining temperature and depth of ancient lakes, inferring types of lake stratification, seasons and causes of fish death, and of estimating oxygen conditions and perhaps even something of the water chemistry. Modern taxonomic and functional analogues may provide interpretations of physical and ecological limiting factors, behaviour and trophic dynamics of fossil fishes, as inferred from presumed functional relationships between morphology and habitats or feeding modes. However, the reliability of such inferences decreases with increasing geological age, and with increasing phyletic distance between subject and analogue.

   The Shanwang Lake Basin was formed during the Miocene, and is renowned for the superb condition of fossil specimens of various groups of animals and plants preserved in its diatomaceous deposits. It is also one of the most abundant and diverse Tertiary fish fossil deposits in China. The fish fossils were first described by Young and Tchang (1936), and later by Zhou (1990) and Chen *et al.* (1999). The structural complexity of an organism is positively correlated with the potential taphonomic information that it can provide, and fish skeletons are thus potentially rich sources of information about temperature, depth, hydraulic energy, scavengers and oxygen concentration of ancient lakes.

   This chapter demonstrates just how much information can be gleaned from such studies of fossil fish faunas. The main taphonomic data used are from experimental studies (Smith and Elder, 1985) and from field observations on fish skeletons recovered during two intensive excavations at no. 2 locality of the Miocene Shanwang Basin during 1996. Using this information, character-istics of death and burial are inferred, and then combined with sedimentary and other palaeontological evidence to interpret aspects of limnology and the environment of the deposition, and to infer the season and possible causes of fish deaths.

A. Geological History of Shanwang Lake

The Miocene Shanwang Lake Basin (36°33′ N, 118°41′ E) is located 22 km east of Linqu County in Shandong Peninsula (Shandong Province), eastern China. It is one of several small basins within the Tertiary lake-basin system, located on the Tancheng–Lujiang fracture zone at the eastern margin of the Eurasia plate. During the 1950s this was a mining district operated for diatomite ore. It was the frequent discovery of high-quality fossil material during the mining operations which prompted the Chinese government, in 1980, to declare a 1.5 km^2 National Preserved Area in the ancient lake basin.

Precambrian schist and Cretaceous pyroclastic rocks can be found near the study area. The lake originally formed on basaltic bedrock due to water gathering on the low-lying basalt surface after eruption and the activities of provincial faults. The lake-basin floor is formed by the Niushan Formation (the Tertiary olivine tholeiitic basalt), directly overlying Palaeogenic shale and Precambrian metamorphic rocks, and several overlying strata are clearly evident. The Shanwang Formation rests on top of the Niushan Formation and consists of breccia, sandstone, diatomite, mudstone, volcanic ash, vivianite nodules and basalt interbeddings. The Yaoshan Formation, consisting of mainly Tertiary alkalic olivine basalt, tops the Tertiary sequence in the study area (Li, 1991). Stratigraphic studies of the Shanwang Formation are still in progress, but it is already clear that it can be divided into six members (Figure 1A). Most of the fossil specimens are found in the lower member of the Shanwang Formation, the laminated diatomaceous units (Figure 1B).

Both palaeontological (Li, 1982; Qiu, 1990) and radiometric evidence (Chen and Peng, 1985; Zhu et al., 1985) has derived a late Early Miocene age (15.5–17 Mya) for the Shanwang fossil biota. Species composition of fossil assemblages indicates the climate to have been warm temperate with mild winters but distinct seasonality; weather similar to that in the present-day lower Yangtze River valley in south-eastern China (Yan et al., 1983). The reconstruction of palaeopole position during the deposition of diatomaceous units in the Shanwang Formation indicate that the latitude of the Shanwang area during the Miocene was 32.4°N (Liu and Shi, 1989), considerably further south than the area is today (36°33′ N), confirming the reconstruction of the Miocene Shanwang climate based on palaeontological evidence.

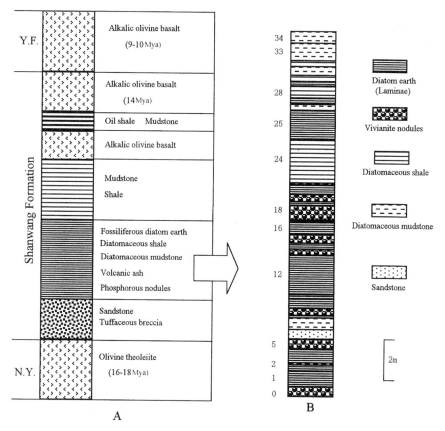

Fig. 1. (A) Generalized stratigraphic sequences of the Shanwang Formation and (B) stratigraphic division of the fossiliferous member of the excavating profile showing the sedimentary deposition cycles at no. 2 locality of the Miocene Shanwang Basin. Y.F., Yaoshan Formation; N.F., Niushan Formation.

III.   FISH TAPHONOMY AND ENVIRONMENTAL INFERENCES

A. Temperature

The most critical process in fish taphonomy is the eventual floating of the carcass by gas in the swimbladder and that produced during decomposition. Experimental observations (Smith and Elder, 1985) have shown that, at temperatures over 15°C, a fish carcass will float for several hours after death until disarticulation (the heavy jaw bones disarticulate first). However, at around 15°C, a fish may take over a week to float, whereas at temperatures below 10°C the fish carcass remains on the bottom for over 60 days and such

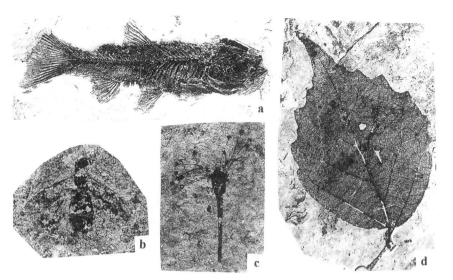

Fig. 2. Typical fossils from thinly laminated diatom shale of the Miocene Shanwang Formation. (a) *Plesioleuciscus miocenicus* (×1.5) preserved in the light-coloured lamina. (b) An insect (×2) preserved in the dark grey lamina. (c) A flower (×2) preserved in the dark grey lamina. (d) A deciduous leaf (×1) preserved in the light-coloured lamina.

carcasses decay without becoming buoyant. Through examination of the state and degree of disarticulation of fossil fish specimens it is thus possible to infer ambient water temperatures at the time of and immediately following that individual's death.

The Shanwang fish specimens from the lower fossiliferous thinly laminated diatom shale are dominated by *Plesioleuciscus* (Leuciscinae). All 556 specimens recovered from 17 to 21 beds (e.g. Figure 2a) in a 2 m × 2 m block were intact and showed no disarticulation, even of the delicate fin and tail rays and other fine bones originally held in place only by tissue (see Figure 2a). Accordingly, these must be interpreted as carcasses that never floated, and this may be considered as unequivocal evidence of decay at temperatures lower than 15°C. In contrast, fish specimens from the upper fossiliferous stratum of diatom-bearing mudstone and shale are mainly composed of *Coreoperca* (order Perciformes), *Lucyprinus* and *Qicyprinus* (order Cypriniformes: subfamily Cyprininae). Of 119 specimens recovered from 27–31 beds in a 2 m × 2 m block, 112 were disarticulated, a feature interpreted as evidence of decay at temperatures higher than 15°C in the hypolimnion.

B. Oxygen Availability

Macroscopic scavengers cause considerable scattering of elements of dead specimens, especially of the component bones of the head and thorax. In comparison, microscopic decay organisms cause separation of connective tissue in such a way that observed movements of bones are clearly attributable to the forces of gravity and to the effects of water current. Therefore, the presence or absence of scavengers can provide information about ambient oxygen levels and other important physical aspects of ancient palaeolakes.

Experiments with macroscopic scavengers (e.g. gastropods; Elder and Smith, 1988), indicate that fish bones are normally moved in all directions and are without preferred orientation. The distance of displacement is not correlated between paired bones, nor does it seem to be correlated with the hydrodynamic character of the bone. The presence of this pattern in the fossil record indicates the effects of scavengers on the carcass, while its absence may indicate conditions unsuitable for scavengers, such as anoxic bottom waters. Fish skeletons in unidirectional flowing waters, i.e. streams and rivers, usually show a more unimodal distribution of dispersal direction, and there is a correlation between distances moved by members of paired bones and between elements having similar shapes. Accordingly, fossil fishes preserved in this pattern can be interpreted as having decomposed in a lotic habitat.

Nearly all of the Shanwang fish skeletons from the fossiliferous thinly laminated diatom shale show no macroscopic scavenger disturbance whatsoever, indicating that macroscopic scavengers were absent between the time of death and burial. From this one can infer either that burial was extremely rapid or that the environment did not usually support the presence of macroscopic scavengers, such as in anoxic bottom waters. Geological evidence presented by the distribution of vivianite nodules in the layers indicates a low sedimentary rate (Zhang and Shan, 1994), effectively excluding the first possibility, and one can therefore interpret the lack of scavengers as evidence of toxic or anaerobic conditions. This interpretation is supported by other evidence for anoxic or toxic conditions, such as the original colours of leaves and insects when they were first unearthed, the absence of benthic invertebrate fossils and the lack of any signs of bioturbation.

In the diatom-bearing mudstone and shale a high percentage of the fish fossils recovered is not perfectly preserved but has undergone varying amounts of disarticulation, especially in the head region and the delicate fin and tail rays. Isolated fish bones are common and the fish bones are dispersed in all directions. The distance of displacement is not correlated between paired bones (Figure 3). This pattern of preservation suggests that the carcasses decomposed in aerobic conditions where they were accessible to scavengers.

C. Water Depth and Thermal Regime

The warm Miocene climate with distinct seasonal fluctuations, the distribution of the fossiliferous thinly laminated diatom shale composed of the light-coloured diatom-rich laminae and dark grey or dark brown organic-rich

Fig. 3. Example of the most common fish species in the diatom-bearing mudstone and shale of the Miocene Shanwang formation, *Coreoperca shandongensis* (×1).

laminae, and the excellent preservation of fossils and fine laminations, all suggest that, during the Miocene, Shanwang Lake was a stratified water body.

Lakes in subtropical areas, as at Shanwang, are more likely to be monomictic (one overturn per year) than dimictic (two overturns per year) (Bradley, 1948). Shanwang Lake had a rounded shape and only covered an area of 1.5 km². The basin floor was at least 70 m lower than the surrounding area. It was sheltered by surrounding small hills and, during the period when the fossiliferous thinly laminated diatom shale was deposited, it was probably deep enough to be stratified with an annual winter circulation. These features are consistent with the idea that the lake was stratified, because depth of mixing depends on wind fetch and velocity (Hutchinson, 1957) and because small, deep lakes are more likely to stratify than their opposites; the maximum depth of mixing in a meromictic lake is proportional to the fourth root of the surface area, or the square root of the radius (Ruttner, 1963).

There is a close relationship between anaerobic conditions and the preservation of laminae in strata. Anaerobic conditions preclude much decomposition and prevent sediment-disturbing factors such as bubbling gasses and benthic organisms. Modern examples of this relationship include Fayetteville Green Lake, New York, where stratification exists for most or all of the year and where fine laminae are preserved (Ludlam, 1969), and the Gulf of California (Calvert, 1964), where laminated diatomaceous sediments are confined to those depths where there is least oxygen. The presence of stratification in the lake during the deposition of the fossiliferous thinly laminated diatom shale allows a minimum depth for the lake to be inferred. In a warm, subtropical climate, such as that at Shanwang Lake during the Miocene, water deeper than 8–12 m is required for the development of a cold, anoxic hypolimnion (Smith and Elder, 1985).

To recap, interpretation of fish preservation patterns and geological features allow reconstruction of the environmental conditions at Shanwang Lake during the Miocene. Evidence indicates that, at the time of deposition of the diatom-bearing mudstone and shale, Shanwang Lake was warm and shallow; the hypolimnion was aerobic and suitable for scavengers, indicating a holomictic lake. One can infer that water depth was less than 8–12 m.

D. Seasonal Environmental Variation

Seasonal variation in the environment of an ancient lake is most readily demonstrated when the lacustrine sediments are varves, i.e. composed of couplets of laminae of different composition, where each couplet represents 1 year of deposition. Through direct analogy with modern lake sediments, the presence of laminated couplets of sediment is sometimes taken to be sufficient evidence that the couplets are varves (Olsen et al., 1978), and probably this conclusion is usually correct.

The fossiliferous thinly laminated diatom shales of Shanwang Lake are couplets, averaging about 0.3 mm in thickness, of light-coloured, less organic laminae, mainly composed of diatoms (up to 90–95%), a few montmorillonite lamellae, plagioclase, quartz and zircon, and dark grey or dark brown organic-rich laminae, which contain more montmorillonite, minor diatoms, a little basalt, plagioclase, quartz and mica detritus, and chlorite lumps. Virtually all of the well-preserved fossils are found in these couplets. Pyrite is found more often in dark laminae than in the light laminae, a feature agreeing with biogenic varves described elsewhere (O'Sullivan, 1983; Anderson and Dean, 1988).

Elemental analysis also reveals significant differences in the mean compositions of light versus dark laminae. Dark laminae have significantly higher amounts of aluminium, magnesium, calcium, potassium, sodium, iron, titanium, sulphur, phosphor, manganese and barium, and significantly lower amounts of silicon. This is due to the higher proportion of clay minerals, such as chlorite and montmorillonite, found in dark laminae than in light laminae, and indicates a seasonal variation in rainfall.

The diatom production in lakes is almost invariably cyclic with a strong annual signal, and the diatom–clay couplets indicate seasonality. The suggested depositional mechanism is a seasonal variation in rainfall with a wetter summer and drier winter. During the spring and summer, the humid–warm climate, high precipitation rate and intermittent water supplies resulted in suspended mud particles and made the lake water turbid. It was during this phase that the dark grey or dark brown organic-rich layers were deposited. During the autumn and winter, however, the lake had little or no surface water supply, and diatoms died *en masse* owing to the changes of insolation affecting primary production. These dead diatoms were sedimented out on the calm lake bottom, forming the light-coloured diatom-rich layers.

The excellent preservation of the animals (e.g. insects and fishes) and of the plants, especially deciduous leaves, permitted a test of the hypothesis that the laminated couplets are varves. If so, fossils representing different parts of the annual cycle would be expected to occur at different frequencies. Therefore, it was recorded whether each fossil specimen was preserved in a light lamina or in a dark lamina. Because the light laminae are much thinner than the dark laminae, a fossil was considered to be preserved in the former only if it was in, immediately on or immediately beneath the light lamina. A fossil was considered to be in a dark lamina if part of the dark lamina could be observed both beneath and above the fossil. The census of macrofossils revealed that 97.5% of deciduous leaves, 73.1% of plant seeds, 80% of fishes and 69.1% of insects are buried in the light laminae (Figure 4). Deciduous leaves and plant seeds normally occur in laminae that formed during autumn and early winter. All of the insects recorded are winged adults (Figure 2b), which are therefore most probably terrestrial in origin and normally disappear in early autumn. The relatively low percentage of insects buried in the light laminae indicates that the transition from dark laminae to light laminae deposition occurred during the period of greatest insect abundance, probably in the late summer.

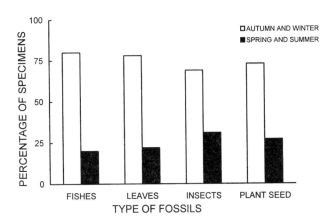

Fig. 4. Proportions of major types of fossils in the Miocene varves of the Shanwang Formation. Black bars represent proportions found in the dark (organic-rich) laminae, interpreted as spring and summer deposits. White bars represent proportions found in the light (diatom-rich) laminae, interpreted as autumn and winter deposits.

E. Possible Causes of Mass Fish Deaths

Under conditions where mortality occurs each year at about the same stage in the annual growth cycle, the size–frequency distribution of fossil fish would show special "year-class" peaks which reflect seasonal growth patterns (Wilson, 1984). Normally, fishes grow more rapidly during summer than winter. If fishes from the population died at a uniform rate and were preserved with equal probability during all seasons of the year, then the slower growth of fish during the winter would result in a slightly greater frequency of fish in the fossil sample having lengths equal to the length attained by winter of each year. In such a case, no clear peaks would be expected in the fossil sample. However, if death occurred each year during a period of several summer months, the rapid summer growth would result in the peaks in the sample being less distinct than they would be if death occurred each year during a period of little or no growth (late autumn and winter). Based on this logic, one can investigate the periodicity and timing of mortalities within the fish populations of Shanwang Lake during the Miocene.

The length–frequency histogram of 474 *Plesioleuciscus miocenicus* (Pisces: Leuciscinae) specimens (the most abundant species in the thinly laminated diatom shale) showed a series of distinct peaks (Figure 5), indicating that fish death mostly occurred at a single time each year, especially during a period of little or no growth (winter). Special year-classes, indicated by body length, are evident, with average peaks at 57.5, 90, 102.5 and 112.5 mm, corresponding to fishes of ages $0+$, $1+$, $2+$ and $3+$, respectively. Figure 6 shows a rapid

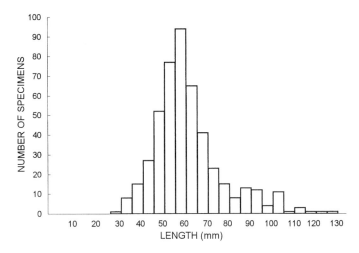

Fig. 5. Length–frequency distribution for *Plesioleuciscus miocenicus* from the thinly laminated diatom shale of the Miocene Shanwang Formation.

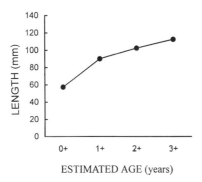

ESTIMATED AGE (years)

Fig. 6. Lengths of the inferred year-class peaks plotted against estimated ages.

decrease in the absolute rate of growth (measured as an increase in length) after the second year (age 1 +), which is consistent with the normal lessening of annual growth increments with age seen in modern fishes (Ricker, 1975).

There are several theories to explain the cause of the fish mortalities in Shanwang Lake, and it would be a mistake to assume that the same type of catastrophe caused all of them. The most plausible scenarios include starvation, noxious algal blooms and stratified water turnover involving an anoxic hypolimnion. The absence of gut contents in the fish fossils indicates that starvation might have been a factor in their death. The modern relatives of the Leuciscinae are omnivorous or herbivorous, and these Miocene fishes therefore probably fed on plankton, aquatic macrophytes and (or) aquatic insects, all of which are seasonal in availability. If winter temperature remained relatively high, the fish may have had to maintain a high metabolic rate, and some may have died of starvation.

Plankton is usually beneficial to the fish community, for all fishes obtain nutrition, either directly (planktivores) or indirectly (via the food chain), from these tiny organisms. However, under certain conditions, one or several species of plankton can multiply with abnormal rapidity, such as *Melosira* (Diatomaceae) in the Miocene Shanwang Lake. During such algal blooms, toxins produced by the algae can reach lethal concentrations and cause mass fish kills.

If the lake were monomictic as stated above, the single annual overturn would have occurred some time during the winter. The abundant evidence for anaerobic conditions suggests that, during overturn, anoxic conditions due to the mixing of hydrogen sulphide-rich water from the hypolimnion with the waters of the epilimnion could have caused the death of the fish. At the same time, by dissipating hydrogen sulphide and excess carbon dioxide, and also by mixing with the more alkaline surface water, the complete circulation of the waters would increase the pH value of the hypolimnion. Large amounts of sediments would be precipitated, which would facilitate the relatively rapid burial of dead fishes. During the stratified periods of summer, the anoxic, hydrogen sulphide-rich hypolimnion would also protect dead animal and plant remains on the lake bottom from attack by most scavengers and decomposition by aerobic bacteria.

Present evidence favours the overturn-death hypothesis as the most likely scenario responsible for the majority of the mortality horizons in the thinly laminated diatomaceous layers. However, other hypotheses remain viable, and some combination of two or more may have been involved in the annual mass kills of fishes detected in the strata of Shanwang Lake.

IV. CONCLUSIONS

Two kinds of fish taphonomical facies are present in the Miocene Shanwang Lake Basin of Shandong, eastern China, a thinly bedded diatom-bearing mudstone and shale typical of the upper part of the section, and a thinly laminated diatom shale typical of the lower part.

A census of the macrofossils in the diatomite couplets showed that significantly higher proportions of deciduous leaves, plant seeds, fishes and insects were buried in the light-coloured laminae, indicating these couplets to be varves, in which the light-coloured laminae are mainly autumn and winter deposits, and the dark grey or dark brown organic-rich laminae are spring and summer deposits.

There is direct evidence of thinly laminated diatom shales composed of couplets of light-coloured diatom-rich laminae and dark grey or dark brown organic-rich laminae, and the excellent preservation of fossils and fine laminations. This, together with the known configuration of the lake basin and evidence of a warm temperate climate, suggests that the Miocene Shanwang Lake was stratified with one overturn per year (monomictic). Because in a warm subtropical climate water deeper than 8–12 m is required for the development of a cold, anoxic hypolimnion, one may infer that the minimum water depth during the deposition of the thinly laminated diatom shale was 8–12 m.

Fish fossils from the thinly bedded diatom-bearing mudstone and shale are dominated by *Lucyprinus, Qicyprinus* (Cyprininae) and *Coreoperca* (Perciformes). A high percentage of these specimens is not perfectly preserved but has undergone varying amounts of disarticulation, especially in the head region and the delicate fin and tail rays. Isolated fish bones are common, the bones around the fish skeleton are scattered in all directions and the distance of displacement is not correlated between paired bones. This pattern of preservation is evidence of decay at temperatures higher than 15°C in the hypolimnion and indicates aerobic conditions suitable for scavengers. The lake during that time is thought to have been a holomictic water body, with a water depth of less than 8–12 m.

Fish fossils from the thinly laminated diatom shale are dominated by *Plesioleuciscus* (Leuciscinae) and are excellently preserved: even the delicate fin and tail rays and other bones originally held in place only by tissue show no scattering, disarticulation or evidence of macroscopic scavenger disturbance. This is unequivocal evidence of decomposition having taken place at temperatures lower than 15°C and in the absence of scavengers. Evidence for a low sedimentary rate is provided by the distribution of vivianite nodules in the layers, and the lack of scavengers can thus be interpreted as evidence for toxic or anaerobic conditions in the hypolimnion.

The length–frequency histogram of 474 *P. miocenicus* specimens showed a series of distinct peaks, indicating that fish death mostly occurred at a single time each year, especially during a period of little or no growth (winter). The most probable causes of the fish mortalities in the thinly laminated diatom shale include starvation, noxious algal bloom and stratified water turnover. Overturn-death is the most favoured hypothesis at the present stage of knowledge.

ACKNOWLEDGEMENTS

The author particularly thanks Professor M.-m. Chang for her comments that helped to improve an earlier draft of the manuscript, Professor J.-j. Zhou, F. Jin, D.-f. Yan and Z.-x. Qiu for generous help and discussion with aspects of this research, and Mr. J.-x. Yan and H. Zhang for assistance in the field. This research was funded by the National Science Foundation of China (grant no. 49832010) and Special Funds for Paleontology and Paleoanthropology (IVPP no. 960102).

REFERENCES

Anderson, R.Y. and Dean, W. E. (1988). Lacustrine varve formation through time. *Paleogeogr. Paleoclimatol. Paleoecol.* **62**, 215–235.

Bradley, W.H. (1948). Limnology and the Eocene lakes of the Rocky Mountain Region. *Bull. Geol. Soc. Am.* **59**, 635–648.

Calvert, S.E. (1964). Factors affecting distribution of laminated diatomaceous sediments in Gulf of California. *Mem. Am. Assoc. Petrol. Geol.* **3**, 311–330.

Chen, D.-g. and Peng, Z.-c. (1985). K–Ar ages and Pb, Sr isotopic characteristics of Cenozoic volcanic rocks in Shandong, China. *Geochemistry* **4**, 303–311 (in Chinese with English abstract).

Chen, P.-f., Liu, H.-z. and Yan, J.-x. (1999). Discovery of fossil *Coreoperca* (Perciformes) in China. *Vertebr. PalAsiat.* **37**, 165–177 (in Chinese with English abstract).

Elder, R.L. and Smith, G.R. (1988). Fish taphonomy and environmental inference in paleolimnology. *Paleogeogr. Paleoclimatol. Paleoecol.* **62**, 577–592.

Hutchinson, G.E. (1957). *A Treatise on Limnology. I. Geography, Physics, and Chemistry.* John Wiley and Sons, New York.

Li, F.-l. (1991). Reconsideration of the Shanwang Formation, Linqu, Shandong. *J. Stratigr.* **15,** 123–129 (in Chinese with English abstract).

Li, H.-m. (1982). The age of the Shanwang flora, Shandong. *Selected Papers from the 12th Annual Convention of Palaeontological Society of China*, pp. 159–162. Scientific Publishing House, Beijing (in Chinese).

Liu, H.-f. and Shi, N. (1989). Paleomagnetic study of Shanwang Formation, Shandong Province. *Acta Sci. Nat. Univ. Pekinensis* **5,** 585–593 (in Chinese with English abstract).

Ludlam, S.D. (1969). Fayetteville Green Lake, New York. III. The laminated sediments. *Limnol. Oceanogr.* **14,** 848–857.

Olsen, P.E., Remington, C.L., Corner, B. and Thomson, K.S. (1978). Cyclic change in Late Triassic lacustrine communities. *Science* **201,** 729–733.

O'Sullivan, P.E. (1983). Annually-laminated lake sediments and the study of Quaternary environmental changes–a review. *Quatern. Sci. Rev.* **1,** 245–313.

Qiu, Z.-x. (1990). The Chinese Neogene mammalian biochronology–its correlation with the European Neogene mammalian zonation. In: *European Neogene Mammal Chronology* (Ed. by E.H. Lindsay, V. Fahlbusch and P. Mein), pp. 527–556. Plenum Press, New York.

Ricker, W.E. (1975). Computation and interpretation of biological statistics of fish populations. *Bull. Fish. Res. Brd Can.* **191,** 1–382.

Ruttner, F. (1963). *Fundamentals of Limnology*. University of Toronto Press, Toronto.

Smith, G.R. and Elder, R.L. (1985). Environmental interpretation of burial and preservation of *Clarkia* fishes. In: *Late Cenozoic History of the Pacific Northwest* (Ed. by C.J. Smiley), pp. 85–93. California Academy of Sciences, San Francisco, CA.

Wilson, M.V.H. (1984). Year classes and sexual dimorphism in the Eocene catostomid fish *Amyzon aggregatum*. *J. Vertebr. Paleontol.* **3,** 137–142.

Yan, D.-f., Qiu, Z-d. and Men, Z.-y. (1983). Miocene stratigraphy and mammals of Shanwang, Shandong. *Vertebr. PalAsiat.* **21,** 211–221 (in Chinese with English abstract).

Young, C.-c. and Tchang, T.-l. (1936). Fossil fishes from the Shanwang Series of Shandong. *Bull. Geol. Soc. China* **15,** 197–207.

Zhang, M.-s. and Shan, L.-f. (1994). *Sedimentary Geology of Shanwang Basin*. Geology Publishing House, Beijing.

Zhou, J.-j. (1990). The Cyprinidae fossils from Middle Miocene of Shanwang Basin. *Vertebr. PalAsiat.* **28,** 95–127 (in Chinese with English abstract).

Zhu, M., Hu, H.-g. and Zhao, D.-z. (1985). Potassium–argon dating of Neogene basalt in Shanwang area, Shandong Province. *Petrol. Res.* **5,** 47–59 (in Chinese with English abstract).

Historical Changes in the Environment of Lake Titicaca: Evidence from Ostracod Ecology and Evolution

P. MOURGUIART

I. SUMMARY

Lake Titicaca is a high-altitude, nearly closed, basin lake in the tropical belt of South America. The lake is subjected to the influence of the intertropical convergence zone (ITCZ) during the summer months, and is characterized by a flora and fauna that, while not as diverse as that of many other ancient lakes, nevertheless has groups such as molluscs and ostracods which show a fascinating ecological and morphological variability (polymorphism). For these and other groups, Lake Titicaca is an ideal natural laboratory for studies of evolution.

This chapter demonstrates how, by applying transfer functions to quantitative information on the response of extant ostracod communities to habitat heterogeneity and environmental variability, past environments of Lake Titicaca can be reconstructed. Results show the recent history of Titicaca (since 8000 ^{14}C years bp) to have been characterized by major lake-level fluctuations which separated the basin into three palaeolakes at various intervals of geological time, and which have possibly influenced population dynamics in the lake for groups such as fishes, molluscs and ostracods. The high-amplitude palaeoenvironmental variations indicated by the transfer function also seem to have had important effects on the dynamics of ostracod populations.

Special attention is given to the formation of habitats and evolution of the lake ecosystem during the past 10 000 years, with particular emphasis on the ostracod crustaceans within the subfamily Limnocytherinae, a key ostracod group in the central Andes, and their highly variable carapace morphology. It is possible to distinguish among several morphs (or subspecies?) within this group, but the overlap is significant and individuals are sometimes difficult to classify. Accordingly, while taxonomic identifications remain based on external carapace morphology only, it is impossible to propose a definitive hypothesis for phylogenetic relations in the Andean limnocytherinid flock and its taxonomic structure. Even so, preliminary surveys of this group beyond the Lake Titicaca basin and adjacent areas suggest that the actual Titicaca fauna is not an isolated lineage, but part of extensive radiations which span the entire Andes. The observed differences between extant ostracod communities of Lake Titicaca and lakes Huaron and Junín can be explained by differential effects of the ITCZ, which produced distinct climatic regimes during the Holocene and resulted in different evolutionary pathways for the respective faunas of these lakes. This evolution appears to be more uniform in the case of Lake Junín, probably because of its closer proximity to the equator (and hence a lower exposure to the influence of the ITCZ) than Lake Titicaca.

II.   INTRODUCTION

Lake Titicaca is located on the Andean high plateau (Figure 1), and is the largest and oldest of a series of high mountain lakes and temporary lakelets scattered from north to south along the Andean Cordilleras. These lacustrine environments are home to abundant and diverse ostracod populations, whose wide geographical distribution and subtle ecological requirements make them ideal ecological indicator organisms. Accordingly, the ostracods are a favoured group used by biologists in the reconstruction of palaeoenvironments, and are especially useful in the reconstruction of the water chemistry, temperature and trophic dynamics of lakes. The first goal of this chapter is to demonstrate how information from transfer functions applied to extant lacustrine ostracod assemblages can be used to infer past limnological changes in Lake Titicaca. The second goal is to demonstrate and examine the ecophenotypic variability of carapace characters within the Limnocytherinae, one of the most abundant ostracod groups in South American inland waters (Martens and Behen, 1994), in relation to environmental changes.

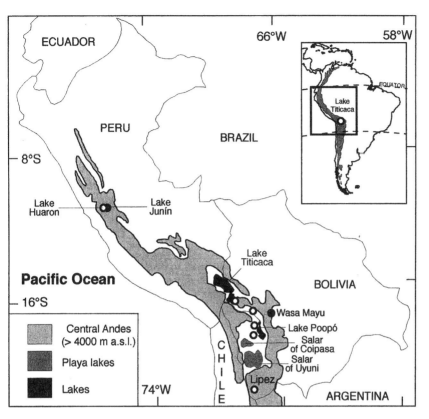

Fig. 1. Map of the central Andes showing localities of ostracod samples referred to in this study. Modern samples (open circles): Lake Huaron, Peru, Lake Titicaca and surroundings, Lake Poopó and adjacent lacustrine systems, South Lipez lakes; and fossil samples (closed circles): Lake Junín, Peru, and Wasa Mayu, Bolivia.

III. LAKE TITICACA: GEOLOGICAL SETTING, REGIONAL CLIMATE AND LIMNOLOGICAL CHARACTERISTICS

Lake Titicaca is a large (c. 8000 km^2), warm, monomictic lake (Lazzaro, 1981) that extends between 16°15′ S and 17°30′ S latitude and 68°30′ W and 70° W longitude, at an altitude of c. 3809 m. It is located in the northern part of the Altiplano, a 200 000 km^2 intermontane endorheic basin which formed during the Pliocene and Early Pleistocene, some 3–2 Mya, in the central Andes of Peru, Bolivia and Argentina (Lavenu, 1992). The Pleistocene was marked by periodic advances and retreats of glaciers, and these resulted in high-amplitude fluctuations of lacustrine basins (Servant and Fontes, 1978; Lavenu et al., 1984; Mourguiart et al., 1997). The present-day altiplanean climate is dominated by typical tropical wet–dry seasons and local orographic effects, induced by the seasonal latitudinal movements of the intertropical convergence zone (ITCZ).

From November to April, the ITCZ envelops the central Andes. During this
period, the warm, moist Amazonian air penetrates the eastern Cordillera from
the north-east, bringing stormy rains and causing a steep rainfall gradient
between the north and the south (Roche et al., 1992).

The lake drains to the south via the Río Desaguadero to Lake Poopó, a saline
lake at an altitude of c. 3686 m (Figure 1), and is divided into three main basins
(Figure 2). The northern basin, Lake Chucuito, has a maximum depth of c. 285 m
and is separated from the two southern basins, which combine to form Lake
Huiñaimarca, by the Tiquina Strait (Figure 2). The total dissolved salt
concentration varies from 0.9 g l^{-1} in Lake Chucuito to 1.2 g l^{-1} in Lake
Huiñaimarca. The waters are dominated by chloride, sulphate and sodium ions,
with mean pH values around 8.5 (Carmouze et al., 1981, 1992). Water clarities are
greater in Lake Chucuito (Secchi disc mean values between 11.3 m and 14.6 m)
than in Lake Huiñaimarca (mean values from 3.2 m to 5.6 m) (Iltis et al., 1992).

Lake Titicaca is characterized by some flat shallow areas (Puno and
Achacachi bays in Lake Chucuito, and Lake Huiñaimarca), steep escarpment
margins (flanks of Lake Chucuito) and flat profundal areas (central part of
Lake Chucuito). These different habitat types support different macrophyte
communities. A depth-related zonation of macrophytes is also present, and this
creates well-defined habitats for benthic communities. According to Collot et
al. (1983), in Lake Huiñaimarca, at depths from 0.2 m to 2.5 m the emergent
Lilaeopsis and Hydrocotyle growth along the beach shore is replaced by an
aquatic macrophyte assemblage of Myriophyllum, Elodea, Ruppia and
Potamogeton. From 2.5 m to c. 4.5 m depth Schoenoplectus tatora (an emergent sedge of the family Cyperaceae) dominates, and from this depth down to
c. 7.5 m, dense mats of Characeae species occur. Beyond this there is a zone of
Potamogeton settlement extending down to 9 m, which is the limit of the photic
zone. These macrophyte belts extend all around Lake Titicaca, but some
bathymetric differences between basins are known, related mainly to bottom
sediment characteristics and light penetration (Iltis and Mourguiart, 1992).

The macrophyte belts provide food and refugia for a variety of invertebrates,
which themselves provide potential food for fishes. In turn, the distribution
and depth zonation of aquatic macrophytes influence the distribution of
associated invertebrate animals. This is illustrated for one group, the ostracods,
in Figure 3. Unfortunately, very few ecological studies have been conducted in
this remarkable ecosystem, the sum of which has been compiled by Dejoux and
Iltis (1992). Rudimentary taxonomic studies have been carried out for several
invertebrate groups (e.g. Amphipoda, Cladocera, Copepoda, Mollusca and
Ostracoda), among which molluscs and amphipods are the most abundant
(Dejoux, 1992), but many taxa remain poorly known or undescribed.

The native fish fauna in the lake is well documented and comprises 26
species, contained within only two genera, Orestias and Trichomycterus
(Lauzanne, 1982, 1992; Parenti, 1984). In the 1940s and the 1950s, two alien
species were introduced into the lake as food fish; the rainbow trout

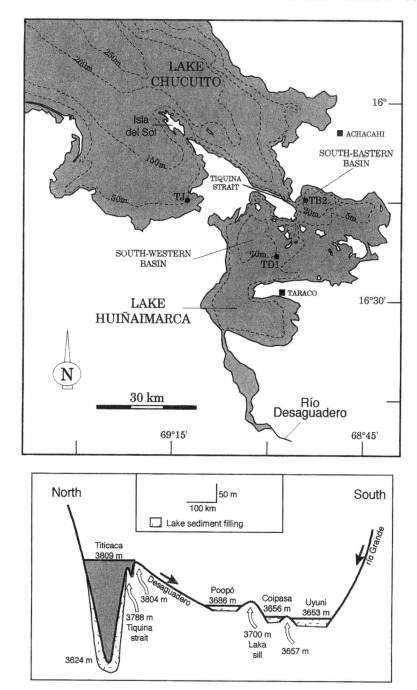

Fig. 2. Bathymetric map of present-day Lake Titicaca, with the three main basins. The lower figure indicates the main communication between basins along the Altiplano.

Oncorhynchus mykiss from North America and the pejerrey *Basilichthys bonariensis* from Argentina. The establishment of these two non-native species has drastically affected the composition and diversity of the natural fish communities. Several endemic species of the genus *Orestias* are now considered endangered or extinct (Lauzanne, 1992), but much more work is needed to determine fully the impact of these introductions on the functioning of Titicaca's ecological communities and ecosystem.

IV. THE OSTRACODA

A. The Ostracod Communities

Ostracods are small, bivalved crustaceans that are abundant in both benthic and periphytic habitats within Lake Titicaca. Their calcified carapaces have an average length of about 1 mm and completely envelop the soft parts. Their carapace valves fossilize readily and constitute common microfossils in cores from lake sediments. Most species are easily identifiable from their carapace and this, combined with the fact that most species are strongly stenotopic, makes the group useful as proxies to reconstruct past environments. In the Altiplano, the large number of aquatic environments, distributed over a hydrologically and climatically diverse area, provides ideal circumstances for comparative studies of ostracod ecology and evolution.

B. Ostracod Distribution Patterns

Field surveys have revealed that modern Lake Titicaca is home to about 50 species of ostracods (Lerner-Seggev, 1973; Mourguiart, 1987, 1992; Carbonel *et al.*, 1990; Mourguiart and Roux, 1990; Mourguiart and Carbonel, 1994), and that the abundance and species composition of extant ostracod assemblages changes from the shallow littoral habitats to the deeper environments (Figure 3). This change is due to both abiotic and biotic factors, such as hydraulic energy level (which acts on the nature of the substrate), oxygen levels at the water–sediment interface, macrophyte cover and food supply. At the landward edge and in the shallower waters of Lake Titicaca (less than 2 m water depth), the ostracod community is dominated by species adapted to unstable conditions. This zone is characterized by maximum physical disturbance (wave action), seasonal water-level fluctuation and wide temperature variation, and here periphytic genera such as *Chlamydotheca* and *Herpetocypris* are always dominant. Within the dense mats of *Schoenoplectus*, which occur at c. 2.5–4.5 m water depth, conditions at the water–sediment interface are generally anoxic and this prevents colonization by benthic ostracod species; periphytic species, however, are present on the *Schoenoplectus* stems. In less dense mats, one can occasionally find species well-adapted to poorly oxygenated waters (e.g.

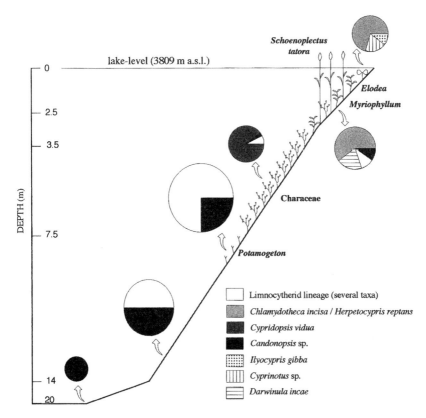

Fig. 3. Ostracod assemblages in relation to macrophyte distribution in Lake Huiñaimarca. The size of the circles is proportional to the density of ostracods. The dominant macrophyte community is indicated in bold type. Modified from Wirrmann and Mourguiart (1995).

Candonopsis sp. and *Limnocythere* sp. A1). At 4.5–7.5 m the ostracod community of Lake Titicaca is poorly diversified: this is in stark contrast to the situation in most lacustrine systems where this bathymetric zone is characterized by high species richness and density. The ostracod assemblages found on and in sediment substrates are dominated by *Candonopsis* sp., *Limnocythere titicaca* and *Darwinula incae*, whereas *Cypridopsis vidua* occasionally occurs at high densities on plant stems. This relatively low species richness is explicable in terms of the low habitat diversity provided by the dense Characeae mats which dominate this depth zone, and the low ambient oxygen concentrations at the water–sediment interface. Just below this zone, species richness and abundance of ostracods increase dramatically down to approximately 10 m. This limit is a function of water clarity and so varies between Lake Chucuito (up to 25–30 m) and Lake Huiñaimarca (*c*. 10 m) (Mourguiart, 1992). Thereafter, at progressively greater depths the diversity and abundance of ostracods decrease slowly, until dropping sharply at the limit of the thermocline, where only low oxygen-tolerant species are present (e.g. *Candonopsis* sp., *Limnocythere* sp. A1 and *L.* sp. A2).

To the south, the effects of the ITCZ are reduced and the lakes become progressively more saline. This is because solute residence time increases as precipitation declines and evaporation increases. In these more saline lakes the most common benthic species are *Cyprideis salebrosa* and *Limnocythere bradburyi* (Mourguiart and Carbonel, 1994; Mourguiart and Corrège, 1998).

This brief section has summarized the extent of the present level of knowledge about this fascinating example of ecological zonation in Lake Titicaca. Clearly, further extensive fieldwork in this and other altiplanean lacustrine ecosystems is merited.

C. Intraspecific Variation

The benthic ostracod community of Lake Titicaca is endemic, whereas periphytic species found there are widespread, occurring also in marshes, springs, or permanent and temporary lakes. Among the species present, the members of the subfamily Limnocytherinae are of particular interest because they exhibit an impressive plasticity in external valve morphology, which is believed to be at least partly environmentally induced (e.g. Mourguiart, 1987; Peypouquet *et al.*, 1987; Carbonel *et al.*, 1990). Furthermore, most of the other species not in this subfamily appear to be more widespread (e.g. *Candonopsis* and *Amphicypris*): in South America, four genera (*Limnocythere, Neolimnocythere, Pampacythere* and *Paracythereis*) and 19 species are presently recognized (Martens and Behen, 1994; Mourguiart and Corrège, 1998).

Because ostracods are very sensitive to strong short-term fluctuations within habitats, they can be used to reconstruct the past limnological conditions of Lake Titicaca during the Holocene epoch (the past 10 000 years). The present-day Titicacan limnocytherinid lineage includes at least four different species, divided into perhaps two distinct genera, and occurring always in discrete microhabitats. The polymorphism displayed by this fauna can be related to ecophenotypic variations under the control of environmental factors such as depth or salinity, and well-defined forms, differing in their carapace shape and ornamentation, are here regarded as "entities". Thus, ecophenotypic morphs, subspecies or species are all considered as entities in the transfer function database, under the criterion that each entity occurs only in a discrete and well-defined habitat type.

D. Methods

Since the pioneering work of Imbrie and Kipp (1971), many attempts to quantify the relationships between biological data and the environment have been made. Researchers have used data on such varied organisms as pollen, beetles, diatoms and molluscs to reconstruct past environments (e.g. Gasse, 1994; Guiot, 1994). Most modern approaches towards reconstructing palaeoenvironments utilize transfer functions, a method whereby values of an environmental variable can be expressed as a function of species data (ter Braak, 1995). In the present study, transfer functions are applied to extant ostracod communities in order to infer past lake levels and climatic changes in Lake Titicaca. Lake-level fluctuations are also reconstructed by subjecting this data to factor analysis of correspondence (FAC).

Fourteen cores were sampled in Lake Titicaca using a Mackereth corer (Barton and Burden, 1979). Radiocarbon dates on carbonated deposits were obtained on six of these cores, either conventionally or by accelerator mass spectrometry. However, three of these cores (TD1, TB2 and TJ) provided records extending back to c. 25 000 years bp (Wirrmann and Mourguiart, 1995) and this chapter will henceforth focus on findings and interpretations from these cores only (Figure 4). The lowest section of the lake sediment cores contains no ostracods; in most cores the faunas do not begin to appear until after about 8000 years bp. The absence of ostracods during isotope stage 2 (the last glacial maximum, LGM) has been interpreted in terms of ambient physicochemical conditions having been unfavourable to calcified organisms: the climate was much colder and water pH was low (Mourguiart et al., 1995). In general, ostracod density was between 100 and 50 000 individuals cm^3 of core sample. Ostracod relative abundances and assemblages, defined by examining at least 300 adult specimens per sample, indicated that members of the Limnocytherinae comprised c. 50%, 20% and 80%, respectively, of the ostracod assemblages in the three most complete cores (Mourguiart, 1987; Carbonel et al., 1990; Mourguiart et al., 1998). It should be noted, however, that most periphytic ostracod species have such thin shells that normally they do not fossilize. A complete listing of all the ostracod species recovered and their ecological requirements is given in Mourguiart and Roux (1990) and Mourguiart and Carbonel (1994). Because the present study considers only the species preserved in sediments, among the 115 modern samples available, only 28 taxa from nine genera are subjected to statistical analysis.

Seventeen "environmental" categories were determined according to the results of a first FAC by combining (i) five ranges of water depth, (ii) four ranges for the total dissolved salts, and (iii) four ranges for the Mg:Ca ratio in the water. A second FAC was then run by incorporating two matrices, which corresponded to the 115 modern samples and the fossil samples for a given core, respectively (Mourguiart and Roux, 1990; Mourguiart et al., 1998). The product of these three matrices was then subjected to reciprocal averaging and a multiple linear regression (MLR) was then run for the three environmental parameters. The MLR produces a mathematical formula linking the environmental parameters to the relative abundances of ostracod species. Results validating the transfer function are discussed in detail elsewhere

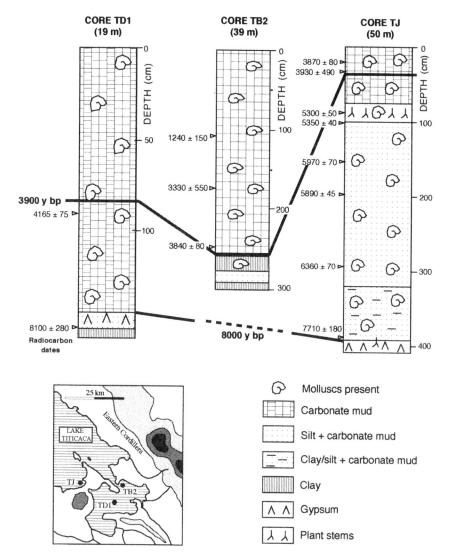

Fig. 4. Principal correlations between cores TD1, TB2 and TJ, based on lithological stratigrahy and radiocarbon dates. Inset map shows the location of the coring sites in the present Lake Titicaca.

(Mourguiart and Roux, 1990; Mourguiart and Carbonel, 1994; Mourguiart *et al.*, 1992, 1997, 1998) and only a summary of the most important findings is presented here (Figure 5).

According to the two first factors, which represent more than 40% of total information or variance (Figure 5b), four separate groups, corresponding to different environment types, can be distinguished (Figure 5a). Plots of residuals given by the MLR results show that the estimates are within ± 0.82 m of the water depths measured in the field, with a corresponding correlation coefficient of 0.98 (Figure 5c). Further, it is important to emphasize that there is good correspondence between the modern samples and the fossil ones. This high

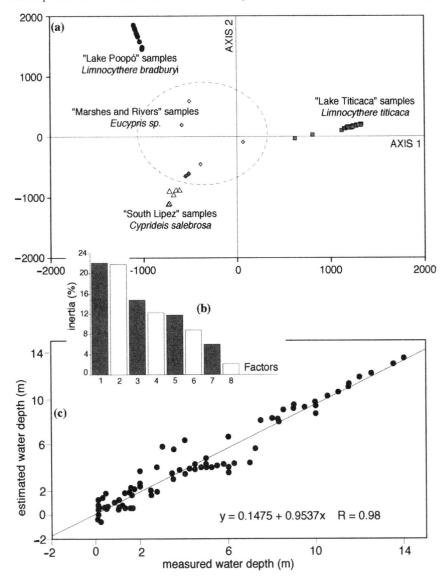

Fig. 5. Main results of the transfer function. (a) Correspondence analysis of modern data as passive elements. For each group, a representative species is given. (b) Factor weightings. (c) Estimated versus measured water depth for the 115 modern samples. Modified from Mourguiart and Carbonel (1994).

reliability of these data means that by applying transfer functions to extant ostracod assemblages at least three ecological variables can be quantitatively reconstructed: (i) depth, (ii) salinity, and (iii) Mg:Ca ratio of water.

The aim here is to advocate, through example, the usefulness of the above approach in reconstructing palaeoenvironments in Lake Titicaca, and perhaps also in other ancient lakes. Therefore, for the purpose of simplicity and clarity, only one factor, water depth, is considered, because it appears to be the most reliable parameter and the easiest one to correlate from core to core. Details and analyses of other factors are given in Mourguiart et al. (1998).

V. RESULTS

Transfer function results from three cores retrieved from the three main basins of Lake Titicaca are presented in Figure 6. Palaeolake levels indicate a complex evolution of the three basins, showing in particular the occasional isolation of separate basins (Figures 6 and 7). Some major phases are also evident. From about 8000 to 3900 ^{14}C years bp, the lake levels fluctuated around a mean position of about 3765 m for Lake Chucuito and about 3790 m for the south-western basin (data obtained from cores TJ and TD1). During this period, in the south-eastern basin (Lake Huiñaimarca, core TB2) there are no ostracod fossils in the sediments, except in one level which contains species such as *Potamocypris, Strandesia* and *Cypridopsis*. This association is found today only in pond habitats, and indicates that between 8000 and 3900 ^{14}C years bp, this part of Lake Huiñaimarca corresponded to a very shallow lacustrine environment (Figure 7). At 3900 ^{14}C years bp, the water levels of the three lakes rose drastically. However, evidence suggests that the present-day lake level was not reached until after 680 ^{14}C years bp (Mourguiart et al., 1998).

Detailed analysis of relative abundance of the subfamily Limnocytherinae from the three primary cores (Figures 8–10) reveals different patterns within each of the three basins. These patterns are not fully explicable in terms of water-depth or water-quality differences, and suggest that the three basins should be considered as having been discrete entities throughout most of the Holocene. The Limnocytherinae recovered from core TJ (Figure 8) are more highly diversified than those found in cores TB2 and TD1 (Figures 9 and 10). Separation of the two lineages of the most common species (*Limnocythere* groups A and B)* is based on carapace shape morphology: *Limnocythere* group

---

*These closely related pairs of species are similar to males (*L.* sp. B1) and females (*L.* sp. A1) of *L. grafi*, a species from a palaeolake in the Cochabamba region of Bolivia, described by Purper and Pinto (1980). The confusion of these latest authors is probably due to the poor quality of the preserved material in the sediments that they examined. It is evident that for further clarification of the presently confused systematics of *Limnocythere*, a new synthesis of data from different analytical methods and sources (e.g. soft parts) is necessary.

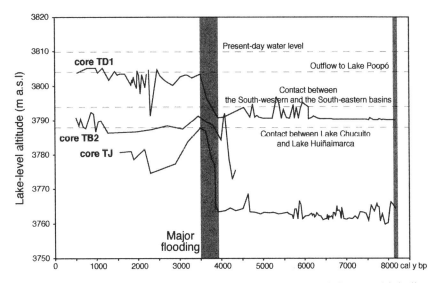

Fig. 6. Lake-level evolutions for the three basins reconstructed from multiple linear regression on the ostracod faunas. All radiocarbon dates were calibrated using the Calib 3.0 program (Stuiver and Reimer, 1993). Between two dated horizons, the ages were calculated by linear interpolation, with the exception of the lowermost and uppermost layers where ages were extrapolated from adjacent intervals. The horizontal broken lines indicate altitudes at which connection was possible between basins. The vertical shaded areas indicate major changes in the evaporation:precipitation ratio. Modified from Martin *et al.* (1997).

B have a more developed flat anterior margin and in dorsal view they appear narrower than do *L.* group A individuals (Carbonel *et al.*, 1990). Both sexes are always present. In Lake Titicaca, these two lineages show an extraordinary morphological variability, and often more than one type or morph is present in a sample. Previous studies (Mourguiart, 1987; Carbonel *et al.*, 1990) have referred to them as "polymorphic" populations, and the same terminology is used here (Figure 11). Sixteen morphs are present in core TJ (four for group A and 12 for group B), whereas only three morphs are present in cores from Lake Huiñaimarca (two for group A and one for group B).

An important aspect of polymorphism concerns the external morphology of the carapace. The observed polymorphism is manifest in differences in reticulation, nodation, spinosity, microconation and size. However, the most spectacular morphological variability seen in this group is in the development of spines. Four morphs belonging to *Limnocythere* "group B" are characterized by having no, two, six or eight spines (Figure 11). The spiny morphs are not found in present-day altiplanean lacustrine environments and can be considered as extinct. In modern Titicaca, another morph belonging to the same lineage (perhaps the same species) is very similar to the six-spined morph,

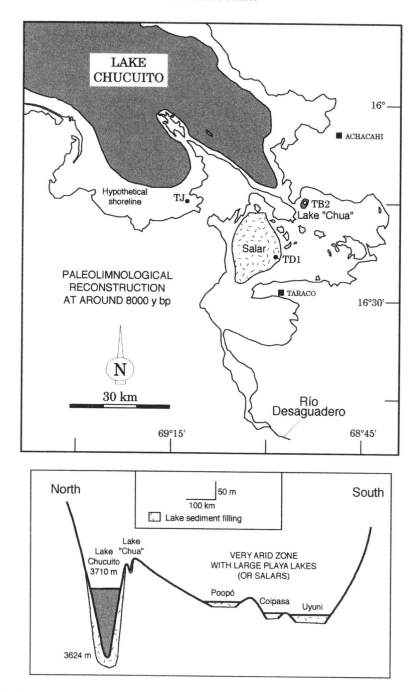

Fig. 7. (Upper) Hypothetical reconstruction of the three basins at 8000 years bp and (lower) corresponding transect along the Altiplano.

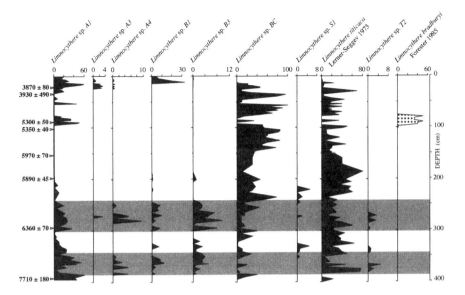

Fig. 8. Distribution of the different Limnocytherinae morphs and species for core TJ. The shaded zones highlight periods of major development of polymorphism. In the figure, the presence of brackish to saline species is indicated by stippled shading.

but displays an atrophy and a different shape of spine (Figure 11). Another very similar species, *Neolimnocythere hexaceros*, has been described from Lake Huaron, Peru (Delachaux, 1928), and in Holocene sediments from Lake Junín (Peru) (De Deckker, 1987; Mourguiart, unpubl.), but this form shows relatively large differences in the length, shape and direction of the spines (cf. Figure 11). These similarities lend support to the hypothesized close phylogenetic affinity between the limnocytherinid species of Lake Titicaca and those of Lakes Huaron and Junín.

VI. DISCUSSION

It is widely accepted that the contemporary biodiversity of South American tropical rainforest has been affected by climatic events during the Pleistocene. In particular, during the LGM, some 20 000–18 000 years bp, the Amazon Basin was believed to have contained widely separated "islands" of forest, which served as refugia for plants and animals (Prance, 1982). The high diversity of the South American rainforest was thus explained by the presence of these numerous perennial refugia. However, recent studies (Ledru *et al.*, 1998) have shown that the Amazon Basin was actually very arid during the LGM and have refuted the refugia hypothesis. In addition, several workers

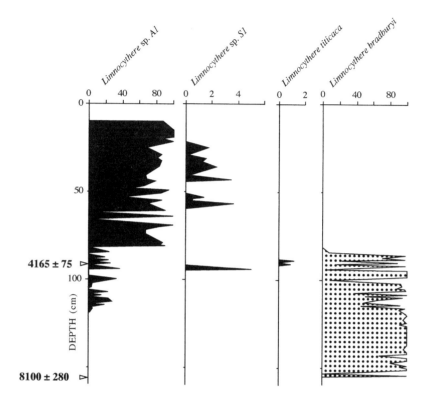

Fig. 9. Distribution of the different limnocytherinid morphs and species for core TB2. Stippled shading denotes the presence of brackish to saline species.

have identified short-term dry events during the Holocene (e.g. Turcq *et al.*, 1998). An alternative hypothesis to explain the high biodiversity of this area is needed. Evidence from reconstruction of ancient ostracod communities offers one such scenario, and shows how the effects of the ITCZ and lake-level fluctuations might have provided a template for speciation within this group.

Analysis of sedimentary archives from Lake Titicaca reveals that ostracod communities in this ancient lake are not rigid entities. Instead, these communities show drastic and relatively short-term (centuries) changes in faunal composition, with certain species disappearing completely (Figure 8). Such species either reappear at later stages of the lake's history, indicating them to have found refugia in other parts of the lake itself or in adjacent water bodies (small marginal lakes or headwaters of rivers) during their absence from the lake, or are not recorded again, indicating their extinction. The highest population densities and degrees of polymorphism occurred during the mid-Holocene epoch (*c.* 7500–6000 years bp; Figure 8). Palaeodata (microcharcoal

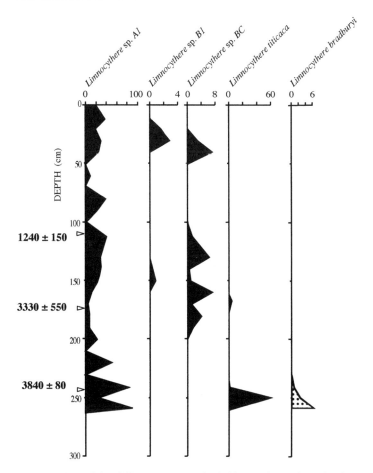

Fig. 10. Distribution of the different limnocytherinid morphs and species for core TD1. Stippled shading denotes the presence of brackish to saline species.

remains in lacustrine sediments) from Amazonian rainforest have shown that short dry-climate episodes frequently occurred during this time (Turcq *et al.*, 1998). When one considers the different ostracod communities in the three basins, it is possible to conclude that the high morphological diversification of the Limnocytherinae in Lake Chucuito (core TJ) is a result of the high habitat diversity and the highly stable environment of the system, which was subject to the effects of the ITCZ. In comparison, in the two cores from Lake Huiñaimarca (cores TB2 and TD1) the ostracod faunas display low morphological and species diversity: in the southern lake basins the effects of the ITCZ were minimal, and data show that during the mid-Holocene these water bodies were temporary and sometimes dried out completely (Hansen *et al.*, 1984; Mourguiart *et al.*, 1997).

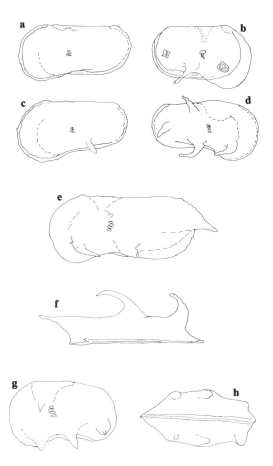

Fig. 11. Examples of morphological variability in ostracods of the subfamily Limnocytherinae. The first four illustrations (a–d) represent morphs from core TJ (Lake Chucuito, sample TJ-295) during the mid-Holocene, respectively characterized by the complete absence of, or the presence of two, six or eight spines on the carapace. The "spiny" morphs are considered as extinct. (a) Smooth morph: *Limnocythere* sp. B1 Mourguiart, 1987; female, lateral view of left valve [length (L) = 0.95 mm]; (b) "biceros" morph: *Limnocythere* sp. B2 Mourguiart, 1987; female, lateral view of left valve (L = 0.91 mm); (c) "hexaceros" morph: *Limnocythere* sp. B4b Mourguiart, 1987, female, lateral view of right valve (L = 0.91 mm); (d) "octaceros" morph: *Limnocythere* sp. B4c Mourguiart, 1987; female, lateral view of right valve (L = 0.95 mm). Parts e and f represent *Neolimnocythere hexaceros* Delachaux, 1928, an ostracod from the Holocene sediments of Lake Junín. (e, f) *Neolimnocythere hexaceros*; male, (e) lateral and (f) ventral views of right valve (L = 1 mm). The final two illustrations show modern morphs found in Lake Chucuito (Isla del Sol). (g, h) "hexaceros" morph: *Limnocythere* sp. B4a Mourguiart, 1987; female, lateral view of left valve (g) and ventral view of carapace (h) (L = 0.85 mm). Differences in spine lengths and shapes are particularly evident between morphs.

Although the effects of climate on the evolutionary history of Titicaca's aquatic fauna are only now beginning to be unravelled, it is apparent that factors outside the lake basin need to be considered (Argollo and Mourguiart, 2000; Lendru and Mourguiart, 2000; Lendru et al., 2000). The humid air carried by the ITCZ to the Altiplano originates from the Amazon Basin, and future studies on the hydrology and limnology of Lake Titicaca should consider the effects of large-scale climate changes within that area.

A. Sympatric and Allopatric Speciation: Climatic and Habitat Effects

The fact that different morphs are found in present-day Lake Titicaca environments, and also in other isolated small lakes (Lakes Huaron and Junín) some hundreds of kilometers north of Lake Titicaca, is particularly significant. It is as yet unclear whether all of these species have a single ancestor or not, but two different evolutionary scenarios can be forwarded: (i) the morphs are different species of allopatric origin, and have evolved in different refuges since the last deglaciation in the Andes, some 12 000 years bp; or (ii) the morphs have a sympatric origin, and developed within each lake system through selection and micropartitioning of diverse stable niches.

The importance of geographical barriers as factors promoting allopatric divergence of species is well recognized. In the high Andean plateau, Pleistocene glaciations could have promoted divergence into closely related pairs of species in different lacustrine systems, such as those found in Lake Titicaca in the south, and Lakes Junín and Huaron in the north, respectively. During glaciations, each unconnected lake basin could thus have provided a refuge, within which species evolved allopatrically. However, the history of these environments is one of drastic climate changes and high-amplitude water-level variations, extending even to complete desiccation in most cases (Hansen et al., 1984; Mourguiart et al., 1997). This begs the question: at times of severe environmental stress or desiccation, where did ostracod communities find conditions favourable to their survival? Possibly, lakes at lower elevations served during Ice Ages as refugia for some lacustrine ostracod taxa. Tentative support for this theory is provided by Wasa Mayu, a lake located at a lower elevation in the Valle Alto near Cochabamba, Bolivia, and where at least three limnocytherinid species lived between c. 35 000 and 15 000 years bp (Purper and Pinto, 1980; Strahl, 1998).

In the author's opinion, the differences in the diversities of Titicacan ostracods can be partly explained by the unusual regional climate, characterized by high-amplitude interannual variations in pluviometry owing to the effects of the ITCZ, especially at its southernmost portion. A combination of environmental stability and habitat diversity seems to have promoted species diversity. The highest species richness occurred during the mid-Holocene, a period characterized by frequent disturbances (wet–dry episodes) and high lacustrine productivity, as evidenced by the very high sedimentation rates of biological remains seen in core TJ (Figures 3, 8). At that time, hydrological conditions in Lake Huiñaimarca (cores TB2 and TD1) were too severe to allow colonization by diverse ostracod communities (Figures 9, 10). A contrasting situation is observed in the north, where the high species diversity in Lake Chucuito (core TJ) is a result of the highly diverse habitat and the relatively stable environment. Lastly, in Lakes Huaron and Junín, climate and habitats (mainly Characeae mats) were more uniform throughout the Holocene (Hansen *et al.*, 1984; Mourguiart *et al.*, 1997), and only a small number of limnocytherinid species colonized these lakes. The low habitat diversity there would not have promoted speciation through partitioning of spatial niches.

B. The Need for a Sound Taxonomy

The fundamental difference between ostracod populations from Lake Titicaca and those from Lake Junín concerns the degree of diversification. The Titicaca "morphs" are examples of an incredible ecophenotypic variation under the control of environmental factors (e.g. water depth), which is quasicontinuous (Mourguiart, 1987). Within the same population (living or fossilized), it is generally possible to observe a series of forms intermediate between smooth and highly spinous morphs. In contrast, in Holocene and present-day environments, *Neolimnocythere hexaceros* from Lakes Huaron and Junín shows no morphological variation. However, it is difficult to make meaningful comparison between the fauna of Lake Titicaca and these two lakes. This is because of the disparity in the present level of taxonomic knowledge of the respective ostracod faunas of these lakes. The limnocytherinid taxa found in Lakes Huaron and Junín have been identified based on both external morphology of the carapace and descriptions of soft parts, e.g. *Limnocythere elongata, Neolimnocythere erinacea* and *Paracythereis impudica* (Delachaux, 1928). The identification and separation of different species is thus much more accurate. For example, using characters of external morphology only, it is difficult to differentiate between *Limnocythere titicaca* of Lake Titicaca and *L. elongata*, or even *P. impudica*, of Lake Huaron, yet these species are easily separated on the basis of soft-part anatomy (Lerner-Seggev, 1971). Clearly, whether the different morphs in Lake Titicaca are indeed different species or subspecies, and their relationship to species in other Andean lakes, needs to be founded upon a sound taxonomy. This can only be achieved by considering characters of the soft-part anatomy.

Further studies are needed to elucidate the processes responsible for the geographical distribution of past and present ostracod diversity in the Andes. However, it is increasingly evident that rather than being species depauperate, the Andean plateau harbours surprisingly high levels of diversity for some aquatic taxa (e.g. ostracods). Unfortunately, many of the questions raised in this chapter can presently be addressed only on the basis of speculation and hypothesis, and there is a clear and urgent need for further studies of the remarkable morphological variation in certain Andean ostracod species. This must be done in tandem with taxonomic studies based on soft parts, especially the morphology of the copulatory processes of the hemipene, which represent more conservative characters to diagnose species accurately, and on genetic material. The application of molecular DNA techniques, as used so fruitfully on other ancient lake faunas by several contributors in this volume, seems certain to play an important role in Lake Titicaca. With a sound taxonomy in place, such studies of the limnocytherinid faunas of these Andean lacustrine environments should provide unique insights into allopatric and sympatric evolutionary processes.

ACKNOWLEDGEMENTS

This work is part of the AIMPACT and PVC programmes of the ME Department (UR1) at IRD (Institut de Recherche pour le Développement, ex. ORSTOM). It was undertaken through a convention between IRD and UMSA (Universidad Mayor de San Andrés, La Paz, Bolivia). The author thanks J. Argollo, P. Baker, K. Martens and G. Seltzer, as well as the editors, A. Rossiter and H. Kawanabe, and two anonymous reviewers for greatly improving the manuscript.

REFERENCES

Argollo, J. and Mourguiart, P. (2000). Late Quaternary climate history of the Bolivian Altiplano. *Quatern. Int.* (in press).

Barton, C.E. and Burden, F.R. (1979). Modifications to the Mackereth corer. *Limnol. Oceanogr.* **24**, 977–983.

Braak, C.J.F. ter (1995). Calibration. In: *Data Analysis in Community and Landscape Ecology* (Ed. by R.H.G. Jongman, C.J.F. ter Braak and O.F.R. van Tongeren). pp 78–90. Cambridge University Press, Cambridge.

Carbonel, P., Mourguiart, P. and Peypouquet, J.-P. (1990). The external mechanism responsible for morphological variability in Recent Ostracoda, seasonality and biotope situation: an example from Lake Titicaca. In: *Ostracoda and Global Events* (Ed. by R. Whatley and C. Maybury), pp. 331–340. Chapman and Hall, London.

Carmouze, J.-P., Arze, C. and Quintanilla, J. (1981). Régulation hydrochimique du lac Titicaca et l'hydrochimie de ses tributaires. *Rev. Hydrobiol. Trop.* **14**, 329–348.

Carmouze, J.-P., Arze, C. and Quintanilla, J. (1992). Hydrochemical regulation of the lake and water chemistry of its inflow rivers. In: *Lake Titicaca: A Synthesis of Limnological Knowledge* (Ed. by C. Dejoux and A. Iltis), pp. 98–112. Kluwer, Dordrecht.

Collot, D., Koriyama, F. and Garcia, E. (1983). Répartitions, biomasses et productions des macrophytes du Lac Titicaca. *Rev. Hydrobiol. Trop.* **16**, 241–261.

De Deckker, P. (1987). *Neolimnocythere hexaceros* Delachaux 1928. *Stereo-Atlas of Ostracod Shells* **14**, Front cover.

Dejoux, C. (1992). The benthic populations: distribution and seasonal variations. In: *Lake Titicaca: A Synthesis of Limnological Knowledge* (Ed. by C. Dejoux and A. Iltis), pp. 383–404. Kluwer, Dordrecht.

Dejoux, C. and Iltis, A. (eds) (1992). *Lake Titicaca: A Synthesis of Limnological Knowledge*. Kluwer, Dordrecht.

Delachaux, T. (1928). Faune invertébrée d'eau douce des hauts plateaux du Pérou. *Bull. Soc. Neuchâtel Sci. Nat.* **52**, 45–77.

Forester, R.M. (1985). *Limnocythere bradburyi* n.sp.: a modern ostracode from central Mexico and possible paleoclimatic indicator. *J. Paleontol.* **59**, 8–20.

Gasse, F. (1994). Lacustrine diatoms for reconstructing past hydrology and climate. In: *Long-term Climatic Variations* (Ed. by J.-C. Duplessy and M.-T. Spyridakis), pp. 335–369. NATO ASI Series, Vol. 122. Springer, Berlin.

Guiot, J. (1994). Statistical analysis of biospherical variability. In: *Long-term Climatic Variations* (Ed. by J.-C. Duplessy and M.-T. Spyridakis), pp. 299–334. NATO ASI Series, Vol. 122. Springer, Berlin.

Hansen, B.C.S., Wright, H.E., Jr and Bradburyi, J.P. (1984). Pollen studies in the Junin area, central Peruvian Andes. *Bull. Soc. Am.* **95**, 1454–1465.

Iltis, A. and Mourguiart, P. (1992). Higher plants: distribution and biomass. In: *Lake Titicaca: A Synthesis of Limnological Knowledge* (Ed. by C. Dejoux and A. Iltis), pp. 241–252. Kluwer, Dordrecht.

Iltis, A., Carmouze, J.-P. and Lemoalle, J. (1992). Physico-chemical properties of the waters. In: *Lake Titicaca: A Synthesis of Limnological Knowledge* (Ed. by C. Dejoux, and A. Iltis), pp. 89–97. Kluwer, Dordrecht.

Imbrie, J. and Kipp, N.G. (1971). A new micropaleontological method for quantitative paleoclimatology: application to a late Pleistocene Caribbean core. In: *The Late Cenozoic Glacial Ages* (Ed. by K.K. Turekian), pp. 71–181. Yale University Press, New Haven, CT.

Lauzanne, L. (1982). Les Orestias (Pisces, Cyprinodontidae) du Petit Lac Titicaca. *Rev. Hydrobiol. Trop.* **15**, 39–70.

Lauzanne, L. (1992). Fish fauna, native species: the Orestias. In: *Lake Titicaca: A Synthesis of Limnological Knowledge* (Ed. by C. Dejoux and A. Iltis), pp. 405–419. Kluwer, Dordrecht.

Lavenu, A. (1992). Formation and geological evolution. In: *Lake Titicaca: A Synthesis of Limnological Knowledge* (Ed. by C. Dejoux and A. Iltis), pp. 3–15. Kluwer, Dordrecht.

Lavenu, A., Fornari, M. and Sebrier, M. (1984). Existence de deux nouveaux épisodes lacustres quaternaires dans l'Altiplano péruvo-bolivien. *Cahiers ORSTOM, Sér. Géol.* **14**, 103–114.

Lazzaro, X. (1981). Biomasses, peuplements phytoplanctoniques et production primaire du lac Titicaca. *Rev. Hydrobiol. Trop.* **14**, 349–380.

Ledru, M.-P., Bertaux, J., Sifeddine, A. and Suguio, K. (1998). Absence of Last Glacial Maximum records in the lowland tropical forests. *Quatern. Res.* **49**, 233–237.

Ledru, M.-P., Campello Cordeiro, R., Landim Dominguez, J.M., Martin, L., Mourguiart, P., Sifeddine, A. and Turcq, B. (2000). Late glacial cooling in Amazonia. *Quatern. Res.* (in press).

Ledru, M.-P. and Mourguiart, P. (2000). Late glacial vegetation records in the Americas and climatic implications. In: *Interhemispheric Climate Linkages* (Ed. by V. Markgraf). Academic Press, London (in press).

Lerner-Seggev, R. (1973). *Limnocythere titicaca*, a new species (Ostracoda) (Cytheridae) from Lake Titicaca, Bolivia. *Crustaceana* **25**, 88–94.

Martens, K. and Behen, F. (1994). A checklist of the recent non-marine ostracods (Crustacea, Ostracoda) from the inland waters of South America and adjacent islands. *Trav. Sci. Mus. Nat. Hist. Nat. Luxemb.* **22**, 84 pp.

Martin, L., Bertaux, J., Corrège, T., Ledru, M.-P., Mourguiart, P., Sifeddine, A., Soubiès, F., Wirrmann, D., Suguio, K. and Turcq, B. (1997). Astronomical forcing of contrasting rainfall changes in tropical South America between 12,400 and 8800 cal yr B.P. *Quatern. Res.* **47**, 117–122.

Mourguiart, P. (1987). Les Ostracodes lacustres de l'Altiplano bolivien. Le polymorphisme, son intérêt dans les reconstitutions paléohydrologiques et paléoclimatiques de l'Holocène. Unpublished PhD Thesis, University of Bordeaux.

Mourguiart, P. (1992). The Ostracoda. In: *Lake Titicaca: A Synthesis of Limnological Knowledge* (Ed. by C. Dejoux and A. Iltis), pp. 337–345. Kluwer, Dordrecht.

Mourguiart, P. and Carbonel, P. (1994). A quantitative method of palaeolake-level reconstruction using ostracod assemblages: an example from the Bolivian Altiplano. *Hydrobiologia* **288**, 183–193.

Mourguiart, P. and Corrège, T. (1998). Ecologie et paléoécologie des ostracodes actuels et holocènes de l'Altiplano bolivien. In: *What about Ostracoda!* (Ed. by S. Crasquin-Soleau, E. Braccini, and F. Lethiers), pp. 103–115. Elf ep-éditions, Mém. 20.

Mourguiart, P. and Roux, M. (1990). Une approche nouvelle du problème posé par les reconstructions des paléoniveaux lacustres: utilisation d'une fonction de transfert basée sur les faunes d'ostracodes. *Géodynamique* **5**, 151–165.

Mourguiart, P., Wirrmann, D., Fournier, M. and Servant, M. (1992). Reconstruction quantitative des niveaux du petit lac Titicaca au cours de l'Holocène. *C. R. Acad. Sci. Paris* **315**, 875–880.

Mourguiart, P., Argollo, J. and Wirrmann, D. (1995). Evolución del lago Titicaca desde 25 000 años BP. In: *Climas cuaternarios en América del Sur* (Ed. by J. Argollo and P. Mourguiart), pp. 157–171. ORSTOM Editions, La Paz.

Mourguiart, P., Argollo, J., Corrège, T., Martin, L., Montenegro, M.E., Sifeddine, A. and Wirrmann, D. (1997). Changements limnologiques et climatologiques dans le bassin du lac Titicaca (Bolivie), depuis 30 000 ans. *C. R. Acad. Sci. Paris* **325**, 139–146.

Mourguiart, P., Corrège, T., Wirrmann, D., Argollo, J., Montenegro, M.E., Pourchet, M. and Carbonel, P. (1998). Holocene palaeohydrology of Lake Titicaca estimated from an ostracod-based transfer function. *Palaeogeogr. Palaeoclimatol. Palaeoecol.* **143**, 51–72.

Parenti, L.R. (1984). A taxonomic revision of the Andean killifish genus *Orestias* (Cyprinodontiformes, Cyprinodontidae). *Bull. Am. Mus. Nat. Hist.* **178**, 107–214.

Peypouquet, J.-P., Carbonel, P., Ducasse, O., Tölderer-Farmer, M. and Lété, C. (1987). Environmentally cued polymorphism of ostracods–a theoretical and practical approach. A contribution to geology and to the understanding of ostracod evolution. In: *Evolutionary Biology of Ostracoda, its Fundamentals and Application* (Ed. by T. Hanai, N. Ikeya and K. Ishizaki), pp. 1003–1019. Elsevier, Amsterdam.

Prance, G.T. (1982). Forest refuges: evidence from woody angiosperms. In: *Biological Diversification in the Tropics* (Ed. by G.T. Prance), pp. 37–158. Columbia University Press, New York.

Purper, I. and Pinto, I.D. (1980). Interglacial ostracodes from Wasa Mayu, Bolivia. *Pesquisas* **13**, 161–184.

Roche, M.A., Fernandez Jauregui, C., Aliaga Rivera, A., Peña Mendez, J., Salas Rada, E. and Montaño Vargas, J. (1992). *Balance hídrico superficial de Bolivia*, pp. 1–29. ORSTOM/UNESCO-ORCYT, Paris.

Servant, M. and Fontes, J.-Ch. (1978). Les lacs quaternaires des hauts plateaux des Andes boliviennes Premières interprétations paléoclimatiques. *Cahiers ORSTOM Sér. Géol.* **10**, 9–23.

Strahl, J. (1998). Palynological study of borehole Cb297 in the Valle Central of Cochabamba, Bolivia. *Bol. Serv. Nac. Geol. Min. (SERGEOMIN)* **14**, 1–77.

Stuiver, M. and Reimer, P.J. (1993). Extended ^{14}C data base and revised CALIB 3.0 ^{14}C age calibration program. *Radiocarbon* **35**, 215–230.

Turcq, B., Sifeddine, A., Martin, L., Absy, M.L., Soubiès, F., Suguio, K. and Volkmer-Ribeiro, C. (1998). Amazonia rainforest fires: a lacustrine record of 7000 years. *Ambio* **27**, 139–142.

Wirrmann, D. and Mourguiart, Ph. (1995). Late Quaternary spatio-temporal variations in the Altiplano of Bolivia and Peru. *Quatern. Res.* **43**, 344–354.

Linking Spatial and Temporal Change in the Diversity Structure of Ancient Lakes: Examples from the Ecology and Palaeoecology of the Tanganyikan Ostracods

A.S. COHEN

I. SUMMARY

The linked questions of what is a community and how does it change through space and time are some of the most long-lasting and vexing issues in ecology. In part, this reflects the different scales of time and space from which researchers draw their data, yet the issue is of great practical importance because it impinges on decision making in conservation biology.

In offering the potential for linking differently scaled studies from both ecology and palaeoecology, ancient lakes are ideal areas in which to investigate this problem. High degrees of endemism and morphological specializations in the faunas of these lakes suggest that ecological interactions among such organisms are also likely to be complex, highly deterministic and stable, as a result of long periods of coevolved interactions. Data from long-term studies of fish ecology in Lake Tanganyika support this notion of highly evolved interactions and community stability, and have served as a general model for understanding ecological interactions in ancient lakes.

However, the highly endemic and diverse ostracods of Lake Tanganyika do not conform to this model. Ostracod species associations are extremely variable in space, probably as a result of local colonization and extinction events within local population patches. Whereas individual species show

ADVANCES IN ECOLOGICAL RESEARCH VOL. 31
ISBN 0-12-013931-6

affinities for particular environments, the assemblage observed in a given locality is highly unpredictable. Palaeoecological analyses of sediment cores also show species composition to have varied greatly over periods of hundreds of years. In some cases, however, this assemblage variability is accompanied by stability in total species richness.

These results suggest that several models of community assembly may be relevant to ancient lake faunas, and that caution must be exercised in applying any particular model of community structure or stability to conservation biology and biodiversity management.

II. INTRODUCTION

For most of the twentieth century ecologists have asked the question "What are communities?" (e.g. Clements, 1916; Gleason, 1926). Are they the result of extensive coevolution of organisms that have coinhabited specific regions or environments on the earth's surface over geologically long time intervals, or are they the chance byproducts of the association of species with similar habitat requirements? Is "diversity structure" (i.e. the component contributions of local α diversity, across-gradient β diversity and regional γ diversity; Whittaker, 1977) highly variable over time, or is the contribution of each component to total diversity within an ecosystem relatively constant? If the latter, what allows for such homeostasis? How important are factors such as predation, competitive interactions or mutualisms in structuring species associations, and are they major forces in driving evolutionary diversification? Are there emergent properties of communities at progressively larger spatial and temporal scales, and to what extent is the perception of "community" influenced by the spatial and temporal scale of sampling and data collection (Haury et al., 1978; Levin, 1992)? These questions are important not only to theoreticians, but also for conservation biologists and resource managers. In this chapter these questions will be examined from the perspective of ancient lakes, their ecosystems and their conservation biology. Reference will be made to the cichlid fishes, and especially to the ostracod crustaceans, of Lake Tanganyika, two groups for which relevant data are available.

As humans manipulate nature, either for intended purposes or incidentally through our own population growth, we have generally ignored Aldo Leopold's well-known dictum of "intelligent tinkering", that we keep all of nature's interactions intact until we understand fully the potential consequences of doing otherwise (Leopold, 1953; Samways, 1996). It is central to the effort of avoiding "unintelligent tinkering", that we understand the relative importance of particular species interactions in these lakes and what the ecosystem-level consequences are of disrupting such species-level interactions.

The theoretical and practical importance of understanding community composition and interactions has led to a variety of approaches to address this question, foremost among which have been ecological, palaeoecological and phylogenetic perspectives. Ecologists have used simulation models, and field and laboratory monitoring and manipulation. Notable within the latter category are long-term ecological research sites (LTER sites), which can provide extraordinarily detailed records of communities and the interactions of organisms. However, even the longest running of these sites have only been monitored for about 100 years and most have been studied for only a few decades or less. So, despite the wealth of information that can be obtained about the structure and composition of these extant species assemblages, from a practical standpoint it may never be possible to uncover the origins and evolution of species associations using this approach.

Palaeoecologists have utilized the fossil record to attempt to reconstruct time-series data on species associations (e.g. Davis, 1986; Valentine and Jablonski, 1993; Brett and Baird, 1995; Pandolfi, 1996). Palaeoecological data have the advantage of covering evolutionarily meaningful time periods, through which the historical background and stability of associations can become evident. However, the temporal resolution of palaeoecological data is frequently poor and often uncertain, and the nature of interactions between species usually can only be inferred.

Phylogenetic reconstructions have the power to illustrate the sequential divergence patterns of species with known modern associations, and can reveal patterns of cospeciation through the demonstrating of matching phylogenies in closely linked groups of species (e.g. Armbruster, 1992; Moran and Baumann, 1994). However, the temporal and spatial resolution of these historical inferences is highly variable, and the results are dependent on the quality of the phylogenetic inference used.

When studying the history and durability of species interactions it is desirable to link the complementary strengths of these differently scaled approaches. Yet this is easier said than done, since not all of these types of data are available for every ecosystem. However, ancient lakes are one setting for which such complementary data are available. Furthermore, many ancient lakes have complex ecosystems with numerous endemic species, and consequently with great potential for phylogenetic reconstructions. In addition, all of these lakes are, or could be, fruitful LTER sites. More importantly, these lakes are self-contained tectonic, depositional and evolutionary systems. Their ages are often known with some certainty and highly resolved records of their long histories are preserved in their sediments.

III. LAKE TANGANYIKA

A. Ecological Interactions

Lake Tanganyika is one such ancient lake, and provides an outstanding setting for drawing links across these scales and understanding the history of species associations. This lake is one of the largest (c. 34 000 km^2) and oldest freshwater lakes in the world (c. 9-12 My; Cohen et al., 1993a). For over 100 years its peculiar, endemic fauna has been a source of interest among naturalists (Smith, 1881; Moore, 1903; Fryer, this volume). Over 1500 species of protists, plants and animals have been described from the lake, perhaps 600 of which are endemic to it, and new species are being described each year (Coulter, 1991, 1994; De Vos and Snoeks, 1994; Martens, 1994; Michel 1994, this volume). Numerous studies document Lake Tanganyika's extraordinary community complexity and species interactions on a local scale (e.g. for fish communities: Yamaoka, 1983; Takamura, 1984; Hori, 1987; Kuwamura, 1987, 1992; Hori et al., 1993) and the implications of these diverse biotic interactions for the evolution of species and communities (Fryer and Iles, 1972; Hecky and Fee, 1981; Hecky and Kling, 1981; West et al., 1991; Rossiter, 1995). Biodiversity is high in Lake Tanganyika at all levels: within habitats, between habitats and between provinces (Coulter, 1991). On a local scale species diversity is also strongly and inversely correlated with sedimentation disturbance, caused primarily by watershed erosion and sediment discharge following deforestation (Alin et al., 1999). This inverse correlation is mostly evident in the northern lake basin, where human population density is high and land use has historically been very intense (Cohen et al., 1993b).

Not only are Lake Tanganyika species diverse, but they are also highly disparate from their sister taxa outside the lake basin, both morphologically and genetically (e.g. Michel et al., 1992; Sturmbauer and Meyer, 1993; West and Cohen, 1996). The extraordinary variability seen in feeding structures, defensive mechanisms and courtship behaviours within many of the faunal groups has been used as evidence for the existence of strong (and often obligate) species interactions within Tanganyikan species assemblages. Hori et al. (1993) attribute resource partitioning among Tanganyikan cichlids to a combination of competitive displacement between feeding guilds for food resources, and commensal and mutual relationships involving various species during foraging. West and Cohen (1994, 1996) have argued that the extraordinary shell sculpturing and thickening observed in endemic Tanganyikan gastropods is probably driven by strong predation pressure from potamonautid crabs (which have also undergone a species radiation in the lake of taxa with unusually large, shell-crushing chelae).

Collectively, this body of work supports the notion that a strong relationship exists between the extraordinary diversity of Lake Tanganyika and the complexity of what appear to be highly deterministic ecological interactions. In this view, Lake Tanganyika species flocks qualify as true adaptive radiations (sensu Simpson, 1949), in which extreme niche specialization and finely tuned adaptation to local environments or resources explain how so many species can be packed into one locality (e.g. Fryer and Iles, 1972; Liem, 1978). Some studies suggest that community integration is very tight in Lake Tanganyika, particularly within littoral, rocky-bottom communities of cichlid fishes (Nakai *et al.*, 1994). At one well-studied site (Mbemba, Democratic Republic of Congo), these workers have documented a relatively stable fish fauna over a time scale of 5–10 years.

B. Ostracod Ecology

The ostracods are another group of Tanganyika fauna whose ecology and evolution have been well studied, but unlike the cichlid fishes, they do not appear to conform to the pattern of community integration described above. The Tanganyikan ostracods are highly diverse: as many as 200 species may exist in the lake, almost all of which are endemic (Martens, 1994). The ostracod "radiation" actually incorporates numerous independent, monophyletic lineages (Martens, 1994; Park and Downing, this volume). Many species are highly derived (numerous endemic genera are recognized), some of which possess the hallmark "thalassoid" features of heavy calcification and spinosity and shell sculpturing well known among the Tanganyikan prosobranch gastropods (West and Michel, this volume). Phylogenetic reconstructions for one clade, the genus *Gomphocythere*, suggest that speciation within at least some endemic Tanganyika lineages has been slow, and that independent lineages of *Gomphocythere* have been accumulating since at least the Pliocene, based on information from fossil sister taxa outside the lake (Park and Downing, this volume). However, a pre-Pleistocene ostracod fossil record from Lake Tanganyika is currently unavailable, since no appropriate age outcrops of lake sediments occur surrounding the lake.

1.   Composition of Ostracod Species Assemblages

Lake Tanganyika ostracods occur on a variety of substrate types, ranging from aquatic macrophytes to rocks on soft bottoms (sand and mud). A factor analysis comparison of endemic species occurring on soft-bottom substrates from 103 sampling localities (350 individuals counted per locality) around Lake Tanganyika shows that species are assorted into assemblages that broadly correspond to two major environmental variables (Figure 1). Factor 1 is closely related to substrate texture (sand versus mud), whereas factor 2 is mainly associated with the degree of sediment disturbance (high versus low sedimentation rates). Both of these variables are probably indirectly related to feeding behaviour (most species are non-selective detritus feeders) and to the ability of individual species to move on top of or interstitially within the sediment. However, when a factor analysis is performed on the similarity between site assemblages, rather than broad species' habitat ranges, assemblages do not cluster along substrate (mud versus sand versus rock) habitat or by depth range, and no clear pattern is evident (Figure 2). Taken together, these data indicate that whereas individual ostracod species show

A.S. COHEN

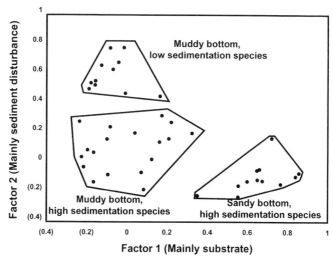

Fig. 1. Rotated factor analysis matrix for common Lake Tanganyika ostracods, by species. Note the significant clustering of species by substrate and sedimentation regime.

some habitat stenotopy, the species composition of any given assemblage is highly unpredictable.

2. *Distribution of Ostracod Populations*

Earlier studies have indicated that ostracod populations within Lake Tanganyika may be very patchy, with species being common in one location, yet rare or absent at a similar site, even within a distance of several hundred metres (Cohen, 1995). This patchiness exists even within areas that superficially appear to be of uniform habitat or substrate type. Most species have been found at only a very limited number of localities, as with many cichlid species from Lakes Malawi and Tanganyika (Brichard, 1989; Ribbink *et al.*, 1983). What is unusual about the distribution pattern of ostracods in Tanganyika (and sets them apart from most observations of cichlids) is the spatial distribution of rare taxa. Rare Malawi and Tanganyika cichlids tend to be clustered in geographically limited areas, often a single island or stretch of rocky coast, as regional endemics. In contrast, rare Tanganyika ostracods (ignoring species that have been found only at a single locality) are often widespread, with sampling localities separated by hundreds of kilometres, and few species qualify as regional endemics (Figure 3). This distribution pattern also supports the notion that geographical (and perhaps habitat) range limitations for these species are more a function of sampling bias against rare species than the actual distribution pattern of these taxa. Species assemblages

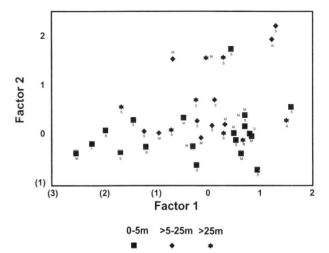

Fig. 2. Rotated factor analysis matrix for Tanganyika ostracod assemblages, by locality. R, rocky habitat; S, sandy habitat; M, muddy habitat. Note that locality assemblages do not cluster by either substrate type or depth.

can change so rapidly even within a habitat that cumulative α diversity (adding adjacent sites within a habitat type) can climb at least as quickly as cumulative β diversity (between-habitat) curves (Figure 4a, b). Ongoing investigations are attempting to define the spatial scale over which this variability exists and the extent to which sampling resolution determines the pattern of patchiness that is inferred (S. Alin, unpubl.).

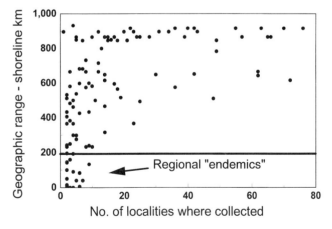

Fig. 3. Geographical range of Lake Tanganyika ostracod species. Distances are along the shoreline of the lake. Regional "endemics" are those species which have been found at multiple sites, all within 200 km (shoreline distance) of one another.

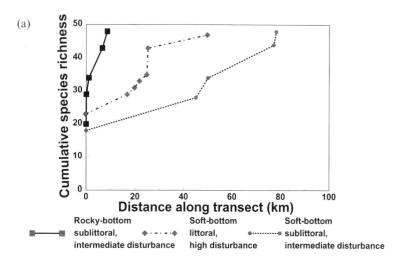

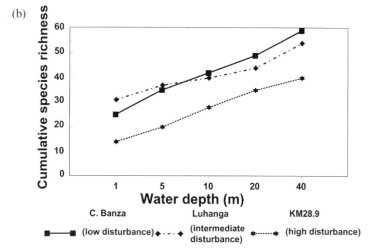

Fig. 4. Cumulative diversity with additional samples within and between habitats for Lake Tanganyika ostracods. (a) Along distance gradients (parallel to shoreline) within a given habitat (α diversity). Sample sites – Cape Banza, Luhanga and KM28.9 – were chosen as representative of low, intermediate and high disturbance regimes, respectively. (b) Along depth gradients between habitats (β diversity). As more samples are added, new species appear in the species pool. New species appear within habitats at approximately the same rate as they appear between habitats.

C. Models of Diversity Maintenance

Models of ancient lake community structure and integration have been strongly influenced by studies of both the morphology and feeding behaviour of cichlids. At an ecosystem level, and particularly for conservation planning purposes, it is important both to verify the hypotheses that have been derived from these studies, and to determine how relevant they are for other clades within ancient lake biotas.

The coexistence of numerous cichlid species with similar resource requirements has been used as evidence for the functionality of hyperspecialization in their pharyngeal jaw and tooth structure (Liem, 1978, 1980). However, this hypothesis has always been troubled by two facts. (i) Although cichlids show dietary differentiation, gut content analyses show that they consume a wide variety of food items, suggesting that these differences are facultative, not obligatory (e.g. Hori et al., 1993). (ii) Furthermore, the number of species that occurs within a given locality greatly exceeds the range of feeding guilds available. Nakai et al. (1994) have argued that these observations can be explained by retaining the hyperspecialization model by also invoking a behavioural component to the specialization (in one well-studied case, foraging behaviours that result in feeding mutualisms). However, the extent to which specific cichlid associations and feeding interactions observed at the primary LTER-type study site for this type of work (Mbemba) occur throughout the ranges of these species in Lake Tanganyika is unknown. Only by documenting the species associations and feeding interactions among these same cichlids at other localities around the lake might it be possible to determine whether they result from short-term contingencies of recruitment and locally learned behaviour, or are the result of coevolution dating to the divergence of the species involved in the association.

Even if feeding and behavioural specialization facilitate the coexistence and maintenance of high diversity in the Tanganyikan cichlids, this mechanism is unlikely to be useful in explaining the high diversity of ostracods in that lake. Tanganyikan ostracods show only modest habitat differentiation and substrate specificity, comparable to that seen in other non-diverse lacustrine faunas. Within-clade morphological specialization in Tanganyikan ostracod feeding and locomotory appendages is insignificant, and behavioural observations, although few, suggest that most species are non-selective detritus feeders. Body size differentiation (often implicated in feeding specialization) is also very limited within these clades.

Elsewhere the author has argued that the organization of the endemic ostracods into numerous metapopulations, rather than hyperspecialization, may be the mechanism for maintaining their high species diversity in Lake Tanganyika. This idea is supported by the frequent sympatry among numerous competitors whose resource requirements are remarkably similar (Cohen, 1994, 1995). Metapopulation dynamics as a mechanism for regenerating local, ephemeral populations is consistent with the highly fragmented nature of the littoral and sublittoral habitats found in the lake, especially coupled with the poor dispersal capabilities of many endemic ostracod species. The relative importance of habitat fragmentation versus dispersal mode in structuring metapopulations would depend on the degree of substrate stenotopy shown by individual species. For species with strong habitat stenotopy (such as many cichlids and molluscs) degree of habitat fragmentation would be more important, whereas for habitat generalists (i.e. many ostracod species), rate of dispersal would be more important. Theoretically, if most species occur as transient populations in a habitat patch, then disequilibrium conditions may disallow the development of competitive dominance and higher local species diversity may result (Hanski, 1983). Under such circumstances local habitat patches may experience a continual flux of species, marked by local extinction and recolonization by a variety of species from adjacent populations.

D. Ostracod Palaeoecology

What do palaeoecological data reveal about the history of community stability and species diversity over longer time scales in Lake Tanganyika? Palaeoecological studies of lake-sediment cores of soft sand and mud substrates from relatively shallow water (40–70 m water depth) are now enabling analysis of community change at comparatively high temporal resolution and over much longer times intervals than ecologists can sample. Given average rates of sedimentation of 1–10 mm y^{-1} at these water depths and bioturbation depths of c. 2–5 cm, one can expect a maximum sampling resolution of between 2 and 50 years (more typically 5–10 years), with sampling intervals measuring in the thousands of years.

Four cores were analysed in detail, chosen originally to represent a spectrum of physical disturbance regimes from river sedimentation. Core 86-DG-32 was collected near the Luamfi River delta in the southern, Tanzanian part of the lake. The Luamfi River watershed has a low human population (5–10 inhabitants km^2) and has experienced very little deforestation, and this core was selected to represent a low disturbance site. Core 86-DG-14 was collected offshore of the Dama River delta, central Burundi coast, in an area of intense, but not complete, deforestation, and represented an intermediate disturbance site. The remaining two cores, Bur-1 (Ruzizi River delta) and Karonge no. 3 (Karonge River delta), were collected from sites adjacent to heavily populated watersheds (c. 350 inhabitants km^2) that have experienced near total deforestation. These two cores represent sites of high physical disturbance. Details of these cores and their faunal lists are presented elsewhere (Wells *et al.*, 1999).

Fossil ostracods present several advantages for detailed study over other taxa of benthic or epibenthic organisms commonly represented in the cores (fish, insects, molluscs, sponges). They are abundant and diverse as fossils, generally well preserved and relatively easy to identify. For these reasons the following palaeoecological analyses and discussion will address only the ostracod fauna. The intention in carrying out these preliminary analyses of fossil ostracod fauna present in the core samples was to ask three questions. (i) What is the historical pattern of diversity difference between sites undergoing different disturbance regimes? (ii) Is an intensification of human disturbance recorded in these cores in the past few hundred years? (iii) How stable or unstable have ostracod species assemblages been over these time intervals?

The Karonge no. 3 core, collected 500 m offshore from the Karonge River delta, showed a clear increase in sedimentation rates over the past few decades and was almost barren of fossil invertebrates (Wells *et al.*, 1999). As such, it was of little use in the present context and is not discussed further.

The other three cores all contained ostracod fossils, and within each core we analysed ostracod populations at approximately decadal resolution to understand patterns of community change and diversity. The data cover the last *c.* 1100 years, during which time independent data suggest that lake levels, water balance and major ion water chemistry have remained relatively stable (Cohen *et al.*, 1997).

Cumulative and standing diversity plots for these three cores are shown in Figure 5. Core BUR-1 has by far the lowest standing diversity of the three sites, and standing diversity varies greatly between sampling horizons (*x*-axis). This difference is well evident before the rapid increase in population pressure and land degeneration of the twentieth century. If human activity is implicated in this difference, the threshold for having an effect on diversity must have been reached long before the twentieth century.

Standing diversities at intermediate and low disturbance site cores (86-DG014 and 86-DG-32, respectively) are much higher than the high-disturbance site (using a standardized sample size), as are all measured diversity indices (Fisher α, Shannon–Weiner and Simpson). Diversity in 86-DG-32 is relatively constant throughout the core. In 86-DG-14 diversity remains relatively constant except for one barren interval near the base of the core and a drop at the top of the core, which may be an anthropogenic signal.

Cumulative diversity curves display relatively rapid saturation of species (*c.* 50–100 years) for both the high and intermediate disturbance sites. In contrast, new species continue to appear at the low disturbance site throughout this *c.* 1100 year record. This continual appearance of new species despite a relatively constant standing species richness is indicative of a continual turnover of taxa. Significant species turnover and change is also indicated by the relatively low Jaccard similarity index values calculated for adjacent samples throughout the 86-DG-32 and 86-DG-14 cores (Wells *et al.*, 1999). At the highly disturbed site (BUR-1) a pattern of co-ordinated local extinction and colonization is evident, with substantial faunal similarity between the species composition of stratigraphically adjacent samples.

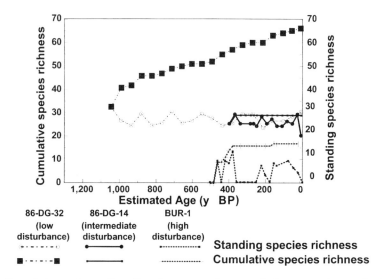

Fig. 5. Cumulative and standing diversity curves of ostracods for cores taken at the three sites discussed in the text.

One explanation for the differences between these cores is that the low disturbance site contains the fossil record of numerous transient metapopulations which colonized the site over time. These species collectively contributed to maintain a high diversity, but individually were continually going extinct and recolonizing the site, without strong linkages to other sympatric species. Adjacent habitat patches both provided new colonists for the core site and were perhaps colonized by individuals from the study site. Conversely, at the more disturbed site, the interconnectedness of metapopulations may have broken down as a result of environmental perturbations becoming too severe. This would have led to fewer species being available to repopulate a locality. Such a scenario would explain the lower overall diversity at this site and, more importantly, why co-ordinated extinction and recolonization was the dominant pattern of faunal turnover. With a restructuring of community functions to allow competition to become important among a reduced but persistent pool of species, the integration of species interactions may have become tighter. A similar pattern of simultaneous local extinction and recolonization events has been observed in palaeolakes that have undergone major environmental fluctuations (salinity or lake level) during the Quaternary (e.g. Palacios-Fest *et al.*, 1993; Magny *et al.*, 1995).

Thus, the preliminary palaeoecological results are consistent with a view of considerable ecological change and an absence of species integration for ostracods over the 10–100 year time scale. However, this study is still in its infancy, and only through the study of many cores and other taxonomic groups can generalizations be made about long-term ecological change in Lake Tanganyika. Exciting prospects for such understanding arise from ongoing efforts to obtain long cores (several hundreds of metres, spanning several million years of lake history) through scientific drilling in several of the African Great Lakes, including Tanganyika. Such cores promise to provide a view of the historical underpinnings of the ostracod radiation in Lake Tanganyika and its ecological dynamics.

IV. IMPLICATIONS FOR CONSERVATION BIOLOGY

It is now recognized that some areas of Lake Tanganyika are threatened by various human impacts. Threats from greatly increased rates of nearshore sedimentation caused by watershed deforestation and subsequent erosion are well documented (e.g. Cohen et al., 1993b), and there is a significant correlation between higher sedimentation rate and reduced diversity among ostracods, molluscs and fishes (Alin et al., 1999). A rapid increase in sedimentation is directly relevant to questions of habitat and population fragmentation because many benthic and epibenthic species of invertebrates and fish are sensitive to sediment loading, as it interferes with feeding and, particularly on rocky habitats, reduces habitat heterogeneity. As such, the extent to which extreme specialization and niche partitioning or continuous disequilibrium/metapopulation fragmentation regulate species diversity is of more than theoretical interest, and impinges upon habitat management and conservation decision making.

Based on the findings from these studies of ostracod assemblages, past and present, and those of cichlid fishes, several informed recommendations can be made. It is clear that reserve designs should incorporate an understanding of how permanent or ephemeral species are at the proposed reserve site, recognizing that stability may vary greatly among various components of the community. For example, if fish communities at a given locality are the sole target for preservation and they prove to be stable entities in Lake Tanganyika, with high persistence, then recruitment from adjacent habitat patches is a less important consideration in determining the area to be covered by a reserve. In such circumstances reserves that cover relatively small areas and that remain undisturbed might be viable. Conversely, conservation efforts might be directed more broadly to encompass both highly variable and stable elements of the biota. If recruitment from adjacent parts of numerous metapopulations is required for periodically rescuing habitat patches (sensu Brown and Kodric-Brown, 1977), then reserves must be designed to link adjacent habitat patches and should be larger. Elimination of adjacent patches (or the deterioration of their quality) could lead to a cascading sequence of extinctions, both directly, through a loss of recolonization potential, and indirectly, by eliminating the very population structure that originally facilitated the high α diversity among competitors.

In the author's view, conservation and biodiversity management for Lake Tanganyika should take a broad, ecosystem-level approach, and avoid focusing only on high-visibility flagship taxa or on those with greatest immediate economic importance (Cohen, 1994). The contrast between cichlid and ostracod diversity structure in space and time suggests that no one taxonomic group is likely to serve as a robust model for how diversity is maintained or how ancient lake communities function as a whole. Only by combining palaeoecological and LTER data for multiple taxonomic groups are we likely to attain sufficient understanding of ecosystem change and species interactions over the 10–100 year time scale required for making informed decisions about conservation management practices in ancient lakes.

ACKNOWLEDGEMENTS

Research was funded by NOAA-NURC-UCAP grant no. UCAP-92-04, and NSF grants BSR 8415289, BSR 869074 and EAR 9627766. The author thanks the governments of Burundi, Tanzania and the Democratic Republic of Congo for providing numerous research permits for this work, as well as S. Alin, D. Dettman, K. Martens, E. Michel, C. O'Reilly, L. Park, T. Wells and K. West, all of whom continue to teach the author new things about Lake Tanganyika. This is publication no. 54 of IDEAL (the International Decade of East African Lakes).

REFERENCES

Alin, S., Cohen, A.S., Bills, R., Gashagaza, M.M., Michel, E., Tiercelin, J.J., Martens, K., Coveliers, P., Mboko, S.K., West, K., Soreghan, M., Kimbadi, S. and Ntakimazi, G. (1999). Rapid faunal assessments for monitoring biodiversity in Lake Tanganyika, East Africa. *Conserv. Biol.* **13**, 1017–1033.
Armbruster, W.S. (1992). Phylogeny and the evolution of plant–animal interactions. *Bioscience* **42**, 12–20.
Brett, C. and Baird, G. (1995). Coordinated stasis and evolutionary ecology of Silurian to Middle Devonian faunas in the Appalachian Basin. In: *New Approaches to Speciation in the Fossil Record* (Ed. by D. Erwin and R. Anstey), pp. 285–315. Columbia University Press, New York.
Brichard, P. (1989). *Pierre Brichard's Guide to the Cichlids and All other Fishes of Lake Tanganyika.* T.F.H. Publications, Neptune City, NJ.
Brown, J. and Kodric-Brown, A. (1977). Turnover rates in insular biogeography: effects of immigration on extinction. *Ecology* **58**, 445–449.
Clements, F.E. (1916). *Plant Succession: An Analysis of the Development of Vegetation.* Carnegie Institute, Washington, DC.
Cohen, A.S. (1994). Extinction in ancient lakes: biodiversity crises and conservation 40 years after J.L. Brooks. *Arch. Hydrobiol. Beiheft. Ergebnisse Limnol.* **44**, 451–479.

Cohen, A.S. (1995). Paleoecological approaches to the conservation biology of benthos in ancient lakes: a case study from Lake Tanganyika. *J. N. Am. Benthol. Soc.* **14**, 654–668.

Cohen, A.S., Soreghan, M. and Scholz, C. (1993a). Estimating the age of formation of lakes: an example from Lake Tanganyika, East African Rift System. *Geology* **21**, 511–514.

Cohen, A.S., Bills, R., Cocquyt, C. and Caljon, A. (1993b). The impact of sediment pollution on biodiversity in Lake Tanganyika. *Conserv. Biol.* **7**, 667–677.

Cohen, A.S., Talbot, M.R., Awramik, S.M., Dettman, D.L. and Abell, P. (1997). Lake level and paleoenvironmental history of Lake Tanganyika, Africa, as inferred from late Holocene and modern stromatolites. *Geol. Soc. Am. Bull.* **109**, 444–460.

Coulter, G.W. (ed.) (1991). *Lake Tanganyika and its Life.* Oxford University Press, London.

Coulter, G.W. (1994). Lake Tanganyika. *Arch. Hydrobiol. Beiheft. Ergebnisse Limnol.* **44**, 127–137.

Davis, M. (1986). Climatic instability, time lags, and community disequilibrium. In: *Community Ecology* (Ed. by J. Diamond and T. Case), pp. 269–295. Harper and Row, New York.

De Vos, L. and Snoeks, J. (1994). The non-cichlid fishes of the Lake Tanganyika basin. *Arch. Hydrobiol. Beiheft. Ergebnisse Limnol.* **44**, 391–405.

Fryer, G. and Iles, T.D. (1972). *The Cichlid Fishes of the Great Lakes of Africa. Their Biology and Evolution.* Oliver and Boyd, Edinburgh.

Gleason, H.A. (1926). The individualistic concept of the plant association. *Bull. Torrey Bot. Club* **53**, 7–26.

Hanski, I. (1983). Coexistence of competitors in patchy environments. *Ecology* **64**, 491–500.

Haury, L.R., McGowan, J.A. and Wiebe, P.H. (1978). Patterns and processes in the time–space scales of plankton distributions. In: *Spatial Pattern in Plankton Communities* (Ed. by J.H. Steele), pp. 277–327. Plenum, New York.

Hecky, R. and Fee, E. (1981). Primary production and rates of algal growth in Lake Tanganyika. *Limnol. Oceanogr.* **26**, 532–547.

Hecky, R. and Kling, H. (1981). The phytoplankton and protozooplankton of the euphotic zone of Lake Tanganyika: species composition, biomass, chlorophyll content and spatiotemporal distribution. *Limnol. Oceanogr.* **26**, 548–564.

Hori, M. (1987). Mutualism and commensalism in the fish community in Lake Tanganyika. In: *Evolution and Coadaptation in Biotic Communities* (Ed. by S. Kawano, J. Connell and T. Hidaka), pp. 219–239. University of Tokyo Press, Tokyo.

Hori, M., Gashagaza, M.M., Nshombo, M. and Kawanabe, H. (1993). Littoral fish communities in Lake Tanganyika: irreplaceable diversity supported by intricate interactions among species. *Conserv. Biol.* **7**, 657–666.

Kuwamura, T. (1987). Distribution of fishes in relation to depth and substrate at Myako, east middle coast of Lake Tanganyika. *Afr. Stud. Monogr.* **7**, 1–14.

Kuwamura, T. (1992). Overlapping territories of *Pseudosimochromis curvifrons* males and other herbivorous fishes in Lake Tanganyika. *Ecol. Res.* **7**, 43–53.

Leopold, A. (1953). *Round River.* Oxford University Press, New York.

Levin, S.A. (1992). The problem of pattern and scale in ecology. *Ecology* **73**, 1943–1967.

Liem, K. (1978). Modulatory multiplicity in the feeding mechanism in cichlid fishes as exemplified by the invertebrate pickers of Lake Tanganyika. *J. Zool. Soc. Lond.* **189**, 93–125.

Liem, K. (1980). Adaptive significance of intra and interspecific differences in the feeding repertoires of cichlid fishes. *Am. Zool.* **20**, 295–314.

Magny, M., Mouthon, J. and Ruffaldi, P. (1995). Late Holocene level fluctuations of Lake Ilay in Jura, France: sediment and mollusc evidence and climatic implications. *J. Paleolimnol.* **13**, 219–229.

Martens, K. (1994). Ostracod speciation in ancient lakes: a review. *Arch. Hydrobiol. Beiheft. Ergebnisse Limnol.* **44**, 203–222.

Michel, A.E. (1994). Why snails radiate: a review of gastropod evolution in long-lived lakes, both recent and fossil. *Arch. Hydrobiol. Beiheft. Ergebnisse Limnol.* **44**, 285–317.

Michel, A.E., Cohen, A.S., West, K., Johnston, M.R. and Kat, P.W. (1992). Large African lakes as natural laboratories for evolution: examples from the endemic gastropod fauna of Lake Tanganyika. *Mitt. Int. Verein. Limnol.* **23**, 85–99.

Moore, J.E.S. (1903). *The Tanganyika Problem.* Hurst and Blackett, London.

Moran, N. and Bauman, P. (1984). Phylogenetics of cytoplasmically inherited microorganisms of arthropods. *Trends Ecol. Evol.* **9**, 15–20.

Nakai, K., Kawanabe, H. and Gashagaza, M.M. (1994). Ecological studies on the littoral cichlid communities of Lake Tanganyika: the coexistence of many endemic species. *Arch. Hydrobiol. Beiheft. Ergebnisse Limnol.* **44**, 373–389.

Palacios-Fest, M., Cohen, A.S., Ruiz, J. and Blank, B. (1993). Comparative paleoclimatic interpretations from nonmarine ostracodes using faunal assemblages, trace element shell chemistry, and stable isotope data. *Geophys. Monogr.* **78**, 179–190.

Pandolfi, J. (1996). Limited membership in Pleistocene reef coral assemblages from the Huon Peninsula, Papua New Guinea: constancy during global change. *Paleobiology* **22**, 152–176.

Ribbink, A., Marsh, B., Marsh, A., Ribbink, A. and Sharp, B. (1983). A preliminary survey of the cichlid fishes of rocky habitats in Lake Malawi. *S. Afr. J. Zool.* **18**, 149–310.

Rossiter, A. (1995). The cichlid fish assemblages of Lake Tanganyika: ecology, behaviour and evolution of its species flocks. *Adv. Ecol. Res.* **26**, 187–252.

Samways, M.J. (1996). The art of unintelligent tinkering. *Conserv. Biol.* **10**, 1307.

Simpson, G.G. (1949). *The Meaning of Evolution.* Yale University Press, New Haven, CT.

Smith, E.A. (1881). On a collection of shells from Lakes Tanganyika and Nyassa and other localities in East Africa. *Proc. Zool. Soc. Lond.* **1881**, 276–300.

Sturmbauer, C. and Meyer, A. (1993). Mitochondrial phylogeny of the endemic mouthbrooding lineages of cichlid fishes from Lake Tanganyika, East Africa. *Mol. Biol. Evol.* **10**, 751–768.

Takamura, K. (1984). Interspecific relationships of Aufwuchs-eating fishes in Lake Tanganyika. *Envir. Biol. Fishes* **10**, 225–241.

Valentine, J. and Jablonski, D. (1993). Fossil communities: compositional variation at many time scales. In: *Species Diversity in Ecological Communities* (Ed. by R. Ricklefs and D. Schluter), pp. 341–349. University of Chicago Press, Chicago, IL.

Wells, T.M., Cohen, A.S., Park, L.E., Dettman, D.L. and McKee, B.A. (submitted). Ostracode stratigraphy and paleoecology from surficial sediments of Lake Tanganyika, Africa. *J. Paleolimnol.* **22**, 259–276.

West, K. and Cohen, A.S. (1994). Predator–prey coevolution as a model for the unusual morphologies of the crabs and gastropods of Lake Tanganyika. *Arch. Hydrobiol. Beiheft. Ergebnisse Limnol.* **44**, 267–283.

West, K. and Cohen, A.S. (1996). Shell microstructure of gastropods from Lake Tanganyika, Africa: adaptation, convergent evolution, and escalation. *Evolution* **50**, 672–681.

West, K., Cohen, A.S. and Baron, M. (1991). Morphology and behavior of crabs and gastropods from Lake Tanganyika, Africa: implications for lacustrine predator–prey coevolution. *Evolution* **45**, 589–607.

Whittaker, R.H. (1977). Evolution of species diversity in land communities. *Evol. Biol.* **10**, 1–67.

Yamaoka, K. (1983). Feeding behavior and dental morphology of algae-scraping cichlids (Pisces:Cichlidae) in Lake Tanganyika. *Afr. Stud. Monogr.* **4**, 77–89.

Conservation of the Endemic Cichlid Fishes of Lake Tanganyika: Implications from Population-level Studies Based on Mitochondrial DNA

E. VERHEYEN and L. RÜBER

I. SUMMARY

Several recent population-level studies of mitochondrial DNA (mtDNA) variation in Tanganyikan cichlid fishes illustrate the strength and utility of mtDNA analysis as a powerful tool in evolutionary biology. This chapter attempts to show that these studies also contain information about the various levels of biodiversity that are relevant for the identification of "evolutionary significant units" (ESUs) and the assessment of conservation priorities of cichlid taxa or areas from an evolutionary perspective. These results based on mtDNA studies on cichlid populations are consistent with the notion that species should never be viewed as undifferentiated monotypic entities. The presence or absence of genetically and geographically distinct populations is an important feature that characterizes each of these species, and is important to consider when arriving at informed conservation or management decisions.

Mitochondrial DNA studies have also provided new information on the distributional boundaries of some genetically divergent cryptic species, a conclusion that, in a number of examples, has been supported by the subsequently discovery of morphological and behavioural differences. It is

ADVANCES IN ECOLOGICAL RESEARCH VOL. 31
ISBN 0–12–013931–6

possible that there may be many more undiscovered cases of allopatrically occurring morphologically similar fishes that are erroneously considered to be conspecific.

The phylogeographical results indicate that closely related cichlids can have markedly different within-lake distribution ranges, a conclusion that has major implications in any efforts towards the conservation of the Tanganyikan rock-dwelling cichlid fauna. Provided that concordant lines of evidence are found for historical centres of biodiversity in Lake Tanganyika, emphasis should be placed on preserving the integrity of such regional biotas by establishing regionally structured protected sanctuaries. Logistic considerations may make it impossible to preserve the genetic diversity in all components of the cichlid fauna by such measures. However, to formulate guidelines for the establishment of protected areas within Lake Tanganyika, and for the management of the faunas that these areas harbour, genetic information can be a very useful complement to zoogeographical data that is based solely on traditional morphologically based species descriptions.

II. INTRODUCTION

With an estimated age of between 9 and 12 My, Lake Tanganyika is by far the oldest East African rift lake (Cohen *et al.*, 1993a). This lake harbours an extraordinary diversity of endemic organisms and its cichlid fish fauna is morphologically, ecologically and behaviourally the most diverse species flock of the African lakes (Fryer and Iles, 1972; Greenwood, 1984; Coulter, 1991; Rossiter, 1995). The rapidly increasing growth of the human population around the lake has meant that this unique freshwater ecosystem is threatened not only by resource overexploitation, such as the depletion of its fish stocks by overfishing, but also by industrial and agricultural pollution (see Cocquyt, this volume) and excessive amounts of sedimentation caused by deforestation (Cohen, 1992; Cohen *et al.*, 1993b). As a result of these environmental changes, water quality may soon deteriorate to a degree that threatens the survival of the unique endemic fish and invertebrate species flocks of the lake, particularly along its extremely species-rich rocky littoral habitats.

Recent studies have shown that the taxonomic knowledge of the Tanganyikan cichlid fauna is less advanced than had been assumed (Snoeks *et al.*, 1994; Snoeks, this volume). This is important, because a sound taxonomic knowledge of these complex species assemblages is an essential prerequisite to any effort to evaluate the effects of environmental changes on the species communities in the littoral habitats of the lake. However, to date, many important facts about these species assemblages are still ignored; for example, many species are described only on the basis of very few specimens, usually collected from only one or two localities. At the same time, it is becoming increasingly evident that an unknown number of species has yet to be discovered and named. This lack of fundamental knowledge about this fauna raises critical questions, such as (i) how much do we really know about the biodiversity in Lake Tanganyika, and (ii) what measures should be taken to conserve the unique fauna of this lake? Some guidelines to assist in selecting which species should be conserved can be drawn from the available, but unfortunately often anecdotal and sometimes incorrect, literature on the distribution ranges of species (Snoeks *et al.*, 1994). However, to select which populations should be conserved is an even more difficult task since very few species have been subjected to any form of systematic investigation on a lake-wide scale. An additional problem is that the cichlid fishes from the rocky littoral habitats, the supposedly better known component of this lacustrine fauna, are characterized by a considerable degree of geographically restricted within-lake endemisms, that until recently were usually studied through the distribution patterns of the often brightly coloured, geographically restricted populations (Konings, 1988). It is also important to note that although several studies have dealt with the possible importance of body colour in driving speciation by sexual selection (e.g. Dominey, 1984), not one study takes a firm position with regard to the taxonomic rank that should be given to these geographically restricted "colour races", subspecies or species (e.g. *Tropheus* spp. or *Petrochromis* spp.). It is only recently that mitochondrial DNA (mtDNA) studies at the population level have shed some light on the within-lake evolution and interrelationships among allopatric rock-dwelling populations of some mouthbrooding cichlid taxa (Sturmbauer and Meyer, 1992; Meyer *et al.*, 1996; Verheyen *et al.*, 1996; Sturmbauer *et al.*, 1997, Rüber *et al.*, 1998). This chapter attempts to illustrate that mtDNA population-level studies on rock-dwelling Tanganyikan cichlid fishes also yield information concerning the true nature of the extant biodiversity, information of fundamental importance to any meaningful efforts directed towards conserving this unique fauna.

III. INTERSPECIFIC AND INTRASPECIFIC PHYLOGENY, MITOCHONDRIAL DNA SEQUENCES AND THE CONSERVATION OF ROCK-DWELLING CICHLID FISHES

The mtDNA molecule offers several advantages for studies that attempt to infer the evolutionary history of cichlid populations. First, it is almost exclusively maternally inherited (Giles *et al.*, 1980) and the analysis of molecular lineages is thus not complicated by the effects of recombination as is the case for most nuclear DNA (e.g. microsatellites). Secondly, the high rate of evolution of the mtDNA permits the accumulation of substantial numbers of substitutions over time periods relevant for the study of population histories (Brown *et al.*, 1979; Meyer, 1993; Sturmbauer *et al.*, 1997). In addition, since each cell contains many more copies of mtDNA than of autosomal DNA, retrieval of mtDNA from old tissue samples is more easily achieved. Consequently, the mtDNA genome is a promising molecule for future studies that may need to examine rare and/or extinct taxa from formalin-fixed museum specimens (O'Rourke et al., 1996).

In order to study the genetic substructure in a particular geographical area, mtDNA sequence data from different localities within this area must be collected and compared with each other. To date, there have been only three published studies of mtDNA diversity in populations of mouthbrooding Tanganyikan cichlid species: Sturmbauer and Meyer (1992), Meyer et al. (1996) and Verheyen et al. (1996). These studies determined the nucleotide sequences of a variable part of the mitochondrial control region (Meyer, 1993) from five genera comprising an uncertain number of species (reviewed in Sturmbauer et al., 1997; Rüber et al., 1998). As the main focus of an attempt to elucidate the evolutionary mechanisms that induce and result in explosive speciation events in rock-dwelling cichlid fishes, these studies examined whether and how recent historical geographical barriers are reflected in mtDNA sequence comparisons among cichlid populations. This approach may permit the evaluation of the importance of ecological specialization and niche partitioning, e.g. habitat specificity, site fidelity or territoriality, homing behaviour and social organization, in the within-lake evolution of these species flocks, as suggested by various authors (e.g. Fryer and Iles, 1972; McKaye and Gray, 1984; Greenwood, 1984; Rossiter and Yamagishi, 1997; Kohda, 1997; Yanagisawa et al., 1997).

This chapter illustrates how population-level mtDNA studies can yield a variety of information on the cichlid biodiversity in Lake Tanganyika. Usually, the conservation of biodiversity is considered at three levels: (i) the ecosystem or landscape level (containing landscapes and habitats and the relationships and processes that occur within them), (ii) the taxonomic or species level (most commonly used), and (iii) the genetic level (the least understood area of biodiversity that addresses the diversity found within and among populations of species). Two examples are presented that illustrate how these mtDNA studies have a bearing on all three levels of biodiversity mentioned above. These two examples concern studies on Tanganyikan endemic mouthbrooding rock-dwelling cichlid fishes that have different life-history strategies (e.g. presence or absence of locally restricted populations characterized by different colours, presence or absence of territoriality, presence or absence of schooling behaviour). Although none of these studies was designed specifically to provide data relevant to the management and conservation of these fishes, the importance and utility of this information for the conservation of these faunas are illustrated. These data sets need to be extended and complemented to answer adequately questions pursuant to the short- and long-term conservation and management of these unique faunas.

A. Genetic Variation and Phylogeography in Two *Simochromis* Species

The mtDNA sequence variation in *Simochromis babaulti* and *S. diagramma* populations (Meyer *et al.*, 1996) reveals that, in contrast to previous findings in the Tanganikan genus *Tropheus* (Sturmbauer and Meyer, 1992), few populations contain more than one haplotype. In addition, several mtDNA haplotypes are shared among geographically distinct populations (separated by 300 km, see Figure 1). Furthermore, no particular region in the lake appears to be genetically depauperate (possess a significantly lower number of mtDNA haplotypes), and genetic and geographic distances do not seem to be closely related. The low sequence divergences between these two species and the absence of spatial genetic differentiation within these species suggests their recent speciation and a probably extensive contemporary gene flow among populations.

These findings demonstrate clearly that even closely related Tanganyika cichlids (*Simochromis* species versus *Tropheus* species) can differ markedly in their capacity for dispersal and in their responses to physical barriers that affect gene flow. Indirectly, this had already been assumed from ecological data (schooling behaviour, different range of habitats, territoriality in males and/or females) and distribution data on colour variation in *Simochromis* and *Tropheus* (e.g. Brichard, 1989; Konings, 1988). Consequently, it seems that

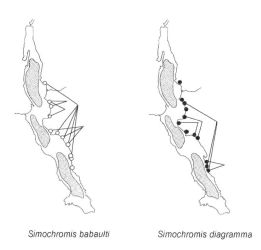

Simochromis babaulti   Simochromis diagramma

Fig. 1. Maps of Lake Tanganyika with localities where the two studied *Simochromis* species were collected (redrawn after Meyer *et al.*, 1996). The connected symbols indicate shared haplotypes among specimens from different localities. Several localities were found to contain specimens with different haplotypes (data not shown). The three shaded areas within the current lake basin indicate three separate palaeolakes that existed during a recent (*c.* 22 000 years bp) major 600 m drop in lake level (Tiercelin and Mondeguer, 1991).

some rock-dwelling cichlids, at least the two *Simochromis* species studied so far, may indeed be truly circumlacustrine taxa.

B. Genetic Variation, Phylogeny and Phylogeography of Three Genera of the Tribe Eretmodini

The mtDNA sequence variation in the four nominal eretmodine species *Eretmodus cyanostictus, Tanganicodus irsacae, Spathodus marlieri* and *S. erythrodon* (Figure 2) agrees to a large extent with previous findings in *Tropheus* (Sturmbauer and Meyer, 1992). Both data sets reveal the existence of several genetically distinct lineages that often have a restricted geographical distribution (i.e. intralacustrine endemism). However, the data set on the eretmodines (Verheyen *et al.*, 1996) contains sequences of specimens collected at more locations than are available for the two studied *Simochromis* species, for which higher numbers of specimens per locality were sequenced (Meyer *et al.*, 1996).

Nevertheless, the data on eretmodines suggest that, unlike in *Simochromis babaulti* and *S. diagramma* discussed above, very few specimens share an identical mtDNA haplotype, and in the rare cases that this was observed it was in specimens from the same or adjacent populations (Figure 3a, b). Interestingly, the within-lake distribution of the different mtDNA lineages that consist of closely related mtDNA haplotypes is restricted to the northern,

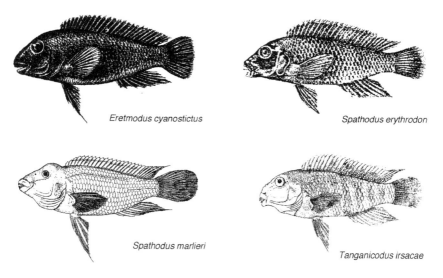

Fig. 2. Illustration of the general morphology of the four currently described species of the tribe Eretmodini, the so-called goby cichlids (Poll, 1950; Boulenger, 1898, redrawn from Poll, 1986).

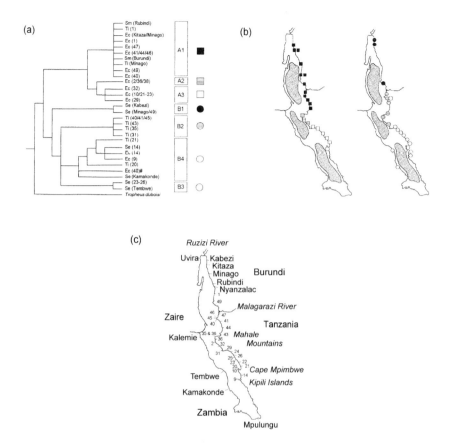

Fig. 3. (a) Strict consensus tree constructed from 275 equally parsimonious trees [tree length of 208 steps, consistency index (CI) 0.62, rescaled consistency index (RC) 0.53, redrawn from Verheyen et al., 1996]. Heuristic search with random addition of taxa (10 replications using PAUP (Swofford, 1993). The species names follow current taxonomic assignments: Ec, *Eretmodus cyanostictus*; Ti, *Tanganicodus irsacae*; Se, *Spathodus erythrodon*; Sm, *Spathodus marlieri*. Locality names and numbers are given in parentheses. Ec (40)# indicates *Eretmodus* species from locality 40 that is morphologically distinct from Ec (40). (b) Maps of Lake Tanganyika showing the mtDNA distribution of the studied Eretmodini (redrawn from Verheyen et al., 1996). The symbols indicate genetically distinct lineages based on phylogenetic analyses (see Figure 3a). The left map shows the intralacustrine distribution of mtDNA haplotypes belonging to lineages A1, A2 and A3, while the map on the right shows the distribution of mtDNA haplotypes belonging to lineages B1, B2, B3 and B4. Both maps show the three separate palaeolakes that follow the present 600 m depth contour. The lake level dropped by almost 600 m below its current level between 220 000 and 75 000 years ago (Tiercelin and Mondeguer, 1991). Localities where two symbols appear together indicate that individuals with different generic assignments have been collected there. (c) Map of Lake Tanganyika showing all of the localities and sample sites mentioned in the text. Lake Tanganyika is about 650 km in length and has a maximum width of 80 km.

the central and the southern parts of the lake, respectively (Figure 3b). This distribution pattern, including the presence of very similar haplotypes on opposite sides of the lake, matches quite closely the now inundated shorelines of the three Lake Tanganyikan palaeolakes (Scholz and Rosendahl, 1988). As observed in *Tropheus* (Sturmbauer and Meyer, 1992) but not in *Simochromis* (Meyer *et al*., 1996), the existence of two inclusive lineages, A and B, living sympatrically along nearly the whole length of the studied area, indicates that at least two consecutive periods of rapid diversification occurred during the within-lake evolution of the Eretmodini (Figure 3a).

The strong phylogeographical structure in the Eretmodini, as well as the virtual absence of shared haplotypes among adjacent populations, as was observed within *Tropheus*, suggests very limited contemporary gene flow. Another important finding is that closely related cichlids can have markedly different ranges of within-lake distribution. Indirectly, this deviation from the traditionally invoked two- or three-basin distribution patterns by "species couples" (Poll, 1956) agrees with more detailed data on cichlid species distribution ranges in Lake Tanganyika (Snoeks *et al*., 1994).

This observation is extremely relevant for the conservation of the Tanganyikan rock-dwelling cichlid fauna. The protection of a particular rocky littoral habitat type in different parts of the lake will not necessarily result in the conservation of the same cichlid taxa. Indeed, the mitochondrially defined clades are in conflict with the current taxonomy and suggest that the northern *Eretmodus* and the southern "true" *E. cyanostictus* (type locality Mpulungu, Zambia, Figure 3c) are not conspecifics (Rüber *et al*., 1999), and that the taxonomy at the genus level of this tribe needs to be revised. The supposed existence of two *Eretmodus*-like species is supported by differences in body coloration and morphology that have long remained undetected (Konings, 1988; Verheyen *et al*., 1996; Rüber *et al*., 1999).

IV. RELEVANCE OF MITOCHONDRIAL DNA SEQUENCES FOR THE MANAGEMENT AND CONSERVATION OF LACUSTRINE BIODIVERSITY

These two studies, as well as the mentioned mtDNA-based population study on *Tropheus* (Sturmbauer and Meyer, 1992), illustrate that mtDNA analysis is a powerful tool in evolutionary biology, and have documented much of what is known of the evolutionary processes that underlie the extant cichlid biodiversity in Lake Tanganyika. However, the results of these studies also contain information about the various levels of biodiversity that are relevant for the identification of evolutionary significant units (ESUs) and the assessment of conservation priorities of taxa or areas from an evolutionary perspective (Ryder, 1986). These findings of mtDNA studies on cichlid populations are also consistent with the generally accepted notion that species should never be viewed as undifferentiated monotypic entities (Avise, 1994). The use of mtDNA and other genetic markers allows the identification of ESUs, which are increasingly being recognized as appropriate units for conservation efforts. Therefore, it is important that future studies should seek to provide information about the presence or absence of genetically and geographically distinct populations for a number of mouthbrooding and substrate-brooding model species that represent the extremely speciose cichlid communities of the rocky littoral habitats of Lake Tanganyika. Only after this type of information becomes available will it be possible to conserve these species communities effectively.

Another important result of these mtDNA studies is that they provide new information on the boundaries of some genetically divergent cryptic species. Too strict an application of mtDNA phylogeographies may be inappropriate to delineate the boundaries among cichlid populations that are not also separated by another, independent, set of characters. However, one example of a taxon that was "discovered" on the basis of its mtDNA sequences is an undescribed *Eretmodus* species, a claim that supports earlier observations on the morphology (Konings, 1988) and recent data on behavioural differences between the "true" southern *E. cyanostictus* and the northern *E.* c.f. *cyanostictus* (K. Yamaoka, unpubl.). *Eretmodus cyanostictus* seems to represent a similar case of taxonomic confusion to *T. moori* (type locality Mpulungu, Zambia) because there are strong indications that morphologically very similar *Tropheus* are not necessarily conspecifics (Sturmbauer and Meyer, 1992; Snoeks *et al.*, 1994). To date, many authors continue mistakenly to consider *T. moori* a circumlacustrine species (e.g. Kohda, 1997), an opinion that does not acknowledge the recent evidence that the taxonomy of these taxa is considerably more complex than was originally thought. Unfortunately, the prevalence of this "one species, many colour races" view in the literature could result in a conservation policy that protects particular rocky littoral habitats in the lake, but will not result in the conservation of the several morphologically very similar, but undescribed, species in other parts of the lake.

The documented examples illustrate how the sometimes complex evolutionary histories of rock-dwelling cichlids results in hierarchically arranged genetic diversity from geographical population structures within species to genetic differences among reproductively isolated taxa that have been phylogenetically separated for different periods of time. Since the conservation of biodiversity implies concerns about the conservation of genetic diversity as well as the preservation of evolutionary processes, it is important that more extensive studies on dynamic evolutionary processes such as speciation, introgression and hybridization should be studied in several taxa.

Clearly, in order to evaluate the possible importance of hybridization events in the evolutionary history of certain cichlid taxa, it will also be necessary to study nuclear markers. In addition, in order to allow testing of the potential importance of introgressive hybridization in the evolution of these fishes, these molecular data sets will have to be supplemented with observations concerning the ecology, ethology and reproductive behaviour of these fishes. The adoption of ESUs (Ryder, 1986) as the focus of conservation efforts would satisfy the generally accepted goal that recognizes the importance of genetic biodiversity (Moritz, 1994). The presence or absence of strong phylogeographical structure in the studied rock-dwelling cichlids, as well as the presence or absence of shared haplotypes among adjacent populations can, if performed on a sufficiently high number of specimens per population, be indicative of the amount of contemporary gene flow. In managing small, genetically distinct populations, precaution should be taken to avoid mixing populations by translocating fishes from populations that are naturally isolated and may be adapted to local environmental conditions. It is clear that such information would facilitate management programmes targeting the preservation of the genetic integrity of each species while maintaining its genetic variability. Phylogeographical and other genetic information will also shed light on the precautions necessary to avoid artificial movements of taxa between areas that contain genetically isolated populations or, worse still, different species that may compete or hybridize.

In this context it is important to note that phylogeographical data seem to indicate that even cichlids within the closely related genera *Tropheus* and *Simochromis* can have markedly different within-lake distribution ranges, a conclusion that has major implications for any efforts at conserving the Tanganyikan rock-dwelling cichlid fauna.

The mtDNA data on the studied cichlid fauna indicate that particular areas in Lake Tanganyika are geographical centres for an important fraction of its regional, intraspecific diversity. Future studies should test whether these areas are comparable with zoogeographical provinces that can be distinguished based on species' distribution limits (Snoeks et al., 1994). This will involve using mtDNA phylogeographical data to define geographical regions for a substantial proportion of species to evaluate whether, or to what extent, they evolved independently from conspecifics in other regions of the lake. The value of this approach is illustrated by comparable studies on other faunas, where congruent genetic breaks along extant or historical geographical barriers have been revealed (e.g. Avise, 1992).

One important consequence of such studies would be that the presence of high numbers of ESUs based on mtDNA data would indicate which regions within Lake Tanganyika merit protection, even if these regions are not exceptional in the number of endemic species that they harbour. Provided that concordant lines of evidence can be found for historical centres of biodiversity in Lake Tanganyika, emphasis should be placed on preserving the integrity of such regional biotas by establishing regionally structured protected sanctuaries. Logistic considerations may make it impossible to preserve the genetic diversity in all of the components of the cichlid fauna by such measures. However, to formulate guidelines for the establishment of protected areas within Lake Tanganyika, and for the management of the faunas that these areas harbour, genetic information can be a very useful complement to zoogeographical data that, thus far, have too often been based solely on species identification in the field and too seldom confirmed by thorough study of appropriate reference specimen collections.

V. CONCLUSIONS AND SUGGESTIONS FOR FUTURE RESEARCH

Mitochodrial DNA-based phylogeographies of some rock-dwelling cichlids indicate that there are markedly different degrees of geographical population subdivisions. Some taxa appear to have lake-wide distribution ranges and experience extensive gene flow, whereas other taxa have restricted distribution patterns that often seem to differ between taxa. However, in many cases it may be possible to characterize regionally restricted populations using mtDNA sequences. In addition, mtDNA sequences have proven useful for the "discovery" of previously unknown cichlid species.

In view of the taxonomic problems characteristic for this fauna and revealed by mtDNA studies, it is recommended that a thorough comparison should be made with the existing type material and adequate specimen collections, with the objective of the eventual re-evaluation of the species identifications in ecological and zoogeographical studies. Although it seems obvious that any study should be based on properly identified specimens, in reality many ecological, and even some taxonomic studies on Tanganyika cichlids have been based on species identifications made in the field or specimens that were studied without reference to type collections.

To be useful, future mtDNA studies will require additional and more fine-grained sampling programmes that complement the already existing specimen and tissue collections. Many of the results discussed above are still preliminary, and more extensive sampling will be required to produce more reliable qualitative and quantitative signals of temporal and spatial population changes. In order to allow the study of within-population variation, which is important but has yet to be addressed, mtDNA studies of the type presented in this chapter will need to be supplemented with additional sequence data. If supported by the presence of data on nuclear genetic variation, the results of mtDNA studies at the population level will provide information of fundamental importance to the short- and long-term management of cichlid populations and species in Lake Tanganyika.

ACKNOWLEDGEMENTS

This study was supported by the Belgian (FWO-V 2.0004.91) and the Swiss (83EU-045301) Science Foundation programmes. The authors thank J.

Snoeks, C. Sturmbauer and N.F. Broccoli for constructive criticism on earlier drafts of this manuscript, and the referees and editors for their valuable comments and suggestions, which significantly improved the manuscript. The original sources of the data used in this chapter are Meyer *et al.* (1996) and Verheyen *et al.* (1996), reviewed in Sturmbauer *et al.* (1997).

REFERENCES

Avise, J.C. (1992). Molecular population structure and the biogeographical history of a regional fauna: a case study with lessons for conservation biology. *Oikos* **63**, 62–76.

Avise, J.C. (1994). *Molecular Markers, Natural History and Evolution.* Chapman and Hall, New York.

Brichard, P. (1989). *Cichlids of Lake Tanganyika.* T.F.H. Publications, Neptune, NJ.

Brown, W.M., George, M. and Wilson, A.C. (1979). Rapid evolution of animal mitochondrial DNA. *Proc. Nat. Acad. Sci. USA* **76**, 1967–1971.

Cohen, A.S. (1992). Criteria for developing viable underwater natural reserves in lake Tanganyika. *Mitt. Int. Verein. Limnol.* **23**, 109–116.

Cohen, A.S., Soreghan, M.J. and Scholz, C.A. (1993a). Estimating the age of formation of lakes, an example from Lake Tanganyika, East-African Rift System. *Geology* **21**, 511–514.

Cohen, A.S., Bills, R., Cocquyt, C. and Caljon, A. (1993b). The impact of sediment pollution on biodiversity in Lake Tanganyika. *Conserv. Biol.* **7**, 667–677.

Coulter, G.W. (1991). The benthic fish community. In: *Lake Tanganyika and its Life* (Ed. by G.W. Coulter), pp. 151–199. Oxford University Press, London.

Dominey, W.J. (1984). Effect of sexual selection and life history on speciation, species flocks in African cichlids and Hawaiian *Drosophila*. In: *Evolution of Fish Species Flocks* (Ed. by A.A. Echelle and I. Kornfield), pp. 231–249. University of Maine at Orono Press, Orono, ME.

Fryer, G. and Iles, T.D. (1972). *The Cichlid Fishes of the Great Lakes of Africa. Their Biology and Evolution.* Oliver and Boyd, Edinburgh.

Giles, R.E., Blanc, H., Cann, H.M. and Wallace, D.C. (1980). Maternal inheritance of human mitochondrial DNA. *Proc. Nat. Acad. Sci. USA* **77**, 6715–6719.

Greenwood, P.H. (1984). African cichlids and evolutionary theories. In: *Evolution of Fish Species Flocks* (Ed. by A.A. Echelle and I. Kornfield), pp. 141–154. University of Maine at Orono Press, Orono, ME.

Kohda, M. (1997). Interspecific society among herbivorous cichlid fishes. In: *Fish Communities in Lake Tanganyika* (Ed. by H. Kawanabe M. Hori and M. Nagoshi), pp. 107–120. Kyoto University Press, Kyoto.

Konings, A. (1988). *Tanganyika Cichlids.* Verduijn Cichlids and Lake Fish Movies, Zevenhuizen, The Netherlands.

McKaye, K.R. and Gray, W.N. (1984). Extrinsic barriers to gene flow in rock-dwelling cichlids of Lake Malawi: macrohabitat heterogeneity and reef colonization. In: *Evolution of Fish Species Flocks* (Ed. by A.A. Echelle and I. Kornfield), pp. 169–183. University of Maine at Orono Press, Orono, ME.

Meyer, A. (1993). Evolution of mitochondrial DNA in fishes. In: *Molecular Biology Frontiers, Biochemistry and Molecular Biology of Fishes* (Ed. by P.W. Hochacka and T.P. Mommsen), pp. 1–38. Elsevier, Amsterdam.

Meyer, A., Knowles, L. and Verheyen, E. (1996). Widespread geographic distribution of mitochondrial haplotypes in Lake Tanganyika rock-dwelling fishes. *Mol. Ecol.* **5**, 341–350.

Moritz, C. (1994). Applications of mitochondrial DNA analysis in conservation: a critical review. *Mol. Ecol.* **3**, 401–411.

O'Rourke, D.H., Carlyle, S.W. and Parr, R.L. (1996). Ancient DNA, methods, progress and perspectives. *Am. J. Hum. Biol.* **8**, 557–571.

Poll, M. (1956). Poissons Cichlidae. Résultats scientifiques de l'exploration hydrobiologique du Lac Tanganika (1946–1947). *Inst. R. Sci. Nat. Belg.* **3**, 1–619.

Poll, M. (1986). Classification des Cichlidae du Lac Tanganika: tribus, genres et espèces. *Acad. R. Belg. Mem. Classe Sci.* **45**, 1–163.

Rossiter, A. (1995). The cichlid fish assemblages of Lake Tanganyika, ecology, behaviour and evolution of its species flock. *Adv. Ecol. Res.* **26**, 187–252.

Rossiter, A. and Yamagishi, S. (1997). Intraspecific plasticity in the social system and mating behaviour of a lek-breeding cichlid fish. In: *Fish Communities in Lake Tanganyika* (Ed. by H. Kawanabe, M. Hori and M. Nagoshi), pp. 193–218. Kyoto University Press, Kyoto.

Rüber, L., Verheyen, E., Sturmbauer, C. and Meyer, A. (1998). Lake level fluctuations and speciation in rock-dwelling cichlid fish in Lake Tanganyika, East Africa. In: *Evolution on Islands* (Ed. by P.R. Grant), pp. 225–241. Oxford University Press, Oxford.

Rüber, L., Verheyen, E. and Meyer, A. (1999). Replicated evolution of trophic specializations in an endemic cichlid fish lineage from Lake Tanganyika. *Proc. Nat. Acad. Sci. USA* **96**, 10230–10235.

Ryder, O.A. (1986). Species conservation and systematics, the dilemma of subspecies. *Trends Ecol. Evol.* **1**, 9–10.

Scholz, C.A. and Rosendahl, B.R. (1988). Low lake stands in lakes Malawi and Tanganyika, delineated with multifold seismic data. *Science* **240**, 1645–1648.

Snoeks, J., Rüber, L. and Verheyen, E. (1994). The Tanganyika problem, taxonomy and distribution of its ichthyofauna. *Arch. Hydrobiol. Beiheft. Ergebnisse Limnol.* **44**, 357–374.

Sturmbauer, C. and Meyer, A. (1992). Genetic divergence, speciation and morphological stasis in a lineage of African cichlid fishes. *Nature* **359**, 578–581.

Sturmbauer, C., Verheyen, E., Rüber, L. and Meyer, A. (1997). Phylogeographic patterns in populations of cichlid fishes from rock habitats in Lake Tanganyika. In: *Molecular Systematics of Fishes* (Ed. by T.D. Kocher and C. Stepien), pp. 93–107. Academic Press, London.

Swofford, D.L. (1993). *Phylogenetic Analysis Using Parsimony (PAUP), Version 3.1.1.* Smithsonian Institution, Washington, DC.

Tiercelin, J.J. and Mondeguer, A. (1991). The geology of the Tanganyika through. In: *Lake Tanganyika and its Life* (Ed. by G.W. Coulter), pp. 7–48. Oxford University Press, London.

Verheyen, E., Rüber, L. Snoeks, J. and Meyer, A. (1996). Mitochondrial phylogeography of rock dwelling cichlid fishes reveals evolutionary influence of historic lake level fluctuations in Lake Tanganyika, Africa. *Phil. Trans. R. Soc. Lond. B* **351**, 797–805.

Yanagisawa, Y., Ochi, H. and Gashagaza, M.M. (1997). Habitat use in cichlid fishes for breeding. In: *Fish Communities in Lake Tanganyika* (Ed. by H. Kawanabe, M. Hori and M. Nagoshi), pp. 149–174. Kyoto University Press, Kyoto.

The Use of Perturbation as a Natural Experiment: Effects of Predator Introduction on the Community Structure of Zooplanktivorous Fish in Lake Victoria

J.H. WANINK and F. WITTE

I. SUMMARY

This chapter illustrates the importance of ancient lakes as natural laboratories for the study of community structure. The major problem in this field, assessing the relative importance of competition and predation, can be successfully attacked by the use of human perturbation as a natural experiment. Using East African Lake Victoria as an example, the effects are reviewed of eutrophication, overfishing and invasive aliens on several communities within this complex ecosystem. The long-term study by the authors' research team on the zooplanktivorous fish community of Lake Victoria provided the opportunity to use the explosive population increase of an introduced predator, the Nile perch *Lates niloticus*, as a natural experiment. Changes in population densities, habitat choice and life histories of the zooplanktivores suggest a shift from a community structured by competition to one structured by predation. Although the complexity of their foodwebs may complicate the

ADVANCES IN ECOLOGICAL RESEARCH VOL. 31
ISBN 0–12–013931–6

separation of the structuring factors, it may be concluded that ancient lakes, which are almost inevitably exposed to human perturbation, offer unique opportunities to study community structure.

II. INTRODUCTION

Since Darwin (1859) elucidated the importance of competition as the driving factor in evolution, ecologists have commonly believed that competition was the principal factor structuring communities (Seifert, 1984). Classical field studies by Lack (1947) and MacArthur (1958) provided the basis for the concepts of resource partitioning and niche segregation, which assume that current community structures are the result of former competition (e.g. Hutchinson, 1959; Schoener, 1974; Pianka, 1981). However, the lack of strong evidence for ongoing competition has prompted many scientists to question this approach, which has been satirically described as studying "the ghost of competition past" (Connell, 1980).

In more recent years, an increasing amount of attention has been directed to two other factors which may be substantial in the structuring of communities: predation (e.g. Hairston et al., 1960; Paine, 1966; Slobodkin et al., 1967) and parasitism (e.g. Anderson and May, 1979; May and Anderson, 1979). The presence of keystone predators was found to facilitate high diversity in some complex foodwebs by reducing the abundance of potential competitors and thus preventing resource limitation (Paine, 1969, 1980). However, the effects of keystone predators were not generalizable and in some situations species diversity was found to decrease instead of increase (Seifert, 1984).

The relative importance of competition and predation in community structure has been heavily disputed, but the prevailing view among ecologists has been described as: "competition is important at times or under certain conditions, and one important mission of community ecology is to determine how, where, and when competition compares in importance relative to other factors" (Matthews, 1998). However, the relative importance of separate factors can only be determined accurately through experimental manipulation (Pianka, 1981; Paine, 1992). Although there are some examples of elegant field experiments (e.g. Werner, 1984), controlled manipulation of natural systems is difficult and, consequently, most experimental studies of competition are performed on simple "communities" under laboratory conditions (Pianka, 1981; Matthews, 1998).

III. ANCIENT LAKES AS NATURAL LABORATORIES

Ancient (> 100 000 years old; Gorthner, 1994) lakes have long been recognized as providing great opportunities for the study of community structure (Coulter et al., 1965). Long-term conditions of stability and isolation have resulted in extremely high levels of speciation and adaptive radiation (Brooks, 1950; Fryer and Iles, 1972; Beadle, 1981), although this may hold only for the littoral and the benthic habitat, as the pelagic foodwebs in ancient lakes are usually simplified (Dumont, 1994). Many years of "environmental calm" are thought to "permit populations to reach densities or assemblages to become saturated to a point at which competition becomes important" (Matthews, 1998).

Large species flocks of cichlid fishes have evolved in the African Great Lakes Tanganyika, Malawi and Victoria (Fryer and Iles, 1972), where virtually 100% of the species are endemic (Greenwood, 1991). In a seminal paper, Lowe-McConnell (1969) explained the success of the cichlids in these lakes by their preadaptations for lacustrine life: (i) their ability to breed in still water, and (ii) their morphological and physiological plasticity, which allowed for the exploitation of a wide array of food types. She also argued that it was probably intraspecific competition that directed evolutionary forces towards producing the observed adaptive radiations.

However, when conditions change and specialized species are at a disadvantage, these same lakes that allow for extreme trophic radiation may act as "evolutionary traps" (Lowe-McConnell, 1969). Although ancient lakes are usually extremely stable habitats, human perturbation can cause dramatic changes in their ecosystems: changes which may be regarded as "natural experiments", an opportunity to study the structuring factors of fish communities (Pitcher, 1995; Wanink, 1998b). Eutrophication can affect the competitive abilities of species through changes in the food stocks and the oxygen regime (Matthews, 1998). Overfishing and species introductions cause the removal or addition of predators or competitors, a basic condition for experimental community research (Pianka, 1981). The introduction of alien species has frequently been used as a natural experiment, since no fewer than 1354 cases of alien fish introductions, involving 237 species, have been documented world-wide since the middle of the nineteenth century (Welcomme, 1988). It is one such introduction, that of the Nile perch *Lates niloticus* into Lake Victoria, and its effects on the native fauna of that lake that are considered here.

IV. NATURAL EXPERIMENTS IN LAKE VICTORIA

A. The Original State

Lake Victoria is the world's largest tropical lake (surface area *c*. 69 000 km^2) and originated about 750 000 years ago (Kendall, 1969). It is relatively young in comparison with neighbouring Lakes Tanganyika (estimated age 9–12 My) (Cohen *et al.*, 1993) and Malawi (*c*. 2 My) (Greenwood, 1974; Pitcher and Hart, 1995). It is debatable whether Lake Victoria should be classified as a true ancient lake, since a recent study has indicated that a temperature rise about

15 000 years ago made the lake dry up entirely until 12 400 years ago (Stager *et al.*, 1986; Johnson *et al.*, 1996; but see the critique by Fryer, 1997). However, irrespective of its age, Lake Victoria offers ample opportunities to study community structure, as a monophyletic flock (Meyer *et al.*, 1990; Lippitsch, 1993) of more than 500 endemic species of haplochromine cichlid fishes has evolved (Greenwood, 1974; Witte *et al.*, 1992b; Seehausen, 1996; Kaufman *et al.*, 1997). In addition to the haplochromines, the lake's natural fish fauna comprised, until recently, two species of tilapiine cichlids and 45 non-cichlid species, belonging to nine families (Greenwood, 1974; van Oijen, 1995).

B. Anthropogenic Perturbation

1. Eutrophication

Agricultural activities along the lake shore have caused increasing eutrophication of Lake Victoria since the 1920s (Hecky, 1993). Since the 1960s a strong increase in phytoplankton production has been observed (Mugidde, 1993), resulting in the frequent occurrence of algal blooms and hypoxia, occasionally accompanied by mass fish kills (Ochumba and Kibaara, 1989; Kaufman, 1992). Although these changes would be expected to have an effect on the fish community structure through positive effects on herbivores and negative effects on oxygen-sensitive species, their impact has not been studied extensively. However, it has recently been demonstrated that the decrease in water transparency (Witte *et al.*, 1999), resulting from the eutrophication may lead to a loss of species diversity among closely related haplochromine cichlids (Seehausen *et al.*, 1997a). The narrowed light spectrum frustrates the colour-based discrimination by females between conspecific and other males, resulting in hybridization (Seehausen *et al.*, 1997a; Seehausen and van Alphen, 1998).

2. Overfishing

Artisanal fisheries did not affect the fish fauna of Lake Victoria until the beginning of the twentieth century (Worthington and Lowe-McConnell, 1994). After the introduction of gillnets in 1905, increased fishing pressure on the two indigenous species of tilapiine cichlids resulted in a strong decline in their stocks during the first half of the century (Graham, 1929; Fryer and Iles, 1972; Fryer, 1973). Commercial bottom trawling for haplochromine cichlids, which began in the mid-1970s in the southern part of the lake, resulted in local overfishing, affecting mainly the larger piscivorous species (Marten, 1979; Witte, 1981; Witte and Goudswaard, 1985; Barel *et al.*, 1991).

3. Invasive Aliens

To compensate for the decreasing catches of tilapiine cichlids, in the 1950s the exotic tilapiines *Tilapia rendalli, T. zilii, Oreocheomis leucostictus* and *O.*

niloticus (Nile tilapia) were introduced into the lake, where they were expected to fill empty niches (Beauchamp, 1958; Welcomme, 1967). Since the tilapiines were all herbivores, there were no predation effects of the introduced species on the fish community. However, the indigenous tilapiines quickly declined as a result of competition for spawning sites and nurseries, and hybridization with the alien species (Welcomme, 1967; Lowe-McConnell, 1987; Ogutu-Ohwayo, 1990b; Twongo, 1995). Eventually, the Nile tilapia, which expanded its diet to include zooplankton and macrobenthos (Gophen *et al.*, 1995; Balirwa, 1998), became the dominant tilapiine species in the lake (Ogutu-Ohwayo, 1990b; Kudhongania and Chitamwebwa, 1995).

In addition to the herbivorous tilapiine cichlids, a large predator, the Nile perch *Lates niloticus*, was introduced into the lake during the 1950s and early 1960s (Hamblyn, 1961; Arunga, 1981; Welcomme, 1988). The intended purpose of this introduction was for the Nile perch to convert the small bony haplochromine cichlids into high-quality table fish. Populations of Nile perch remained low until the end of the 1970s, when its abundance started to increase explosively (Barel *et al.*, 1985, 1991). Within a few years after the start of the Nile perch boom, about 200 offshore species of haplochromines probably became extinct and many inshore species decreased in abundance (Witte *et al.*, 1992b; Seehausen *et al.*, 1997b). In the offshore waters, the habitat of adult Nile perch, virtually no haplochromines survived and most of the non-cichlid fishes showed a strong population decline (Ogutu-Ohwayo, 1990b; Witte *et al.*, 1992b; Goudswaard and Witte, 1997).

Although Nile perch seemed to be non-selective feeders within the haplochromine cichlids, they showed a strong preference for haplochromines as a group, and switched to the zooplanktivorous cyprinid *Rastrineobola argentea* (locally called dagaa), the prawn *Caridina nilotica* and juvenile Nile perch only after the collapse of the haplochromines (Acere, 1985; Hughes, 1986; Ogutu-Ohwayo, 1990a; Mkumbo and Ligtvoet, 1992). Nile perch were able to switch to prawns because the abundance of the latter had explosively increased after the collapse of the haplochromines, a good example of the top–down effects that cascaded through the foodweb after the Nile perch boom (Goldschmidt *et al.*, 1993; Witte *et al.*, 1995). The complexity of the trophic cascade and the interference with bottom-up effects, caused by eutrophication, limit the possibilities to study the effects of Nile perch introduction on the fish community (Wanink, 1998b).

At the end of the 1980s, the water hyacinth *Eichhornia crassipes*, a South American species brought to Africa by humans as an ornamental plant, found its way into Lake Victoria and rapidly spread to form vast mats covering most of the bays (Njuguna, 1991). The low ambient levels of oxygen found under the hyacinth mats are feared to reduce strongly the habitat of the inshore fishes that have survived Nile perch predation relatively well (Njuguna, 1991; Witte *et al.*, 1992a, 1995). Conversely, the hypoxic water beneath the mats may act as a refuge for haplochromines that are less sensitive to low oxygen levels than Nile

perch (Ogutu-Ohwayo, 1995; Wanink, 1998b). Recent observations in the
Kenyan waters of the lake have indicated a recovery of the catfish *Clarias
gariepinus* and the lungfish *Protopterus aethiopicus* in areas covered by water
hyacinth (Bugenyi and van der Knaap, 1998).

In the remainder of this chapter some results from the research on
zooplanktivorous fish conducted between 1977 and 1990 in the southern part
of the lake by the Haplochromis Ecology Survey Team (HEST) and the
Tanzania Fisheries Research Institute (TAFIRI) are used to illustrate the use
of natural experiments in the analysis of community structures. The
zooplanktivores consisted of several haplochromine species and the cyprinid
dagaa, of which the former were preferred by Nile perch. In addition to the
different impact of Nile perch predation, the haplochromines and dagaa were
both confronted with an increasing fishing pressure and were presumably
affected in the same way by eutrophication. These factors made the
zooplanktivores a promising group in which to study the relative importance
of competition and predation as factors structuring their community.

V. THE ZOOPLANKTIVOROUS FISH COMMUNITY

A. The Original State

Until the mid-1970s dagaa was believed to be the only zooplanktivorous fish
species in Lake Victoria (Greenwood, 1974). However, since HEST started to
sample the pelagic zone in 1977, it became clear that at least 24 species of
zooplanktivorous haplochromines existed in the lake (van Oijen *et al.*, 1981;
Seehausen, 1991). Zooplanktivorous haplochromines constituted more than
20% of the commercial bottom trawl catches made near the Tanzanian town of
Mwanza (Witte, 1981; Witte and Goudswaard, 1985). Although the
zooplanktivores occurred in all areas of the lake, they were most common
over mud bottoms in the sublittoral (6–20 m deep) waters (Witte and van
Oijen, 1990). In the Mwanza Gulf, where 15 species of haplochromine
zooplanktivores and dagaa used to coexist in a relatively small area, the highest
abundance was found at the 14 m deep station G, the master sampling station
of HEST (Witte, 1984; Goldschmidt and Witte, 1990; Goldschmidt *et al.*, 1990;
Wanink, 1998b). At this station, six species of haplochromine zooplanktivores
and dagaa were regularly caught (Witte, 1984; Goldschmidt, 1989).

If competition for food or space was likely to occur anywhere in the
haplochromine species flock of Lake Victoria, this was between the
zooplanktivores, which largely overlap in gross morphological characteristics.
Therefore, the distribution and foraging patterns of the six most common
species at station G were studied intensively. Most species appeared to occupy
the lower half of the water column by day, taking mainly cyclopoid and
calanoid copepods and to a lesser extent cladocerans, while they moved to the

upper half of the column at night, switching to chaoborid larvae and pupae (Witte, 1984; Goldschmidt, 1989; Goldschmidt et al., 1990). These general patterns still allow for the existence of present competition. However, detailed observations on vertical and horizontal migration, diets and reproductive strategies showed that even these relatively young species were already ecologically segregated (Witte, 1984; Goldschmidt, 1989; Goldschmidt and Witte, 1990; Goldschmidt et al., 1990).

Until recently, little was known about the ecology of dagaa (Wanink, 1999). This fish was regarded as a strict surface dweller and zooplanktivore (Greenwood, 1974), although there were indications that it occasionally took emerging lakeflies, the common name used for several species of chaoborid and chironomid midges (Corbet, 1961; Hoogenboezem, 1985). Since the lakeflies swarm at night, usually around the new moon (MacDonald, 1956; Wanink and Goudswaard, 2000; see Irvine, this volume), dagaa hunting for adult lakeflies at the water surface may have interfered with the nocturnal foraging activities of the zooplanktivorous haplochromines. However, the haplochromines probably restricted their exploitation of lakeflies to the larvae and pupae of chaoborids, which leave the sediment at night and concentrate in the upper part of the water column. Like the haplochromine zooplanktivores, dagaa took mainly cyclopoid copepods by day (Hoogenboezem, 1985). Although dagaa appeared not to be the strict surface dweller as assumed, the species concentrated in midwater during daytime and did not use the bottom area, where most of the haplochromines were concentrated during the day (Wanink, 1998b).

Before the population boom of the introduced Nile perch, the predation pressure on the zooplanktivores was probably low, owing to a very low piscivore biomass in spite of the existence of many species of haplochromine piscivores (van Oijen, 1982). Therefore, it may be assumed that competition has been more important than predation in the structuring of the original community of zooplanktivorous fish in Lake Victoria.

B. Effects of the Nile Perch Boom

The effects of the Nile perch boom on the population densities, the niche characteristics and the life histories of haplochromine zooplanktivores and dagaa are summarized in Table 1. Nearly all data presented in Table 1 originate from station G in the Mwanza Gulf, where the Nile perch population started to rise in 1983 and increased explosively during 1984 and 1985 (Barel et al., 1991). Data representing the periods before and after the Nile perch boom were collected in 1983 and 1987–1988, respectively, unless stated otherwise.

C. Population Densities

During 1988, the formerly dominant haplochromines were no longer caught at station G, whereas the catch rate for dagaa increased to levels much higher

Table 1
Population, niche and life-history parameters of zooplanktivorous haplochromines and
dagaa, before (1983) and after (1987–1988) the Nile perch boom

	Haplochromines		Dagaa	
	Before	After	Before	After
Average trawl catch (nh^{-1})	293	0	216	5574
Habitat (day)	Bottom	–	Midwater	Bottom
Habitat (night)	Surface	–	Surface	Surface
Diet (day)	Zooplankton	–	Zooplankton	Zooplankton, benthos
Diet (night)	Lakefly larvae	–	(Lakeflies)	Lakeflies
Average number of gill filaments	–	–	42.3	43.9
Average number of gill rakers	–	–	17.5	16.8
Average length ripe ♀♀ (cm)	6.4–6.4	7.4–7.1	6.3	4.5
Absolute fecundity (egg number)	15–28	40–65	2117	748
Relative fecundity (egg number cm^{-3})	0.06–0.11	0.10–0.18	8.47	8.21
Growth constant K ($year^{-1}$)	–	–	1.42	3.00
Asymptotic length L_∞ (cm)	–	–	6.1	5.3
Size at maturity ♂ (cm)	4.7–4.7	–	5.2	4.6
Size at maturity ♀ (cm)	4.3–4.4	–	4.4	3.3
Age at maturity ♂ (years)	1–(1)	(1)–1	1.35	0.67
Age at maturity ♀ (years)	1–(1)	(1)–1	0.90	0.32

All differences mentioned in the text are statistically significant (details in the original
sources). Data in parentheses are uncertain. Sources: populations: Wanink (1998b);
niche, haplochromines: Witte (1984), Goldschmidt (1989), Goldschmidt et al. (1990);
niche, dagaa: Hoogenboezem (1985), Wanink (1992, 1998b), Wanink et al. (1999); life
history, haplochromines (data for two species: H. (Y.) laparogramma–H. (Y.)
pyrrhocephalus): Goldschmidt and Witte (1990; maturity data), Wanink, (1991); life
history, dagaa: Okedi (1974; NB. data from 1970–1974), Wanink (1998b). Minimum
values are given for size and age at maturity.

than those seen in 1983. Because the values for average trawl catch were
obtained from standardized trawls by research vessels, these data could be used
as an estimate for fish abundance. Since different gear was used to sample the
haplochromine zooplanktivores (bottom trawls during daytime) and dagaa
(surface trawls at night), the data for haplochromines and dagaa are not
directly comparable. It should be noted that, although the catches of
haplochromines at station G remained at very low levels during several years

after the Nile perch boom, a recovery of two zooplanktivorous species, *Haplochromis (Yssichromis) laparogramma* and *H. (Y.) pyrrhocephalus*, has been observed during 1997 (Witte *et al.*, submitted).

D. Niche Shift of Dagaa

As mentioned above, most of the haplochromine zooplanktivores used the bottom area by day and the surface area at night. Although there were too few animals caught after the Nile perch boom to determine whether or not this pattern had changed, there were no indications of any change during the years in which the population collapsed (Wanink, 1998b). Dagaa moved between midwater and the surface layer before the Nile perch boom, but gradually invaded the lower part of the water column by day during the decline of the haplochromines (Wanink, 1998b). After the Nile perch boom dagaa concentrated near the bottom by day and near the surface at night.

Concomitant with the habitat shift of dagaa, this species started to exploit the increased stocks of benthic invertebrates, such as the shrimp *Caridina nilotica* and chironomid larvae. Indications from the pre-Nile-perch period, that dagaa might occasionally take emerging lakeflies at night, were confirmed. In fact, after the Nile perch boom emerging lakeflies were commonly taken whenever available. During the shift towards a more benthic habitat and diet, two morphological changes in the gill apparatus of dagaa became apparent: an increase in the number of filaments and a decrease in the number of rakers.

The increased number of gill filaments was probably an adaptation to the relatively poor oxygen conditions in the new habitat. Before the ecological changes in the lake, hypoxic conditions occasionally occurred in the lower half of the water column at the 14 m sampling station G during the rainy season (van Oijen *et al.*, 1981). Even these short periods of hypoxia were reflected in a different number of gill filaments in two closely related haplochromines from the area, of which the species with the higher number occurred mainly below, and the other above a depth of 8 m (Hoogerhoud *et al.*, 1983). Although the recently observed increase in the occurrence of hypoxia is most severe in the offshore waters (Kaufman, 1992), extended hypoxic periods have also been reported from station G (Wanink *et al.*, 2000). It should be noted that the increase in hypoxia is restricted to the lower part of the water column, i.e. the new habitat of dagaa while, in contrast, the oxygen levels in the superficial layers have increased (Hecky *et al.*, 1994; Wanink and Kashindye, 1998).

The decrease in the number of gill rakers, as depicted in Table 1, continued after 1988 (Wanink and Witte, 2000). The reduction in numbers, which is assumed to imply an increase in the distance between the rakers, may be an adaptation to the shift towards benthic feeding of dagaa. In Lake Michigan the same phenomenon has been observed in the cisco *Coregonus hoyi*, which was forced to switch from zooplankton to benthic prey after the invasion of the competitively superior alewife *Alosa pseudoharengus* (Crowder, 1984).

E. Life-history Tactics

Assuming that Nile perch predation caused an increase in the ratio of adult
over juvenile mortality in the zooplanktivores (Wanink, 1991), one or more of
the following reproductive tactics may be predicted by life-history theory:
increased fecundity, earlier maturation, faster growth, shorter lifespan and, as
a result of allocating more energy to reproduction at the cost of somatic
growth, dwarfed fish (Noakes and Balon, 1982; Roff, 1992; Stearns, 1992).
Unfortunately, most of the haplochromines disappeared so rapidly after the
Nile perch boom that it could not be determined whether or not they had
adapted their life history. Although dwarfing was observed in some species,
there were indications for slower instead of faster growth rates (Witte and
Witte-Maas, 1987). Only for *Haplochromis (Yssichromis) laparogramma* and
H. (Y.) pyrrhocephalus, the two longest surviving haplochromines at station
G, which eventually even showed a strong recovery (Witte *et al.*, unpubl. data),
are some data on size and fecundity available from both before and after the
Nile perch boom. Of these species, the average size of ripe females initially
declined, but increased strongly after 1986, together with the number of eggs
produced per clutch (Wanink, 1991). The increase in absolute fecundity with
increasing body size was to be expected from the positive relationship observed
between these parameters for both species (Goldschmidt and Witte, 1990).
However, the increase in their relative fecundities (egg number divided by the
third power of body length) suggests that *H. (Y.) laparogramma* and *H. (Y.)
pyrrhocephalus* increased their reproductive effort as a response to the high
predation pressure by Nile perch. In addition to size and fecundity, the life-
history parameters after the Nile perch boom include only rough estimates for
age at maturity. The age at maturity has probably not changed, because the
haplochromine zooplanktivores used to show a breeding peak in the dry
season, when zooplankton abundance was high. This timing might not be
absolute, however, since ripe/brooding females were present throughout the
year and, especially in *H. (Y.) laparogramma*, the annual peak was not very
clear (Goldschmidt and Witte, 1990).

In contrast to the haplochromine zooplanktivores, dagaa showed a strong
reduction in body size after the Nile perch boom. The positive correlation between
body size and egg numbers (Okedi, 1974; Wanink, 1991, 1998b) therefore meant
that there was a concomitant decrease in the absolute fecundity of dagaa during
this time. Since this correlation did not change after the Nile perch boom, the
relative fecundity of dagaa remained constant. It has been shown, however, that
dagaa, in spite of its reduced absolute fecundity per clutch, has probably increased
its annual egg production, assuming that the species is able to breed continuously
(Wanink, 1998b). The increase in annual egg production was achieved by a strong
increase in growth rate (which ceases earlier, resulting in a lower asymptotic
length) and maturation at a smaller size, which resulted in a reduction in the
generation time by 50% in males and 64% in females.

VI. CONCLUSIONS

A. Structuring Factors

1. Competition

Although the zooplanktivorous haplochromines were studied intensively before the Nile perch boom, the hypothesis that competition has been an important factor structuring this community could not be tested. In spite of the coexistence of several closely related, only recently developed species, detailed research showed that they were already ecologically segregated (Witte, 1984; Goldschmidt and Witte, 1990; Goldschmidt et al., 1990). The only support for the competition hypothesis came from a quasi-experimental use of observational data on prey densities (Goldschmidt, 1989). Goldschmidt selected two nights from his sampling programme at station G on which he observed, respectively, the lowest and the highest densities of Chaoborus larvae in the surface layer. From his stomach analysis of the zooplanktivores caught in the top metre of the water column on those nights, it appeared that Haplochromis (?) "argens" was highly selective for the large fourth instar larvae during both nights, while H. (Yssichromis) heusinkveldi, H. ((Y.) laparogramma and H. (Y.) pyrrhocephalus were only so during the night on which prey densities were highest. Since H. (?) "argens" was the only species caught in its own microhabitat, Goldschmidt suggested that, in that habitat, it might be competitively superior.

The observed niche shift and the strong population increase of dagaa after the collapse of the haplochromines suggests that the haplochromines were competitively superior to dagaa (Table 1; Wanink, 1998b, 1999). Therefore, under the pre-Nile-perch conditions, competition with the haplochromines probably kept dagaa at relatively low population levels, as the fish were unable to exploit the bottom area. At that time, they experienced poorer feeding conditions in their former midwater to surface habitat during the main foraging periods than currently, when they can also access the bottom layer. Remarkably, the explosive population increase of Nile perch, the natural experiment that allowed the presence of competition in the zooplanktivorous fish community to be detected, simultaneously almost removed this structuring factor from the system, although the recovery of two haplochromine species may yet change the situation again.

2. Predation

Although the importance of predation in structuring the original zooplanktivorous fish community could not be assessed, it probably played only a minor role, owing to the relatively low piscivore biomass in the lake (van Oijen, 1982). However, the virtually complete removal of all the haplochromine zooplanktivores from station G by the introduced Nile perch clearly illustrates the importance of predation under the new circumstances. The success of dagaa after the Nile perch boom could be explained by a combination of competitive release and life-history tactics which facilitated coexistence with the new predator. Although there were some implications for reproductive tactics in *H. (Y.) laparogramma* and *H. (Y.) pyrrhocephalus* (Table 1), these species were almost completely absent from station G for several years preceding their recent recovery (Witte *et al.*, submitted). Since the mouthbrooding habit of the haplochromines limits the number of eggs per clutch, the potential for increasing their fecundity is only slight. Furthermore, a species utilizing mouthbrooding behaviour is likely to be much more vulnerable to predation than is one using a reproductive system without parental care, such as dagaa. It is not yet clear whether the two recovering haplochromine species, which may have survived in shallow refuge areas, have continued to modify their reproductive tactics or whether the risk of Nile perch predation at station G has decreased over the years.

B. The Need for Experiments

This work on the zooplanktivorous fish community of Lake Victoria supports the view that the relative importance of competition and predation in community structure can only be determined by experimental manipulation (Pianka, 1981; Paine, 1992). In spite of intensive research on the most promising group for detecting current competition, this was found only after using a quasi-experimental approach with regard to food densities (Gold-schmidt, 1989) or the lake-wide natural experiment of the Nile perch boom (Wanink, 1998b). Although large-scale planned experiments in ancient lakes are both difficult to organize and ethically undesirable, the intensive human activities on and around most such lakes will often result in perturbation, which may be used as natural experiments by ecologists.

The recent recovery of two zooplanktivorous haplochromines in Lake Victoria offers a further opportunity for a natural experiment. Assuming that these species are still competitively superior to dagaa, a shift of dagaa back to its original habitat would be expected, together with a decrease in population size. It should be noted, however, that the system has definitely changed in comparison with the period before the Nile perch, and that predation is now an important factor structuring the community. Although at first sight the foodweb seems to have become much simpler after the disappearance of many cichlid species from several trophic groups, a complicating factor has arisen in the form of increased intraguild predation (Wanink, 1998a). This phenomenon of predators eating potential competitors (Polis and Holt, 1992) will make it more difficult to separate the factors contributing to the community structure of the zooplanktivores. However, the ongoing changes in Lake Victoria offer such unique opportunities for quasi-experimental research that any ecologist should meet the challenge of using even this complex ecosystem as a natural laboratory.

ACKNOWLEDGEMENTS

The authors thank their colleagues from HEST and TAFIRI for support and co-operation. Portia Chifamba critically read the draft and made it possible to complete the manuscript in a stimulating tropical environment. The research by HEST was financially supported by The Netherlands Foundation for the Advancement of Tropical Research (WOTRO; grants W87-129, W87-161, W87-189) and by the Section for Research and Technology of The Netherlands Minister of Development Cooperation.

REFERENCES

Acere, T.O. (1985). Observations on the biology of the Nile perch *L. niloticus* (Linne) and growth of its fishery in the northern waters of Lake Victoria. *FAO Fish. Rep.* **335**, 42–61.

Anderson, R.M. and May, R.M. (1979). Population biology of infectious diseases: Part I. *Nature* **280**, 361–367.

Arunga, J.O. (1981). A case study of the Lake Victoria Nile perch *Lates niloticus* (mbuta) fishery. In: *Proceedings of the Workshop of the Kenya Marine and Fisheries Research Institution on Aquatic Resources of Kenya, Mombasa, 13–19 July 1981*, pp. 165–184. KMFRI and the Kenyan National Academy for the Advancement of Arts and Science, Nairobi.

Balirwa, J.S. (1998). *Lake Victoria Wetlands and the Ecology of the Nile Tilapia, Oreochromis niloticus Linné*. Balkema, Rotterdam.

Barel, C.D.N., Dorit, R., Greenwood, P.H., Fryer, G., Hughes, N., Jackson, P.B.N., Kanawabe, H., Lowe-McConnel, R.H., Witte, F. and Yamaoka, K. (1985). Destruction of fisheries in Africa's lakes. *Nature* **315**, 19–20.

Barel, C.D.N., Ligtvoet, W., Goldschmidt, T., Witte, F. and Goudswaard, P.C. (1991). The haplochromine cichlids of Lake Victoria: an assessment of biological and fisheries interest. In: *Cichlid Fishes: Behaviour, Ecology and Evolution* (Ed. by M.H.A. Keenleyside), pp. 258–279. Chapman and Hall, London.

Beadle, L.C. (1981). *The Inland Waters of Tropical Africa: An Introduction to Tropical Limnology*, 2nd ed. Longman, New York.

Beauchamp, R.S.A. (1958). Utilising the natural resources of Lake Victoria for the benefit of fisheries and agriculture. *Nature* **181**, 1634–1636.

Brooks, J.L. (1950). Speciation in ancient lakes. *Q. R. Biol.* **25**, 30–36, 131–176.

Bugenyi, F.W.B. and Knaap, M. van der (1998). Lake Victoria: an example of ACP-EU fisheries research. *EC Fish. Coop. Bull.* **10**, 20–21.

Cohen, A.S., Soreghan, M.J. and Scholz, C.A. (1993). Estimating the age of formation of lakes, an example from Lake Tanganyika, East-African Rift System. *Geology* **21**, 511–514.

Connell, J.H. (1980). Diversity and the coevolution of competitors, or the ghost of competition past. *Oikos* **35**, 131–138.

566   J.H. WANINK and F. WITTE

Corbet, P.S. (1961). The food of non-cichlid fishes in the Lake Victoria basin, with remarks on their evolution and adaptation to lacustrine conditions. *Proc. Zool. Soc. Lond.* **136**, 1–101.

Coulter, G.W., Harding, D., Eccles, D.H. and Bell-Cross, G. (1965). Unique opportunities for research in the Great Lakes of Central Africa. *Nature* **206**, 4–6.

Crowder, L.B. (1984). Character displacement and habitat shift in a native cisco in southeastern Lake Michigan: evidence for competition? *Copeia* **1984**, 878–883.

Darwin, C. (1859). *The Origin of Species by Means of Natural Selection.* Murray, London.

Dumont, H.J. (1994). Ancient lakes have simplified pelagic food webs. *Arch. Hydrobiol. Beiheft. Ergebnisse Limnol.* **44**, 223–234.

Fryer, G. (1973). The Lake Victoria fisheries: some facts and fallacies. *Biol. Cons.* **5**, 304–308.

Fryer, G. (1997). Biological implications of a suggested Late Pleistocene desiccation of Lake Victoria. *Hydrobiologia* **354**, 177–182.

Fryer, G. and Iles, T.D. (1972). *The Cichlid Fishes of the Great Lakes of Africa.* Oliver and Boyd, Edinburgh.

Goldschmidt, T. (1989). Reproductive strategies, subtrophic niche differentiation and the role of competition for food in haplochromine cichlids (Pisces) from Lake Victoria, Tanzania. *Ann. Mus. R. Afr. Centrale Sci. (Zool.)* **257**, 119–132.

Goldschmidt, T. and Witte, F. (1990). Reproductive strategies of zooplanktivorous haplochromine species (Pisces, Cichlidae) of Lake Victoria before the Nile perch boom. *Oikos* **58**, 356–368.

Goldschmidt, T., Witte, F. and Visser, J. de (1990). Ecological segregation of zooplanktivorous haplochromines (Pisces, Cichlidae) from Lake Victoria. *Oikos* **58**, 343–355.

Goldschmidt, T., Witte, F. and Wanink, J. (1993). Cascading effects of the introduced Nile perch on the detritivorous/phytoplanktivorous species in the sublittoral areas of Lake Victoria. *Cons. Biol.* **7**, 686–700.

Gophen, M., Ochumba, P.B.O. and Kaufman, L.S. (1995). Some aspects of perturbation in the structure and biodiversity of the ecosystem of Lake Victoria (East Africa). *Aquat. Liv. Res.* **8**, 27–41.

Gorthner, A. (1994). What is an ancient lake? *Arch. Hydrobiol. Beiheft. Ergebnisse Limnol.* **44**, 97–100.

Goudswaard, P.C. and Witte, F. (1997). The catfish fauna of Lake Victoria after the Nile perch upsurge. *Envir. Biol. Fishes* **49**, 21–43.

Graham, M. (1929). *The Victoria Nyanza and its Fisheries.* Waterlow and Sons, London.

Greenwood, P.H. (1974). The cichlid fishes of Lake Victoria, East Africa: the biology and evolution of a species flock. *Bull. Br. Mus. (Nat. Hist.), Zool.* **6**, Suppl., 1–13.

Greenwood, P.H. (1991). Speciation. In: *Cichlid Fishes: Behaviour, Ecology and Evolution* (Ed. by M.H.A. Keenleyside), pp. 86–102. Chapman and Hall, London.

Hairston, N.G., Smith, F.E. and Slobodkin, L.B. (1960). Community structure, population control, and competition. *Am. Nat.* **94**, 421–425.

Hamblyn, E.L. (1961). The Nile perch project. In: *Annual Report 1960*, pp. 26–32. EAFRO, Jinja.

Hecky, R.E. (1993). The eutrophication of Lake Victoria. *Verh. Int. Verein. Limnol.* **25**, 39–48.

Hecky, R.E., Bugenyi, F.W.B., Ochumba, P., Talling, J.F., Mugidde, R., Gophen, M. and Kaufman, L. (1994). Deoxygenation of the deep water of Lake Victoria, East Africa. *Limnol. Oceanogr.* **39**, 1476–1481.

Hoogenboezem, W. (1985). *The Zooplankton in the Mwanza Gulf (Lake Victoria) and the Food of a Zooplanktivorous Cyprinid* (Rastrineobola argentea). HEST Report, University of Leiden.

Hoogerhoud, R.J.C., Witte, F. and Barel, C.D.N. (1983). The ecological differentiation of two closely resembling *Haplochromis* species from Lake Victoria (*H. iris* and *H. hiatus*; Pisces, Cichlidae). *Neth. J. Zool.* **33**, 283–305.

Hughes, N.F. (1986). Changes in the feeding biology of the Nile perch, *Lates niloticus* (L.) (Pisces:Centropomidae), in Lake Victoria, East Africa since its introduction in 1960, and its impact on native fish community of the Nyanza Gulf. *J. Fish Biol.* **29**, 541–548.

Hutchinson, G.E. (1959). Homage to Santa Rosalia, *or* why are there so many kinds of animals? *Am. Nat.* **93**, 145–159.

Johnson, T.C., Scholz, C.A., Talbot, M.R., Kelts, K., Ricketts, R.D., Ngobi, G., Beuning, K., Ssemmanda, I. and McGill, J.W. (1996). Late Pleistocene desiccation of Lake Victoria and rapid evolution of cichlid fishes. *Science* **273**, 1091–1093.

Kaufman, L. (1992). Catastrophic change in species-rich freshwater ecosystems: the lessons of Lake Victoria. *BioScience* **42**, 846–858.

Kaufman, L.S., Chapman, L.J. and Chapman, C.A. (1997). Evolution fast forward: haplochromine fishes of the Lake Victoria region. *Endeavour* **21**, 23–30.

Kendall, R.L. (1969). An ecological history of the Lake Victoria Basin. *Ecol. Monog.* **39**, 121–176.

Kudhongania, A.W. and Chitamwebwa, D.B.R. (1995). Impact of environmental change, species introductions and ecological interactions on the fish stocks of Lake Victoria. In: *The Impact of Species Changes in African Lakes* (Ed. by T.J. Pitcher and P.J.B. Hart), pp. 19–32. Chapman and Hall, London.

Lack, D. (1947). *Darwin's Finches*. Cambridge University Press, Cambridge.

Lippitsch, E. (1993). A phyletic study on haplochromine fishes (Perciformes, Cichlidae) of East Africa, based on scale and squamation characters. *J. Fish Biol.* **42**, 903–946.

Lowe-McConnell, R.H. (1969). Speciation in tropical freshwater fishes. *Biol. J. Linn. Soc. Lond.* **1**, 51–75.

Lowe-McConnell, R.H. (1987). *Ecological Studies in Tropical Fish Communities*. Cambridge University Press, Cambridge.

MacArthur, R.H. (1958). Population ecology of some warblers of northeastern coniferous forests. *Ecology* **39**, 599–619.

MacDonald, W.W. (1956). Observations on the biology of chaoborids and chironomids in Lake Victoria and on the feeding habits of the "elephant-snout fish" *Mormyrus kannume* (Forsk). *J. Anim. Ecol.* **25**, 36–53.

Marten, G.G. (1979). Impact of fishing on the inshore fishery of Lake Victoria (East Africa). *J. Fish. Res. Brd Can.* **36**, 891–900.

Matthews, W.J. (1998). *Patterns in Freshwater Fish Ecology*. Chapman and Hall, New York.

May, R.M. and Anderson, R.M. (1979). Population biology of infectious diseases: Part II. *Nature* **280**, 455–461.

Meyer, A., Kocher, T.D., Basasibwaki, P. and Wilson, A.C. (1990). Monophyletic origin of Lake Victoria cichlid fishes, suggested by mitochondrial DNA sequences. *Nature* **347**, 550–553.

Mkumbo, O.C. and Ligtvoet, W. (1992). Changes in the diet of Nile perch, *Lates niloticus* (L), in the Mwanza Gulf, Lake Victoria. *Hydrobiologia* **232**, 79–83.

Mugidde, R. (1993). The increase in phytoplankton primary productivity and biomass in Lake Victoria (Uganda). *Verh. Int. Verein. Limnol.* **25**, 846–849.

Njuguna, S.G. (1991). Water hyacinth: the world's worst aquatic weed infests lakes Naivasha and Victoria. *Swara* **14** (6), 8–10.

Noakes, D.L.G. and Balon, E.K. (1982). Life histories of tilapias: an evolutionary perspective. In: *The Biology and Culture of Tilapias*. (Ed. by R.S.V. Pullin and R.H. Lowe-McConnell), pp. 61–82. ICLARM, Manila.

Ochumba, P.B.O. and Kibaara, D.I. (1989). Observations on blue–green algal blooms in the open waters of Lake Victoria, Kenya. *Afr. J. Ecol.* **27**, 23–34.

Ogutu-Ohwayo, R. (1990a). Changes in the prey ingested and the variations in the Nile perch and other fish stocks of Lake Kyoga and the northern waters of Lake Victoria (Uganda). *J. Fish Biol.* **37**, 55–63.

Ogutu-Ohwayo, R. (1990b). The decline of the native fishes of Lakes Victoria and Kyoga (East Africa) and the impact of introduced species, especially the Nile perch, *Lates niloticus* and the Nile tilapia, *Oreochromis niloticus*. *Envir. Biol. Fishes* **27**, 81–96.

Ogutu-Ohwayo, R. (1995). Diversity and stability of fish stocks in Lakes Victoria, Kyoga and Nabugabo after establishment of introduced species. In: *The Impact of Species Changes in African Lakes* (Ed. by T.J. Pitcher and P.J.B. Hart), pp. 59–81. Chapman and Hall, London.

Oijen, M.J.P. van (1982). Ecological differentiation among the piscivorous haplochromine cichlids of Lake Victoria (East Africa). *Neth. J. Zool.* **32**, 336–363.

Oijen, M.J.P. van (1995). Key to Lake Victoria fishes other than haplochromine cichlids: In: *Fish Stocks and Fisheries of Lake Victoria: A Handbook for Field Observations* (Ed. by F. Witte and W.L.T. van Densen), pp. 209–300. Samara Publishing, Cardigan.

Oijen, M.J.P. van, Witte, F. and Witte-Maas, E.L.M. (1981). An introduction to ecological and taxonomic investigations on the haplochromine cichlids from the Mwanza Gulf of Lake Victoria. *Neth. J. Zool.* **31**, 149–174.

Okedi, J. (1974). Preliminary observations on *Engraulicypris argenteus* (Pellegrin) 1904 from Lake Victoria. In: *Annual Report 1973*, pp. 39–42. EAFFRO, Jinja.

Paine, R.T. (1966). Food web complexity and species diversity. *Am. Nat.* **100**, 65–75.

Paine, R.T. (1969). The *Pisaster–Tegula* interaction: prey patches, predator food preference, and intertidal community structure. *Ecology* **50**, 950–961.

Paine, R.T. (1980). Food webs: linkage, interaction strength and community infrastructure. *J. Anim. Ecol.* **49**, 667–685.

Paine, R.T. (1992). Food-web analysis through field measurement of per capita interaction strength. *Nature* **355**, 73–75.

Pianka, E.R. (1981). Competition and niche theory. In: *Theoretical Ecology: Principles and Applications*, 2nd edn (Ed. by R.M. May), pp. 167–196. Blackwell Scientific Publications, Oxford.

Pitcher, T.J. (1995). Species changes and fisheries in African lakes: outline of the issues. In: *The Impact of Species Changes in African Lakes* (Ed. by T.J. Pitcher and P.J.B. Hart), pp. 1–16. Chapman and Hall, London.

Pitcher, T.J. and Hart, P.J.B. (eds) (1995). *The Impact of Species Changes in African Lakes*. Chapman and Hall, London.

Polis, G.A. and Holt, R.D. (1992). Intraguild predation: the dynamics of complex trophic interactions. *Trends Ecol. Evol.* **7**, 151–154.

Roff, D.A. (1992). *The Evolution of Life Histories; Theory and Analysis*. Chapman and Hall, New York.

Schoener, T.W. (1974). Resource partitioning in ecological communities. *Science* **185**, 27–39.

Seehausen, O. (1991). Die zooplanktivoren Cichliden des Victoriasees. *Deutsche Aquar. Terrari. Z.* **44**, 715–721.

Seehausen, O. (1996). *Lake Victoria Rock Cichlids: Taxonomy, Ecology, and Distribution*. Verduijn Cichlids, Zevenhuizen.

Seehausen, O. and Alphen, J.J.M. van (1998). The effect of male coloration on female mate choice in closely related Lake Victoria cichlids (*Haplochromis nyererei* complex). *Behav. Ecol. Sociobiol.* **42**, 1–8.

Seehausen, O., Alphen, J.J.M. van and Witte, F. (1997a). Cichlid fish diversity threatened by eutrophication that curbs sexual selection. *Science* **227**, 1808–1811.

Seehausen, O., Witte, F., Katunzi, E.F., Smits, J. and Bouton, N. (1997b). Patterns of the remnant cichlid fauna in southern Lake Victoria. *Conserv. Biol.* **11**, 890–904.

Seifert, R.P. (1984). Does competition structure communities? Field studies on neotropical *Heliconia* insect communities. In: *Ecological Communities: Conceptual Issues and the Evidence* (Ed. by D.R. Strong, Jr, D. Simberloff, L.G. Abele and A.B. Thistle), pp. 54–63. Princeton University Press, Princeton.

Slobodkin, L.B., Smith, F.E. and Hairston, N.G. (1967). Regulation in terrestrial ecosystems, and the implied balance of nature. *Am. Nat.* **101**, 109–124.

Stager, J.C., Reinthal, P.N. and Livingstone, D.A. (1986). A 25,000-year history for Lake Victoria, East Africa, and some comments on its significance for the evolution of cichlid fishes. *Freshwat. Biol.* **16**, 15–19.

Stearns, S.C. (1992). *The Evolution of Life Histories*. Oxford University Press, Oxford.

Twongo, T. (1995). Impact of fish species introductions on the tilapias of Lakes Victoria and Kyoga. In: *The Impact of Species Changes in African Lakes* (Ed. by T.J. Pitcher and P.J.B. Hart), pp. 45–57. Chapman and Hall, London.

Wanink, J.H. (1991). Survival in a perturbed environment: the effects of Nile perch introduction on the zooplanktivorous fish community of Lake Victoria. In: *Terrestrial and Aquatic Ecosystems: Perturbation and Recovery* (Ed. by O. Ravera), pp. 269–275. Ellis Horwood, Chichester.

Wanink, J.H. (1992). The pied kingfisher *Ceryle rudis* and dagaa *Rastrineobola argentea*: estimating the food-intake of a prudent predator. In: *Proceedings of the VIIth Pan-African Ornithological Congress, Nairobi, Kenya, 28 August–5 September 1988* (Ed. by L. Bennun), pp. 403–411. PAOCC, Nairobi.

Wanink, J.H. (1998a). How predictable were the effects of Nile perch introduction on the Lake Victoria ecosystem? In: *Proceedings of the ULKRS Seminar Series* (Ed. by P.C. Chifamba), University of Zimbabwe. *ULKRS Bull.* **1/96**, 35–44.

Wanink, J.H. (1998b). The pelagic cyprinid *Rastrineobola argentea* as a crucial link in the disrupted ecosystem of Lake Victoria: dwarfs and giants–African adventures. Unpublished PhD Thesis, Rijksuniversiteit, Leiden.

Wanink, J.H. (1999). Prospects for the fishery on the small *Rastrineobola argentea* in Lake Victoria. *Hydrobiologia* **407**, 189–195.

Wanink, J.H. and Goudswaard, P.C. (2000). The impact of Lake Victoria's lakefly abundance on Palearctic passerines. *Ostrich.* **71** (in press).

Wanink, J.H. and Kashindye, J.J. (1998). Short-term variations in pelagic photosynthesis demand well-timed sampling to monitor long-term limnological changes in Lake Victoria. *Hydrobiologia* **377**, 177–181.

Wanink, J.H. and Witte, F. (2000). Rapid morphological changes following niche shift in the zooplanktivorous cyprinid *Rastrineobola argentea* from Lake Victoria. *Neth. J. Zool.* **50** (in press).

Wanink, J.H., Goudswaard, P.C. and Berger, M.R. (1999). *Rastrineobola argentea*, a major resource in the ecosystem of Lake Victoria. In: *Fish and Fisheries of Lakes and Reservoirs in Southeast Asia and Africa* (Ed. by W.L.T. van Densen and M.J. Morris), pp. 295–309. Westbury Publishing, Otley.

Wanink, J.H., Kashindye, J.J., Goudswaard, P.C. and Witte, F. (2000). Dwelling at the oxycline: does increased stratification provide a predation refugium for the Lake Victoria sardine *Rastrineobola argentea*? *Freshwat. Biol.* (in press).

Welcomme, R.L. (1967). Observations on the biology of the introduced species of *Tilapia* in Lake Victoria. *Rev. Zool. Bot. Afr.* **76**, 249–279.

Welcomme, R.L. (1988). International introductions of inland fish species. *FAO Fish. Tech. Paper* **294**, 1–318.

Werner, E.E. (1984). The mechanisms of species interactions and community organization in fish. In: *Ecological Communities: Conceptual Issues and the Evidence* (Ed. by D.R. Strong, Jr, D. Simberloff, L.G. Abele and A.B. Thistle), pp. 360–382. Princeton University Press, Princeton.

Witte, F. (1981). Initial results of the ecological survey of the haplochromine cichlid fishes from the Mwanza Gulf of Lake Victoria, Tanzania: breeding patterns, trophic and species distribution. *Neth. J. Zool.* **31**, 175–202.

Witte, F. (1984). Ecological differentiation in Lake Victoria haplochromines: comparison of cichlid species flocks in African lakes. In: *Evolution of Fish Species Flocks* (Ed. by A.A. Echelle and I. Kornfield), pp. 155–167. University of Maine at Orono Press, Orono, ME.

Witte, F. and Goudswaard, P.C. (1985). Aspects of the haplochromine fishery in southern Lake Victoria. *FAO Fish. Rep.* **335**, 81–88.

Witte, F. and Oijen, M.J.P. van (1990). Taxonomy, ecology and fishery of Lake Victoria haplochromine trophic groups. *Zool. Verh. Leiden* **262**, 1–47.

Witte, F. and Witte-Maas, E.L.M. (1987). Implications for taxonomy and functional morphology of intraspecific variation in haplochromine cichlids of Lake Victoria. In: From form to fishery: an ecological and taxonomical contribution to morphology and fishery of Lake Victoria cichlids, pp. 1–83. Witte, F., Unpublished PhD Thesis, Rijksuniversiteit Leiden.

Witte, F., Goldschmidt, T., Goudswaard, P.C., Ligtvoet, W., Oijen, M.J.P. van and Wanink, J.H. (1992a). Species extinction and concomitant ecological changes in Lake Victoria. *Neth. J. Zool.* **42**, 214–232.

Witte, F., Goldschmidt, T., Wanink, J., Oijen, M. van Goudswaard, K., Witte-Maas, E. and Bouton, N. (1992b). The destruction of an endemic species flock: quantitative data on the decline of the haplochromine cichlids of Lake Victoria. *Envir. Biol. Fishes* **34**, 1–28.

Witte, F., Goldschmidt, T. and Wanink, J.H. (1995). Dynamics of the haplochromine cichlid fauna and other ecological changes in the Mwanza Gulf of Lake Victoria. In: *The Impact of Species Changes in African Lakes* (Ed. by T.J. Pitcher and P.J.B. Hart), pp. 83–110. Chapman and Hall, London.

Witte, F., Goudswaard, P.C., Katunzi, E.F.B., Mkumbo, O.C., Seehausen, O. and Wanink, J.H. (1999). Lake Victoria's ecological changes and their relationships with the riparian societies. In: *Ancient Lakes: Their Cultural and Biological Diversities* (Ed. by H. Kawanabe, G. Coulter and A. Roosevelt), pp. 189–202. Kenobi Productions, Ghent.

Worthington, E.B. and Lowe-McConnell, R.H. (1994). African lakes reviewed: creation and destruction of biodiversity. *Environ. Conserv.* **21**, 199–213.

Lake Biwa as a Topical Ancient Lake

A. ROSSITER

I. SUMMARY

Present knowledge concerning the Japanese ancient Lake Biwa is briefly summarized, and extant and potential threats to its well-being are documented. From this example of Lake Biwa, strengths and weaknesses in our knowledge of other ancient lakes and the problems that they face were identified. Several areas of concern were evident.

Knowledge of the geological history of ancient lakes is uneven, as is the accuracy and degree of detail of geochronological information available. This greatly complicates meaningful results from studies of palaeohydrology, palaeobiology and palaeoecology. In all ancient lakes, future research must place greater emphasis on these areas. Such studies can both unravel and facilitate an understanding of the history of the lake and the reconstruction of evolutionary scenarios.

Ancient lakes are recognized as hotspots of diversity and centres of endemism, but in no ancient lake has an adequate inventory of biodiversity been compiled to test hypotheses concerning patterns of diversity or its regulation. More effort is needed in this area. The importance of such baseline data is exemplified by the case of Lake Biwa, where the absence of quantitative studies of the native fish communities before, during or after the introduction

ADVANCES IN ECOLOGICAL RESEARCH VOL. 31
ISBN 0–12–013931–6

and subsequent population explosions of largemouth bass and bluegill has prevented any meaningful evaluation of the effects of these alien fishes on the native fauna.

Lake Biwa has had more ecological insults inflicted upon it than any of the other ancient lakes of the world; physical, chemical and biotic perturbations, including alien introductions and overfishing, have all taken their toll on the lake ecosystem and continue to do so. Measures to ameliorate some of these perturbations are suggested, some of which will require heightened ecological awareness and brave decision making on the part of the authorities who manage this ecological treasure. It is hoped that the organizations and governments responsible for the well-being of the other ancient lakes of the world will heed the example of Lake Biwa when arriving at management decisions concerning the unique biotas and resources of which they have charge.

II. INTRODUCTION

The main objective of this chapter is to summarize the present level of knowledge concerning the Japanese ancient Lake Biwa, and to highlight extant and possible threats to its well-being. A second aim is to use Lake Biwa as an example to identify strengths and weaknesses in the knowledge of ancient lakes and the problems that they face.

A. Why Lake Biwa?

Lake Biwa is the only ancient lake situated in a developed country amidst a highly industrialized and intensive agricultural setting. Its biological uniqueness and importance have long been recognized by Japanese researchers, and numerous faunal, geological and, especially, limnological studies have been made there over a long period. In addition, because of its accessibility and relatively small size, investigations of Lake Biwa should have been free of the logistic difficulties and problems to comprehensive study and sampling presented by the larger ancient lakes. This long history of continuous research and the many and extensive research efforts conducted on Biwa are perhaps equalled only by those on the much larger Lake Baikal. This concatenation of positive factors suggests that a reasonably comprehensive knowledge of the ecology and an almost complete inventory of the biota of Lake Biwa should have been achieved. Lake Biwa should thus be at the forefront of what is known, and what needs to be known, about ancient lakes *per se*.

A second reason to highlight Lake Biwa comes from its negative perspectives. The lake has been exposed to the effects of high human population pressure, and to those of industrialization, water resources development and intensive agriculture within the lake basin. Eutrophication has advanced rapidly since the 1950s and native taxa, including some biologically and commercially important endemic forms, have shown alarming declines in population numbers. Introductions of alien species have further perturbed the lake ecosystem. As such, Lake Biwa is a microcosm of many of the problems experienced by several ancient lakes. An examination of Lake Biwa may also offer a pointer towards potential threats facing other ancient lakes, and perhaps also a lesson in how to, or how not to, proceed to alleviate these problems should they arise elsewhere.

III. PHYSICAL AND LIMNOLOGICAL FEATURES OF LAKE BIWA

The modern-day Lake Biwa is a monomictic lake situated in Honshu Island, Japan, and experiences a temperate climate. Its ancient name was "niho-no-umi" or "Lake of the grebe", but its modern name is derived from its perceived resemblance in shape to the lute-like Japanese musical instrument, the biwa. Accordingly, Lake Biwa has a larger north basin and a smaller south basin, joined by a narrow neck. The south basin is only 16 km long along its main axis, with a surface area of 58 km^2 and a maximum depth of 8 m, but is mostly shallower, with a mean depth of only 4.0 m. The north basin is 48 km long, with a surface area of 616 km^2, and mean and maximum depths of 43 m and 103.6 m, respectively. The lake is 63.5 km long along its major axis (NNE–SSW), only 1.35 km wide at the neck between the two basins, and has a surface area of c. 671 km^2 and a volume of 27.5 km^3 (Shiga Prefectural Government, 1997). Over 400 rivers and streams flow into Lake Biwa, but there is only one natural outlet, the Seta River, at the extreme southern end of the lake. The residence time of the lake's water is 19 years (Shiga Prefectural Government, 1997). Water temperatures at the epilimnion can exceed 30°C in the summer, but a strong thermocline means that waters of the hypolimnion remain at around 5–10°C. This cold-water refuge has implications for the presence and evolution of some of the extant fauna.

IV. GEOLOGICAL HISTORY

The history of Lake Biwa is long and complicated, but is reasonably well understood, based mainly on findings from geological surveys at various locations within the Omi basin, and especially on information from two cores (to 200 m, in 1971–1972, and to 1400 m, in 1982–1983) taken at the middle of the north basin (Horie, 1984, 1987). From such sources, the lake's history is understood from about 4 Mya to the present (e.g. Yokoyama, 1984; Kawabe, 1989, 1994; Meyers et al., 1993; Nakajima and Nakai, 1994; Research Group for Natural History of Lake Biwa, 1994, hereinafter RGNHLB; Nishino and Watanabe, this volume). During the Pliocene and Pleistocene there were at least eight phases in the history of Lake Biwa: five major lacustrine phases of different ages, collectively known as Palaeo-Lake Biwa, which were precursors to the present lake, two fluvial periods, during when the lacustrine environment was replaced by a riverine one, and finally, the modern Lake Biwa. The present Lake Biwa is *c*. 430 000 years old (Meyers *et al*., 1993), and although a relatively young member of the group, it still exceeds comfortably the 100 000 year age criterion (Gorthner, 1994) to qualify as an ancient lake.

A. The Importance of an Accurate Geological History

For ancient lake studies in particular, knowledge of the geological history provides a canvas upon which climatic and physical events during its formation, and the subsequent colonization and evolution of its biota, can be drafted. Accurate ageing of the lake and of events during its history provide invaluable guidelines; a framework over which the skein of evolutionary scenarios of the resident organisms, whether extant or fossil, can be drawn and better understood.

A taxonomically diverse and rich array of fossils has been recovered from the Palaeo-Lake Biwa sediments (RGNHLB, 1986, 1994), and these have contributed greatly towards envisioning the faunal and climatic changes which occurred throughout the lake's existence. For example, each phase in the lake's complex history is characterized by a different cyprinid fish fauna. The xenocypridine and cultrinine fish faunas flourished until the mid-Quaternary but then disappeared, an event attributed to the loss of large and shallow lakes (high primary productivity for these filter feeders) and the presence of torrential rivers (unsuitable habitat) during the Island Arc Movement of the Middle to Late Quaternary (Nakajima, 1986, 1994). These dramatic environmental changes are also reflected in the history of the lake's mollusc faunas, which have been completely extirpated at least four times, at 3.1, 2.5–2.4, 1.9–1.8 and 0.4–0.3 Mya (Matsuoka, 1987).

In contrast to the molluscs, none of which survived throughout the entire history of Palaeo-Lake Biwa, fossils closely resembling those of the extant endemic catfish *Silurus biwaensis* have been recovered from the Palaeo-Lake Ohyamada (Kobayakawa and Okuyama, 1994), indicating that the ancestor of this species was already present some 3.2 Mya. The only major difference between this ancestor and its extant descendant seems to be that of size, the modern species being much larger (Nishino and Watanabe, this volume).

As demonstrated elsewhere in this volume (Geary *et al*.; Chen; Mourguiart, this volume) fossil evidence can be used to infer features of the palaeoenvironment. Modern Lake Biwa was formed by the gradual expansion of the last palaeolake phase, Palaeo-Lake Katata. In the upper part of the Katata Formation, mollusc fossils increase in abundance and diversity, which suggests that the lake gradually became an open lake at the later stages of Palaeo-Lake Katata. At *c*. 0.8 Mya the southern part of the present north basin filled with water, whereas the northern, deep basin has existed continuously for 430 000 years (Meyers *et al.*, 1993). Further evidence for environmental changes in the modern lake comes from the diatom flora (Tanaka *et al.*, 1984; Tanaka and Matsuoka, 1985); the presence of certain *Melanosira* spp. and *Stephanodiscus* spp. in sediments from around 400 000 years ago suggests that the deep oligotrophic character of Lake Biwa was present at that time and has continued until the present day.

From this variety of data sources, a history and an ecological and evolutionary picture of modern Lake Biwa can be proposed. During the Middle to Lake Pleistocene (0.8–0.5 Mya) tectonic faulting produced the deep bottom plain of the present lake (Itihara, 1982). This resulted in discrete open-water, deep offshore and rocky littoral habitats in Lake Biwa, which offered potential opportunities for speciation into a diverse array of vacant ecological niches. The relatively deep waters and large volume of the lake may have served as a refuge during the interglacials, buffering against the thermal stresses and desiccation which eliminated faunas in other shallow lakes in Japan at that time. Extant Biwa taxa to which this apply include the gastropod *Valvata biwaensis* Preston, whose ancestors were believed to have been widely distributed but were exterminated in other lakes (Kawanabe, 1978), the bivalve *Pisidium* (*Eupisidium*) *kawamurai* and the turbellarian *Bdellocephala annandalei* Ijima et Kaburaki (Annandale, 1922).

In several other ancient lakes, palaeolimnological studies (Davis, 1990; Johnson and Odada, 1996) have also allowed for partial reconstruction of palaeoenvironments, and as shown by Geary *et al.*, Chan, Mourguiart and Cohen (this volume), this approach has much to offer in achieving an understanding of the evolutionary processes which have acted to shape the faunas of ancient lakes seen today. Similarly, knowledge of the ecology of the fossil fauna and flora can provide much information on ancient lake palaeoenvironments. However, to reconstruct species evolutionary histories and to evaluate hypotheses for speciation within this framework, an accurate age estimation of geological events is essential. Inaccurate age estimation can act to confuse or obfuscate elucidation of evolutionary scenarios.

Such inaccuracies derive from two main sources. First, the citing of incorrect data and information. Although well-documented and despite the many publications (in English) detailing the history and age of Lake Biwa (e.g. Horie 1987; Nakajima and Nakai, 1994), several recent publications contain erroneous information on this topic. For example, Martin (1996) states that the present Lake Biwa has twice filled up with sediments. This error is presumably based on a misunderstanding of the filling up of two different palaeolakes during the lake's complicated early history. The age of Lake Biwa being incorrectly given as 0.1 My (Morino and Miyazaki, 1994) or as 2 My (Noakes, 1998) is less easy to explain.

The second source is non-awareness of revised age estimates from new studies. Recent geological surveys, utilizing techniques and equipment undreamed of even 20 years ago, have refined our knowledge of the histories of many ancient lakes, and have also increased the precision of their age estimation and perhaps also its accuracy. These new techniques often produce a different age estimate to that obtained by older methodologies, and generally the age of the lake is modified with each successive study. Citing of older, obsolete, estimates (or inaccurate recent ones) can be the source of much confusion. For example, age estimates for most ancient lakes have been revised repeatedly during recent decades, and even today, age estimates have large ranges and for many ancient lakes there is still no firm concensus on just how old they really are (Table 1).

This situation partly reflects the fact that for several ancient lakes (e.g. Lake Malawi; Ribbink, 1994) geological history is poorly understood, and for many their isolation, and also in the case of the deeper lakes, logistic difficulties in obtaining long cores, mean that their palaeohydrology and palaeobiology are still insufficiently known. When the difference between two estimates is several million years, as is sometimes the case in ancient lakes, details of the evolutionary time frames of the biota are that much more difficult to unravel.

Modern geophysical surveys, using such tools as new seismic, radar, gravity and magnetic data to map each ancient lake and its bathymetry, and long sediment cores to reveal their geological and limnological history, are needed. Palaeobiological data can complement these studies and assist in the interpretation of the lake's climatic and limnological history. The information derived from such a comprehensive approach is essential in furthering our understanding of evolutionary and speciation processes of ancient lake biotas.

V. SPECIES COUNTS: BIODIVERSITY

Surveys of biodiversity are fundamental to ecological studies of ancient lakes. But how much do we actually know and how accurate are our assessments? Lake Biwa has been studied intensively for several decades, and six research institutes, including both university and governmental bodies, are engaged in research on the lake environment (International Lake Environment Committee, hereinafter ILEC, 1993). One would therefore expect the fauna and flora of the lake to have been fully documented, and its ecology to be well understood.

Yet, despite this long history of research on this biologically important ancient lake, there exists no comprehensive inventory of its flora and fauna. The incomplete lists of Mori (1970) and Mori and Miura (1980, 1990) are the best that are available, but in view of the lack of original sources for species descriptions, an incomplete survey of older references, and the citing of obsolete or misspelled taxa (O.A. Timoshkin; M. Nishino, pers. comm.) are of limited use only. Given this incomplete knowledge of the biota of the lake and the inadequacies of the existing inventory, the list of taxa recorded from Lake Biwa is certain to be revised.

Table 1
Examples of differences in the estimated ages or time of origin for several ancient lakes
or their rift zones, listed in chronological order of publication

	Estimated age or time of origin	Source where cited
Lake Baikal		
	c. 1.6 My	Dumitrashko (1952)
	c. 26–38 My	Sarkisyan (1955)
	1.6–65 My (Tertiary)	Lamakin (1955)
Southern depression	10–30 My	Mats (1993)
Northern deep basin	3.5–10 My	Logatchev (1993)
Present lake	0.8–3.3 My	
Protolakes, which may not have been persistent	60 Mya	Mats (1993)
One or more lakes existed continuously	> 28 My	Mats (1993)
Formation of Baikal rift zone	35 Mya	Martin (1994)
Formation of Baikal rift zone	70 Mya	Sherbakov (1999)
Lakes present in the Baikal rift zone	*c.* 60 Mya	Sherbakov (1999)
Lake Malawi		
	1–2 My	Fryer (1959)
	100 000 years	Cromie (1982)
	1.5 My	Greenwood (1984)
	< 2 My	Mayr (1984)
	> 5 My	Johnson and Ng'ang'a (1990)
	c. 2 My, maybe more	Ribbink (1994)
Lake Lanao		
	10 000 years	Myers (1951, 1960)
	8000–11 000 years	Wood and Wood (1963)
	*c.*1.6–26 My (late Tertiary)	Frey (1969)
	3.6–5.51 My	Lewis (1978)
Lake Tanganyika		
Tanganyika rift zone	18.6 Mya	Patterson (1983)
	c. 2 My	Greenwood (1984)
	> 2 My	Mayr (1984)
Tanganyika rift zone	13–20 Mya	Burgess (1985), Morley (1988)

Table 1 *continued*

	Estimated age or time of origin	Source where cited
Tanganyika rift zone	5–26 Mya (Miocene)	Haberyan and Hecky (1987)
Protolakes	20 My	Tiercelin and Mondeguer (1991)
Isolated lake	14 My	Coulter (1994)
	9–12 My	Cohen *et al.* (1993)
Lake Victoria		
	> 750 000 years	Bishop and Trendall (1967)
	c. 800 000 years (mid-Pleistocene)	Temple (1969)
	c. 14 000 years	Kendall (1969)
	c. 14 000 years	Livingstone (1975)
	14 000 years	Livingstone (1980)
	750 000 years	Greenwood (1984)
	100 000–750 000 years	Mayr (1984)
	25 000 years	Stager *et al.* (1986)
	14 000–750 000 years	Greenwood (1994)
	12 400 years	Johnson *et al.* (1996)

Not all cited references are primary sources of the age estimate.

For example, the fish fauna of the lake is its most conspicuous biota and also perhaps its best known, yet even now, new fishes are being discovered there (e.g. a description of a new endemic *Rhinogobius* species is currently awaiting publication: S. Takahashi, pers. comm.) and the presence of other undescribed or misidentified fish taxa is suspected. The lake's invertebrate fauna is less well known, but recent efforts, especially those of M. Nishino and collaborators, have improved matters greatly (e.g. Nishino, 1991, 1992, 1993; Kawakatsu and Nishino, 1993, 1994). However, a literature survey and sampling regime of benthic invertebrates in Lake Biwa conducted recently by O.A. Timoshkin and colleagues resulted in over 170 taxa newly recorded from the lake, including several newly described, and probably endemic, species. This inventory, which the authors emphasize is incomplete, lists over 1070 animal species and 534 plant species from Lake Biwa: the authors predict that detailed surveys and taxonomic studies will result in a further 50% increase in the number of taxa described from the lake (O.A. Timoshkin, pers. comm.).

Clearly, even for this small, supposedly well-known lake, knowledge of the biota is incomplete and leaves one wondering what undocumented biological diversity the other less well-studied, but already highly species-rich ancient lakes might hold. The invertebrate fauna of many ancient lakes is incompletely or poorly known, but as highlighted by several of the contributions in this volume, recent investigations have revealed many new taxa and a higher than expected diversity even in supposedly well-known groups. This incomplete knowledge means that biodiversity inventories are constantly being revised as taxonomic studies proceed. Once a reasonably complete data set has been assembled, however, ancient lakes will be ideal sites in which to explore whether any particular mechanism is the dominant explanation for a specific distribution pattern, and whether there are any general rules about which community assembly rules are likely to be important under particular environmental conditions, among specific groups of organisms, or at particular temporal or spatial scales. In this regard, despite present deficiencies, given careful planning, clear objectives and a quantitative research approach, future studies in Lake Biwa could provide a model for this type of research in ancient lakes.

VI. THREATS TO THE LAKE BIWA ECOSYSTEM

The human population of Shiga Prefecture, within which Lake Biwa lies, has increased rapidly in recent decades, from c. 850 000 in 1970 to c. 1.4 million today. Over this same period Japan has experienced an economic upsurge, reflected in Shiga by a six-fold increase in industrial production (Shiga Prefectural Government, 1997). These and other factors have conspired to produce a dramatic increase in eutrophication and a decline in the water quality and ecological well-being of Lake Biwa, most clearly evinced by algal blooms, foul-smelling drinking water and population declines in many of the lake's flora and fauna. However, the number of people relying on Lake Biwa as a source of drinking water has increased from 5.5 million in 1955 to over 14 million today. From both biological and applied perspectives, the environmental condition of Lake Biwa is thus of great concern.

In this section some of the dangers posed to the ecology of Lake Biwa and their sources are catalogued. In step with modern conservation perspectives, rehabilitation of the ecosystem, and not individual species, will be addressed. For convenience these threats are considered under four headings–pollution, habitat destruction and modification, alien introductions, and overharvesting– although in some instances these categories overlap.

A. Pollution

1. Entry of Toxins into the Lake Waters

Despite the seemingly clear cause (modern industrialization and increased human population pressure) and effect (increased industrial and domestic waste entering the lake) outlined above, pollution of Lake Biwa is not an exclusively modern phenomenon. The earliest documented case of industrial pollution that the author was able to locate took place in 1928, when effluent from a rayon factory in Otsu City caused mass mortalities of fishes and shellfish in the lake's southern basin, and severely impacted upon the commercial fisheries of that area. Another major incident involving industrial pollution occurred in the 1960s, when the southern basin and the Seta River were heavily contaminated by polychlorinated biphenyls (PCBs) in the untreated effluent from a condenser plant in Kusatsu City. Subsequently, high PCB concentrations were detected in the lake's fishes and in lake sediments, and in the late 1960s the local government responded by dredging and removing contaminated mud.

Other cases of severe perturbation have occurred when long-term accumulations of non-biodegradable compounds, mainly those of agricultural origin, in lake sediments have achieved toxic levels. The cause and effect of several such events is well correlated. For example, after the 1950s, large areas of wetland around the lake were converted into rice paddies (see below). This increased agricultural usage of the lake basin was accompanied by an increased use of fertilizers and pesticides, especially chlorinated hydrocarbons. Seepage, runoff and, perhaps, overapplication of these chemicals and their entry into the lake resulted in several instances of mass poisoning of fishes and molluscs in the lake during the 1960s and 1970s. Sublethal, but equally severe effects on the fish communities were also apparent, such as the offshore migration of fish when agricultural chemicals were applied to the rice paddies (T. Chatani, cited in Kada, 1999).

It is debatable whether the situation has improved or not during the 20 year interim since these well-publicized events. For example, in 1993, concern was expressed over the levels of the organic herbicide CNP which had accumulated in the lake sediments and certain aquatic fauna since its first use by farmers in rice paddies during the early 1980s (Ishida, 1986) In some cities in Shiga, trichloroethylene contamination of the groundwater has been noted (ILEC, 1995) and in mid-1999, alarm was raised over the level of dioxins in the lake sediments and in aquatic organisms.

2.   Eutrophication

Perhaps the most insidious form of pollution of Lake Biwa comes through excessive nitrogen and phosphorus loads. These nutrients have accumulated in the lake, have resulted in increased eutrophication, and are the major causes of algal blooms and foul water in Lake Biwa.

Public awareness of the long-term eutrophication of Lake Biwa was triggered in 1969 when residents of Kyoto City complained about the smell of their tapwater, which originates from Lake Biwa. Studies revealed the smell to be due to the generation of geosmin associated with a bloom of *Phormidea*, *Anabaena* and other phytoplankters, caused by severe contamination of the lake water. Thereafter, the quality of the lake water has continued to degrade rapidly. Prior to *c.* 1965 the water of the north basin was oligotrophic, but is now mesotrophic, while the previously mesotrophic south basin has degraded to eutrophic.

Again, the deterioration of the lake ecosystem is not an exclusively modern phenomenon. The mean annual transparency in the central part of the north basin has declined steadily since 1920, from over 10 m to only 5.3 m in 1995 (the most recent published data). In the south basin, transparency is extremely poor, with a 1995 mean value of only 1.7 m (Shiga Prefectural Government, 1997). Changes in transparency need to be interpreted carefully, as reduced transparency can occur through an increased productivity of phytoplankton, most often as a response to increased nutrient loadings, or through increased suspension of organic or inorganic particulates. However, the almost 10-fold increase in plankton biomass which occurred between 1950 and the early 1980s (ILEC, 1993) leaves little room for ambiguity on this matter. These increases in phytoplankton productivity have occurred in tandem with increased nutrient inputs into the lake, for example, between 1960 and 1980 nitrogen input into the lake increased from 5461 to 9196 t y^{-1} (Inoue *et al.*, 1981). Since 1965, as eutrophication has proceeded, algal blooms have occurred during the summer and autumn months. In 1977 the first "red tide" was recorded and these blooms of *Uroglena americana, Perinidium* spp. and *Anabaena* spp. have since occurred annually, except during 1984 (Shiga Prefectural Government, 1997). In 1983, blooms of the planktonic algae *Anabaena* spp. and *Microcystis* spp. caused the first "green tide", which was so serious that many beaches were closed. Green tides have occurred every year since, except during 1984. The pollutant loads which trigger these plankton blooms are circulated northwards along the east shore of the north basin by gyres (Nakanishi *et al.*, 1986) and produce algal blooms in areas far from the point sources of the pollution.

From where does this nutrient loading originate? The amounts of nitrogen, phosphorus and coarse organic detritis (COD) entering Lake Biwa each day from within the lake watershed are given in Table 2 in relation to four major sources (Shiga Prefectural Government, 1997) and it is clear that by far the greatest amount of nutrients has an anthropogenic origin.

Table 2
Sources and amounts of nutrient loads to Lake Biwa (all values given as tonnes d^{-1})

Source	COD	Total nitrogen	Total phosphorus
Natural	22.21	9.11	0.29
Domestic	18.41	5.81	0.59
Industrial	15.61	4.80	0.69
Agricultural	10.47	4.39	0.16
Total	66.70	24.10	1.73

3. Attempts at Reducing Nutrient Loadings

The problems of excessive nutrient inputs into the lake have long been recognized, and measures to reduce these loadings have included a 1973 local government ordinance limiting industrial pollution rates and another in 1979 regulating effluent nitrogen and phosphorus. This Prevention of Pollution Ordinance included the prohibition of the sale and use of phosphorus-based domestic detergents, and regulations to control nitrogen and phosphorus inputs from industrial and agricultural sources (ILEC, 1995). Efforts have also been made to improve the treatment of sewage. [The reader may be surprised to learn that only 40% of homes in Shiga are covered by the sewage system: in areas without any sewage system, domestic effluent, including human waste, discharged from each house flows directly into public water areas (Shiga Prefectural Government, 1997). Traditionally, human waste and domestic effluent have been utilized as fertilizers. These are clearly areas of strong concern.*]

However, despite these and other regulatory measures, there has been no significant reduction in eutrophication, and the end result has been an inability to meet even self-imposed targets to reduce COD in the lake. In 1986 a 5-year project (1986–1990) was begun to reduce COD levels to 2.5 mg l^{-1} in the south basin and 1.8 mg l^{-1} in the north basin, but in 1990 COD levels were 3.3 mg l^{-1} and 2.3 mg l^{-1}, respectively. A second 5-year project (1991–1995) aimed at reducing levels to 3.3 mg l^{-1} in the south basin and 2.5 mg l^{-1} in the north basin, but 1995 levels were again higher than the targets. A third 5-year project is now underway (1996–2000), and while the results are awaited with interest, the overall prognosis for success does not seem high: recently, an extensive growth of the sulphur-oxidizing bacterium *Thioploca*, indicative of anaerobic conditions, was discovered in the deep zone of the lake (Nishino *et al.*, 1998).

Why have these efforts failed, despite intensive long-term studies, recognition of the causal factors and implementation of some remedial measures? One plausible possibility is that by addressing the eutrophication problem solely in terms of physicochemical properties and ignoring the importance of the biology of the whole lake and its basin, the authorities have misunderstood the problem. This ignoring of biology is most evident in construction projects which, considered from the perspective of eutrophication control, have replaced beneficial natural systems with deleterious artificial constructs.

B. Ecosystem Destruction and Habitat Modification

The riparian habitat of Lake Biwa and its inflowing rivers has been severely impacted by the destruction of ecosystems and by habitat modification. Many

*In addition to the effects of increased eutrophication, it should be noted that increasing nitrate concentrations can cause health problems in humans (Moss, 1988). The health risks inherent in using human waste as a fertilizer on leaf and other vegetables destined for human consumption need not be stated.

of these efforts appear little more than attempts at dominating nature and resculpting natural environments. In some cases "environmental improvement" projects seem to have been primarily cosmetically orientated and their results purely aesthetic. However, in some instances, these projects might have exacerbated the eutrophication and environmental degradation of the lake.

1. The Importance of Streams, Rivers and Wetlands, and the Effects of Modification

a. Rivers and Streams. Lake Biwa is fed by over 400 streams and rivers, and these contribute to nutrient transport and flushing of lake waters. They cannot be divorced from either the adjacent riparian zone or the surrounding biotic and geomorphic catchment that funnels water into the lake. The lower reaches of these inflows are utilized by both lacustrine and riverine faunas and the nature of the downstream assemblage, and ultimately that of the lake ecosystem, is thus inextricably linked with processes occurring upstream. This interconnectedness means that any meaningful consideration of lake ecology requires inclusion of the entire watershed.

This is also true with regard to eutrophication, yet attempts at controlling eutrophication in Lake Biwa since the 1970s have been approached as a physicochemical problem and considered from a black-box perspective. The important implications of biology and its potential usefulness in providing solutions to the eutrophication problem have not been fully considered and, as a result, ecologically unwise construction practices and policies have been effected. Some of these are likely to have inadvertently hampered other efforts at controlling nutrient inputs and might even have accelerated the eutrophication of the lake. This is especially true of the modification of rivers, streams and wetland ecotones which has taken place around Lake Biwa.

Under natural conditions, aquatic macrophytes in streams and rivers remove dissolved nutrients through a process called spiralling (Elwood *et al.*, 1983; Ward, 1988). In addition, filter-feeding invertebrates remove fine and coarse organic particulates from the stream water, and detritivorous invertebrates consume the larger particulates deposited on the stream bottom. The stream biota thus helps to purify stream water as it travels downstream and before it enters a lake.

Unfortunately, very few of Lake Biwa's inflowing streams and rivers remain in near natural conditions and retain their original flora and fauna. Instead, despite having long been known to be an ecologically deleterious practice (e.g. Emerson, 1971), extensive channelization of these streams and rivers has been undertaken. As would be predicted (e.g. Bayless and Smith, 1967), ecologically the effect has been to reduce drastically the numbers, diversity and productivity of the stream biota. The practical implications for eutrophication are equally stark. The depauperate fauna and flora which now inhabit these straight-walled concrete channels have but minimal biotic impact: little or no spiralling or biological processing takes place, and instead nitrates, phosphates and particulates in the stream are rushed straight into the lake. That many of these channels run through the middle of paddy field complexes or urban areas serves only to amplify the negative effect that they have in loading the lake with chemicals and nutrients.

Ostensibly, the reasons for channelization were flood prevention and the supply of irrigation water to paddy fields, but the extent of channelization–in some places almost inaccessible mountain streams several centimeters deep and barely a metre wide have been concreted–hints at overzealousness. The low priority afforded to biological concerns and the natural environment can be gauged from the portion of the total budget of the 25-year-long Comprehensive Development Project of Lake Biwa (Shiga Prefectural Government, 1983). Although one of its touted aims was to "conserve the natural environment of Lake Biwa and its surrounding areas", less than 1% of the total budget was allocated towards this purpose, compared with over 14% for "improvement" of paddy irrigation systems (Shiga Prefectural Government, 1997), i.e. mainly channelization-related work.

Further ecological problems have been caused by the water-regulation aspects of this development project, whose objective was to utilize Lake Biwa as a water resource by manipulating water level from $-1.5\,m$ to $+1.5\,m$ in accord with water supply needs, and which has resulted in dam and lock systems being installed on many of these streams and rivers. Under this scheme, many rivers and streams are sealed off and allowed to run dry for large parts of the year, and mature aquatic communities thus have no opportunity to develop.

b. Wetland and Reedbed Areas. Large parts of the Biwa shoreline formerly comprised large wetland expanses of reedbeds and satellite lakes, often fed by rivers and steams. These lakes were an important habitat whose shallow waters warmed throughout the summer, and where a high macrophyte productivity supported a rich invertebrate community and provided a spawning and nursery habitat for many fish species.

Wetland ecotones such as these comprise both aquatic and terrestrial components of production and so have the highest biological activity and productivity in lake ecosystems (Jorgensen and Vollenweider, 1988). Because reedbeds and wetlands retain sediments and nutrients, the former by sedimentation and the latter through uptake by macrophytes and bacteria (Novitski, 1978), they also have a practical application in controlling eutrophication. Furthermore, the aquatic macrophytes in satellite lakes utilize dissolved nitrates and phosphates, and around Lake Biwa were themselves used by farmers as fertilizer for their fields (Ogawa et al., 1997). These properties can improve the chemical quality and clarity of the downstream water which has percolated through these habitats, and the practical usefulness of wetlands as natural filters has long been recognized by civil engineers (e.g. Kadlek and Tilton, 1979; Gersberg et al., 1983); but not, apparently, by those responsible for the extensive filling in of satellite lakes and cutting down of reedbeds which has taken place around Lake Biwa since the 1960s. Since World War II more than 2500 ha of satellite lakes and reedbeds have been reclaimed for rice production (Kada, 1999), resulting in a loss of fish-spawning habitat, reedbeds and large areas of littoral aquatic macrophytes (Hanafusa, 2000).

This loss of spawning habitat has severely impacted upon several fish species, including the endemic crucian carp *Carassius c. grandoculis*. This commercially important species spawns mainly in aquatic macrophyte zones in satellite lakes. However, by the early 1960s a shortage of available satellite lakes meant that *c.* 12% of the entire spawning population of *C. c. grandoculis* was using a single satellite lake, with an area of less than 100 ha (Kawanabe, 1999). However, despite its obvious importance to the well-being of this species, this lake was also reclaimed, starting in 1964. The population and catches of *C. c. grandoculis* have subsequently undergone a catastrophic decline.

Decimation of reedbed habitat has been equally severe. In 1953 there were 219 ha of reedbed around the Lake Biwa shoreline, but by 1980 this had been reduced to 163 ha, and to 130 ha by 1992 (ILEC, 1994; Shiga Prefectural Government, 1997). In addition to the loss of habitat for a diverse fauna, the removal of reedbeds (and with them their filtering capacity) will have meant an increased nutrient load entering the lake. The situation was further compounded by the initiation, in 1972, of a 25 year project to construct roads and dams around the lake. The natural coastline comprised 48.6% of the lake in 1979, but was reduced to 44% by 1985, and to 40.8% by 1991.

To summarize, the effects of these construction practices around the lake in increasing eutrophication have been manifold.

(1) Channelization of inflowing streams and rivers has resulted in nutrient loads being transported directly into the lake before spiralling within the biotic community can reduce them.
(2) Many channels run directly through urban and agricultural areas, and so the nutrient and pollutant loads they carry are prone to be abnormally high.
(3) The removal of reedbeds took away a natural filter which would normally have reduced some of the nutrient load. It also removed an important biotope for many of the lake organisms.
(4) The filling in of satellite lakes and their replacement with rice paddies was doubly inadvisable:
    (a) these areas were important and irreplaceable spawning habitats for many fishes, and for productivity and growth of a wide variety of other taxa. The rich aquatic macrophyte community there absorbed dissolved nitrates and phosphates;
    (b) replacing satelite lakes with rice paddies served to increase the phosphate and nitrate loading from fertilizers and chemicals from pesticides to the lake, carried there directly by channels and swiftly by virtue of their lakeshore location.

C. Alien Introductions

The long history of introductions of alien species into Lake Biwa can be traced back to 1883 when the lake was stocked with salmon *Onchorhynchus keta*. Although this species failed to become established, other alien introductions thereafter have been many, and have met with varying degrees of success. Yet other non-native taxa have been introduced accidentally or have migrated to the lake after being introduced to nearby waterbodies. A list of non-native taxa recorded in the lake is given in Table 3.

Two of the most successful and the most visible of these aliens are the North American aquatic macrophytes *Elodea nuttallii* and *Egeria densa*. *Elodea nuttallii* was first noted in the lake in 1961 and *E. densa* in 1969 (Miura, 1980). *Elodea nuttallii* is the dominant macrophyte in the lake (Hamabata, 1991) and together these species now dominate the aquatic vegetation of the lake's littoral zone. It is unknown how they arrived in Lake Biwa, but as both are popular aquarium and ornamental pond plants, it is probable that they were introduced by aquarists. The South American water hyacinth, *Eichhornia crassipes*, is also now in the lake, and probably arrived by the same means.

Five intentionally introduced taxa have flourished in the lake: the Japanese smelt *Hypomesus transpacificus nipponensis*, introduced in the 1920s and 1940s; the North American bluegill sunfish *Lepomis macrochirus*, introduced officially in 1967, but illegally a few years prior; the North American largemouth bass *Micropterus salmoides*, introduced illegally by anglers, probably around 1974 but certainly several times thereafter (Nakai, 1999); the Japanese long-armed shrimp *Macrobrachium nipponense*, from Lake Kasumi ga ura, Ibaraki Prefecture, Japan, in the 1920s; and the North American bullfrog *Rana catesbeiana* in 1918. One (probably) accidentally introduced taxon, the Chinese mussel *Limnoperna fortunei*, has successfully colonised the lake, where its effects parallel those of the early stages of the zebra mussel infestation of the Laurentian Great Lakes. The North American crayfish *Procambarus clarkii*, believed introduced by anglers in 1930, is also abundant. Because of their biological and commercial impact, the discussion hereafter will focus on the introduced fishes of Lake Biwa.

1.   The Non-native Fishes of Lake Biwa

a. Japanese Smelt. The Japanese smelt was introduced into Lake Biwa in large numbers in the early 1900s and again throughout the 1940s, but although smelt were captured in small quantities then and thereafter, these introductions were considered to have failed. Large numbers have been present since a 1994 population explosion (Ide, 1996a). The source of this population has been attributed to individuals introduced elsewhere in the lake subsequent to these earlier failed introductions. However, it seems equally parsimonious to suggest

Table 3
Non-native species recorded from Lake Biwa: some are solitary records only, and not all
have self-sustaining populations in the lake

Aquatic macrophytes	Pond weed, *Elodea nuttallii*
	Pond weed, *Egeria densa*
	Water hyacinth, *Eichhornia crassipes*
Molluscs	*Quadrula bagini*
	Lampsilis luteola
	Anodonta woodiana
	Margaritifera margaritifera
	Chinese mussel, *Limnoperna fortunei*
Crustaceans	Opposum shrimp, *Mysis* sp.
	Japanese long-armed shrimp, *Macrobrachium nipponense*
	Malaysian giant prawn, *Macrobrachium rosenbergii*
	Crayfish, *Pacifastacus leniusculus*
	Crayfish, *Procambarus clarkii*
Fishes	Sturgeon *Acipenser* hybrid form
	Atlantic sturgeon, *Acipenser oxyrhynchus*
	Alligator gar, *Atractosteus spatula*
	Shortnose gar, *Lepistosteus platostomus*
	Japanese eel, *Anguilla japonica*
	Asian snakehead, *Channa argus*
	Brook charr, *Salvelinus fontinalis*
	Japanese nikou charr *Salvelinus leucomaenis* f. *pluvius*
	Rainbow trout, *Onchorhynchus mykiss*
	Pink salmon, *Onchorhynchus gorbuscha*
	Sockeye salmon, *Onchorhynchus nerka*
	Chum salmon, *Onchorhynchus keta*
	Lake whitefish, *Coregonus* (*olbus* = *albus*?) *clupeaformis*
	European whitefish, *Coregonus lavaretus baeri* and *C. l. maraeniodes*
	Pond smelt, *Hypomesus olidus*
	Common carp, *Cyprinus carpio*?
	Rudd, *Scardinus erythrophthalamus*
	Goldfish, *Carassius auratus*
	Chinese grass carp, *Ctenopharyngodon idella*
	Bighead carp, *Hypopthalmichthys molitrix*
	Rosy bitterling, *Rhodeus ocellatus ocellatus*
	Chinese false gudgeon, *Abbotina rivularis*
	Black tetra, *Gymnocorhymbus ternetzi*

Table 3 *continued*

Fishes	Halfbeak, *Hyporhamphus intermedius*
	Japanese smelt, *Hypomesus transpacificus nipponensis*
	Bluegill sunfish, *Lepomis macrochirus*
	White crappie, *Pomoxis annularis*
	Largemouth bass, *Micropterus salmoides*
	Smallmouth bass, *Micropterus dolomieu dolomieu*
	Oscar cichlid, *Astronotus ocellatus*
	Various *Plecostomus* catfishes
	Red-tailed catfish, *Phractocephalus hemrioliopterus*
	Oxydoras catfish
	Royal knifefish, *Chitala branchi*
	Japanese short-spined trident goby, *Tridentiger brevis*
	Mullet, *Mugil* sp.
Amphibians	North American bullfrog, *Rana catesbeiana*
	Axolotl, *Ambystoma mexicanum*
Reptiles	Common snapping turtle, *Chelydra serpentina*
	Alligator snapping turtle, *Macroclemys temmincki*
	Western painted turtle, *Chrysemys picta*
	Red-eared slider, *Trachemys scripta elegans*

that conditions in the lake changed such that a residual population of the original stocks experienced an ecological release, and an increased growth rate, fecundity and survival. The Japanese smelt is an offshore forager of zooplankters, and it may be utilizing an increased zooplankter population blossoming on phytoplankton blooms driven by the increased nitrate and phosphate loads in the lake. Mass spawning is now observed every spring at the mouths of streams and lower reaches of affluent rivers, and the fish is now targeted by commercial fishery interests.

b. Bluegill Sunfish and Largemouth Bass. Bluegill were presented to the then Crown Prince of Japan (the present Emperor) as a gift by an American delegation. These fish were given to the Shiga Prefecture Fisheries Experimental Station, who bred them and, starting in 1967, introduced them to many satellite lakes of Lake Biwa, ostensibly as a host for the parasitic glochidium larvae of freshwater mussels, at that time the hub of a thriving Biwa freshwater pearl industry. Bluegill flourished and by the early 1970s had achieved a lake-wide distribution (Terashima, 1980). The population then exploded and has remained high ever since.

Largemouth bass were first recorded in the lake in 1974, after having been illegally introduced by anglers for sport purposes. Introductions probably occurred prior to this and thereafter for several years. The population showed an explosive increase and by the early 1980s, largemouth bass were extremely abundant throughout Lake Biwa, where they are the top aquatic predator.

Bluegill and largemouth bass now dominate the littoral fish communities, and their arrival has coincided with a perceived decline in the populations of many of the native cyprinid fauna. However, it has also been suggested that the native fish faunas were already in decline before the explosion in bluegill and largemouth bass populations (Nakajima and Nakai, 1994) and that the largemouth bass population has decreased since the 1980s (Maehata, 1990). Unfortunately, the absence of quantitative ecological studies of fish community composition before and during this process mean that baseline data necessary for an accurate perspective on temporal flux in the densities of native species are unavailable. It is thus impossible to quantify this as a cause-and-effect process or to study its dynamics. Similarly, the dynamics of both the largemouth bass and bluegill booms were, for whatever reason, undocumented. As such, the complete absence of any statistically investigable data makes statements concerning the effects of bluegill and largemouth bass introductions (e.g. Maehata, 1990; Nakajima and Nakai, 1994) impossible to verify and renders them anecdotal only.

Bluegill are omnivorous, biased towards carnivory, and largemouth bass are exclusively predatory, and both species are probably causing major perturbations to the structure and function of the native Biwa fish community. In addition to dominating the littoral habitat with their breeding arenas, bluegill prey on the eggs and young of native fishes and compete with them for invertebrate food items. However, while it seems intuitively obvious that the effects of these fishes are deleterious to the native fishes, the absence of any quantitative studies mean that such statements must be cloaked in caveats. It should be noted that the presence of alternative prey (native fishes) reduces predation pressure on bluegill by largemouth bass, and hence bluegill may ultimately prove the major problem species in the lake.

Bluegill seemingly offer little or no economic potential in Lake Biwa as the Japanese consider them unpalatable and they are of limited interest to sports anglers. The economic benefits accrued from largemouth bass have not been assessed, but are likely to be great. The lake is now a popular venue for sports anglers seeking this fish, and apart from fishing tackle suppliers and bait shops, specialist "bass boat" manufacturers and sellers have also established themselves around the lake. Local communities also derive financial benefits in terms of accommodation, meals and other incidentals provided to visiting anglers. Negative impacts are also evident, in that the high numbers of bluegill mean that they clog the nets of commercial inshore fishermen targeting other species. Another concern is that many of the visiting anglers show scant regard for nature, and litter the shoreline with wrappers and discarded fishing line. The former is unsightly and the latter is a danger to wildfowl, waders and other shoreline birds, which can become entangled and die.

c. Other Alien Fishes. Several other alien fish species are also present in Lake Biwa. The Japanese short-spined trident goby *Tridentiger brevis* was first recorded in the lake in the north basin in 1989 (Takahashi, 1990), but is now extremely abundant at rocky shores throughout the lake, where it competes with naturally occurring fishes. The Asian snakehead *Channa argus* is present and breeding, and the North American smallmouth bass *Micropterus dolomieu dolomieu* has been recorded in Biwa since 1995 (Ide, 1996b). The goldfish *Carassius auratus* is also present, and may pose a problem through potential hybridization with native and endemic carassids. The Chinese grass carp *Ctenopharyngodon idella* and bighead carp *Hypopthalmichthys molitrix* occur in the lake, but are not believed to be breeding there, as they require a riverine habitat for successful hatching of eggs (Lever, 1996). The rosy bitterling *Rhodeus ocellatus ocellatus* and the Chinese false gudgeon *Abbotina rivularis* occur in other waters in Japan, but arrived in Lake Biwa only in the late 1960s (Miura, 1971).

More recently, several exotic fish species, many originating in the tropics of Asia and South America, have been recorded, such as the oscar cichlid *Astronotus ocellatus*, various *Plecostomus* catfishes, the black tetra *Gymnocorhymbus ternetzi*, the red-tailed catfish *Phractocephalus hemioliopterus*, the royal knife fish *Chitala branchi, Oxydoras* catfish, the North American alligator gar *Lepisoteus spatula* and shortnose gar *L. platostomus*, and sturgeon (Acipenseridae). These fishes are believed to have been released by aquarists. The shortnose gar and sturgeon could probably survive through the temperate Biwa winter, but in normal circumstances the tropical species will not be able to survive the winter temperatures. However, at some localities warm water discharged from factories or domestic areas might permit survival. A regulation forbids the release of fishes into the lake, but since its introduction in the 1970s no one has been charged. Regulations mean nothing if they cannot be monitored and enforced.

D. Overharvesting

The statistics for fish catches from Lake Biwa over the past 30 years make for sombre reading. Almost without exception, dramatic declines have taken place in almost all target species. For example, in the mid-1960s annual catches of the endemic goby *Chaenogobius isaza* totalled 300–400 t, but had crashed to only 0.1 t by the mid-1990s (Ueda, 1998). Catches of carp were 180 t annually around the early to mid-1970s but thereafter have declined continuously and stand at *c.* 50 t today. Shellfish catches have also shown serious declines, for example, the recent catches of the setashijimi clam *Corbicula sandai* are only 2.5% of their former levels during the 1960s.

The source of these figures is official fisheries data and they are therefore of limited value for ecological analysis. However, two aspects are noteworthy: first, the decrease in catches is of such magnitude that clearly all is not well in these fish and shellfish populations, and secondly, the continued fishing of *Chaenogobius isaza* and *Corbicula sandai* to such low levels suggests that conservation and sustainable yield are alien concepts to the fishermen of this lake. As well as being victims of the ecological perturbations caused by alien introductions and pollution, they may have partly contributed to the decline in fish and shellfish populations by overharvesting. This is perhaps best illustrated by the example of the ikechogai *Hyriopsis schlegeli*, an endemic mussel formerly important in Biwa freshwater pearl cultivation. Catches of this species plummeted from 102 t in 1962 to 2 kg in 1991. Since then none have been caught and it is feared extinct (Ueda, 1999). Further examples of declining catches of several commercially important species are given in Nakai (1999).

Habitat rehabilitation and ecological conservation measures must meet with the approval of socioeconomic pressures, but Lake Biwa does not have the excuse for overutilization that many of the other ancient lakes have to offer, that of being located in countries where they are the sole source of protein (fish) for millions of people, and which do not have the finances, technology or infrastructure to regulate effectively what goes into (pollution) and comes out of (fishes, shellfish, plants) the lake.

Clearly, Lake Biwa is facing a multitude of problems, the scope and magnitude of some of which suggest that complete recovery is improbable and only partial reclamation of the natural system might be possible. It might therefore be optimal to focus remedial efforts on those areas where solutions may be most easily attained. For example, despite its present dire situation, examples from other lakes offer hope that it might still be possible to halt or perhaps reverse the eutrophication of Lake Biwa. Measures to reduce phosphate loadings in lakes were widely known, and already being applied successfully in North America and Europe as early as the 1960s; for example, the implementation of such measures in Lake Washington, USA, resulted in a reversal of eutrophication within a few years (Edmonson, 1970). More efficient methods are available today. However, implementation of realistic measures to control eutrophication of Lake Biwa will necessitate an increased awareness and understanding of ecological principles, and must consider the problem in a basin-wide context. If, through wisdom and vision, this challenge is accepted, a real opportunity will exist for a reduction of eutrophication via ecological rehabilitation measures, such as the recovery and reconstruction of natural habitats, i.e. ecosystem restoration with both biological and practical   benefits.

VII.   CONCLUSIONS

Studies in Lake Biwa typify those in other ancient lakes and can be used to identify key areas of concern.

Knowledge of the geological history of ancient lakes is uneven. For some lakes much is known, but for others very little. The accuracy and degree of detail of geochronology also vary among lakes, greatly complicating the reconstruction of evolutionary scenarios.

In all ancient lakes, future research on palaeohydrology and palaeobiology is recommended. This can unravel both the history of the lake and thus the template in which speciation has occurred. It can also contribute towards the resolution of such instances when the biological data are in conflict with the geological age estimate (e.g. Johnson *et al.*, 1996; Fryer, 1997). Geochemical investigations are also encouraged.

Biodiversity inventories have been instrumental in the identification of ancient lakes as hotspots of diversity and centres of endemism, but in no ancient lake has an adequate inventory of biodiversity been compiled (Martens *et al.*, 1994). Even in the intensively studied and relatively small ancient Lake Biwa the level of knowledge of species richness is insufficient to test hypotheses concerning patterns of diversity or its regulation. More effort is needed in this area.

Efforts at conservation should be directed towards protecting the existing state from anthropogenically generated acceleration, with the priority on habitat and ecosystem protection, rather than on cataloguing efforts or on individual species protection. Unfortunately, while such recommendations are common parlance in the international conservation community, they do not seem to have been applied in Lake Biwa, where the authorities responsible for the management of the lake seem to have busied themselves with unending limnological water chemistry-type surveys over the past 30 years. Meanwhile, the ecology of the lake is poorly known, as is the biology of a large portion of its fauna. For example, even fundamental data such as length-at-age, longevity, fecundity, habitat preference and life history are wanting for most of the fish species. This is especially apparent and debilitating in view of the invasion of bluegill and largemouth bass: the effects of these fishes on the native fish communities cannot be assessed accurately because little quantitative information exists. In this respect, research in Lake Biwa lags behind the kind of detailed ecological studies done on several other ancient lakes. Lake Victoria, with its Nile perch problem paralleling the bass/bluegill of Biwa, comes immediately to mind, but meaningful studies of the ecology of fishes and other faunas have also been conducted in Lakes Tanganyika, Malawi, Titicaca and Baikal.

It is clear that Lake Biwa has had more ecological insults inflicted upon it than any other ancient lake; among the ancient lakes of the world, only Lake Victoria equals the magnitude and scope of the problems experienced by Lake Biwa. In Lake Victoria, human pressures and alien introductions have been to blame, but in Biwa it is these same factors plus a policy of environmentally unfriendly large-scale changes in the natural habitat which have contributed to the situation in various ways; physical, chemical and biotic perturbations, including alien introductions and overfishing, have all taken their toll on the lake ecosystem, and continue to do so.

Dangers to ancient lakes are often not single, but come from a variety of sources, and the effects of these operating together can be greater than the sum of their independent effects. Lakes are extremely susceptible to disruption by introduced species (Moyle, 1993), but it is also true that inflowing rivers that are dammed, channelized, polluted or otherwise modified also pose a danger. Species and ecosystems rarely decline from single causes or events and, most often, introducing an alien species is but the final blow to an ecosystem already disturbed by human influences. In this regard the demise of the cyprinid species flock of Lake Lanao is salutary. Fifteen of the 18 purported endemic species within this species flock went extinct in less than 25 years, as a result of overexploitation, anthropogenic pollution stemming from human population pressure, and the deliberate and accidental introduction of alien species into the lake (Kornfield, 1982; Kornfield and Carpenter, 1984). The parallels with Lake Biwa and the portent for its future seem obvious.

Some of the factors affecting Lake Biwa are unlikely to improve, e.g. largemouth bass and bluegill populations and *Elodea* and *Egeria* populations are unlikely to recede, but other goals are attainable and not excessively Utopian, e.g. decreased pesticide and fertilizer runoff into the lake, implementation of an ecologically aware river and stream management policy (ideally one which includes such measures as habitat restoration, decreased pollution, controlled fishing quotas), and restoration of reedbeds and littoral and satellite lake areas. Similarly, Lake Biwa has an advantage over most other ancient lakes in that fishing is not essential: fish is not the sole source of protein for millions of lakeshore people, and no one will starve to death if fishing were controlled.

Taking such measures will require heightened ecological awareness and brave decision making on the part of the authorities who manage this ecological treasure, but unless this is done the prognosis for Lake Biwa seems poor. Perhaps those responsible for the well-being of the other ancient lakes of the world can take heed of the example of Lake Biwa, and act early and decisively to protect the unique biotas and resources of which they have charge.

ACKNOWLEDGEMENTS

It is a pleasure to thank Y. Rossiter for her support and for her assistance in translating several of the cited publications into English, Y. Kada for her perceptive comments on an earlier draft of this manuscript, Mrs N. Nakayama for locating some of the reference sources and G. Patterson for advice and assistance. N.F. Broccoli provided inspiration, provoked perspiration and invoked aspiration.

REFERENCES

Annandale, T.N. (1922). The macroscopic fauna of Lake Biwa. *Annot. Zool. Jpn* **10**, 127–153.

Bayless, J. and Smith, W.B. (1967). The effects of channelization upon the fish populations of lotic waters in eastern North Carolina. *Proc. Annu. Conf. Southeast Assoc. Game Fish Commiss.* **18,** 230–238.

Bishop, W.W. and Trendall, A.F. (1967). Erosion surface, tectonics, and volcanic activity in Uganda. *Q. J. Geol. Soc. Lond.* **122,** 385–420.

Burgess, C.F. (1985). The structural and stratigraphic evolution of Lake Tanganyika: a case study of continental rifting. Unpublished MSc Thesis, Duke University, USA.

Cohen, A.S., Soreghan, M.J. and Scholz, C.A. (1993). Estimating the age of formation of lakes: an example from Lake Tanganyika, East African Rift system. *Geology* **21,** 511–514.

Coulter, G.W. (1994). Lake Tanganyika. *Arch. Hydrobiol. Beiheft. Ergebnisse Limnol.* **44,** 13–18.

Cromie, W.J. (1982). Beneath the African lakes. Mosaic **13,** 2–6.

Davis, R.B. (ed.) (1990). *Paleolimnology and the Reconstruction of Ancient Environments.* Kluwer, Dordrecht.

Dumitrashko, N.V. (1952). The geomorphology and paleography of the Baikal Mountain Region. *Trudy Geogr. Inst. Acad. Sci. USSR* **55,** 3–189 (in Russian).

Edmonson, W.T. (1970). Phosphorus, nitrogen and algae in Lake Washington after diversion of sewage. *Science* **169,** 690–691.

Elwood, J.W., Newbold, J.D., O'Neill, R.V. and Winkle, W. van (1983). Resource spiralling: an operational paradigm for analyzing lotic ecosystems. In: *Dynamics of Lotic Ecosystems* (Ed. by T.D. Fontaine and S.M. Bartell). Ann Arbor Science Publishers, Ann Arbor, MI.

Emerson, J.W. (1971). Chanellization. A case study. *Science* **173,** 325–326.

Frey, D.G. (1969). A limnological reconnaisance of Lake Lanao. *Verh. Int. Verein. Limnol.* **17,** 1090–1102.

Fryer, G. (1959). The trophic interrelationships and ecology of some littoral communities of Lake Nyasa with especial reference to the fishes, and a discussion of the evolution of a group of rock-frequenting Cichlidae. *Proc. Zool. Soc. Lond.* **132,** 153–281.

Fryer, G. (1997). Biological implications of a suggested Lake Pleistocene desiccation of Lake Victoria. *Hydrobiologia* **354,** 177–182.

Gersberg, R.M., Elkins, B.W. and Goldman, C.R. (1983). Nitrogen removal in artificial wetlands. *Wat. Res.* **17,** 1009–1014.

Gorthner, A. (1994). What is an ancient lake? *Arch. Hydrobiol. Beiheft. Ergebnisse Limnol.* **44,** 97–100.

Greenwood, P.H. (1984). African cichlids and evolutionary theories. In: *Evolution of Fish Species Flocks* (Ed. by A.A. Echelle and I. Kornfield), pp. 141–154. University of Maine at Orono Press, Orono, ME.

Greenwood, P.H. (1994). Lake Victoria. *Arch. Hydrobiol. Beiheft. Ergebnisse Limnol.* **44,** 19–26.

Haberyan, K.A. and Hecky, R.E. (1987). The late Pleistocene and Holocene stratigraphy and paleolimnology of Lake Kivu and Tanganyika. *Palaeogeogr. Palaeoclimatol. Palaeoecol.* **61,** 169–197.

Hamabata, E. (1991). Studies of aquatic macrophyte communities in Lake Biwa (2). Distribution of species and communities – results from on-board surveys with sonars and sample collection. In: *Landscape and Environment of Shiga; Scientific Studies of Shiga Prefecture, Japan,* pp. 1295–1310. The Foundation of Nature Conservation in Shiga Prefecture, Otsu, Japan (in Japanese).

Hanafusa, Y. (2000). Wetlands of Lake Biwa: Their history, significance and fate. *Lakes Reservoirs, Res. Manage.* **5,** 1–3.

Horie, S. (ed.) (1984). *Lake Biwa. Monographiae Biologicae* Vol. 54. Dr W. Junk Publishers, The Hague.

Horie, S. (1987). *The History of Lake Biwa.* Kyoto University, Takashima, Japan.

Ide, M. (1996a). A preliminary summary of information on Japanese smelt in Lake Biwa. In: *A Report on Ecological Investigations of Fishes in Lake Biwa and its Rivers,* pp. 136–137. Shiga Prefecture Fisheries Experimental Station, Hikone, Japan (in Japanese).

Ide, M. (1996b). Catch records of non-native fishes in Lake Biwa. In: *A Report on Ecological Investigations of Fishes in Lake Biwa and its Rivers,* pp. 134–135. Shiga Prefecture Fisheries Experimental Station, Hikone, Japan (in Japanese).

Inouc, Y., Iwai, S., Ikeda, S. and Kunimatsu, T. (1981). Eutrophication of Lake Biwa – nutrient loading and ecological modelling. *Verh. Int. Verein. Limnol.* **21,** 248–255.

International Lake Environment Committee (ILEC). (1993). *Data Book of World Lake Environments,* Vol. 1. International Lake Environment Committee/United Nations Environment Programme, Kusatsu, Japan.

International Lake Environment Committee (ILEC). (1994). *Data Book of World Lake Environments,* Vol. 2. International Lake Environment Committee/United Nations Environment Programme, Kusatsu, Japan.

International Lake Environment Committee (ILEC). (1995). *Data Book of World Lake Environments,* Vol. 2. International Lake Environment Committee/United Nations Environment Programme, Kusatsu, Japan.

Isiida, N. (1986). Present status of toxic organic contaminants in the Lake Biwa/Yodo River system. In: *Toxic Organic Compounds in Inland Waters,* pp. 54–66. Lake Biwa Research Institute, Otsu, Japan (in Japanese).

Itihara, M. (1982). The bottom surface of Lake Biwa and its relation to the climax of the Rokko Movements. *Monogr. Assoc. Geol. Collab. Jpn* **24,** 229–233 (in Japanese).

Johnson, T.C. and Ng'ang'a, P. (1990). Reflections on a rift lake. *Am. Assoc. Petrol. Geol. Mem.* **50,** 113–135.

Johnson, T.C. and Odada, E.O. (eds) (1996). *The Limnology, Climatology and Paleoclimatology of the East African Lakes.* Gordon and Breach, Amsterdam.

Johnson, T.C., Scholtz, C.A., Talbot, M.R., Kelts, K., Ricketts, R.D., Ngobi, G., Beuning, K., Ssemmanda, I. and McGill, J.W. (1996). Late Pleistocene desiccation of Lake Victoria and rapid evolution of cichlid fishes. *Science* **273,** 1091–1093.

Jorgensen, S.E. and Vollenweider, R.A. (1988). *Guidelines of Lake Management.* International Lake Environment Committee and UNEP Publication, Otsu, Japan.

Kada, Y. (1999). Socio-ecological changes around Lake Biwa: how have the local people experienced the rapid modernization? In: *Ancient Lakes: Their Cultural and Biological Diversity* (Ed. by H. Kawanabe, G.W. Coulter and A.C. Roosevelt), pp. 243–260. Kenobi Productions, Ghent.

Kadlek, R.H. and Tilton, D.L. (1979). The use of freshwater wetlands as a tertiary wastewater treatment alternative. *Crit. Rev. Envir. Control* **9,** 185–212.

Kawabe, T. (1989). Stratigraphy of the lower part of the Kobiwako Group around the Ueno Basin, Kinki District, Japan. *J. Geosci. Osaka City Univ.* **32,** 39–90.

Kawabe, T. (1994). The history of Lake Biwa. In: *The Natural History of Lake Biwa* (Ed. by the Research Group for the Natural History of Lake Biwa), pp. 25–72. Yasaka Shobo, Tokyo (in Japanese).

Kawakatsu, M. and Nishino, M. (1993). A list of publications on turbellarians recorded from Lake Biwa-ko, Honshu, Japan. *Bull. Fuji Women's Coll.* **31,** 87–102.

Kawakatsu, M. and Nishino, M. (1994). A list of publications on turbellarians recorded from Lake Biwa-ko, Honshu, Japan. Addendum 1. A supplemental list of publications and a revision of the section Platyhelminthes in the papers by Mori (1970) and Mori and Miura (1980; 1990). *Bull. Fuji Women's Coll.* **32**, 87–103.

Kawanabe, H. (1978). Some biological problems. *Int. Assoc. Theor. Appl. Limnol.* **20**, 2674–2677.

Kawanabe, H. (1999). Biological and cultural diversities in Lake Biwa. In: *Ancient Lakes: Their Cultural and Biological Diversity* (Ed. by H. Kawanabe, G.W. Coulter and A.C. Roosevelt), pp. 17–41. Kenobi Productions, Ghent, Belgium.

Kendall, R.L. (1969). The ecological history of the Lake Victoria basin. *Ecol. Monogr.* **39**, 121–176.

Kobayakawa, M. and Okuyama, S. (1994). Catfish fossils from the sediments of ancient Lake Biwa. In: *Arch. Hydrobiol. Beiheft. Ergebnisse Limnol.* **44**, 425–431.

Kornfield, I. (1982). Report from Mindanao. *Copeia* **1982**, 493–495.

Kornfield I. and Carpenter, K.E. (1984). Cyprinids of Lake Lanao, Philippines: taxonomic validity, evolutionary rates and speciation scenarios. In: *Evolution of Fish Species Flocks* (Ed. by A.A. Echelle and I. Kornfield), pp. 69–84. University of Maine at Orono Press, Orono, Maine.

Lamakin, V.V. (1955). The Obruchev fault in the Baikal depression. *Voprosy Geol. Azii* **2**, 448–478.

Lever, C. (1996). *Naturalized Fishes of the World.* Academic Press, London.

Lewis, W.M. (1978). A compositional, phytogeographical and elementary structural analysis of the phytoplankton in a tropical lake: Lake Lanao, Philippines. *J. Ecol.* **66**, 213–266.

Livingstone, D.A. (1975). Late Quaternary climatic change in Africa. *Ann. Rev. Ecol. Syst.* **6**, 249–280.

Livingstone, D.A. (1980). Environmental change in the Nile headwaters. In: *Sahara and the Nile* (Ed. by M.A.J. Williams and H. Faure), pp. 339–359. Blakema, Rotterdam.

Logatchev, N.A. (1993). History and geodynamics of the Lake Baikal rift in the context of the East Siberia Rift system: a review. *Bull. Central Rech. Expl. Prod. Elf Aquitaine* **17**, 353–370.

Maehata, M. (1990). The continuing tale of black bass in Lake Biwa. *Conserv. Freshwat. Fishes* **3**, 125–128 (in Japanese).

Martens, K., Goddeeris, B.R. and Coulter, G.W. (eds) (1994). Speciation in ancient lakes. *Arch. Hydrobiol. Beiheft. Ergebnisse Limnol.* **44**, 508.

Martin, P. (1994). Lake Baikal. *Arch. Hydrobiol. Beiheft. Ergebnisse Limnol.* **44**, 3–11.

Martin, P. (1996). Oligochaeta and Aphanoneura in ancient lakes: a review. *Hydrobiologia* **334**, 63–72.

Mats, V.D. (1993). The structure and development of the Baikal rift depression. *Earth Sci. Rev.* **34**, 81–118.

Matsuoka, K. (1987). Malacofaunal succession in Pliocene to Pleistocene non-marine sediments in the Omi and Ueno Basins, Central Japan. *J. Earth Sci. Nagoya Univ.* **35**, 23–115.

Mayr, E. (1984). Evolution of fish species flocks: a commentary. In: *Evolution of Fish Species Flocks* (Ed. by A.A. Echelle and I. Kornfield), pp. 3–11. University of Maine at Orono Press, Orono, ME.

Meyers, P.A., Takemura, K. and Horie, S. (1993). Reinterpretation of late Quaternary sediment chronology of Lake Biwa, Japan, from correlation with marine glacial-interglacial cycles. *Quatern. Res.* **39**, 154–162.

Miura, T. (1971). Fishes of Lake Biwa. In: *A Report on Scientific Investigations in Lake Biwa Quasi-national Park*, pp. 313–330. Scientific Investigation Group of Lake Biwa Quasi-national Park (in Japanese).

Miura, T. (1980). Primary producers. In: *An Introduction to the limnology of Lake Biwa* (Ed. by S. Mori), pp. 44–46. Kyoto, Japan.

Mori, S. (1970). List of plant and animal species living in Lake Biwa. *Mem. Fac. Sci. Kyoto Univ. Ser. B* **3**, 22–46.

Mori, S. and Miura, T. (1980). List of plant and animal species living in Lake Biwa (revised). *Mem. Fac. Sci. Kyoto Univ. Ser. B* **8**, 1–33.

Mori, S. and Miura, T. (1990). List of plant and animal species living in Lake Biwa (corrected third edition). *Mem. Fac. Sci. Kyoto Univ. Ser. B* **14**, 14–22.

Morino, H. and Miyazaki, N. (1994). *Lake Baikal – Field Science of an Ancient Lake.* University of Tokyo Press, Tokyo.

Morley, C.K. (1988). Variable extension in Lake Tanganyika. *Tectonics* **7**, 785–801.

Moss, B. (1988). *Ecology of Freshwaters: Man and Medium.* Blackwell Scientific Publications, Oxford.

Moyle, P.B. (1993). *Fish. An Enthusiast's Guide.* University of California Press, Berkeley, CA.

Myers, G.S. (1951). Fresh-water fishes and East Indian zoogeography. *Stanford Icthyol. Bull.* **4**, 11–21.

Myers, G.S. (1960). The endemic fish fauna of Lake Lanao, and the evolution of higher taxonomic categories. *Evolution* **14**, 323–333.

Nakai, K. (1999). Recent faunal changes in Lake Biwa, with particular reference to the bass fishing boom in Japan. In: *Ancient Lakes: Their Cultural and Biological Diversity* (Ed. by H. Kawanabe G.W. Coulter and A.C. Roosevelt), pp. 227–241. Kenobi Productions, Ghent.

Nakajima, T. (1986). Fossil pharyngeal teeth of cyprinids from the Kusuhara Formation of the Iga Group – a comparison to the cyprinid fauna of the Kobiwako Group. *Bull. Mizunami Fossil Mus.* **13**, 105–114.

Nakajima, T. (1994). Succession of cyprinid fauna in Paleo-lake Biwa. In: *Arch. Hydrobiol. Beiheft. Ergebnisse Limnol.* **44**, 433–439.

Nakajima, T. and Nakai, K. (1994). Lake Biwa. In: *Arch. Hydrobiol. Beiheft. Ergebnisse Limnol.* **44**, 43–54.

Nakanishi, M., Narita, T., Mitamura, O., Suzuki, N. and Okamoto, K. (1986). Horizontal distribution and seasonal change of chlorophyll a concentrations in the north basin of Lake Biwa. *Jpn. J. Limnol.* **47**, 155–164.

Nishino, M. (ed.) (1991). *Handbook of Zoobenthos in Lake Biwa*, Vol. I, *Molluscs.* Lake Biwa Research Institute, Otsu (in Japanese).

Nishino, M. (ed.) (1992). *Handbook of Zoobenthos in Lake Biwa*, Vol. II, *Aquatic Insects.* Lake Biwa Research Institute, Otsu (in Japanese).

Nishino, M. (ed.) (1993). *Handbook of Zoobenthos in Lake Biwa*, Vol. III, *Porifera, Platyhelminthes, Annelida, Tenticulata and Crustacea.* Lake Biwa Research Institute, Otsu (in Japanese).

Nishino, M., Fukui, M. and Nakajima, T. (1998). Dense mats of *Thioploca*, gliding filamentous sulfur-oxidising bacteria in Lake Biwa, central Japan. *Wat. Res.* **32**, 953–957.

Noakes, D.L.G. (1998). A new perspective on lakes: Kawanabe's latest achievements. *Environ. Biol. Fishes* **52**, 391–394.

Novitski, R.P. (1978). Hydrologic characteristics of Wisconsin's wetlands and their influence on floods, stream flow and sediment. *Wetland Functions and Values: The State of our Understanding.* American Water Resources Association, Minneapolis, MN.

Ogawa, S., Kawanabe, H. and Kada, Y. (1997). Change of fisheries and other things in Oki Island. *Umido* **3**, 2–3 (in Japanese).

Patterson, M.B. (1983). Structure and acoustic stratigraphy of the Lake Tanganyika Rift Valley: a single-channel seismic survey of the Lake north of Kalemie, Zaire. Unpublished MSc Thesis, Duke University, USA.

Research Group for the Natural History of Lake Biwa. (1986). Freshwater fossil assemblages from the Pleistocene Kobiwako Group on the southwest side of Lake Biwa. *Bull. Mizunami Fossil Mus.* **13**, 55–100 (in Japanese).

Research Group for the Natural History of Lake Biwa. (1994). *The Natural History of Lake Biwa.* Yasaka Shobo, Tokyo (in Japanese).

Ribbink, A.J. (1994). Lake Malawi. In: *Arch. Hydrobiol. Beiheft. Ergebnisse Limnol.* **44**, 27–33.

Sarkisyan, S.G. (1955). *Baikal.* Geografizdat, Moscow (in Russian).

Sherbakov, S.Y. (1999). Molecular phylogenetic studies on the origin of biodiversity in Lake Baikal. *Trends Ecol. Evol.* **14**, 92–95.

Shiga Prefectural Government (1983). *Lake Biwa Comprehensive Development Project.* Shiga Prefectural Government, Otsu, Japan.

Shiga Prefectural Government. (1997). *Lake Biwa. Conservation of Aquatic Environments.* Shiga Prefectural Government, Otsu, Japan.

Stager, J.C., Reinthal, P.N. and Livingstone, D.A. (1986). A 25,000 year history for Lake Victoria, East Africa, and some comments on its significance for the evolution of cichlid fishes. *Freshwat. Biol.* **16**, 15–19.

Takahashi, S. (1990). The first record of the goby *Tridentiger kuroiwae brevispinis* in Lake Biwa. *Annu. Rep. Lake Biwa Cult. Cent.* **8**, 7 (in Japanese).

Tanaka, M. and Matsuoka, K. (1985). Pliocene freshwater diatoms from the Koka and Ayama areas in Shiga Prefecture, Central Japan. *Bull. Mizunami Fossil Mus.* **12**, 57–70 (in Japanese).

Tanaka, M., Matsuoka, K. and Takagi, Y. (1984). The genus *Melosira* (Bacillariophyceae) from the Pliocene Iga Formation of the Kobiwako Group in Mie Prefecture, Central Japan. *Bull. Mizunami Fossil Mus.* **11**, 55–68.

Temple, P.H. (1969). Some biological implications of a revised geological history for Lake Victoria. *Biol. J. Linn. Soc. Lond.* **1**, 363–371.

Terashima, A. (1980). Bluegill, occupying an empty niche in Lake Biwa. In: *Japanese Freshwater Organisms: The Ecology of Invasion and Disturbance* (Ed. by H. Kawanabe and N. Mizuno), pp. 63–70. Tokai University Press, Tokyo (in Japanese).

Tiercelin, J.-J. and Mondeguer, A. (1991). The geology of the Tanganyika trough. In: *Lake Tanganyika and its Life* (Ed. by G.W. Coulter), pp. 7–48. Oxford University Press, Oxford.

Ueda, Y. (1999). Shiga Government fighting to rescue "dying" Lake Biwa. *Yomiuri Shimbun*, **14**.

Ward, J.V. (1988). Riverine–wetland interactions. In: *Freshwater Wetlands and Wildlife* (Ed. by R.R. Scharitz and J.W. Gibbons). US Department of Energy, Oak Ridge, TN.

Wood, C.E. and Wood, J.C. (1963). A monograph of the fishes of Lake Lanao. Unpublished manuscript, cited in Kornfied and Carpenter (1984).

Yokoyama, T. (1984). Stratigraphy of the Quaternary systema round Lake Biwa and geohistory of the ancient Lake Biwa. In: *Lake Biwa* (Ed. by S. Horie), pp. 43–138. Dr W. Junk Publishers, The Hague.

Long-term Changes in Lake Balaton and its Fish Populations

P. BÍRÓ

I. SUMMARY

Ancient lakes harbour unique biotas and offer biologists unrivalled opportunities for studies of ecology and evolution. They are often referred to as "natural laboratories", yet for many ancient lakes and for many faunal groups meaningful research is still in its infancy. The aim of this chapter is to illustrate, through the example of a "modern" lake, the Hungarian Lake Balaton, how detailed and long-term studies are fundamental to achieving an understanding of both physical and biotic processes in a lake's ecosystem. Long-term biotic changes resulting from environmental perturbations are highlighted and examples given of how the fish biota has responded to human-induced changes in this highly eutrophic shallow lake. It is hoped that such long-term research perspectives will also be applied to the ancient lakes of the world.

   A comprehensive and systematic study of Lake Balaton began in the late nineteenth century. Changes in the water (environment) quality of Lake Balaton have been monitored using research data on geological history, limnology and water chemistry, phytoplankton and primary productivity. Studies on the zooplankton and benthic invertebrate communities have been conducted, and fish population dynamics, stock-recruitment relationships and trophic interactions have been widely documented. Together, these studies have revealed profound anthropogenic changes in the ecology of the lake since the 1960s. Furthermore, reports on the distribution, introductions and immigration of several fish species showed human-induced changes in species

ADVANCES IN ECOLOGICAL RESEARCH VOL. 31
ISBN 0–12–013931–6

diversity and demonstrated the ecologically deleterious effects of dredging of wetlands. More recently, restoration measures implemented since the late 1970s have begun to take effect, and recent years have seen unusually low algal and benthic biomass values.

II.   HISTORICAL AND ANTHROPOGENIC CHANGES IN LAKE BALATON

A. Geological History and Climate

Geologically, Lake Balaton is a relatively young glacial formation, and pollen stratigraphic analyses of sediments indicate the present bed profile to have existed for some 10 500–12 000 years (Zólyomi, 1995). Lake Balaton clearly does not qualify as an ancient lake (> 100 000 years old; Gorthner, 1994), but is included in this volume for two reasons: (i) as an exemplar of detailed, comprehensive and long-term research, studies of Lake Balaton illustrate the benefits and insights obtained through such a research perspective; and (ii) as an example of the deleterious and complex effects of perturbation on a lake ecosystem, and of remedial measures to counter them, Lake Balaton studies can serve as a surrogate from which to extrapolate the potential dangers from anthropogenic factors facing ancient lakes today.

Lake Balaton (> 300 000 km^2) is located in the Carpathian basin, a closed oval-shaped area approximately 1000 km long, which lies in central Europe (44°30′–49°30′ N, 15°–26°30′ E) (Figure 1). Historical changes in the area are summarized in Table 1. The Carpathian basin is the buffer zone between three European climate areas: the Atlantic, the Mediterranean and the Continental. The annual mean air temperature of the area is about 10°C. The lake is normally covered by ice by the last week of December. Total snow cover is common, but all snow has usually melted by mid-March. Annual precipitation varies between 600 and 700 mm (Baranyi, 1975). Since the early 1960s, both ice-on and ice-off dates have become later, and ice-duration periods have become shorter, showing some relationship to El Niño years. This ice-phenology pattern may be related to global climatic changes, but its impact on the lake's biota is not yet known.

B. Limnology, Cultural History and Eutrophication

The eutrophic–hypertrophic Lake Balaton is one of the largest and probably most studied shallow lakes (mean depth > 3.0 m) in central Europe (Figure 1). Despite having a large surface area of 593 km^2 the lake's volume is only 1.8 km^3. Historical records (1900–1998) of algal (> 2000 species), invertebrate (> 1200 species) and fish communities (47 native and introduced species), and the abundance of individual taxa, illustrate dramatic changes in the ecology of

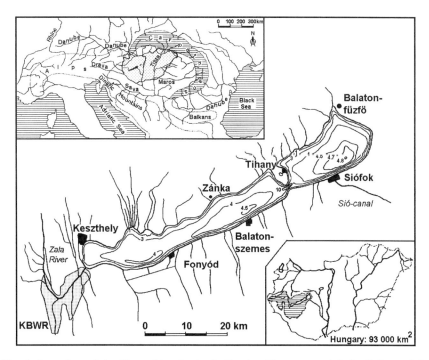

Fig. 1. Location of the Carpathian basin, indicating Hungary and Lake Balaton (after Bíró, 1984) (top left) and the present bed profile of Lake Balaton, the position of Kis-Balaton water reservoir (KBWR) (centre) and the drainage area of the lake (lower right).

the lake. From about 1860 water volume began to decline as the Sió canal, shores and water levels were regulated, and wetlands and flooded areas were disconnected (Tölg, 1961; Bíró, 1984) (Figures 2 and 3). Early signs of human-induced eutrophication were already apparent in the 1930s, but it is only since the 1970s that the lake's biota has become heavily affected by anthropogenic activities (Table 2). In the 1950s, there was a rapid increase in the application of fertilizers, insecticides and pesticides in the agricultural surroundings of the lake (Sebestyén, 1967). Through runoff, seepage and entry into the ground-water system, these compounds have entered the lake ecosystem. Several cyanobacterial blooms in the lake, and concomitant massive fishkills and alterations of the nutrient supply and budget, have resulted (Bíró, 1984; Bíró and Vörös, 1990; Padisák and Reynolds, 1998).

The water of Lake Balaton is highly mineralized and has a high buffering capacity (pH 7.8–9.0) (Bíró, 1984). Chemical characteristics include unequal distributions of chlorophyll *a*, organic carbon, total phosphorus (TP) and nitrogen (TN) along the longitudinal axis of the lake. During 1975–1982, the total annual phosphorus load was estimated at 314 t and that of nitrogen at

Table 1
Historical changes in Lake Balaton and its surroundings

Geological era (period)	Time (years bp)	Formation of the area
Late Pliocene:		The shallow Pannonian lake covers the area
Levantine epoch	1.5–2 My	Pannonian lake becomes silted up
Pliocene–Pleistocene	1 My to 700 000 y	Ancient River Danube meanders along the south-western edge of the Pliocene–Pleistocene tectonic depression
Pleistocene:		
"Danube" or Günz glacial stage	650–600 000 y	Translocation of River Danube, appearance of Zala River
Mindel	450 000 y	Emergence of Bakony Mountains, formation of Balaton uplands
Riss (dry, periglacial period)	230 000 y	Crustal movements, alterations in the shape, extension and location of ancient Lake Balaton
Würm glacial stage	115 000 y	Greatest extension of Lake Balaton
Early Holocene	10 500–12 000 y	Silting, crustal movements: formation of the present bed profile of Lake Balaton

Data based on Hankó (1931), Mike (1976), Rónai (1969), Zólyomi (1995).

3148 t. Annually, the Zala River transported 84 t of TP and 916 t of TN into the lake. The phosphorus content of the lake water from the north-east to south-west basins varied from 30 to 95 mg m^{-3} and the nitrogen content from 761 to 1653 mg m^{-3}, respectively (Herodek, 1988).

Bacteria may play an important role in phosphorus cycling within the lake sediment (25%) and at the sediment–water interface (Istvánovics, 1993). In the 1970s the average turnover time of PO_4–P was 2.5 days, but in recent years this has decreased to only a few minutes. In the winter of 1974–1975, an intensive bloom of *Nitzschia acicularis* W. Smith developed and comprised over 50% of the extremely high total algal biomass (9.8 g m^{-3}) (Herodek, 1977). The result was a massive fishkill (*c.* 70 t). The first large invasion of the alga *Cylindrospermopsis* (*Anabaenopsis*) *raciborskii* (Wolosz.) Seenayya and Subba Raju occurred in 1982, and was followed in the 1990s by an increase in both the relative and absolute importance of nitrogen-fixing cyanobacteria. After the second half of August 1994, an explosive bloom of *Cylindrospermopsis raciborskii* covered the entire lake surface (biomass 70 mg m^{-3}; L. Vörös, pers. comm.), endangering both fish populations and human health. Its toxic effect

Fig. 2. Initial lakes and main tectonic lines around Lake Balaton (Rónai, 1969) (top), and the ancient flood-prone and marshy areas prior to dredging (Erdös, 1898, modified from Tölg, 1961) (bottom). The "original" Kis Balaton covered an area several times larger than that of the reservoir under reconstruction (see Figure 1).

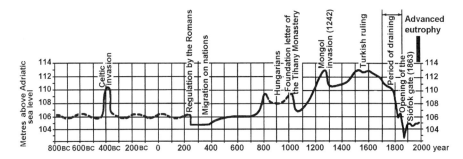

Fig. 3. Historical changes in the water level of Lake Balaton (after Bendefy and Nagy, 1969).

resulted in an inhibition of metabolism and liver functions in the majority of cyprinids (Bíró, 1997).

C. Variations in Fish Populations and Their Biomass

Regulations of water level and the disconnection (draining) of the marshy areas (Nagyberek, Kis-Balaton) started in the nineteenth century and reduced the area and the diversity of habitats of the lake (Figure 2) (Tölg, 1961). Later, shoreline building resulted in huge losses of natural spawning grounds and

Table 2

Anthropogenic changes in Lake Balaton based on literature records

Period	Cultural effects
1950s to 1960s	Intensified agriculture on the drainage area, shoreline regulations and building resulted in progressive eutrophication and the first massive fishkill, due to chlorinated hydrocarbons, in 1965 (500 t)
1970s to 1980s	Two- to eight-fold increase in primary production from 100 to 830 g C m^{-2} y^{-1}, regular algal blooms, second massive fishkill in 1975 (70 t), thereafter subsequent fishkills. High biomass of macrobenthos (10–110 g m^{-2}) in 1982–1983
1980s to present	Lake-management initiatives, restoration of water quality (P removal from treated sewage; diversion of treated sewage, reduction of external and internal P–loading, restoration of Kis-Balaton reservoir (1985–); appearance of toxic cyanoprokaryotes (1982–); first kill of eel (1991), second massive mortality of eel (1995)
1995 to 1999	Low algal biomass (< 5–10 mg l^{-1}), drop in biomass of macrobenthic communities (from 16.1 to 2.4 g m^{-2} during 1995–1997)

nursery areas. The parallels here with the effects of shoreline modification of ancient Lake Biwa are striking (see Kawanabe, 1996; Rossiter, this volume).

Since the early twentieth century, a total of 47 fish species has been recorded in the lake and its drainage area (Répássy, 1909). However, only 31 species have been recorded during the last three decades. During the twentieth century 13 non-native species have become established. Although 20–24 species are common in the lake, the commercial fish catch comprises only 15–17 species (Bíró, 1994). The lake is dominated by cyprinids: bream *Abramis brama* L., Danubian bream *A. sapa* Pallas, blue bream or zope *A. ballerus* L., silver bream *Blicca bjoerkna* L., roach *Rutilus rutilus* L., rudd *Scardinius erythrophthalmus* L., bleak *Alburnus alburnus* L. and common carp *Cyprinus carpio* L. The dominant predators are the zander or pikeperch *Stizostedion lucioperca* L., Volga zander *S. volgensis* Gmelin, asp *Aspius aspius* L. and wels catfish *Silurus glanis* L. Species rare or believed to have been extirpated in the lake, e.g. the introduced black bass *Micropterus salmoides* Lacepède and the native perch *Perca fluviatilis* L., disappeared from the lake during the 1910s and 1970s, respectively. However, self-sustaining stocks currently persist in northern and southern streams and in the reconstructed Kis-Balaton water reservoir (Przybylski *et al.*, 1991; Bíró and Paulovits, 1994). Following the impoundment of the Kis-Balaton reservoir in 1985, the goby *Neogobius fluviatilis* Pallas, first recorded in 1970 (Bíró, 1972), and the gibel or Prussian carp *Carassius auratus gibelio* Bloch have invaded the entire littoral zone (Bíró and Paulovits, 1994). The common eel *Anguilla anguilla* L. and silver carp *Hypophthalmichthys molitrix* Cuv. et Val. have been stocked annually since 1961 and 1972, respectively, and both have reached fairly high densities (Bíró, 1992, 1994, 1997).

In the early twentieth century (1902–1920), the commercial catch varied between 500 and 800 t y^{-1} (Répássy, 1909; Hankó, 1931). During the 1950s, this catch increased to 1960 t y^{-1}, but it decreased in the 1960s for a variety of reasons: mass mortalities, progressive eutrophication, introductions, invasions, disappearance, competition and the balanced/unbalanced nature of populations (Bíró, 1994) (Figure 4). During the 1970s, a *c.* 50% reduction in fishing intensity (selectivity) further decreased the catch. In contrast, from 1950 to 1988, recreational or sport fishing catches have increased from 50 to 550 t y^{-1} (Bíró, 1977, 1997).

The density and biomass of fish populations (70–300 kg ha^{-1}) seem to reflect the trophic gradient along the major longitudinal axis of the lake, e.g. the biomass of fish is four to six times higher in the western than in the eastern basin (Bíró and Vörös, 1990). Catches decreased from 1972 to 1992 in different fishing areas. However, since the mid-1980s increased catches in the north-east basin have occurred as a result of intensive gillnetting of silver carp and eel trapping. Between 1980 and 1995, exotic species (silver carp, crucian carp, eel, bighead carp *Aristichthys nobilis* Rich., grass carp *Ctenopharyngodon idella* Val.) comprised 60% of the annual commercial catch (Figure 5). However, following the mass mortality of eels in 1991 (due to hyperinfestation by the

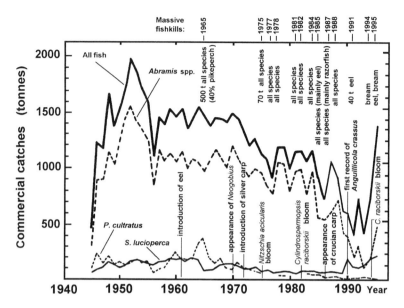

Fig. 4. Annual commercial catches in Lake Balaton during 1945–1995 (modified from Bíró, 1997). Massive fishkills recorded between 1965 and 1995, either for all or for several species, are indicated in the upper section. The appearance or introduction of certain species and algal blooms is indicated in the lower section.

parasite *Anguillicola crassus* Kuwahara, Niimi et Itagaki), introduction of eels was prohibited and the stock has since been fished to extinction (Bíró, 1992).

Analysis of the commercial fish catches (kg ha^{-1} y^{-1}) (1980–1995) in different areas reveals an overall fluctuation and decrease of native species. Between 1980 and 1985, total annual catches increased from 85 to 160 kg ha^{-1}, then decreased by 50–75% between 1985 and 1995 (Figure 5).

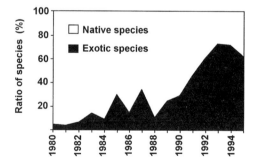

Fig. 5. Relative proportion, by weight, of native and exotic fish species in the commercial catches in Lake Balaton between 1980 and 1995.

The four breams, *Abramis brama, A. sapa, A. ballerus* and *B. bjoerkna*, are the dominant fish species showing generalized or mixed feeding patterns (Specziár *et al.*, 1997). Total annual catches in the period 1980–1995 varied from 1.5 to 16.1 kg ha^{-1} y^{-1}, but the overall trend has been a decrease of approximately 0.9 kg ha^{-1} y^{-1}. Bream stocks increased with eutrophication and reduced fishing effort, and since the 1980s the relative contribution of older age groups (over 6 years) to population composition has increased by 10–15%. However, because the stock increased in both body size and density, numbers subsequently decreased and the body size became negatively density dependent (Figure 6) (Bíró and Garádi, 1974; Dauba and Bíró, 1992). In the summer of 1994, cyanobacteria toxins caused a massive mortality of mature breams, indicating the need for increased fishing selectivity for the removal of older age groups. During 1973–1983, the estimated biomass of breams increased from 160 to 180 kg ha^{-1} and started to oscillate thereafter. The ratio of production to biomass (P/B) decreased from 72.6% to 70.5%, and both the coefficients of growth in weight (from 0.47 to 0.25) and total mortality (from 0.99 to 1.04) exhibited significant changes (Bíró, 1997).

Common carp, the most specialized feeder, has also been stocked for angling purposes, and maximum commercial catches occur in cyclic periods of 3–5 years ($<$ 0.5 kg ha^{-1}), but the long-term trend shows a decrease. Tench (*Tinca tinca* L.) prefers the highly eutrophic parts of the lake, but yields very low catches of only 0.02–0.002 kg ha^{-1}. During the late 1980s and early 1990s, this species practically disappeared from catches. Yields of asp *Aspius aspius* L., an open-water predator, showed great fluctuations ($<$ 0.43 kg ha^{-1}), with an overall decreasing trend. In the 1950s and 1960s, total annual commercial catches of the pelagic ziege or razorfish *Pelecus cultratus* L. varied from 100 to 150 t y^{-1}, but from 1980 to 1995 these values decreased to $<$ 2.8 kg ha^{-1} y^{-1}. In the period 1970–1980, the stock of ziege fell to 10% of its former levels, but recovered slowly in the years thereafter. Its growth rate first increased during the 1970s and then decreased during the 1980s (Figure 6) (Bíró, 1982, 1997).

The bleak is the most important prey for fish predators. It feeds mainly on periphytic crustaceans. The age structure of this species also shows changes and the annual mortality of the stock varies from 38% to 72%. In contrast to the dramatic and continuous decrease in its stock size, its growth rate has become very slow (Figure 6) (Bíró, 1990a; Bíró, and Muskó, 1995). An assessed biomass of 36.7–59.4 kg ha^{-1} in 1986 had, by 1991–1992, decreased to 4.8–12.4 kg ha^{-1} (Bíró, 1997).

The pikeperch *Stizostedion lucioperca* responded sensitively to eutrophication and reduced fishing intensity. In the early 1980s, the yield of pikeperch varied between 0.4 and 2.0 kg ha^{-1}, but these values dropped to 0.01–0.6 kg ha^{-1} in the 1990s, with an annual decrease of 0.04 kg ha^{-1} y^{-1}. Maximum catches of pikeperch seem to occur in 6-year cyclic periods. Between 1970 and 1980, the pikeperch stock decreased to only 15–50% of its previous norm. Numbers recovered slowly in the years thereafter; however, the population is a

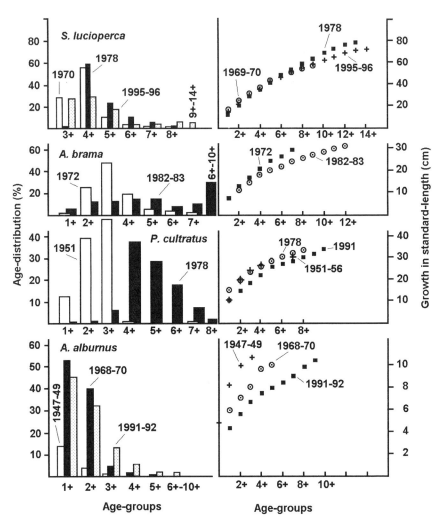

Fig. 6. Variations in the age distribution (left) and annual growth in standard length (right) of pikeperch, common bream, razor fish and bleak in Lake Balaton over a long time scale. Year and source of data for pikeperch: 1970 (Bíró, 1970), 1978 (Bíró, 1985), 1995–1996 (Bíró, unpubl.); common bream: 1972 (Bíró and Garádi, 1974), 1982–1983 (Dauba and Bíró, 1992); razorfish: 1951–1956 (Entz and Lukacsovics, 1957), 1978 (Bíró, 1982), 1991 (Perényi, 1991); bleak: 1947–1949 (Entz, 1950), 1968–1970 (Bíró, 1975), 1991–1992 (Bíró and Muskó, 1995).

self-regulated one (through cannibalism). In the late 1970s, the number of fishes in older age groups increased in the south-west basin and, owing to lower ambient population density, individual growth rates (length and weight) increased, but these have declined again recently (Figure 6) (Bíró, 1985, 1990a, b). In 1973, the average biomass of pikeperch was estimated at 9.7 kg ha^{-1}, but this had decreased to 1.6–4.9 kg ha^{-1} by 1984. During the 1980s, average annual mortality varied from 30% to 71% (Bíró, 1997). Based on its stock-recruitment relationship, the pikeperch population is inherently oscillatory. Volga pikeperch *S. volgensis*, pike *Esox lucius* and wels catfish *Silurus glanis* are also captured in low quantities (0.01–0.3 kg ha^{-1} y^{-1}).

The introduced silver carp inhabits mainly the north-east and central basins and has a moderately high standing crop in the lake (specimens of 25–30 kg body weight are common). Catch data indicate a 0.07 kg ha^{-1} increase in biomass per year. Bighead carp, introduced accidentally in the early 1960s, has an insignificant stock ($<$ 0.2 kg ha^{-1}). Grass carp, also introduced accidentally in the 1960s, produced relatively low but slowly increasing yields ($<$ 0.07 kg ha^{-1} y^{-1}). Stocking of eel was done annually from 1961 to 1991, but was prohibited after the mass mortality of eels which occurred in 1991 (Bíró, 1992). An intensified commercial exploitation of the stock (10–20 kg ha^{-1} y^{-1}) in 1993–1994 resulted in a sharply increasing catch (0.3 kg ha^{-1} y^{-1}). The population density of crucian carp *Carassius carassius* L. is low. Giebel, or Prussian carp, appeared in the lake in 1986, following the impoundment of the Kis-Balaton reservoir. These two carp species are mainly caught by the eel traps at the mouth of the Sió canal.

Since the 1970s the age structure and growth rates of several fish species have varied greatly. These changes reflect the effects of environmental perturbations (massive mortalities due to pesticides, parasites and cyanobacteria toxins), selective or non-selective fisheries practices and variations in the stock-recruitment relationships.

D. Trophic Relationships

Cultural eutrophication of the shallow Lake Balaton since the 1960s has resulted in profound changes in both environmental conditions and fish communities. From the 1960s to the 1990s, a two- to eightfold increase in primary production (from 100 to 182–830 g C m^{-2} y^{-1}) has provided more favourable conditions for the common cyprinids. Primary production of the lake increased substantially from 7.2–17.1 kJ m^{-2} d^{-1} (open water, central basin, 1960) to 7615–6456 kJ m^{-2} y^{-1} (open water, north-east basin, 1970–1990), or to 33 890–30 264 kJ m^{-2} y^{-1} (open water, south-west basin, 1970–1990) (Herodek, 1988; G.-Tóth, 1992). If these figures are considered as representing 100% of total production, the secondary production (plankton plus benthos) in the open water from north-east to south-west basins (170–400 kJ m^{-2} y^{-1}; Ponyi, 1992) transforms about 1.0–2.3% of primary energy. In the littoral zone, the production of zooplankton from north-east to south-west varied from 61.8 to 618.0 kJ m^{-2} y^{-1} and the transfer efficiency from 0.8% to 1.7% (Simonian *et al.*, 1995). Along the longitudinal axis of the lake the efficiency of energy transfer to fish varies between 0.05% and 0.1% in open water (Bíró and Vörös, 1990) and between 0.1 and 0.4% in the littoral zone (Bíró, 1984; Simonian et al., 1995).

Comparative studies on the trophic relationships of periphyton–zooplankton–fish and periphyton–zoobenthos–fish revealed significant differences between the littoral zone and the open water areas. The periphyton biomass (dry weight/substratum area) ranged from 4.1 to 36.4 g m^{-2}, the zooplankton biomass from 0.5 to 1.9 mg wet weight l^{-1} mo^{-1} and the fish biomass from 71.6 to 90.4 kg ha^{-1} (Simonian et al., 1995). The biomass of zoobenthos varied from 0.1 to 20 g m^{-2} wet weight. Measured as kilojoules, biomass and production values suggest that the transfer of energy from primary producers to fish is more efficient in the littoral zone than in the open-water areas.

Trophic relationships (both planktonic and benthic foodwebs) respond sensitively to environmental and fish community changes. In the littoral zone, a significant amount of energy is channelled through the periphyton–zooplankton/zoobenthos–bleak–pikeperch food chain (Bíró, 1995a, b, 1997). As a rough estimate, from April to October the bleak stock (4.8–12.4 kg ha^{-1}) consumes about 223–297 kJ ha^{-1} in the north-east basin and about 1236–1649 kJ ha^{-1} in the south-west basin (4.5–33.0 kJ m^{-2}). In the north-east basin 11% of energy may originate from zooplankton and c. 89% from benthos. In the central area of the lake, equal amounts (c. 50%) of energy may be channelled through periphytic–zooplanktonic and benthic foodwebs; in the south-west basin these figures are about 14% and 86%, respectively.

Since pikeperch, the top predator in Lake Balaton, consumes about 2.2–6.7 kJ m^{-2} y^{-1} and its food consists of bleak (c. 90–98%), it seems evident that bleak has an important role in energy mediation through the littoral foodwebs linked to the periphytic–planktonic–benthic communities (Bíró and Muskó, 1995). However, the energy dissipates significantly in both the littoral and the open-water areas, and the energy flowing through benthos and periphyton seems to be more important than that of the plankton community (Bíró and Vörös, 1990). The various bream species consume about 93–141 kJ m^{-2} y^{-1} of food energy, contributing about 2–10% of food energy to the pikeperch (Bíró and Vörös, 1990).

III. MANAGEMENT OF WATER QUALITY OF LAKE BALATON

A wide-scale programme for the restoration of targeted mesotrophic water quality, as well as engineering and pollution control on Lake Balaton, was begun in the late 1970s (Herodek, 1988). The main steps include the radical reduction of external nutrient loading from agricultural and inhabited areas, the reconstruction of the Kis-Balaton water quality protection reservoir (213.7 km^2) for phosphorus retention, the construction of a 190 km long sewage transfer pipeline around the lake, and phosphorus removal (by precipitation) from the treated sewage. The internal phosphorus storage in the sediments (Total phosphorus 500–841 μg g^{-1} dry sediment) will probably be sufficient to supply the primary producers with this limiting nutrient for many years (Herodek and Istvánovics, 1986; Istánovics, 1993). Regulation of tourism and water use, as well as the development of infrastructure in densely inhabited shorelines and drainage areas, will further contribute to the effective rehabilitation of the lake and its biota. Management strategies for the rehabilitation of fish populations include habitat restoration, water-quality improvement, regulation of fisheries (both sport and commercial) and increased stocking of native predators. All of these measures can also produce concomitant and unexpected changes in the biota.

During 1997 and 1998, an unusually low algal biomass (5–10 mg l^{-1}) was present throughout the lake, the characteristic longitudinal trophic gradient disappeared and, consequently, both primary and benthic productivity decreased significantly. As a long-term trend, trophic differences between eastern and western lake basins have started to equilibrate in a way which corresponds to the supposition that most of the external phosphorus load to the eastern areas of the lake emanates from the more eutrophic western basin (Padisák and Reynolds, 1998). This peculiar behaviour of Lake Balaton has never been described before, and may result in the formation of new foodwebs and fish species assemblages. Clearly, even after a century of study, much remains to be learned from Lake Balaton.

It is only through comprehensive and long-term studies that we have begun to unravel and understand some of the interplay among trophic relations, community interactions, and population dynamics and their effects on the ecosystem of this lake. Detailed studies of this nature should also be conducted on all ancient lakes, to understand fully the complex interactions within each of these unique habitats.

ACKNOWLEDGEMENTS

These studies were supported financially by the National Research Fund (OTKA) Project no. T016012 and by the Balaton Secretariat of the Prime Minester's Office.

REFERENCES

Baranyi, S. (1975). A Balaton hidrológiai jellemzői 1921–70. *Vízgazdálkodási Tudományos Kutató Intézet, Tanulmányok Kutatási Eredmények* **45,** 1–140.
Bendefy, L. and Nagy, I. (1969). *Historical Changes of the Shoreline of Lake Balaton.* Műszaki Könyvkiadó, Budapest (in Hungarian).
Bíró, P. (1970). Investigation of growth of pike-perch (*Lucioperca lucioperca* L.) in Lake Balaton. *Ann. Inst. Biol. (Tihany) Hung. Acad. Sci.* **37,** 145–164.

Bíró, P. (1972). *Neogobius fluviatilis* in Lake Balaton–a Ponto-Caspian goby new to the fauna of central Europe. *J. Fish Biol.* **4**, 249–255.

Bíró, P. (1975). The growth of bleak (*Alburnus alburnus* L.) (Pisces, Cyprinidae) in Lake Balaton and the assessment of mortality and production rate. *Ann. Inst. Biol. (Tihany) Hung. Acad. Sci.* **42**, 139–156.

Bíró, P. (1977). Effects of exploitation, introductions, and eutrophication on percids in Lake Balaton. *J. Fish. Res. Brd Can.* **34**, 1678–1683.

Bíró, P. (1982). Growth, mortality, P/B-ratio and yield of ziege (*Pelecus cultratus* L.) in Lake Balaton. *Aquacult. Hung. (Szarvas)* **3**, 181–200.

Bíró, P. (1984). Lake Balaton: a shallow Pannonian water in the Carpathian Basin. In: *Lakes and Reservoirs. Ecosystems of the World 23* (Ed. by F.B. Taub), pp. 231–245. Elsevier, Amsterdam.

Bíró, P. (1985). Dynamics of pike-perch, *Stizostedion lucioperca* (L.) in Lake Balaton. *Int. Rev. Gesamten Hydrobiol.* **70**, 471–490.

Bíró, P. (1990a). Population structure, growth, P/B-ratio and egg-production of bleak (*Alburnus alburnus* L.) in Lake Balaton. *Aquacult. Hung. (Szarvas)* **6**, 105–118.

Bíró, P. (1990b). Population parameters and yield-per-recruit estimates for pikeperch (*Stizostedion lucioperca* L.) in Lake Balaton. In: *Management of Freshwater Fisheries* (Ed. by W.L.T. van Densen, B. Steinmetz and R.H. Hughes), pp. 248–261. Pudoc, Wageningen.

Bíró, P. (1992). Die Geschichte des Alas (*Anguilla anguilla* L.) im Plattensee (Balaton). *Österreichs Fisch.* **45**, 197–207 (in German).

Bíró, P. (1994). Fish production of Lake Balaton–past, present and future. *Halászat (Tudomány)* **87**, 180–186 (in Hungarian).

Bíró, P. (1995a). Management of the pond ecosystems and trophic webs. *Aquaculture* **129**, 373–386.

Bíró, P. (1995b). Management of eutrophication of lakes to enhance fish production. In: *Guidelines of Lake Management*, Vol. 7, *Biomanipulation in Lakes and Reservoirs Management* (Ed. by R. de Bernardi and G. Giussani), pp. 81–96. ILEC/UNEP, Otsu, Japan.

Bíró, P. (1997). Temporal variations in Lake Balaton and its fish populations. *Ecol. Freshwat. Fishes* **6**, 196–216.

Bíró, P. and Garádi, P. (1974). Investigations on the growth and population structure of bream (*Abramis brama* L.) at different areas of Lake Balaton, the assessment of mortality and production. *Ann. Inst. Biol. (Tihany) Hung. Acad. Sci.* **41**, 153–179.

Bíró, P. and Muskó, I.B. (1995). Population dynamics and food of bleak (*Alburnus alburnus* L.) in the littoral zone of Lake Balaton. *Hydrobiologia* **310**, 139–149.

Bíró, P. and Paulovits, G. (1994). Evolution of fish fauna in Little Balaton Water Reservoir. *Verh. Int. Verein. Limnol.* **25**, 2164–2168.

Bíró, P. and Vörös, L. (1990). Trophic relationships between primary producers and fish yield in Lake Balaton. *Hydrobiologia* **191**, 213–221.

Dauba, F. and Bíró, P. (1992). Growth of bream, *Abramis brama* L., in two outside basins of different trophic state of Lake Balaton. *Int. Rev. Gesamten Hydrobiol.* **77**, 225–235.

Entz, B. (1950). Autumn and winter shoals of fish in the shore zones of Lake Balaton in 1947–1949. *Ann. Inst. Biol. (Tihany) Hung. Acad. Sci.* **19**, 83–94.

Entz, B. and Lukacsovics, F. (1957). Investigations during the winter months on some Balaton fishes as to their feeding, growing and propagation conditions. *Ann. Inst. Biol. (Tihany) Hung. Acad. Sci.* **24**, 71–86 (in Hungarian).

G.-Tóth, L. (1992). The activity of the terminal electron transport system (ETS) of the plankton and sediment in Lake Balaton. *Hydrobiologia* **243/244**, 157–167.

Gorthner, A. (1994). What is an ancient lake? *Arch. Hydrobiol. Beiheft. Ergebnisse Limnol.* **44**, 97–100.

Hankó, B. (1931). Ursprung und Verbreitung der Fischfauna Ungarns. *Arch. Hydrobiol.* **23**, 520–556 (in German).

Herodek, S. (1977). Recent results of phytoplankton research in Lake Balaton. *Ann. Inst. Biol. (Tihany) Hung. Acad. Sci.* **44**, 181–198.

Herodek, S. (1988). Limnology of Lake Balaton. Eutrophication of Lake Balaton. In: *Lake Balaton Research and Management* (Ed. by Misley K), pp. 9–58. Nexus, Budapest.

Herodek, S. and Istvánovics, V. (1986). Mobility of phosphorous fractions in the sediments of Lake Balaton. *Hydrobiologia* **135**, 149–154.

Istvánovics, V. (1993). Transformations between organic and inorganic sediment phosphorus in Lake Balaton. *Hydrobiologia* **253**, 193–206.

Kawanabe, H. (1996). Asian great lakes, especially Lake Biwa. *Envir. Biol. Fishes* **47**, 219–234.

Mike, K. (1976). A Balaton vizgyűjtőjének geomorfológiája. 2. A Balaton kialakulása és fejlődése. In: *Vituki, Vízrajzi Atlasz sorozat* 21. *Balaton I. Hidrográfia, Geomorfológia*, pp. 30–39. Vituki, Budapest (in Hungarian).

Padisák, J. and Reynolds, C.S. (1998). Selection of phytoplankton associations in Lake Balaton, Hungary, in response to eutrophication and restoration measures, with special reference to the cyanoprokaryotes. *Hydrobiologia* **384**, 41–53.

Perényi, M. (1991). Population dynamics and food-consumption of razor fish (*Pelecus cultratus* L.) in Lake Balaton. Unpublished CSc Thesis, Tihany (in Hungarian).

Ponyi, J. (1992). One century of research of the invertebrate fauna. In: *100 Years of Balaton Research* (Ed. by P. Bíró), pp. 77–84. Reproprint, Nemesvámos, Hungary (in Hungarian).

Przybylski, M., Bíró, P., Zalewski, M., Tátrai, I. and Frankiewicz, P. (1991). The structure of fish communities in streams of the northern part of the catchment area of Lake Balaton (Hungary). *Acta Hydrobiol.* **33**, 135–148.

Répássy, M. (1909). Fisheries of Lake Balaton. In: *Fisheries and Fish Farming in Fresh Waters* (Ed. by I. Darányi), pp. 421–431. Pallas Rt., Budapest (in Hungarian).

Rónai, A. (1969). The geology of Lake Balaton and surroundings. *Mitt. Int. Verein. Limnol.* **17**, 275–281.

Sebestyén, O. (1967). A kemizáció kihatása vízi ekoszisztémákban. *Magyar Tudományos Akadémia V. Osztályának Közleményei* **18**, 389–391 (in Hungarian).

Simonian, A., Tátrai, I., Bíró, P., Paulovits, G., Tóth, L.G. and Lakatos, Gy. (1995). Biomass of planktonic crustaceans and the food of young cyprinids in the littoral zone of Lake Balaton. *Hydrobiologia* **303**, 39–48.

Specziár, A., Tölg, L. and Bíró, P. (1997). Feeding strategy and growth of cyprinids in the littoral zone of Lake Balaton. *J. Fish Biol.* **51**, 1109–1124.

Tölg, I. (1961). Über die Ursache des Nahrungsmangels der Balaton-Zander (*Lucioperca lucioperca* L.) und Begründung des Nahrungsersatzplanes. *Ann. Inst. Biol. (Tihany) Hung. Acad. Sci.* **28**, 179–195 (in German).

Zólyomi, B. (1995). Opportunities for pollen stratigraphic analysis of shallow lake sediments: the example of Lake Balaton. *GeoJournal* **36**, 237–241.

Advances in Ecological Research
Volumes 1–31

Cumulative List of Titles

Aerial heavy metal pollution and terrestrial ecosystems, **11**, 218

Age determination and growth of Baikal seals (*Phoca sibirica*), **31**, 449

Age-related decline in forest productivity: pattern and process, **27**, 213

Analysis of processes involved in the natural control of insects, **2**, 1

Ancient Lake Pannon and its endemic molluscan fauna (Central Europe; Mio-Pliocene), **31**, 463

Ant-plant-homopteran interactions, **16**, 53

The benthic invertebrates of Lake Khubsugul, Mongolia, **31**, 97

Biogeography and species diversity of diatoms in the northern basin of Lake Tanganyika, **31**, 115

Biological strategies of nutrient cycling in soil systems, **13**, 1

Bray-Curtis ordination: an effective strategy for analysis of multivariate ecological data, **14**, 1

Can a general hypothesis explain population cycles of forest lepidoptera?, **18**, 179

Carbon allocation in trees: a review of concepts for modelling, **25**, 60

Catchment properties and the transport of major elements to estuaries, **29**, 1

Coevolution of mycorrhizal symbionts and their hosts to metal-contaminated environments, **30**, 69

Conservation of the endemic cichlid fishes of Lake Tanganyika: implications from population-level studies based on mitochondrial DNA, **31**, 539

The cost of living: field metabolic rates of small mammals, **30**, 177

A century of evolution in *Spartina anglica*, **21**, 1

The climatic response to greenhouse gases, **22**, 1

Communities of parasitoids associated with leafhoppers and planthoppers in Europe, **17**, 282

Community structure and interaction webs in shallow marine hardbottom communities: tests of an environmental stress model, **19**, 189

The decomposition of emergent macrophytes in fresh water, **14**, 115

Delays, demography and cycles: a forensic study, **28**, 127

Dendroecology: a tool for evaluating variations in past and present forest environments, **19**, 111

The development of regional climate scenarios and the ecological impact of greenhouse gas warming, **22**, 33

Developments in ecophysiological research on soil invertebrates, **16**, 175

The direct effects of increase in the global atmospheric CO_2 concentration on natural and commercial temperate trees and forests, **19**, 2

The distribution and abundance of lakedwelling Triclads—towards a hypothesis, **3**, 1

Index